# Methods in Enzymology

Volume 396
NITRIC OXIDE
Part E

# METHODS IN ENZYMOLOGY

## EDITORS-IN-CHIEF

### John N. Abelson     Melvin I. Simon

DIVISION OF BIOLOGY
CALIFORNIA INSTITUTE OF TECHNOLOGY
PASADENA, CALIFORNIA

## FOUNDING EDITORS

### Sidney P. Colowick and Nathan O. Kaplan

*Methods in Enzymology*

*Volume 396*

# Nitric Oxide

*Part E*

EDITED BY

*Lester Packer*

*Enrique Cadenas*

UNIVERSITY OF SOUTHERN CALIFORNIA
DEPARTMENT OF MOLECULAR PHARMACOLOGY AND TOXICOLOGY
SCHOOL OF PHARMACY
LOS ANGELES, CALIFORNIA

**ELSEVIER**
ACADEMIC
PRESS

AMSTERDAM • BOSTON • HEIDELBERG • LONDON
NEW YORK • OXFORD • PARIS • SAN DIEGO
SAN FRANCISCO • SINGAPORE • SYDNEY • TOKYO

Elsevier Academic Press
525 B Street, Suite 1900, San Diego, California 92101-4495, USA
84 Theobald's Road, London WC1X 8RR, UK

This book is printed on acid-free paper.

Permissions may be sought directly from Elsevier's Science & Technology Rights
Department in Oxford, UK: phone: (+44) 1865 843830, fax: (+44) 1865 853333,
E-mail: permissions@elsevier.co.uk. You may also complete your request on-line
via the Elsevier homepage (http://elsevier.com), by selecting
"Customer Support" and then "Obtaining Permissions."

For all information on all Elsevier Academic Press publications
visit our Web site at www.books.elsevier.com

ISBN-13: 978-0-12-182801-1
ISBN-10: 0-12-182801-8

PRINTED IN THE UNITED STATES OF AMERICA
05   06   07   08   09   9   8   7   6   5   4   3   2   1

# Table of Contents

## Section I. Biochemical, Molecular, and Real-Time Detection of Nitric Oxide

## Section II. Nitration and S-Nitrosylation

## Section III. Peroxynitrite

## Section IV. Signaling and Gene Expression

## Section V. Cell Biology and Physiology

# Contributors to Volume 396

Article numbers are in parentheses and following the names of contributors.
Affiliations listed are current.

GAMAL EL-DIN ALI AHMED HASSAN ABUO-RAHMA (2), *Medicinal Chemistry Department, Faculty of Pharmacy, Minia University, Minia, 61519, Egypt*

BARRY W. ALLEN (7), *Department of Anesthesiology and Center for Hyperbaric Medicine and Environmental Physiology Duke University Medical Center, Durham, North Carolina 27710*

K. KRISTOFFER ANDERSSON (38), *Department of Molecular Bioscience, University of Oslo, N-0316 Oslo, Norway*

ROBERTO J. ARAI (29), *Department of Biochemistry/Molecular Biology, CINTER-GEN—Universidade Federal São Paolo, Escola Paulista de Medicina, Rua Mirassol, 020 São Paulo, Brazil*

MELINDA L. ASBURY (36), *Department of Pharmacology, Joan C. Edward School of Medicine, Marshall University, Huntington, West Virgina, 25704*

JOSEPH S. BECKMAN (19), *Department of Biochemistry and Biophysics, Oregon State University, Environmental Health Science Center, The Linus Pauling Institute, Corvallis, Oregon 97331*

VLADIMIR BERKA (40), *Division of Hematology, Department of Internal Medicine, UT Health Science Center at Houston Medical School, Houston, Texas 77030*

MAURO BERTOTTI (4), *Instituto de Química-USP, Universitdade de São Paulo, Lineu Prestes 748, 05508-900 São Paulo, Brazil*

LISARDO BOSCÁ (49), *Instituto de Bioquimica, Centro Mixto CSIC-UCM, Facultad de Farmacia, Universidad Complutense, 28760 Madrid, Spain*

BETH M. BOULDEN (42), *Wallace H. Coulter Department of Biomedical Engineering, Georgia Institute of Technology, Atlanta, Georgia 30033*

ALBERTO BOVERIS (37), *Laboratory of Free Radical Biology, School of Pharmacy and Biochemistry, University of Buenos Aires, 1113 Buenos Aires, Argentina*

DANIEL A. BRAZEAU (33), *Department of Pharmaceutical Sciences, School of Pharmacy and Pharmaceutical Sciences, University at Buffalo, State University of New York, Buffalo, New York 14260-1200*

HOLLY BROWN-BORG (24), *University of North Dakota, School of Medicine and Health Sciences, Grand Forks, North Dakota, 58202-9037*

YUANLIN CAO (8, 9), *Institute of Biophysics, Academia Sinica, Beijing 100101, China*

MARIA CECILIA CARRERAS (34), *Laboratory of Oxygen Metabolism, University Hospital and School of Pharmacy and Biochemistry, University of Buenos Aires, 1120 Buenos Aires, Argentina*

JEANNEAN CARVER (10), *Department of Pediatrics, Critical Care Medicine, University of Virginia School of Medicine, Charlottesville, Virginia 22908*

YOUNG-NAM CHA (31, 35), *Department of Pharmacology and Toxicology, Medicinal Toxicology Center, College of Medicine, Inha University, Inchon, 402-751, South Korea*

KENNY K. K. CHUNG (14), *Institute for Cell Engineering, Departments of Neurology and Neuroscience, Johns Hopkins University School of Medicine, Baltimore, Maryland 21205*

DANIELA P. CONVERSO (34), *Laboratory of Oxygen Metabolism, University Hospital and School of Pharmacy and Biochemistry, University of Buenos Aires, 1120 Buenos Aires, Argentina*

JACK H. CRAWFORD (47), *Department of Pathology, Unviersity of Alabama at Birmington, Biomedical Research Building II, Birmingham, Alabama 35294-2180*

CLAUDETTE M. ST. CROIX (26), *Department of Environmental and Occupational Heath, University of Pittsburgh Graduate School Public Health, Pittsburgh, Pennsylvania 15261*

MARLI F. CURCIO (29), *Department of Biochemistry/Molecular Biology, CINTER-GEN—Universidade Federal Sâo Paolo, Escola Paulista de Medicina, Rua Mirassol, 020 São Paulo, Brazil*

TAYFUN DALBASTI (50), *Department of Neurosurgery, University of Ege, School of Medicine, Bornova, 35100 Izmir, Turkey*

TED M. DAWSON (14, 30), *Institute for Cell Engineering, Departments of Neurology and Neuroscience, Johns Hopkins University School of Medicine, Baltimore, Maryland 21205*

VALINA L. DAWSON (14, 30), *Institute for Cell Engineering, Departments of Neurology and Neuroscience, Johns Hopkins University School of Medicine, Baltimore, Maryland 21205*

SERGEY DIKALOV (52), *Division of Cardiology, Emory University School of Medicine, Suite 319, Woodruff Memorial Research Building, 101 Woodruff Circle, Atlanta, Gainesville 30022*

SERGEY I. DIKALOV (42), *Department of Medicine, Emory University, Atlanta, Georgia 30322*

ALLAN DOCTOR (10), *Department of Pediatrics, Critical Care Medicine, University of Virginia School of Medicine, Charlottesville, Virginia 22908*

SAMUEL C. DUDLEY, JR. (42), *Department of Medicine, Emory University, Atlanta, Georgia 30322*

ANDREW S. DUTTON (3), *Department of Chemistry and Biochemistry, University of California, Los Angeles, Los Angeles, California 90095-1569*

MANUCHAIR EBADI (24), *School of Medicine and Health Sciences, University of North Dakota, Grand Forks, North Dakota, 58202-9037*

ALI EL-EMAN (2), *Department of Pharmaceutical Chemistry, College of Pharmacy, King Saud University, Riyadh 11451, Saudi Arabia*

MICHAEL G. ESPEY (27), *Laboratory of Human Carcinogenesis, National Cancer Institute, Bethesda, Maryland 20892-4255*

BRUNO FINK (52), *NOxygen Science Transfer & Diagnostics, Lindenmatte 42, 79215 Elzach, Germany*

PAOLA FINOCCHIETO (34), *Laboratory of Oxygen Metabolism, University Hospital and School of Pharmacy and Biochemistry, University of Buenos Aires, 1120 Buenos Aires, Argentina*

ROBERT A. FLOYE (45), *Oklahoma Medical Research Foundation, Free Radical Biology and Aging Research Program, Oklahoma City, Oklahoma 73104*

PETER C. FORD (1), *Department of Chemistry and Biochemistry, University of California, Santa Barbara, Santa Barbara, California 93106-9510*

DOUGLAS WAGNER FRANCO (4), *Universidade de São Paulo, Instituto de Química de São Carlos, Departamento de Química e Fisica Molecular, Av. Trabalhador Saocariense, 400, Centro - Cx Postal 780, CEP 13566-590 - São Carlos - SP, Brazil*

MARIA CLARA FRANCO (34), *Laboratory of Oxygen Metabolism, University Hospital and School of Pharmacy and Biochemistry, University of Buenos Aires, 1120 Buenos Aires, Argentina*

XIAMING FU (22), *Center for Cardiovascular Diagnostics and Prevention, Cleveland Clinic Foundation, Department of Cell Biology, Cleveland, Ohio 44195*

JON M. FUKUTO (3, 25), *Department of Chemistry & Biochemistry, University of California, Los Angeles, Los Angeles, California 90095-1569*

HO-LEUNG FUNG (33), *Department of Pharmaceutical Sciences, School of Pharmacy and Pharmaceutical Sciences, University at Buffalo, State University of New York, Buffalo, New York 14260-1200*

RALPH GÄBLER (5, 51), *In Vivo GMBH – Institute for Trace Gas Technology, Brueghel Strasse 4, Sankt Augustin, D-53757, Germany*

SOLEDAD GALLI (34), *Laboratory of Oxygen Metabolism, University Hospital and School of Pharmacy and Biochemistry, University of Buenos Aires, 1120 Buenos Aires, Argentina*

BENJAMIN GASTON (10), *Department of Pediatrics, University of Virginia Health System, Charlottesville, Virginia 22908*

PEDRAM GHAFOURIFAR (24, 36), *University of North Dakota, School of Medicine and Health Sciences, Grand Forks, North Dakota, 58202-9037*

ANTONIUS C. F. GORREN (38), *Department of pharmacology and Toxicology, Karl-Franzens-University Graz, A-8010 Graz, Austria*

MATTHEW B. GRISHAM (12), *Department of Molecular and Cellular Physiology, Louisiana State University Health Sciences Center, Shreveport, Louisiana 71130*

PING GUO (8), *Institute of Biophysics, Academia Sinica, Beijing 100101, China*

CURTIS C. HARRIS (27), *Laboratory of Human Carcinogenesis, National Cancer Institute, Bethesda, Maryland 20892-4255*

C. MICHAEL HART (42), *Department of Medicine, Emory University, Atlanta VA Medical Center, Decatur, Georgia 30033*

STANLEY L. HAZEN (22), *Center for Cardiovascular Diagnostics and Prevention, Cleveland Clinic Foundation, Department of Cell Biology, Cleveland, Ohio 44195*

HARMUT HELLER (51), *Physiologisches Institut I, Bonn Universitat, Nussallee 11, 53115 Bonn, Germany*

KARSTEN HEMMRICH (39), *Department of Plastic Surgery, Hand Surgery, Burn Center, Univeristy Hospital of the Aachen, University of Technology, Pauwelsstr. 30, D-52057 Aachen, Germany*

KENNETH HENSLEY (17), *Oklahoma Medical Research Foundation, Free Radical Biology and Aging Research Program, Oklahoma City, Oklahoma 73104*

LORNE J. HOFSETH (27), *College of Pharmacy, CLS 109, University of South Carolina, Columbia, South Carolina 29208*

SUK J. HONG (30), *Institute for Cell Engineering, Department of Neurology, Johns Hopkins University School of Medicine, Baltimore, Maryland 21205*

SONSOLES HORTELANO (49), *Institute de Bioquimica, Centro Mixto CSIC-UCM, Facultad de Farmacia, Universidad Complutense, 28760 Madrid, Spain*

K. N. HOUK (3), *Department of Chemistry and Biochemistry, University of California, Los Angeles, Los Angeles, California 90095-1569*

JOHN HUGHES (12), *Center for Cardiovascular Sciences, Albany Medical College, Albany, New York 12208*

T. SCOTT ISBELL (47), *Department of Pathology, University of Alabama at Birmington, Biomedical Research Building II, Birmingham, Alabama 35294-2180*

MALCOLM J. JACKSON (43), *Department of Metabolic and Cellular Medicine, School of Clinical Sciences, University of Liverpool, Liverpool L69 3GA, United Kingdom*

SAMIE R. JAFFREY (11), *Department. of Pharmacology, Weill Medical College, Cornell University, New York, New York 10021*

JOY JOSEPH (18), *Department of Biophysics, Free Radical Research Center, Medical College of Wisconsin, Milwaukee, Wisconsin 53226-0509*

SANDEEP S. JOSHI (36), *Department of Pharmacology, Joan C. Edward School of Medicine, Marshall University, Huntington, West Virginia 25704*

DAVID JOURD'HEUIL (12), *Center for Cardiovascular Sciences (MC8), Albany Medical College, Albany, New York 12208*

FRANCES L. JOURD'HEUIL (12), *Center for Cardiovascular Sciences (MC8), Albany Medical College, Albany, New York 12208*

SHASI KALIVENDI (44), *Department of Biophysics, Free Radical Research Center, Medical College of Wisconsin, Milwaukee, Wisconsin 53226-0509*

B. KALYANARAMAN (18, 44), *Department of Biophysics, Free Radical Research Center, Medical College of Wisconsin, Milwaukee, Wisconsin 53226-0509*

JAROSLAW KANSKI (16), *Department of Pharmaceutical Chemistry, University of Kansas, Lawrence, Kansas 66047*

EMRAH KILINC (50), *Department of Analytical Chemistry, School of Pharmacy, University of Ege, Bornova, Izmir, Turkey*

CHAEKYUN KIM (31, 35), *Department of Pharmacology and Toxicology, Medicinal Toxicology Center, College of Medicine, Inha University, Inchon, 402-751, South Korea*

SANG GEON KIM (28, 32), *College of Pharmacy, Seoul National University, Sillimdong, Kwanak-gu, Seoul 151-742, South Korea*

ERIC D. KINCAID (36), *Department of Pharmacology, Joan C. Edward School of Medicine, Marshall University, Huntington, West Virginia 25704*

HIDEKI KISHIDA (45), *Oklahoma Medical Research Foundation, Free Radical Biology and Aging Research Program, Oklahoma City, Oklahoma 73104*

REINHARD KISSNER (6), *Laboratory F. Anorg. Chemie, CHI H 211, ETH Hönggersberg, CH-8093 Zürich, Switzerland*

DEAN J. KLEINHENZ (42), *Department of Medicine, Emory University, Atlanta VA Medical Center, Atlanta, Gainesville 30033*

JEFFREY R. KOENITZER (47), *Department of Biology, Unviersity of Alabama at Birmington, Birmingham, Alabama 35294-2180*

DORIS KOESLING (41), *Institut für Pharmakologie und Toxi Kologie, Ruhr-Universität Bochum, Medizinische Fakultät MA N1, 44780 Bochum, Germany*

VICTORIA KOLB-BACHOFEN (39, 48), *Research Group Immunobiology, Heinrich-Heine-University of Düesseldorf, D-40001 Düsseldorf, NRW, Germany*

JOERG KONTER (2), *Department of Pharmaceutical Medicinal Chemistry, Institute of Pharmacy, Friedrich-Schiller-University, Philosophenweg 14, D-07743 Jena, Germany*

WILLEM H. KOPPENOL (6), *Laboratory F. Anorg. Chemie, CHI H 211, ETH Hönggersberg, CH-8093 Zürich, Switzerland*

YASHIGE KOTAKE (45), *Oklahoma Medical Research Foundation, Free Radical Biology and Aging Research Program, Oklahoma City, Oklahoma 73104*

SRIGIRIDHAR KOTAMRAJU (44), *Department of Biophysics, Free Radical Research Center, Medical College of Wisconsin, Milwaukee, Wisconsin 53226-0509*

DAVID W. KRAUS (47), *Department of Pathology, University of Alabama at Birmington, Biomedical Research Building II, Birmingham, Alabama 35294-2180*

SANTIAGO LAMAS (13), *Centro de Investigaciones Biologicas, Consejo Superior de Investiaciones Cientificas and Instituto Reina Sofia de Investigaciones Nefrologicas, Ramiro de Maeztu, 9, Madrid E-28040*

REINHARD LANGE (38), *Institut für Pharmakologie und Toxikologie, Karl-Franzens-Universität Graz, Universitätsplatz 2, A-8010 Graz, Austria*

CHANG HO LEE (32), *Department of Pharmacology and Institute of Biomedical Science, College of Medicine, Hanyang University, Seoul 133-791, South Korea*

JOCHEN LEHMANN (2, 5), *Department of Pharamaceutical X Medicinal Chemistry, Institute of Pharmacy, Friedrich-Schiller-University, Philosophenweg 14, D-07743 Jena, Germany*

BRUCE S. LEVISON (22), *Center for Cardiovascular Diagnostics and Prevention, Cleveland Clinic Foundation, Department of Cell Biology, Cleveland, Ohio 44195*

MARK D. LIM (1), *Department of Chemistry and Biochemistry, University of California, Santa Barbara, Santa Barbara, California 93106-9510*

JIE LIU (7), *Department of Chemistry, Duke University, Durham, North Carolina 27708*

BRENDA E. LOPEZ (25), *Department of Pharmacology, School of Medicine, Center for the Health Sciences, University of California, Los Angeles, Los Angeles, California 90095-1772*

IVAN M. LORKOVIĆ (1), *Department of Chemistry and Biochemistry, University of California, Santa Barbara, Santa Barbara, California 93106-9510*

ANTHONY M. LOWERY (12), *Center for Cardiovascular Sciences, Albany Medical College, Albany, New York 12208*

CINZIA MALLOZZI (20), *Department of Cell Biology and Neuroscience, Unit of Free Radical Pathophysiology, Istituto Superiori di Sanita, Viale Regina Elena, 299-00161, Rome, Italy*

ALI R. MANI (15), *Centre for Hepatology, Department of Medicine, Royal Free Campus, University College London, London NW3 2PF United Kingdom*

STEPHANE MARCHAL (38), *INSERM U 710*, *Université Montpellier, 34095 Montpellier Cédex 5, France*

EMIL MARTIN (40), *Department of Integrative Biology and Pharmacology, UT Health Science Center at Houston Medical School, Houston, Texas 77030*

ANTONIO MARTÍNEZ-RUIZ (13), *Centro de Investigaciones Biologicas, Consejo Superior de Investiaciones Cientificas and Instituto Reina Sofia de Investigaciones Nefrologicas, Ramiro de Maeztu, 9, Madrid E-28040*

BERND MAYER (38), *Institut für Pharmakologie und Toxikologie, Karl-Franzens-Universität Graz, Universitätsplatz 2, A-8010 Graz, Austria*

CRAIG J. MCMACKIN (46), *Evans Department of Medicine and Whitaker Cardiovascular Institute, Boston University School of Medicine, Boston, Massachusetts 02118*

MAURIZIO MINETTI (20), *Department of Cell Biology and Neuroscience, Unit of Free Radical Pathophysiology, Istituto Superiori di Sanita, Viale Regina Elena, 299-00161, Rome, Italy*

HUGO P. MONTIERO (29), *Department of Biochemistry/Molecular Biology, CINTERGEN—Universidade Federal Sâo Paolo, Escola Paulista de Medicina, Rua Mirassol, 020 São Paulo, Brazil*

KEVIN P. MOORE (15), *Centre for Hepatology, Department of Medicine, Royal Free Campus, University College London, London NW3 2PF United Kingdom*

MIRIAM S. MORAES (29), *Department of Biochemistry/Molecular Biology, CINTERGEN—Universidade Federal Sâo Paolo, Escola Paulista de Medicina, Rua Mirassol, 020 São Paulo, Brazil*

VÂNIA MORI (4), *Instituto de Química, Universitdade de São Paulo, Lineu Prestes 748, 05508-900 São Paulo, Brazil*

FERID MURAD (23, 40), *Department of Integrative Biology and Pharmacology, UT Health Science Center at Houston Medical School, Houston, Texas 77030*

DAI NAKAE (45), *Oklahoma Medical Research Foundation, Free Radical Biology and Aging Research Program, Oklahoma City, Oklahoma 73104*

STEPHEN J. NICHOLLS (22), *Center for Cardiovascular Diagnostics and Prevention, Cleveland Clinic Foundation, Department of Cell Biology, Cleveland, Ohio 44195*

CARLOS J. ROCHA OLIVEIRA (29), *Department of Biochemistry/Molecular Biology, CINTERGEN—Universidade Federal Sâo Paolo, Escola Paulista de Medicina, Rua Mirassol, 020 São Paulo, Brazil*

EUN YOUNG PARK (28), *College of Pharmacy, Seoul National University, Sillim-dong, Kwanak-gu, Seoul 151-742, South Korea*

RAKESH P. PATEL (47), *Department of Pathology, University of Alabama at Birmington, Biomedical Research Building II, Birmingham, Alabama 35294-2180*

ADNANA PAUNEL (48), *Research Group Immunobiology, Institute of Molecular Medicine, Heinrich-Heine-University of Düesseldorf, D-40001 Düsseldorf, NRW, Germany*

CLAUDE A. PIANTADOSI (7), *Department of Medicine & Anesthesiology, Duke University Medical Center, Durham, North Carolina 27710*

BRUCE R. PITT (26), *Department of Environmental and Occupational Heath, University of Pittsburgh Graduate School Public Health, Pittsburgh, Pennsylvania 15261*

JUAN JOSÉ PODEROSO (34), *Laboratory of Oxygen Metabolism, University Hospital, and School of Pharmacy and Medicine, University of Buenos Aires, 1120 Buenos Aires, Argentina*

RAFAEL RADI (21), *Departamento de Bioquímica and Center for Free Radical and Biomedical Research, Facultad de Medicina, Universidad de la Repblica Montevideo, Uruguay, Avda. Gral. Flores 2125, 11800 Montevideo, Uruguay*

H. EL RAFAEY (24), *University of North Dakota, School of Medicine and Health Sciences, Grand Forks, North Dakola, 58202-9037*

KRISTINE M. ROBINSON (19), *Department. of Biochemistry and Biophysics, Oregon State University, Environmental Health Science Center, The Linus Pauling Institute, Corvallis, Oregon 97331*

ANA I. ROBLES (27), *Laboratory of Human Carcinogenesis, National Cancer Institute, Bethesda, Maryland 20892-4255*

CHESTER E. RODRIGUEZ (25), *Department of Pharmacology, School of Medicine, Center for the Health Sciences, University of California, Los Angeles, Los Angeles, California 90095-1772*

NATALIA ROMERO (21), *Departamento de Bioquímica and Center for Free Radical and Biomedical Research, Facultad de Medicina, Universidad de la Repblica, Avda. Gral. Flores 2125, 11800 Montevideo, Uruguay*

MICHAEL RUSSWUM (41), *Institut Für Pharmakologie und Toxikologie, Ruhr-Universität Bochum, Medizinische Fakultät MA N1, 44780 Bochum, Germany*

CHRISTIAN SCHÖNEICH (16), *Department of Pharmaceutical Chemistry, University of Kansas, Lawrence, Kanas 66047*

K.-D. SCHUSTER (51), *Physiologisches Institut I, Bonn Universitat, Nussallee 11, 53115 Bonn, Germany*

MAURO SERAFINI (20), *Antioxidant Research Laboratory, Unit of Human Nutrition, National Institute for Food and Nutrition Research, Via Ardeatine, 546, 00178 Rome, Italy*

TIESONG SHANG (44), *Department of Biophysics, Free Radical Research Center, Medical College of Wisconsin, Milwaukee, Wisconsin 53226-0509*

SUSHIL K. SHARMA (24), *University of North Dakota, School of Medicine and Health Sciences, Grand Forks, North Dakola, 58202-9037*

MASARU SHINYASHIKI (25), *Department of Pharmacology, School of Medicine, Center for the Health Sciences, University of California, Los Angeles, Los Angeles, California 90095-1772*

ZHONGZHOU SHEN (22), *Center for Cardiovascular Diagnostics and Prevention, Cleveland Clinic Foundation, Department of Cell Biology, Cleveland, Ohio 44195*

MORTEN SORLIE (38), *Department of Chemistry of Biotechnology, Agricultural University of Norway, N-1432Ås, Norway*

KLAOKWAN SRISOOK (31, 35), *Department of Pharmacology and Toxicology, Medicinal Toxicology Center, College of Medicine, Inha University, Inchon, 400-103, South Korea*

ANNA MARIA MICHELA DI STASI (20), *Department of Cell Biology and Neuroscience, Unit of Free Radical Pathophysiology, Istituto Superiori di Sanita, Viale Regina Elena, 299-00161, Rome, Italy*

MOLLY S. STITT (26), *Department of Environmental and Occupational Heath, University of Pittsburgh Graduate School Public Health, Pittsburgh, Pennsylvania 15261*

CHRISTOPH V. SUSCHEK (39, 48), *Institute of Biochemistry and Molecular Biology II, Heinrich-Heine-University of Düsseldorf, D-40001 Düsseldorf, NRW, Germany*

YI TAO (9), *Institute of Biophysics, Academia Sinica, Beijing 100101, China*

JOSE CARLOS TOLEDO (4), *Universidade de Sâo Pâulo, Instituto de Química de Sao Carlos, Departamento de Química e Fisica Molecular, Saocariense, 400, Centro - Cx Postal 780, CEP 13566-590 - Sâo Carlos - SP, Brazil*

DOANH C. TRAN (33), *Department of Pharmaceutical Sciences, School of Pharmacy and Pharmaceutical Sciences, University at Buffalo, State University of New York, Buffalo, New York 14260-1200*

PAQUI G. TRAVÉS (49), *Instituto de Bioquimica, Centro Mixto CSIC-UCM, Facultad de Farmacia, Universidad Complutense, 28760 Madrid, Spain*

AH-LIM TSAI (40), *Division of Hematology, Department of Internal Medicine, UT Health Science Center at Houston Medical School, Houston, Texas 77030*

ILLARION V. TURKO (23), *Department of Integrative Biology and Pharmacology, UT Health Science Center at Houston Medical School, Houston, Texas 77030*

LAURA B. VALDEZ (37), *Laboratory of Free Radical Biology, School of Pharmacy and Biochemistry, University of Buenos Aires, 1113 Buenos Aires, Argentina*

JOSEPH A. VITA (46), *Evans Department of Medicine and Whitaker Cardiovascular Institute, Boston University School of Medicine, Boston, Masschusetts 02118*

SIMON C. WATKINS (26), *Center for Biological Imaging, Department of Cell Biology and Physiology, University of Pittsburgh, Pittsburgh, Pennsylvania 15261*

C. R. WHITE (47), *Department of Medicine, Center for Free Radical Biology, University of Alabama at Birmington, Birmingham, Alabama 35294-2180*

KELLY S. WILLIAMSON (17), *Oklahoma Medical Research Foundation, Free Radical Biology and Aging Research Program, Oklahoma City, Oklahoma 73104*

YANGCANG XU (8, 9), *Institute of Biophysics, Academia Sinica, Beijing 100101, China*

KHALEQUZ ZAMAN (10), *Department of Pediatrics, Critical Care Medicine, University of Virginia School of Medicine, Charlottesville, Virginia 22908*

PATRICIA ZANICHELLI (4), *Universidade de Sâo Pâulo, Instituto de Química de Sao Carlos, Departamento de Química e Fisica Molecular, Saocariense, 400, Centro - Cx Postal 780, CEP 13566-590 - Sâo Carlos - SP, Brazil*

TAMARA ZAOBORNYJ (37), *Laboratory of Free Radical Biology, School of Pharmacy and Biochemistry, University of Buenos Aires, 1113 Buenos Aires, Argentina*

MIRIAM ZEINI (49), *Instituto de Bioquimica, Centro Mixto CSIC-UCM, Facultad de Farmacia, Universidad Complutense, 28760 Madrid, Spain*

HAO ZHANG (18), *Department of Biophysics, Free Radical Research Center, Medical College of Wisconsin, Milwaukee, Wisconsin 53226-0509*

BAOLU ZHAO (8, 9), *Institute of Biophysics, Academia Sinica, Beijing 100101, China*

# Preface

The discovery that nitrogen monoxide or nitric oxide (NO) is a free radical formed in a variety of cell types by nitric oxide synthase and is involved in a wide array of physiological and pathophysiological phenomena has ignited enormous interest in the scientific community. One of the unique features of nitric oxide is its function as an intercellular messenger and, in this capacity, its involvement in the modulation of cell signaling and mitochondrial respiration. Nitric oxide metabolism and the interactions of this molecule with multiple cellular targets are currently areas of intensive research and have important pharmacological implications for health and disease.

Accurately assessing the generation, action, and regulation of nitric oxide in biological systems has required the development of new analytical methods at the molecular, cellular, tissue, and organismal levels. This was the impetus for Methods in Enzymology Volumes 268, 269, 301, and 359 Nitric Oxide Parts A, B, C, and D, respectively. Only a few years later, this new Volume 396 reflects the development of new and important tools for the assessment of nitric oxide action. Nitric Oxide, Part E contains five major sections: Biochemical, Molecular, and Real-Time Detection of Nitric Oxide, (II) Nitration and S-Nitrosylation, (III) Peroxynitrite, (IV) Signaling and Gene Expression, and (V) Cell Biology and Physiology.

In bringing this volume to fruition, credit must be given to the experts in various specialized fields of nitric oxide research who have contributed outstanding chapters to these sections on nitric oxide methodology. To these colleagues, we extend our sincere thanks and most grateful appreciation.

LESTER PACKER
ENRIQUE CADENAS

# METHODS IN ENZYMOLOGY

VOLUME LV. Biomembranes (Part F: Bioenergetics)
*Edited by* SIDNEY FLEISCHER AND LESTER PACKER

VOLUME LVI. Biomembranes (Part G: Bioenergetics)
*Edited by* SIDNEY FLEISCHER AND LESTER PACKER

VOLUME LVII. Bioluminescence and Chemiluminescence
*Edited by* MARLENE A. DELUCA

VOLUME LVIII. Cell Culture
*Edited by* WILLIAM B. JAKOBY AND IRA PASTAN

VOLUME LIX. Nucleic Acids and Protein Synthesis (Part G)
*Edited by* KIVIE MOLDAVE AND LAWRENCE GROSSMAN

VOLUME LX. Nucleic Acids and Protein Synthesis (Part H)
*Edited by* KIVIE MOLDAVE AND LAWRENCE GROSSMAN

VOLUME 61. Enzyme Structure (Part H)
*Edited by* C. H. W. HIRS AND SERGE N. TIMASHEFF

VOLUME 62. Vitamins and Coenzymes (Part D)
*Edited by* DONALD B. MCCORMICK AND LEMUEL D. WRIGHT

VOLUME 63. Enzyme Kinetics and Mechanism (Part A: Initial Rate and
Inhibitor Methods)
*Edited by* DANIEL L. PURICH

VOLUME 64. Enzyme Kinetics and Mechanism (Part B: Isotopic Probes and
Complex Enzyme Systems)
*Edited by* DANIEL L. PURICH

VOLUME 65. Nucleic Acids (Part I)
*Edited by* LAWRENCE GROSSMAN AND KIVIE MOLDAVE

VOLUME 66. Vitamins and Coenzymes (Part E)
*Edited by* DONALD B. MCCORMICK AND LEMUEL D. WRIGHT

VOLUME 67. Vitamins and Coenzymes (Part F)
*Edited by* DONALD B. MCCORMICK AND LEMUEL D. WRIGHT

VOLUME 68. Recombinant DNA
*Edited by* RAY WU

VOLUME 69. Photosynthesis and Nitrogen Fixation (Part C)
*Edited by* ANTHONY SAN PIETRO

VOLUME 70. Immunochemical Techniques (Part A)
*Edited by* HELEN VAN VUNAKIS AND JOHN J. LANGONE

VOLUME 71. Lipids (Part C)
*Edited by* JOHN M. LOWENSTEIN

VOLUME 72. Lipids (Part D)
*Edited by* JOHN M. LOWENSTEIN

# Section I

# Biochemical, Molecular, and Real-Time Detection of Nitric Oxide

# [1]  The Preparation of Anaerobic Nitric Oxide Solutions for the Study of Heme Model Systems in Aqueous and Nonaqueous Media: Some Consequences of NO$_x$ Impurities

*By* MARK D. LIM, IVAN M. LORKOVIĆ, and PETER C. FORD

Abstract

The reactions of nitric oxide (NO) have been a subject of broad interest among biochemists and chemists. This report discusses several quantitative techniques to handle and use NO gas that has been delivered from compressed cylinders. The focus is on techniques that minimize and avoid the presence of oxygen impurities, which when present, result in the generation of other nitrogen oxides (NO$_x$). These NO$_x$ species typically exhibit different reactivity, and unless removed or quantified, their presence will complicate studies focusing on the reactions caused by NO itself.

Introduction

The chemistry and biochemistry of nitric oxide (NO, also known as *nitrogen monoxide*) has seen an enormous surge in interest over the past 15 years. This interest derives from the recognition that NO and its chemical derivatives have important roles in a number of functions in mammalian physiology and are generated endogenously by a cast of enzymes categorized as NO synthases (NOSs) (Ignarro *et al.*, 1987; Palmer *et al.*, 1987). In the laboratory, commonly there is the need to generate or add NO during *in vitro* experiments, and various techniques can be used. For example, certain NO donors are commercially available, and a widely used class of these are diazeniumdiolates, which are often referred to as "NONOates." An example is DEA/NO, the diethylamide adduct of NO with the chemical formula Na[Et$_2$NN(O)NO] (this topic has been reviewed by Hrabie and Keefer (2002) and Wang *et al.* (2002)). Another technique is to add NO to an experiment as a gas or as a solution aliquot in which gaseous NO has been dissolved. In such experiments, the common source is compressed NO gas available in tanks of various sizes from several commercial sources. The purpose of this chapter is to describe techniques that have been developed in this laboratory to purify and to deliver quantitatively NO gas to aqueous and nonaqueous systems. Our focus is the investigation of the thermodynamics and kinetics of biochemical and chemical reactions.

METHODS IN ENZYMOLOGY, VOL. 396                                          0076-6879/05 $35.00
Copyright 2005, Elsevier Inc. All rights reserved.                    DOI: 10.1016/S0076-6879(05)96001-1

Although many biochemical experiments with NO are carried out under ambient conditions (*i.e.*, in aerated media), it is important to ensure that the NO source is free of reactive $NO_x$ impurities. Such impurities are inherent to the compressed gas sources and can easily be generated during the handling of NO gas. It is for this reason that the somewhat laborious procedures described here were developed. Furthermore, the ability to deliver pure NO under deaerated conditions allows one to study the chemistry inherent to NO itself.

NO Impurities

*Disproportionation and Autoxidation*

When working with NO gas, one must account for the common $NO_x$ impurities that are derived from several reactions. Those found in NO sources and the kinetics of certain reactions relevant to their formation are summarized in Table I. One type of reaction leading to the formation of $NO_x$ is the "disproportionation" of NO [Eq. (1)], which leads to $N_2O$ (nitrous oxide) and $NO_2$ (nitrogen dioxide) products (Melia, 1965). The rate of this reaction is dependent on pressure and temperature and has been shown to be catalyzed by copper. Commercial NO gas sources (high-pressure tanks) commonly have $N_2O$ and $N_2O_3$ impurities resulting from such disproportionation [Eqs. (1) and (2)]. NO also reacts with dioxygen (autoxidation) to form $NO_2$ in the gas phase or aprotic medium [Eq. (3)], or $NO_2^-$ [nitrite, Eq. (4)] in aqueous medium (Ford *et al.*, 1993; Wink *et al.*, 1993). Although the precise mechanism of this reaction remains controversial, it is clear that reactive intermediates formed during NO autoxidation are powerful oxidizing and nitrosating agents. Nitrogen dioxide can further dimerize to form $N_2O_4$ [Eq. (5)] or, under conditions of excess NO, will form $N_2O_3$ [Eq. (2)]. The species formed in Eqs. (1)–(5) can further undergo disproportionation or hydrolysis reactions. For example, in acidic solution, nitrite ions disproportionate to NO and $NO_2$, via Eqs. (6) and (2). Thus, to understand the specific reactivity of NO itself, one must have a well-defined system in which all key parameters are understood.

$$3NO \rightleftharpoons N_2O + NO_2 \qquad (1)$$

$$NO + NO_2 \rightleftharpoons N_2O_3 \qquad (2)$$

$$2NO + O_2 \rightleftharpoons 2NO_2 \qquad (3)$$

$$4NO + O_2 + 2H_2O \rightleftharpoons 4NO_2^- + 4H^+ \qquad (4)$$

TABLE I
REACTION OF NO

| | Rate law | $k$ (T = 293 K) | References |
|---|---|---|---|
| $3NO \rightleftharpoons N_2O + NO_2$ | $-\dfrac{d[NO]}{dt} = k[NO][(NO)_2]$ | $7.2 \times 10^{-1} \ M^{-1} \ s^{-1}$ | Melia, 1965 |
| $2NO + O_2 \rightleftharpoons 2NO_2$ gas phase | $-\dfrac{d[NO]}{dt} = 2k[NO]^2[O_2]$ | $1.42 \times 10^4 \ M^{-2} \ s^{-1}$ | Hisatsune and Zafonte, 1969 |
| $2NO + O_2 \rightleftharpoons 2NO_2$ hydrophobic medium | $-\dfrac{d[NO]}{dt} = k[NO]^2[O_2]$ | $2.8 \times 10^6 \ M^{-2} \ s^{-1}$ | Nottingham and Sutter, 1986 |
| $4NO + O_2 + 2H_2O \rightleftharpoons 4NO_2^- + 4H^+$ | $-\dfrac{d[NO]}{dt} = 4k[NO]^2[O_2]$ | $2 \times 10^6 \ M^{-2} \ s^{-1}$ | Ford et al., 1993 |
| $NO + NO_2 \rightleftharpoons N_2O_3$ | $-\dfrac{d[NO]}{dt} = k[NO][NO_2]$ | $1.1 \times 10^9 \ M^{-1} \ s^{-1}$ | Grätzel et al., 1969 |
| $2NO_2 \rightleftharpoons N_2O_4$ | $-\dfrac{d[NO_2]}{dt} = k[NO_2]^2$ | $4.5 \times 10^8 \ M^{-1} \ s^{-1}$ | Grätzel et al., 1969 |

$$2NO_2 \rightleftharpoons N_2O_4 \qquad (5)$$

$$2NO_2^- + 2H^+ \rightarrow 2HNO_2 \rightarrow N_2O_3 + H_2O \qquad (6)$$

*Methods of Impurity Removal from Nitric Oxide Gas*

$NO_x$ impurities can be removed by taking advantage of their different boiling points and reactivity. As shown in Table II, NO has a boiling point at 121 K, whereas the other $NO_x$ have much higher boiling points. This allows for the selective removal of $NO_x$ simply by distilling NO from a cell into which NO and its impurities from a source have been condensed. The impurities $NO_2$, $N_2O_3$, and $N_2O_4$ can also be trapped out of NO gas streams by bubbling through an alkaline solution; the overall reactions are shown in Eqs. (7) and (8).

$$2NO_2 + 2OH^- \rightleftharpoons NO_2^- + NO_3^- + H_2O \qquad (7)$$

$$N_2O_3 + 2OH^- \rightleftharpoons 2NO_2^- + H_2O \qquad (8)$$

It is more difficult to remove the common impurity, $N_2O$, which is less reactive. This is best done by passing the NO gas stream through a column of activated silica at low temperature (196K) to absorb the $N_2O$.

*Example of Impurity Effects.* An important feature of $NO_x$ impurities is that they are more soluble in most solvents than is NO itself. Thus, whenever there is a gas–solution interface, these impurities will partition selectively into the solution, and the net effect of the impurities on the solution environment is considerably amplified. For example, the reaction of the heme model Fe(TPP)(NO) (**I**, TPP = tetraphenylporphyrin) with NO was reported to form Fe(TPP)(NO_2)(NO) (**II**), by an NO disproportionation pathway similar to Eq. (1) but mediated by the metal center (Yoshimura, 1984). However, in subsequent studies, where exceeding care was taken to remove the higher nitrogen oxides from the NO supply, it was shown that the ultraviolet (UV)–visible (vis) and infrared spectrum of **I** do not change when solutions of this material were exposed to NO at room

TABLE II
PROPERTIES OF $NO_x^a$

|         | Boiling points | Melting points |
|---------|----------------|----------------|
| NO      | 121 K          | 109 K          |
| $N_2O_3$ | 275 K          | 171 K          |
| $N_2O$   | 185 K          | 182 K          |
| $N_2O_4$ | 262 K          | 263 K          |

$^a$ From Laane and Ohlsen (1980).

temperature (Lorkovic and Ford, 2000a). A different product was seen when the temperature of this solution was lowered to 179 K. The bisnitrosyl $Fe(TPP)(NO)_2$ (**III**) formed in a reversible fashion [Eq. (9)] and was characterized spectroscopically (Lorkovic and Ford, 2000a). However, when traces of air were added to ambient temperature solutions of **I** under NO, the formation of **II** was rapid. Thus, it is clear that the $Fe(TPP)(NO_2)$ (NO) is formed from the presence of impurities in the gas source. Such impurities were probably the result of adventitious $O_2$ impurities leading to NO autoxidation and the formation of $N_2O_3$ [Eqs. (2) and (3)]. The reaction of $N_2O_3$ with **I** [Eq. (10)] has an equilibrium constant of 160 (toluene, 298 K) that was measured to quantify the system (Lorkovic and Ford, 2000b).

$$Fe(TPP)(NO) + NO \rightleftharpoons Fe(TPP)(NO)(NO) \tag{9}$$

$$Fe(TPP)(NO) + N_2O_3 \rightleftharpoons Fe(TPP)(NO_2)(NO) + NO \tag{10}$$

Mechanistic studies of model complexes may give insight toward the functions of higher nitrogen oxides that are exogenously or endogenously produced. Another $NO_x$ product of NO autoxidation is nitrite ion, $NO_2^-$, which is formed in aqueous solution [Eq. (4)] (Ford *et al.*, 1993), and recently nitrite has drawn considerable attention as a possible vasodilator (Cosby *et al.*, 2003; Gladwin *et al.*, 2000). In addition, a study of the reductive nitrosylation of ferriheme models and proteins by NO [Eq. (11)] demonstrated that $NO_2^-$ catalyzes this reaction with the possible intermediacy of $N_2O_3$, a powerful nitrosating agent (Fernandez and Ford, 2003; Fernandez *et al.*, 2003).

$$Fe(III)(P) + 2NO + H_2O \rightleftharpoons Fe(II)(P)(NO) + 2H^+ + NO_2^- \tag{11}$$

In the course of our mechanistic investigations of the reactions between NO and metal complexes, we have found it necessary to develop techniques to prepare NO solutions that are clean and well defined. This chapter addresses issues relevant to preparing such solutions from gaseous NO measured and delivered using vacuum line techniques that have been designed to minimize such impurities. In addition, we discuss methods for evaluating the presence of other $NO_x$ impurities.

## Method

### Preparing Solutions to Have a Specified NO Concentration

Preparing solutions containing well-defined concentrations of NO is straightforward but requires careful attention to details. The techniques described here were developed to meet our specific requirements but carry over to broader applications. By using the equipment available in a typical chemistry laboratory, pure and well-defined aqueous and nonaqueous solutions of NO may be prepared, if certain criteria are followed in the preparation. The basis for preparing solutions of dissolved gas follows Henry's Law, which states that the amount of gas dissolved in solution is proportional to the partial pressure of that gas. The handling and preparation of solutions of gases in liquids have been nicely reviewed and provide a good reference point for those working with such systems (Wilhelm *et al.*, 1977).

The solubility of NO in different solvents at various temperatures has been compiled by the International Union of Pure and Applied Chemistry (IUPAC) as part of its Solubility Data Series (Young, 1981). Table III summarizes the measured solubility of NO in various solvents at 298 K. The solubility data are commonly reported either in units of molar concentration per atmosphere of NO or as the Ostwald coefficient ($L$). The latter is a ratio that describes the headspace–liquid partitioning of the gas as a function of volume of the liquid, as shown in Eq. (12). This can be converted simply to molar concentration using the molar volume of NO (~24.5 L/mol at 298 K).

$$L = \frac{\text{Volume (gas absorbed)}}{\text{Volume (liquid)}} \qquad (12)$$

*Apparatus*

*The Vacuum Line and Purification Column.* As shown in Fig. 1, a cylinder of nitric oxide is connected by stainless-steel tubing and Swagelock fittings to a custom-made stainless-steel purification column (Fig. 2). It is important not to use copper tubing or connectors, as copper is known to catalyze the disproportionation reaction shown in Eq. (1). The purification column (No. 2 in Fig. 1, expanded in Fig. 2) is packed with Ascarite II

TABLE III
SOLUBILITY OF NO IN VARIOUS SOLVENTS[a]

| Solvent | Solubility at 298 K ($M$/atm) | Measured Ostwald coefficient[a] |
|---|---|---|
| Water | $1.8 \times 10^{-3}$ | $4.77 \times 10^{-2}$ (298 K) |
| Methanol | $14.5 \times 10^{-3}$ | $3.46–3.55 \times 10^{-1}$ (293.2 K) |
| Acetonitrile | $14.1 \times 10^{-3}$ | N.R.[b] |
| Carbon tetrachloride | $14.6 \times 10^{-3}$ | $3.38–3.55 \times 10^{-1}$ (292.8 K) |
| Toluene | $11.9 \times 10^{-3}$ | N.R. |

[a] From Young (1981).
[b] N.R., not reported.

(NaOH on a silicate carrier) that has been heated under vacuum at 400 K overnight. This column will remove $NO_2$, $N_2O_3$, and $N_2O_4$ from the NO stream at this point. The stainless-steel tubing is attached via a Viton O-ring seal to the greaseless vacuum line, or it can be streamed to a custom-made silica-packed column (No. 3 in Fig. 1) that has been activated by heating under vacuum at 400 K overnight. The silica column, when cooled to acetone/dry ice acetone temperatures (T = 196 K), can be used to trap the nitrous oxide impurity ($N_2O$) commonly formed in NO gas sources before vacuum line. The vacuum line (No. 4 in Fig. 1) is constantly maintained at reduced pressure. It is constructed with greaseless Teflon-to-glass stopcocks and is connected to a vacuum pump (No. 8 in Fig. 1) guarded by a liquid nitrogen trap (No. 7 in Fig. 1) on one end and to a manometer gauge (No. 5 in Fig. 1) on the other. All glassware is connected to the vacuum line with a greaseless Viton O-ring seal.

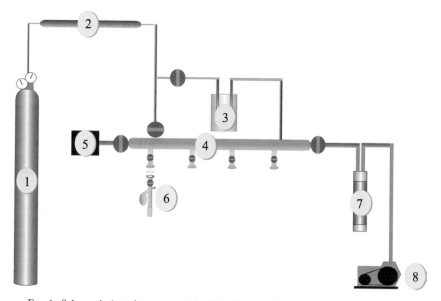

Fig. 1. Schematic (not drawn to scale) of the line used for the purification, measurement, and delivery of nitric oxide (NO). A commercially available cylinder of NO (1) is connected by stainless-steel tubing to an Ascarite containing column (2; see Fig. 2), which will remove $NO_2$, $N_2O_3$, and $N_2O_4$ from the supply gas. This is connected by stainless-steel tubing coupled with Swagelock fittings to the glass vacuum line (4), and delivery of the gas is regulated by a Teflon-to-glass stopcock. NO can also be diverted to a custom-made silica coil (3), which at 196 K will trap out $N_2O$. The all-glass vacuum line is composed of greaseless Teflon-to-glass stopcocks, a vacuum pump (8), and a liquid nitrogen trap (7). Pressures of NO can be measured directly in the line by an attached manometric gauge (5) and into the appropriate flask (6; see Fig. 3), which is connected by a greaseless O-ring (Viton) connection.

Fig. 2. Custom stainless-steel column used for scrubbing nitric oxide (NO) gas stream (No. 2 in Fig. 1). Stainless-steel tubing leading from the gas cylinder (a) is connected by high-pressure Swagelock connections to a Teflon-lined screw-top lid (b). Ascarite or sodium hydroxide pellets (c) are hand-packed into the column, which are supported by a stainless-steel frit (d). The packing has been activated and dried by heating to 400 K overnight under vacuum. This column is connected by stainless-steel tubing and Swagelock fittings to a T-joint (e), where the gas is controlled by needle valves to an additional silica coil for the removal of $N_2O$ (f) or sent directly to the Schlenk line (g).

*Glassware.* As shown in Fig. 3, reaction flasks are equipped with Teflon-to-glass stopcocks and have a bulb that can fit inside a cryogenic dewar for freezing the NO (b.p. = 121 K, m.p. = 109 K) using liquid nitrogen ($LN_2$). The flasks are connected to the vacuum line by greaseless O-ring joints.

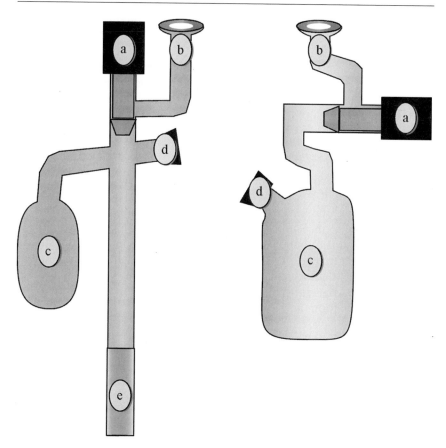

FIG. 3. Greaseless anaerobic glass flasks are used for handling nitric oxide (NO) solutions (No. 6 in Fig. 1). Basic components are a Chemglass Teflon-to-glass stopcock (a) to seal from the outside environment, an O-ring (Viton) fitting (b) for connection to the vacuum line, and a glass bulb (c) that can fit inside a liquid $N_2$ dewar for freezing in known $P_{NO}$. Additional accessories include an injection port (electrode adapter, d) sealed by a degassed three-layer rubber-Teflon-rubber septum for air titrations or solution injections, and a quartz cuvette connected by a quartz- to Pyrex-graded seal for spectrophotometric measurements and photolysis experiments (e).

Additional accessories include a four-sided UV–vis quartz cuvette that is permanently attached using a quartz- to Pyrex-graded seal to allow for spectrophotometric measurements or photolysis experiments. A small injection port (electrode adapter) sealed by a deoxygenated three-layer septum can also be attached for either mixing of solutions or titrating air (to account for $N_2O_3$ reactivity, as discussed later) into a flask containing

NO. It is important to deoxygenate the septum completely before use (storage in an inert glovebox or under vacuum) and to dispose of the septum after each experiment, especially once it has been punctured or exposed to organic solvents. A cell equipped with the electrode adapter should not be used for slow reactions with NO, because this is a potential source of leaks over long periods (hours). Techniques used for syringing solutions are described later.

### Transferring Known Quantities of NO

As already described, it is important to fully deoxygenate all solutions, cells, stopcocks, and septa before introducing NO. This is accomplished by storage under vacuum or in an inert atmosphere glovebox. Several methods can be used to measure a known pressure of NO into a reaction flask. The first method is to bubble NO directly through a solution from an NO line at a predetermined partial pressure, $P_{NO}$. The main disadvantage of this procedure is that because the higher nitrogen oxides are more soluble than NO itself, the entraining process will result in concentrating the impurities in solution, as volumes of NO much greater than those of the solution and headspace are needed to accomplish this task. This method is also wasteful of the NO gas and will require an additional step to contain and quench unreacted NO.

A second method is to add NO directly to an evacuated flask containing a solution sample that is being stirred or agitated to facilitate gas–liquid equilibration. The stopcocks leading to the flask are open to the line, and the NO partial pressure is measured directly for the volume of the line and cell connected to the manometer. The main disadvantages are the requirement to account for solvent vapor pressure of the solution and the need to contain and quench unreacted NO, once the gas–solution equilibration is complete. This may be the preferred method for those working with temperature-sensitive compounds such as proteins. To account for the vapor pressure of the solvent, one can use Eq. (13), where $[NO]_p$ is the concentration of dissolved NO, $[NO]_{sol}$ is the solubility of 1 atm of NO (from previous measured values), $P_{solvent}$ is the vapor pressure of the solvent, and P is the total pressure ($P_{tot} = P_{NO} + P_{solvent}$) of the system measured at the manometer.

$$[NO]_p = \left(\frac{[NO]_{sol}}{P_{bar} - P_{solvent}}\right)(P_{tot} - P_{solvent}), \tag{13}$$

where

$$P_{total} = P_{NO} + P_{solvent}$$

$P_{bar}$ = barometic pressure
$P_{solvent}$ = vapor pressure of the solvent
$P_{NO}$ = partial pressure of NO
$[NO]_p$ = concentration of NO dissolved in solvent
$[NO]_{sol}$ = solubility of 1 atm of NO over solvent

A third method is to condense a known quantity of NO directly into an evacuated flask at $LN_2$ temperatures. The stopcock is then closed after complete evacuation of the reaction flask, and the pressure of NO is measured over the line only. The reaction flask is frozen to $LN_2$ temperature (77 K), and the stopcock opened to condense the NO from the line to the flask.

A related procedure is to measure the $P_{NO}$ over the total line and small flask of known volume at ambient temperature. The cell is then closed off to the line by a stopcock, and the line evacuated. A reaction cell that has already been attached to the line and evacuated (but isolated by a stopcock) is cooled to $LN_2$ temperature. The stopcocks to both the cell and the flask are then opened, and the NO is distilled from the original flask to the final cell.

The quantitative accuracy of these methods based on condensing NO into the reaction flask depends on the accuracy of the $P_{NO}$ measurements and knowing the quantitative volumes of the relevant flasks and vacuum line. In addition, this procedure introduces a potential hazard: *It is extremely important not to condense the NO from an open line to the source or to condense NO from a larger volume at a certain value of $P_{NO}$ into a smaller volume, which when warmed to ambient temperature, cannot contain the resulting pressure. Under such conditions, especially if condensation occurs from an open NO line, the resulting excessive pressure in the reaction flask may create an explosion hazard!*

It may be necessary to transfer solutions of NO from one flask to another by syringe. All samples should be transferred using gas-tight syringes equipped with small-diameter non-coring needles and a valve to isolate the contents of the barrel from the outside atmosphere. The syringe barrel and needle should be deoxygenated by purging with an inert gas such as argon, and particular care should be taken to ensure that the septum on the flask properly reseals after being punctured.

*Other Precautions When Working with NO*

*Distillation of NO.* To minimize the effects of trace leaks that may have occurred during handling of NO gas on the vacuum line, one can use an additional vacuum distillation technique to remove $NO_x$ impurities. NO is frozen directly into an Ascarite-containing flask, which is connected to another anaerobic flask (or NMR tube) by a T-shaped glass connector that is sealed from the vacuum line. The Ascarite flask is then slowly warmed as NO is condensed to a second evacuated flask that has been cooled to $LN_2$ temperature.

*Accounting for Impurities.* As a control, it is important to account for the reactions caused by impurities. NO is colorless as a gas at room temperature and as a solid at $LN_2$ temperatures. Any color in the head-space or solution is a simple indicator of contamination. In non-protic medium, $N_2O_3$ has a broad UV absorption band (assigned as $\pi\pi$ transition) centered at 350 nm ($\varepsilon = 1060\ M^{-1}\ cm^{-1}$) and a weakly absorbing band at 700 nm. Such solutions appear green-blue in color. In aqueous systems, $N_2O_3$ will be quickly hydrolyzed to form $HNO_2$ [Eq. (14)], which will have sharp absorption bands at 336 ($\varepsilon = 17\ M^{-1}/cm^{-1}$), 346 (26), 358 (34), 371 (34), and 385 nm (19) (Fig. 4).

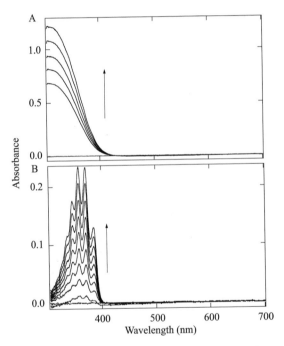

FIG. 4. Ultraviolet–visible (UV–vis) spectra tracking the changes that occur during the titration of air to a cell (shown in Fig. 3) containing nitric oxide (NO) at 273K. (A) The changes in the spectrum of a 5-ml toluene solution containing 6.3 m$M$ NO following an initial injection of 100 $\mu$l of air ($3.9 \times 10^{-4}\ M\ N_2O_3$), followed by 25-$\mu$l aliquots ($4.9 \times 10^{-4}$, $5.9 \times 10^{-4}$, $6.9 \times 10^{-4}$, and $7.9 \times 10^{-4}\ M\ N_2O_3$ total). (B) The changes in the spectrum of an unbuffered aqueous solution (5ml) containing 1.0m$M$ NO followed by 100-$\mu$l injections of air ($\Delta[NO_2^-] = 7.8 \times 10^{-4}\ M$ each time). The dominant $NO_x$ formed under these conditions is $HNO_2$

TABLE IV
EXAMPLES OF SOURCES OF PARTS AND EQUIPMENT MENTIONED IN THIS CHAPTER

| Item | Vendor | Special notes |
|------|--------|---------------|
| Gas cylinder (NO or $N_2O$) | Air Liquide | 99% purity |
| Stainless-Steel tubing and fittings | Swagelock | |
| Ascarite column | See Figure 2 | |
| Ascarite | Thomas Scientific | Heat under vacuum overnight before use |
| Silica column | Alltech Associates | Custom made |
| Vacuum line | Ace Glass | Glassblower required to convert stopcocks |
| Joint, O-ring seal | Ace Glass | Need clamp no. 7669 |
| Viton O-rings | Ace Glass | |
| Chem-Cap high-vacuum stopcocks | Chemglass | |
| Electrode adapter (injection port) | Ace Glass | |
| Three-layer septum for injection port | Alltech Associates | |
| UV–vis cuvette with Pyrex- to quartz-graded stem | NSG Precision Cells | |
| 1010 Series gas-tight syringes | Hamilton | With noncoring needles and valve |

$$N_2O_3 + H_2O \rightleftharpoons 2HNO_2 \qquad (14)$$

The simplest method to differentiate reactions involving NO from those involving $NO_x$ is to contaminate the experimental conditions deliberately. Using a sealed injection port, pure dioxygen or laboratory air can be titrated with a gas-tight syringe to an NO-containing reaction to account for $NO_2$, $N_2O_3$, and $N_2O_4$ in organic media, or for experiments in protic medium, nitrite salts can be added. The UV–vis spectra shown in Fig. 4 display the results of air titrations to toluene (spectrum A) and water (spectrum B) solutions containing NO. Commercially available $N_2O$ can also be added to a reaction mixture to account for its reactivity.

Sources of Equipment and Parts Described in This Chapter

Table IV contains a nonexclusive list of some sources from which equipment and parts described here are available.

## References

Cosby, K., Partovi, K. S., Crawford, J. H., Patel, R. P., Reiter, C. D., Martyr, S., Yang, B. K., Waclawiw, M. A., Zalos, G., Xu, X., Huang, K. T., Shields, H., Kim-Shapiro, D. B., Schechter, A. N., Cannon, R. O., and Gladwin, M. T. (2003). Nitrite reduction to nitric oxide by deoxyhemoglobin vasodilates the human circulation. *Nat. Med.* **9,** 1498–1505.

Fernandez, B. O., and Ford, P. C. (2003). Nitrite catalyzes ferriheme protein reductive nitrosylation. *J. Am. Chem. Soc.* **125,** 10510–10511.

Fernandez, B. O., Lorkovic, I. M., and Ford, P. C. (2003). Nitrite catalyzes reductive nitrosylation of the water-soluble ferri-heme model $Fe^{III}(TPPS)$ to $Fe^{II}(TPPS)(NO)$. *Inorg. Chem.* **42,** 2–4.

Ford, P. C., Wink, D. A., and Stanbury, D. M. (1993). Autoxidation of NO in aqueous nitric oxide. *FEBS Lett.* **326,** 1–3.

Grätzel, M., Henglein, A., Lilie, J., and Beck, G. (1969). Pulsradiolytische untersuchung einiger elementarprozesse der oxidation und reduktion des nitritions. *Ber. Bunsenges. Phys. Chem.* **73,** 646–653.

Gladwin, M. T., Shelhamer, J. H., Schechter, A. N., Pease-Fye, M. E., Waclawiw, M. A., Panza, J. A., Ognibene, F. P., and Cannon, R. O. (2000). Role of circulating nitrite and S-nitrosohemoglobin in the regulation of regional blood flow in humans. *Proc. Natl. Acad. Sci. USA* **97,** 11482–11487.

Hrabie, J. A., and Keefer, L. K. (2002). Chemistry of the nitric oxide–releasing diazeniumdiolate functional group and its oxygen-substituted derivatives. *Chem. Rev.* **102,** 1135–1154.

Hisatsune, I. C., and Zafonte, L. (1969). A kinetic study of some third-order reactions of nitric oxide. *J. Phys. Chem.* **73,** 2980–2989.

Ignarro, L. J., Buga, G. M., Wood, K. S., Byrns, R. E., and Chaudhuri, G. (1987). Endothelium-derived relaxation factor produced and released from artery and vein is nitric oxide. *Proc. Natl. Acad. Sci. USA* **84,** 9265–9269.

Laane, J., and Ohlsen, J. R. (1980). Characterization of nitrogen oxides by vibrational spectroscopy. *Prog. Inorg. Chem.* **27,** 465–513.

Lorkovic, I. M., and Ford, P. C. (2000a). Nitric Oxide Addition to the Ferrous Nitrosyl Porphyrins Fe(P)(NO) Gives *Trans*-Fe(P)(NO)$_2$ in Low-Temperature Solutions. *J. Am. Chem. Soc.* **122,** 6516–6517.

Lorkovic, I. M., and Ford, P. C. (2000b). Reactivity of the Iron Porphyrin Fe(TPP)(NO) with Excess NO. Formation of Fe(TPP)(NO)(NO$_2$) Occurs via Reaction with Trace NO$_2$. *Inorg. Chem.* **39,** 632–633.

Melia, T. P. (1965). Decomposition of nitric oxide at elevated pressures. *J. Inorg. Nucl. Chem.* **27,** 95–98.

Nottingham, W. C., and Sutter, J. R. (1986). Kinetics of the oxidation of nitric oxide by chlorine and oxygen in nonaqueous media. *Int. J. Chem. Kin.* **18,** 1289–1302.

Palmer, R. M. J., Ferrige, A. G., and Moncada, S. (1987). Nitric oxide release accounts for the biological activity of endothelium-derived relaxation factor. *Nature* **327,** 447–540.

Wang, P. G., Xian, M., Tang, X., Wu, X., Wen, Z., Cai, T., and Janczuk, A. J. (2002). Nitric Oxide Donors: Chemical Activities and Biological Applications. *Chem. Rev.* **102,** 1091–1134.

Wilhelm, E., Battino, R., and Wilcock, R. J. (1977). Low-pressure solubility of gases in liquid water. *Chem. Rev.* **77,** 219–262.

Wink, D. A., Darbyshire, J. F., Nims, R. W., Saavedra, J. E., and Ford, P. C. (1993). Reactions of the bioregulatory agent nitric oxide in oxygenated aqueous media: Determination of the kinetics for oxidation and nitrosation by intermediates generated in the nitric oxide/oxygen reaction. *Chem. Res. Toxicol.* **6,** 23–27.

Yoshimura, T. (1984). Reaction of nitrosylporphyrinatoiron (II) with nitrogen oxide. *Inorg. Chim. Acta* **83,** 17–21.

Young, C. L. (1981). Oxides of nitrogen. *In* "Solubility Data Series" (R. Battino, H. L. Clever, and C. L. Young, eds.) Vol. 8. Pergamon Press, New York.

## [2] The NOtizer—A Device for the Convenient Preparation of Diazen-1-ium-1,2-diolates

*By* JOERG KONTER, GAMAL EL-DIN ALI AHMED
HASSAN ABUO-RAHMA, ALI EL-EMAM, and JOCHEN LEHMANN

### Abstract

*N*-bound diazen-1-ium-1,2-diolates, also known as *NONOates* or "solid nitric oxide" (NO), have become popular tools in biomedical research since the discovery of NO as a very important multifunctional endogenous messenger. In contrast to other well-known NO donors, NONOates are capable of releasing NO spontaneously in aqueous media. The rate of NO liberation is determined by the molecular structure of the diazeniumdiolate and the pH value and temperature of the medium in which it is dissolved.

In this chapter, we introduce a novel device (the NOtizer) for simple and convenient preparation of diazeniumdiolates. It not only enables the user to provide all the necessary conditions for reliable synthesis such as anaerobic conditions and high pressure of NO gas in the translucent reaction chamber but also includes software that records the course of pressure and temperature online and calculates the consumption of NO by the reaction. The plot of the pressure decay shows the user completion of the reaction and allows the user to study kinetic characteristics from synthesis of different NONOates. A brief guide for the synthesis of PYRRO/NO, DEA/NO, PAPA/NO, SPER/NO, and DETA/NO, which are the most widely applied diazeniumdiolates, is presented in this chapter. Finally, characteristics of NONOates that need to be considered concerning analytics and storage are mentioned.

### Introduction

1-Substituted diazen-1-ium-1,2-diolates, also known as *NONOates,* are compounds of the general structure X-[N(O)NO]$^-$. According to the atom X, diazeniumdiolates can be classified in three major classes: *C*-, *O*-/*S*-, and *N*-bound diazeniumdiolates. Because the *N*-bound NONOates can directly release the endogenous mediator nitric oxide (NO), they are interesting

METHODS IN ENZYMOLOGY, VOL. 396                    0076-6879/05 $35.00
DOI: 10.1016/S0076-6879(05)96002-3

compounds from a biomedical point of view. One mole of diazeniumdiolate is capable of generating 2 mol of NO by acid-catalyzed dissociation according to Eq. (1) (Keefer *et al.*, 1996).

$$
\begin{array}{c}
R \\
\diagdown \\
\quad N-N^{+} \underset{\diagup}{\overset{\diagup N-O^-}{=}} \\
R \diagup \quad O^-
\end{array}
\xrightarrow{H^+}
\begin{array}{c}
R \\
\diagdown \\
\quad NH \\
R \diagup
\end{array}
+ \quad 2NO
\qquad (1)
$$

Organic nitrates are generally declared to be "NO donors," but in contrast to diazeniumdiolates, they need to be bioactivated. The corresponding bioactivation processes are not yet completely understood, and the resulting bioactive species are still under discussion (Chen *et al.*, 2002; Chung and Fung, 1990). Nitrosothiols [e.g., *S*-nitroso-*N*-acetyl-penicillamine (SNAP)] can be considered $NO^+$ donors rather than NO donors. In contrast to these compounds, *N*-diazeniumdiolates have proven to be excellent tools to demonstrate the physiological properties of NO *in vitro* and have gained importance as potential drugs for *in vivo* application. Innovative work on the development of novel NONOates and NONOate prodrugs, such as the liver-selective NO donor prodrug V-PYRRO/NO, has been performed by Saavedra *et al.* (1997) and others have reported on the development of thromboresistant NO-releasing polymeric coatings (Batchelor *et al.*, 2003; Keefer, 1998; Maragos *et al.*, 1991, 1993; Saavedra *et al.*, 1992, 1997, 2001; Woodward and Wintner, 1969; Zhang *et al.*, 2003).

The most useful preparation of *N*-diazeniumdiolates is the direct exposure of a secondary amine, acting as a nucleophile, to several atmospheres of gaseous NO under anaerobic conditions in a suitable solvent and added base to keep the product in a stable anionic form (Keefer and Hrabie, 2002; Longhi *et al.*, 1962). We now introduce a novel device (NOtizer) that allows the convenient preparation of diazeniumdiolates, the monitoring of NO consumption, temperature pattern, and thus kinetic studies on the NONOate formation.

## NONOates as NO Donors in Biological Systems

The rate of NO release is determined by the molecular structure of the respective NONOate, pH value, and temperature of the solution in which the compound is dissolved. Concerning pH, the decomposition of diazeniumdiolates proceeds extremely slowly at a pH value of more than 9, at a moderate rate at physiological pH values, and almost instantaneously at acetic pH values (Davies *et al.*, 2001; Keefer *et al.*, 1996). Depending on the structure of the parent nucleophile, the half-life of diazeniumdiolates ranges from a few seconds to several days, which grants access to versatile sources of NO adjustable to the requirements of the experiment (Lehmann *et al.*, 2002; Thomas *et al.*, 2002). The structures of the most intensively described diazeniumdiolates and their corresponding half-lives are given in Table I (Hrabie *et al.*, 1993; Keefer *et al.*, 1996; Maragos *et al.*, 1991; Saavedra *et al.*, 1997; Thomas *et al.*, 2002). Because the NO releasing rate of diazeniumdiolates may vary because of different conditions of the experiment, one should measure the NO release under study rather than rely on half-lives given in literature (Keefer *et al.*, 1996; Kröncke and Kolb-Bachofen, 1999).

TABLE I
STRUCTURE AND HALF-LIVES OF THE MOST ESTABLISHED NONOATES

| Name | Corresponding amine | Structure | Half-life (37°, pH 7.4) |
|---|---|---|---|
| PYRRO/NO | Pyrrolidine | | 3 s |
| DEA/NO | Diethylamine | | 2–4 min |
| PAPA/NO | N-propyl-1,3-propane-diamine | | 15 min |
| SPER/NO | Spermine | | 40 min |
| DETA/NO | Diethylene-triamine | | 20 h |

In addition to other advantages of diazeniumdiolates as direct NO donors ("solid NO"), NO production *in vitro* is not influenced by catalytic copper or iron ions, which guarantees the predictability and reproducibility of NO release. Besides delivering NO, NONOates themselves interact very little with other compounds in media, but nevertheless pharmacological investigations should go along with control experiments, using the amines resulting by cleavage of the respective diazeniumdiolates (Keefer *et al.*, 1996; Kröncke and Kolb-Bachofen, 1999; Thomas *et al.*, 2002). For example, the physiologically important polyamine spermine is the resulting nucleophile generated by dissociation of SPER/NO (Keefer *et al.*, 1996; Maragos *et al.*, 1993).

## Synthesis of Diazen-1-ium-1,2-diolates Using the NOtizer

In the 1960s, Drago and Paulik (1960) and Drago and Karstetter (1960) reported on the reaction of NO with anhydrous diethylamine dissolved in ether at a temperature of $-78°$ using a three-necked flask and passing NO through the reaction mixture. In the 1990s, when the role of endogenous NO was recognized, interest in NO donors increased and the synthesis of diazeniumdiolates was reinvestigated and improved, mainly by Keefer, Saavedra, Hrabie, and others. These authors used a glass bottle placed in a standard hydrogenation apparatus with a pressure of up to 5 atm NO at room temperature (Lehmann, 2000; Maragos *et al.*, 1991).

To shift the equilibrium toward the product and to keep the evolving diazeniumdiolate in the stable anionic form, the presence of base is necessary (Keefer *et al.*, 1996). Usually, equimolar quantities of sodium or potassium methylate are added. Dry methanol or mixtures of methanol and ether can be considered solvents of first choice. It is important to consider that ethanol (specifically ethoxide), under alkaline conditions, reacts with NO to yield undesired products by Traube's degeneration (Keefer *et al.*, 2001; Traube, 1898; Woodward and Wintner, 1969). A study by Arnold *et al.* (2002) shows that commonly used acetonitrile is not recommendable, because it reacts with NO, yielding a C-bound polydiazeniumdiolate as byproduct. Because oxygen or water will lead to the corresponding ammonium nitrite salts or even toxic nitrosamines, dry solvents and anaerobic conditions are obligatory (Bonner, 1996; Hansen *et al.*, 1982; Keefer, 1998; Masuda *et al.*, 2000; Srinivasan *et al.*, 2001).

Together with the Hyscho company, Bonn, Germany, we developed the NOtizer (Fig. 1), which is a modified hydrogenation apparatus.

As illustrated in Fig. 1, the NOtizer consists of a translucent pressure vessel as reaction chamber, stirrer, protective grating, regulator valves, indicators for pressure and temperature, and a computer with the

CPU for online monitoring

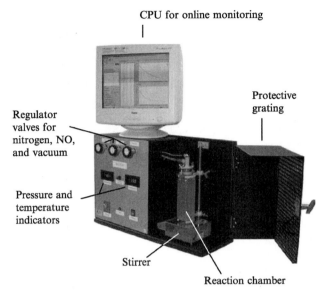

Regulator
valves for
nitrogen, NO,
and vacuum

Protective
grating

Pressure and
temperature
indicators

Stirrer

Reaction chamber

Fig. 1. The NOtizer for the preparation of diazeniumdiolates. (See color insert.)

appropriate software to record the data obtained. The computer can plot the numerical values of pressure and temperature changes and calculate the consumption of NO in the course of the synthesis. Particularly, the pressure decay curves are of interest, because they inform about the progression of the synthesis and can be fitted with appropriate software. The fit of these curves permits the user to gain information about the reaction rate, the kinetic rate constant, and the kinetic order of the reaction.

To demonstrate the benefits of the NOtizer, we have synthesized the diazeniumdiolates displayed in Table I. Exactly 0.05 mol of the respective amines was assorted with dry methanol to give a final volume of 100 ml. Before adding the solvent to the amine, 0.05 mol of elementary sodium was dissolved in the methanol to produce the necessary base. The vessel was placed in the NOtizer, evacuated, and filled with nitrogen four consecutive times, and then it was evacuated again and NO was administered to give a pressure of 4 bar. The recording of the reaction was ended when no more decrease of pressure occurred. All reactions were performed at room temperature, and the (either immediately or after concentration) precipitated diazeniumdiolates were obtained by suction filtration and washed with ether. To compare the reaction rate of the respective amines, one can superpose the pressure decay curves (Fig. 2).

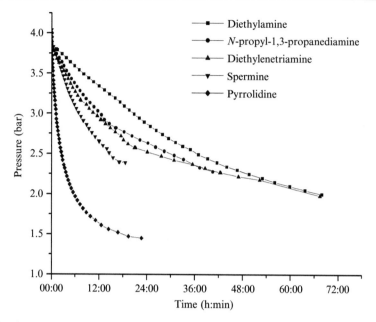

FIG. 2. Pressure decay curves for the reaction of diethylamine, N-propyl-1,3-propanedia-mine, diethylenetriamine, spermine, and pyrrolidine with nitric oxide.

The pressure decay curves in Fig. 2 show that steric and electronic effects have great influence on the reaction rate. For example, in reflection to the course of pressure decline, diethylamine shows a much slower reaction in comparison to pyrrolidine. These attained curves can be fitted with the appropriate computer software to reveal clues about the kinetic properties of the respective reaction.

It is noteworthy that if only formation of the product is desired, the addition of NO to give a pressure of 4–5 atm should be repeated until no appreciable amount of NO is consumed by the reaction, thus resulting in a higher yield. Figure 3 demonstrates how the consumption of NO decreases each time the pressure is raised back to 4.0 bar.

Because the software of the NOtizer also records the temperature pattern, we could demonstrate that formation of diazeniumdiolates is an exothermic reaction. Furthermore, it is possible to analyze the changes in reaction rate that occur with changes in temperature. Figure 4 shows the superposed pressure decay curves for the synthesis of PYRRO/NO at room temperature and under cooled conditions to point out this aspect.

As is expected from the collision theory, under ice-cold conditions the reaction rate slows and the yield of the product is slightly lower. It is

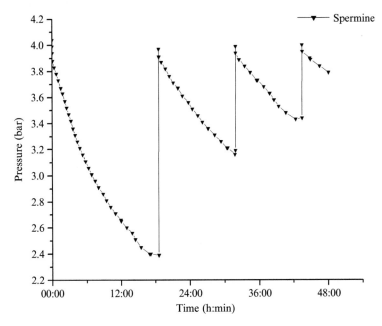

FIG. 3. Pressure decay curve for the synthesis of SPER/NO when inflowing NO repeatedly.

noteworthy that no improvement in quality or purity was attained at low temperature.

### Characteristics of Diazeniumdiolates to be Considered

To confirm quality and purity of the synthesized NONOates, nuclear magnetic resonance (NMR) spectroscopy and elemental analysis are the most reliable techniques. A suitable solvent for the NMR spectroscopic analysis is $D_2O$ with NaOD for alkaline conditions in which the compounds are kept in their stable anionic form. Using [1]H-NMR spectroscopy, the signal of the protons bound to the alpha-C-atoms shows a downfield shift usually about 0.4–0.5 ppm compared to the amine. Another criterion for the quality of the formed product is, of course, that NO is released by the compound when dissolved in aqueous media. A commercially available NO-selective sensor can be used to determine the amount of NO being released. If the compound is pure, 1 mol of NONOate will generate 2 mol of NO. Because of the volatility of NO from solution, this method is not very convincing for the evaluation of purity (Thomas *et al.*, 2002). A superior method for quantitative acquisition of NO from solution is laser magnetic

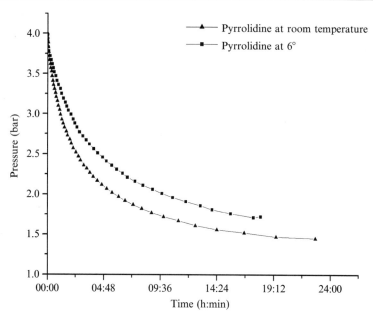

FIG. 4. Pressure decay curves when synthesizing PYRRO/NO at room temperature and under cooling.

resonance spectroscopy (LMRS) because NO is detected in its gaseous form and accumulatively over longer periods (Lehmann *et al.*, 2002).

Concerning the storage of diazeniumdiolates, Keefer *et al.* (1996) recommend storage in a refrigerator at 4° under dry nitrogen or argon atmosphere protected from moisture. Initial indications for the decomposition are agglutination and especially change in color.

## References

Arnold, E. V., Citro, M. L., Keefer, L. K., and Hrabie, J. A. (2002). A nitric oxide–releasing polydiazeniumdiolate derived from acetonitrile. *Org. Lett.* **4**(8), 1323–1325.

Batchelor, M. M., Reoma, S. L., Fleser, P. S., Nuthakki, V. K., Callahan, R. E., Shanley, C. J., Politis, J. K., Elmore, J., Merz, S. I., and Meyerhoff, M. E. (2003). More lipophilic dialkyldiamine-based diazeniumdiolates: Synthesis, characterization, and application in preparing thromboresistant nitric oxide release polymeric coatings. *J. Med. Chem.* **46,** 5153–5161.

Bonner, F. T. (1996). Nitric oxide gas. *Methods Enzymol.* **268,** 50–57.

Chen, Z., Zang, J., and Stamler, J. S. (2002). Identification of the enzymatic mechanism of the nitroglycerin bioactivation. *Proc. Natl. Acad. Sci. USA* **99,** 8306–8311.

Chung, S. J., and Fung, H. L. (1990). Identification of the subcellular site for nitroglycerin metabolism to nitric oxide in bovine coronary smooth muscle cells. *J. Pharmacol. Exp. Ther.* **253,** 614–619.

Davies, K. M., Wink, D. A., Saavedra, J. E., and Keefer, L. K. (2001). Chemistry of the diazeniumdiolates, 2: Kinetics and mechanism of dissociation to nitric oxide in aqueous solution. *J. Am. Chem. Soc.* **123,** 5473–5481.

Drago, R. S., and Karstetter, B. R. (1960). The reactions of nitrogen(II) oxide with various primary and secondary amines. *J. Am. Chem. Soc.* **83,** 1819–1822.

Drago, R. S., and Paulik, F. E. (1960). The reaction of nitrogen(II) oxide with diethylamine. *J. Am. Chem. Soc.* **82,** 96–98.

Hansen, T. J., Croisy, A. F., and Keefer, L. K. (1982). N-nitrosation of secondary amines by nitric oxide via the "Drago complex." *IARC Sci. Publ.* **41,** 21–29.

Hrabie, J. A., Klose, J. R., Wink, D. A., and Keefer, L. K. (1993). New nitric oxide–releasing zwitterions derived from polyamines. *J. Org. Chem.* **58,** 1472–1476.

Keefer, L. K. (1998). Nitric oxide-releasing compounds: From basic research to promising drugs. *Chemtech.* **28**(8), 30–35.

Keefer, L. K., Flippen-Anderson, J. L., George, C., Shanklin, A. P., Dunams, T. M., Christodoulou, D., Saavedra, J. E., Sagan, E. S., and Bohle, D. S. (2001). Chemistry of the diazeniumdiolates, I: Structural and spectral characteristics of the [N(O)NO]-functional group. *Nitric Oxide* **5**(4), 377–394.

Keefer, L. K., and Hrabie, J. A. (2002). Chemistry of the nitric oxide–releasing diazeniumdiolate ("nitrosohydroxylamine") functional group and its oxygen-substituted derivatives. *Chem. Rev.* **102,** 1135–1154.

Keefer, L. K., Nims, R. W., Davies, K. M., and Wink, D. A. (1996). "NONOates" (1-substituted diazen-1-ium-1,2-diolates) as nitric oxide donors: Convenient nitric oxide dosage forms. *Methods Enzymol.* **268,** 281–293.

Kröncke, K. D., and Kolb-Bachofen, V. (1999). Measurement of nitric oxide–mediated effects on zinc homeostasis and zinc finger transcription factors. *Methods Enzymol.* **301,** 126–135.

Lehmann, J. (2000). Nitric oxide donors—Current trends in therapeutic applications. *Expert Opin. Ther. Pat.* **10**(5), 559–574.

Lehmann, J., Horstmann, A., Menzel, L., Gäbler, R., Jentsch, A., and Urban, W. (2002). Release of nitric oxide from novel diazeniumdiolates monitored by laser magnetic resonance spectroscopy. *Nitric Oxide Biol. Ch.* **6**(2), 135–141.

Longhi, R., Ragsdale, R. O., and Drago, R. S. (1962). Reactions of nitrogen(II) oxide with miscellaneous Lewis bases. *Inorg. Chem.* **1,** 768–770.

Maragos, C. M., Morley, D., Wink, D. A., Dunams, T. M., Saavedra, J. E., Hoffman, A., Bove, A. A., Isaac, L., Hrabie, J. A., and Keefer, L. K. (1991). Complexes of NO with nucleophiles as agents for the controlled biological release of nitric oxide. Vasorelaxant effects. *J. Med. Chem.* **34**(11), 3242–3247.

Maragos, C. M., Wang, J. M., Hrabie, J. A., Oppenheim, J. J., and Keefer, L. K. (1993). Nitric oxide/nucleophile complexes inhibit the *in vitro* proliferation of A375 melanoma cells via nitric oxide release. *Cancer Res.* **53,** 564–568.

Masuda, M., Mower, H. F., Pignatelli, B., Celan, I., Friesen, M. D., Nishino, H., and Ohshima, H. (2000). Formation of N-nitrosamines and N-nitramines by the reaction of secondary amines with peroxynitrite and other reactive nitrogen species: Comparison with nitrotyrosine formation. *Chem. Res. Toxicol.* **13**(4), 301–308.

Saavedra, J. E., Billiar, T. R., Williams, D. L., Kim, Y.-M., Watkins, S. C., and Keefer, L. K. (1997). Targeting nitric oxide (NO) delivery *in vivo*. Design of a liver-selective NO donor prodrug that blocks tumor necrosis factor-α–induced apoptosis and toxicity in the liver. *J. Med. Chem.* **40,** 1947–1954.

Saavedra, J. E., Dunams, T. M., Flippen-Anderson, J. L., and Keefer, L. K. (1992). Secondary amine/nitric oxide complex ions, R2N[N(O)NO]-. O-Functionalization chemistry. *J. Org. Chem.* **57**(23), 6134–6138.

Saavedra, J. E., Srinivasan, A., Bonifant, C. L., Chu, J., Shanklin, A. P., Flippen-Anderson, J. L., Rice, W. G., Turpin, J. A., Davies, K. M., and Keefer, L. K. (2001). The secondary amine/nitric oxide complex ion R(2)N[N(O)NO](−) as nucleophile and leaving group in S$_N$Ar reactions. *J. Org. Chem.* **66**(9), 3090–3098.

Srinivasan, A., Kebede, N., Saavedra, J. E., Nikolaitchik, A. V., Brady, D. A., Yourd, E., Davies, K. M., Keefer, L. K., and Toscano, J. P. (2001). Chemistry of the diazeniumdiolates, 3: Photoreactivity. *J. Am. Chem. Soc.* **123**(23), 5465–5472.

Thomas, D. D., Miranda, K. M., Espey, M. G., Citrin, D., Jourd'Heuil, D., Paolocci, N., Hewett, S. J., Colton, C. A., Grisham, M. B., Feelisch, M., and Wink, D. A. (2002). Guide for the use of nitric oxide (NO) donors as probes of the chemistry of NO and related redox species in biological systems. *Methods Enzymol.* **359**, 84–105.

Traube, W. (1898). Über synthesen stickstoffhaltiger verbindungen mit hilfe des stickoxyds. *Liebigs Ann. Chem.* **300**, 81–128.

Woodward, R. B., and Wintner, C. (1969). The methoxazonyl group. *Tetrahedron Lett.* **32**, 2689–2692.

Zhang, H., Annich, G. M., Miskulin, J., Stankiewicz, K., Osterholzer, K., Merz, S. I., Bartlett, R. H., and Meyerhoff, M. E. (2003). Nitric oxide–releasing fumed silica particles: Synthesis, characterization, and biomedical application. *J. Am. Chem. Soc.* **125**(17), 5015–5024.

# [3] Quantum Mechanical Determinations of Reaction Mechanisms, Acid Base, and Redox Properties of Nitrogen Oxides and Their Donors

*By* ANDREW S. DUTTON, JON M. FUKUTO, and K. N. HOUK

## Abstract

This chapter reviews computational methods based on quantum mechanics and commonly used commercial programs for the exploration of chemical phenomena, particularly in the field of nitrogen oxides. Examples from the literature are then used to demonstrate the application of these methods to the chemistry and biochemistry of various nitrogen oxides. These examples include determining reaction mechanisms using computed reaction energies, predicting rates of reactions using transition state theory, and determining chemical properties such as hydration equilibria, pK$_a$s, and reduction potentials.

## Introduction

Nitrogen oxides have a rich history in the study of atmospheric and biological chemistry. The studies of Paul Crutzen and others demonstrate that nitrogen oxides could catalyze the decomposition of ozone. This work led to the Nobel Prize in Chemistry for Crutzen in 1995. Three years later,

METHODS IN ENZYMOLOGY, VOL. 396
Copyright 2005, Elsevier Inc. All rights reserved.

0076-6879/05 $35.00
DOI: 10.1016/S0076-6879(05)96003-5

research on nitrogen oxides, specifically nitric oxide (NO), again received the attention of the Royal Swedish Academy of Sciences. The discovery that the simple gaseous molecule, NO, was a signaling molecule in the cardiovascular system resulted in the Nobel Prize in Physiology or Medicine in 1998 to Louis Ignarro, Robert F. Furchgott, and Ferid Murad.

These discoveries demonstrate the importance of the chemistry and biology of nitrogen oxides. The nitrogen oxide nitroxyl (HNO) has garnered interest in both chemistry and biology (Fukuto et al., 2004). Nitrogen oxides are known for the many reactions they undergo, often involving fleeting intermediates that are difficult to study experimentally. Both NO and HNO can be quantified experimentally and are sometimes measured by detecting stable end products such as nitrate ($NO_3^-$), or nitrite ($NO_2^-$), or in the case of HNO, $N_2O$ (Grisham et al., 1996). The rate constants for dimerization of HNO and reactions of HNO with various agents were determined using competition studies (Miranda et al., 2003). The rate for HNO dimerization followed by dehydration to yield water and $N_2O$ is very rapid, on the order of $10^6\ M^{-1}\ s^{-1}$ (Miranda et al., 2003; Shafirovich and Lymar, 2002). A detailed understanding of the chemistry of these reactive intermediates is required in the exploration of the biological significance of various nitrogen oxides. A practical approach is to use theoretical methods to predict chemical properties. Computational chemistry allows for an exhaustive study of the pathways and properties that are currently impossible or very difficult to study experimentally. This chapter summarizes the state of the art of computational chemistry for the study of acid–base properties of nitroxyl, the reaction mechanism of various nitrogen oxides, and the redox proprieties of NO and HNO.

## Computational Methodology

A brief description of the available computational models and their accuracies is presented here. This is not a thorough review on the subject but serves as a guide to the basic molecular quantum mechanical (QM) methods and the computational programs used for such calculations. There are many introductions to this subject (Cramer, 2002; Goodman, 1998; Leach, 2001; Young, 2001).

QM methods can be very accurate, to within ±1 kcal/mol of experimental energies, but to achieve this level of accuracy, significant computational resources are required. All QM methods are approximations to the exact solution of the Schrödinger equation, which can only be solved analytically for the hydrogen atom. Nevertheless, methods have been developed to solve the Schrödinger equation by iterative methods to any accuracy desired. These methods can be broken down into groups based on the approximations that are used: semi-empirical methods [currently, Austin model 1 (AM1) and parameterized method 3 (PM3) are most frequently used], *ab initio* methods [Hartree-Fock (HF), configuration interaction (CI), Møller-Plesset (MPn), and multiconfigurational self-consistent field (MCSCF) methods], density functional methods (*e.g.*, B3LYP, BPW91, and MPW1K), and extrapolative methods (*e.g.*, CBS, Gn, and Wn).

The simplest and fastest QM methods are the semi-empirical methods. Semi-empirical methods use numerical approximations to the values of some of the complex integrals found in quantum mechanics. Various parameters are determined by fitting the parameters so that the computed energies and geometries of a test set of molecules will reproduce experimental data. Semi-empirical methods differ by which integrals are approximated or neglected, as well as the choice of the experimental data used for parameterization. The most commonly used semi-empirical methods are AM1 and PM3. AM1 was developed by Dewar *et al.* (1985) as an advance over an already existing modified neglect of diatomic overlap (MNDO) method. Stewart (1989) developed a very similar method and used a larger range of molecules for parameterization; he called this method parameterized method 3 (PM3). Repasky *et al.* (2002) have reported an improvement of the PM3 and MNDO methods, called PDDG/PM3 and PDDG/MNDO.

*Ab initio* ("from the beginning") methods use the principles of quantum mechanics without any empirical parameters. These methods are generally the most accurate but require significant computational resources. The HF theory (Roothaan, 1951) is the basic *ab initio* method; it is essentially molecular orbital (MO) theory with no parameters. However, it may not be highly accurate, for two reasons. (1) Electron–electron repulsion is treated as the repulsion between average positions of the electrons, rather than the instantaneous positions. This is called the *neglect of correlation,* because electron motions are in reality correlated. This approximation allows the separation of the multi-electron wavefunction into individual one-electron wavefunctions that provide an averaged electron repulsion. (2) All QM calculations involve a second approximation, the use of relatively simple mathematical functions centered on atoms (although plane waves are sometimes used) to represent atomic orbitals. This set of functions is known as the *basis set.* Simple exponential or Gaussian functions are not sufficiently flexible to give correct electron distributions; a large number of functions with many different scale factors and exponents need to be used to achieve the actual electron distribution in molecules. Nowadays, a linear combination of Gaussian functions is usually used to approximate the Slater functions that closely resemble hydrogenic orbitals. The determination of the orbital coefficients involves an iterative process used to achieve a self-consistent field (SCF) wavefunction. The HF theory gives good results for geometries and for reaction energies when the change in electron correlation energy is small. The HF theory is a variational method: HF theory–computed energies are always greater than the actual energy. By increasing the basis set size and extrapolating to a complete basis set, the HF limit is achieved. The difference between the exact energy of a system and the HF limit is defined as the correlation energy.

The MPn perturbation theory, where $n$ corresponds to the order of correction (Møller and Plesset, 1934), is a method that corrects the HF energy to include correlation energy. Because this method is not a variational method, it is possible to overestimate the correlation energy, which sometimes happens with the most commonly used perturbation method, MP2. MP3 and MP4 calculations are generally quite accurate, but the computational cost is very high, and these are not used routinely.

CI methods involve a linear combination of HF wavefunctions with different orbital occupations. CI calculations are named according to the type of electronic excitations used in the calculation; for example, CI singles (CISs) calculations involve all possible single excitations from the lowest energy configuration (all orbitals doubly occupied or vacant, and filled according to the Aufbau principle); CISD contains all single- and double-electron excitations. CISDT calculations are possible, but the triple's contribution is usually estimated perturbatively, leading to a method called *CISD(T)*. The computer time required for CI calculations is large and is possible only on relatively small systems.

MCSCF calculations involve simultaneous optimization of orbital coefficients and configuration coefficients of CI calculations. This method is more complex than CI methods, because the computational chemist has to choose which orbital configurations will be used. The orbitals for which different occupations are allowed constitute the active space. Calculations that allow for all possible combinations of electronic configurations in the active space are called complete active space SCF (CASSCF) calculations. These calculations are computationally quite expensive but yield accurate results for open-shell systems, that is, systems with partially occupied orbitals or unpaired electrons.

The density functional theory (DFT) (Hohenberg and Kohn, 1964; Kohn and Sham, 1965) builds on the theorems of Hohenberg, Kohn, and Sham, which show that the energy of a molecule can be determined from the electron density. Walter Kohn shared the Nobel Prize in Chemistry, with John Pople, in 1998, for the development of computational methods and programs. DFT methods include electron correlation energy directly. However, the exact form of the functional that relates energy to electron density is not known, and many approaches have been taken in an attempt to find an accurate functional. Gradient-corrected methods use gradients of the electron density, and hybrid methods use both DFT and HF exact exchange. The hybrid methods [*e.g.*, B3LYP (Becke, 1993), BPW91 (Perdew *et al.*, 1996), and MPW1K (Adamo and Barone, 1998)] have been used extensively for organic and inorganic systems. The letters generally refer to the initials of the creators of the functional (B = Becke, P = Perdew, etc.).

The *ab initio* and DFT methods described previously all require the computational chemist to specify the basis set. In general, the accuracy of a calculation increases with an increasing size of the basis set used. The total energy of a molecule reaches an asymptote upon systematic increases of the basis set size. This has led to the development of highly accurate extrapolative methods that systematically increase the level of correlation energy or the basis set size to extrapolate to solutions that would be equivalent to full CI with a complete basis set. These extrapolative methods [CBS (Montgomery *et al.*, 1994), Gn (Curtiss *et al.*, 1991, 1998; Pople *et al.*, 1989), and Wn (Martin and de Oliveira, 1999), where *n* corresponds to the version of the method] provide very accurate energies, usually with errors less than 1 kcal/mol when compared to experimental data. An error of 1 kcal/mol is considered equal to the error of most experimentally determined values of energies such as heats of atomization or heats of formation.

These methods are readily available in software packages that have now been adapted to run on most computers. Software packages run the gamut from freely downloadable software involving a few methods to software that has all of the methods described earlier plus graphical interfaces and extensive wavefunction analysis capabilities. These can be purchased from a number of commercial vendors. The emphasis of these programs is on molecular mechanics, semi-empirical methods, and various types of *ab initio* and DFT methods. The principal uses of these programs are for the computation of an optimized geometry of a molecule or of all the conformations of a molecule, computing the energies of specific molecular geometries, computing molecular frequencies, optimizing transition state structures and energies, and computing molecular properties such as nuclear magnetic resonance (NMR) and infrared (IR) spectra. The computed energies can be electronic energies (E) for hypothetical fixed geometries, or by using the frequency data, the enthalpy of formation (H), the entropy (S), and the Gibbs free energy (G). Rate constants for a reaction can be computed from computed activated free energies using transition state theory.

Programs containing a large variety of high-level methods, such as Gaussian (Frisch *et al.*, 1998, 2003), Jaguar (Schrodinger, Inc., Portland,

OR), NWChem (Straatsma *et al.*, 2004), and QChem (Kong *et al.*, 2000), are generally used by computational chemists, and SPARTAN (Wavefunction, Inc., 2004) and HyperChem (Hypercube, Inc., Gainesville, FL), have been especially useful as teaching aides. These programs include the current state-of-the-art methodologies and solvation models, molecular dynamics, and other advanced functions. These programs are routinely upgraded as newer and better computational strategies are reported. These programs will compute all of the properties discussed in the previous paragraph, as well as electronically excited energy state geometries, energies, and properties.

Most of the work in our laboratories on nitrogen oxide chemistry involves a DFT method, B3LYP, and the extrapolative method, CBS-QB3, computed with Gaussian 98 (Frisch *et al.*, 1998) or 03 (Frisch *et al.*, 2003).

## Quantum Mechanical Determinations of Reaction Mechanism

The usual strategy for determining reaction mechanisms computationally involves the exploration of the potential energy surfaces for potential mechanisms and, using the tenets of transition state theory, finding the transition states that give the lowest energy process. Transition states are stationary points with one negative force constant corresponding to motion along the reaction coordinate. QM calculations can be used to optimize geometries and energies, enthalpies, and free energies associated with the reactants, transition states, and products. Because only a limited number of structures considered by the theoretical chemist are computed and are used to determine the reaction mechanism, a mechanism may be overlooked inadvertently. If a transition structure that is not considered has a lower energy than the transition state of the proposed mechanism, the prediction will be in error. As many plausible structures as possible must be computed to ensure that no low-energy mechanism has been overlooked.

In the following section, examples are presented to illustrate the QM methods that have been used to determine reaction mechanisms of nitrogen oxides and their precursors.

## N-*Hydroxyguanidine Oxidative Decomposition*

*N*-Hydroxy-L-arginine (NOHA) is an intermediate in the enzymatic production of NO from L-arginine and possesses the *N*-hydroxyguanidine functional group. It has been reported that NOHA is capable of escaping from the enzyme, NO synthase (NOS) (Hecker *et al.*, 1995; Wingand *et al.*, 1997). Cho *et al.* (2003) studied the oxidation of *N*-hydroxyguanidines to understand the biological chemistry and decomposition of NOHA.

They found that the *N*-hydroxyguanidine functional group readily reduced Cu(II) to Cu(I) and yielded an iminoxyl intermediate. Cho *et al.* (2003) then proposed that the iminoxyl radical has a finite lifetime in solution and can either react with added NO, dimerize, or be further oxidized by a second Cu(II). Finally, these products decompose to the resulting cyanamide, *N*-nitrosoguanidine, and $N_2O$ (Cho *et al.*, 2003). They proposed a mechanism to account for the formation of the observed products, noting that the intermediates could not be detected. Their mechanism was then tested theoretically (the results are shown in Fig. 1). Computations corroborated the proposed mechanism, as all steps are thermodynamically feasible. This suggests that these are reasonable reactions and the oxidative chemistry of the *N*-hydroxyguanidine functional group should be considered biologically relevant in systems where NOHA is being generated.

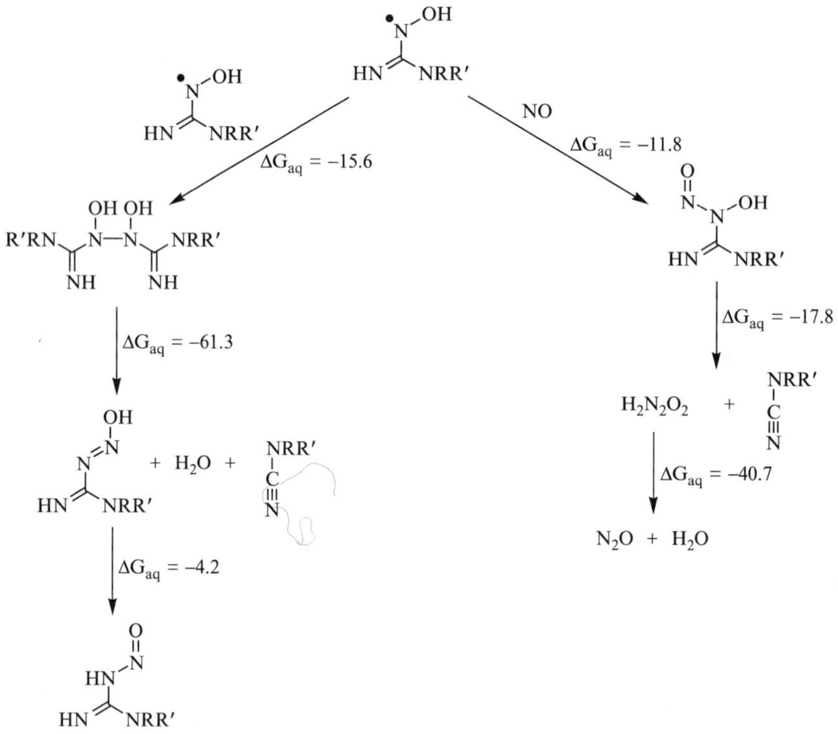

FIG. 1. The free energy changes for the proposed reactions of the iminoxyl intermediate; all energies are in kcal/mol. [CBS-QB3, PCM (B3LYP/6-311 + G[d])] S—N bond dissociation energies of *S*-nitrosothiols and transnitrosation reaction energies.

### S–N Bond Dissociation Energies of S-Nitrosothiols and Transnitrosation Reaction Energies

The stabilities of S-nitrosothiols (R-SNOs) and the mechanisms of decomposition are important for understanding the roles of these species as NO carriers and as therapeutic agents (Bartgerger et al., 2001a). The reported stabilities of R-SNOs vary greatly, with primary and secondary R-SNOs being generally unstable and tertiary R-SNOs being stable (Cranahan et al., 1978; Roy et al., 1994). Bartberger et al. (2001a) used the highly accurate CBS-QB3 method to compute the bond dissociation enthalpies (BDEs) and the free energy barriers for dissociation of a representative set of R-SNOs to test whether the mechanism of R-SNO decomposition involved simple S–N bond homolysis. The theoretical work was compared with experimentally determined BDEs (Bartberger et al., 2001a). They found that the thermal homolysis energy of the S–N bond is quite high, approximately 30 kcal/mol, which makes the process improbable under physiological conditions. The computed BDEs of alkyl nitrosothiols have a narrow range, 31–32 kcal/mol, and these values are nearly independent of alkyl substitution (Bartberger et al., 2001a). The computed free energy of activation for $CH_3SNO$ SN homolysis is 29.5 kcal/mol, which agrees reasonably well with the measured activation-free energies of 26–27 kcal/mol for various R-SNOs, when R is a saturated alkyl group (Bartberger et al., 2001a).

R-SNOs can also act as biological signaling agents through transnitrosation reactions [Eq. (1)]. This process, which formally transfers an $NO^+$ equivalent from one thiol to another thiol, can selectively modify protein function.

$$R\text{-}SH + R'\text{-}SNO \longrightarrow R\text{-}SNO + R'\text{-}SH \tag{1}$$

Experimental evidence suggests that this process proceeds through nucleophilic attack of a thiol/thiolate on the nitrogen of the R-SNO (Houk et al., 2003). Houk et al. (2003) used the simple model transnitrosation of S-nitrosomethane with methane thiolate to determine the energetics and mechanism of Eq. (1) theoretically. The CBS-QB3 method along with the PCM solvation model was used to determine the reaction energetics (Fig. 2).

The mechanism for transnitrosation involves the formation of an interesting intermediate, the nitroxyl disulfide, which lies 15.2 kcal/mol above the starting materials in water (Houk et al., 2003). The reaction barrier to form the nitroxyl disulfide in water was computed to be 17.4 kcal/mol. Mass spectroscopy (MS) was used to verify experimentally the existence of the nitroxyl disulfide (Houk et al., 2003). Analysis of the electrospray mass

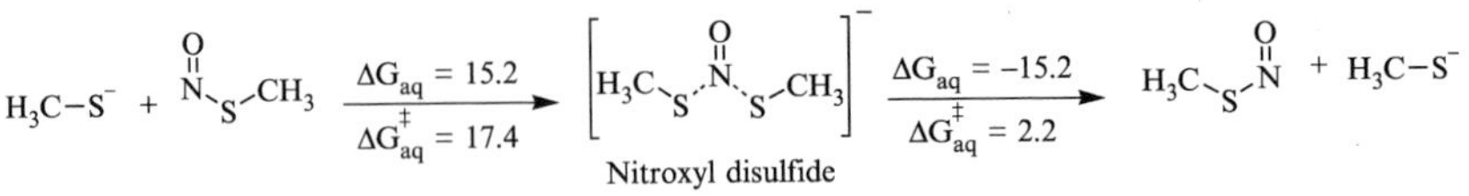

$$H_3C-S^- + \overset{O}{\underset{\underset{S}{\|}}{N}}\diagdown CH_3 \quad \xrightarrow[\Delta G^{\ddagger}_{aq} = 17.4]{\Delta G_{aq} = 15.2} \quad \left[ H_3C\diagdown_S\cdot\overset{O}{\overset{\|}{N}}\diagdown_S\diagdown CH_3 \right]^- \quad \xrightarrow[\Delta G^{\ddagger}_{aq} = 2.2]{\Delta G_{aq} = -15.2} \quad H_3C\diagdown_S\diagdown\overset{O}{\overset{\|}{N}} + H_3C-S^-$$

Nitroxyl disulfide

FIG. 2. Computed mechanism and energetics for transnitrosation via the intermediate nitroxyl disulfide. The changes in free energy are reported in kcal/mol (from Houk et al., 2003). [CBS-QB3, PCM B3LYP/6-311 + G(d)].

spectra of $S$-nitroso-$N$-acetyl-penicillamine (SNAP) in acetonitrile provided a strong peak at $m/e$ of 410. This mass-to-charge ratio matches the expected value for the monoanionic nitroxyl disulfide. Further MS/MS analysis of the $m/e$ 410 ion revealed two peaks at an $m/e$ of 219 and 190 corresponding to SNAP and $N$-acetylpenicillamine carboxylate, respectively.

## Comparing the Reactivities of Peroxynitrate and Peroxynitrite

Olsen et al. (2003) investigated the reactivity of the important nitrogen oxides peroxynitrate (PNA) and peroxynitrite (PNI). The CBS-QB3 method was used to investigate the propensities of O-atom transfer from PNA and PNI to $NH_3$, $H_2S$, and $H_2CCH_2$, as well as the BDEs for PNA, PNI, $O_2NOOCO_2^-$, and $ONOOCO_2^-$ (Olsen et al., 2003). Interestingly, they found that the barrier for O-atom transfer from PNA and PNI was approximately equal (Fig. 3), implying similar two-electron oxidation behavior.

In contrast to the similar two-electron oxidation chemistry of PNI and PNA, the BDEs for PNI and PNA, as well as the carbon dioxide adducts, differ drastically (Fig. 4). The BDEs for PNI are significantly lower than those for PNA. This difference in BDEs can largely be attributed to the higher reorganization energy of $NO_2$ radical versus $NO_3$ radical (Olsen et al., 2003). $CO_2$ accelerates the decomposition of PNI much more than for PNA, via homolytic cleavage of the peroxo OO bond. The principal decomposition pathway of PNI involves rapid formation of $ONOOCO_2^-$, which rapidly homolyzes to $NO_2$ and $CO_3^-$. Although PNA also reacts readily with $CO_2$, the product, $O_2NOOCO_2^-$, does not undergo homolysis readily.

## NONOate Decomposition Mechanisms

The compounds referred to as NONOates all possess the $[NONO]^-$ functional group. These compounds, also called diazeniumdiolates, have the ability to release various nitrogen oxides in a pH-dependent decomposition (Thomas et al., 2002). Dutton et al. used the hybrid density functional B3LYP, the extrapolative CBS-QB3 method, and the continuum solvation model PCM to investigate the decomposition mechanisms for dialkylamine NONOates (2004a) and Angeli's salt (2004b).

| Reaction | Computed energy barriers (kcal/mol) |
|----------|-------------------------------------|

PNI     $ONOO^- + H_2C{=}CH_2 \xrightarrow{\Delta E^{\ddagger} = 14.7} (CH_2)_2O + NO_2^-$

PNA     $O_2NOO^- + H_2C{=}CH_2 \xrightarrow{\Delta E^{\ddagger} = 13.6} (CH_2)_2O + NO_3^-$

PNI     $ONOOH + H_2C{=}CH_2 \xrightarrow{\Delta E^{\ddagger} = 17.1} (CH_2)_2O + HNO_2$

PNA     $O_2NOOH + H_2C{=}CH_2 \xrightarrow{\Delta E^{\ddagger} = 19.6} (CH_2)_2O + HNO_3$

PNI     $ONOOH + H_2S \xrightarrow{\Delta E^{\ddagger} = 20.9} H_2SO + HNO_2$

PNA     $O_2NOOH + H_2S \xrightarrow{\Delta E^{\ddagger} = 20.6} H_2SO + HNO_3$

PNI     $ONOOH + H_3N \xrightarrow{\Delta E^{\ddagger} = 20.6} H_3NO + HNO_2$

PNA     $O_2NOOH + H_3N \xrightarrow{\Delta E^{\ddagger} = 23.5} H_3NO + HNO_3$

Fig. 3. Computed activation energy barriers, in kcal/mol, for oxygen atom transfer reactions of PNI and PNA (from Olsen et al., 2003) (CBS-QB3).

Diethylamino NONOate (the diethylamide adduct of NO, or DEA/NO) is a commonly used NO source for *in vivo* and *in vitro* experiments. There have been many explorations of the mechanism of its decomposition (Dutton et al., 2004a; Davies et al., 2001; Hrabie et al., 1993; Taylor et al., 1995). Using a simplified model compound, dimethylamino NONOate (DMA/NO), Dutton et al. (2004a) proposed a mechanism based on the computed relative energies of the products of DMA/NO protonation (Fig. 5). Using the $pK_a$ and rate data collected by Davies et al. (2001), Dutton et al. (2004a) were able to fit a kinetic expression derived from the proposed mechanism to the observed pH-dependent rate of decomposition. The constants from the fitted rate data allowed for the calculation of the $pK_a$, $-6.9$ for N(1) protonation and the rate constant for decomposition of the N(1) protonated species, $k = 7.2 \times 10^{11} \text{ s}^{-1}$ (Dutton et al., 2004a).

Angeli's salt decomposes in a pH-dependent manner to generate HNO in the pH range of 10–4 and produces NO when the pH is less than 4 (Angeli, 1986; Hughes and Wimbledon, 1976). Dutton et al. (2004b) used the relative energies of the protonated species to propose a mechanism that explained the generation of both HNO and NO (Fig. 6). The generation of HNO results from the protonation of N(2) followed by cleavage of the N(1)–N(2) bond. Protonation a second time will result in the production of

| Reaction | | Computed BDE (kcal/mol) |
|---|---|---|

PNI    ONOOH    $\xrightarrow{\Delta E_{rxn} = 19.7}$    $NO_2^\bullet$ + $^\bullet OH$

PNA    $O_2NOOH$    $\xrightarrow{\Delta E_{rxn} = 39.0}$    $NO_3^\bullet$ + $^\bullet OH$

PNI    ONOOH    $\xrightarrow{\Delta E_{rxn} = 27.9}$    $NO^\bullet$ + $^\bullet OOH$

PNA    $O_2NOOH$    $\xrightarrow{\Delta E_{rxn} = 24.8}$    $NO_2^\bullet$ + $^\bullet OOH$

PNI    $ONOOCO_2^-$    $\xrightarrow{\Delta E_{rxn} = 7.5}$    $NO_2^\bullet$ + $CO_3^{-\bullet}$

PNA    $O_2NOOCO_2^-$    $\xrightarrow{\Delta E_{rxn} = 33.2}$    $NO_3^\bullet$ + $CO_3^{-\bullet}$

PNI    $ONOOCO_2^-$    $\xrightarrow{\Delta E_{rxn} = 28.5}$    $NO^\bullet$ + $^\bullet OOCO_2^-$

PNA    $O_2NOOCO_2^-$    $\xrightarrow{\Delta E_{rxn} = 31.8}$    $NO_2^\bullet$ + $^\bullet OOCO_2^-$

PNI    $ONOOCO_2^-$    $\xrightarrow{\Delta E_{rxn} = 18.3}$    $ONOO^-$ + $CO_2$

PNA    $O_2NOOCO_2^-$    $\xrightarrow{\Delta E_{rxn} = 17.7}$    $O_2NOO^-$ + $CO_2$

FIG. 4. Computed bond dissociation enthalpies in kcal/mol (from Olsen *et al.*, 2003) (CBS-QB3).

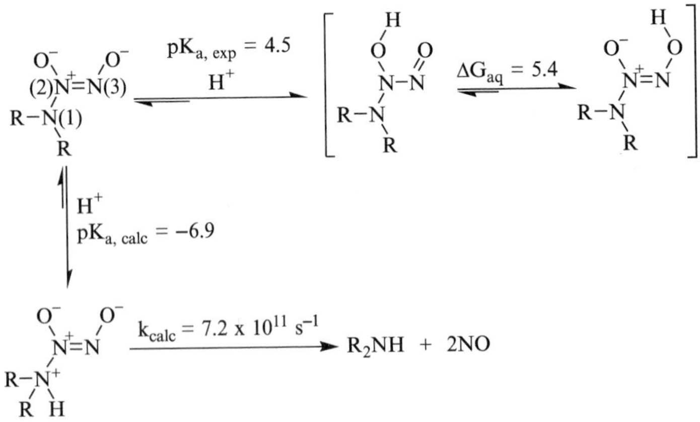

FIG. 5. The proposed mechanism for the decomposition of dialkyamine NONOates based on theoretical calculations (from Dutton *et al.*, 2004a). Changes in aqueous free energy are in kcal/mol [CBS-QB3, PCM B3LYP/6-311 + G(d)].

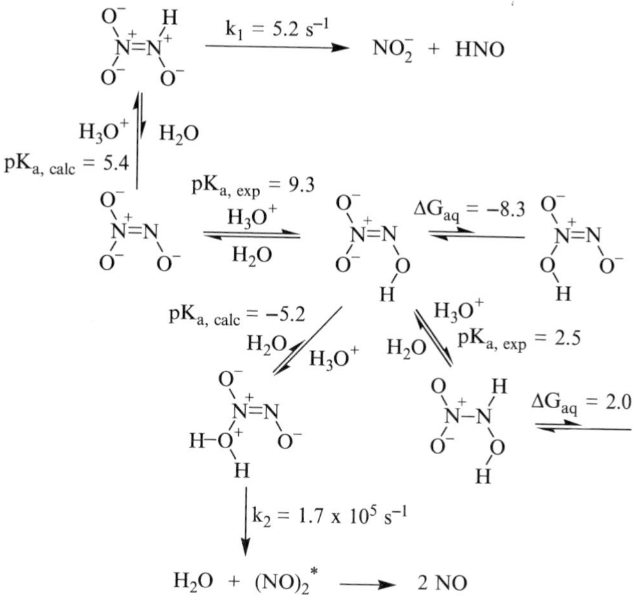

$$k_1 = 5.2 \ s^{-1} \longrightarrow NO_2^- + HNO$$

$$pK_{a, \, calc} = 5.4$$

$$H_3O^+ \quad H_2O$$

$$pK_{a, \, exp} = 9.3$$

$$H_3O^+ / H_2O$$

$$\Delta G_{aq} = -8.3$$

$$pK_{a, \, calc} = -5.2$$

$$H_2O / H_3O^+$$

$$H_3O^+ \quad pK_{a, \, exp} = 2.5$$

$$H_2O$$

$$\Delta G_{aq} = 2.0$$

$$k_2 = 1.7 \times 10^5 \ s^{-1}$$

$$H_2O + (NO)_2^* \longrightarrow 2 \ NO$$

FIG. 6. The proposed mechanism for the decomposition of Angeli's salt based on theoretical calculations (from Dutton et al., 2004b). Computed aqueous changes in free energy are reported in kcal/mol [CBS-QB3, PCM B3LYP/6-311 + G(d)].

two molecules of NO via the diprotonated species at O(2). A kinetic expression was derived from the proposed mechanism (Dutton et al., 2004b). This expression was fit to the experimentally (Hughes and Wimbledon, 1976) observed pH-dependent rate of decomposition. Using the experimentally determined $pK_a$s (Hughes and Wimbledon, 1976) of 9.3 and 2.5 and the constants from the fitted expression, rate constants for the decomposition, $k_1$ and $k_2$, of the protonated species were determined to be 5.2 $s^{-1}$ and $1.7 \times 10^5 \ s^{-1}$, respectively (Dutton et al., 2004b).

## Determining Chemical Properties: HNO Hydration Equilibria, HNO pK_a, and Various Reduction Potentials

### The Hydration $K_{eq}$ of HNO

Bartberger et al. (2001b) computed the energetics of hydration of eight ketones and aldehydes and found a good linear correlation with experimental data (Fig. 7). They used this correlation and the computed

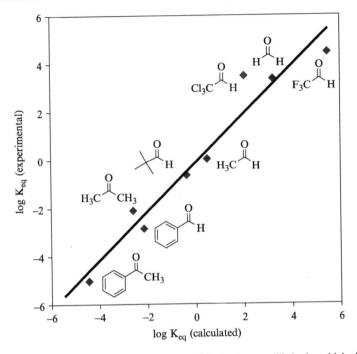

FIG. 7. Plot of calculated versus experimental hydration equilibria for aldehydes and ketones (from Bartberger *et al.*, 2001b).

hydration energy of HNO to predict the $K_{eq}$ for HNO hydration. The hydration reaction [Eq. (2)] is used to compute the change in free energy, $\Delta E_{rxn, \, PCM}$, which can be used to determine the $K_{eq}$.

$$
\underset{R}{\overset{O}{\underset{\parallel}{C}}}{}_{R'(H)} \;+\; H_2O \;\xrightarrow{\;\Delta E_{rxn,PCM} \text{ or } K_{eq}\;}\; R-\underset{\underset{OH}{|}}{\overset{\overset{OH}{|}}{C}}-R'(H) \qquad (2)
$$

Plotting the eight computed $K_{eq}$ versus the experimental $K_{eq}$ gives a linear correlation with an $R^2$ value of 0.73 and the equation of log $K_{eq, \, exp}$ = 1.01(log $K_{eq, \, calc}$) − 0.13. This equation predicts a $K_{eq}$ for HNO of $6.9 \times 10^{-5}$ (Bartberger *et al.*, 2001b). This predicted value implies that essentially all HNO in aqueous solution will exist as HNO and not as the hydrate.

*Computing the $pK_a$ of HNO and Reduction Potential of Nitric Oxide*

The NO reduction potential and the HNO $pK_a$ are related and therefore discussed in this section together. The aqueous NO reduction potential and the HNO $pK_a$ are related to the aqueous free energies of the reactions in Eqs. (3) and (4), respectively.

$$NO + e^- \longrightarrow {}^3NO^- \tag{3}$$

$$HNO \longrightarrow H^+ + {}^3NO^- \tag{4}$$

In 1970, Grätzel *et al.* (1970) published the widely cited value of 4.7 for the $pK_a$ of HNO. This value of 4.7 has since been revised upward by Bartberger *et al.* (2001b, 2002) and Shafirovich and Lymar (2002), converging on an HNO $pK_a$ of about 11.6. Initially, Bartberger *et al.* (2001b) revised the HNO $pK_a$ to 7.2 by using the CBS-QB3 method combined with the PCM aqueous solvation model to compute the solvated proton affinities, $PA_{solv}$, for 30 experimentally known acids, spanning a $pK_a$ range of 0.2–33.5. A plot of the calculated proton affinities in aqueous solution ($PA_{solv}$) versus the experimental $pK_a$ provided a linear correlation: $pK_a = 0.549(PA_{solv}) - 139.8$, and a correlation constant of 0.95 (Bartberger *et al.*, 2001b). Using this equation, a value of 7.2 is obtained, indicating that at neutral pH, the HNO is approximately half in the protonated form and half deprotonated as ${}^3NO^-$ (Bartberger *et al.*, 2001b). However, a year later Bartberger *et al.* (2002) and Shafirovich and Lymar (2002) independently revised the $pK_a$ for HNO to about 11.6. Shafirovich and Lymar (2002) used experimental data, collected themselves along with the reduction potential of −0.81 V versus NHE determined by Benderskii *et al.* (1989) and estimated values to derive the $\Delta_f G°$ in aqueous solutions for the species involved in Eqs. (3) and (4) (Shafirovich and Lymar, 2002). Bartberger *et al.* (2002) used the accurately measured $O_2/O_2^-$ reduction potential (Woods, 1987) of −0.16 V versus NHE as a reference point to calibrate their computed reduction potential. Using the CBS-QB3 method combined with the PCM model of water, Bartberger *et al.* (2002) computed the solvated electron affinity, $EA_{solv}$, for the $O_2/O_2^-$ and $NO/{}^3NO^-$ reaction couples. By adjusting the $EA_{solv}$ of NO by the difference between the computed and experimental values of $O_2$, an estimated reduction potential for $NO/{}^3NO^-$ of −0.76 V versus NHE is obtained (Bartberger *et al.*, 2002). This theoretically derived value agrees well with the value measured by Benderskii *et al.* (1989), and subsequently the value used by Shafirovich and Lymar (2002). In similar fashion to Shafirovich and Lymar (2002), Bartberger *et al.* (2002) used their estimated value of the $NO/{}^3NO^-$ reduction potential to determine the HNO $pK_a$ of 11.6. The earlier value was in error primarily because of problems in computing aqueous solvation energies.

### Theoretically Derived Reduction Potentials of HNO and Angeli's Salt

Building on the methodology developed by *Bartberger et al.*, Dutton *et al.* (in press) and Miranda *et al.* (2005) estimated the reduction potentials of HNO and Angeli's salt, a commonly used HNO and NO donor. Using the CBS-QB3 method with the PCM and CPCM aqueous solvation models, the reduction potential for 36 experimentally known compounds with a range of 2.65 to −1.8 V versus NHE was computed. A plot of the computed versus experimental reduction potentials, shown in Fig. 8, provided a linear correlation of $E^{o}_{exp} = 0.84(E^{o}_{calc}) + 0.03$, with a correlation constant of 0.97 (Dutton *et al.*, submitted). Using this equation, it is then possible to predict theoretically the reduction potentials for nitrogen oxides that are difficult to measure experimentally. The estimated reduction potentials for HNO and Angeli's salt using this method are shown in Table I (Dutton *et al.*, in press; Miranda *et al.*, 2005).

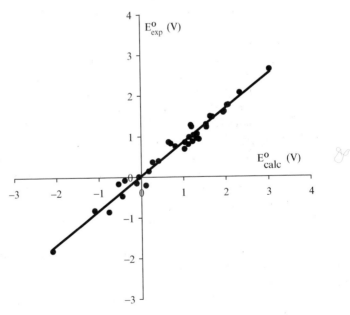

FIG. 8. Plot of 36 standard state experimental reduction potentials versus the calculated reduction potentials using the PCM solvation model.

TABLE I
COMPUTED REDUCTION POTENTIALS FOR HNO, $H_2NO$, AND ANGELI'S SALT $vs.$ NHE

| Reaction couple | $E^o_{pred}(V)$ | $E^o_{exp}(V)$ |
|---|---|---|
| $HNO + H_3O^+ + e^-/H_2NO + H_2O$ | $0.6 \pm 0.1$ | |
| $H_2NO + H_3O^+ + e^-/H_2NOH + H_2O$ | $0.9 \pm 0.1$ | |
| $HNO + 2H_3O^+ + 2e^-/H_2NOH + 2H_2O$ | $0.8 \pm 0.1$ | $0.7^a$ |
| $HNO + e^-/HNO^-$ | $-0.7 \pm 0.1$ | |
| $N_2O_3^- + e^-/N_2O_3^{2-}$ | $0.5 \pm 0.1$ | |
| $HN_2O_3 + e^-/HN_2O_3^-$ | $0.9 \pm 0.1$ | |

$^a$ From Shafirovich and Lymar (2002).

## Conclusion

These examples demonstrate of the use of computational methods for the study of reactive nitrogen oxide intermediates. Computational chemistry can provide insights into the properties of highly reactive molecules that are difficult to observe or study experimentally. Reactive intermediates involved in a reaction mechanism can be computed to determine chemical properties such as $pK_a$s, redox potentials, and hydration equilibria. Furthermore, in most biological systems, there are a wide variety of possible reactions that nitrogen oxides may undergo, and computational chemistry allows for a thorough investigation of an individual reaction pathway. Although experiments are necessary to establish what occurs under physiological conditions, theory can predict which of myriad possibilities are thermodynamically and kinetically feasible.

## Acknowledgments

We are grateful to the National Institute of General Medical Sciences, the National Institutes of Health (NIH) for financial support of this research. A. S. D. acknowledges the support of the NIH, Chemistry and Biology Interface Training Grant from UCLA.

## References

Adamo, C., and Barone, V. (1998). Exchange functionals with improved long-range behavior and adiabatic connection methods without adjustable parameters: The $m$PW and $m$PW1PW models. *J. Chem. Phys.* **108,** 664.

Angeli, A. (1986). Sopra il nitroidrosilammina. *Gazz. Chim. Ital.* **26,** 17.

Bartberger, M. D., Mannion, J. D., Powell, S. C., Stamler, J. S., Houk, K. N., and Toone, E. J. (2001a). S-N dissociation energies of S-nitrosothiols: On the origins of nitrosothiol decomposition rates. *J. Am. Chem. Soc.* **123,** 8868.

Bartberger, M. D., Fukuto, J. M., and Houk, K. N. (2001b). On the acidity and reactivity of HNO in aqueous solution and biological systems. *Proc. Natl. Acad. Sci. USA* **98,** 2194.

Bartberger, M. D., Liu, W., Ford, E., Miranda, K. M., Switzer, C., Fukuto, J. M., Farmer, P. J., Wink, D. A., and Houk, K. N. (2002). The reduction potential of nitric oxide (NO) and its importance to NO biochemistry. *Proc. Natl. Acad. Sci. USA* **99**, 10958.

Becke, A. D. (1993). Density-functional thermochemistry. III. The role of exact exchange. *J. Chem. Phys.* **98**, 5648.

Benderskii, V. A., Krivenko, A. G., and Ponomarev, E. A. (1989). *Soviet Electrochem.* **25**, 154.

Cho, J. Y., Dutton, A., Miller, T., Houk, K. N., and Fukuto, J. M. (2003). Oxidation of N-hydroxyguanidines by copper(II): Model systems for elucidating the physiological chemistry of the nitric oxide biosynthetic intermediate N-hydroxyl-L-arginine. *Arch. Biochem. Biophys.* **417**, 65.

Cranahan, G. E., Lenhart, P. G., and Ravichandran, R. (1978). *S*-Nitroso-*N*-acetyl-DL-penicillamine. *Acta Crystallogr.* **B34**, 2645.

Cramer, C. J. (2002). "Essentials of Computational Chemistry." John Wiley & Sons, New York.

Curtiss, L. A., Raghavachari, K., Trucks, G. W., and Pople, J. A. (1991). Gaussian-2 theory for molecular energies of first- and second-row compounds. *J. Chem. Phys.* **94**, 7221.

Curtiss, L. A., Raghavachari, K., Redfern, P. C., Rassolov, V., and Pople, J. A. (1998). Gaussian-3 (G3) theory for molecules containing first and second-row atoms. *J. Chem. Phys.* **109**, 7764.

Davies, K. M., Wink, D. A., Saavedra, J. E., and Keefer, L. K. (2001). Chemistry of the Diazeniumdiolates. 2. Kinetics and Mechanism of Dissociation to Nitric Oxide in Aqueous Solution. *J. Am. Chem. Soc.* **123**, 5473.

Dewar, M. J. S., Zoebisch, E. G., Healy, E. F., and Stewart, J. J. P. (1985). Development and use of quantum mechanical molecular models. 76. AM1: A new general purpose quantum mechanical molecular model. *J. Am. Chem. Soc.* **107**, 3902.

Dutton, A. S., Fukuto, J. M., and Houk, K. N. (2004a). The mechanism of NO formation from the decomposition of dialkylamino diazeniumdiolates: Density functional theory and CBS-QB3 predictions. *Inorg. Chem.* **43**, 1039.

Dutton, A. S., Fukuto, J. M., and Houk, K. N. (2004b). Mechanisms of HNO and NO production from Angeli's salt: Density functional and CBS-QB3 theory predictions. *J. Am. Chem. Soc.* **126**, 3795.

Dutton, A. S., Fukuto, J. M., and Houk, K. N. Theoretical reduction potentials for nitrogen oxides from CBS-QB3 energetics and (C)PCM solvation calculations. *Inorg. Chem.* In press.

Frisch, M. J., Trucks, G. W., Schlegel, H. B., Scuseria, G. E., Robb, M. A., Cheeseman, J. R., Zakrzewski, V. G., Montgomery, J. A., Jr., Stratmann, R. E., Burant, J. C., Dapprich, S., Millam, J. M., Daniels, A. D., Kudin, K. N., Strain, M. C., Farkas, O., Tomasi, J., Barone, V., Cossi, M., Cammi, R., Mennucci, B., Pomelli, C., Adamo, C., Clifford, S., Ochterski, J., Petersson, G. A., Ayala, P. Y., Cui, Q., Morokuma, K., Malick, D. K., Rabuck, A. D., Raghavachari, K., Foresman, J. B., Cioslowski, J., Ortiz, J. V., Stefanov, B. B., Liu, G., Liashenko, A., Piskorz, P., Komaromi, I., Gomperts, R., Martin, R. L., Fox, D. J., Keith, T., Al-Laham, M. A., Peng, C. Y., Nanayakkara, A., Gonzalez, C., Challacombe, M., Gill, P. M. W., Johnson, B. G., Chen, W., Wong, M. W., Andres, J. L., Head-Gordon, M., Replogle, E. S., and Pople, J. A. (1998). "Gaussian 98," Rev. A.1. Gaussian, Inc., Pittsburgh PA.

Frisch, M. J., Trucks, G. W., Schlegel, H. B., Scuseria, G. E., Robb, M. A., Cheeseman, J. R., Montgomery, J. A., Jr., Vreven, T., Kudin, K. N., Burant, J. C., Millam, J. M., Iyengar, S. S., Tomasi, J., Barone, V., Mennucci, B., Cossi, M., Scalmani, G., Rega, N., Petersson, G. A., Nakatsuji, H., Hada, M., Ehara, M., Toyota, K., Fukuda, R., Hasegawa, J., Ishida, M., Nakajima, T., Honda, Y., Kitao, O., Nakai, H., Klene, M., Li, X., Knox, J. E., Hratchian, H. P., Cross, J. B., Adamo, C., Jaramillo, J., Gomperts, R., Stratmann, R. E., Yazyev, O., Austin, A. J., Cammi, R., Pomelli, C., Ochterski, J. W., Ayala, P. Y.,

Morokuma, K., Voth, G. A., Salvador, P., Dannenberg, J. J., Zakrzewski, V. G., Dapprich, S., Daniels, A. D., Strain, M. C., Farkas, O., Malick, D. K., Rabuck, A. D., Raghavachari, K., Foresman, J. B., Ortiz, J. V., Cui, Q., Baboul, A. G., Clifford, S., Cioslowski, J., Stefanov, B. B., Liu, G., Liashenko, A., Piskorz, P., Komaromi, I., Martin, R. L., Fox, D. J., Keith, T., Al-Laham, M. A., Peng, C. Y., Nanayakkara, A., Challacombe, M., Gill, P. M. W., Johnson, B., Chen, W., Wong, M. W., Gonzalez, C., and Pople, J. A. (2003). "Gaussian 03," Rev. B.04. Gaussian, Inc., Pittsburgh PA.

Fukuto, M., Dutton, A. S., and Houk, K. N. (2004). The chemistry and biology of nitroxyl (HNO): A chemically unique species with novel and important biological activity. *ChemBioChem.* http://www3.interscience.wiley.com/cgi-bin/abstract/109860812/ABSTRACT December 27, 2004.

Goodman, J. M. (1998). "Chemical Applications of Molecular Modeling." Royal Society of Chemistry, Cambridge, UK.

Grätzel, M, Taniguchi, S., and Henglein, A. (1970). Pulsradiolytische untersuchung kurzlebiger zwischenprodukte der NO-reduktion in wässriger lösung. *Ber. Bunsenges. Phys. Chem.* **74**, 1003.

Grisham, M. B., Johnson, G. G., and Lancaster, J. R., Jr. (1996). Quantitation of nitrate and nitrite in extracellular fluids. *Methods Enzymol.* **268**, 237.

Hecker, M., Schott, C., Bucher, B., Busse, R., and Stoclet, J.-C. (1995). Increase in serum NG-hydroxy-L-arginine in rats treated with bacterial lipopolysaccharide. *Eur. J. Pharmacol.* **275**, R1.

Hohenberg, P., and Kohn, W. (1964). Inhomogeneous Electron Gas. *Phys. Rev.* **136**, B864.

Houk, K. N., Hietbrink, B. N., Bartberger, M. D., McCarren, P. R., Choi, B. Y., Voyksner, R. D., Stamler, J. S., and Toone, E. J. (2003). Nitroxyl disulfides, novel intermediates in transnitrosation reactions. *J. Am. Chem. Soc.* **125**, 6972.

Hrabie, J. A., Klose, J. R., Wink, D. A., and Keefer, L. K. (1993). New nitric oxide-releasing zwitterions derived from polyamines. *J. Org. Chem.* **58**, 1472.

Hughes, M. N., and Wimbledon, P. E. (1976). The chemisry of trioxodinitrates. Part I. Decompostion of sodium trioxodinitrate (Angeli's salt) in aqueous solution. *J. Chem. Soc. Dalton* 703.

*HyperChem*. Hypercube, Inc., Gainesville, FL.

*Jaguar*, v. 5.5. (2003). Schrodinger, Inc., Portland, OR.

Kohn, W., and Sham, L. J. (1965). Self-Consistent Equations Including Exchange and Correlation Effects. *Phys. Rev.* **140**, A1133.

Kong, J., White, C. A., Krylov, A. I., Sherrill, C. D., Adamson, R. D., Furlani, T. R., Lee, M. S., Lee, A. M., Gwaltney, S. R., Adams, T. R., Ochsenfeld, C., Gilbert, A. T. B., Kedziora, G. S., Rassolov, V. A., Maurice, D. R., Nair, N., Shao, Y., Besley, N. A., Maslen, P. E., Dombroski, J. P., Daschel, H., Zhang, W., Korambath, P. P., Baker, J., Byrd, E. F. C., Van Voorhis, T., Oumi, M., Hirata, S., Hsu, C.-P., Ishikawa, N., Florian, J., Warshel, A., Johnson, B. G., Gill, P. M. W., Head-Gordon, M., and Pople, J. A. (2000). Q-Chem 2.0: A high-performance *ab initio* electronic structure program. *J. Comput. Chem.* **21**, 1532–1548.

Leach, A. R. (2001). "Molecular Modeling," 2nd Ed. Prentice Hall, New York.

Martin, J. M. L., and de Oliveira, G. (1999). Towards standard methods for benchmark quality *ab initio* thermochemistry—W1 and W2 theory. *J. Chem. Phys.* **111**, 1843.

Miranda, K. M., Dutton, A. S., Ridnour, L., Foreman, C. A., Ford, E., Paolocci, N., Katori, T., Mancardi, D., Thomas, D. D., Espey, M. G., Houk, K. N., Fukuto, J. M., and Wink, D. A. (2005). Mechanism of aerobic decomposition of Angeli's salt (sodium trioxodinitrate) at physiological pH. *J. Am. Chem. Soc.* **127**, 722.

Miranda, K. M., Paolocci, N., Katori, T., Thomas, D. D., Ford, E., Bartberger, M. D., Espey, M. G., Kass, D. A., Fukuto, J. M., and Wink, D. A. (2003). A biochemical rationale for the discrete behavior of nitroxyl and nitric oxide in the cardiovascular system. *Proc. Natl. Acad. Sci. USA* **100**, 9197.

Møller, C., and Plesset, M. S. (1934). Note on an Approximation Treatment for Many-Electron Systems. *Phys. Rev.* **46**, 618.

Montgomery, J. A., Ochterski, J. W., and Petersson, G. A. (1994). A complete basis set model chemistry. IV. An improved atomic pair natural orbital method. *J. Chem. Phys.* **101**, 5900.

Olsen, L. P., Bartberger, M. D., and Houk, K. N. (2003). Peroxynitrate and peroxynitrite: A complete basis set investigation of similarities and differences between these NOx species. *J. Am. Chem. Soc.* **125**, 3999.

Perdew, J. P., Burke, K., and Wang, Y. (1996). Generalized gradient approximation for the exchange-correlation hole of a many-electron system. *Phys. Rev.* **B54**, 16533.

Pople, J. A., Head-Gordon, M., Fox, D. J., Raghavachari, K., and Curtiss, L. A. (1989). Gaussian-1 theory: A general procedure for prediction of molecular energies. *J. Chem. Phys.* **90**, 5622.

Repasky, M. P., Chandrasekhar, J., and Jorgensen, W. L. (2002). PDDG/PM3 and PDDG/MNDO: Improved semiempirical methods. *J. Comput. Chem.* **23**, 1601.

Roothaan, C. C. J. (1951). New Developments in Molecular Orbital Theory. *Rev. Mod. Phys.* **23**, 69.

Roy, B., d'Hardemare, A. M., and Fontcave, M. (1994). New Thionitrites: Synthesis, Stability, and Nitric Oxide Generation. *J. Org. Chem.* **59**, 7019.

*SPARTAN '04.* (2004). Wavefunction, Inc., Irvine, CA.

Shafirovich, V., and Lymar, S. V. (2002). Nitroxyl and its anion in aqueous solutions: Spin states, protic equilibria, and reactivities toward oxygen and nitric oxide. *Proc. Natl. Acad. Sci. USA* **99**, 7340.

Stewart, J. J. P. (1989). Optimization of parameters for semiempirical methods I. Method. *J. Comput. Chem.* **10**, 209.

Straatsma, T. P., Aprà, E., Windus, T. L., Bylaska, E. J., de Jong, W., Hirata, S., Valiev, M., Hackler, M. T., Pollack, L., Harrison, R. J., Dupuis, M., Smith, D. M. A., Nieplocha, J., Tipparaju, V., Krishnan, M., Auer, A. A., Brown, E., Cisneros, G., Fann, G. I., Fruchtl, H., Garza, J., Hirao, K., Kendall, R., Nichols, J., Tsemekhman, K., Wolinski, K., Anchell, J., Bernholdt, D., Borowski, P., Clark, T., Clerc, D., Dachsel, H., Deegan, M., Dyall, K., Elwood, D., Glendening, E., Gutowski, M., Hess, A., Jaffe, J., Johnson, B., Ju, J., Kobayashi, R., Kutteh, R., Lin, Z., Littlefield, R., Long, X., Meng, B., Nakajima, T., Niu, S., Rosing, M., Sandrone, G., Stave, M., Taylor, H., Thomas, G., van Lenthe, J., Wong, A., and Zhang, Z. (2004). "NWChem, A Computational Chemistry Package for Parallel Computers," v. 4.6. Pacific Northwest National Laboratory, Richland, WA.

Taylor, D. K., Bytheway, I., Barton, D. H. R., Bayse, C. A., and Hall, M. B. (1995). Toward the Generation of NO in Biological Systems Theoretical Studies of the N2O2 Grouping. *J. Org. Chem.* **60**, 435.

Thomas, D. D., Miranda, K. M., Espey, M. G., Citrin, D., Jourd'heuil, D., Paolocci, N., Hewett, S. J., Colton, C. A., Grisham, M. B., Feelisch, M., and Wink, D. A. (2002). Guide for the use of nitric oxide (NO) donors as probes of the chemistry of NO and related redox species in biological systems. *Methods Enzymol.* **359**, 84.

Wigand, R., Meyer, J., Busse, R., and Hecker, M. (1997). Increased serum NG-hydroxy-L-arginine in patients with rheumatoid arthritis and systemic lupus erythematosus as an index of an increased nitric oxide synthase activity. *Ann. Rheum. Dis.* **56**, 30.

Woods, P. M. (1987). The two redox potentials for oxygen reduction to superoxide. *Trends Biochem. Sci.* **12**, 250.

Young, D. (2001). "Computational Chemistry." Wiley-Interscience, New York.

[4]   Electrochemical and Spectrophotometric Methods
for Evaluation of NO Dissociation Rate Constants from
Nitrosyl Metal Complexes Activated through Reduction

*By* Mauro Bertotti, Vânia Mori, Patrícia Zanichelli,
José Carlos Toledo, and Douglas Wagner Franco

## Abstract

Electrochemical and spectrophotometric methods are described for measuring the rate of nitric oxide (NO) dissociation ($k_{NO}$) from coordination compounds. Electrochemical methods based on double-potential step chronoamperometry and rotating ring-disc electrode voltammetry techniques proved to be suitable for measuring NO dissociation from 0.03 to 4.0 s$^{-1}$. The spectrophotometric method using an ancillary ligand as a colorimetric indicator is illustrated on measuring $k_{-NO} = 0.002$ s$^{-1}$. This methodology is limited only by the rate of the ancillary ligand substitution.

## Introduction

Nitrosyl complexes have been considered an alternative for nitric oxide NO carrier drugs (Clarke, 2002). For example, sodium nitroprusside ($Na_2[Fe(CN)_5(NO)]$) has been largely employed in clinical practices and in biological and chemical studies (Clarke, 2002; Tfouni *et al.*, 2003).

This contribution is intended to present compounds type the *trans*-$[Ru(NH_3)_4(L)(NO)]^{3+}$ (where L = py, pz, L-His, 4-pic, 4-Clpy, nic, isn, imN) and $[Ru(Hedta)NO]$ as NO donors for chemical or biological *in vivo* studies. We also present two different methods, electrochemical and spectrophotometric, to calculate a specific rate constant for NO dissociation from nitrosyl complex reductively activated.

### Nitrosyl Complexes as NO Donor

*Chemical Properties.* The formal oxidation state of $[Ru^{II}NO^+]$ has been assigned to the *trans*-$[RuII(NH_3)_4(L)(NO)]^{3+}$ and $[Ru(Hedta)NO]^{3+}$ species (Borges *et al.*, 1998; Clarke, 2002; Tfouni *et al.*, 2003; Zanichelli *et al.*, 2004), which is consistent with their diamagnetism.

DFT calculations for *trans*-$[Ru^{II}(NH_3)_4(L)(NO^+)]^{3+}$ (L = py and pz) indicate that their LUMO orbital is predominantly on the $\pi^*$ of NO (68–70%) (Toledo *et al.*, 2004). Thus, the one-electron reduction of

METHODS IN ENZYMOLOGY, VOL. 396
0076-6879/05 $35.00
DOI: 10.1016/S0076-6879(05)96004-7

these complexes is expected to generate $[Ru^{II}NO^0]$ species (Reaction 1) (Borges *et al.*, 1998; Toledo *et al.*, 2004), as confirmed by electron paramagnetic resonance spectroscopy for the reduced form of the *trans*-$[Ru(NH_3)_4(H_2O)(NO)]^{3+}$ (Tfouni *et al.*, 2003) and $[Ru(Hedta)NO]$ (Zanichelli *et al.*, 2004).

The reduced species, *trans*-$[Ru(NH_3)_4(L)(NO)]^{2+}$, releases $NO^0$ (Reaction 2) at a rate constant dependent on the electronic characteristics of the ligand *trans*-positioned to NO (Tfouni *et al.*, 2003).

$$\text{\textit{trans}-}[Ru^{II}(NH_3)_4(L)(NO^+)]^{3+} + e^- \longrightarrow$$
$$\text{\textit{trans}-}[Ru^{II}(NH_3)_4(L)(NO^0)]^{2+} \quad (1)$$

$$\text{\textit{trans}-}[Ru^{II}(NH_3)_4(L)(NO^0)]^{2+} + H_2O \xrightarrow{k_{-NO}}$$
$$\text{\textit{trans}-}[Ru^{II}(NH_3)_4(L)(H_2O)]^{2+} + NO \quad (2)$$

These compounds have a potential applicability as NO donors, because their synthetic procedure is well established, they are water soluble and inert regarding substitution reactions in solution, and they can keep their integrity for months if stored appropriately under vacuum or inert atmosphere and in the absence of light (Borges *et al.*, 1998; Tfouni *et al.*, 2003). The labilization of NO through reduction of *trans*-$[Ru(NH_3)_4 L(NO)]^{3+}$ species proceeds according to the equations (1 and 2) presented above.

The $[Ru^{II}NO^+/Ru^{II}NO^0]$ redox potential for the nitrosyl complexes presented here ranges from $-0.118$ (L = ImN) to $0.072$ (L = nic) V *vs.* NHE (Tfouni *et al.*, 2003) and is $-0.1$ V *vs.* NHE for $[Ru(Hedta)NO]$ complex (Zanichelli *et al.*, 2004), suggesting that all of them could be reduced by biological reducing agents such as NADH, flavin coenzymes $FADH_2$ and $FMNH_2$, and iron sulfur proteins.

The *trans*-$[Ru(NH_3)_4P(OC_2H_5)_4(NO)]^{3+}$ nitrosyl complex may be reduced by mitochondrial components and exhibits an hypotensive effect in mice (Torsoni *et al.*, 2002). Its biological activity has been attributed to the complex NO donor ability.

Thus, the *trans*-$[Ru(NH_3)_4(L)(NO)]^{3+}$ and $[Ru(Hedta)NO]$ nitrosyl complexes can be directly employed as an NO donor in most biological assays, which contain reducing agents. For chemical studies, the NO release can be assessed through the use of a one-electron reductor specie such as $Eu^{II}$ ions.

*Electrochemical Methods.* Electrochemistry (EC) is a very powerful tool to determine rate constants associated with this system, classically denominated as an EC process (a chemical reaction following an electrochemical step: Reactions 1 and 2). Two main electrochemical techniques are suitable to provide this kinetic information once the mechanism of the electrode process is well understood, as in the present situation: double-potential step chronoamperometry and voltammetry with rotating ring-disc electrode (RRDE). Both electrochemical techniques have been used to evaluate the influence of the *trans*-ligand (L) on the kinetics of the labilization of the NO molecule after the cathodic reduction of $[Ru^{II}(NH_3)_4(L)(NO^+)]^{3+}$ complexes.

*Double-Potential Step Chronoamperometry.* An experiment with double-potential step chronoamperometry (Bard and Faukner, 1980) is carried out by suddenly changing the potential of the working electrode from a value at which no faradaic process occurs (electron transfer between the investigated species and the electrode) to a potential at which the kinetics of the electrochemical process is very fast, hence no material can coexist with the electrode surface. In this experimental condition, the resulting current-time dependence (chronoamperogram) exhibits an exponential decrease of current as a function of time, which is justified by the decrease in the rate at which the species can reach the electrode by diffusion. If after a period ($\tau$) the potential is stepped back to the initial value, the reverse electrode reaction proceeds, producing a current of inverse signal. It is then possible to calculate the amount of material lost by the chemical change and accordingly; the faster the coupled chemistry reaction, the less current will be obtained in the second part of the experiment. The determination of rate constant values is performed by measuring the ratio between currents measured on both potential steps at determined times and comparing those with working curves described in the literature (Bard and Faukner, 1980).

*Rotating Ring-Disc Electrode Voltammetry.* Voltammetry carried out with RRDEs (Bard and Faukner, 1980; Pletcher, 1991) can also be conveniently used to obtain kinetic information on EC processes. The design of such a device consists of a conducting ring concentric to a conventional disc electrode that rotates at controlled speeds, both electrodes being separated by a thin insulating gap. Separate electrode connections allow electrodes to be polarized at different potentials simultaneously; hence, a substance generated at the disc surface can be monitored adequately at the ring. The amount of material collected at the ring is dependent on the distance between electrodes and the chemistry of the intermediate formed.

## Materials and Methods

The nitrosyl complexes of the class of *trans*-$[Ru(NH_3)_4(L)(NO)](BF_4)_3$ [L = py, pz, 4-pic, imN, nic, $H_2O$, L-His, 4-Clpy, isn and imC (bounded by carbon)] are prepared reacting the respective aqua complexes, *trans*-$[Ru(NH_3)_4(L)(H_2O)]^{2+}$, with excess of sodium nitrite in an acidic solution (2.0 mol/L$^{-1}$ HBF$_4$). The synthetic procedure for *trans*-$[Ru(NH_3)_4(L)(H_2O)]^{2+}$ and $[Ru(Hedta)NO]^{3+}$ complex synthesis can be found in the literature (Borges *et al.*, 1998; Tfouni *et al.*, 2003; Zanichelli *et al.*, in press).

An Autolab PGSTAT 30 (Eco Chemie) bipotentiostat with data acquisition software made available by the manufacturer (GPES version 4.8) was used for electrochemical measurements. Experiments were done in a conventional electrochemical cell with a Ag/AgCl (saturated KCl) electrode and a platinum wire being used as reference and counter electrodes, respectively. Voltammetry with a gold RRDE was carried out using an analytical rotator (AFMSRX) connected to the bipotentiostat. The chronoamperometric experiments were performed using a glassy carbon disc electrode as a working electrode. The dissociation of NO from $[Ru^{II}(Hedta)NO^+]$ was monitored through ultraviolet–visible (UV-vis) measurements after the reduction of $[Ru^{II}(Hedta)NO^+]$ with Eu(II) ions, which was previously generated in solution by the reduction with Zn(Hg) amalgam. The solutions of $[Ru^{II}(Hedta)NO^+]$ and Eu(II) ions were separately deoxygenated by argon bubbling. The electronic spectra were recorded with a Hewlett-Packard 8451A instrument.

Results and Discussion

Figure 1 shows a typical chronoamperogram of a solution containing the *trans*-$[Ru^{II}(NH_3)_4(PZ)(NO^+)]^{3+}$ complex. By measuring $I_a$:$I_c$ values (ratio between anodic and cathodic currents measured at times t and t $-$ $\tau$, respectively) and by using working curves described in the literature (Bard and Faukner, 1980) (plots of $I_a$:$I_c$ as a function of kt for different (t $-$ $\tau$)/$\tau$ values, where k = kinetic constant and $\tau$ = time of potential reversal), rate constant values for a series of compounds were determined as shown in Table I. Values for $k_{-NO}$ are in agreement with the electronic properties of the *trans* ligand coordinated to the ruthenium complex.

The rate of loss of the NO molecule from the reduced material is too slow to be monitored in a fast electrochemical experiment such as voltammetry with RDDE. Accordingly, for this technique to yield reliable results in this chemical system, temperatures around 40° were employed to accelerate the release of the NO molecule. Figure 2 shows a series of voltammograms recorded with an RRDE for a solution of *trans*-$[Ru^{II}(NH_3)_4 (4\text{-pic})(NO^+)]^{3+}$. By measuring the ratio between limiting currents at the ring and the disc (collection efficiency, $N_k$) at different rotation rates ($\omega$), the rate constant was conveniently calculated using equations described in the literature (Albery and Bruckenstein, 1966). Briefly, the kinetic constant is obtained by plotting

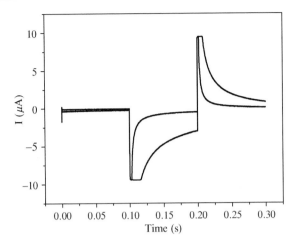

Fig. 1. Double-potential step chronoamperogram for *trans*-[Ru$^{II}$(NH$_3$)$_4$(pz)NO]$^{+/0}$ in CF$_3$COOH solution (pH = 2) at T = 25°. Potential limits: 0.10 V and −0.20 V.

TABLE I
RATE CONSTANTS VALUES (k$_{NO}$) OF
*TRANS*-[Ru$^{II}$(NH$_3$)$_4$(L)NO]$^{+/0}$ COMPLEXES$^a$

| Ligand | k$_{NO}$/s$^{-1}$ |
|---|---|
| H$_2$O | 0.04 |
| ImN | 0.16 |
| L-hist | 0.14 |
| 4-pic | 0.09 |
| py | 0.06 |
| 4-Clpy | 0.03 |
| nic | 0.025 |
| isn | 0.043 |
| ImC | 4.0 |
| pz | 0.34 |

$^a$ k$_{NO+}$/ss$^{-1-1}$, T = (25.0 ± 0.2°)$^0$, pH = 2 (CF$_3$COOH).

collection efficiency values (N$_k$) as a function of $\omega^{-1}$, the slope of the straight line containing the kinetic parameter.

Table II shows k$_{-NO}$ values calculated for some complexes at 40°, as well as comparative data obtained by using the chronoamperometric technique. The good agreement between the data ensures that both techniques are suitable to obtain kinetic information of this system. It is important to

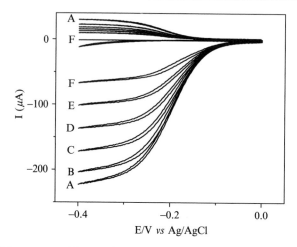

FIG. 2. Voltammetry with rotating ring-disc electrode for a solution containing *trans*-[Ru(NH$_3$)$_4$(4-pic)NO]$^{3+}$ in CF$_3$COOH, pH = 2 ($\mu$ = 0.10 mol L$^{-1}$). Scan rate = 20 mV s$^{-1}$; rotation rates: 400 (F), 900 (E), 1600 (D), 2500 (C), 3600 (B), e 4900 (A). Potential at the ring maintained at 0 V. T = 40°.

TABLE II
SPECIFIC RATE CONSTANTS (k$_{-NO}$) FOR
*TRANS*-[Ru$^{II}$(NH$_3$)$_4$(L)NO]$^+$/0AT 40.0 ± 0.2°

| Ligand | k$_{-NO}$/s$^{-1}$ Ring-disc | k$_{-NO}$/s$^{-1}$ Chronoamperometry |
|--------|------|------|
| 4-pic | 2.3 | 2.2 |
| Py | 2.1 | 1.9 |
| Nic | 1.9 | 1.7 |

notice that the time windows of chronoamperometric methods usually are extended in the 0.001–50 s range, and in the case of voltametry with rotating electrodes, a shorter range is reported (0.001–0.3 s) (Bard and Faukner, 1980), justifying the use of both techniques to measure rate constants associated with the release of NO upon electroreduction of these ruthenium compounds.

## Spectrophotometric Method

The specific rate constants for NO dissociation from nitrosyl complexes can also be determined using spectrophotometric methods, mainly when k$_{-NO}$ < 10$^{-2}$ s$^{-1}$, and diffusing of the reduced species generated at the electrode could occur, yielding miscellany data.

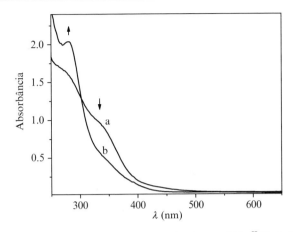

FIG. 3. Successive electronic spectra showing the decay of $[Ru^{II}(Hedta)(NO^0)]^-$ after reduction of $[Ru^{II}(Hedta)(NO^+)]$ with Eu(II) ions. $(I = 0.2\ M$, NaCF$_3$COO/CF$_3$COOH, pH 1.2, at 25°, [Ru] $= 1.8 \times 10^{-4}\ M$, [Eu] $= 1.8 \times 10^{-3}\ M$. (a) 90 s, (b) 1830 s after Eu addition.

When $[Ru^{II}(Hedta)NO^+]$ was mixed with excess of Eu(II) ions, the electronic absorption band at 370 nm ($\varepsilon = 170\ mol^{-1}\ L\ cm^{-1}$) disappeared and a pale yellow solution resulted, with a new absorbance at 340 nm and a shoulder at 280 nm. An isosbestic point is noticed at 300 nm (Fig. 3). The absorbance maximum at 340 nm was attributed to the reduced species $[Ru(Hedta)NO^0]^-$ formation, and the further decrease is due to NO dissociation (Zanichelli *et al.*, 2004). The $[Ru(Hedta)H_2O]^-$ has absorbance maximum at $\lambda_{max} = 280$ nm ($\varepsilon = 2900\ mol^{-1}\ cm^{-1}\ L$) and 427 nm ($\varepsilon = 260\ mol^{-1}\ cm^{-1}\ L$), so the increase of the absorbance maximum at 280 nm is attributed to the aquaspecies formation. When the ruthenium complex was in excess regarding Eu (II), only the band at 340 nm was used to probe the reaction.

Pseudo–first-order rate constants ($k_{obs}$) were determined from the slope of $\ln(A_\infty - A_t)$ against time by a standard linear least-square method. Values that differed less than 8% were averaged.

In addition, the fast replacement of H$_2$O ligand by pyrazine (pz) in the $[Ru(edta)(H_2O)]$ ($k_f = 2.0 \times 10^5\ M^{-1}\ s^{-1}$, pH $= 5.1$), the complex $[Ru(edta)pz]^{2-}$ ($\lambda_{max} = 463$ nm, $\varepsilon = 11,600 \pm 1000\ mol^{-1}\ cm^{-1}\ L$), is quite convenient to follow NO dissociation. In acid medium, this maximum shifts to 545 nm. Therefore, pz behaved as an auxiliary ligand to trap the initial product of the dissociation process (Fig. 4).

Table III summarizes the kinetic data obtained under the different experimental conditions and strongly suggests that NO aquates from $[Ru^{II}(Hedta)NO]^-$ and $[Ru^{II}(edta)(NO)]^{2-}$ at similar rates. Overall, the results support the release of NO to the medium, according to the following scheme:

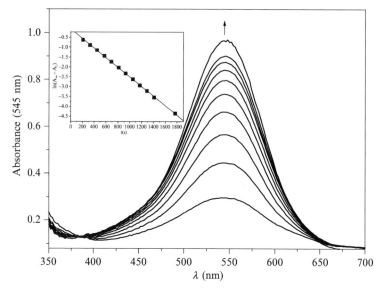

FIG. 4. Successive electronic spectra showing the formation of the $[Ru^{II}(Hedta)pz]^-$ complex at 545 nm, after reducing $[Ru^{II}(Hedta)NO^+]$ with Eu(II) ions in the presence of pyrazine. ($I = 0.2$ M, NaCF$_3$COO/CF$_3$COOH, pH 1.00, at 25°; [Ru] = $1.8 \times 10^{-4}$ M, [Eu] = $1.8 \times 10^{-3}$ M, [pz] = $3.6 \times 10^{-1}$ M. Inset: Plot of $\ln(A_\infty - A_t)$ versus time at 545 nm. Reprinted with permission from Zanichelli *et al.*, 2004.

TABLE III
OBSERVED RATE CONSTANTS ($k_{obs}$) FOR THE AQUATION REACTION OF
$[Ru^{II}(Hedta)NO^+]$ COMPLEX

| $C_{[Ru(Hedta)NO]}$ ($\times 10^3$ M) | $C^{II}_{Eu}$ ($\times 10^3$ M) | $C_{pz}$ ($\times 10^2$ M) | ($k_{obsd}$), s$^{-1}$ ($\times 10^3$) |
|---|---|---|---|
| 5.00 | 0.50 | | 1.7 |
| 5.00 | 1.50 | | 2.1 |
| 5.00 | 2.50 | | 2.1 |
| 1.00 | 2.00 | | 2.5 |
| 0.20 | 2.00 | 4.0 | 2.1 |
| 0.18 | 1.8 | 3.6 | 2.4 |
| 0.14 | 1.80 | 2.8 | 2.2 |
| 0.10 | 1.80 | 2.0 | 2.5 |
| 5.00* | 0.50 | | 2.1 |

$C_{[Ru(Hedta)NO]}$, concentration of the ruthenium complex; $C^{II}_{Eu}$, concentration of the reductor; $C_{pz}$, concentration of the pyrazine.

* Aquation in phosphate buffer, pH 7.4; all other data refer to pH 1.0.

Reprinted with permission from Zanichelli *et al.*, 2004.

$$[Ru^{II}(Hedta)NO^+] \xrightarrow[\text{fast}]{Eu(II)} [Ru^{II}(Hedta)NO^0]^-$$

$$+H_2O \downarrow \text{slow}$$

$$[Ru^{II}(Hedta)(H_2O)]^- + NO^0$$

$$+pz \downarrow \text{fast}$$

$$[Ru^{II}(Hedta)pz]$$

## Conclusion

Both electrochemical and spectrophotometric methods have been demonstrated to be useful for measuring these kinetic rate constants. In comparison to $[Ru(NH_3)L(NO)]$ complexes (range $10^{-2}$ to $10^{-1}$ s$^{-1}$), the release of NO from [Ru(Hedta)NO] is slower ($\sim 10^{-3}$ s$^{-1}$). In this particular case, spectrophotometric methods constitute a suitable tool for measuring the kinetic rate constant.

## References

Albery, W. J., and Bruckenstein, S. (1966). Ring-disc electrodes. Part 5—First-order kinetic collection efficiencies at the ring electrode. *Trans. Faraday Soc.* **62,** 1946–1954.

Bard, A. J., and Faukner, L. R. (1980). "Electrochemical Methods: Fundamentals and Applications." Wiley & Sons, New York.

Borges, S. S., Davanzo, C. V., Castelano, E. E., K-Schepector, J., Silva, S. L., and Franco, D. W. (1998). Ruthenium nitrosyl complexes with *N*-heterocyclic ligands. *Inorg. Chem.* **37,** 2670–2677.

Clarke, M. J. (2002). Ruthenium metallopharmaceuticals. *Coord. Chem. Rev.* **232,** 69–93.

Pletcher, D. (1991). "A First Course in Electrode Process." The Electrochemical Consultancy, Romsey.

Tfouni, E., Krieger, M., McGarvey, B. R., and Franco, D. W. (2003). Structure, chemical and photochemical reactivity and biological activity of some ruthenium amine nitrosyl complexes. *Coord. Chem. Rev.* **236,** 57–69.

Toledo, J. C., Mori, V., Silva, H. A. S., Scarpellini, Camargo, A. J., Bertoti, M., and Franco, D. W. (2004). Ruthenium tetramines as a model of nitric oxide donor compounds. *Eur. J. Inorg. Chem.* **53,** 1879–1885.

Torsoni, A. S., Barros, B. F., Toledo, J. C., Haun, M., Krieger, M. H., Tfouni, E., and Franco, D. W. (2002). Hypotensive properties and acute toxicity of *trans*-[Ru(NH_3)_4(POEt)_3(NO)] (PF_6)_3, a new nitric oxide donor. *Nitric Oxide Biol. Chem.* **6,** 247–254.

Zanichelli, P. G., Miotto, A. M., Estrela, H. F. G., Soares, F. R., Kassisse, D. M. G., Bratfisch, R. C. S., Castellano, E. E., Roncaroli, F., Parise, A. R., Olabe, J. A., Brito, A. R. M. S., and Franco, D. W. (2004). The [Ru(Hedta)NO]$^{0,1-}$ system. Structure, chemical reactivity and biological assays. *J. Inorg. Biochem.* **98**(11), 1921–1932.

## [5]   Sensitive and Isotope Selective ($^{14}$NO/$^{15}$NO) Online Detection of Nitric Oxide by Faraday–Laser Magnetic Resonance Spectroscopy

By RALPH GÄBLER and JOCHEN LEHMANN

### Abstract

The monitoring of trace amounts of nitric oxide (NO) is of great interest for biomedical applications. High-resolution infrared spectroscopy, like laser magnetic resonance spectroscopy (LMRS), is a versatile technique for the quantitative and isotope-selective analysis of low concentrations of NO. The ability to distinguish between different NO isotopomeres is of special interest for pharmaceutical and biomedical applications when tracer investigations are performed. With LMRS using Faraday modulation, a sensitivity of 1–2 ppbV for $^{15}$NO can be achieved. With further improvements and development of the sensor, we expect a sub-ppbV sensitivity.

### Introduction

Nitric oxide (NO) is a highly reactive molecule that is synthesized by various biological tissues. It plays a major role in the regulation of blood pressure, in nerve cell communication, and numerous physiological processes. The need to understand how and when this molecule participates in a chemical pathway *in vivo* has made it necessary to develop methods for its detection in biological matrices and fluids.

The capability of distinguishing between different biogenic NO sources by using isotope-labeled samples and precursors could spur major advances in the field of NO biology and chemistry. The production rate of biogenic NO is generally very low, so it is very difficult to carry out precise quantitative online measurements. Commonly used NO sensors based on chemiluminescence and electrodes show the ppbV sensitivity and below for NO required for biomedical applications but fail to distinguish between different isotopomeres of NO.

Spectroscopic techniques using monochromatic laser light are the most promising techniques to achieve both a high sensitivity and a high selectivity for isotopomeres.

METHODS IN ENZYMOLOGY, VOL. 396
Copyright 2005, Elsevier Inc. All rights reserved.

0076-6879/05 $35.00
DOI: 10.1016/S0076-6879(05)96005-9

## Spectroscopic Detection of Nitric Oxide

Spectroscopic sensing of NO is preferably carried out in the 5-$\mu$m wavelength region, where the fundamental vibrational band of NO is located. Various spectroscopic techniques have been demonstrated for NO sensing (Menzel *et al.*, 2001; Roller *et al.*, 2002), but most approaches are based on absorption measurements. Such measurements often require extra efforts to avoid interference from water and other molecules that are present in complex gas mixtures obtained from biological samples.

In spectroscopic trace gas detection, the optical medium consists of molecules, and the interaction between incoming light and the optical medium reveals the gas concentration. Monochromatic light is only influenced near discrete transition frequencies of the molecules. Far away from the resonance frequencies (*i.e.*, in most parts of the spectrum), the optical medium is completely transparent and does not show any refractive properties. The absorption of light is described by the Lambert-Beer law [Eq. (1)] with the absorption coefficient $\alpha$ and the length of the sample cell z.

$$I(z) = I_0 * \exp(-\alpha z) \tag{1}$$

The attenuation of the incoming light $I_0$ can be used to determine the absorption $\alpha$; in other words, the molecule concentration in the optical medium. In or near resonance, the medium not only absorbs the incoming light but also changes the phase velocity of light (dispersion). The dispersion of light is typically described by the refractive index n. Both effects are strictly connected, although most times just the absorption is used in trace gas detection.

Physical properties of open shell molecules (*e.g.*, radicals) like NO offer an alternative spectroscopic approach for the trace gas detection using the dispersion of light (Evenson *et al.*, 1980; McKellar, 1981). The absorption frequency of an open shell molecule can be tuned by an external magnetic field (Koch *et al.*, 1997), resulting in a Zeeman splitting (Fig. 1) of the magnetic substates.

The key to the detection of changes of the refractive index is the polarization of the incoming light. Linearly polarized light is a superposition of two circular components with reverse sense. Because magnetic substates of, for example, NO just interact with and attenuate one circular component of the light, the polarization of linearly polarized light is changed (Fig. 2).

Figure 2 is a schematic drawing of the physical changes that occur when a linearly polarized light passes an optical medium. The selective absorption of the second circular component of the incoming light (Fig. 2A) is indicated by a smaller diameter of the component. The selective dispersion

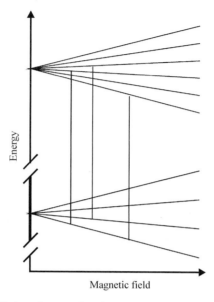

FIG. 1. Zeeman splitting of magnetic substates in an external magnetic field. At zero magnetic field, the substates have identical energy.

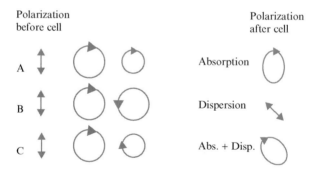

FIG. 2. Changes of the polarization of light. (A) Pure absorption; (B) pure dispersion; (C) combined absorption and dispersion.

of the second circular component (Fig. 2B) is indicated by a change of the position of the arrow. Because absorption and dispersion are strictly connected, Fig. 2C shows the behavior of the light intensity after the detection cell.

The dispersion of light can be easily detected by introducing an optical element with a well-defined axis of polarization, a linear polarizer detuned by 90 degrees to the polarization of the incoming light. In linear approximation (for small angles), the rotated intensity passing the polarizer is described by Eq. (2) with the rotation angle of the polarization $\Delta\phi$ and an offset angle $\delta$:

$$I(\delta) = I_0 * (\delta^2 + 2\delta * \Delta\phi) \tag{2}$$

Signal Generation in LMRS

To understand the signal generation of molecules (Ganser *et al.*, 2003) with a magnetic moment, we look at the simplest molecular transition, a Q (1/2) transition. The molecular transition occurs between the two states that are doubly degenerate and have equal magnetic moments.

Applying an external magnetic field results in the so-called Zeeman splitting of the molecular states (Fig. 1). A straightforward sinusoidal Zeeman modulation in combination with a linearly polarized beam does not produce a net signal at zero external field. The alternating modulation field defines the axis of quantization for the two circular components of the laser light. The magnetic substates coincide when the flux density $B_0$ is zero.

The allowed transitions in the experimental setup of Faraday-LMRS (Figs. 3 and 4) are $\Delta M = +1$ or $\Delta M = -1$. When the field is scanned through zero, one type shows increasing and the other one decreasing frequency. For linearly polarized light, both components give signals of opposite sign and, therefore, cancel exactly to zero.

To obtain a nonzero signal, an asymmetry for the two coinciding transitions must be introduced. Either a static external magnetic field that is used for Zeeman tuning of the molecular transitions or an optical analyzer, placed in front of the detector, lifts the degeneracy. This analyzer is offset at a small angle $\delta$ from the 90-degree crossed position for optimum sensitivity.

Without degeneracy, only one circular component of the incoming light interacts with the selected magnetic substate (either $\Delta M = +1$ or $\Delta M = -1$). Only one circular component is attenuated, which results in a rotation of the polarization and a nonzero signal.

Carbon Monoxide Laser–Based Faraday-LMRS

The central part of a Faraday-LMRS (Fig. 3) is the source of linearly polarized light. For detection of NO, a carbon monoxide (CO) laser (Mürtz *et al.*, 1999) in the midinfrared is a suitable and powerful light source. The resonance between the $X^2\Pi_{3/2}$ R(1.5) transition of the $^{14}N^{16}O$ molecule can be tuned into resonance (Zeeman tuning) with one the specific laser

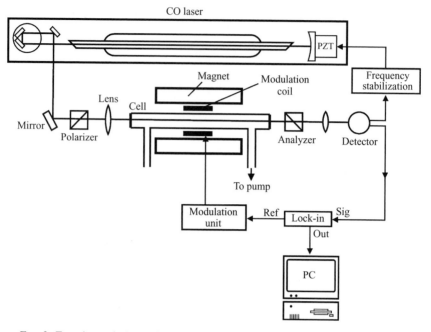

FIG. 3. Experimental setup of Faraday-laser magnetic resonance spectroscopy using a carbon monoxide laser.

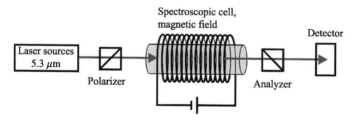

FIG. 4. Linear setup of new Faraday-laser magnetic resonance spectroscopy using a quantum cascade laser.

transition P(13) $\nu = 9 \rightarrow 8$ (1,884,349 cm$^{-1}$, 400-mW laser power, single mode) at a magnetic flux density of 0.1490 Tesla. For the $^{15}N^{16}O$, a resonance between the $X^2\Pi_{3/2}$ Q(1.5) transition and the CO laser transition P(17) $\nu = 10 \rightarrow 9$ (1,842,816 cm$^{-1}$, 200 mW, single mode) at a magnetic flux density of 0.1440 Tesla is used. A reproducible sensitivity of about 1 ppbV could be achieved in daily measurements for both isotopomeres.

The CO laser is stepwise tunable, running on single molecular transitions of the CO molecule. The flow laser normally is liquid nitrogen–cooled stabilized on the gain maximum of a particular laser line by means of a standard stabilization technique (frequency modulation). The laser output is deflected into the magnet (length of detection zone, 200 mm; it is surrounded by a small modulation coil for Zeeman modulation at 8 kHz). A polarizer in front of the magnet entry ensures that the incoming laser beam is well polarized. The polarization of the emerging beam is analyzed by a Rochon prism (analyzer) that is almost crossed with respect to the polarization of the incoming laser beam. The infrared radiation is detected by a liquid nitrogen–cooled InSb detector and fed into an amplifier. The analogue signal output is converted and processed by a personal computer.

The flow rate of the gas stream through the sensor can be varied with a needle valve and a pressure controller at the exit of the detection cell. For time-resolved measurements of breath maneuvers, a flow of 800 ml/min is sufficient. Depending on the application [*e.g.*, lung function tests (Heller *et al.*, 2004)], much smaller flows can be chosen (40 ml/min). The working pressure inside the cell is 25 mbar.

The drawbacks of the CO laser–based Faraday-LMRS are the bulky and cumbersome experimental setup of the liquid nitrogen–cooled laser and the liquid helium–cooled magnet. You need an optical table with dimensions of 200 × 120 cm to run the experiment. Furthermore, with a CO laser, the detection of NO cannot be performed at the optimal wavelength, because the molecular transitions of CO running on the laser are not in resonance with the molecular transitions of NO.

### Quantum Cascade Laser–Based Faraday-LMRS

Using a tunable quantum cascade laser (QCL) for the detection, the optimal wavelength (1875.8 cm$^{-1}$) for the detection of $^{15}$N$^{16}$O can be chosen (Ganser *et al.*, 2004). With the laser light in resonance with the molecular transition of NO, the static external magnetic field for Zeeman tuning of the molecular transition can be omitted. Therefore, the setup of the Faraday-LMRS can be simplified and miniaturized (Fig. 4). The new linear setup of the Faraday-LMRS has the dimensions of 120 × 30 × 45 cm. Within first measurements of the sensitivity, a value of 1–2 ppbV for the $^{15}$N$^{16}$O isotopomere with a laser intensity of 20 mW can be achieved. With further optimization, we expect a sub-ppbV sensitivity for the labeled $^{15}$N$^{16}$O isotopomere.

Within the tunability range of the QCL, a molecular transition of the normal $^{14}$N$^{16}$O isotopomere can be reached by changing the laser current and temperature. The modulation strength of this transition is about

10 times smaller than the strength of the optimal $^{15}N^{16}O$ transition. For biomedical applications, these transitions are ideal for tracer investigations because the abundance of $^{15}N^{16}O$ is just 0.36% of the normal isotopomere. A fast switching between these two transitions measures nearly simultaneously the concentrations of both isotopomeres.

The volume of the detection cell is about 15 ml. With a working pressure inside the cell of 25 mbar, this corresponds to a volume of about 0.4 ml at ambient pressure. The flow through the cell can be selected with a needle valve at the entrance and a pressure controller at the exit of the cell. It covers a wide range from about 20 ml/min to 1 liter/min.

## References

Evenson, K. M., Saykally, R. J., Jennings, D. A., Curl, R. F., and Brown, J. M. (1980). *In* "Chemical and Biochemical Application of Lasers" (C. B. Moore, ed.). Academic, New York.

Ganser, H., Urban, W., and Brown, J. M. (2003). The sensitive detection of NO by Faraday modulation spectroscopy with a quantum cascade laser. *Mol. Phys.* **101**(4–5), 545–550.

Ganser, H., Horstjann, M., Suschek, C. V., Hering, P., and Mürtz, M. (2004). Online monitoring of biogenic nitric oxide with a QC laser-based Faraday modulation technique. *Appl. Phys. B* **78**(3-4), 513–517.

Heller, H., Korbmacher, N., Gäbler, R., Brandt, S., Breitbach, T., Jürgens, U., Grohe, C., and Schuster, K.-D. (2004). Pulmonary $^{15}NO$ uptake in interstitial lung disease. *Nitric Oxide* **10**, 229–232.

Koch, M., Luo, X., Mürtz, P., Urban, W., and Mörike, K. (1997). Detection of small traces of $^{15}NO$ and $^{14}NO$ by Faraday LMR spectroscopy of the corresponding isotopomeres of nitric oxide. *Appl. Phys. B* **64**, 683.

McKellar, A. R. W. (1981). Mid-infrared laser magnetic resonance spectroscopy. *Faraday Discuss. Chem. Soc.* **71**, 63.

Menzel, L., Kosterev, A. A., Curl, R. F., Tittel, F. K., Gmachl, C., Capasso, F., Sivco, D. L., Bailargeon, J. N., Hutchinson, A. L., Cho, A. Y., and Urban, W. (2001). Spectroscopic detection of biological NO with a quantum cascade laser. *Appl. Phys. B* **72**, 859.

Mürtz, P., Menzel, L., Bloch, W., Hess, A., Michel, O., and Urban, W. (1999). LMR spectroscopy: A new sensitive method for on-line recording of nitric oxide in breath. *J. Appl. Physiol.* **86**(3), 1075–1080.

Roller, C., Namjou, K., Jeffers, J. D., Camp, M., Mock, A., McCann, P. J., and Grego, J. (2002). Nitric oxide breath testing by tunable diode laser absorption spectroscopy: Application in monitoring respiratory inflammation. *Appl. Opt.* **41**, 6018.

# [6]   Qualitative and Quantitative Determination of Nitrite and Nitrate with Ion Chromatography

By Reinhard Kissner and Willem H. Koppenol

## Abstract

Reactions of nitrogen monoxide and peroxynitrite often yield nitrite and nitrate as stable end products. The simultaneous detection of these two ions by anion chromatography with conductivity detection is described. The chromatographic system used is similar to conventional isocratic or gradient high-performance liquid chromatography equipment. The columns are packed with ion-exchanging resins instead of silica-derived adsorbents. Conductivity, though inherently nonspecific, has the advantage of covering a linear dynamic signal range of five orders of magnitude, which is far better than spectroscopic techniques; these are generally limited to two orders of magnitude. Typical run times per chromatogram are 15–30 min, and sample and standard concentrations can be between 100 n$M$ and 10 m$M$. Injection volumes vary from 5 to 200 $\mu$l. Unlike with the Griess method, which determines only nitrite, a true mass balance from independent signals can be obtained if nitrate and nitrite are the only products.

## Introduction

In nitrogen monoxide–related biochemistry, it is often useful to determine quantitatively the reaction products to deduce the reaction mechanisms and stoichiometries. Among the commonly expected substances are nitrite and nitrate. The ratio at which they are produced is often a telltale for the reaction pathways (Kissner and Koppenol, 2002). The classic method for nitrite identification and quantitative determination is the Griess reaction (Granger et al., 1996; Griess, 1864; Nims et al., 1996; Wishnok et al., 1996; Yokoi et al., 1996), which uses diazotation and coupling to form a purple dye. Nitrate cannot be determined directly this way, because only the sum of nitrite and nitrate is accessible after reduction of the nitrate. The determination by the difference of two analytical results suffers the drawback of increased error.

Ion chromatography with ultraviolet (UV) and electrochemical detection has been used to determine nitrite and nitrate (Stratford, 1999). This technique is an instrumental method that has been used in trace analytical and environmental chemistry for quite a while, especially for the

METHODS IN ENZYMOLOGY, VOL. 396
0076-6879/05 $35.00
DOI: 10.1016/S0076-6879(05)96006-0

determination of a huge variety of anions, which are not selectively detectable by spectroscopic methods. The chromatographic system used is similar to conventional isocratic or gradient high-performance liquid chromatography (HPLC) equipment. The columns, however, are packed with ion-exchanging resins instead of silica-derived adsorbents. Depending on the application, cations or anions can be determined; there even exist automated systems that split the sample simultaneously onto a cation and anion exchanging column. The detection usually is based on changes in conductivity, a nonspecific but rather sensitive technique. Conductivity has the advantage that it covers an enormous linear dynamic signal range of five orders of magnitude, which is far better than spectroscopic techniques; these are generally limited to two orders of magnitude. Typical run times per chromatogram are 15–30 min, and sample and standard concentrations can be between 100 and 10 m$M$. Injection volumes vary from 5–200 $\mu$l. Any existing conventional HPLC system can be converted to ion chromatography by introducing a suitable column and adding the conductivity detector. The pump, the sampler, and the signal processing systems can be used without modifications. Such an extension allows for detection with electronic background conductivity suppression (Gjerde et al., 1979, 1980). The conductivity caused by the eluent alone is subtracted electronically from the total signal, which requires amplifiers designed for the accurate processing of small signal changes on a large but steady background signal. The more classic ion chromatography design circumvents this problem by eliminating the eluent ions before detection, a method called *chemical suppression* (Small et al., 1975). It works well for two popular eluents, namely carbonate and hydroxide solutions, which are simply neutralized by a proton-loaded ion-exchange column or in a small membrane reactor between column and detector. Modern hollow-fiber membrane reactors generate the acid by water electrolysis and do not require a regeneration cycle. With its low background conductivity, chemical suppression allows for the detection of nanomolar concentrations. The drawback of chemical suppression in anion chromatography is the restricted eluent choice; in particular, the use of acidic eluents is not possible. The reverse is true for cation chromatography, in which an eluent with a low pH can be neutralized directly by water electrolysis in a membrane reactor or a hydroxide-loaded ion-exchange column.

## Procedure

The choice of the column is crucial with respect to the composition of the sample. For low ionic strength samples, standard anion exchange columns like the Hamilton PRP-X100 or the Dionex IonPac AS18 can be used. However, if the sample contains high concentrations of buffer ions, special high-capacity columns (e.g., Dionex IonPac AS11-HC, Metrohm Metrosep A Supp 1, Phenomenex Star-Ion A300 HC) should be used. The reason is that because the ion-exchanging polymer can be saturated with high concentrations of buffer ions even if they bind only weakly, retention times will be shortened for the ions of interest, and peak resolution will be impaired. With high-capacity columns, the overload problem is reduced at the cost of longer retention times. In case of a huge overload, one should consider dilution of the sample, which improves resolution and retention times while the loss in signal intensity is partially compensated by the high sensitivity of the conductometer. Furthermore, the more dilute sample buffer influences the eluent pH less, the stability of which is essential for constant background conductivity. The weaker signal is often better resolved on a reduced and less distorted background.

## Examples

Some examples of nitrite and nitrate chromatograms obtained with acidic and alkaline eluents are discussed here. The time axis unit in chromatography is minutes in most cases. The units of the signal axis are usually not indicated, because peak area and height are determined by calibration with standard samples. Most detectors do not indicate conductance values but only a proportional voltage.

The graphs in Figs. 1, 2, and 3 were obtained with acidic eluent and electronic background suppression. Figure 1 shows the chromatogram of a standard solution, 20 $\mu M$ in each nitrite and nitrate, with 1 m$M$ phosphate buffer. Peak 1 represents nitrite, and Peak 2 nitrate. The sharp dip in front of Peak 1 stems from dihydrogen phosphate, which has a lower specific conductance than the eluent electrolyte. At the very beginning, a negative peak is obtained because the sample has a lower total ionic strength and conductance than the pure eluent, which is, therefore, diluted. The positive peak immediately following the negative peak is caused by cations in the sample, which are not or hardly retained. The large peak at 14 min corresponds to the eluent anions displaced by the sample anions when they entered the column.

Figure 2 shows the chromatogram of a 20-fold diluted sample from a 100-$\mu M$ peroxynitrite solution freshly decomposed at pH 7.8 in 100 m$M$ phosphate buffer. Peaks 1 and 2 are caused by nitrite and nitrate again. Because the buffer concentration (5 m$M$) is higher than that of the eluent, there is a positive deflection at the beginning instead of a dilution dip. Peak 3 is chloride and peak 4 is sulfate, both impurities in the phosphates.

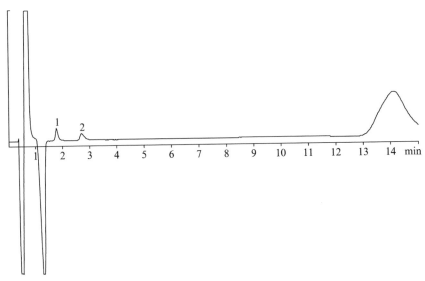

FIG. 1. Acidic eluent: phthalate buffer 2 m*M*, pH 5.0 set with NaOH, 7.6% acetone, column: Hamilton PRP-X100, 125 × 4 mm, polystyrene/divinylbenzene copolymer with quaternary ammonium functions. Sample: standard, 20 $\mu M$ nitrite, 20 $\mu M$ nitrate in 1 m*M* phosphate buffer, pH = 5.6, time axis in min, flow: 2 mL/min.

Figure 3 is a chromatogram obtained from a 1-m*M* sodium hydroxide solution used to collect the nitrogen monoxide gas that evolved from a sample of nitrate that was reduced by boiling it with copper metal powder and sulfuric acid. It can be neatly seen that the aerobic oxidation of nitrogen monoxide in alkaline solution leads preferably to nitrite, because the nitrate Peak 2 is close to the detection limit. Peak 1 is nitrite. Peak 4 is sulfate, as in the previous example. Peak 3, however, is not chloride, which corresponds to the tiny peak between Peaks 3 and 1. It was finally identified as fluoride, which was present as an impurity in the sulfuric acid used and was enriched in the distillation because hydrogen fluoride is rather volatile.

Figure 4 shows the separation capability of a column with higher capacity, together with the advantages of chemical suppression in a system with alkaline eluent. Peaks 1 and 2 represent nitrite and nitrate, each 1 $\mu M$, and the huge peak is due to 10 m*M* phosphate buffer. A similar problem is solved in the chromatogram in Fig. 5. Here, 20 $\mu M$ peroxynitrite was reduced with a large excess of iodide. Again, Peaks 1 and 2 are the clearly separated nitrite and nitrate. The small later peaks were not identified, but

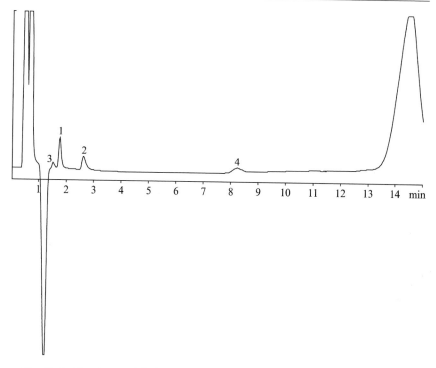

Fig. 2. Acidic eluent: phthalate buffer 2 m$M$, pH 5.0 set with NaOH, 7.6% acetone, column: Hamilton PRP-X100, 125 × 4 mm, polystyrene/divinylbenzene copolymer with quaternary ammonium functions. Sample: nitrite and nitrate, concentrations sum 5 $\mu M$ in 5m$M$ phosphate buffer, pH = 7.8, time axis in min, flow: 2 mL/min.

they could be iodine species of higher oxidation states. The huge peak that elutes after 13 min stems from 1 m$M$ iodide. The 0.1 m$M$ perchlorate also present is obscured by the iodide.

Interferences

With regard to the nitrite determination, the most serious interference is chloride, which elutes just before nitrite. A large excess of chloride can be removed chemically by precipitation before chromatography. Smaller chloride excesses can be handled with an eluent gradient system that "stretches" the initial part of the chromatogram. The problem of high chloride concentrations applies to, for example, biological samples

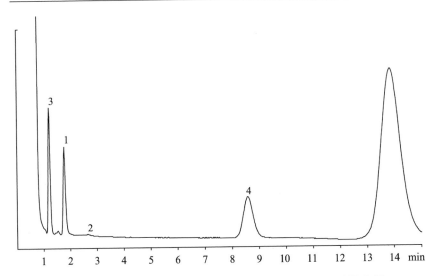

FIG. 3. Acidic eluent: phthalate buffer 2 m$M$, pH 5.0 set with NaOH, 7.6% acetone, column: Hamilton PRP-X100, 125 × 4 mm, polystyrene/divinylbenzene copolymer with quaternary ammonium functions, flow: 2 mL/min. Sample: nitrite and nitrate after absorption of nitrogen monoxide in 2 ml air-saturated 1 m$M$ NaOH, time axis in min. Nitrogen monoxide was obtained from the reduction of 10 ml 20 $\mu M$ nitrate with copper metal.

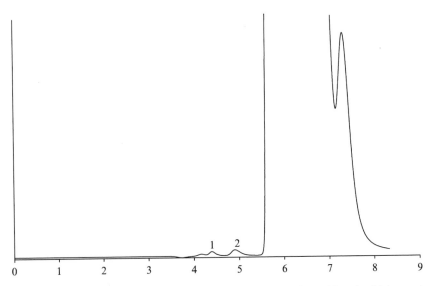

FIG. 4. Alkaline eluent: 1.7 m$M$ NaHCO$_3$/1.8 m$M$ Na$_2$CO$_3$, column: Metrohm Metrosep A Supp 3, 250 × 4.6 mm, polystyrene/divinylbenzene copolymer. Sample: standard 1 $\mu M$ nitrite, 1 $\mu M$ nitrate in 10 m$M$ phosphate buffer, pH = 5.6, time axis in min, flow: 2 mL/min.

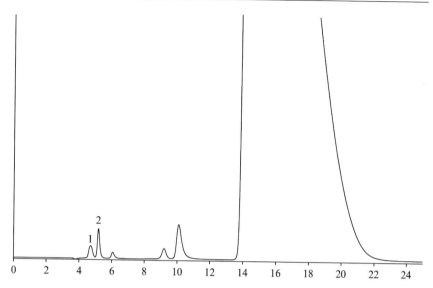

FIG. 5. Alkaline eluent: 1.7 m$M$ NaHCO$_3$/1.8 m$M$ Na$_2$CO$_3$, column: Metrohm Metrosep A Supp 3, 250 × 4.6 mm, polystyrene/divinylbenzene copolymer. Sample: nitrite and nitrate from the reaction of 20 $\mu M$ peroxynitrite with 1 m$M$ sodium iodide in 0.1 m$M$ HClO$_4$, time axis in min, flow: 2 mL/min.

(Stratford, 1999). In the case of a complex mixture, one should apply a highly specific method to determine nitrite, such as nitrosation of diphenylamine, followed by polarographic detection (Chang *et al.*, 1977). High-pressure side-eluent gradients typically require at least two pumps to generate a gradient. If the eluent is hydroxide, the gradient can be produced elegantly with only one pump, because it is generated by water electrolysis between the pump and the injection valve. Ultrafiltration of all samples is recommended, as for all HPLC experiments.

Conclusions

Because nitrite and nitrate have different retention times for all working ion chromatographic column/eluent combinations, they can be determined directly and simultaneously. In contrast to the Griess method, a true mass balance from independent signals is obtained if nitrate and nitrite are the only products. A lack of mass balance allows one to conclude that nitrogen has been lost in the form of gaseous products, such as nitrogen monoxide.

References

Chang, S.-K., Kozenlauskas, R., and Harrington, G. W. (1977). Determination of nitrite ion using differential pulse polarography. *Anal. Chem.* **49,** 2272.
Gjerde, D. T., Fritz, J. S., and Schmuckler, G. (1979). Anion chromatography with low-conductivity eluents. *J. Chromatogr.* **186,** 509.
Gjerde, D. T., Schmuckler, G., and Fritz, J. S. (1980). Anion chromatography with low-conductivity eluents. II. *J. Chromatogr.* **187,** 35.
Granger, D. L., Taintor, R. R., Boockvar, K. S., and Hibbs, J. B., Jr. (1996). Measurement of nitrate and nitrite in biological samples using nitrate reductase and Griess reaction. *Methods Enzymol.* **268,** 142.
Griess, P. (1864). On a new series of bodies in which nitrogen is substituted for hydrogen. *Philos. Trans. R Soc. London* **154,** 667.
Kissner, R., and Koppenol, W. H. (2002). Product distribution of peroxynitrite decay as a function of pH, temperature, and concentration. *J. Am. Chem. Soc.* **124,** 234.
Nims, R. W., Cook, J. C., Krishna, M. C., Christodoulou, D., Poore, C. M. B., Miles, A. M., Grisham, M. B., and Wink, D. A. (1996). Colorimetric assays for nitric oxide and nitrogen oxide species formed from nitric oxide stock solutions and donor compounds. *Methods Enzymol.* **268,** 93.
Small, H., Stevens, T. S., and Bauman, W. C. (1975). Novel ion exchange chromatographic method using conductimetric detection. *Anal. Chem.* **47,** 1801.
Stratford, M. R. L. (1999). Measurement of nitrite and nitrate by high-performance ion chromatograpy. *Methods Enzymol.* **301,** 259.
Wishnok, J. S., Glogowski, J. A., and Tannenbaum, S. R. (1996). Quantitation of nitrate, nitrite, and nitrosating agents. *Methods Enzymol.* **268,** 130.
Yokoi, I., Habu, H., Kabuto, H., and Mori, A. (1996). Analysis of nitrite, nitrate, and nitric oxide synthase activity in brain tissue by automated flow injection technique. *Methods. Enzymol.* **286,** 152.

# [7]  Electrochemical Detection of Nitric Oxide in Biological Fluids

*By* BARRY W. ALLEN, JIE LIU, and CLAUDE A. PIANTADOSI

Abstract

The challenges that must be overcome in order to detect nitric oxide (NO) in biological fluids include its low physiological concentration ($\bar{1}nM$) and its short half-life (a few seconds or less). Electrochemistry is capable of making such measurements, if certain principles, both biological and electrochemical, are kept in mind. We discuss these principles and demonstrate an example of practical measurement by detecting NO release in a drop of blood suspended within the reference electrode of an electrochemical cell. We elicit the NO release by decreasing the oxygen concentration in the gaseous atmosphere surrounding the drop.

METHODS IN ENZYMOLOGY, VOL. 396                                      0076-6879/05 $35.00
                     DOI: 10.1016/S0076-6879(05)96007-2

## Introduction

Electrochemistry is well suited to measuring nitric oxide (NO) in biological fluids because it can detect authentic NO in real time and *in situ*. In addition, NO electrodes can be made small enough to be used for many applications *in vivo*. However, there are pitfalls to making reliable electrochemical measurements in any system, especially in complex biological systems. In a previous volume of *Methods in Enzymology,* we discussed some of the difficulties involved in making such measurements along with strategies for overcoming those difficulties, and we gave the example of measuring NO in circulating blood *in vivo* (Allen *et al.*, 2002). In this chapter, we discuss the somewhat different challenges involved in measuring NO in biological fluids that are maintained in contact with a gaseous environment, in order to study such processes as the sequestration or release of NO from blood cells as they move between regions of high and low $PO_2$ levels.

## Nitric Oxide in Blood

The interactions of blood with NO are complex, partly reflecting the nature of blood itself—variously a solution, suspension, and emulsion of cells, lipids, proteins, gases, electrolytes and water—and partly because of the facility with which NO, as a free radical, can coordinate, bond, or react with many other chemical species, particularly with those containing oxygen, transition metals, or sulfur. The physiological consequences of these interactions are also diverse and span extremes: Blood can either destroy or preserve NO biological activity. Thus, a concentrated hemoglobin solution can rapidly extinguish NO biological activity (Lancaster, 2000). However, within the supporting chemical milieu of the red blood cell, hemoglobin can bind NO and then release it at another time and place, with bioactivity intact (McMahon *et al.*, 2000; Pawloski *et al.*, 2001). In addition, NO can react with components of the blood to produce other species that have biological activities of their own, such as peroxynitrite and nitrotyrosine (Beckman *et al.*, 1997; Crow and Beckman, 1995). Elucidation of the mechanisms by which the fate of NO in the blood is determined is under investigation in a number of laboratories. Of particular interest is the question of how the interaction of red blood cells with NO might be modulated by changes in oxygen content of the blood.

As the blood circulates between the meshwork of capillaries surrounding an alveolus and the capillary bed of intensely metabolizing tissue, its oxygen content may fall fivefold. Typically, the vessels from which gas exchange takes place between the blood and its immediate environment are a few hundred microns in diameter. To illustrate the principles and problems involved in detecting the status of NO in the blood *in vitro*, we present experimental data from preparations in which an NO-selective electrode was immersed in a drop of heparinized mammalian blood suspended within a flowing and humid gas mixture, which keeps the blood fluid for several hours and allows NO measurements to be made under a wide range of experimental conditions.

## Materials and Methods

### New Experimental Approach

Small aliquots of blood or other biological fluids can be suspended from an electrode array by means of capillarity. For example, we suspended a 20-$\mu$l drop of rabbit aortic blood restrained by surface tension within a helix of silver wire 1.85 mm in internal diameter. The helix was coated with silver chloride so it could function as the electrically stable reference electrode of a two-electrode electrochemical cell. Centered within the drop and coaxial with the reference electrode was an NO–specific electrode, approximately 100 $\mu M$ in diameter and 3 mm long. The drop and the associated electrodes were housed within a water-jacketed 4-ml airspace, which was maintained at 30° (Fig. 1). Humidified gas mixtures were

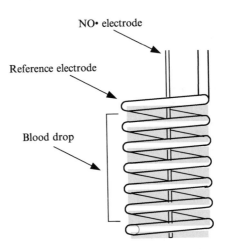

NO• electrode

Reference electrode

Blood drop

FIG. 1. Blood drop experiments were conducted by suspending approximately 20 $\mu$l of rabbit arterial blood within the helical reference electrode. The blood drop was held in place by capillarity and kept fluid by a continuous flow of warm humid gas.

allowed to flow past the drop and were alternated between an air–$CO_2$ mixture (20% $O_2$, 5% $CO_2$, 75% $N_2$) and a $CO_2$–nitrogen mixture (5% $CO_2$, 95% $N_2$). Gas flow was maintained at constant rate.

## Electrodes

NO-specific electrodes were fabricated according to a procedure described in detail by Wang *et al.* (submitted for publication). Briefly, platinum wires, approximately 100 $\mu$ in diameter, were coated with aligned, multiwalled carbon nanotubes using the fluoropolymer Nafion as adhesive. The nanotubes were subsequently coated with ruthenium, using chemical vapor deposition (Fig. 2). Finally, the electrodes were coated with five additional coats of Nafion, which excludes anions and thereby rejects many substances that could be oxidized on the electrode and produce a signal that could confound that of NO (Bedioui *et al.*, 1994). The combination of high specific surface area provided by the nanotubes with the ability of ruthenium to catalyze the electrochemical oxidation of NO (Allen *et al.*, 2000) was used to increase the response of the electrode to NO, because the free NO concentration in blood is expected to be very small (Allen *et al.*, 2002).

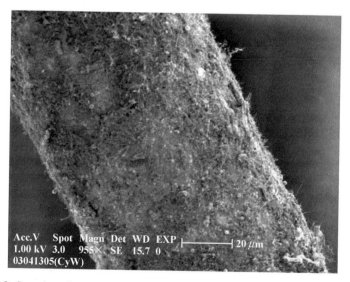

FIG. 2. Scanning electron micrography (955×) reveals surface characteristics of the platinum filament electrode coated with carbon nanotubes over which a film of ruthenium was deposited using chemical vapor deposition.

## Blood Samples and Chemical Reagents

Rabbit aortic blood was withdrawn into 3-ml syringes containing 7 units of lyophilized heparin, according to procedures approved by the Duke University Institutional Review Board, and kept on ice for up to 30 min before use.

Although Nafion is well known to confer a high degree of specificity to NO electrodes (Pariente *et al.*, 1994), we tested the electrode for its response to a small group of relevant substances that could be expected to be present in significant concentrations in blood. Thus, we prepared 100-$\mu M$ solutions in deionized water of the following: ascorbate (Sigma 99%); L-cystine (Fluka, >99.5%); 2,3-diphospho-D-glyceric acid (DPG) (Sigma, 98%); sodium nitrite (Sigma, 95.5%); and sodium nitrate (Sigma, 95.5%).

## Electrochemical Methods

Electrochemical measurements were performed using amperometry, with a constant potential of +675 mV (vs. Ag/AgCl,) maintained at the working electrode, with current the measured variable. Immediately before NO measurements, the electrodes were activated electrochemically by applying alternating potentials of 200 and 800 mV for 250 ms each at 500-ms intervals for a total of 120 s, a procedure modified from a method we have published (Allen and Piantadosi, 2003). Before and after each experiment, the composite resistance of the electrochemical cell was measured three times, and if the final average resistance was more than 10% greater than the initial average, it is assumed that the electrode was fouled or that the blood drop had dried, and data from that experiment were not used. In addition, the electrode assembly was always examined visually under 30× magnification after each experiment, and if the blood was not fluid, or if the drop did not fill the helix, the data were not used. Electrochemical activation and measurements of NO oxidation and electrical resistance can be performed on any suitable potentiostat. For example, we used a BAS 100 B/W potentiostat equipped with a low-current module.

## Results

### Selectivity of the Sensor for Nitric Oxide

The amperometric responses (nA/$\mu M$) of the electrode for ascorbate, L-cysteine, DPG, sodium nitrite, and sodium nitrate were normalized to the response for NO and are shown in Table I.

TABLE I
SELECTIVITY OF SENSOR FOR NITRIC OXIDE

| | | |
|---|---|---|
| NO | $21.14 \pm 6.35$ nA/$\mu M$ | 100.00% |
| Ascorbate | $0.44 \pm 0.03$ nA/$\mu M$ | 2.08% |
| Cysteine | $0.17 \pm 0.04$ nA/$\mu M$ | 0.80% |
| DPG | $0.05 \pm 0.01$ nA/$\mu M$ | 0.24% |
| Nitrate | $0.10 \pm 0.01$ nA/$\mu M$ | 0.47% |
| Nitrite | $0.09 \pm 0.02$ nA/$\mu M$ | 0.43% |

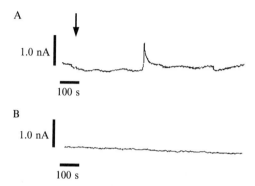

FIG. 3. Amperometric recordings of blood drop experiments. (A) The composition of the gas mixture flowing slowly past a blood drop was changed (at arrow) from an oxygen-containing mixture to an oxygen-free mixture. Approximately 350 s later, a clear signal appears that is consistent with nitric oxide (NO) oxidation. (Note that the standard electrochemical convention in which oxidation is shown in the downward direction has been reversed to demonstrate the presumed increase in NO.) (B) In a control experiment in which the blood was only exposed to the oxygen-containing mixture, there is a complete absence of an oxidation signal.

### Responses to Changing Gas Mixtures

Signals consistent with NO oxidation were seen in blood-drop experiments involving gentle deoxygenation. In the system of Fig. 1, these apparent NO oxidation signals were first detected from 200 to 400 s after the flowing gas was changed from the oxygen-containing mixture to the oxygen-free mixture. An initial oxidation spike was followed by a continuous signal of 1–2 nA, which would correspond to nitric oxide in the tens of nM (Fig. 3A).

Such signals were not seen when blood was exposed for the same length of time to only the oxygen-containing mixture (Fig. 3B).

Discussion

The blood-drop preparation described here may represent a useful approach for further investigation of the response of NO levels in various biological fluids to alterations in concentration of other dissolved gases, such as oxygen. However, at this initial stage, care must be used in interpreting data. For example, electrochemistry can detect the transfer of electrons between electrodes but cannot definitively identify the specific source of those electrons. Various procedures that increase the specificity of a particular electrode, by limiting the potential to that at which a particular substance is oxidized or reduced or by applying coatings to the electrode that exclude species that have certain characteristics of charge or size, only increase the probability that the signal measured is proportional to the expected analyte. It is always useful to confirm any experimental result by using other nonelectrochemical methods, such as spectroscopy, until confidence can be established concerning the source of the electrochemical signal. This is particularly true with the biological chemistry of NO in blood, given the present state of this field in which our curiosity exceeds the sensitivity and specificity of our experimental tools.

However, in general, guidelines for applying electrochemistry to the study of NO biology are straightforward and have been discussed previously in greater detail (Allen *et al.*, 2002). These guidelines include the following:

1. *Control for interventional artifacts*: All interventions used to elicit changes in NO electrochemical activity should themselves be eliminated as sources of artifacts that could mimic an authentic NO signal. This requires careful design of control experiments.

2. *Control for changes in fluid path or fluid composition between electrodes*: The current measured between the electrodes in an electrochemical cell—consisting of two or more electrodes and an electrolyte solution—is a function of the substances being electrolyzed *and* of the electrical resistance between electrodes. This means that anything that changes the electrical resistance, such as movement of the electrodes relative to each other, or that changes the resistivity of the electrolyte itself, such as evaporation of solvent, will alter the measured current and, therefore, could be confounded with a change in analyte concentration.

3. *Control for changes in electrochemical background oxidation or reduction*: Many analytes that are present in biological systems in low concentrations (low $\mu M$ and below) are detected on top of a steady background provided by the oxidation or reduction of substances such as $Cl^-$ that are present in much higher concentrations. Ignoring this

important fact can lead to serious misinterpretation of data. For example, if an NO electrode is inserted inside an excitable cell (a neuron or a muscle cell) and the reference electrode is left in the fluid bathing the cell, then the two electrodes are separated by a membrane that changes its ionic permeability when the cell is stimulated. The resulting "signal" may simply be a rediscovery of the action potential, work first done decades ago.

4. *Calibrate according to experimental conditions*: The electrode should be calibrated in materials and under conditions as close as possible to those in which biological NO is to be detected.

5. *Control for changes in temperature*: Since rates of chemical reaction change with temperature, either constant temperature must be maintained during an electrochemical experiment or the electrode's response to the changing temperatures must be determined.

6. *Control for changes in hydrodynamics*: Reaction rates also change with alterations in mass transport of the reactants or products. Therefore, fluid flow must be held constant, or its effects must be accounted for.

7. *Select an appropriate time constant*: All experimental apparatus, both mechanical and electronic, have time constants that impose a delay between a stimulus and its signal. Because NO can be very fugitive in biological systems, the composite time constant of all components of the experimental system must be kept below the expected time course of signal.

8. *Avoid mistaking NO for oxygen when using a polarographic oxygen electrode*: NO and oxygen are oxidized at different potentials but are electrochemically reduced at the same potential. The practical consequences of this are that NO electrodes, which are normally operated at oxidizing potentials, do not detect oxygen. But oxygen electrodes, which are normally operated at reducing potentials, do detect both NO and oxygen. Therefore, when NO and oxygen are to be measured simultaneously using electrochemistry, the response of the oxygen electrode to NO must be known so it can be corrected using readings from the NO-specific electrode.

Conclusions

A highly sensitive electrochemical system can be designed to detect n$M$ concentrations of NO activity. In such a system, we consistently observed a clear oxidation signal at an NO-specific electrode, from 200 to 400 s after a suspended drop of rabbit arterial blood was exposed to a decrease in ambient P$O_2$. The interval of hundreds of seconds between the onset of hypoxia and the appearance of the oxidation signal appears to reflect, at least in part, the time it takes for oxygenated rabbit hemoglobin to desaturate under these conditions. It is worth noting that the oxygen saturation curve for rabbit hemoglobin is shifted to the left compared to human hemoglobin (Holland and Calvert, 1995), and therefore, a lower $PO_2$ is required to unload its oxygen. Also, because we did not measure the change in either ambient $PO_2$ or blood-drop $PO_2$ in this preliminary work, we do not yet know how long it takes for either of these oxygen levels to reach critical values in this kind of experiment. Because gas flows must be deliberately kept low in order to prevent drying of the blood drop, $PO_2$ will change slowly.

Our data are consistent with the ability to detect specific NO release from biological fluids at n$M$ concentrations using carefully configured NO-sensitive electrochemical systems. Our preliminary blood-drop experiments implicate the transformation of bound or sequestered NO to a form that is free or that is at least electrochemically active. And this transformation is correlated with an experimental condition—hypoxia—in which hypoxemia could be induced in the captured blood drop. At this point, we cannot assign this release to a particular source in the blood, for example, the red cells or the plasma, although the ability of the red blood cells to dilate blood vessels by means of release of NO bioactivity has been demonstrated in several laboratories (James *et al.*, 2004; Stamler *et al.*, 1997).

## Acknowledgments

This work was supported by the Office of Naval Research. We also acknowledge the excellent technical assistance of Dr. Cuiying Wang in preparing the electrodes and for providing the scanning electron micrograph. We are grateful to Dr. Shaoming Huang for supplying the multiwalled carbon nanotubes.

## References

Allen, B. W., Coury, L. A., and Piantadosi, C. A. (2002). Electrochemical detection of physiological nitric oxide: Materials and methods. *Methods Enzymol.* **359**, 125–134.

Allen, B. W., and Piantadosi, C. A. (2003). Electrochemical activation of electrodes for amperometric detection of nitric oxide. *Nitric Oxide* **8**, 243–252.

Allen, B. W., Piantadosi, C. A., and Coury, L. A., Jr. (2000). Electrode materials for nitric oxide detection. *Nitric Oxide* **4**(1), 75–84.

Beckman, J. S., Estevez, A. G., Spear, N., and Crow, J. P. (1997). "Interactions of Nitric Oxide, Superoxide and Peroxynitrite with Superoxide Dismutase in Neurodegeneration. Nitric Oxide and Other Diffusible Signals in Brain Development, Plasticity, and Disease, Conference Proceedings." Louisiana State University Medical Center, New Orleans, LA.

Bedioui, F., Trevin, S., and Devynck, J. (1994). The use of gold electrodes in the electrochemical detection of nitric oxide in aqueous solution. *J. Electroanal. Chem.* **377**(1–2), 295–298.

Crow, J. P., and Beckman, J. S. (1995). Quantitation of protein tyrosine, 3-nitrotyrosine, and 3-aminotyrosine utilizing HPLC and intrinsic ultraviolet absorbance. *Methods* **7**(1), 116–120.

Holland, R. A. B., and Calvert, S. J. (1995). Oxygen transport by rabbit embryonic blood: High cooperativity of hemoglobin-oxygen binding. *Respir. Physiol.* **99**(1), 157–164.

James, P. E., Lang, D., Tufnell-Barret, T., Milsom, A. B., and Frenneaux, M. P. (2004). Vasorelaxation by red blood cells and impairment in diabetes: Reduced nitric oxide and oxygen delivery by glycated hemoglobin. *Circ. Res.* **94**(7), 976–983.

Lancaster, J. R. (2000). The physical properties of nitric oxide, determinants of the dynamics of nitric oxide in tissue. *In* "Nitric Oxide: Biology and Pathobiology" (L. J. Ignarro, ed.), pp. 209–224. Academic Press, San Diego.

McMahon, T. J., Exton Stone, A., Bonaventura, J., Singel, D. J., and Solomon Stamler, J. (2000). Functional coupling of oxygen binding and vasoactivity in S-nitrosohemoglobin. *J. Biol. Chem.* **275**(22), 16738–16745.

Pariente, F., Alonso, J. L., and Abruña, H. D. (1994). Chemically modified electrode for the selective and sensitive determination of nitric oxide (NO) *in vitro* and in biological systems. *J. Electroanal. Chem.* **379**(1–2), 191–197.

Pawloski, J. R., Hess, D. T., and Stamler, J. S. (2001). Export by red blood cells of nitric oxide bioactivity. *Nature* **409**, 622–626.

Stamler, J., Jia, L., Eu, J. P., McMahon, T. J., Demchenko, I. T., Bonaventura, J., Gernert, K., and Piantadosi, C. A. (1997). Blood flow regulation by S-nitrosohemoglobin in the physiological oxygen gradient. *Science* **276**(27 June 1977).

Wang, C., Piantadosi, C., Liu, J., and Allen, B. W. (Submitted). Novel electrochemical nitric oxide sensor using ruthenium-modified aligned carbon nanotubes. Submitted for publication.

# [8] Simultaneous Detection of NO and ROS by ESR in Biological Systems

*By* Yuanlin Cao, Pink Guo, Yangcang Xu, and Baolu Zhao

## Abstract

A large body of evidence shows that the generation of nitric oxide (NO) and reactive oxygen species (ROS) and the rate of ROS/NO play an important role in the biological system. We developed a method to simultaneously detect NO free radical and ROS in biological systems using ERS spin trapping technique. The adduct $(DETC)_2$-$Fe^{2+}$-NO and *N-tert-* butyl-$\alpha$-phenylnitrone (PBN)–ROS in biological systems can be extracted by organic solvent and then measured on an electron spin resonance (ESR) spectrometer at room temperature because the $g = 2.035$ of $(DETC)_2$-$Fe^{2+}$-NO is different from that of PBN-ROS ($g = 2.005$) and their ESR signals can be separated clearly. Using this method, we measured the production of NO and ROS in plant and animal systems.

METHODS IN ENZYMOLOGY, VOL. 396
0076-6879/05 $35.00
DOI: 10.1016/S0076-6879(05)96008-4

## Introduction

Living systems have developed a mechanism of defense and regulation that uses reactive oxygen species (ROS) and nitric oxide (NO) during the process of growth, development, and death. Of these, the formation of hydrogen peroxide, superoxide, and NO is ubiquitous in plant systems. They are often formed as byproducts of normal metabolism as a result of leaky electron transport systems. Several sources are known to exist for the generation of ROS. These include plasma membrane localized NADPH and NADH oxidases, apoplastic peroxidases, amine oxidase, and oxidases, as well as protoplastic sources from mitochondria, chloroplasts, and peroxisomes. The presence and synthesis of NO in plant cells are undisputed. NO can be generated as a byproduct of denitrification, nitrogen fixation, and/or respiration. In most cases, NO production in plant tissues has been linked to the accumulation of $NO_2$. NO is produced from $NO_2$ nonenzymatically through NADPH nitrate reductases. NO emission from plants occurs under stress situations, such as herbicide treatment, as well as under normal growth conditions. Guo *et al.* (2003) found that there is NO synthase (NOS) in plants as in animals, but the structure is different.

A large body of evidence shows that the generation of NO and ROS plays an important role in the biological system (Bolwell, 1999; Cao *et al.*, 2002; Guo *et al.*, 2003; Nathan, 1992; Zhang *et al.*, 2002; Zhao and Chen, 1993). Hypersensitive cell death for plant defense needs not only ROS and NO but also a certain rate of ROS/NO. To study the law of NO and ROS during the process of biological development and regulation, it is essential to set up an effective method for simultaneous detection of NO and ROS.

Electron spin resonance (ESR) is the most effective and direct method for detecting free radicals. Not only is the life of oxygen-free radical and nitrogen-free radicals very short (e.g., the life of the hydroxyl radicals is about $10^{-6}$ s), but also the concentration of the free radicals is very low in biological systems. A spin trapping technique has to be used with ESR to detect the short-lived free radical in this case (Capani *et al.*, 2001; Mordvintcev *et al.*, 1991; Zhang *et al.*, 2001; Zhou *et al.*, 1999). We have developed a method of simultaneously detecting NO and ROS by ESR in the biological system and have measured the relative signal strength in soybean and rat.

## Principal

NO can be trapped by $(DETC)_2$ Fe ,and the signal of $(DETC)_2$ FeNO has a triplet at g = 2.035 with $a_N$ = 13.5G; superoxide, hydroxyl, and lipid free radicals and other reactive oxygen free radicals can be trapped by *N-tert*-butyl-α-phenylnitrone (PBN); and the signal of PBN-ROS has a triplet to hexplet at g = 2.005 with $a_N$ = 15.0G (Capani *et al.*, 2001; Mordvintcev *et al.*, 1991; Zhang *et al.*, 2001; Zhou *et al.*, 1999). The signal of $(DETC)_2$-FeNO is between two and three peaks of copper, and the signal of PBN-ROS is between three and four peaks of copper, so there is no overlap between the signals of NO and ROS, and they can be measured in one ESR spectrum. The complexes of $(DETC)_2$-FeNO and PBN-ROS are stable enough for different treatment, so they can be used to detect NO and ROS *in vivo*.

## Reagents

Diethyldithiocarbamate (DETC), PBN, diethylenetriaminepentaacetic acid (DPTA), ethyl acetate, $Na_2S_2O_4$, $FeSO_4$, $CaCl_2$ sucrose, HEPES, Tween 80, PBS.

## Materials

Soy bean, rat.

## Solutions

Solution A: 0.1 *M* phosphate buffer, pH 7.4, containing 0.32 *M* sucrose, 10 m*M* HEPES, 10 m*M* PBN, 2 m*M* DPTA, 0.05% Tween 80, 5 mM thioaethylenglycol
Solution B: 0.5 *M* $Na_2S_2O_4$, 0.3 *M* $FeSO_4$
Solution C: 0.6 *M* DETC

## Procedure

We ground 0.25 g of sample in Solution A at ice water in the presence of quartz sand. The mixture was centrifuged at 13,201g for 20 min, and 450 μl of the supernatant was added 10 μl Solution B and Solution C, respectively, which was maintained at 37° for 1 h. ESR spectra were measured at room temperature with a Bruker ER200D-SRC spectrometer. The conditions for ESR were as follows: X-band, 100 kHz modulation with 3.2-G amplitude; microwave power 20 mW, central magnetic field 3385 G with scan 400 G, scan time 200 s. The mean height of the three peaks in each signal is taken as the relative intensity of the NO and ROS signal.

Statistical Analysis

Each experiment was performed at least three times, and results are presented as the mean ± SEM. Statistical analyses were performed using one-way analysis of variance (ANOVA), and a $p$ value $<.05$ was considered significantly different.

Results

### Measurement of the NO and ROS Level in Various Organs of Soybean

Figure 1 is the ESR spectrum of NO and ROS generated in soybean leaves and trapped by $(DETC)_2$-Fe and PBN. The signal of $(DETC)_2$-FeNO has a triplet at $g = 2.035$ with $aN = 13.5$ G. The signal of PBN-ROS has a triplet to hexplet at $g = 2.005$ with $aN = 15.0$ G. Figure 1 and Table I show that the level of NO and ROS in various organs of soybean is different, with the signal intensity of euphylla the highest among the organs of soybean.

### Detection of NO with ROS Production in Various Organs of Rat

NO and ROS have widespread tissue distribution and biological activity. With the ESR technique described in this chapter, we measured the NO and ROS levels in liver, heart, lung, and kidney of normal rats. From Table II, we can see that different organs have different levels of NO and ROS in the rats.

### Impacts of pH to Detection of NO and ROS in Plants

The pH dependence of trapping NO and ROS was performed in phosphate buffer. At acidic case pH 6.5, the signal strength of NO and ROS was higher than that of those at alkaline pH 8.5. It is interesting to note that at a more neutral condition of pH 7.4, we detected a higher signal in the reaction mixture of soybean leaves.

### Impacts of Detergents to Detection of NO and ROS in Plants

In the biological systems, the production of NO and ROS is involved in membrane protein. Different detergents have different efficiency of extraction during the reaction of enzyme. In our experiment, we found that 0.05% Tween 80 was suitable for extraction of the complex.

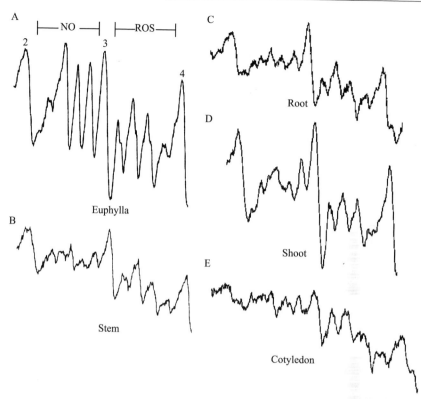

FIG. 1. Electron spin resonance (ESR) spectra of NO and ROS generated from the organs of soybean and trapped by of DETC-Fe and PBN, respectively. ESR condition: The spectrum was recorded at a Bruker ER200D-SRC ESR spectrometer. X-band, 100 KHz modulation with amplitude 3.2 G, microwave power 1 mW, time constant 200 s, central magnetic field 3446 G, swept width 200 G.

TABLE I

NO AND ROS LEVELS IN VARIOUS ORGANS OF SOYBEAN (RELATIVE CONCENTRATION) (14 DAYS)

|  | Root | Stem | Cotyledon | Euphylla | Shoot |
|---|---|---|---|---|---|
| NO level | $5.67 \pm 0.67$ | $6.17 \pm 0.60$ | $9.17 \pm 0.70^{*}$ | $27.58 \pm 1.13^{**}$ | $7.00 \pm 0.86$ |
| ROS level | $9.92 \pm 1.28$ | $9.83 \pm 0.62$ | $12.67 \pm 1.36^{*}$ | $20.42 \pm 1.91^{**}$ | $12.17 \pm 1.30^{*}$ |

Data are expressed as mean $\pm$ SEM for n = 6.
*Denotes that the value has $p \leq .01$ compared to the case with the root.
**Denotes that the value has $p \leq .05$ compared to the case with the root.

TABLE II
NO AND ROS LEVELS IN VARIOUS INTERNAL ORGANS OF RATS
(RELATIVE CONCENTRATION)

|  | Liver | Heart | Lung | Kidney |
|---|---|---|---|---|
| NO level | $3.17 \pm 0.40$ | $6.75 \pm 0.63^*$ | $6.47 \pm 0.87^*$ | $4.92 \pm 0.33^*$ |
| ROS level | $4.17 \pm 0.31$ | $6.17 \pm 0.48^*$ | $6.50 \pm 0.89^*$ | $6.92 \pm 0.27^*$ |

Data are expressed as mean $\pm$ SEM of three times.
*Denotes that the value has $p \leq .05$, compared to the case with liver.

### Stability of the $(DETC)_2$-$Fe^{2+}$-NO with PBN-ROS Complex in Ethyl Acetate

*Stability to Temperature.* The extractions of the $(DETC)_2$-$Fe^{2+}$-NO and PBN-ROS complex by ethyl acetate from the mixture of plant homogenates were immediately detected by ESR, and then were kept in the dark at $-20$, $-4$, or $25°$ and measured after 24 h. The results showed that the signal decreased about 7.5% and 31% after extracts for 24 h at $-20°$ and $25°$, respectively; however, at $0–4°$, the signal decreased only about 1.6%. The results indicate that $0–4°$ is the optimum temperature to keep the extractions of the $(DETC)_2$-$Fe^{2+}$-NO and PBN-ROS complex in the ethyl acetate.

*Stability to Light.* The extraction of the $(DETC)_2$-$Fe^{2+}$-NO and PBN-ROS complex by ethyl acetate from the mixture of plant homogenates was exposed to sunlight for different times, and then the extraction was measured by ESR. It was found that the ESR signal intensity of the $(DETC)_2$-$Fe^{2+}$-NO and PBN-ROS complex in ethyl acetate decreased quickly in light, and after 10 min of exposure, the intensity decreased about 50%,

After 2-h exposure, the signal disappeared completely. The signal of PBN-ROS was more sensitive to light than that of the $(DETC)_2$-$Fe^{2+}$-NO. This indicates that the $(DETC)_2$-$Fe^{2+}$-NO and PBN-ROS complex are sensitive to light.

*Stability to Time.* The extraction of the $(DETC)_2$-$Fe^{2+}$-NO and PBN-ROS complex by ethyl acetate from the mixture of plant homogenates was kept in the dark at $0–4°$ for different times, and then the extraction was measured by ESR. It was found that there was little change for NO when it was kept in the dark at $0–4°$ for 24 h, implying that the $(DETC)_2$-$Fe^{2+}$-NO complex in ethyl acetate was stable in this condition.

### Conclusions

1. The signal strength of NO and ROS in one biological system can be determined simultaneously with the methods described in this chapter. It

can be investigated by the technique in different biological materials, including plants and animals.

2. The signals in the acetate at 4° in the dark are rather stable during 24 h, so the method can be used to deal with a mass of biological materials.

3. Ethyl acetate can extract NO and ROS complex from low concentration in big volume to small volume, and the extracts can be investigated at room temperature, so this is a more sensitive, economical, and convenient method.

## Acknowledgment

This work was supported by a grant from the National Science Foundation of China (30070196, 30370369).

## References

Bolwell, G. P. (1999). Role of active oxygen species and NO in plant defence responses. *Curr. Opin. Plant Biol.* **2,** 287–284.

Cao, Y., Niu, Y., and Zhao, B. (2002). Study of effect of NO on wheat stripe rust by ESR. *Free Radical Biol. Med.* **33,** S74.

Capani, F., Loidl, C. F., Aguirre, F., Piehl, L., Facorro, G., Hagger, A., De Paoli, T., Farach, H., and Pecci-Saavedra, J. (2001). Changes in reactive oxygen species (ROS) production in rat brain during global perinatal asphyxia: An ESR study. *Brain Res.* **914,** 204–207.

Guo, F.-Q., Okamoto, M., and Crawford, N. M. (2003). Identification of a plant nitric oxide synthase gene involved in hormonal signaling. *Science* **302,** 100–103.

Mordvintcev, P., Mulsch, A., Busse, R., and Vanin, A. (1991). On-line detection of nitric oxide formation in liquid aqueous phase by electron paramagnetic resonance spectroscopy. *Anal. Biochem.* **199,** 142–146.

Nathan, C. (1992). Nitric oxide as a secretory product of mammalian cells. *FASEB J.* **6,** 3051–3064.

Zhang, D., Xiong, J., Hu, J., Li, Y., and Zhao, B. (2001). Improved method to detect nitric oxide in biological syste. *Appl. Magn. Reson.* **20,** 345–356.

Zhang, Y. T., Zhang, D. L., Cao, Y. L., and Zhao, B. L. (2002). Developmental expression and activity variation of nitric oxide synthase in the brain of golden hamster. *Brain Res. Bull.* **58,** 385–389.

Zhao, B. L., and Chen, W. C. (1993). Properties and biological function of NO free radicals. *Prog. Biochem. Biophys.* **20,** 409–411.

Zhou, G. Y., Zhao, B. L., Hou, J. W., Li, M. F., Wan, Q., and Xin, W. J. (1999). Detection of nitric oxide by spin trapping EPR spectroscopy and triacetylglycerol extraction. *Biotechnol. Tech.* **13,** 507–511.

# [9]   The ESR Method to Determine Nitric Oxide in Plants

*By* Yangcang Xu, Yuanlin Cao, Yi Tao, and Baolu Zhao

## Abstract

Nitric oxide (NO) plays an important role not only in animal system but also in plant system. We developed a method to detect NO free radical in plant system using electron spin resonance (ERS) spin trapping technique. The adduct $(DETC)_2$-$Fe^{2+}$-NO in plant can be extracted by organic solvent and then measured on an ESR spectrometer at room temperature (indirect method) or can be measured in live plant on an ESR spectrometer, but the water in the plant has to be partially dehydrated (direct method).

## Introduction

Nitric oxide (NO) is a small uncharged free radical containing one unpaired electron. NO exists in space as an interstellar molecule and in the atmosphere of Venus and Mars. On earth, NO has been recognized as an atmospheric pollutant and a potential health hazard (Nagano and Yoshimura, 2002). However, in 1987 it was reported that NO is identical to endothelium-derived relaxing factor (EDRF), which is biosynthesized in the living body (Palmer *et al.*, 1987). Now NO has been found in a wide variety of organisms ranging from mammals to invertebrates, bacteria, and plants, and many important physiological functions were discovered (Beligni and Lamattina, 2000; Mata and Lamattina, 2001; Moilanen and Vapaatalo, 1995).

Therefore, the direct detection of NO has become attractive, especially in biological models. Electron spin resonance (ESR) spectroscopy is a specific technique to detect and characterize molecules with unpaired electron(s). In the last 2 decades, the ESR technique has attracted increasing attention from both biologists and medicinal scientists because of increasing interest in reactive oxygen radicals in biological systems. Since NO was added as a member of the group of endogenously produced radicals (bioradicals) in 1987, endogenous NO has provided a challenging target for ESR spectrometric detection. Although NO is detectable with an ESR nitroxide spin trapping technique in chemical systems, the ESR signal of NO in biological systems is hardly detectable with this technique,

METHODS IN ENZYMOLOGY, VOL. 396
0076-6879/05 $35.00
DOI: 10.1016/S0076-6879(05)96009-6

probably because of the short lifetime of the spin trapping adduct. The dithiocarbamate (DTC) derivatives-Fe spin trapping technique was applied to overcome these difficulties. This technique enables one to conduct *in vitro* and *in vivo* NO measurements in biological systems. Several spin trapping compounds (spin traps) that selectively react with and trap NO have been developed and applied to biological NO measurements, such as iron complexes with DTC derivatives (Fe-DTC complexes). Among NO trapping reagents, Fe-DTC complex alone has been applied to measurements of NO *in vivo* (Nagano and Yoshimura, 2002). *N,N*-Diethyldithiocarbamate (DETC) is one of the dithiocarbamate derivatives, and Fe-DETC complex has been widely used for determination of NO in animals by ESR spectroscopy at either 77 K or room temperature (Mordvintcev *et al.*, 1991; Ryan *et al.*, 1993; Zhang *et al.*, 2001).

Reports on measuring NO by ESR in plants are very few. Mathieu *et al.* (1998) detected the spectra of leghemoglobin-NO using ESR at 77 K in soybean nodules. In this study, we reported the method of detecting NO in plant tissue extraction and the methods of detecting NO directly and indirectly in live plants using ESR at room temperature.

## Principal

ESR spectroscopy is a specific technique to detect and characterize molecules with unpaired electron(s). The electronic configuration of NO, with 11 valence electrons, is $(K^2K^2)(2s\sigma b)^2(2s\sigma*)^2(2\rho\pi b)^4(2\rho\sigma b)^2(2\rho\pi*)$ (Nagano and Yoshimura, 2002). Thus, NO is a free radical with one unpaired electron in the antibonding $\pi$ orbital and should be detected by ESR spectroscopy. However, the spin trapping technique must be applied in detecting NO because of the interaction of spin and orbital momentums. $Fe^{2+}$ and DETC easily react to form Fe-DETC, which is one of the effective NO trapping reagents, but the adduct $(DETC)_2$-$Fe^{2+}$-NOsignal in water solution is difficult to be detected by ESR spectroscopy at room temperature because of the low concentration and water absorbing microwaves. The adduct $(DETC)_2$-$Fe^{2+}$-NO is lipid soluble, so it can be extracted by organic solvent and then measured on ESR spectrometer at room temperature. If $Fe^{2+}$ and DETC are absorbed by different parts such as roots and leaves, they can form Fe-DETC complex either in intracellular or in extracellular spaces and then trap NO in the situation. Because Fe-DETC is lipid soluble, the complex can not only permeate through a membrane but also trap NO in the membrane. Therefore, after the plant absorbs iron salt and DETC, the detected ESR signal represents whole NO in physiological condition. Because the resultant adduct $(DETC)_2$-$Fe^{2+}$-NO is relatively stable and lipid soluble, the

$(DETC)_2$-$Fe^{2+}$-NO in the leaf can also be extracted by organic solvent and then determined on ESR spectrometer. Alternatively, it can be directly detected on ESR spectrometer after decreasing water content in leaf.

## Reagents

DETC, 2-phenyl-4,4,5,5,-tetramethylimidazoline-1-oxyl 3-oxide (PTIO), ethyl acetate, $FeSO_4$, sucrose, EDTA, thioaethylenglycol, $Na_2S_2O_4$, $KH_2PO_4$, $K_2HPO_4$, bovine serum albumin.

## Materials

Wheat (*Triticum aestivum* L.), Orchid (*Malaxis monophyllos* L.)

## Procedure

1. *Trapping and detecting NO in plants (wheat or orchid seedling extract) indirectly:* Fresh leaf tissue samples (0.3 g) were ground in a chilled mortar with 0.8 ml 100 m$M$ phosphate buffer solution containing 0.32 $M$ sucrose, 0.1 mM EDTA and 5 mM thioaethylenglycol (pH 7.4). The homogenate was centrifuged at 13,201$g$ for 20 min (4°). The supernate containing 5.7 mg protein was incubated with spin trapping reagent (7.5 mM $FeSO_4$ and 25 mM DETC with 0.5 $M$ $Na_2S_2O_4$) for 60 min. Then 250 $\mu$l ethyl acetate was added to the mixture, which was shaken for 3 min and centrifuged at 13,201$g$ for 6 min. The organic solvent layer was used to determine NO on ESR spectrometer.

2. *Trapping NO and detecting NO in plants (wheat or orchid seedling extract) directly:* Wheat or orchid seedlings were treated according to the following procedure: Seedlings were transferred to a culture medium containing 10 mM $FeSO_4$ and 0.5 $M$ $Na_2S_2O_4$, and the leaves were brushed with 30 mM DETC; 24 h later, the leaf discs were divided into two parts. Part one: 0.3 g leaf tissue was ground in a chilled mortar with 0.8 ml 100 mM phosphate buffer solution (pH 7.4). After adding 250 $\mu$l ethyl acetate to the homogenate, the mixture was shaken, centrifuged, and separated as in the procedure described earlier. The organic solvent layer was used to determine NO on the ESR spectrometer (indirect method). Part two: 0.25 g of leaves were treated in a 700-W microwave for 30 s to inactivate the enzymes in the leaves and then kept at 40° to partially dehydrate. Thereafter, the leaf discs were inserted directly into a quartz tube to determine NO on an ESR spectrometer. The lost water ratio was calculated by measuring the weight of the leaves before and after the drying treatment.

3. *Standard NO spectrum preparation:* NO gas was prepared as described by Zhao *et al.* (1995), and its saturated concentration in water was calculated as described by Borutaite and Brown (1996). The reaction mixture (638 $\mu$l) contained 100 m$M$ phosphate buffer solution (pH 7.4), 5.7 mg bovine serum albumin, 7.5 m$M$ $FeSO_4$, 0.5 $M$ $Na_2S_2O_4$, 25 m$M$ DETC and $H_2O$, or various concentrations of NO solution. Then the procedures of incubation, shaking, separation, and measurement were done as described earlier, and an NO standard spectrum was prepared.

4. *ESR operation:* ESR experiments were performed on a Bruker 200D-SRC spectrometer (Germany). The $(DETC)_2$-$Fe^{2+}$-NOcomplex both in solution and in intact leaves was measured in a 2.5-mm (for solution) or 3.5-mm (for leaf discs), respectively, internal diameter quartz tube at 25°. For a control of $(DETC)_2$-$Fe^{2+}$-NOcomplex in frozen solution, the ESR spectrum was measured at 100 K (Fig. 3A). The volume of organic solvent was 120 $\mu$l, and the height of leaf discs was 2 cm. The conditions for measurement were: X-band; 100 kHz modulation with 3.2-G amplitude; microwave power, 20 mW; central magnetic field 3385 G, scan width 400 G, time constant 0.3 s, scan time 4 min.

Statistical Analysis

Data were analyzed by analysis of variance and Student's $t$ test. Values are means of three independent experiments; standard deviations are indicated by error bars. At least 10 plant seedlings were used per experiment.

Results

*Detection of NO in Plant (Wheat or Orchid) Tissue Extract*

Figure 1A presents the typical ESR spectrum of $(DETC)_2$-$Fe^{2+}$-NO complex in ethyl acetate extracted from plant samples. The spectrum, recorded in solution at room temperature, has an isotropic spectrum as a result of rapid molecular motion, is an axial signal with an easily recognized hyperfine structure (hfs) triplet at $g^1 = 2.035$ with $a_N = 12.5$ G. And this triplet hfs is due to the interaction between the unpaired electron and the nitrogen nucleus of NO. This triplet hfs was similar to that of control (Fig. 1B), where authentic NO gas was prepared by chemical process. This triplet hfs also is consistent with that reported in animal samples (Zhang *et al.*, 2001). It indicated that the signal came from $(DETC)_2$-$Fe^{2+}$-NO in plant tissue. In addition to the ESR signal due to the $(DETC)_2$-$Fe^{2+}$-NO,

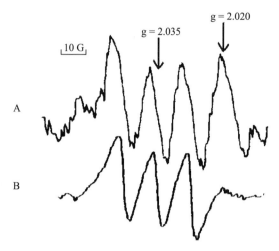

FIG. 1. The electron spin resonance (ESR) spectra of $(DETC)_2$-$Fe^{2+}$-NO in orchid tissue extract. (A) The $(DETC)_2$-$Fe^{2+}$-NO complex in the leaf homogenate was extracted by ethyl acetate and then measured on an ESR spectrometer at room temperature. (B) The $(DETC)_2$-$Fe^{2+}$-NO complex in standard NO solution system was extracted by ethyl acetate and then measured on an ESR spectrometer at room temperature.

the spectrum in Fig. 1A also shows the fourth peak at g = 2.02, which is attributed to the complex of DETC and copper (Kubrina *et al.*, 1992; Tominaga *et al.*, 1993; Zhou *et al.*, 1999), and this copper certainly came from plant tissue extract.

### Indirect Detection of NO in Live Plant (Wheat or Orchid)

After DETC and $FeSO_4$ were absorbed by the leaves and roots of plant seedlings, respectively, they trapped NO in plant and formed $(DETC)_2$-$Fe^{2+}$-NOcomplex, which was extracted by ethyl acetate and then measured on ESR spectrometer. Figure 2A presents the spectrum that shows the hfs triplet at g = 2.035 with $a_N$ = 12.5 G and the fourth peak at g = 2.02. The hfs triplet was similar to that of standard NO spectrum (Fig. 2B), and the fourth peak at g = 2.02 was similar to that of plant tissue extract. It indicated that the hfs triplet detected by ESR spectrometer came from $(DETC)_2$-$Fe^{2+}$-NOin plant. To further ascertain the source of the signal, PTIO (scavenger of NO⋅) experiment was conducted. When 1.5 m$M$ PTIO was provided for 24 h before uptake of trapping reagent, almost no signal could be detected (Fig. 2C). These results offer proof that the signals detected by ESR spectrometer came from NO in the plant.

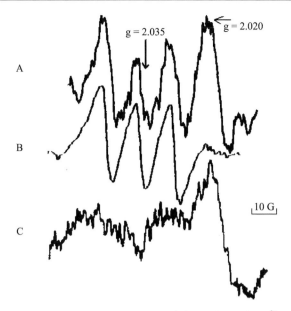

FIG. 2. Indirectly detected NO in live orchid leaves. (A) The $(DETC)_2$-$Fe^{2+}$-NO complex in live orchid leaves was extracted by ethyl acetate and then measured on an ESR spectrometer at room temperature. (B) The $(DETC)_2$-$Fe^{2+}$-NO complex in standard NO solution system was conducted as Fig. 1B. (C) 1.5 m$M$ PTIO (2-phenyl-4,4,5,5,-tetramethylimidazoline-1-oxyl 3-oxide) was supplied for 24 h before providing trapping reagent to orchid seedlings, The $(DETC)_2$-$Fe^{2+}$-NO signal was measured as (A).

### Direct Detection of NO in Live Plant (Wheat or Orchid)

As mentioned earlier, water in biological samples restricted ESR detection in certain cases. After DETC and $FeSO_4$ was absorbed and the $(DETC)_2$-$Fe^{2+}$-NO complex accumulated to a specific level in the leaf, the intact leaf was heated by microwave for 30 s, to inactivate enzyme activity *in vivo*, and then the leaves were kept at 40° to partially dehydrate them. Because the molecular motion of $(DETC)_2$-$Fe^{2+}$-NO in dehydrated leaf was restricted, the spectrum detected by ESR spectrometer is different than that recorded in solution at room temperature but is similar to that recorded in frozen solution. Figure 3B shows the spectrum in intact leaves when the leaves lost water equal to about 40% of their total weight. The spectrum has similar spectrum parameters to that shown in Fig. 3A in frozen solution at 100 K (g = 2.035 with $a_N$ = 12.5 G) and to the report by Goodman *et al.* (1969). In addition, when the seedlings were treated by PTIO as described earlier, the ESR signal in the intact leaf was almost

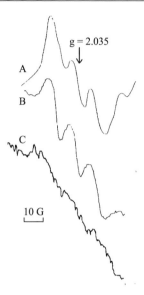

Fig. 3. Directly detected NO in live orchid leaves. (A) $(DETC)_2\text{-}Fe^{2+}\text{-}NO$ complex, which was obtained by the reaction of $NO^-$ solution with $(DETC)_2\text{-}Fe^{2+}$, as described in the sections "Materials" and "Methods," then 120 $\mu$l mixture solution was introduced into quartz tube and put into liquid nitrogen immediately, was measured on an ESR spectrometer at 100 K. (B) The live orchid leaves (0.25 g) containing $(DETC)_2\text{-}Fe^{2+}\text{-}NO$ complex were dried to lose 40% water of total weight, then the leaf discs were inserted directly into a quartz tube of 3.5-mm internal diameter packed 2 cm in height and measured $(DETC)_2\text{-}Fe^{2+}\text{-}NO$ signal on an ESR spectrometer. (C) 1.5 m$M$ PTIO was supplied for 24 h before providing trapping reagent to orchid seedlings. The $(DETC)_2\text{-}Fe^{2+}\text{-}NO$ signal was measured as (B).

eliminated (Fig. 3C). This indicates that the three peaks were also ascribed to $(DETC)_2\text{-}Fe^{2+}\text{-}NO$. The baseline of the spectrum was not parallel with the x axis, but similar to what had been found with that of NO-leghemoglobin complex in intact soybean nodules Mathieu *et al.*, 1998). Probably, $Fe^{3+}$ (e.g., as ferritin) and $Cu^{2+}$ (as mononuclear complexes) produce absorptions in this region.

Conclusions

1. NO in plant tissue extract was trapped by spin trapping reagent DETC-Fe and formed $(DETC)_2\text{-}Fe^{2+}\text{-}NO$complex. The adduct $(DETC)_2\text{-}Fe^{2+}\text{-}NO$ is lipid soluble, so it can be extracted by organic solvent and then measured on ESR spectrometer at room temperature.

Because cell structure was destroyed during the extraction, NO detected using this method represents NO content produced in the plant tissue extract during the incubating time with the spin trapping reagent.

2. To detect NO in live plants, the trapping reagent must enter into the foliage. However, DETC easily combines with $FeSO_4$ to form hydrophobic complexes, which was difficult for plant absorption. Therefore, DETC and $FeSO_4$ must be absorbed by different parts of a plant, such as roots and leaves. When $(DETC)_2$-$Fe^{2+}$-NO accumulates in the leaf, the complex can be extracted by ethyl acetate and then measured on ESR spectrometer (indirect method) or can be measured directly on ESR spectrometer (direct method).

3. As water in the leaf absorbs microwaves, the $(DETC)_2$-$Fe^{2+}$-NO signal in fresh leaves cannot be detected by the direct method at room temperature. Therefore, part of the water in fresh leaves must be eliminated. And this make the spectrum similar to that recorded in frozen solution.

## Acknowledgment

This work was supported by a grant from the National Science Foundation of China (grant no. 30070196).

## References

Beligni, M. V., and Lamattina, L. (2000). Nitric oxide induces seed germination and de-etiolation, and inhibits hypocotyl elongation, three light-inducible responses in plants. *Planta* **210**, 215–221.

Borutaite, V., and Brown, G. C. (1996). Rapid reduction of nitric oxide by mitochondria, and reversible inhibition of mitochondrial respiration by nitric oxide. *Biochem. J.* **315**, 295–299.

Guo, P., Cao, Y.-L., Li, Z.-Q., and Zhao, B.-L. (2004). Role of an endogenous nitric oxide burst in the resistance of wheat to stripe rust. *Plant Cell Environ.* **27**, 473–477.

Kubrina, L. N., Caldwell, W. S., Mordvintcev, P. I., Malenkova, I. V., and Vanin, A. F. (1992). EPR evidence for nitric oxide production from guanidino nitrogens of L-arginine in animal tissues *in vivo. Biochim. Biophys. Acta* **1099**, 233–237.

Mata, C. G., and Lamattina, L. (2001). Nitric oxide induces stomatal closure and enhances the adaptive plant responses against drought stress. *Plant Physiol.* **126**, 1196–1204.

Mathieu, C., Moreau, S., Frendo, P., Puppo, A., and Davies, M. J. (1998). Direct detection of radicals in intact soybean nodules: Presence of nitric oxide-leghemoglobin complexes. *Free Radic. Biol. Med.* **24**, 1242–1249.

Moilanen, E., and Vapaatalo, H. (1995). Nitric oxide in inflammation and immune response. *Ann. Med.* **27**, 359–367.

Mordvintcev, P., Mulsch, A., Busse, R., and Vanin, A. (1991). On-line detection of nitric oxide formation in liquid aqueous phase by electron paramagnetic resonance spectroscopy. *Anal. Biochem.* **199**, 142–146.

Nagano, T., and Yoshimura, T. (2002). Bioimaging of nitric oxide. *Chem. Rev.* **102,** 1235–1269.

Palmer, R. M. J., Ferrige, A. G., and Moncada, S. (1987). Nitric oxide release accounts for the biological activity of endothelium-derived relaxing factor. *Nature* **327,** 524–526.

Ryan, T. P., Miller, D. M., and Aust, S. D. (1993). The role of metals in the enzymatic and nonenzymatic oxidation of epinephrine. *J. Biochem. Toxicol.* **8,** 33–39.

Tominaga, T., Sato, S., Ohnishi, T., and Ohnishi, S. T. (1993). Potentiation of nitric oxide formation following bilateral carotid occlusion and focal cerebral ischemia in the rat: *In vivo* detection of the nitric oxide radical by electron paramagnetic resonance spin trapping. *Brain Res.* **614,** 342–346.

Zhang, D., Xiong, J., Hu, J., and Zhao, B. (2001). Improved method to detect nitric oxide in biology systems. *Appl. Magn. Reson.* **20,** 345–356.

Zhao, B. L., Shen, J. G., Li, M., Li, M. F., Wan, Q., and Xin, W. J. (1995). Scanvenging effect of chinonin on NO and oxygen free radicals and its protective effect on the mycardium from the injury of ischemia-reperfussion. *Biochim. Biophys. Acta* **1315,** 131–137.

Zhou, G., Zhao, B., Hou, J., Li, M., Chen, C., and Xin, W. (1999). Detection of nitric oxide in tissue by spin trapping EPR spectropy and triacetylglycerol extraction. *Biotechnol. Tech.* **13,** 507–511.

# Section II

## Nitration and S-Nitrosylation

# [10]  S-Nitrosothiol Formation

By Jeannean Carver, Allan Doctor, Khalequz Zaman, and
Benjamin Gaston

## Abstract

Protein and peptide S-nitrosothiols (SNOs) are involved in guanylate cyclase–independent signaling associated with nitric oxide synthase (NOS) activation. As a general rule, SNO formation requires the presence of an electron acceptor such as $Cu^{2+}$. Various proteins have been identified that catalyze SNO formation, including NOS itself, ceruloplasmin, and hemoglobin. Biochemical evidence suggests the existence of other SNO synthases and NOS-associated proteins involved in SNO formation following NOS activation. Indeed, both hydrophilic and hydrophobic consensus motifs have been identified that favor protein S-nitrosylation. Inorganic SNO formation appears also to occur in biological systems at low pH levels and/or in membranes. Once formed, SNOs localized to specific cellular compartments signal specific effects, ranging from gene regulation to ion channel gating. Indeed, the number of cellular and physiological functions appreciated to be regulated through SNO synthesis, localization, and catabolism is increasing. Although research into SNO biosynthesis is in its infancy, the importance of this field of biochemistry has been confirmed repeatedly by investigators from a broad spectrum of disciplines.

## Introduction

Many cellular processes signaled by nitric oxide synthase (NOS) are independent of the classical NO-mediated guanylate cyclase pathway (Gaston *et al.*, 2003). These cyclic guanosine monophosphate (cGMP)–independent NOS signals often involve thiol S-nitrosylation reactions. In general, these reactions require an electron acceptor; NO is oxidized to a complex containing a nitrosonium ($NO^+$) equivalent, which reacts with a cysteine thiolate to form an S-nitrosothiol (SNO) bond (Arnelle and Stamler, 1995; Gaston *et al.*, 2003). Target thiols include protein or peptide cysteine residues and cysteine itself. Products of these S-nitrosylation reactions, SNOs, are present endogenously (Gaston *et al.*, 1993; Gow *et al.*, 2002; Kluge *et al.*, 1997), and SNO concentrations in cells and tissues increase with NOS activation (Gow *et al.*, 1995; Zhang and Hogg, 2004b). The synthesis, storage, and catabolism of SNOs

METHODS IN ENZYMOLOGY, VOL. 396                                    0076-6879/05 $35.00
DOI: 10.1016/S0076-6879(05)96010-2

appear to be regulated by specific proteins and by regional cellular chemistry (Arnelle and Stamler, 1995; Gaston, 1999; Gaston *et al.*, 2003). However, the science of SNO signaling and metabolism is in its infancy. This chapter presents methods of biochemical and inorganic SNO synthesis.

In general, it is a mistake to view SNOs as "NO• donors." The extent to which a particular SNO is biologically active usually decreases with the extent to which the SNO bond breaks homolytically to release NO• (Gaston *et al.*, 1994, 2003; Lipton *et al.*, 2001). It has been common practice to design experiments using SNO concentrations that are 100–1000 times higher than physiological levels, based on the anticipated NO• concentration in solution. Often the biological activities reported using these high concentrations of SNOs—whether with regard to ion channel regulation, gene regulation, posttranslational protein modification, cell survival, or other effects of interest—have been the opposite of the effects caused by S-nitrosylation reactions associated with physiological SNO levels (Fig. 1) (Gaston *et al.*, 2003; Zaman *et al.*, 2004). Indeed, SNOs are generally formed and activated through covalent interactions, commonly involving transfer of an $NO^+$ equivalent to or from the cysteine thiolate anion. Occasionally, transfer of a nitroxyl ($NO^-$) equivalent may also be involved (Arnelle and Stamler, 1995).

SNOs can be challenging to isolate from cells and to assay. This is particularly true for SNO proteins, because even the smallest protein modification caused by isolation and handling can result in artifactual loss

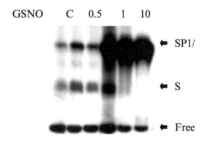

FIG. 1. Effect of S-nitrosoglutathione (GSNO) dose on Sp3/Sp1–DNA-binding activity in A549 cells. This figure illustrates the importance of appropriate dosing of the nitrosothiol in the physiological range. It demonstrates an electromobility shift assay using a consensus Sp3/Sp1 oligonucleotide as probe and nuclear extracts from control (Lane 1) and GSNO-treated cells at concentrations of 0.5–500.0 $\mu M$ (Lanes 2–6). Physiological GSNO concentrations increased Sp3/Sp1–DNA binding, whereas supraphysiological GSNO concentrations inhibited Sp3 binding but augmented Sp1 binding. Reproduced, with permission, from Zaman *et al.* (2004).

or formation of an SNO bond. Stringent criteria have been proposed by which a specific SNO signaling reaction can be shown to be relevant to a specific signaling process (Table I) (Lancaster and Gaston, 2004). Of note, it is unlikely that any one manuscript will be able to address all of these criteria for a specific process.

As noted, SNO formation ordinarily requires the availability of an electron acceptor. $NO^{\bullet}$ does not react with thiols except in the unusual circumstance in which thiyl radical has been formed (Arnelle and Stamler, 1995; Gaston, 1999; Gaston et al., 2003). A variety of electron acceptors catalyze the oxidation of $NO^{\bullet}$ to an $NO^{+}$ equivalent—including oxygen, $Cu^{2+}$, $Fe^{2+}$, and $NAD^{+}$ (Arnelle and Stamler, 1995; Gow and Stamler, 1998; Gow et al., 1997; Hunt et al., 2000; Rafikov et al., 2002; Vanin et al., 1997). These kinds of reactions may be relevant both to in vitro synthesis of SNO standards and to endogenous SNO formation in cells and tissues.

TABLE I

PROPOSED CRITERIA TO ESTABLISH THAT A SPECIFIC BIOACTIVITY IS ASSOCIATED WITH S-NITROSATION OR DENITROSATION OF A SPECIFIC PROTEIN[a]

1. The altered bioactivity of the target protein is associated with increased (or decreased) activity of a nitric oxide synthase (NOS) isoform.
2. S-Nitrosation of the target protein isolated from cells (and/or in situ) following NOS activation can be demonstrated by more than one independent assay (ideally three assays). In addition, the extent of nitrosation from endogenous nitric oxide formation is sufficient in magnitude to affect the activity of the protein, and nitrosation/denitrosation occurs rapidly enough to account for regulated changes in activity.
3. Mutation of a specific cysteine in the target protein results in (1) loss of the NOS-responsive bioactivity and (2) inability to identify the S-nitrosated protein following NOS activation.
4. In association with termination of the bioactivity, loss of the cellular protein S-nitrosation is demonstrated.
5. Alteration of the function of the purified protein can be demonstrated in association with S-nitrosation under conditions relevant to the protein's cellular environment (i.e., in the presence of a nitrogen oxide at concentrations measured in the specifically relevant tissue or cell compartment at the appropriate $pO_2$, $H^+$, etc.).
6. Pharmacological experiments demonstrate that cyclic guanosine monophosphate is not exclusively involved in mediating the bioactivity.
7. Pharmacological experiments suggest that a thiol modification is involved (i.e., the altered bioactivity is blocked by pretreatment with N-ethylmaleimide and/or reversed by excess DTT).
8. Pharmacological modifications of specific SNO metabolic enzymes relevant to the putative signaling process (such as $\gamma$-glutamyl transpeptidase or glutathione-dependent formaldehyde dehydrogenase) appropriately alter the bioactivity, and/or the bioactivity is caused by S-nitroso-L-cysteine–containing compounds but not S-nitroso-D-cysteine–containing compounds.

[a]From Lancaster and Gaston (2004).

Inorganic Synthesis of SNOs *In Vitro*

It is useful to synthesize SNOs, both to determine whether these compounds mimic the effects of NOS activation and/or to use the synthesized SNO in assay standards (Table I). However, two caveats apply:

1. SNOs are not interchangeable in standard curves, because the stability of the S-NO bond—as well as autocapture of the nitrogen oxide released for measurement—varies substantially from compound to compound. For example, depending on the assay and the preassay preparatory conditions, 100 nM S-nitrosoglutathione (GSNO) may or may not give the same signal as 100 nM S-nitroso-hemoglobin (Doctor *et al.*, 2005).

2. All SNOs are not equal with regard to bioactivity. Perhaps the most striking example is the difference between the L and the D-isomers of S-nitrosocysteine (CSNO). In many bioassays, the L-isomer of CSNO is active while the D-isomer is inactive (Davisson *et al.*, 1996; Lipton *et al.*, 2001; Ohta *et al.*, 1997). This difference appears to be associated with stereoselective cell membrane receptors and/or transport mechanisms (Gaston *et al.*, 2003; Zhang and Hogg, 2004a). Additionally, GSNO may require bioactivation by $\gamma$-glutamyl transpeptidase (GGT) to be as active as S-nitrosocysteinyl glycine or CSNO (Table I) (Lipton *et al.*, 2001; Zaman *et al.* 2004).

## Inorganic Synthesis of Low-Mass SNOs in Acid

Most low-mass SNOs can be prepared from the corresponding reduced thiol by reaction with protonated nitrous acid (as $H_2ONO^+$); reactions uniformly go virtually to completion in less that 5 s (Stamler and Feelisch, 1996; Stamler *et al.*, 1992). The method involves the following stage. The reduced thiol is assayed for purity and dissolved (generally to 100 mM) in 1 N HCl. An equal concentration of $NaNO_2$ is prepared in water. If a metal chelator such as EDTA will not interfere with subsequent assays or experiments, it is often advisable to prepare the $NaNO_2$ in water containing 200 $\mu M$ EDTA because transition metals (in particular, copper) can substantially decrease the stability of SNOs (Williams, 1996). It is best not to dissolve the $NaNO_2$ in acid because NO is evolved from HONO (pKa $\sim$ 3.6) (Hunt *et al.*, 2000). The thiol and nitrite solutions are then mixed in equal volumes in a fume hood. For most SNOs, a pink-orange solution will result (absorbance $\sim$340 nm). Tertiary thiols will produce a green SNO solution. Generally, it will be advisable to add 3 volumes of buffer and 1 volume of 0.5 normal NAOH (both containing 100 $\mu M$ EDTA) to neutralize the pH before use in cell systems. It is important to avoid overshooting the titration, because some SNOs are less stable at alkaline pH. The reaction should be performed in dark Eppendorf or foil-covered tubes at 4° to minimize photolytic and thermal SNO decomposition (Arnelle and Stamler, 1995; Stamler et al., 1992). SNOs can be lost with freeze-thawing and will gradually break down, even at −80°. Concentrations should be confirmed by one of several SNO assays (Fang et al., 1998; Stamler and Feelisch, 1996; Williams, 1996; Wink et al., 1999).

*Alternative Methods for Inorganic Synthesis of Low-Mass SNOs*

Alternative synthetic reactions are not efficient. Low-mass SNOs may also be prepared by transnitrosylation reactions from ethyl nitrite (EtNO) in aqueous solution or *tert*-butyl nitrite (tBNO) in chloroform (Doyle et al., 1983; Moya et al., 2002; Stamler and Feelisch, 1996; Welch et al., 1999). One hundred millimoles of thiol is dissolved in chloroform or acetone:water (15:1, v:v), depending on thiol solubility. Two milliliters of EtNO or tBNO is added, and the organic solvent is evaporated. High-intensity ultraviolet (UV) light can form CSNO from cysteine and nitrate (Dejam et al., 2003). However, in our hands, UV light always eliminates SNOs detected by nonphotolytic methods (such as mass spectrometry and reductive chemiluminescence), and UV light–based assays with appropriate controls are sensitive and specific for biological SNOs (Gaston et al., 1993; Lipton et al., 2001; Mannick et al., 2001).

Synthesis of SNO Proteins *In Vitro*

*General Comments*

Nitrous acid, as described earlier, can be used to S-nitrosylate protein-reduced thiols (Stamler et al., 1992). However, the acid conditions can denature the protein, resulting in S-nitrosylation of thiols that would not be modified under physiological conditions and/or loss of protein structural constraints that favor S-nitrosylation. Therefore, transnitrosylation reactions (Meyer et al., 1994; Scharfstein et al., 1994) are conventionally favored as a mechanism for preparing most SNO proteins. By this method, the protein of interest is isolated and purified. It is mixed with a 10- to 50-fold excess of concentration of CSNO or S-nitroso-*N*-acetyl-cysteine (SNOAC) and incubated at 4° in the dark for 30 min. The CSNO is then removed by overnight dialysis and/or by washing across a Sephadex G25 or similar column. Particularly if dialysis is used, it is important to know the stability of the SNO protein to be certain that it will not be lost during the overnight incubation. Also, it is important to be certain that all of the low-mass SNO has been eliminated. CSNO and SNOAC can readily be detected by high-performance liquid chromatography (HPLC) as previously described (Fang *et al.*, 1998). If the co-incubation is done with CSNO in the absence of EDTA, CSNO generally will have been lost within 2 h, leaving the SNO-protein as the only SNO species. It is also important to recognize that some proteins can have more than one target site thiol, depending on redox condition. Therefore, excess CSNO (or other more stable SNO) may be required. In each case, validation of the protein concentration and protein SNO concentration is critical.

## Synthesis of SNO-Hemoglobin *In Vitro*

SNO-hemoglobin (SNOHb) can be synthesized in both the free form of hemoglobin and the intact erythrocyte. Synthesis of free SNOHb is accomplished using purified human Hb (HbA$_0$) (Doctor *et al.*, 2005; Gow *et al.*, 1999; Jia *et al.*, 1996; Pawloski *et al.*, 2001) that has been dialyzed overnight against 2% aerated borate, pH 9.2, then S-nitrosylated by incubation for 5 min at room temperature with 10-fold molar excess of freshly made 0.5 *M* CSNO. The CSNO is prepared as described earlier with an important modification, that of reacting 1 *M* NaNO$_2$ (H$_2$O) with 1.1 *M* L-cysteine (in 0.5 *M* HCl); it is important to have cysteine in slight molar excess to nitrite to limit subsequent formation of methemoglobin. The reaction is quenched on a Sephadex G25 column equilibrated with phosphate-buffered saline (PBS) (Doctor *et al.*, 2005; Gow *et al.*, 1999); care should be taken to remove all residual CSNO from the Hb preparation. All solutions should contain 0.5 m*M* EDTA. The samples must be protected from light at all times and stored at −80°.

A method for synthesis of SNOHb in the intact erythrocyte has also been developed. In experiments designed to test the relation of Hb conformational state to SNO bioactivities, erythrocytic SNOHb appears to have more physiological relevance than cell-free SNOHb (p50 = 5–10 Torr) because it will not be possible to induce relevant R ↔ T state shifts (and deploy SNO bioactivities) without depressing pO$_2$ to a nonphysiological extreme (Doctor *et al.*, 2005; Jia *et al.*, 1996). In the first deoxygenation step, fresh erythrocytes washed in PBS are placed in a gastight vial under pure Argon until pO$_2$ is less than 10 Torr and Hb is fully desaturated. The deoxygenated erythrocytes are then exposed for 5 min to physiological amounts of NO• added as aliquots of oxygen-purged NO•-saturated PBS, yielding erythrocytes containing HbFe[II]NO. The saturated aqueous NO• for this step is prepared by cleansing 99% pure NO• gas of higher oxides by serial passage through gas washing vessels containing 1 *M* NaOH and deoxygenated deionized distilled water (Beckman *et al.*, 1996). Purified NO$^{\bullet}$ gas is then bubbled for 10 min through 10 m$M$ PBS (pH 7.4) in deionized distilled water that has been thoroughly deoxygenated by alternate exposure to vacuum and vigorous bubbling with ultrapure Argon (scrubbed to remove any residual $O_2$). Saturated aqueous NO$^{\bullet}$ can be stored in septated airtight vials. The second step of reoxygenation is done by simple exposure to room air, resulting in the intramolecular transfer of NO from heme to $\beta$cys$^{93}$, and yielding the desired SNOHb (Doctor *et al.*, 2005).

## Biochemical Mechanisms of SNO Formation *In Vivo*

Enzyme systems have been identified or proposed as SNO synthases. However, many unanswered questions remain regarding the metabolism and regulation of these SNO proteins. These biochemical SNO synthetic pathways are not yet conventionally used for SNO synthesis *in vitro*.

Activation or upregulation of each NOS isoform has been associated with increased production of cellular SNOs, as detected by several methodologies (Gow *et al.*, 2004; Haendeler *et al.*, 2004; Liu *et al.*, 2001; Mannick *et al.*, 1999; Zhang and Hogg, 2004b). Different mechanisms have been proposed for this NOS-associated SNO formation, such as diffusion of NO away from the enzyme to form intermediate $N_2O_3$. However, the details of these mechanisms have not been worked out satisfactorily. Spatial constraint of SNO formation by NOS isoforms has been controversial (Lancaster and Gaston, 2004) but is suggested by localized SNO immunostaining, by localized SNO fluorescent staining, and by differential SNO localization in organelles (Ckless *et al.*, 2003; Gow *et al.*, 2002; Mannick *et al.*, 2001).

Ceruloplasmin catalyzes GSNO formation (Inoue *et al.*, 1999). In this reaction sequence, NO radical reacts with $Cu^{2+}$ to form an NO$^{+}$-Cu$^{+}$ complex. NO$^{+}$ is transferred to GSH with release of a proton. The electron on $Cu^{+}$ is transferred through the copper systems of the enzyme to oxygen, forming water.

Hemoglobin functions as an SNO synthase with the NO radical initially binding to ferrous iron or (perhaps as nitroxyl) to the ferric iron of methemoglobin to form an iron nitrosyl species (Gow and Stamler, 1998; Gow *et al.*, 1999). NO in this Fe-NO complex may be moved to the $\beta$cys$^{93}$ thiol during hemoglobin oxygenation (switch from T to R state) as noted earlier in this chapter. With subsequent deoxygenation, however, this NO$^{+}$ equivalent can then be transferred to additional thiols, including both cytosolic glutathione and anion exchange protein 1 (AE1) on the erythrocyte membrane. In this sense, membrane-associated hemoglobin can be thought of as an SNO kinase for AE1 (Pawloski et al., 2001).

Enzymatic SNO synthesis has been proposed to occur in the mitochondria; this mechanism is believed to involve a flavin (Foster and Stamler, 2004). Additionally, eNOS-dependent SNO synthases have been proposed in endothelial cells, and evidence from our laboratory suggests that N-acetyl-cysteine can "steal" NO, dysregulating eNOS-dependent SNO synthesis and/or SNO hemoglobin to permit the development of pulmonary hypertension (unpublished observations). However, mechanistic details for transfer of NO from NOS to target protein thiols have not been worked out. It may be that proteomic analysis will facilitate more rapid progress in this field (Jaffrey et al., 2001).

Additional inorganic reactions discussed previously that lead to SNO formation in vitro may be relevant to SNO formation in vivo. These include (1) the slow oxidation (third-order rate constant) of $NO^\bullet$ by oxygen— which is somewhat faster in membranes than in aqueous solution— followed by thiolate nitrosation by $N_2O_3$; and (2) an SNO formation from iron nitrosyl intermediates (Arnelle and Stamler, 1995; Vanin et al., 1997). However, these reactions are not conventionally used for SNO synthesis in vitro.

### SNO Bioactivities

GSNO is the classic SNO identified in tissues and synthesized exogenously for experimental models. Normal extracellular concentrations are on the order of 500 n$M$ to 1 $\mu M$ in certain tissues but tend to be undetectable in cytosolic extracts (Gaston et al., 1993; Kluge et al., 1997; Liu et al., 2001). Applied extracellularly, GSNO signals a broad spectrum of bioactivities. Importantly, however, physiological levels and supraphysiological (nitrosative stress levels) have diametrically opposed effects on many processes, ranging from cell survival to gene regulation (Mannick et al., 1999; Zaman et al., 2004). For example, low micromolar concentrations of GSNO increase Sp3 binding, increasing transcription of genes such as cftr, whereas higher concentrations have the opposite effect (Fig. 1) (Zaman et al., 2004). This gene regulatory activity of GSNO appears to depend on expression and activity of γ-glutamyl transpeptidase (Zaman et al., 2004). Another classic SNO bioactivity experiment involves the use of L-CSNO with the D-isomer serving as a control. L-CSNO, for example, affects heart rate, blood pressure, and minute ventilation at the level of the nucleus tractus solitarius, whereas the D-isomer is inactive (Davisson et al., 1996; Lipton et al., 2001; Ohta et al., 1997).

## Summary

S-nitrosothiols should not generally be thought of as NO• donors. They signal cGMP-independent bioactivities mediated by cysteine transnitrosylation reactions rather than by homolytic cleavage and NO• reactions. Physiological SNO concentrations tend to be in the low micromolar range (or picomoles per milligram of protein in cell extracts) (Gaston *et al.*, 1993; Kluge *et al.*, 1997; Mannick *et al.*, 2001; Zhang and Hogg, 2004b). Enzyme systems are being identified that catalyze endogenous formation of SNOs in biological systems, but these enzymes are not classically used for SNO synthesis *in vitro*. Acid nitrosation, often coupled with transnitrosylation reactions to form SNO proteins, remains the most commonly used method for synthesizing SNOs in the laboratory. More involved procedures are also now available to synthesize SNO-hemoglobin from both isolated and intraerythrocyte hemoglobin *in vitro*.

## References

Arnelle, D. R., and Stamler, J. S. (1995). NO$^+$, NO, and NO$^-$ donation by S-nitrosothiols: Implications for regulation of physiological functions by S-nitrosylation and acceleration of disulfide formation. *Arch. Biochem. Biophys.* **318,** 279–285.

Beckman, J., Wink, D., and Crow, J. (1996). Nitric oxide and peroxynitrite. *In* "Methods in Nitric Oxide Research" (M. Feelisch and J. Stamler, eds.). John Wiley & Sons, West Sussex, England.

Ckless, K., Reynaert, N. L., Taatjes, D. J., Lounsbury, K. M., van derVliet, A., and Janssen-Heininger, Y. (2003). S-nitrosoproteins can be detected via chemical derivitization and immunocytochemistry in intact lung epithelial cells. *Am. J. Respir. Crit. Care Med.* **167,** A52.

Davisson, R. L., Travis, M. D., Bates, J. N., and Lewis, S. J. (1996). Hemodynamic effects of L- and D-S-nitrosocysteine in the rat. *Circ. Res.* **79,** 256–262.

Dejam, A., Kleinbongard, P., Rassaf, T., Hamada, S., Gharini, P., Rodriguez, J., Feelisch, M., and Kelm, M. (2003). Thiols enhance NO formation from nitrate photolysis. *Free Radic. Biol. Med.* **35,** 1551–1559.

Doctor, A., Platt, R., Sheram, M. L., Eischeid, A., McMahon, T., Maxey, T., Doherty, J., Axelrod, M., Kline, J., Gurka, M., Gow, A., and Gaston, B. (2005). Hemoglobin conformation couples S-nitrosothiol content in erythrocytes to oxygen gradients. *Proc. Natl. Acad. Sci. USA* **102,** 5709–5714.

Doyle, M. P., Terpstra, J. W., Pickering, R. A., and LePoire, D. M. (1983). Hydrolysis, nitrosyl exchange and synthesis of alkyl nitrites. *J. Org. Chem.* **48,** 3379–3382.

Fang, K., Ragsdale, N. V., Carey, R. M., MacDonald, T., and Gaston, B. (1998). Reductive assays for S-nitrosothiols: Implications for measurements in biological systems. *Biochem. Biophys. Res. Commun.* **252,** 535–540.

Foster, M. W., and Stamler, J. S. (2004). New insights into protein S-nitrosylation: Mitochondria model system. *J. Biol. Chem.* **279,** 25891–25897.

Gaston, B. (1999). Nitric oxide and thiol groups. *Biochim. Biophys. Acta* **1411,** 323–333.

Gaston, B. M., Carver, D. J., Doctor, A., and Palmer, L. A. (2003). S-nitrosylation signaling in cell biology. *Mol. Interv.* **3,** 253–263.

Gaston, B., Drazen, J. M., Jansen, A., Sugarbaker, D. A., Loscalzo, J., Richards, W., and Stamler, J. S. (1994). Relaxation of human bronchial smooth muscle by S-nitrosothiols *in vitro. J. Pharmacol. Exp. Ther.* **268,** 978–984.

Gaston, B., Reilly, J., Drazen, J. M., Fackler, J., Ramdev, P., Arnelle, D., Mullins, M. E., Sugarbaker, D. J., Chee, C., Singel, D. J., Loscalzo, J., and Stamler, J. S. (1993). Endogenous nitrogen oxides and bronchodilator S-nitrosothiols in human airways. *Proc. Natl. Acad. Sci. USA* **90,** 10957–10961.

Gow, A. J., Chen, Q., Hess, D. T., Day, B. J., Ischiropoulos, H., and Stamler, J. S. (2002). Basal and stimulated protein S-nitrosylation in multiple cell types and tissues. *J. Biol. Chem.* **277,** 9637–9640.

Gow, A. J., Buerk, D. G., and Ischiropoulos, H. (1997). A novel reaction mechanism for the formation of S-nitrosothiol *in vivo. J. Biol. Chem.* **272,** 2841–2845.

Gow, A. J., Luchsinger, B. P., Pawloski, J. R., Singel, D. J., and Stamler, J. S. (1999). The oxyhemoglobin reaction of nitric oxide. *PNAS USA* **96,** 9027–9032.

Gow, A. J., and Stamler, J. S. (1998). Reactions between nitric oxide and haemoglobin under physiological conditions. *Nature* **391,** 169–173.

Haendeler, J., Hoffmann, J., Zeiher, A. M., and Dimmeler, S. (2004). Antioxidant effects of statins via S-nitrosylation and activation of thioredoxin in endothelial cells: A novel vasculoprotective function of statins. *Circulation* **110,** 856–861.

Hunt, J. F., Fang, K., Malik, R., Snyder, A., Malhotra, N., Platts-Mills, T. A., and Gaston, B. (2000). Endogenous airway acidification: Implications for asthma pathophysiology. *Am. J. Respir. Crit. Care Med.* **161,** 694–699.

Inoue, K., Akaike, T., Miyamoto, Y., Okamoto, T., Sawa, T., Otagiri, M., Suzuki, S., Yoshimura, T., and Maeda, H. (1999). Nitrosothiol formation catalyzed by ceruloplasmin. Implication for cytoprotective mechanism *in vivo. J. Biol. Chem.* **274,** 27069–27075.

Jaffrey, S. R., Erdjument-Bromage, H., Ferris, C. D., Tempst, P., and Snyder, S. H. (2001). Protein S-nitrosylation: A physiological signal for neuronal nitric oxide. *Nat. Cell Biol.* **3,** 193–197.

Jia, L., Bonaventura, C., Bonaventura, J., and Stamler, J. S. (1996). S-Nitrosohaemoglobin: A dynamic activity of blood involved in vascular control. *Nature* **380,** 221–226.

Kluge, I., Gutteck-Amsler, U., Zollinger, M., and Do, K. Q. (1997). S-Nitrosoglutathione in rat cerebellum: Identification and quantification by liquid chromatography-mass spectrometry. *J. Neurochem.* **69,** 2599–2607.

Lancaster, J. R., Jr., and Gaston, B. (2004). NO and nitrosothiols: Spatial confinement and free diffusion. *Am. J. Physiol. Lung Cell Mol. Physiol.* **287**(3), 465–566.

Lipton, A. J., Johnson, M. A., Macdonald, T., Lieberman, M. W., Gozal, D., and Gaston, B. (2001). S-Nitrosothiols signal the ventilatory response to hypoxia. *Nature* **413,** 171–174.

Liu, L., Hausladen, A., Zeng, M., Que, L., Heitman, J., and Stamler, J. S. (2001). A metabolic enzyme for S-nitrosothiol conserved from bacteria to humans. *Nature* **410,** 490.

Mannick, J. B., Hausladen, A., Liu, L., Hess, D. T., Zeng, M., Miao, Q. X., Kane, L. S., Gow, A. J., and Stamler, J. S. (1999). Fas-induced caspase denitrosylation. *Science* **284,** 651–654.

Mannick, J. B., Schonhoff, C., Papeta, N., Ghafourifar, P., Szibor, M., Fang, K., and Gaston, B. (2001). S-Nitrosylation of mitochondrial caspases. *J. Cell Biol.* **154,** 1111–1116.

Meyer, D. J., Kramer, H., Ozer, N., Coles, B., and Ketterer, B. (1994). Kinetics and equilibria of S-nitrosothiol-thiol exchange between glutathione, cysteine, penicillamines, and serum albumin. *FEBS Lett.* **345,** 177–180.

Moya, M. P., Gow, A. J., Califf, R. M., Goldberg, R. N., and Stamler, J. S. (2002). Inhaled ethyl nitrite gas for persistent pulmonary hypertension of the newborn. *Lancet* **360,** 141–143.

Ohta, H., Bates, J. N., Lewis, S. J., and Talman, W. T. (1997). Actions of S-nitrosocysteine in the nucleus tractus solitarii are unrelated to release of nitric oxide. *Brain Res* **746,** 98–104.

Pawloski, J. R., Hess, D. T., and Stalmer, J. S. (2001). Export by red blood cells of nitric oxide bioactivity. *Nature* **409,** 622–626.

Rafikov, O., Rafikov, R., and Nudler, E. (2002). Catalysis of S-nitrosothiols formation by serum albumin: The mechanism and implication in vascular control. *Proc. Natl. Acad. Sci. USA* **99,** 5913–5918.

Scharfstein, J. S., Keaney, J. F., Jr., Slivka, A., Welch, G. N., Vita, J. A., Stamler, J. S., and Loscalzo, J. (1994). *In vitro* transfer of nitric oxide between a plasma protein-bound reservoir and low molecular weight thiols. *J. Clin. Invest.* **94,** 1432–1439.

Stamler, J. S., and Feelisch, M. (eds.) (1996). "Preparation and Detection of S-Nitrosothiols, Methods in Nitric Oxide Research." John Wiley & Sons, Ltd.

Stamler, J. S., Jaraki, O., Osborne, J., Simon, D. I., Keaney, J., Vita, J., Singel, D., Valeri, C. R., and Loscalzo, J. (1992). Nitric oxide circulates in mammalian plasma primarily as an S-nitroso adduct of serum albumin. *Proc. Natl. Acad. Sci. USA* **89,** 7674–7677.

Vanin, A. F., Malenkova, I. V., and Serezhenkov, V. A. (1997). Iron catalyzes both decomposition and synthesis of S-nitrosothiols: Optical and electron paramagnetic resonance studies. *Nitric Oxide. Biol. Chem.* **1,** 191–203.

Welch, G. N., Upchurch, G. R., Jr., and Loscalzo, J. (1996). S-Nitrosothiol detection. *Methods Enzymol.* **268,** 293–298.

Williams, D. L. (1996). S-Nitrosothiols and role of metal ions in decomposition to nitric oxide. *Methods Enzymol.* **268,** 299–308.

Wink, D. A., Kim, S., Coffin, D., Cook, J. C., Vodovotz, Y., Chistodoulou, D., Jourd'heuil, D., and Grisham, M. B. (1999). Detection of S-nitrosothiols by fluorometric and colorimetric methods. *Methods Enzymol.* **301,** 201–211.

Zaman, K., Palmer, L. A., Doctor, A., Hunt, J. F., and Gaston, B. (2004). Concentration-dependent effects of endogenous S-nitrosoglutathione on gene regulation by specificity proteins Sp3 and Sp1. *Biochem. J.* **380**(Pt. 1), 67–74.

Zhang, Y., and Hogg, N. (2004a). The mechanism of transmembrane S-nitrosothiol transport. *PNAS* **101,** 7891–7896.

Zhang, Y., and Hogg, N. (2004b). Formation and stability of S-nitrosothiols in RAW 264.7 cells. *Am. J. Physiol. Lung Cell Mol. Physiol.* **287,** 467–474.

# [11]  Detection and Characterization of Protein Nitrosothiols

*By* SAMIE R. JAFFREY

## Abstract

Protein S-nitrosylation is a post-translational modification of cysteine residues elicited by nitric oxide (NO). Detection and quantification of protein nitrosothiols remains a challenge because of the lability of the nitrosothiol moiety. Here, we describe approaches for labeling S-nitrosylated proteins with affinity and radioactive tags to facilitate their detection, purification, and identification.

METHODS IN ENZYMOLOGY, VOL. 396
0076-6879/05 $35.00
DOI: 10.1016/S0076-6879(05)96011-4

Introduction

A reaction of nitric oxide (NO) in biological systems is the formation of nitrosothiols (Gaston *et al.*, 1993; Stamler *et al.*, 1992), a modification of the sulfur atom on cysteine, also called *S*-nitrosation or *S*-nitrosylation. NO reacts with $O_2$ to yield $NO_x$ such as $NO_2$ and $N_2O_3$, powerful electrophiles that *S*-nitrosylate cysteines to form nitrosothiols (S-NO) (Fig. 1A). Although reactions of $NO_x$ with lysine are possible, lysine nitrosation (or diazotization) is not favored because of the low nucleophilicity of amines at neutral pH levels, as well as spontaneous deamination following its reaction with $NO_x$ (Wink *et al.*, 1991). Indeed S-NO can be detected in biological tissue up to the mid-micromolar range (Gaston *et al.*, 1993; Kluge *et al.*, 1997), but N-NO and C-NO are virtually undetectable (Kluge *et al.*, 1997; Simon *et al.*, 1996).

Detection of *S*-nitrosylated proteins is complicated by the lability of nitrosothiols. Nitrosothiols are sensitive to light and undergo photolytic decomposition to NO and thiyl radical (Barrett *et al.*, 1966; Josephy *et al.*, 1984; Singh *et al.*, 1996). Further, cytosolic reducing agents such as ascorbate (Aquart and Dasgupta, 2004; Holmes and Williams, 1998; Smith and Dasgupta, 2000) and reduced metals, especially Cu(I) (Dicks and Williams, 1996), readily reduce the nitrosothiol bond. Additionally, glutathione, a thiol-containing tripeptide present at concentrations as high as about 5 m$M$ in cellular cytosol (Hwang *et al.*, 1992), reduces nitrosothiols by a transnitrosation reaction in which the glutathione thiolate attacks the nitrogen of the nitrosothiol resulting in the transfer of the nitroso moiety to the

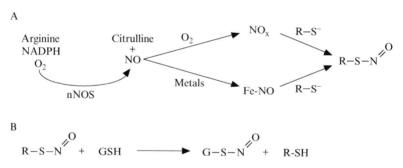

FIG. 1. Schematic diagram of the primary reaction pathways for the generation of *S*-nitrosothiols. (A) Nitric oxide (NO) is synthesized from arginine, NADPH, and $O_2$, leading to the formation of NO and citrulline. The redox requirements for the formation of nitrosating species are provided by transition metals, such as iron, or $O_2$. The interactions of NO and $O_2$ are potentiated by their mutual enrichment in hydrophobic compartments. (B) Transnitrosylation between glutathione (GSH) and R-SNO is also depicted, leading to the formation of *S*-nitrosoglutathione (GSNO).

glutathione thiolate (Fig. 1B). Indeed, experiments with model small-molecule nitrosothiols show that the half-life of NO ranges from 1 to 5 min in kidney and liver lysates (Kashiba-Iwatsuki *et al.*, 1997). Importantly, the *in vivo* half-life of protein nitrosothiols has not been determined.

Although nitrosothiols are difficult to detect, their presence is often invoked to account for various NO-related phenomena. Thus, when NO is found to have an effect in an *in vitro* cellular or behavioral process, and when involvement of guanylyl cyclase can be excluded, perhaps by the absence of effect of guanylyl cyclase inhibitors, nitrosothiol formation can be reasonably inferred as the likely NO effector mechanism.

The reversible regulation of protein function by *S*-nitrosylation has led to suggestions that nitrosothiols function as post-translational modifications analogous to phosphorylation or acetylation (Lane *et al.*, 2001; Stamler *et al.*, 1997). The principal methods used to study phosphorylation utilize radiolabeled precursors, such as $[^{32}P]$ adenosine triphosphate (ATP) or $[^{32}P]PO_4$. In the case of signaling by *S*-nitrosylation, however, radioactive isotopes of nitrogen or oxygen (the atoms that comprise the adduct) are not available, necessitating the development of alternative sensitive methods to detect the nitrosothiol moiety.

One major approach for nitrosothiol detection is the photolytic-chemiluminescence methodology. Using this approach, a number of authors have provided evidence that certain proteins, including the ryanodine receptor (Xu *et al.*, 1998), caspase-3 (Mannick *et al.*, 1999), and albumin (Stamler *et al.*, 1992), possess nitrosothiol moieties in their endogenous state. This assay involves laser-mediated decomposition of the nitrosothiol and detection of released NO by a chemiluminescent reaction with ozone. This method is appropriate to determine and quantitate nitrosothiols in a protein sample. It is not used to identify *S*-nitrosylated proteins present in a pool of unnitrosylated proteins.

Here, we describe a procedure by which *S*-nitrosylated proteins can be labeled with biotin or other tags, thus permitting these proteins to be detected by routine Western blotting procedures or autoradiography or purified by avidin-affinity chromatography. We describe the use of this procedure to detect *S*-nitrosylation of proteins exposed to NO donors *in vitro* and to detect endogenously *S*-nitrosylated proteins.

### Principle of Assay System

We have developed a method to label *S*-nitrosylated proteins with a biotin moiety specifically on *S*-nitrosylated cysteines (Jaffrey *et al.*, 2001) (Fig. 2). This method consists of three steps: In the first step, free thiols in a protein or a protein mixture are "blocked" by incubation with the

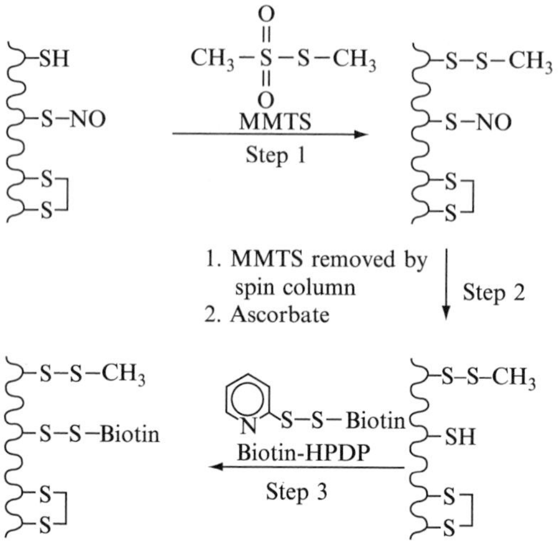

Fig. 2. Schematic diagram of the S-nitrosylation assay. A protein is indicated schematically with cysteines in either the thiol, disulfide, or nitrosothiol state. In Step 1, free thiols are made unreactive by methylthiolation with methyl methanethiosulfonate (MMTS). MMTS is removed in Step 2 by either passing the protein sample through a desalting (spin) column or acetone precipitation. In the final step, nitrosothiols are selectively reduced with ascorbate to regenerate the thiol, which is then reacted with the thiol-modifying reagent biotin-HPDP, which is shown schematically.

thiol-specific methylthiolating agent methyl methanethiosulfonate (MMTS) (Kenyon and Bruice, 1977). Sodium dodecylsulfate (SDS) is used as a protein denaturant to ensure access of MMTS to buried cysteines. Under the conditions used, MMTS does not react with nitrosothiols (S. Jaffrey, unpublished data) or preexisting disulfide bonds (Kenyon and Bruice, 1977). The procedures are performed in the dark to minimize photolysis of the nitrosothiol bonds.

In Step 2, nitrosothiol bonds are selectively decomposed with ascorbate, which results in the reduction of nitrosothiols to thiols (Singh et al., 1996; Smith and Dasgupta, 2000). In Step 3, the newly formed thiols are reacted with N-[6-(biotinamido)hexyl]-3'-(2'-pyridyldithio)propionamide (biotin-HPDP), a sulfhydryl-specific biotinylating reagent. Because MMTS can compete with biotin-HPDP for thiol groups, residual MMTS is removed by desalting on a spin column or by acetone precipitation before treatment with ascorbate.

By the end of Step 3, S-nitrosylated proteins are labeled with biotin specifically on cysteine residues that were once S-nitrosylated. Biotinylated proteins can easily be detected by anti-biotin Western blotting. Because the cysteine biotinylation in this assay is reversible, SDS–polyacrylamide gel electrophoresis (PAGE) is performed without reducing agents, and samples are not boiled, to prevent nonspecific reactions of biotin-HPDP.

Using this procedure, one can detect several types of protein nitrosothiols by anti-biotin Western blotting. First, proteins that have been S-nitrosylated *in vitro* by NO donors can be detected, and these proteins may be either purified recombinant proteins or protein lysates derived from cells or tissues. Second, proteins derived from lysates can be examined for endogenous nitrosothiols. In this case, the amount of S-nitrosylation is likely to be considerably less than that seen after *in vitro* S-nitrosylation.

In some cases, purification of the biotinylated proteins may be desired. Biotinylated proteins can be purified using immobilized streptavidin, a biotin-binding protein, and eluted with 2-mercaptoethanol or dithiothreitol, which reduces the disulfide linkage that couples the biotin moiety to the protein. Importantly, proteins that contain biotin bound covalently as a cofactor, such as pyruvate carboxylase, will not be eluted from the resin because the biotinylation of these proteins is not reversible by reducing agents. The eluted proteins can be identified by protein sequencing using conventional Edman degradation or mass spectrometry (MS)–based techniques.

Also, using this procedure, other types of tags can be incorporated into proteins. Thiol-specific radiolabeling agents can be used to radiolabel S-nitrosylated proteins. Radioactive labeling of protein S-nitrosothiols can be used to facilitate stoichiometric quantitation of S-nitrosylation or for peptide mapping studies.

## Methods

### Materials

Milli-Q water is used for all experiments. $N$-[6-(biotinamido)hexyl]-3′-(2′-pyridyldithio)propionamide is from Pierce (Rockford, IL), 1-[2-(2-aminoethyl)-$N$-(2-ammonioethyl)amino]diazen-1-ium-1,2-diolate (DETA NONOate) is from Alexis Chemicals (San Diego, CA), and glutathione, S-nitrosoglutathione, S-nitroso-$N$-acetylpenicillamine (SNAP), sodium ascorbate, 5,5′-dithio-bis-(2-nitrobenzoic acid) (DNTB), MMTS, neocuproine, and anti-biotin antibodies (clone BN34) are from Sigma (St. Louis, MO). MicroBioSpin6 columns are from Bio-Rad (Hercules, CA).

### Preparation of Protein Samples

Protein samples should be free of exogenous reducing agents, such as dithiothreitol or 2-mercaptoethanol, which might reduce physiological disulfides and create artifactual thiols that may confound analysis of NO donor–induced S-nitrosylation Additionally, dithiothreitol and 2-mercaptoethanol scavenge endogenous nitrosothiols or deactivate exogenous NO donors. Similarly, samples should generally be protected from light, which induces thiyl radical formation from nitrosothiols. Tissue samples are homogenized in 20 volumes (weight of tissue per volume) 0.1X HEN buffer (25 m$M$ HEPES-NaOH, pH 7.7, 0.1 m$M$ EDTA, and 1 $\mu M$ neocuproine). Divalent metal chelation by EDTA and Cu(I) chelation by neocuproine serves to reduce metal-catalyzed denitrosylation (Askew et al., 1995; Dicks and Williams, 1996; Gordge et al., 1995; Singh et al., 1996). If endogenous S-nitrosylation, rather than in vitro S-nitrosylation, is being assessed, thiol-modifying agents, such as 5 m$M$ N-ethylmaleimide or 1 m$M$ MMTS can be included in the homogenization buffer. Blockade of thiols during homogenization will prevent ex vivo transnitrosation of thiols that are not physiologically S-nitrosylated. Protein levels should not exceed 0.8 mg/ml, with lower concentrations of protein producing lower background signals in the assay (see later discussion). Protein concentration can be determined by the Biuret or other assay, and samples can be diluted to the appropriate protein concentration with 0.1X HEN buffer.

Purified proteins can also be prepared. Typically these can be resuspended in 0.1X HEN buffer at a concentration of 0.8 $\mu$g/ml or less. GST-fusion proteins can be used for these experiments, but glutathione must be removed by dialysis; alternatively, GST-fusion proteins can be eluted from glutathione agarose with S-methyl glutathione, a thiol-inactive form of glutathione capable of eluting GST-fusion proteins from glutathione agarose (S. R. J., unpublished observations). In many protocols for GST-fusion protein preparation, Triton X-100 is used in the elution step to enhance elution (Smith and Johnson, 1988). Because Triton X-100 interferes with SDS (used later in this protocol), and because it cannot be readily removed by dialysis, it should not be used for GST-fusion protein elution. Octyl glucoside (2%) is a suitable alternative.

## In Vitro S-*Nitrosylation*

NO donors, in principle, can modify all thiols in a protein. Thus, the ability of an NO donor to modify a given cysteine or a given protein is not inherently proof that the protein is S-nitrosylated in vivo. Typically, the question is whether a given protein or a given thiol exhibits exquisite sensitivity to being S-nitrosylated. Thus, in vitro S-nitrosylation should be performed with low concentrations of NO donors (e.g., <50 $\mu M$ and preferably 1 $\mu M$ or less). Additionally, the S-nitrosylation of a candidate protein should be compared to a known NO target, such as H-ras, which is readily S-nitrosylated on a single Cys (Lander et al., 1997). H-ras can be prepared recombinantly (Fang *et al.*, 2000). An NO donor (e.g., GSNO) or inactive donor control is added to the protein preparation to achieve the desired final concentration. When setting up the assay, it is useful to also have a sample treated with excess NO donor, at concentrations such as 400 $\mu M$. Under these conditions, the protein will likely be $S$-nitrosylated superstoichiometrically and should be readily detectable in the assay. *In vitro* $S$-nitrosylation should be complete in 20 min at 25°. NO donors should be protected from light and frozen at −80° when not in use. Other NO donors are acceptable, such as SNAP or DETA NONOate, and are used at similar concentrations as GSNO.

After $S$-nitrosylation, it is advisable to remove the NO donor from protein samples because the donor may nitrosylate cysteines that are otherwise inaccessible after the proteins are denatured in subsequent steps. To remove the NO donor, acetone precipitation can be used, followed by resuspension of the pellet in the starting volume of HENS buffer (0.1X HEN buffer adjusted to 1% SDS with a 25% SDS (w/v) stock). For dilute samples or samples of less than 100 $\mu l$, desalting (spin) columns, such as the MicroBioSpin6 column can be used in place of acetone precipitation. The column preparation and use are performed as recommended by the manufacturer.

### Blockade of Protein Thiols

Labeling of $S$-nitrosylated thiols relies on (1) the ability to chemically inactivate all the thiols in protein samples and (2) the selective reduction of nitrosothiols. Once nitrosothiols have been reduced to thiols, these thiols can be labeled with thiol-specific labeling reagents. Several criteria have to be achieved in order for this labeling protocol to work properly. First, all the thiols in a protein sample have to be blocked to essentially 100% completion. If, for example, the blocking was only 99% complete, 1% of protein thiols would remain chemically reactive and capable of reacting with the thiol-specific labeling agents used in subsequent steps. If nitrosothiols represent a tiny fraction of all the cysteines in a cellular lysate, the nonspecific signal derived from labeling unblocked cysteines would dwarf the signal from labeling the $S$-nitrosylated cysteines. Second, the blocking step has to be completed rapidly because of the lability of nitrosothiols. Nitrosothiols spontaneously decompose, even if they are protected from light and metals. Third, the reduction of the nitrosothiol bond needs to be selective for the nitrosothiol moiety and not other oxidation states of cysteine, in particular disulfide bonds. This selectivity needs to be exquisite because the amount of disulfides in the protein sample will be exceedingly high, so that even if a few of these are reduced, the resulting signal from these cysteines after labeling with the thiol-specific labeling agent may overwhelm the specific signal.

Several classes of thiol-modifying agents can be used to block cysteine residues. Alkylating agents, such as iodoacetate, iodoacetamide, and N-ethyl maleimide, exhibit reactivity toward both amines and thiols, with weak reactivity toward hydroxyls even at neutral pH (Boja and Fales, 2001; Partis et al., 1983). Another drawback of these reagents is the relatively slow rate of thiol alkylation compared to activated disulfides or alkyl alkane thiolsulfonates, as has been reported in several enzyme inactivation studies (Bednar, 1990; George and Turner, 1989; Pettigrew, 1986). Alkyl alkane thiolsulfonates are the fastest thiol-specific modifying reagents with reactivities on the order of $10^5 \ M^{-1} \ s^{-1}$ (Stauffer and Karlin, 1994). These compounds are not alkylating agents; they do not add alkyl groups to thiols, as does, for example, iodoacetic acid. Rather, these are methylthiolating reagents. They exhibit greater specificity for thiols than alkylating agents in aqueous buffers at physiological pH levels. Additionally, their small size and neutral charge may assist in facilitating complete blockade of thiols. A characteristic of these agents is that the modification of the thiol (i.e., the formation of the disulfide) is reversible by reducing agents, unlike the thioethers formed by alkylation. The rapidity and the possibility of modifying thiols to virtual completion make MMTS particularly useful in this assay.

Thiols are blocked by adding four volumes of blocking buffer [9 volumes of HEN buffer, 1 volume of SDS buffer (25% w/v in $H_2O$), 0.1 volumes 2 $M$ MMTS (from a 2-$M$ stock in $N,N$-dimethylformamide)]. This solution should be prepared fresh before use. Blocking is conducted at 50° for 20 min with frequent vortexing.

After thiol blockade, MMTS is removed. If the sample volume is sufficiently small, desalting columns, such as MicroBioSpin6 columns can be used. Practical experience with desalting columns has shown that they do not completely "desalt" proteins—typically 1% or so of the low-molecular-weight material can still remain in the protein fraction. Thus, to remove MMTS completely, additional rounds of desalting are required. Typically, three rounds are sufficient. To confirm that the MMTS is removed, a colorimetric thiol-modifying agent probe can be used. 5-Mer-capto-2-nitrobenzoic acid (MNB) (50 m$M$ HEPES, pH 8.0, 2 m$M$ DNTB, 0.05 m$M$ dithiothreitol) is yellow when the MNB thiol is in the unmodified form. To determine if residual MMTS is present in the desalted solution, 50 $\mu$l of the desalted protein solution can be added to 50 $\mu$l MNB. The absence of a color change, as visualized by measuring the absorbance at 412 nm, confirms that MMTS has been substantially removed.

Alternatively, MMTS can be removed by precipitation of the protein with acetone. MMTS does not precipitate under these conditions. The sample is precipitated by adding 2 volumes of acetone, prechilled to $-20°$, and then incubated for 20 min at $-20°$, followed by centrifugation at least 2000g for 10 min at $4°$. Residual MMTS on the pellet and the sides of the tubes can be removed by rinsing the tube and pellet with acetone, followed by recentrifugation. The pellet is resuspended in 0.1 ml HENS buffer/mg of protein in the starting sample. Again, the removal of MMTS can be confirmed with MNB, as noted earlier.

### Labeling of S-Nitrosothiols

Nitrosothiols can be labeled with any thiol-labeling agent. Thiol-labeling agents of the activated mixed disulfide class, such as biotin-HPDP, have the advantage of being virtually thiol specific in aqueous buffers in the physiological pH range. To biotinylate nitrosothiols, the nitrosothiols need to be reduced, and the subsequently formed thiols are labeled with thiol-specific modifying agents. For biotinylation, the protein sample is labeled by adding biotin-HPDP (4 m$M$ solution, prepared fresh by diluting a 50 m$M$ stock in DMSO) to the protein solution to a final concentration of 1 m$M$. The solution is then adjusted to 1 m$M$ sodium ascorbate, using a 50-m$M$ stock prepared in deionized water. Ascorbate does not reduce disulfides unless divalent cations, such as copper(II), are present. EDTA and neocuproine, present in the protein solution, prevent this nonspecific disulfide reduction. In many cases, spontaneous denitrosylation is sufficient to obtain labeling, even without exogenous ascorbate. Nitrosothiols are simultaneously reduced and labeled with biotin-HPDP at $25°$ for 1 h.

Other activated mixed disulfides can be used to label nitrosothiols as well. For example, nitrosothiols can be labeled with radioactive probes (Jaffrey et al., 2002). Radiolabeling can facilitate quantitation of the stoichiometry of S-nitrosylation or can be used to perform peptide mapping experiments (Jaffrey et al., 2002). To perform radiolabeling, 2-[$^{35}$S]amino-3-(2-pyridiyldithio)-propionic acid (APDP) can be used in place of biotin-HPDP. Both these compounds have a reactive mixed disulfide, activated by a 2-mercaptopyridine moiety. APDP is prepared by adding 10 $\mu$l [$^{35}$S] cysteine (Perkin-Elmer, Boston, MA, 11 mCi/ml, 1075 Ci/mmol) containing 10 m$M$ DTT (already present in the commercial preparation) to 10 $\mu$l of 50 m$M$ 2,2'-dithiodipyridine in methanol. To this solution is added 10 $\mu$l phosphate-buffered saline (pH 7.4). This reaction is incubated for 60 min at $25°$. To this reaction is added 10 $\mu$l 10:1 methanol:HOAc, and the mixture is spotted onto an analytical silica TLC plate and resolved using 12:6:1:1 (CHCl$_3$:MeOH:HOAc:H$_2$O). APDP is localized by apposing the TLC plate to autoradiography film for 5 s at room temperature and aligning the resulting spot on the film to the TLC plate. APDP migrates to an R$_f$ of

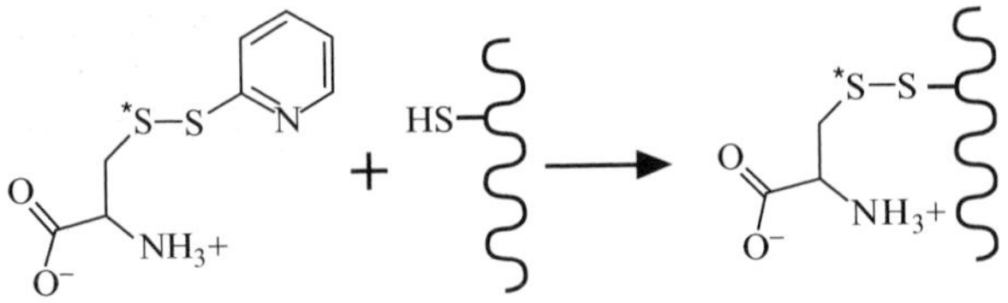

FIG. 3. Radiolabeling of proteins with [$^{35}$S]APDP. A theoretical protein containing a single thiol after ascorbate-mediated denitrosylation is depicted. After reaction with [$^{35}$S]APDP, the protein is labeled with the indicated radiolabeled adduct. The asterisk is used to designate the radioactive sulfur.

about 0.4 and was clearly separated from reduced and oxidized DTT, which is included in the commercial cysteine stock, as well as from the unreacted 2,2′-dithiodipyridine and stoichiometrically produced 2-mercaptopyridine. To purify the APDP, the region of the plate containing the radioactive material is scraped, and the silica is resuspended in 75 $\mu$l resuspension buffer (25 m$M$ HEPES 7.7; 0.1 m$M$ EDTA, 0.01 m$M$ neocuproine). After incubation at 25° for 10 min with frequent vortexing, the silica slurry is clarified by centrifugation. The resulting supernatant typically contains about 80% of the APDP. Typically, about 100,000 cpm of [$^{35}$S]APDP is used in labeling reactions (Fig. 3).

### Detection and Purification of Labeled Proteins

Biotinylated proteins can be detected by SDS-PAGE and immunoblotting with biotin-specific antibodies, or by using an avidin-conjugate such as avidin–horse radish peroxidase. Care should be taken to avoid using reducing agents in the SDS-PAGE as the biotin tag can be cleaved with reducing agents. For radiolabeled proteins, gels can be dried and apposed to autoradiography film for detection.

Alternatively, proteins can be purified for Edman degradation or MS. To purify these proteins, avidin-agarose or related biotin-binding immobilized proteins can be used. Because the labeling reaction contains excess biotin-HPDP, this needs to be removed by dialysis or acetone precipitation before purification on avidin-agarose. To remove biotin-HPDP, precipitate the sample with acetone as described earlier for removing MMTS. Resuspend the protein in 0.1 ml HENS buffer. SDS in HENS in neutralized by adding neutralization buffer (20 m$M$ HEPES-NaOH, pH 7.7, 100 m$M$ NaCl, 1 m$M$ EDTA, 0.5% Triton X-100). To purify the biotinylated proteins, add 15 $\mu$l of packed streptavidin-agarose/mg of protein used in the initial protein sample to purify biotinylated proteins and incubate the protein with the resin for 1 h at 25°. The resin is washed extensively with wash buffer (50 m$M$ HEPES, pH 7.7, 1 $M$ NaCl, 1 m$M$ EDTA). Centrifuge the resin at 200$g$ for 5 s at room temperature between each wash. Bound proteins are eluted in elution buffer (20 m$M$ HEPES-NaOH, pH 7.7, 100 m$M$ NaCl, 1 m$M$ EDTA, 100 m$M$ 2-ME) at 25° for 20 min. Eluted proteins can then be digested for peptide mass fingerprinting or liquid chromatography–MS (LC-MS)/MS. If desired, proteins can be digested with trypsin before incubation of the mixture with immobilized streptavidin. Under these conditions, only the biotinylated peptides bind to the resin. After elution, MS/MS sequencing of these peptides identifies the S-nitrosylated protein. When the peptide only has one cysteine, MS/MS will also identify the specific S-nitrosylated cysteine residue.


Troubleshooting

*False-Positive Signals*

This S-nitrosylation assay is based on the assumption that only an S-nitrosylated cysteine residue will become biotinylated following the various treatments described earlier. However, if all cysteines have not been successfully blocked with the methylthiolating reagent, MMTS, then nonnitrosylated cysteines may produce a signal in this assay. SDS is included to promote denaturation, which should ensure accessibility of MMTS to each thiol. A cysteine that is inaccessible to MMTS may become accessible to biotin-HPDP during the labeling step if time-dependent or DMSO-dependent denaturation occurs and results in the unhindered exposure of a thiol. To ensure the maximal accessibility of cysteines to MMTS, a minimum ratio of SDS to protein is essential to ensure maximal protein denaturation. Thus, protein samples that exceed 0.8 $\mu$g/$\mu$l are more prone to be incompletely blocked by MMTS. If biotinylation is seen in samples that are not treated with NO donors, this signal may be due to endogenous nitrosothiols, due to endogenous biotinylation, or alternatively due to incomplete thiol blockade. Reduction of protein concentration in the sample typically improves thiol blockade.


*Negative Controls for* In Vitro *Nitrosylation*

The best way to ensure that a signal is due to S-nitrosylation is to include an inactive NO donor control. Bands that are present in the NO donor lane, but not in lanes in which control compounds were used, represent proteins that are S-nitrosylated.

*Negative Controls for Detection of* In Vivo *Nitrosylation*

Because the nitrosylation assay includes biotin immunoblotting, proteins that are endogenously biotinylated can be a source of significant background. These enzymes, which perform carboxytransferase reactions, are found at different levels in different animals. For example, the brains of C57/BL6 mice contain these endogenously biotinylated proteins at levels 20 times higher than that found in Norwegian white rats. The presence of endogenously biotinylated proteins can actually be beneficial, because they can serve as loading controls. That is, their uniform intensity in different lanes can ensure that equivalent amounts of protein were utilized in different experiments. A key control in the identification of these proteins is the use of a DMSO vehicle control, instead of biotin-HPDP, in the biotinylation labeling reaction. Bands seen in this sample reflect endogenously biotinylated proteins.

This protocol outlines the methods for the detection of both *in vitro* S-nitrosylated proteins, by biotin immunoblotting, and endogenously biotinylated proteins, by purification of S-nitrosylated proteins by streptavidin-affinity chromatography followed by candidate-specific immunoblotting. In this latter method, some of the same troubleshooting issues arise, but the required controls are different. To ensure that a signal is due to NOS activity and not due to incomplete blocking, tissue samples that are devoid of NOS are ideal. Thus, protein samples prepared from $NOS^{-/-}$ mice provide ideal controls. Alternatively, if NOS-transfected tissue culture cells are used, mock transfected cells would be useful. If these types of controls are unavailable, then samples prepared with NOS inhibitors, such as nitroarginine, can also be used as negative controls.

## Acknowledgments

This work was supported by grants from the National Institutes of Mental Health (MH066204) and the National Alliance for Research on Schizophrenia and Affective Disorders (NARSAD).

## References

Aquart, D. V., and Dasgupta, T. P. (2004). Dynamics of interaction of vitamin C with some potent nitrovasodilators, S-nitroso-N-acetyl-*D,L*-penicillamine (SNAP) and S-nitrosocaptopril (Snocap), in aqueous solution. *Biophys. Chem.* **2,** 117–131.

Askew, S. C., Butler, A. R., Flitney, F. W., Kemp, G. D., and Megson, I. L. (1995). Chemical mechanisms underlying the vasodilator and platelet anti-aggregating properties of S-nitroso-N-acetyl-Dl-penicillamine and S-nitrosoglutathione. *Bioorganic Med. Chem.* **3,** 1–9.

Barrett, J., Fitygibbones, L. J., Glauser, J., Still, R. H., and Young, P. N. (1966). Phytochemistry of the S-nitroso derivatives of hexane-1-thiol and hexane-1,6-dithiol. *Nature* **211,** 848.

Bednar, R. A. (1990). Reactivity and Ph dependence of thiol conjugation to N-ethylmaleimide: Detection of a conformational change in chalcone isomerase. *Biochemistry* **29**, 3684–3690.

Boja, E. S., and Fales, H. M. (2001). Overalkylation of a protein digest with iodoacetamide. *Anal. Chem.* **73**, 3576–3582.

Dicks, A. P., and Williams, D. L. (1996). Generation of nitric oxide from S-nitrosothiols using protein-bound $Cu^{2+}$ sources. *Chem. Biol.* **3**, 655–659.

Fang, M., Jaffrey, S. R., Sawa, A., Ye, K., Luo, X., and Snyder, S. H. (2000). Dexras1: A G protein specifically coupled to neuronal nitric oxide synthase via capon. *Neuron* **28**, 183–193.

Gaston, B., Reilly, J., Drazen, J. M., Fackler, J., Ramdev, P., Arnelle, D., Mullins, M. E., Sugarbaker, D. J., Chee, C., Singel, D. J., Loscalzo, J., and Stamler, J. S. (1993). Endogenous nitrogen oxides and bronchodilator S-nitrosothiols in human airways. *Proc. Natl. Acad. Sci. USA* **90**, 10957–10961.

George, J. N., and Turner, R. J. (1989). Inactivation of the rabbit parotid Na/K/Cl cotransporter by N-ethylmaleimide. *J. Membrane Biol.* **112**, 51–58.

Gordge, M. P., Meyer, D. J., Hothersall, J., Neild, G. H., Payne, N. N., and Noronha-Dutra, A. (1995). Copper chelation-induced reduction of the biological activity of S-nitrosothiols. *Br. J. Pharmacol.* **114**, 1083–1089.

Holmes, A. J., and Williams, D. L. H. (1998). Reactions of S-nitrosothiols with ascorbate: Clear evidence of two reactions. *Chem. Commun.* **16**, 1711–1712.

Hwang, C., Sinskey, A. J., and Lodish, H. F. (1992). Oxidized redox state of glutathione in the endoplasmic reticulum. *Science* **257**, 1496–1502.

Jaffrey, S. R., Erdjument-Bromage, H., Ferris, C. D., Tempst, P., and Snyder, S. H. (2001). Protein S-nitrosylation: A physiological signal for neuronal nitric oxide. *Nat. Cell Biol.* **3**, 193–197.

Jaffrey, S. R., Fang, M., and Snyder, S. H. (2002). Nitrosopeptide mapping: A novel methodology reveals S-nitrosylation of dexras1 on a single cysteine residue. *Chem. Biol.* **9**, 1329–1335.

Josephy, P. D., Rehorek, D., and Janzen, E. D. (1984). Electron-spin resonance spin trapping of thiyl radicals from the decomposition of thionitrites. *Tetrahedron Lett.* **25**, 1685–1688.

Kashiba-Iwatsuki, M., Kitoh, K., Kasahara, E., Yu, H., Nisikawa, M., Matsuo, M., and Inoue, M. (1997). Ascorbic acid and reducing agents regulate the fates and functions of S-nitrosothiols. *J. Biochem. (Tokyo)* **122**, 1208–1214.

Kenyon, G. L., and Bruice, T. W. (1977). Novel sulfhydryl reagents. *Methods Enzymol.* **47**, 407–430.

Kluge, I., Gutteck-Amsler, U., Zollinger, M., and Do, K. Q. (1997). S-Nitrosoglutathione in rat cerebellum: Identification and quantification by liquid chromatography–mass spectrometry. *J. Neurochem.* **69**, 2599–2607.

Lander, H. M., Hajjar, D. P., Hempstead, B. L., Mirza, U. A., Chait, B. T., Campbell, S., and Quilliam, L. A. (1997). A molecular redox switch on P21(Ras). Structural basis for the nitric oxide-P21(Ras) interaction. *J. Biol. Chem.* **272**, 4323–4326.

Lane, P., Hao, G., and Gross, S. (2001). S-Nitrosylation is a specific and fundamental post-translational modification of mammalian proteins: Head-to-head comparison with O-phosphorylation. *Science's STKE* http://stke.sciencemag.org/cgi/content/full/OC_;sig-trans;2001/87/re1.

Mannick, J. B., Hausladen, A., Liu, L., Hess, D. T., Zeng, M., Miao, Q. X., Kane, L. S., Gow, A. J., and Stamler, J. S. (1999). Fas-induced caspase denitrosation. *Science* **284**, 651–654.

Partis, M. D., Griffiths, D. G., Roberts, G. C., and Beechey, R. B. (1983). Crosslinking of protein by W-maleimido alkanoyl N-hydroxysuccinimido esters. *J. Protein Chem.* **2**, 263–277.

Pettigrew, D. W. (1986). Inactivation of *Escherichia coli* glycerol kinase by 5,5′-dithiobis(2-nitrobenzoic acid) and N-ethylmaleimide: Evidence for nucleotide regulatory binding sites. *Biochemistry* **25**, 4711–4718.

Simon, D. I., Mullins, M. E., Jia, L., Gaston, B., Singel, D. J., and Stamler, J. S. (1996). Polynitrosylated proteins: Characterization, bioactivity, and functional consequences. *Proc. Natl. Acad. Sci. USA* **93**, 4736–4741.

Singh, R. J., Hogg, N., Joseph, J., and Kalyanaraman, B. (1996). Mechanism of nitric oxide release from S-nitrosothiols. *J. Biol. Chem.* **271**, 18596–18603.

Smith, D. B., and Johnson, K. S. (1988). Single-step purification of polypeptides expressed in *Escherichia coli* as fusions with glutathione S-transferase. *Gene* **67**, 31–40.

Smith, J. N., and Dasgupta, T. P. (2000). Kinetics and mechanism of the decomposition of S-nitrosoglutathione by L-ascorbic acid and copper ions in aqueous solution to produce nitric oxide. *Nitric Oxide* **4**, 57–66.

Stamler, J. S., Jaraki, O., Osborne, J., Simon, D. I., Keaney, J., Vita, J., Singel, D., Valeri, C. R., and Loscalzo, J. (1992). Nitric oxide circulates in mammalian plasma primarily as an S-nitroso adduct of serum albumin. *Proc. Natl. Acad. Sci. USA* **89**, 7674–7677.

Stamler, J. S., Simon, D. I., Osborne, J. A., Mullins, M. E., Jaraki, O., Michel, T., Singel, D. J., and Loscalzo, J. (1992). S-nitrosylation of proteins with nitric oxide: Synthesis and characterization of biologically active compounds. *Proc. Natl. Acad. Sci. USA* **89**, 444–448.

Stamler, J. S., Toone, E. J., Lipton, S. A., and Sucher, N. J. (1997). (S)No signals: Translocation, regulation, and a consensus motif. *Neuron* **18**, 691–696.

Stauffer, D. A., and Karlin, A. (1994). Electrostatic potential of the acetylcholine binding sites in the nicotinic receptor probed by reactions of binding-site cysteines with charged methanethiosulfonates. *Biochemistry* **33**, 6840–6849.

Wink, D. A., Kasprzak, K. S., Maragos, C. M., Elespuru, R. K., Misra, M., Dunams, T. M., Cebula, T. A., Koch, W. H., Andrews, A. W., Allen, J. S., and Keefer, L. K. (1991). DNA deaminating ability and genotoxicity of nitric oxide and its progenitors. *Science* **254**, 1001–1003.

Xu, L., Eu, J. P., Meissner, G., and Stamler, J. S. (1998). Activation of the cardiac calcium release channel (ryanodine receptor) by poly-S-nitrosylation. *Science* **279**, 234–237.

# [12] Detection of Nitrosothiols and Other Nitroso Species *In Vitro* and in Cells

*By* David Jourd'heuil, Frances L. Jourd'heuil, Anthony M. Lowery, John Hughes, and Matthew B. Grisham

## Abstract

The nitric oxide (NO)–mediated nitrosation of peptides and proteins may play important roles in normobiology and pathobiology. With the realization that S-nitrosothiols (RSNOs) participate in the transport, storage, and delivery of NO, as well as posttranslational modifications in cell signaling and inflammatory processes, there is an increasing need for the detection of nitrosothiols (RSNOs) and other nitroso species in cells and tissues. In this chapter, we describe the utilization of a gas phase

METHODS IN ENZYMOLOGY, VOL. 396
0076-6879/05 $35.00
DOI: 10.1016/S0076-6879(05)96012-6

chemiluminescence-based assay and "biotin switch" method for the detection of nitroso species in cells. These methods are sensitive enough to quantify and contrast the different pools of nitroso species that may coexist under physiologically relevant conditions. They also provide the means to characterize and identify proteins that may represent specific targets for nitrosation reactions.

Introduction

It is becoming increasingly appreciated that $S$-nitrosothiols (RSNOs) play important roles in the physiological chemistry of nitric oxide (NO). For example, RSNOs have been suggested to participate in the transport, storage, and delivery of NO and posttranslational modifications in cell signaling and inflammatory processes. Historically, RSNOs have been shown to be formed from the autoxidation of NO:

$$2NO + O_2^{\cdot} \rightarrow 2^{\cdot}NO_2$$
$$2NO_2 + 2NO \rightarrow 2^{\cdot}N_2O_3$$
$$2N_2O_2 + 2RSH \rightarrow 2^{\cdot}RSNO + 2NO_2^- + 2H^+$$

This reaction is second order with respect to NO and first order with respect to $O_2$. Although this mechanism is well known to occur in the gas phase under conditions of large NO and $O_2$ concentrations, more recent data suggest that RSNO formation by this mechanism in most cells and tissues would be limited. It is known, for example, that oxygen tension within most tissues (except for the lung) would be close to venous $pO_2$ or approximately 20–40 mmHg. Furthermore, steady state levels of NO are known to approximate no more than nanomolar amounts (Lancaster, 1994). In addition to the small amounts of NO and $O_2$ within cells, millimolar levels of GSH exist, suggesting that a reevaluation of the mechanisms responsible for RSNO formation in the presence of large amounts of RSH. Recent data by several laboratories (Jourd'heuil *et al.*, 2003; Schrammel *et al.*, 2003) suggest that a more physiological mechanism to account for RSNO formation *in vivo* under reduced fluxes of $O_2$ and NO but in the presence of millimolar amounts of thiols would be

$$2NO + O_2^{\cdot} \rightarrow {}^{\cdot}2NO_2$$
$$2NO_2 + 2RSH \rightarrow {}^{\cdot}2NO_2 + 2RS^{\cdot}$$
$$2RS^{\cdot} + 2NO \rightarrow 2^{\cdot\cdot}RSNO$$

Because the rate constant for the interaction between $NO_2$ and RSH is two orders of magnitude faster than that for the interaction between $N_2O_3$ and RSH, this mechanism has been suggested as the major pathway for

RSNO *in vivo* (Jourd'heuil *et al.*, 2003). Critical to our understanding of how RSNOs are formed and what role they play in normophysiology and pathophysiology is the ability to specifically quantify RSNOs under physiological conditions. Here, we describe two methods to detect nitroso derivatives *in vitro* and cell systems.

## Chemiluminescence Detection

### *Principles and General Considerations*

Chemiluminescence detection is based on the reductive decomposition of nitroso species by an iodine/triiodide mixture to release NO, which is subsequently measured by gas phase chemiluminescence upon reaction with ozone (Feelisch *et al.*, 2002). RSNOs are sensitive to mercury-induced decomposition, whereas other nitroso species including nitrosamines (RNNOs) and nitrosyl hemes are not (Table I). Thus, aliquots of the same sample are routinely pretreated in parallel with or without mercuric chloride ($HgCl_2$) to determine the contribution of RSNOs to the overall signal. It is important to stress that this method detects NO generated upon reduction of NO carriers. The nature of the NO carrier is then inferred by comparison with known standards relative to their sensitivity to specific decomposition pathways. Clearly, results should be interpreted within this context, and when possible, they should be confirmed using alternative methods.

The presence of nitrite ($NO_2^-$) in biological fluids, tissue extracts, media, and chemicals in amounts that usually exceed the nitrosated biomolecules is the principal limitation to the accurate determination of nitroso compounds using this technique. This is because the iodine/triiodide mixture used to reduce nitrosated molecules also effectively reduces $NO_2^-$ to NO. The obligatory removal of $NO_2^-$ is achieved by treatment of samples with excess sulfanilamide before injection of the sample in the reaction

TABLE I
SENSITIVITY OF RELEVANT NITROSO SPECIES TO MERCURIC CHLORIDE ($HgCl_2$)

| Compound | Abbreviation | Resistance to $HgCl_2$ (%) |
|---|---|---|
| Nitrite | $NO_2^-$ | 100 |
| S-nitrosoglutathione | GSNO | 0 |
| S-nitrosoalbumin | SNOAlb | 0 |
| Nitrosylhemoglobin | HbNO | 97 |
| N-acetyl-nitrosotryptophan | N-AcNOW | 96 |

vessel containing iodine and triiodide. It is recommended that the concentration of sulfanilamide (as well as NEM) should be adjusted to the specific conditions of the experiment. For example, the utilization of an NO donor may lead to the accumulation of $NO_2^-$ in amounts that may exceed the scavenging capacity of standard concentrations of sulfanilamide. Controls containing equivalent amounts of $NO_2^-$ to that of the approximate amount of NO generated in the system should be run in parallel to confirm the complete removal of $NO_2^-$. It is also important to note that any sample processing preceding the addition of sulfanilamide that may involve acidification should be avoided because acidification of $NO_2^-$ is one of the most facile means to nitrosate. This could clearly occur during acidic precipitation of proteins, a step that may be taken to determine low-molecular-weight RSNOs in a sample.

Lastly, a general concern is the inherent instability of RSNOs. The primary mechanisms of decomposition are light (Singh *et al.*, 1996b) and metal mediated (Dicks *et al.*, 1996), so the experiment and analysis of the samples should be performed under minimal light exposure and in the presence of a metal chelator such as ethylenediaminetetraacetic acid (EDTA) or diethylenetetraminepentaacetic acid (DTPA). Excess thiols (Singh *et al.*, 1996a) and ascorbate (Dasgupta and Smith, 2002) may also lead to the unintended decomposition of certain nitroso species, which is prevented by pretreating the samples with *N*-ethylmaleimide (NEM).

## Detection of Nitroso Species in Cells

*General Protocol.* Typically, adherent cells are washed twice with cold phosphate-buffered saline (PBS) containing 100 $\mu M$ DTPA. The cells are detached with trypsin-EDTA, collected by centrifugation, counted, and resuspended in 1 ml of 4 m$M$ phosphate buffer containing 100 $\mu M$ DTPA. Cell suspensions are then homogenized using a 2-ml Dounce homogenizer. Eight hundred microliters of each homogenate are transferred to a glass tube containing 100 $\mu l$ of 100 m$M$ NEM. The samples are kept on ice and in the dark for 15 min before addition of 100 $\mu l$ of 100 m$M$ sulfanilamide in 1 $M$ HCl and incubation for another 15 min on ice. RSNO formation is evaluated by measuring the amount of •NO liberated after injection of the sample into the purge vessel (Feelisch *et al.*, 2002). The purge vessel contains 4.5 ml of glacial acetic acid and 500 $\mu l$ of an aqueous mixture containing 450 m$M$ potassium iodide and 100 m$M$ iodine. The vessel is kept at 70° via a water jacket, and the solution is constantly purged with nitrogen and changed every four injections. The amount of NO evolving from the purge vessel is quantified by gas phase chemiluminescence (NOA 280: Sievers Instruments, Boulder, CO). Peak integration is performed, and results are converted to NO concentrations using authentic NO as a standard. To ascertain that the nitrosation sites are thiols, aliquots of the same sample are pretreated with $HgCl_2$ (5 m$M$ final concentration) and sulfanilamide.

*Chemiluminescence Detection of* S-*Nitrosoglutathione.* Calculated NO concentrations were further validated using $S$-nitrosoglutathione (GSNO) standards. This RSNO is synthesized by co-incubation of glutathione (GSH) with excess acidified $NO_2^-$, as previously described (Wink *et al.*, 1999). Figure 1A shows a typical chemiluminescence detector response after duplicate injection of increasing amounts of GSNO standards. The linearity of the dose response was established by integration of the peak areas (Fig. 1B). Neither nitrite nor GSH—both in excess relative to GSNO—impacted on the signal obtained from 100 pmol of GSNO (Fig. 1D). There was complete recovery of the chemiluminescence signal obtained from 10 pmol of GSNO spiked in cell lysates compared to GSNO in PBS.

*Cell-Associated Nitroso Species from Exogenous Exposure to NO.* Figure 2A shows a typical chemiluminescence detector response from duplicate injection of lysates obtained from cells treated with increasing concentrations of the NO donor spermine NONOate (Sp/NO). For these experiments, the mouse fibroblast cell line NIH 3T3 was maintained in 75-cm$^2$ culture flasks in DMEM supplemented with 10% (v/v) calf serum, penicillin (100 units/ml), and streptomycin (100 $\mu$g/ml). For experimentation, NIH 3T3 cells were seeded in 25-cm$^2$ flasks and grown to 80% confluence in DMEM with 10% (v/v) calf serum, penicillin (100 units/ml), and streptomycin (100 $\mu$g/ml). Before incubation with the NO donor, the cell culture medium was replaced with fresh DMEM containing 10% serum, and the flasks were preincubated for 2 h in humidified atmosphere equilibrated with an $O_2$/nitrogen/$CO_2$ gas mixture, resulting in final $O_2$ tension of 21%. The cells were then incubated with argon-purged Sp/NO for 1 h at 37°. Cell viability was not affected by this treatment as tested by trypan blue exclusion (data not shown). The different Sp/NO concentrations used (10–100 $\mu M$) provided initial rates of NO release, ranging from about 0.1 to 1 $\mu M$/min, well within the physiological range.

*S-Nitrosothiol Formation in Fibroblasts Expressing Human Inducible Nitric Oxide Synthase.* The formation of cell-associated nitroso species from endogenous sources of NO may also be examined by gas phase chemiluminescence. Figure 2B illustrates the formation of RSNOs from DFGiNOS NIH 3T3 cells, a mouse fibroblasts cell line that constitutively expresses recombinant human inducible nitric oxide synthase (iNOS) (Tzeng *et al.*, 1995). These cells are incapable of *de novo* biosynthesis of tetrahydrobiopterin ($H_4B$), a cofactor necessary for the assembly of iNOS subunits into the active dimer. Thus, NO production can be stimulated by

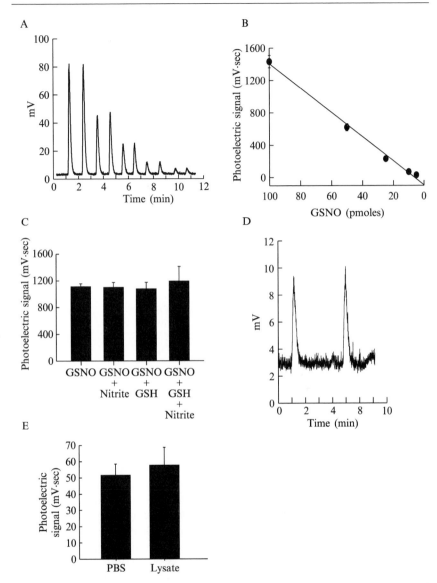

FIG. 1. (A) Chemiluminescence detector response of S-nitrosoglutathione (GSNO) standards in phosphate-buffered saline (PBS). (B) Linear regression of the peak area of the signal and amounts injected (n = 3). (C) Effect of nitrite (NO$_2^-$, 100$\mu$M) and glutathione (GSH, 1mM) on the peak area obtained from 100 pmol of GSNO in PBS (n = 3). (D) Recovery of signal after injection of 10 pmol of GSNO in cell lysate and PBS. (C) Quantification of results described in (D) (n = 3).

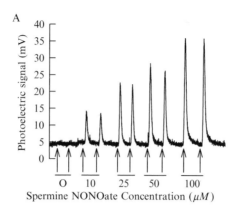

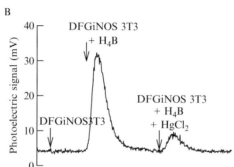

FIG. 2. Representative chemiluminescence signals for nitroso species determination. RSNO detection from NIH 3T3 exposed to exogenous NO (A) or overexpressing the human isoform of nitric oxide synthase (B) (see text for details).

addition of $H_4B$ in the cell culture media. The use of this cell line may represent a simpler alternative to the upregulation of iNOS and concomitant increase in NO production in cells upon cytokine stimulation. This is because cytokine stimulation is associated with changes in the expression and/or activity of many genes such as superoxide dismutase 2 (SOD2) and NADPH oxidases (NOXs) that may have confounding effects on RSNO formation. In the experiment illustrated in Fig. 2B, DFGiNOS cells were incubated for 12 h with 1 $\mu M$ $H_4B$, and the cells were then treated for the determination of nitroso species, as described earlier. Production of NO was confirmed by measurement of $NO_2^-/NO_3^-$ in the cell media (data not shown). To ascertain that NO reaction sites were mostly thiols, samples were examined for their sensitivity to $HgCl_2$, and the results indicate that more than 75% of the signal obtained from the cells stimulated with $H_4B$ represented RSNOs.

*Isolation of Nitrosated Proteins Using Gas-Phase Chemiluminescence as the Detection Assay.* In model systems using cells in suspension or adherent, RSNOs are consistently associated with the high-molecular-weight fraction, indicating that proteins may represent the primary site of *S*-nitrosation in cells (Eu *et al.*, 2000; Hoffmann *et al.*, 2001; Jourd'heuil *et al.*, 2003). There is also a large body of literature indicating that the *S*-nitrosation of cysteine residues may represent an important posttranslational modification, which may regulate protein function and signal transduction pathways (Foster *et al.*, 2003). Thus, it is becoming essential to identify nitrosated proteins in cells in order to understand the molecular basis for nitrosation reactions and to examine the impact of nitrosation on cell function. This cannot be accomplished directly by analysis of nitrosated pools using gas phase chemiluminescence. However, nitrosated proteins may be purified by standard chromatography using the gas phase chemiluminescence method as the detection assay. To demonstrate the feasibility of this method, subconfluent mouse fibroblasts were incubated for 1 h at 37° with the NO donor Sp/NO (100 $\mu M$) in DMEM supplemented with 10% calf serum. After trypsinization, cells were resuspended to a final concentration of $10^7$ cells/ml into a lysis buffer containing 20 m$M$ triethanolamine (TEA), 100 $\mu M$ DTPA, and 0.1% Triton X-100. The cells were then incubated at room temperature for 15 min and centrifuged at 10,000$g$, and the supernatant was collected for immediate injection. The amount of nitroso species recovered in the supernatant represented approximately 50% of the initial amount present in the lysate.

The chromatography for purification was carried out using a Biologic Workstation FPLC instrument (Bio-Rad Laboratories, Hercules, CA) at room temperature, and 1 ml eluent fractions were collected into glass tubes containing the appropriate amount of NEM on ice and protected from light. The fractions were then processed for injection into the reaction vessel of the chemiluminescence detector as described above. For the representative elution profile shown in Fig. 3, the supernatant was chromatographed in one run of 1 ml on a Mono Q HR 5/5 (Amersham Pharmacia, Piscataway, NJ) at a flow rate of 1 ml/min. A programmed gradient was run from 0.05 to 1.0 $M$ NaCl to elute nitrosated fractions. It was evident that multiple fractions contained nitrosated proteins and that further purification was necessary for specific protein identification. We have found that the nitroso signal decays in the collected fraction over a period of 3–5 h allowing further purification by sequential anion exchange and gel filtration chromatography. Another approach is to use the initial anion exchange as a prepurification step for further analysis using the "biotin switch" method described in the following section.

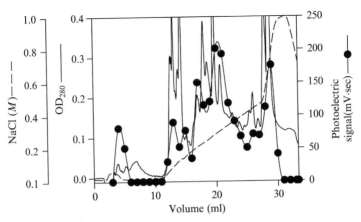

FIG. 3. Representative elution profile for the purification of nitrosated proteins using Mono Q anion exchange chromatography of cell lysates. See text for details.

## Biotin Switch Assay

The biotin switch method developed by Jaffrey *et al.* (2001) has received increasing attention because of its apparent specificity for *S*-nitrosated proteins and its sensitivity when combined with immunoblotting. It offers the exciting advantage to provide a profile of nitrosated proteins from cell preparations when combined with proteomics. The assay is based on the exchange of the NO moiety from the RSNO with a biotin group, which serves as a convenient tag for protein characterization and identification. After preparation of a cell extract under non-reducing conditions, the assay requires blockade of free thiols with the thiol-specific methyl thiolating agent methyl methanethiosulfonate (MMTS). RSNOs are then reduced back to their free thiols by incubation with a large excess of ascorbate. The reduced sulfhydryl groups are conjugated with a biotin derivative, *N*-[6-(biotinamido)hexyl]-3′-(2′-pyridyldithio)propionamide (biotin-HPDP) and the biotinylated proteins are identified by immunoblotting or purified by avidin-affinity chromatography for identification by mass spectrometry. Controls are performed to test for the presence of endogenous biotinylated proteins, to determine potential contamination from thiolated proteins, and to test for artefactual formation of RSNOs during the procedure.

The protocol described below is routinely used in our laboratory for the study of nitrosative pathways in cells. For all experiments described in Fig. 4, cell lysates were prepared from NIH 3T3 cells and diluted to a final concentration of 1 mg/ml in either 100 m$M$ phosphate buffer and 0.1 m$M$

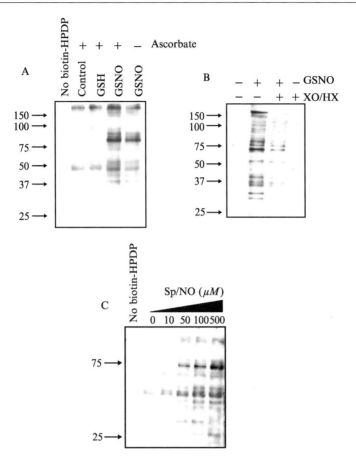

FIG. 4. (A) Representative Western blots of $S$-nitrosated proteins from cell lysates (1 mg/ml) exposed to GSNO (50 $\mu M$) for 1 h at 37° and derivatized using the biotin switch method. (B) Effect of hypoxanthine (HX) (250 $\mu M$) and xanthine oxidase (1 mU/ml) on protein nitrosation of cell lysates exposed to GSNO (50 $\mu M$). See text for details. (C) Effect of increasing concentrations of the NO donor spermine/NO on the nitrosation profile of cell lysates (biotin switch method). See text for details.

DTPA (pH 7.4) or HED buffer containing 250 m$M$ HEPES, 1 m$M$ EDTA, and 0.1 m$M$ DTPA (pH 7.7). After treatment with the NO donor, 75 $\mu$l of the lysate was loaded onto a MicroBioSpin6 column (Bio-Rad, Hercules, CA) equilibrated with HED and the column was centrifuged at 1000$g$ for 5 min. Seven microliters of 25% sodium dodecylsulfate (SDS) and 1.5 $\mu$l of 20% MMTS in DMF were added to the 75 $\mu$l eluate. The sample was then incubated at 50° in a water bath for 20 min with frequent vortexing. The sample was then desalted to remove MMTS and 8 $\mu$l of 2 m$M$ biotin-HPDP in dimethylsulfoxide (DMSO) with 4 $\mu$l of 100 m$M$ ascorbate was then added. The sample was incubated for 60 min before desalting four times and final resuspension into HED containing 0.5% Triton X-100. After SDS-polyacrylamide gel electrophoresis (PAGE) biotinylated proteins were detected by immunoblotting using an anti-biotin antibody (Bethyl Laboratories, Montgomery, TX) as a primary, an anti-rabbit HRP antibody (Amersham Pharmacia, Piscataway, NJ) as a secondary, and an HRP detection kit (Pierce, Rockford, IL).

Figure 4A shows a representative immunoblot of $S$-nitrosated proteins obtained from NIH 3T3 cell lysates treated with 50 $\mu M$ GSNO for 1 h at 37°. Control samples revealed the presence of endogenously biotinylated proteins, although detection may vary, as illustrated in Fig. 4B. Incubation of the cell lysates with GSH did not result in appreciable changes compared to control, whereas incubation with GSNO resulted in the appearance of multiple bands indicative of protein $S$-nitrosation. The intensity of the bands was diminished but not eliminated upon omission of ascorbate during the biotinylation step, suggesting significant spontaneous decomposition of RSNOs during derivatization. Co-addition of equimolar amounts of $NO_2^-$ and GSH did not result in any changes compared to control. As shown in Fig. 4B, the biotin switch assay may represent a useful tool to explore nitrosative pathways in complex biological matrices. In this specific example, co-incubation of cell lysates with 50 $\mu M$ GSNO, 250 $\mu M$ hypoxanthine (HX), and 1 mU/ml xanthine oxidase (XO) resulted in a mark decrease in nitrosated proteins compared to GSNO alone. These results indicated that superoxide ($O_2^{\cdot-}$) and/or hydrogen peroxide ($H_2O_2$) generated by HX/XO may limit RSNO formation in cells, in accordance with previous studies showing an inhibitory effect of XO and $O_2^{\cdot-}$ on the formation of RSNOs in solution (Aleryani et al., 1998; Jourd'heuil et al., 1999; Trujillo et al., 1999). Finally, incubation of cell lysates with increasing concentrations of the NO donor Sp/NO resulted in a dose-dependent increase in $S$-nitrosated proteins (Fig. 4C) in agreement with the chemiluminescence data (Fig. 3A). As a final remark, it is important to stress that the biotin switch assay only probes for $S$-nitrosated proteins. In contrast, the chemiluminescence-based assay allows for the quantitative determination of NO among different carriers including RSNOs, RNNOs, and nitrosyl hemes.

## Summary and Conclusion

The procedures described in this chapter represent important tools to examine nitrosation pathways in cells and tissues. They are sensitive enough to quantify and contrast the different pools of nitroso species that may coexist under physiologically relevant conditions (Bryan *et al.*, 2004). They provide the means to characterize and identify proteins that may represent specific targets for nitrosation reactions. Most importantly, these techniques allow us to examine the relevance of the chemical pathways identified in solution using model compounds, an issue that could not be evaluated earlier with techniques that lacked in sensitivity and specificity. For example, we used this approach to test for the effect of oxygen tension on the formation of cell-associated nitroso species (Jourd'heuil *et al.*, 2003). Under these conditions, maximum nitrosation occurred upon incubation of the cells with 3% $O_2$, and it was partially inhibited upon co-incubation with a thiyl radical scavenger. These results were consistent with those obtained in solution with GSH, and two principal nitrosation pathways explained our results in cells and in solution. First, thiols (RSH, Fig. 5) react with

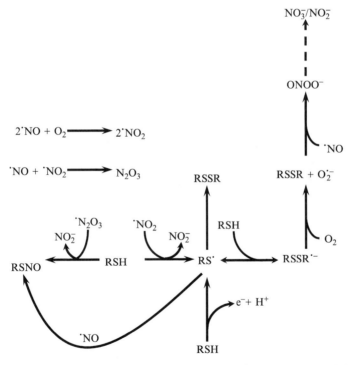

FIG. 5. Reaction of nitric oxide (NO) with thiols (RSH) in oxygenated solution. The reaction of NO with excess RSH yields RSNOs and disulfides (RSSR). RSSR is formed from the one electron oxidation by nitrogen dioxide ($^\bullet NO_2$) of RSH to yield the thiyl radical RS$^\bullet$. NO may be consumed by reaction with superoxide ($O_2^{\bullet-}$), formed from the reduction of molecular oxygen ($O_2$) by the disulfide anion radical (RSSR$^{\bullet-}$). These reactions are described in detail for glutathione (GSH) elsewhere (Jourd'heuil *et al.*, 2003).

dinitrogen trioxide ($N_2O_3$) formed from NO autoxidation to yield $NO_2^-$ and RSNOs. Second, RSNOs may form through the intermediate oxidation of thiols by nitrogen dioxide ($^{\cdot}NO_2$) to generate ($NO_2^-$) and thiyl radicals (RS), the latter either yield oxidized thiol products or combine with NO to form RSNOs. Because $O_2$ drives the removal of thiyl radicals and formation of the NO scavenger superoxide ($O_2^{\cdot-}$), an increase in oxygen tension favors the oxidation of thiols over their nitrosation.

Most often, changes in protein function are seen within the context of a specific modification such as the nitration or nitrosation of an amino acid residue. A major conclusion from studies, including the one briefly described in this chapter, is that both oxidative and nitrosative reactions may always overlap. An important challenge is to identify the specificity, if any, of the nitrosative versus oxidative pathways as chemical modifiers of protein structure and function. Ultimately, it is possible that in many cases, the functional outcome of nitrosative stress may in fact represent an aggregate of multiple modifications on one or several proteins. Advances in proteomics allowing for the simultaneous detection of multiple modifications such as nitrosation, nitration, oxidation, and thiolation will provide the means to tackle this issue.

### Acknowledgements

This work was supported by grants CA89366 (to D. J.) and DK 64023 and DK 47663 (M. B. G.) from the National Institutes of Health.

### References

Aleryani, S., Milo, E., Rose, Y., and Kostka, P. (1998). Superoxide-mediated decomposition of biological S-nitrosothiols. J. Biol. Chem. 273, 6041–6045.

Bryan, N. S., Rassaf, T., Maloney, R. E., Rodriguez, C. M., Saijo, F., Rodriguez, J. R., and Feelisch, M. (2004). Cellular targets and mechanisms of nitros(yl)ation: An insight into their nature and kinetics in vivo. Proc. Natl. Acad. Sci. 101, 4308–4313.

Dasgupta, T. P., and Smith, J. N. (2002). Reactions of S-nitrosothiols with L-ascorbic acid in aqueous solution. Methods Enzymol. 358, 219–229.

Dicks, A. P., Swift, H. R., Williams, D. L. H., Butler, A. R., Al-Sa'doni, H. H., and Cox, B. G. (1996). Identification of Cu+ as the effective reagent in nitric oxide formation from S-nitrosothiols (RSNO). J. Chem. Soc. Perkin Trans. 2, 481–487.

Eu, J. P., Liu, L., Zeng, M., and Stamler, J. S. (2000). An apoptotic model for nitrosative stress. Biochemistry 39, 1040–1047.

Feelisch, M., Rassaf, T., Mnaimneh, S., Singh, N., Bryan, N. S., Jourd'heuil, D., and Kelm, M. (2002). Concomitant S-, N-, and heme-nitros(yl)ation in biological tissues and fluids: Implications for the fate of NO in vivo. FASEB J. 16, 1775–1785.

Foster, M. W., McMahon, T. J., and Stamler, J. S. (2003). S-nitrosylation in health and disease. Trends Mol. Med. 9, 160–168.

Hoffmann, J., Haendeler, J., Zeiher, A. M., and Dimmeler, S. (2001). TNFα and oxLDL reduce protein S-nitrosylation in endothelial cells. *J. Biol. Chem.* **44,** 41383–41387.

Jaffrey, S. R., Erdjument-Bromage, H., Ferris, C. D., Tempst, P., and Snyder, S. H. (2001). Protein S-nitrosylation: A physiological signal for neuronal nitric oxide. *Nat. Cell Biol.* **3,** 193–197.

Jourd'heuil, D., Jourd'heuil, F. L., and Feelisch, M. (2003). Oxidation and nitrosation of thiols at low micromolar exposure to nitric oxide. Evidence for a free radical mechanism. *J. Biol. Chem.* **278,** 15720–15726.

Jourd'heuil, D., Laroux, S. F., Miles, A. M., Wink, D. A., and Grisham, M. B. (1999). Effect of superoxide dismutase on the stability of S-nitrosothiols. *Arch. Biochem. Biophys.* **361,** 323–330.

Lancaster, J., Jr. (1994). Simulation of the diffusion and reaction of endogenously produced nitric oxide. *Proc. Natl. Acad. Sci. USA* **91,** 8137–8141.

Schrammel, A., Gorren, A. C., Schmidt, K., Pfeiffer, S., and Mayer, B. (2003). S-nitrosation of glutathione by nitric oxide, peroxynitrite, and $NO/O_2^-$. *Free Radic. Biol. Med.* **34,** 1078–1088.

Singh, R. J., Hogg, N., Joseph, J., and Kalyanaraman, B. (1996b). Mechanism of nitric oxide release from S-nitrosothiols. *J. Biol. Chem.* **271,** 18596–18603.

Singh, S. P., Wishnok, J. S., Keshive, M., Deen, W. M., and Tannenbaum, S. R. (1996a). The chemistry of the S-nitrosoglutathione/glutathione system. *Proc. Natl. Acad. Sci. USA* **93,** 14428–14433.

Trujillo, M., Alvarez, M. N., Peluffo, G., Freeman, B. A., and Radi, R. (1999). Xanthine oxidase–mediated decomposition of S-nitrosothiols. *J. Biol. Chem.* **273,** 7828–7834.

Tzeng, E., Billiar, T. R., Robbins, P. D., Loftus, M., and Stuehr, D. J. (1995). Expression of human inducible nitric oxide synthase in a tetrahydrobiopterin ($H_4B$)–deficient cell line: $H_4B$ promotes assembly of enzyme subunits into an active dimer. *Proc. Natl. Acad. Sci. USA* **92,** 11771–11775.

Wink, D. A., Kim, M. S., Coffin, D., Cook, J. C., Vodovotz, Y., Chistodoulou, D., Jourd'heuil, D., and Grisham, M. B. (1999). Detection of S-nitrosothiols by fluorometric and colorometric methods. *Methods Enzymol.* **301,** 201–211.

# [13]  Detection and Identification of S-Nitrosylated Proteins in Endothelial Cells

*By* ANTONIO MARTÍNEZ-RUIZ and SANTIAGO LAMAS

## Abstract

Nitric oxide performs some of its direct effects by the induction of various protein posttranslational modifications. Among them, S-nitrosylation of cysteines has gained increasing attention and has been postulated as a signaling mechanism. However, there are still many technical limitations for the detection and identification of this posttranslational modification in cell proteins, and some of them are directly related to the lability of the nitrosothiol bond. We describe protocols for applying a derivatization

METHODS IN ENZYMOLOGY, VOL. 396
0076-6879/05 $35.00
DOI: 10.1016/S0076-6879(05)96013-8

technique, the biotin switch, which substitutes the *S*-nitrosylation for a biotinylation in the detection and proteomic identification of *S*-nitrosylated proteins in intact cells.

Introduction

One of the posttranslational modifications that nitric oxide (NO) can induce in proteins is *S*-nitrosylation, also called *S*-nitrosation; for a discussion of the terminology, see Forman *et al.* (2004) and Martínez-Ruiz and Lamas (2004a). It refers to the formation of a thionitrite ($-S-N=O$) group in cysteine residues, which is not just substitution of the thiol ($-S-H$) hydrogen atom by NO, as it requires loss of an electron. Thus, although the *S*-nitrosylating mechanisms are under discussion, it is generally assumed that NO itself does not directly nitrosylate the cysteine, and it requires previous formation of other chemical species, especially $N_2O_3$. Increasing attention is being paid to this modification as an alternative mode of action of NO, with arguments in favor of a role in cellular signaling, and others supporting that it is formed only in situations of "nitrosative stress," related to some pathophysiological environments.

An increasing number of proteins have been shown to be susceptible of becoming *S*-nitrosylated in specific cysteines, in many cases altering their function (Stamler *et al.*, 2001). However, technical limitations, especially due to the lability of the bond, have hampered the study of the relevance of this modification inside cells and tissues, and studies on individual proteins have been conducted mainly with purified proteins. We have previously reviewed in brief some of the current methodologies for the detection of *S*-nitrosylated proteins (Martínez-Ruiz and Lamas, 2004a). Now we describe one of the methodologies we use, which has allowed us to proteomically identify *S*-nitrosylated proteins in intact endothelial cells (Martínez-Ruiz and Lamas, 2004b).

The Biotin Switch Method

In 2001, Jaffrey *et al.* published a method to specifically derivatize the thiols that were nitrosylated to biotinylated residues; they called this methodology the "biotin switch." It consists of three chemical steps (Fig. 1):

- Blocking of free thiols, using methyl methanethiosulfonate (MMTS)
- Specific reduction of nitrosothiols, using ascorbate
- Labeling of newly formed thiols with a biotinylating agent, using *N*-[6-(biotinamido)hexyl]-3′-(2′-pyridyldithio)propionamide (biotin-HPDP), which incorporates to the thiol by forming a disulfide bridge.

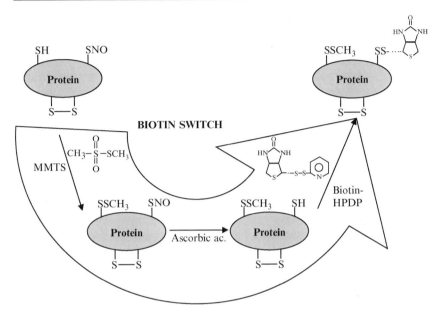

FIG. 1. Schematic representation of the three chemical steps involved in the biotin switch technique that allow an *S*-nitrosylated protein to become specifically biotinylated. The spacer arm included in the biotinylating reagent, biotin-HPDP, is depicted with a dotted line.

In the original report, the authors checked that these reactions did not derivatize cysteines modified in other ways (Jaffrey *et al.*, 2001). A similar methodology has been developed for the case of *S*-glutathionylated proteins, with different chemical reagents for each step, and which uses a modified glutaredoxin to deglutathionylate the protein cysteines (Lind *et al.*, 2002).

In our opinion, specificity of the technique is acceptable, provided that all the adequate controls are performed. However, sensitivity can be a problem for many applications, especially proteomic identification, as we comment later.

Applications of this methodology are summarized in Fig. 2. Biotinylation of purified proteins or extracts that have been derivatized can be detected using a Western blot with avidin or an anti-biotin antibody. As this will detect also endogenously biotinylated proteins, experimental control conditions subjected to the same process adding just the vehicle of biotin-HPDP must be performed. Special care must be taken to use nonreducing buffer and electrophoresis conditions, as the biotin labeling is incorporated via a disulfide bond and is thus reversed by reduction.

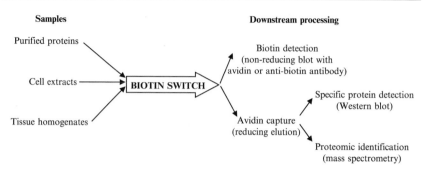

FIG. 2. Overview of the applications of the biotin switch technique for the detection and identification of S-nitrosylated proteins.

Biotinylated proteins can also be purified by capturing them with resins of immobilized avidin. In this case, the incorporation of the biotin via the disulfide bond is an advantage, because it allows an easy elution by incubation with a reducing agent, allowing also to eliminate the endogenously biotinylated proteins, which remain bound to the avidin. However, this has the disadvantage of losing the label from the protein. After avidin capture, specific protein detection can be performed by means of Western blot with an antibody directed against the protein.

Avidin capture can also be used for the proteomic identification of the purified proteins, allowing the description of new proteins that can be targets for S-nitrosylation, although there are still few reports. Most of them added NO donors or nitrosothiols to protein extracts obtained from either murine brains (Jaffrey et al., 2001), murine mesangial cells (Kuncewicz et al., 2003), or rat liver mitochondria (Foster and Stamler, 2004). We have been able to identify the proteins that were S-nitrosylated in intact human endothelial cells, by using S-nitroso-L-cysteine, a nitrosothiol that is transported to the cell cytoplasm (Martínez-Ruiz and Lamas, 2004b). In all cases, however, S-nitrosylation can be considered to have been forced by the addition of high amounts of S-nitrosylating agents; also, milligram to gram quantities of protein extracts were used as the starting material for the biotin switch assay. Thus, direct identification of the proteins that are S-nitrosylated in physiological, or even pathophysiological conditions, would require much more sensitivity and improvement of the technique.

## Preparation of S-Nitrosylated Extracts

### Synthesis of S-nitroso-L-cysteine

S-nitroso-L-cysteine is an unstable nitrosothiol and is not available commercially. We use it as a nitrosylating agent because it is a naturally occurring nitrosothiol, and we and others have seen that it is able to raise the nitrosothiol content in the cell (Zhang and Hogg, 2004), because it is transported across the plasma membrane, and it transnitrosates other low-molecular-weight and protein thiols. We follow a protocol previously described (Jourd'heuil et al., 2000).

1. Prepare 1 volume of L-cysteine 200 m$M$ in HCl 1 $M$. When storing L-cysteine, we treat the bottle with $N_2$ to prevent oxidation.
2. Mix with 1 volume of $NaNO_2$ 200 m$M$ in water. The solution gets red almost immediately. From now on, we protect the tube from light to diminish destruction of the nitrosothiol bond by light.
3. Incubate 30 min at room temperature.
4. Add 2 volumes (respect to L-cysteine) of potassium phosphate 1 $M$, pH 7.4. Immediately place on ice.
5. Aliquot conveniently and store at $-80°$. We thaw aliquots just once.
6. With the remaining solution, make a 1/100 dilution and perform a wave scan. Determine S-nitroso-L-cysteine from absorbance at 338 nm, using the molar absorption coefficient $\varepsilon_{338} = 900\ M^{-1}\ cm^{-1}$ (DeMaster et al., 1995).

We obtain S-nitroso-L-cysteine solutions between 30 and 40 m$M$, which represents a yield between 60 and 80%.

### Treatment of Cells with S-nitroso-L-cysteine and Extract Preparation

We have used the EA.hy926 cell line (kindly provided by Dr. Cora-Jean S. Edgell, University of North Carolina), derived from human endothelial cells (Brown et al., 1993), for the proteomic studies, as high quantities of starting material are needed, and the human origin helps in the proteomic identification. It is maintained in DMEM medium with HAT supplement, 20% fetal bovine serum (FBS), 100 U/ml penicillin, 100 $\mu$g/ml streptomycin, and 5 $\mu$g/ml gentamicin.

1. Culture cells in Petri dishes until they are just confluent.
2. Wash cells with PBS and add RPMI medium, without serum, or PBS.

3. Add S-nitroso-L-cysteine to 1 m$M$ concentration. As a control, we incubate cells with just L-cysteine 1 m$M$. Incubate at 37° in the cell incubator for 15 min. After this step, try to protect the samples from the light, as it decomposes the S-nitrosothiols.

4. After treatment, wash cells with PBS and place on ice. Add non-denaturing lysis solution (50 m$M$ Tris–HCl, pH 7.4, 300 m$M$ NaCl, 5 m$M$ EDTA, 0.1 m$M$ neocuproine, and 1% Triton X-100 plus protease inhibitors). We add 1 ml to a 100-mm diameter dish.

5. Scrape cells, harvest them, and incubate for 15 min in ice.

6. Centrifuge at 10,000g, 4° for 15 min and collect supernatant.

7. Quantify protein concentration. Adjust to 0.5 mg/ml with lysis solution, if necessary.

8. Protein extracts can be processed immediately or stored at −80°, although we try not to maintain them frozen for many days.


Treatment for the Proteomic Identification of S-Nitrosylated Proteins

*Biotin Switch Treatment*

The starting material is the protein extract obtained as explained, at a concentration of 0.5 mg/ml or less. For the proteomic identification, several milligrams of protein extracts will be needed. If the downstream application is the biotin detection, the amount of protein can be significantly reduced. In the case of specific protein detection after avidin capture, the amount can also be reduced, depending on the sensitivity of the specific antibody.

1. Add 4 volumes of blocking buffer [225 m$M$ HEPES, pH 7.7, 0.9 m$M$ EDTA, 90 $\mu M$ neocuproine, 2.5% sodium dodecylsulfate (SDS), 20 m$M$ MMTS], and incubate at 50° for 20 min, with occasional or constant agitation.

2. Precipitate with acetone. Add 2 volumes of acetone stored at −20°. Leave at −20° for more than 10 min. Centrifuge at 2000g at 4° for 5 min. Discard the supernatant. Wash with cold acetone and centrifuge again. Discard all the supernatant without affecting the pellet, and let it dry.

3. Resuspend in 0.1 ml of HENS buffer (250 mM HEPES, pH 7.7, 1 m$M$ EDTA, 0.1 m$M$ neocuproine, 1% SDS) per milligram of protein in the initial sample.

4. Add 1/3 volume of biotin-HPDP solution [4 m$M$ in $N,N$-dimethylformamide (DMF)] and 1/100 volume of 100 m$M$ L-ascorbic acid. A control for determining endogenously biotinylated proteins or

proteins that are basally purified or detected is treating half of the sample with just the vehicle, DMF. After this stage, it is not necessary to protect samples from light.
5. Incubate 1 h at room temperature.
6. Precipitate with acetone as above.
7. Resuspend in HENS buffer as above.

At this point, the biotin switch treatment is complete, and previously S-nitrosylated proteins are now biotinylated. An aliquot may be analyzed to determine the degree of biotinylation by biotin detection with avidin or anti-biotin antibody.

*Avidin Capture*

1. Add 2 volumes of neutralization buffer (20 m$M$ HEPES, pH 7.7, 100 m$M$ NaCl, 1 m$M$ EDTA, 0.5% Triton X-100).
2. At this point it is critical to be sure that the pellet from the acetone precipitation has been completely resuspended. If not, proteins that were not resuspended can be found in the elution fraction, even if they were not biotinylated. More neutralization buffer can be added, and the samples must be centrifuged at high speed (15,000$g$) for 1 min, discarding the pellet.
3. Add the avidin resin previously washed in neutralization buffer without Triton X-100. We use neutravidin-agarose from Pierce, adding 15 $\mu$l of resin/mg of protein.
4. Incubate for 1 h at room temperature with agitation.
5. Centrifuge at 400$g$ for 2 min. Discard the supernatant containing the unbound proteins.
6. Wash the resin five times with washing buffer (20 m$M$ HEPES, pH 7.7, 600 m$M$ NaCl, 1 m$M$ EDTA, 0.5% Triton X-100), centrifuging as above.
7. Add 1 volume of elution buffer (20 m$M$ HEPES, pH 7.7, 100 m$M$ NaCl, 1 m$M$ EDTA, 100 m$M$ 2-mercaptoethanol) and incubate for 20 min at 37°. We mix the suspension by pipetting up and down before, in the middle of, and after the incubation.
8. Centrifuge 1 min at high speed and recover the supernatant, containing the eluted proteins that were formerly biotinylated.
9. A second elution can be performed to increase the recovery, and it can be added to the first one.
10. Add sample buffer for SDS–polyacrylamide gel electrophoresis (PAGE), including reducing agent (2-mercaptoethanol or DTT), and boil as usual.

At this point, standard protocols for SDS-PAGE can be followed. The bands can be detected with Coomassie blue or silver staining, and the proteins can be extracted, digested, and identified by mass spectrometry as usual in proteomic studies. Two-dimensional PAGE can be performed instead of SDS-PAGE, if the number of proteins purified is high, although the yield can be reduced.

### Critical Steps and Anticipated Problems

In addition to some critical steps that have been described earlier, we summarize here some of the problems that can be found using this methodology.

When performing biotin detection, as this is fairly sensible, a smearing of high-signal bands can be obtained. The control without biotin-HPDP can help to discriminate if it is a failure in the blocking step (if the signal is only in biotin-HPDP–treated samples), or if it is a nonspecific detection. If the blocking step is failing, try to reduce the protein concentration, be sure that MMTS is used fresh every time, or even reduce the scaling of the assay. If there is a nonspecific detection, be careful to eliminate all the biotin-HPDP reagent, perform the electrophoresis just after the biotin switch protocol, and separate during the electrophoresis, even in different gels, the samples that were not treated with biotin-HPDP, as this reagent may diffuse.

If you do not obtain any signal in the biotin detection, you can include a positive control treating the extract, without blocking, with biotin-HPDP, as this will biotinylate the free thiol cysteines, which are much more abundant than the S-nitrosylated cysteines.

Finally, as a rule of thumb, try to be always more specific that sensitive. For example, when separating two phases, avoid maintaining traces from the discarded fraction, even if you lose part of your sample.

### References

Brown, K. A., Vora, A., Biggerstaff, J., Edgell, C.-J. S., Oikle, S., Mazure, G., Taub, N., Meager, A., Hill, T., Watson, C., and Dumonde, D. C. (1993). Application of an immortalized human endothelial cell line to the leucocyte: Endothelial adherence assay. *J. Immunol. Methods* **163,** 13–22.

DeMaster, E. G., Quast, B. J., Redfern, B., and Nagasawa, H. T. (1995). Reaction of nitric oxide with the free sulfhydryl group of human serum albumin yields a sulfenic acid and nitrous oxide. *Biochemistry* **34,** 11494–11499.

Forman, H. J., Fukuto, J. M., and Torres, M. (2004). Redox signaling: Thiol chemistry defines which reactive oxygen and nitrogen species can act as second messengers. *Am. J. Physiol. Cell Physiol.* **287,** C246–C256.

Foster, M. W., and Stamler, J. S. (2004). New insights into protein *S*-nitrosylation: Mitochondria as a model system. *J. Biol. Chem.* **279,** 25891–25897.

Jaffrey, S. R., Erdjument-Bromage, H., Ferris, C. D., Tempst, P., and Snyder, S. H. (2001). Protein *S*-nitrosylation: A physiological signal for neuronal nitric oxide. *Nat. Cell Biol.* **3,** 193–197.

Jourd'heuil, D., Gray, L., and Grisham, M. B. (2000). *S*-nitrosothiol formation in blood of lipopolysaccharide-treated rats. *Biochem. Biophys. Res. Commun.* **273,** 22–26.

Kuncewicz, T., Sheta, E. A., Goldknopf, I. L., and Kone, B. C. (2003). Proteomic analysis of *S*-nitrosylated proteins in mesangial cells. *Mol. Cell Proteomics* **2,** 156–163.

Lind, C., Gerdes, R., Hamnell, Y., Schuppe-Koistinen, I., von Löwenhielm, H. B., Holmgren, A., and Cotgreave, I. A. (2002). Identification of *S*-glutathionylated cellular proteins during oxidative stress and constitutive metabolism by affinity purification and proteomic analysis. *Arch. Biochem. Biophys.* **406,** 229–240.

Martínez-Ruiz, A., and Lamas, S. (2004a). *S*-nitrosylation: A potential new paradigm in signal transduction. *Cardiovasc. Res.* **62,** 43–52.

Martínez-Ruiz, A., and Lamas, S. (2004b). Detection and proteomic identification of *S*-nitrosylated proteins in endothelial cells. *Arch. Biochem. Biophys.* **423,** 192–199.

Stamler, J. S., Lamas, S., and Fang, F. C. (2001). Nitrosylation: The prototypic redox-based signaling mechanism. *Cell* **106,** 675–683.

Zhang, Y., and Hogg, N. (2004). The mechanism of transmembrane *S*-nitrosothiol transport. *Proc. Natl. Acad. Sci. USA* **101,** 7891–7896.

## [14]    S-Nitrosylation in Parkinson's Disease and Related Neurodegenerative Disorders

*By* Kenny K. K. Chung, Valina L. Dawson, and Ted M. Dawson

### Abstract

Parkinson's disease (PD) is a common neurodegenerative disorder characterized by impairment in motor function. PD is mostly sporadic, but rare familial cases are also found. The exact pathogenic mechanism is not fully understood, but both genetic and environmental factors are known to be important contributors. In particular, oxidative stress mediated through nitric oxide (NO) is believed to be a prime suspect in the development of PD. NO can exert its effect by modifying different biological molecules, and one of these modifications is through *S*-nitrosylation. Because of the liable nature of *S*-nitrosylation, a number of methods are often used to study this modification. We have successfully employed some of these methods and showed that a familial related protein, parkin, can be *S*-nitrosylated and provide a common pathogenic mechanism for sporadic and familial PD.

METHODS IN ENZYMOLOGY, VOL. 396
0076-6879/05 $35.00
DOI: 10.1016/S0076-6879(05)96014-X

## Introduction

Parkinson's disease (PD) is one of the most common neurodegenerative diseases that is characterized by motor dysfunction such as bradykinesia, rest tremor, gait abnormalities, and postural instability. James Parkinson first described the disease in his "An Essay on the Shaking Palsy" published in 1817. There are at least 500,000 people affected by PD with increasing new cases being reported annually in United States. PD is mostly sporadic, but genetically linked cases are also found. For instance, mutations in $\alpha$-synuclein, parkin, DJ-1, and PINK1 have been linked to rare familial forms of PD (Greenamyre and Hastings, 2004). Pathological studies have revealed that a selective loss of dopaminergic neurons and the presence of Lewy bodies (proteinaceous cytoplasmic inclusions) in the substantia nigra pars compacta (SNc) are common in patients with PD. The exact pathogenic mechanism of PD is not fully understood, but both genetic and environmental factors are believed to contribute to the disease. The contribution of environmental factors is best demonstrated by the finding that some drug addicts who had accidentally injected an impurity of "synthetic heroin" later developed severe parkinsonism (Langston et al., 1983, 1999). The impurity that caused PD in those people was later found to be 1-methyl-4-phenyl-1,2,3,6-tetrahydropyridine (MPTP) (Langston et al., 1983, 1999). Studies have shown that oxidative free radicals generated in the mitochondria by MPP+ (metabolite of MPTP) are responsible for dopaminergic neuronal toxicity (Zhang et al., 2000b). In fact, oxidative stress has long been considered a prime suspect in the pathogenesis of PD. This notion is supported by postmortem pathological studies that mitochondria complex I dysfunction, increased nitrotyrosine immunoreactivity, reduced glutathione and ferritin levels, increased lipid peroxidation, and increased levels of iron are common in patients with PD (Chung et al., 2003). However, the underlying detailed mechanism of how oxidative stress can cause the selective dopaminergic cell death still remains to be elucidated.

## Nitrosative Stress and Parkinson's Disease

One of the major contributors of oxidative stress in PD is believed to be nitric oxide (NO) (Ischiropoulos and Beckman, 2003; Jenner, 2003). NO is the most studied bioactive gas molecule and is involved in a wide spectrum of biological process. Excessive production of NO leads to cellular toxicity (Dawson and Dawson, 1998). In PD, various studies have pointed to the involvement of NO in the pathogenesis (Ischiropoulos and Beckman, 2003;

Jenner, 2003). For instance, increased nitrotyrosine immunoreactivity and nitrated α-synuclein in the Lewy bodies of patients with PD are prominent in postmortem studies (Duda *et al.*, 2000; Giasson *et al.*, 2000). Moreover, both inducible nitric oxide synthase (iNOS) and nNOS knockout mice are resistant to MPTP toxicity (Chung *et al.*, 2003). The exact mechanism of NO-mediated dopaminergic neuronal toxicity in PD is not fully understood, but it is possible that a number of pathways are involved. Particular interest has been elicited in the modification of cysteine residues in proteins by NO, which appears to be another important signaling pathway. The attachment of NO to protein cysteine moieties is termed *S*-nitrosation or *S*-nitrosylation. It has been suggested that nitrosylation may represent a redox sensing and/or signaling mechanism that may be comparable to phosphorylation (Mannick and Schonhoff, 2002; Stamler *et al.*, 2001). For instance, different components of *N*-methyl-D-aspartate (NMDA) receptor complex are modulated by NO through nitrosylation (Stamler *et al.*, 2001). Nitrosothiols have also been linked to hyperventilatory response in respiration (Lipton *et al.*, 2001). In fact more and more proteins are found to be modulated by NO through nitrosylation, and this modification might contribute to a number of human diseases (Foster *et al.*, 2003; Jaffrey *et al.*, 2001; Stamler *et al.*, 2001). One of the important features of *S*-nitrosylation is that the modification is reversible and a specific acid-basic motif seems to be necessary (Stamler *et al.*, 1997, 2001). The modification is also selective, as usually, only specific cysteines are nitrosylated in a protein (Stamler *et al.*, 1997, 2001). Although nitrosative stress is implicated in the pathogenesis of PD, the role of nitrosylation in PD is still largely unknown. One of the obstacles in this field is the availability of sensitive methods that can be used to detect the modification endogenously. There are different methods that can be used to measure nitrosothiols in biological samples, and each has its own advantages and disadvantages.

## Methods Used in the Detection of Nitrosothiols

### Photolysis Chemiluminescence

The method was first developed to measure *S*-nitrosothiols in mammalian plasma (Stamler *et al.*, 1992) and later shown to be able to distinguish different biologically relevant derivatives of NO species (Alpert *et al.*, 1997). The basic theory in this method is to use ultraviolet (UV) radiation to cleave the NO from the S-NO bonds, and the released NO is measured using chemiluminescence method. The system consists of a high-performance liquid chromatographic (HPLC)/GC system, a photolysis chamber (e.g., Nitrolite photolysis unit, Thermo Orion), a cold trap system, and a chemiluminescence spectrometer (e.g., TEA 510 analyzer, Thermo Orion) (Foster and Stamler, 2004; Stamler et al., 1992) (Fig. 1). During measurement, the sample will be injected into the HPLC system and then enter the photolysis chamber together with a steady stream of helium. Inside the photolysis chamber, the sample will enter a quartz coil that coils around a mercury arc lamp. The UV radiation will cleave the NO from the nitrosothiols in the sample and the helium will function as an inert gas carrier of NO. The mixture will then pass through a series of cold traps, which will condense the liquid phase of the sample. The released NO will then enter the chemiluminescence spectrometer and react with ozone gas from an ozone generator. The reaction will produce the resultant chemiluminescent product, $NO_2^*$, which is detected by the spectrometer. This method is one of the most sensitive and at the same time can differentiate types of NO modification in the sample. However, this method requires substantial investment in instrumentation as custom-built components (e.g., photolysis unit and cold traps) are required for assembly of this system.

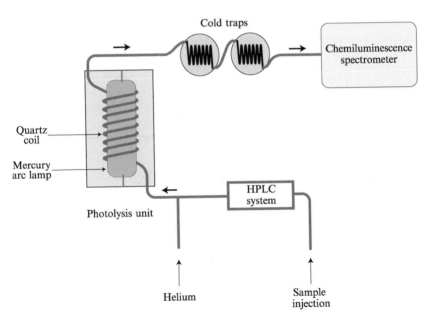

FIG. 1. Schematic diagram showing the photolysis chemiluminescence setup used to measure nitrosothiols in biological samples.

## Chemical-Released Chemiluminescence

The basic theory in this method is similar to the photolysis-chemiluminescence except that NO is released by chemical reagents instead of UV radiation. In this method, nitrosothiols are released by $I_3^-$ reagent, and pretreatment of the samples with different chemicals can be used to distinguish *S*-nitrosothiols, nitrite, and iron nitrosyl (Yang *et al.*, 2003). For example, pretreatment of samples with $HgCl_2$ (5 m$M$) can be used to distinguish *S*-nitrosothiols from nitrite and iron nitrosyl (Yang *et al.*, 2003). The system basically consists of a purge vessel (Purge System, Sievers) and a chemiluminescence spectrometer (NOA 280I, Sievers). During measurement, the sample is injected into the purge vessel, which contain the $I_3^-$. After reaction with the sample, NO will be released and in turn will be carried by a stream of helium gas to the chemiluminescence spectrometer. The released NO will then be reacted with ozone gas, producing a chemiluminescent product, $NO_2^*$, which can be detected by the spectrometer.

## Anti-Nitrosocysteine Antibody

A polyclonal antiserum that was raised in rabbit by using *S*-nitrosocysteine conjugated to bovine serum albumin (BSA) was available commercially from Calbiochem, but it has been discontinued. A similar monoclonal antibody is available commercially from A.G. Scientific. Anti-*S*-nitrosocysteine antibodies have been used in immunohistochemistry or Western blot analysis (Gow *et al.*, 2002; Haendeler *et al.*, 2002; Lorch *et al.*, 2000). However, it is important to have both positive and negative controls, as these antibodies tend to cross-react with unmodified cysteine residues. Pretreatment of samples with NO donors (e.g., 1 m$M$ *S*-nitrosoglutathione) can be used as a positive control, whereas pretreatment of samples with 0.2% $HgCl_2$ or 3.5 m$M$ *p*-chloromercuribenzenesulfonate (Sigma), which releases NO from S-NO bonds, can be used as a negative control (Gow *et al.*, 2002).

## Mass Spectrometry

The development of mass spectrometry for the effective detection of nitrosylation remains one of the methods in this field that requires additional optimization. Mass spectrometry can provide important information of the number and the sites of NO modifications in the molecule. However, up to now, it has only been shown that electrospray ionization mass spectrometry (ESI-MS) can be used in the detection and mapping of *S*-nitrosylation *in vitro* (Kaneko and Wada, 2003; Knipp *et al.*, 2003; Mirza *et al.*, 1995). The more widely used matrix-assisted laser desorption/ionization

MS (MALDI-MS) was found to destroy the modification during the process of ionization (Kaneko and Wada, 2003). Although ESI-MS can provide detailed information about nitrosylation, purified protein or peptide is needed for the analysis (Kaneko and Wada, 2003; Knipp *et al.*, 2003; Mirza *et al.*, 1995).

*Colorimetric and Fluorometric Methods*

Colorimetric and fluorometric assays remain the most popular methods in the detection of nitrosothiols (Chung *et al.*, 2004; Haendeler *et al.*, 2002; Matsushita *et al.*, 2003). In these assays, samples with nitrosothiols are usually treated with $Hg^{2+}$ or $Cu^{2+}$ ions to release the NO that can subsequently react with reagents to form a colored or fluorescent compound (Cook *et al.*, 1996). For example, colorimetric detection of nitrosothiols can be achieved using neutral Griess reagent as follows.

*Chemicals*

Sulfanilamide (SULF) (Sigma)
*N*-(1-naphthyl)ethylenediamine dihydrochloride (NEDD) (Sigma)
*S*-nitrosoglutathione (GSNO) (Calbiochem)
$HgCl_2$ (Sigma)
DMSO (Sigma)

*Protocol*

1. Prepare neutral Griess reagent (57 m$M$ SULF and 1.2 m$M$ NEDD in PBS) by dissolving 0.25 g of SULF and 7.5 mg of NEDD in 25 ml of PBS.
2. Prepare 10 m$M$ of $HgCl_2$ (Sigma) in DMSO as stock solution.
3. Prepare desired concentrations of GSNO as standards.
4. Equal volume of samples or standards can then be mixed with neutral Griess reagent with the addition of $HgCl_2$ in a final concentration of 100 $\mu M$ and incubate at room temperature for 20 min.
5. Measure the absorbance of the mixture at 496 nm using a spectrophotometer.

To achieve higher sensitivity, fluorometric methods that use 2,3-diaminonaphthalene (DAN) can be employed as follows.

*Chemicals*

$HgCl_2$ (Sigma)
DMSO (Sigma)
DAN (Sigma)
*S*-Nitrosoglutathione (GSNO) (Calbiochem)

*Protocol*

1. Prepare 10 m*M* of HgCl$_2$ and 10 m*M* of DAN in DMSO separately as stock solutions.
2. Prepare desired concentrations of GSNO as standards.
3. Mix standards or samples with DAN at a final concentration of 300 $\mu M$ in the presence of 100 $\mu M$ of HgCl$_2$.
4. Incubate the mixture in darkness at room temperature for 1 h.
5. Measure the generated fluorescent compound 2,3-napththyltrazole (NAT) with excitation wavelength of 375 nm and an emission wavelength of 450 nm.

It has been reported that this approach can detect GSNO as low as 50 n*M*. However, the sensitivity and the accuracy of both colorimetric and fluorometric methods are susceptible to interference from protein or biological molecules in the samples (Cook *et al.*, 1996).

*Biotin Switch Method*

This method was developed by Jaffrey *et al.* (2001), and it labels *S*-nitrosylated protein with biotin (Jaffrey and Snyder, 2001; Jaffrey *et al.*, 2001). With proper controls, the method can use to detect *in vivo* *S*-nitrosylated proteins. The basic idea is to block the free cysteine residues in a protein with methanethiosulfonate (MMTS), and then NO is released from S-NO group with ascorbate. The newly formed free cysteine group is in turn labeled with biotin by reacting with biotin HPDP (Fig. 2).

*Materials*

Neocuproine (Sigma)
*S*-nitrosoglutathione (GSNO) (Calbiochem)
Glutathione (GSH) (Calbiochem)
MMTS (Pierce)
Biotin-HPDP (Pierce)
Ascorbate (Sigma)
Dimethylformamide (DMF) (Sigma)
Spin columns (Bio-Rad)
Streptavidin agarose (Pierce)

*Protocol*

1. Prepare samples in HENS buffer [250 mM HEPES-NaOH pH 7.7, 1 m*M* EDTA, 0.1 m*M* neocuproine, 1% sodium dodecylsulfate (SDS)].
2. Incubate samples with GSNO or GSH as control in darkness for 20 min at room temperature. GSNO in the micromolar range can be

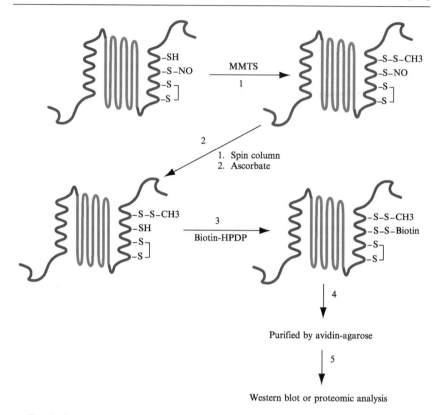

FIG. 2. Schematic diagram showing a theoretical protein with S-nitrosylation is analyzed by the biotin switch method. Free cysteine residue is first block by methyl methanethiosulfonate (MMTS) through sulfonylation. Excess MMTS is removed by spin column or acetone precipitation. Nitric oxide is then released from the nitrosothiol with the treatment of ascorbic acid. The thiol group is subsequently labeled with biotin-HPDP. The labeled protein can then be purified by avidin-agarose and analyzed by Western blot or proceed for proteomic analysis.

used, but lower concentration levels are more likely to represent physiological conditions.

3. Eliminate GSNO by spin column and then incubate the samples with MMTS in a final concentration of 4 m$M$ at 50° for 20 min.

4. Eliminate MMTS by spin column. Alternatively, MMTS can be eliminated by precipitation with 10 volumes of −20° acetone.

5. Incubate samples with biotin-HPDP and ascorbate in a final concentration of 1 m$M$ for 2 h at room temperature.

6. Eliminate biotin-HPDP in the samples by spin column or acetone precipitation.
7. Purify the biotinylated proteins by streptavidin agarose and then proceed for Western blot analysis.

For the *in vivo* detection of *S*-nitrosylation, the sample can be prepared in HENS buffer with the presence of 4 m$M$ MMTS and then follow the same procedure as in the *in vitro* detection of nitrosothiols without the incubation of NO donors. This method has proven to be effective to detect a wide range of proteins that can be *S*-nitrosylated both *in vitro* and *in vivo*, but the sensitivity is not as high as that of other methods (Mannick and Schonhoff, 2002).

It is important to note that during the study of nitrosylation of a protein, different methods should be employed to prevent false-positive or false-negative results. For instance, all the methods described in this chapter have their advantages and disadvantages. The effective use of a combination of them will ease the study of how a selective pathway might be modulated by nitrosylation.

S-Nitrosylation and PD

Because of the strong evidence that NO is one of the major players in the pathogenesis of PD, we investigated the possible involvement of *S*-nitrosylation in regulating the function of parkin, a familial PD–related protein, with a combination of the methods described earlier. Parkin is an E3 ligase in the ubiquitin pathway, which selectively attaches ubiquitin on specific substrates (Chung *et al.*, 2003; Shimura *et al.*, 2000; Zhang *et al.*, 2000a), and it is linked to autosomal recessive juvenile parkinsonism (Kitada *et al.*, 1998). Parkin belongs to a protein family with two interesting new gene (RING) domains and an in-between RINGs (IBR) domain (Capili *et al.*, 2004). Parkin has also been touted as a multipurpose neuroprotectant that is dependent on its E3 ubiquitin ligase activity (Feany and Pallanck, 2003). A number of cysteine moieties cluster together to form the RING finger and IBR motif. Due to the importance of these cysteine moieties in E3 ligase activity, we reasoned that it was possible that some of these cysteines could be modulated by NO through nitrosylation leading to alterations in parkin's function. We first employed both the biotin switch and the fluorometric method to show that parkin can be nitrosylated *in vitro* (Chung *et al.*, 2004). We then further showed that *S*-nitrosylation of parkin impairs its E3 ligase activity and compromises its protective function. We also used the colorimetric method to show that increased nitrosothiols are present in patients with PD/diffuse Lewy body (DLB) disease (Chung *et al.*, 2004). More importantly, we found that parkin is nitrosylated in mouse MPTP model of PD and in brains of patients with PD and DLB disease (Chung *et al.*, 2004). Similar results have also been reported by Yao *et al.* (2004) that parkin can be *S*-nitrosylated under different conditions. They also employed a combination of fluorometric, biotin switch, anti-*S*-nitrosylated protein antibody, and ESI-MS methods for the study of nitrosylation in parkin (Yao *et al.*, 2004). It was shown by ESI-MS that the site of modifications are within the RING1 and IBR domain of parkin (Yao *et al.*, 2004). The findings that parkin can be nitrosylated and that nitrosylation affects its E3 ligase activity provide additional insight into the mechanisms of PD pathogenesis and potential therapeutic targets in PD.

## Concluding Remarks

The understanding of the pathogenic mechanism of PD has improved dramatically in the past decade. The rapid advancement has been fueled by investigating the mechanisms of MPTP dopaminergic toxicity and advances in human genetics, which enabled the identification genes that are linked to familial PD. One of the fundamental questions is whether the pathogenesis of the more common sporadic form of PD converges in a common pathway with MPTP-induced parkinsonism and the familial linked PD. Evidence so far suggests that oxidative stress, mitochondrial dysfunction, and abnormal protein metabolism are important pathogenic mechanisms in PD. Our study further supports the notion that oxidative stress and abnormal protein metabolism are prime suspects in the pathogenesis of PD.

## References

Alpert, C., Ramdev, N., George, D., and Loscalzo, J. (1997). Detection of S-nitrosothiols and other nitric oxide derivatives by photolysis-chemiluminescence spectrometry. *Anal. Biochem.* **245**, 1–7.

Capili, A. D., Edghill, E. L., Wu, K., and Borden, K. L. (2004). Structure of the C-terminal RING finger from a RING-IBR-RING/TRIAD motif reveals a novel zinc-binding domain distinct from a RING. *J. Mol. Biol.* **340**, 1117–1129.

Chung, K. K., Dawson, V. L., and Dawson, T. M. (2003). New insights into Parkinson's disease. *J. Neurol.* **250**(Suppl. 3), III15–III24.

Chung, K. K., Thomas, B., Li, X., Pletnikova, O., Troncoso, J. C., Marsh, L., Dawson, V. L., and Dawson, T. M. (2004). S-nitrosylation of parkin regulates ubiquitination and compromises parkin's protective function. *Science* **304**, 1328–1331.

Cook, J. A., Kim, S. Y., Teague, D., Krishna, M. C., Pacelli, R., Mitchell, J. B., Vodovotz, Y., Nims, R. W., Christodoulou, D., Miles, A. M., Grisham, M. B., and Wink, D. A. (1996).

Convenient colorimetric and fluorometric assays for S-nitrosothiols. *Anal. Biochem.* **238,** 150–158.

Dawson, V. L., and Dawson, T. M. (1998). Nitric oxide in neurodegeneration. *Prog. Brain Res.* **118,** 215–229.

Duda, J. E., Giasson, B. I., Chen, Q., Gur, T. L., Hurtig, H. I., Stern, M. B., Gollomp, S. M., Ischiropoulos, H., Lee, V. M., and Trojanowski, J. Q. (2000). Widespread nitration of pathological inclusions in neurodegenerative synucleinopathies. *Am. J. Pathol.* **157,** 1439–1445.

Feany, M. B., and Pallanck, L. J. (2003). Parkin: A multipurpose neuroprotective agent? *Neuron* **38,** 13–16.

Foster, M. W., McMahon, T. J., and Stamler, J. S. (2003). S-nitrosylation in health and disease. *Trends Mol. Med.* **9,** 160–168.

Foster, M. W., and Stamler, J. S. (2004). New insights into protein S-nitrosylation. Mitochondria as a model system. *J. Biol. Chem.* **279,** 25891–25897.

Giasson, B. I., Duda, J. E., Murray, I. V., Chen, Q., Souza, J. M., Hurtig, H. I., Ischiropoulos, H., Trojanowski, J. Q., and Lee, V. M. (2000). Oxidative damage linked to neurodegeneration by selective alpha-synuclein nitration in synucleinopathy lesions. *Science* **290,** 985–989.

Gow, A. J., Chen, Q., Hess, D. T., Day, B. J., Ischiropoulos, H., and Stamler, J. S. (2002). Basal and stimulated protein S-nitrosylation in multiple cell types and tissues. *J. Biol. Chem.* **277,** 9637–9640.

Greenamyre, J. T., and Hastings, T. G. (2004). Parkinson's–divergent causes, convergent mechanisms. *Science* **304,** 1120–1122.

Haendeler, J., Hoffmann, J., Tischler, V., Berk, B. C., Zeiher, A. M., and Dimmeler, S. (2002). Redox regulatory and anti-apoptotic functions of thioredoxin depend on S-nitrosylation at cysteine 69. *Nat. Cell Biol.* **4,** 743–749.

Ischiropoulos, H., and Beckman, J. S. (2003). Oxidative stress and nitration in neurodegeneration: Cause, effect, or association? *J. Clin. Invest.* **111,** 163–169.

Jaffrey, S. R., Erdjument-Bromage, H., Ferris, C. D., Tempst, P., and Snyder, S. H. (2001). Protein S-nitrosylation: A physiological signal for neuronal nitric oxide. *Nat. Cell Biol.* **3,** 193–197.

Jaffrey, S. R., and Snyder, S. H. (2001). The biotin switch method for the detection of S-nitrosylated proteins. *Sci. STKE.* PL1.

Jenner, P. (2003). Oxidative stress in Parkinson's disease. *Ann. Neurol.* **53**(Suppl. 3), S26–S38.

Kaneko, R., and Wada, Y. (2003). Decomposition of protein nitrosothiols in matrix-assisted laser desorption/ionization and electrospray ionization mass spectrometry. *J. Mass. Spectrom.* **38,** 526–530.

Kitada, T., Asakawa, S., Hattori, N., Matsumine, H., Yamamura, Y., Minoshima, S., Yokochi, M., Mizuno, Y., and Shimizu, N. (1998). Mutations in the parkin gene cause autosomal recessive juvenile parkinsonism. *Nature* **392,** 605–608.

Knipp, M., Braun, O., Gehrig, P. M., Sack, R., and Vasak, M. (2003). Zn(II)-free dimethylargininase-1 (DDAH-1) is inhibited upon specific Cys-S-nitrosylation. *J. Biol. Chem.* **278,** 3410–3416.

Kuncewicz, T., Sheta, E. A., Goldknopf, I. L., and Kone, B. C. (2003). Proteomic analysis of s-nitrosylated proteins in mesangial cells. *Mol. Cell Proteomics* **2,** 156–163.

Langston, J. W., Ballard, P., Tetrud, J. W., and Irwin, I. (1983). Chronic Parkinsonism in humans due to a product of meperidine-analog synthesis. *Science* **219,** 979–980.

Langston, J. W., Forno, L. S., Tetrud, J., Reeves, A. G., Kaplan, J. A., and Karluk, D. (1999). Evidence of active nerve cell degeneration in the substantia nigra of humans years after 1-methyl-4-phenyl-1,2,3,6-tetrahydropyridine exposure. *Ann. Neurol.* **46,** 598–605.

Lipton, A. J., Johnson, M. A., Macdonald, T., Lieberman, M. W., Gozal, D., and Gaston, B. (2001). S-nitrosothiols signal the ventilatory response to hypoxia. *Nature* **413,** 171–174.

Lorch, S. A., Foust, R., III, Gow, A., Arkovitz, M., Salzman, A. L., Szabo, C., Vayert, B., Geffard, M., and Ischiropoulos, H. (2000). Immunohistochemical localization of protein 3-nitrotyrosine and S-nitrosocysteine in a murine model of inhaled nitric oxide therapy. *Pediatr. Res.* **47,** 798–805.

Mannick, J. B., and Schonhoff, C. M. (2002). Nitrosylation: The next phosphorylation? *Arch. Biochem. Biophys.* **408,** 1–6.

Martinez-Ruiz, A., and Lamas, S. (2004). Detection and proteomic identification of S-nitrosylated proteins in endothelial cells. *Arch. Biochem. Biophys.* **423,** 192–199.

Matsushita, K., Morrell, C. N., Cambien, B., Yang, S. X., Yamakuchi, M., Bao, C., Hara, M. R., Quick, R. A., Cao, W., O'Rourke, B., Lowenstein, J. M., Pevsner, J., Wagner, D. D., and Lowenstein, C. J. (2003). Nitric oxide regulates exocytosis by S-nitrosylation of N-ethylmaleimide-sensitive factor. *Cell* **115,** 139–150.

Mirza, U. A., Chait, B. T., and Lander, H. M. (1995). Monitoring reactions of nitric oxide with peptides and proteins by electrospray ionization-mass spectrometry. *J. Biol. Chem.* **270,** 17185–17188.

Shimura, H., Hattori, N., Kubo, S., Mizuno, Y., Asakawa, S., Minoshima, S., Shimizu, N., Iwai, K., Chiba, T., Tanaka, K., and Suzuki, T. (2000). Familial Parkinson disease gene product, parkin, is a ubiquitin-protein ligase. *Nat. Genet.* **25,** 302–305.

Stamler, J. S., Jaraki, O., Osborne, J., Simon, D. I., Keaney, J., Vita, J., Singel, D., Valeri, C. R., and Loscalzo, J. (1992). Nitric oxide circulates in mammalian plasma primarily as an S-nitroso adduct of serum albumin. *Proc. Natl. Acad. Sci. USA* **89,** 7674–7677.

Stamler, J. S., Lamas, S., and Fang, F. C. (2001). Nitrosylation. the prototypic redox-based signaling mechanism. *Cell* **106,** 675–683.

Stamler, J. S., Toone, E. J., Lipton, S. A., and Sucher, N. J. (1997). (S)NO signals: Translocation, regulation, and a consensus motif. *Neuron* **18,** 691–696.

Yang, B. K., Vivas, E. X., Reiter, C. D., and Gladwin, M. T. (2003). Methodologies for the sensitive and specific measurement of S-nitrosothiols, iron-nitrosyls, and nitrite in biological samples. *Free Radic. Res.* **37,** 1–10.

Yao, D., Gu, Z., Nakamura, T., Shi, Z. Q., Ma, Y., Gaston, B., Palmer, L. A., Rockenstein, E. M., Zhang, Z., Masliah, E., Uehara, T., and Lipton, S. A. (2004). Nitrosative stress linked to sporadic Parkinson's disease: S-nitrosylation of parkin regulates its E3 ubiquitin ligase activity. *Proc. Natl. Acad. Sci. USA* **101,** 10810–10814.

Zhang, Y., Dawson, V. L., and Dawson, T. M. (2000b). Oxidative stress and genetics in the pathogenesis of Parkinson's disease. *Neurobiol. Dis.* **7,** 240–250.

Zhang, Y., Gao, J., Chung, K. K., Huang, H., Dawson, V. L., and Dawson, T. M. (2000a). Parkin functions as an E2-dependent ubiquitin-protein ligase and promotes the degradation of the synaptic vesicle-associated protein, CDCrel-1. *Proc. Natl. Acad. Sci. USA* **97,** 13354–13359.

# [15]   Dynamic Assessment of Nitration Reactions *In Vivo*

*By* ALI R. MANI and KEVIN P. MOORE

## Abstract

Assessment of protein nitration is commonly used as a footprint for the formation of reactive nitrogen species *in vivo*. However, one of the major disadvantages of measuring nitrotyrosine in proteins is that nitrated proteins are broken down at variable rates, and the resulting free nitrotyrosine is taken up by cells, metabolized, and excreted. We have discovered a biochemical pathway in which circulating *para*-hydroxyphenylacetic acid (PHPA) undergoes nitration to form 3-nitro-4-hydroxyphenylacetic acid (NHPA), which is rapidly excreted in the urine. Using various animal models, we have shown that measurement of urinary NHPA can be used to assess the formation of reactive nitrogen species *in vivo*.

## Introduction

Nitric oxide (NO) has a wide variety of functions in physiology and pathology. The pathological actions of NO are rooted in its ability to react with other radicals to form reactive nitrogen species (RNSs), which can modify cell function through nitration, nitrosation, or oxidation of cell components, and which is sometimes termed *nitroxidation*. In proteins, most nitration reactions (addition of $-NO_2$) occur on tyrosine residues to form nitrotyrosine. Current data suggest that nitrated proteins are targeted for ubiquitination and catabolism (Souza *et al.*, 2000), and the nitrotyrosine moiety then is metabolized in the liver to 3-nitro-4-hydroxyphenylacetic acid (NHPA) (Ohshima *et al.*, 1990).

Measurement or assessment of protein nitration is commonly used as a footprint for the formation of RNSs *in vivo* and *in vitro*; however, one of the major disadvantages of measuring nitrotyrosine in proteins is that nitrated proteins are broken down at variable rates, and the resulting free nitrotyrosine is taken up by cells, metabolized, and excreted (Souza *et al.*, 2000). Thus, the concept that measurement of NHPA, as the major urinary metabolite of nitrotyrosine, is attractive, because it can provide a time-integrated index of nitrotyrosine formation *in vivo*. We developed a highly sensitive method for measurement of NHPA based on mass spectrometry (MS) (Mani *et al.*, 2003). During extensive validation of the method, we

METHODS IN ENZYMOLOGY, VOL. 396
0076-6879/05 $35.00
DOI: 10.1016/S0076-6879(05)96015-1

discovered that NHPA is not formed exclusively from nitrotyrosine metabolism but is also formed by nitration of circulating *para*-hydroxyphenylacetic acid (PHPA), a metabolite of tyrosine, which is present at high concentrations in biological fluids (Fig. 1).

To study the contribution of circulating PHPA to the formation of NHPA *in vivo*, we synthesized deuterium-labeled PHPA ($D_6$-PHPA), which was infused intravenously (Mani *et al.*, 2003). This enabled us to demonstrate unequivocal nitration of infused $D_6$-PHPA to form $D_5$-NHPA (nitration of $D_6$-PHPA produces $D_5$-NHPA as a result of replacement of a single deuterium of the aromatic ring by the $NO_2$ group). Based on these studies, we estimate that most of the urinary NHPA ($\sim$85%) is formed by nitration of PHPA *in vivo* in the rat. This occurs because PHPA is present in plasma at a concentration of approximately 2.7 $\mu$mol/l, whereas free nitrotyrosine is only present at approximately 4.2 nmol/l. One other important observation that was made is that more than about 95% of infused $D_6$-PHPA is excreted unchanged rapidly (Mani

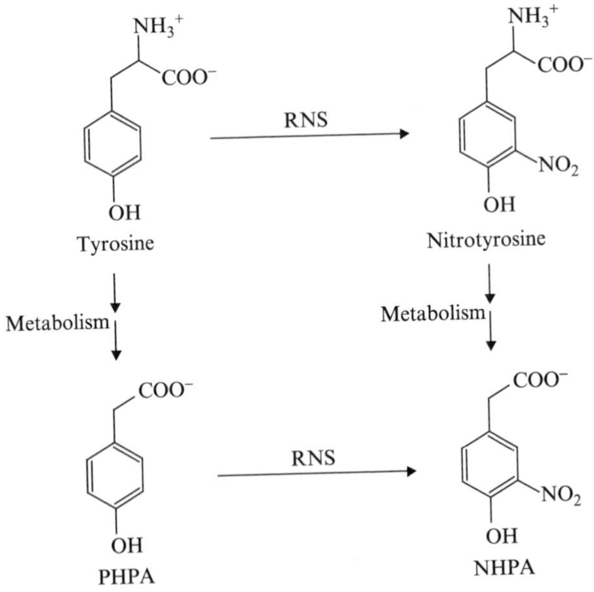

FIG. 1. There are two pathways by which 3-nitro-4-hydroxyphenylacetic (NHPA) can be formed *in vivo*. NHPA can be formed by either nitration of PHPA or metabolism of free nitrotyrosine. Based on studies in the rat, we estimate that at least 85% of NHPA excreted in the urine is formed by nitration of circulating *para*-hydroxyphenylacetic acid (PHPA) *in vivo* (from Mani *et al.*, 2003). Thus, measurement of urinary NHPA may provide an index for the formation of reactive nitrogen species *in vivo*.

*et al.*, 2003). These data suggest that PHPA does not undergo significant metabolism, and the rapidity of excretion suggests that PHPA is not taken up by renal cells before excretion and is probably not reabsorbed from the renal tubules.

The advantage of using deuterium-labeled PHPA (or other isotopes) as a probe to study nitration reactions is rooted in its chemical structure, which is based on a phenolic ring, with a good degree of similarity to tyrosine, and yet is neither taken up by cells or metabolized in the liver. Based on a plasma concentration of PHPA of $2.7 \pm 1.1 \mu mol/l$, and a urinary excretion rate of PHPA at $2.3 \pm 0.7$ nmol/min in rats, the renal clearance of PHPA is estimated to be approximately 0.85 ml/min. This value is very close to the glomerular filtration rate of laboratory rats ($\sim 1.0 \pm 0.1$ ml/min) (Anand *et al.*, 2002) and suggests that PHPA is simply filtered and is not reabsorbed by the kidneys.

To test this novel approach, the effect of systemic inflammation on nitration of PHPA was studied. Injection of bacterial lipopolysaccharide (LPS) leads to increased formation and excretion of NHPA in the urine (Fig. 2), consistent with increased generation of RNSs. To confirm that the increased formation of urinary NHPA arose from nitration of systemic or circulating PHPA, we also measured the urinary excretion of $D_5$-NHPA after infusing $D_6$-PHPA. As shown in Fig. 3, endotoxemia causes a significant increase in both $D_5$-NHPA and NHPA excretion (Fig. 3) and supports the validity of this novel approach. Moreover, the observation that the ratio of nitration of $D_5$-NHPA from infused $D_6$-PHPA was almost identical to that of endogenous NHPA to PHPA suggests that simple measurement

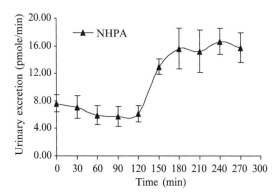

Fig. 2. Urinary excretion of 3-nitro-4-hydroxyphenylacetic acid (NHPA) increases after injection of lipopolysaccharide (LPS) (5 mg/kg intravenously). LPS was injected in anesthetized rats at time 0 and urine samples were collected through a bladder catheter. NHPA excretion showed a significant increased 150-min post-LPS infusion.

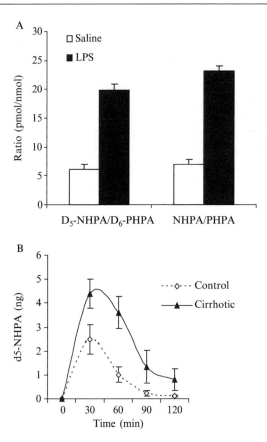

FIG. 3. (A) Nitration of infused deuterated *para*-hydroxyphenylacetic acid (PHPA). Rats were injected with 250 nmol of deuterated PHPA ($D_6$-PHPA). The rate of nitration of $D_6$-PHPA to $D_5$-3-nitro-4-hydroxyphenylacetic acid (NHPA) was determined by gas chromatography (GC)/mass spectrometry (MS) to assess the nitration reactions dynamically. Under basal conditions, the ratio of NHPA/PHPA gives an index of the rate of basal nitration of PHPA, and metabolism of nitrotyrosine. Injection of LPS (endotoxin) increases the ratio of endogenous NHPA/PHPA as well as that of $D_5$-NHPA/$D_6$-PHPA, consistent with increased formation of reactive nitrogen species. (B) Nitration of deuterated PHPA is increased in cirrhosis. Injection of $D_6$-PHPA into rats with biliary cirrhosis leads to increased nitration of PHPA to $D_5$-NHPA. These data suggest that there is increased formation of reactive nitrogen species in cirrhosis and is consistent with data demonstrating increased nitration of tyrosine in liver and plasma of patients with cirrhosis or other models of liver disease.

of the NHPA/PHPA ratio is equally useful to assess nitration reactions *in vivo* (Fig. 3). We have also used this method to examine nitration reactions in health and disease. To test the hypothesis that there is increased formation of RNSs in cirrhosis, we measured the formation of deuterated NHPA from infused deuterated PHPA in rats with biliary cirrhosis. This model is known to be associated with increased formation of NO and nitrotyrosine (Ottesen *et al.*, 2001). As shown in Fig. 3, there is increased formation of deuterated NHPA in cirrhotic rats compared with controls.

## Methods

### Synthesis of Internal Standards

$^{13}C_8$-PHPA is synthesized following the deamination and decarboxylation of $^{13}C_9$-tyrosine using Taiwan cobra (*Naja naja atra*) venom as a catalyzing enzyme (Mani *et al.*, 2003; Nucaro *et al.*, 1998). In brief, $^{13}C_9$-tyrosine is added to 50 m$M$ ammonium formate buffer (pH 7.4) at a concentration of 0.2 mg/ml, and venom is added at a concentration of 1 mg/ml before incubation at 37° for approximately 2 h. $^{13}C_8$-PHPA is extracted into ethyl acetate, dried under nitrogen, and is then further purified by high-pressure liquid chromatography (HPLC) on a Techsphere $C_{18}$ column (25 × 4.6 m$M$). This employs a gradient of water containing 0.1% (v/v) trifluoroacetic acid (TFA) (Solution A) and 0.1% (v/v) TFA/acetonitrile (Solution B). Initial condition is 100% Solution A, changing to 9:1 Solution A/Solution B over 15 min, then to 1:1 Solution A/Solution B from 15 to 30 min. The fractions containing $^{13}C_8$-PHPA can be identified by their retention times and characteristic ultraviolet (UV) spectra with a photodiode array system. Fractions containing $^{13}C_8$-PHPA are pooled and freeze-dried under vacuum. $^{13}C_8$-NHPA is synthesized after exposure of $^{13}C_8$-PHPA to acidified nitrite (1 $M$ HCl and sodium nitrite, 200 m$M$) followed by ethyl acetate extraction, and drying under nitrogen.

### Synthesis of Probe: Deuterium-Labeled PHPA

$D_6$-PHPA ([2,2,2′,3′,5′,6′-$^2$H]PHPA) is synthesized by deuterium exchange (Mani *et al.*, 2003; Shimamura *et al.*, 1986). In brief, 50 mg of PHPA is dissolved in a mixture of $^2H_4$-acetic acid (0.4 ml) and $^2H_2O$ (0.5 ml), and the solvent is evaporated in a stream of nitrogen at 90°. This procedure is repeated twice to remove active protons as completely as possible. The resulting residue is dissolved in a mixture of $^2H_4$-acetic acid

(0.3 ml), $^2H_2O$ (0.3 ml), and $^2HCl$ (37% in $^2H_2O$, 0.8 ml). The solution is sealed in an acid-digestion bomb and heated in an autoclave at 190° for 8 h. The product is extracted with ethyl acetate and washed with water. After drying of the organic phase, the resulting materials are dissolved in 0.1% (v/v) TFA/water (adjusted to pH 5.0 with ammonia solution) and are then applied to $LC_{18}$ reverse-phase columns (Supelclean SPE tubes; Sigma) that had been prewashed with 2 ml of methanol and 5 ml of 0.1% (v/v) TFA/water (pH 5.0). The column is washed with 2 ml of water and $D_6$-PHPA is eluted with 4 ml of 25% (v/v) methanol in water.

The concentrations of $^{13}C$-labeled and deuterated standards are determined against known amounts of unlabeled standards using the method described below.

### Sample Purification and Derivatization for Urinary NHPA and PHPA

$^{13}C_8$-NHPA (10 ng) or $^{13}C_8$-PHPA (100 ng) is added to 100 $\mu l$ of human urine, or 20 $\mu l$ of rat urine, and is diluted to 1 ml with deionized water. Following the addition of 1 ml of ethyl acetate and vortex mixing, the organic phase is removed, and the sample is evaporated under a stream of nitrogen. The residue is reconstituted in 30 $\mu l$ of acetonitrile, applied to a silica TLC plate (LK6 60 A°, 250 $\mu m$ layer thickness, 5 × 20 cm; Whatman, Clifton, NJ), and is run to the top of the plate in chloroform/methanol (80:20, v/v). Compounds migrating in the region of $R_F$ of approximately 0.22 for NHPA and at 0.20 for PHPA, and 1 cm above and below, are scraped from the plate and extracted in 1 ml of ethyl acetate/ethanol (50:50, v/v), dried under nitrogen, and derivatized to the pentafluorobenzyl ester, by the addition of 20 $\mu l$ of 10% (v/v) di-isopropyl ethylamine in acetonitrile and 40 $\mu l$ of 10% (v/v) pentafluorobenzyl bromide in acetonitrile for 1 h at room temperature (21–23°), dried under nitrogen, and redissolved in 20 $\mu l$ of $n$-undecane for analysis of NHPA or 50 $\mu l$ of $n$-undecane for analysis of PHPA.

### Measurement of Plasma Concentrations of NHPA and PHPA

To measure plasma levels of NHPA and PHPA, blood is collected from rats into tubes containing ethylenediaminetetraacetic acid (EDTA), which is centrifuged at 2300$g$ for 30 min at 4°. The internal standards (10 ng of $^{13}C_8$-NHPA or 100 ng of $^{13}C_8$-PHPA) are added to 1 ml of plasma, which is immediately filtered by centrifugation at 9000$g$ in a Microfuge through a 30-kDa molecular-mass cutoff centrifugal membrane (Ultrafree; Millipore, Bedford, MA) to remove high-molecular-mass proteins. The filtrate is applied to an $LC_{18}$ reverse-phase column that had been prewashed with

2 ml of methanol and 5 ml of 0.1% (v/v) TFA/water (pH 5.0). The column is washed with 2 ml of water. Then NHPA and PHPA are eluted with 4 ml of 25% (v/v) methanol in water and dried under vacuum before TLC extraction as described earlier. The samples are derivatized to the pentafluorobenzyl ester by the addition of 20 $\mu$l of 10% (v/v) di-isopropyl ethylamine in acetonitrile and 40 $\mu$l of 10% (v/v) pentafluorobenzyl bromide in acetonitrile for 1 h at room temperature, dried under nitrogen, and redissolved in 20 $\mu$l of *n*-undecane before GC (gas chromatography)/MS analysis.

## GC/MS Analysis of Derivatized Samples

Samples are analyzed on a GC equipped with a 15 m DB-1701 (J&W Scientific, Folsom, CA) capillary column (0.25-mm internal diameter, 0.25 $\mu$m film thickness) interfaced with a mass spectrometer (Trio 1000; Fisons Instruments, Beverly, MA). The ion source and interface temperature are set at 200° and 300°, respectively. Samples are analyzed in negative-ion chemical ionization (NICI) mode with ammonia as the reagent gas, using 1 $\mu$l of each sample for injection. The initial column temperature is maintained at 150° for 1 min increasing to 300° at 20°/min. Samples are quantified by isotope-dilution GC/MS, and ions are monitored at 376 and 384 mass units for NHPA and at 311 and 319 mass units for PHPA with single-ion monitoring. The concentrations are calculated from the known $^{13}C_8$-labeled internal standards, which are 8 mass units heavier than the authentic NHPA and PHPA.

## Nitration of D$_6$-PHPA *In Vivo*

D$_6$-PHPA was injected intravenously as infusion or bolus injection (250 nmol, dissolved in 1 ml of phosphate-buffered saline) into anesthetized rats after the initial basal urine collection. Urine samples can be collected through a bladder catheter every 30 min for 5 h. Nitration of D$_6$-PHPA produces D$_5$-NHPA as a result of replacement of a single deuterium of the aromatic ring by the $-NO_2$ group. Therefore, for measurement of the deuterium-labeled compounds, ion masses ($m/z$) of 381 and 337 are selected in selective-ion monitoring, which are 5 and 6 mass units heavier, respectively, than the authentic NHPA and PHPA (Fig. 4).

## Conclusion

One of the major advantages of this method is that it can employ either measurement of nitrated endogenous PHPA or nitration of infused deuterated PHPA. The latter approach has the advantage in that one

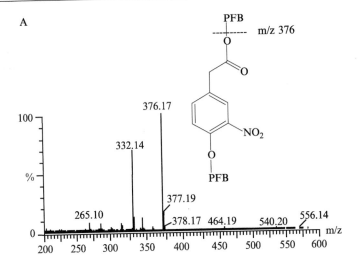

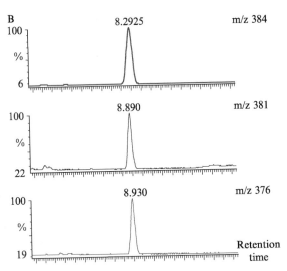

FIG. 4. (A) Structure and mass spectrum of the pentafluorobenzyl ester derivatives of 3-nitro-4-hydroxyphenylacetic acid (NHPA). (B) Selective ion chromatogram of a urine sample obtained from a rat injected with $D_6$-PHPA. *In vivo* formation of $D_5$-NHPA, represented as a distinguished peak, was detected with m/z 381. The values of m/z 376 and m/z 384 represent authentic NHPA and $^{13}C_8$-labeled internal standard, respectively.

can be sure that the excreted deuterated NHPA represents nitration of infused deuterated PHPA. However, we believe that with time, it will be shown that simple measurement of urinary NHPA and its ratio to urinary PHPA will provide a simple reliable index for nitration reactions *in vivo*. Whether these reactions are confined to the systemic circulation is unknown, but the rapid clearance and excretion of PHPA makes this seem likely. The other advantage of this method is that urine is formed from glomerular filtrate collected over hours, and therefore, measurement of metabolites in urine can represent a time-integrated index of *in vivo* formation, or in this case nitration of circulating PHPA or metabolism of nitrotyrosine.

## References

Anand, R., Harry, D., Holt, S., Milner, P., Dashwood, M., Goodier, D., Jarmulowicz, M., and Moore, K. (2002). Endothelin is an important determinant of renal function in a rat model of acute liver and renal failure. *Gut* **50,** 111–117.

Mani, A. R., Pannala, A. S., Orie, N. N., Ollosson, R., Harry, D., Rice-Evans, C. A., and Moore, K. P. (2003). Nitration of endogenous para-hydroxyphenylacetic acid and the metabolism of nitrotyrosine. *Biochem. J.* **374,** 521–527.

Nucaro, E., Jodra, M., Russell, E., Anderson, L., Dennison, P., and Dufton, M. (1998). Conversion of tyrosine to phenolic derivatives by Taiwan cobra venom. *Toxicon* **36,** 1173–1187.

Ohshima, H., Friesen, M., Brouet, I., and Bartsch, H. (1990). Nitrotyrosine as a new marker for endogenous nitrosation and nitration of proteins. *Food Chem. Toxicol.* **28,** 647.

Ottesen, L. H., Harry, D., Frost, M., Davies, S., Khan, K., Halliwell, B., and Moore, K. (2001). Increased formation of *S*-nitrosothiols and nitrotyrosine in cirrhotic rats during endotoxemia. *Free Radic. Biol. Med.* **31,** 790–798.

Shimamura, M., Kamada, S., Hayashi, T., and Naruse, H. (1986). Sensitive determination of tyrosine metabolites, *p*-hydroxyphenylacetic acid, 4-hydroxy-3-methoxyphenyl-acetic acid and 4-hydroxy-3-methoxymandelic acid, by gas chromatography-negative-ion chemical-ionization mass spectrometry. Application to a stable isotope-labelled tracer experiment to investigate their metabolism in man. *J. Chromatogr.* **374,** 17–26.

Souza, J. M., Choi, I., Chen, Q., Weisse, M., Daikhin, E., Yudkoff, M., Obin, M., Ara, J., Horwitz, J., and Ischiropoulos, H. (2000). Proteolytic degradation of tyrosine nitrated proteins. *Arch. Biochem. Biophys.* **380,** 360–366.

## [16]  Protein Nitration in Biological Aging: Proteomic and Tandem Mass Spectrometric Characterization of Nitrated Sites

*By* JAROSLAW KANSKI and CHRISTIAN SCHÖNEICH

### Abstract

Proteomic techniques for the identification of 3-nitrotyrosine–containing proteins in various biological systems are described with emphasis on the direct mass spectrometric detection and sequencing of 3-nitrotyrosine–containing peptides. Strengths and weaknesses of various separation and mass spectrometric techniques are discussed. Some examples for the MS/MS analysis of nitrated peptides obtained from aging rat heart and skeletal muscle are provided, such as nitration of $Tyr^{105}$ of the mitochondrial electron-transfer flavoprotein and $Tyr^{14}$ of creatine kinase.

### Introduction

The nitration of tyrosine to 3-nitrotyrosine (3-NT) represents a common consequence of oxidative stress associated with various pathological conditions and natural aging (Beal, 2002; Gow *et al.*, 2004; Ischiropoulos and Beckman, 2003; Schopfer *et al.*, 2003; Turko and Murrad, 2002). The presence of 3-NT can compromise the function and structure of proteins and affect their biological half-life (Greenacre and Ischiropoulos, 2001). To correlate protein nitration with any biological dysfunction, tools are needed to identify the protein targets of nitration *in vivo* and to characterize the location and extent of this modification. Proteomic approaches can provide such answers, offering a global characterization of the "nitroproteome" of a given system. In general, this technique first separates the proteins of interest by various complementary methods followed by mass spectrometry (MS) sequencing for protein modification. Various articles on the proteomic analysis of 3-NT–containing proteins have appeared and are summarized in Table I.

The purpose of this chapter is to provide a brief description of the various proteomic strategies and discuss their respective advantages and possible shortcomings. We specifically address the difficulties associated with the MS/MS analysis of small quantities of 3-NT–containing peptides from *in vivo* sources and provide some successful examples for the sequencing

METHODS IN ENZYMOLOGY, VOL. 396
Copyright 2005, Elsevier Inc. All rights reserved.
0076-6879/05 $35.00
DOI: 10.1016/S0076-6879(05)96016-3

TABLE I

RECENT PROTEOMIC REPORTS PERTAINING TO THE DETECTION OF 3-NT–CONTAINING PROTEINS IN VARIOUS MODELS

| Reference | System investigated | Protein separation technique used | Mass detection | 3-NT MS/MS |
|---|---|---|---|---|
| Aulak et al., 2001 | In vivo inflammation, rat lungs, liver | 2DE | MALDI-TOF | Not presented |
| Aulak et al., 2004a | Cell lysates, mitochondria from rat livers | 2DE | MALDI-TOF | Not presented |
| Castegna et al., 2003 | Alzheimer's disease brain | 2DE | MALDI-TOF, ESI-MS/MS | Not presented |
| Elfering et al., 2004 | In vitro NO stimulation of rat liver mitochondria | 2DE | MALDI-TOF | Yes |
| Kanski et al., 2003 | In vivo aged rat skeletal and cardiac muslce | 2DE | ESI-MS/MS | No |
| Kanski et al., 2005 | In vivo aged rat skeletal muscle | 2DE, IP | ESI-MS/MS | Yes |
| Koeck et al., 2004a | In vitro rat liver mitochondria | 2DE | MALDI-TOF | Not presented |
| Koeck et al., 2004b | Human skin fiboblasts | 2DE | MALDI-TOF | Not presented |
| Murray et al., 2003 | In vitro mitochodrial complex I | 2DE, 1D | MALDI-TOF, ESI-MS/MS | Yes |
| Miyagi et al., 2002 | In vivo ratina | 2DE | MALDI-TOF, ESI-MS/MS | No |
| Turko et al., 2003 | In vitro diabetic mouse cardiac mitochondria | 2DE | MALDI-TOF | Yes |

of 3-NT–containing peptides from aging tissues. No attempt has been made here to provide detailed technical descriptions of all the respective methods because there is a wealth of information in the literature pertaining to the specific instrumental and fundamental issues associated with each technique.

## Methods for Protein Separation

### 2D-Gel Electrophoresis

Two-dimensional (2D) gel electrophoresis (2DE) has been used most commonly for the separation of proteins. Proteins are separated based on their pI by isoelectrofocusing (IEF) on commercially available immobilized pH gradient (IPG) strips in the first dimension and based on their apparent molecular weight (MW) in the second dimension. This technique offers acceptable resolution, especially when the IEF step is performed within narrow pH ranges, yielding discrete and fairly reproducible protein maps. In addition, 2DE is compatible with transfer protocols, allowing for Western blot analysis of 3-NT–containing proteins by one of the several commercially available antibodies. A recent comparison of various anti-3-NT antibodies was provided by Franze *et al.* (2003). An overlay of a Western blot and a stained gel identifies proteins of interest, which are submitted to MS and MS/MS analysis. Realistically, positive identification of nitrated proteins is possible only if the nitrated peptides of the respective proteins can be confirmed by the MS/MS analysis. Otherwise, the results can be compromised by comigration of more than one protein within the same spot on the 2D map. Our analysis of cardiac proteins from young and old Fisher 344/Brown Norway F1 rats by 2DE has yielded MS/MS confirmation of 3-NT at position 105 in the mitochondrial electron transfer flavoprotein (Kanski *et al.*, 2005). 2DE gels can be analyzed by specialized software for the quantitation of the resolved proteins. Alternatively, differential display using fluorescent dyes can be used for the relative quantification of proteins originating from different tissues.

The main disadvantage of the 2DE technique is its bias against very hydrophobic proteins that are generally underrepresented or missing on the 2D maps. In addition, incomplete focusing resulting in the loss of resolution may occur, especially in the basic region of the 2D gel. Furthermore, the first dimension of 2DE requires special sample preparation to eliminate any artefact associated with the presence of impurities, protein aggregates, high salt contents, and so on, which may compromise the quality of the 2DE gel maps. Efforts have been made to overcome these shortcomings by use of special detergents and lysis conditions (Molloy, 2000).

*Antibody Specificity*

Commonly, the mouse monoclonal 1A6 anti-3-NT antibody has been used to detect protein nitration in Western blot analysis (Beckman *et al.*, 1994). Despite that this is generally considered a specific and reliable antibody, chances are that it can show nonspecific behavior. In order to control for nonspecific binding, experiments need to be performed involving the treatment of the proteins on the PVDF membrane with dithionite ($Na_2S_2O_4$) to reduce 3-NT to 3-aminotyrosine (3-AT) before Western blot analysis (Aulak *et al.*, 2004b). Generally, this approach works well; however, 3-AT oxidizes back to 3-NT with time, potentially resulting in weak signals even on dithionite-treated membranes. Another approach includes the preincubation of the antibody solution with 3-NT, saturating the antibody with the antigen, just before analysis. Likewise, no signal should be observed when membranes are treated with such antigen-competed antibodies. Often, the use of an *in vitro* nitrated protein standard, such as albumin, can be beneficial to control for antibody response during specific nitrating conditions.

*Immunoprecipitation*

Immunoprecipitation (IP) with the anti-3-NT antibody (MacMillan-Crow and Thompson, 1999) allows for the enrichment of 3-NT–containing proteins. Commercially available anti-3-NT antibodies covalently coupled to agarose beads can be used to isolate the antigens, which are then applied directly to one-dimensional (1D) sodium dodecylsulfate polyacrylamide gel electrophoresis (SDS-PAGE) separation. Provided that the antibody–antigen interaction is specific, complete resolution of proteins by 2DE is not required because even multiple proteins in a given gel band will easily be identified by MS and MS/MS analysis. An interesting feature of our recent comparison of nitrated proteins in aging heart is that only a subset of proteins was identified by both methods (Kanski *et al.*, 2005), whereas specific proteins were only detected by either 2DE or IP-1DE. Such results can be a function of the limitations of the 2DE method discussed earlier. Specifically, some proteins identified in the immunoprecipitation experiment are very large (*e.g.*, myoglobin heavy chain, MW 223 kDa) or are characterized by very basic pI values (GERp95, ubiquinol-cytochrome *c* reductase, solute carrier family 25, and $H^+$-transporting ATP synthase with pI values of 9.3, 9.2, 9.8, and 10.03, respectively).

*Affinity Enrichment after Covalent Attachment of a Biotin-Affinity Tag*

Nitrated proteins and peptides can be selectively enriched after reduction of 3-NT to 3-AT and coupling of sulfosuccinimidyl-2-(biotinamido) ethyl-1,3-dithiopropionate (sulfo-NHS-SS-biotin), which contains a cleavable biotin-affinity tag (Nikov *et al.*, 2003). The biotin-reacted peptides are then retained by streptavidin immobilized on beads or an affinity column. After reductive release of the peptides, identification is achieved by MS and MS/MS analysis, accounting for the mass difference introduced by the derivatization chemistry (Nikov *et al.*, 2003). In principle, this approach will only yield the derivatized peptides, thus greatly simplifying the liquid chromatography (LC)–MS/MS runs and data analysis. However, the reaction conditions and pH must be strictly optimized to eliminate any side-reaction with free aliphatic amines of the proteins and/or the peptides.

## Multidimensional HPLC Analysis

Multidimensional high-performance liquid chromatography (HPLC) separation using orthogonal modes can completely eliminate the need for 2DE to separate proteins and peptides of interest. Either proteins can be separated before proteolytic digestion or a whole protein mixture can be digested into the respective peptides before HPLC analysis. The latter strategy has been termed multidimensional protein identification technology (MudPIT) (Washburn *et al.*, 2001). Orthogonal chromatographic modes include size exclusion, ion exchange, and reverse phase. Such tasks may be performed in an automated fashion. The main advantage of this approach is the elimination of time and effort associated with running 2DE and artefacts associated with that technique. For example, as the nitrated tyrosine residue becomes more hydrophobic than tyrosine, the possibility exists that 3-NT–containing peptides may not be completely recovered from the SDS-PAGE gel matrix and, therefore, be absent in the in-gel digest mixture. Moreover, this technique is amenable for direct quantitation by use of isotopically labeled affinity tags (ICATs) (Han *et al.*, 2001). As a potential disadvantage, data analysis can become a very complex and time-consuming task; moreover, the identity of a given peptide cannot necessarily be translated into the presence of a protein because the proteolytic digest precedes the separation step. For example, a peptide of interest may not result from a complete protein but from a protein fragment produced *in vivo*. This is especially important for identifications based on the detection of only one or two peptides.

## Combination of HPLC and SDS-PAGE

Combination HPLC and SDS-PAGE relies on the separation of proteins by any chromatographic mode before SDS-PAGE. It has been applied, for example, to the proteomic characterization of the nuclear pore complex (Rout *et al.*, 2000). Collected chromatographic fractions of interest are applied to SDS-PAGE gel analysis, after which proteins can be stained or transferred for Western blot analysis, as in the case of the 2DE approach. The use of C-4 columns is especially beneficial for large and hydrophobic proteins that are notoriously more difficult to resolve by 2DE. Dissolution and separation of membrane proteins usually requires a high content of organic modifiers, which necessitates solvent evaporation after fraction collection. Solvent evaporation may lead to aggregation of the membrane proteins. However, redissolution of the dried fractions can be achieved by up to 4% SDS. The SDS-PAGE step after HPLC separation represents a good approach to present membrane proteins for proteolytic digestion. In our analysis of a membrane protein, the sarcoplasmic/endoplasmic reticulum CaATPase (SERCA), we have coupled reverse-phase chromatography on a C-4 column with SDS-PAGE, resulting in sufficient quantities of protein purified from skeletal muscle and aorta for the analysis of 3-NT and several nitrosated and *S*-glutathiolated Cys residues (Adachi *et al.*, 2002, 2004; Sharov *et al.*, 2002).

*Solution Isoelectrofocusing*

The principle of IEF in solution is similar to that of IEF in the gel, with the exception of the medium. Proteins are separated across a pH gradient, typically between pH 3 and 10, generated by carrier ampholytes and the applied current. This technique eliminates artefacts usually associated with the use of IPG strips (Lubman *et al.*, 2002). Furthermore, solution IEF can be used on the semipreparative and preparative scale with large amounts of material. The latter can be beneficial for the MS/MS analysis of nitrated peptides, which are usually present *in vivo* in relatively low yields. After focusing is complete, the fractions are collected and either can be applied to 1D SDS-PAGE or 2DE or can be directly applied to MS and MS/MS analysis. Because there are no IPG strips involved, hydrophobic, small, and large proteins are abundant in this type of analysis. The disadvantage of this technique is a general lack of resolution compared to the 2DE method, manifested in the presence of proteins of interest across several collected fractions. This can, however, be overcome by the use of more narrow pH gradients. Using solution electrofocusing, sufficient yields of 3-NT–containing peptides were obtained from skeletal muscle of 34-month-old Fisher 344/Brown Norway F1 rats. Figure 1 displays a representative MS/MS spectrum showing a 3-NT–containing peptide of creatine kinase present in the homogenate of rat skeletal muscle.

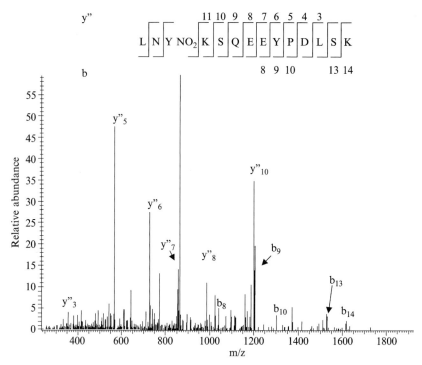

FIG. 1. MS/MS spectrum of the tryptic peptide [12]LNY(NO$_2$)KSQEEYPDLSK[25] of creatine kinase obtained during the proteomic analysis of skeletal muscle homogenate from a 34-month-old Fisher 344/Brown Norway F1 rat. The fragmentation pattern reveals 3-NT at position 14. Protein separation was achieved by solution isoelectrofocusing and sodium dodecylsulfate polyacrylamide gel electrophoresis. MS/MS analysis was achieved by nanoelectrospray MS/MS on an ion-trap mass spectrometer.

## Mass Spectrometry

Mass spectrometric analysis is the ultimate step in determining the protein identity (Aebersold and Goodlett, 2001). Alternative tools such as Edman sequencing or Western blotting potentially suffer from inherent technical problems or nonspecificity. In principle, mass spectrometric identification can be achieved on a whole protein, but this requires a sophisticated high-resolution mass analyzer, such as a Fourier transform-ion cyclotron resonance (FTICR) mass spectrometer, and the single mass itself does not guarantee the protein identity. Instead, the use of proteolysis is preferred. In this approach, the protein of interest is submitted to digestion by known proteolytic enzymes such as trypsin, chymotrypsin, or endo Lys-C. The resulting peptide mixture is then analyzed and compared to existing protein databases. Based on the quality of the fits and the sequence coverage, a score is assigned. Furthermore, by observing mass shifts for the peptides of interest (Aebersold and Goodlett, 2001), one can elucidate any possible posttranslational modifications.

## MALDI and ESI Ionization Techniques

For proteomic studies, there are two common ways to introduce the sample into the mass analyzer, one relying on matrix-assisted laser desorption ionization (MALDI) and the other on electrospray ionization (ESI). MALDI is frequently coupled to a time-of-flight (TOF) analyzer, whereas ESI can be interfaced to various analyzers including the ion-trap, triple quadruple, or the hybrid quadruple–time of flight (Q-TOF). Because of the soft ionization, MALDI and ESI techniques preserve the masses of the peptides or proteins introduced into the source. Under certain circumstances, even noncovalent interactions such as metal complexes can be maintained (Zhu et al., 2003). MALDI TOF spectra are simple to interpret because of the propensity of the method to generate predominantly singly charged ions. Additionally, the MALDI method is relatively resistant to interferences from matrices commonly used in protein chemistry. Traditionally, sequence analysis in MALDI TOF spectra has been achieved by the use of the post–source decay (PSD) technique. Although it is possible to obtain excellent PSD spectra, these spectra are frequently incomplete, requiring the collection of multiple spectra at different instrument settings. The recently developed and commercialized MALDI TOF/TOF and MALDI-Q/TOF instruments can deliver high-quality MS/MS spectra. However, one problem associated with the MALDI TOF analysis of 3-NT–containing peptides is the photochemical net loss of one and two oxygen atoms of the nitro group, resulting in ions that are +29 atomic mass units (a.m.u) and +13 a.m.u, heavier compared to native tyrosine (for comparison, the introduction of a nitro group into tyrosine is associated with a gain of +45 a.m.u.). These photochemical fragmentations may complicate the MALDI TOF fingerprint mass spectrum (Sarver et al., 2001).

ESI represents a powerful alternative to MALDI. Its primary advantage comes from the ability to couple ESI to HPLC separation, allowing for the direct introduction of the sample eluate into the mass analyzer. This offers a tremendous advantage in simplifying the sample matrix, and the introduction of the LC step allows for the direct comparison of chromatographic traces in the search for similarities and differences between samples. In the case of 3-NT–containing peptides, one expects larger retention times compared to the native, unmodified peptides during reverse-phase separation. The development of special stationary phases, including affinity columns, optimized for certain types of analytes, further enhances the selectivity and robustness of the LC-MS/MS technique. Two commonly used mass analyzers for proteomic experiments are the quadrupole-ion trap (IT) and Q-TOF. The ion traps are especially useful for their ability to obtain large amounts of MS/MS data and to perform $MS^n$ sequencing. However, the Q-TOF analyzer provides superior mass resolution and accuracy.

Especially in the case of limited sample amounts, the sensitivity of mass spectrometric analysis can be enhanced by using nano-electrospray ionization (NSI). This technique operates at sub–microliter-per-minute flow rates, offering a far greater sensitivity than the regular ESI source with lower amounts of material consumed. As in the case of ESI, the NSI source can be coupled online to capillary columns of 50–75 $\mu M$ inner diameters.

## Mass Spectrometric Data Analysis for 3-NT

The nitration of a tyrosine residue results in the addition of +45 a.m.u. to the native tyrosine-containing peptide. However, under some circumstances, the nitration of tryptophan was observed (Alvarez et al., 1996; Herold, 2004). Hence, the addition of +45 a.m.u to a peptide per se is no evidence for tyrosine nitration if the same peptides also contains tryptophan, and mass fingerprint analysis alone is not sufficient for a positive detection of 3-NT. Moreover, the possibility exists that a completely unrelated peptide of similar mass, which lacks 3-NT, is present in the mixture.

Sequencing the peptides by MS/MS generally overcomes the limitation of mass fingerprint assignment. In addition, the MS/MS pattern is the only way to assign the location of 3-NT if the original peptide contained more than one tyrosine residue. An alternative approach is to monitor the specific immonium ion of 3-NT (181 a.m.u), which can be traced back to its precursor peptide(s). This approach, followed by analysis by MS/MS, can be a reliable and fast method to identify nitrated peptides in complex peptide mass maps. Such an approach is limited, because of hardware characteristics, to the TOF-TOF, Q-TOF, and triple-quadrupole analyzers.

## Proteomic Studies Providing MS/MS Characterization of
### 3-NT–Containing Peptides

Table I summarizes reports on the proteomic identification of 3-NT–containing proteins as a result of various physiological and pathological conditions. Our data have established an age-dependent increase in protein nitration in both cardiac and skeletal muscle (Kanski *et al.*, 2003, 2005). Several of the proteins identified are associated with energy production and metabolism, providing a correlation between biological aging and loss of motor functions. Using 2DE separation followed by MS/MS analysis, we were able to demonstrate the formation of 3-NT at position 105 on the cardiac mitochondrial electron transfer flavoprotein (Kanski *et al.*, 2005). By employing the solution IEF approach, we have identified 3-NT at position 14 of creatine kinase obtained from *in vivo* aged skeletal muscle (Fig. 1). Assignment of the fragments was done according to the nomenclature given by Roepstorff and Fohlman (1984). Several groups were successful at locating sites of tyrosine nitration during the *in vitro* stimulation of cell culture systems and specific tissues (Table I). In the targeted analysis of a specific protein, Aslan *et al.* (2003) identified sites of nitration of $\beta$-actin in sickle cell disease models.

## Acknowledgments

Support for our research was provided by the National Institutes of Health (grants PO1AG12993, CA072987, and AG023551) and the Center for Bioanalytical Research at the University of Kansas.

## References

Adachi, T., Matsui, R., Xu, S., Kirber, M., Lazar, H. L., Sharov, V. S., Schöneich, C., and Cohen, R. (2002). Antioxidant improves smooth muscle sarco/endoplasmatic reticulum $Ca^{2+}$–ATPase function and lowers tyrosine nitration in hypercholesterolemia and improves nitric oxide–induced relaxation. *Circ. Res.* **90**, 1114–1121.

Adachi, T., Weisbrod, R. M., Pimentel, D. R., Ying, J., Sharov, V., Schöneich, C., and Cohen, R. (2004). *S*-Glutathiolation by peroxynitrite activates SERCA during arterial relaxation by nitric oxide. *Nat. Med.* **10**, 1200–1207.

Aebersold, R., and Goodlett, D. R. (2001). Mass spectrometry in proteomics. *Chem. Rev.* **101**, 269–295.

Alvarez, B., Rubbo, H., Kirk, M., Barnes, S., Freeman, B. A., and Radi, R. (1996). Peroxynitrite-dependent tryptophan nitration. *Chem. Res. Toxicol.* **9**, 390–396.

Aslan, M., Ryan, T. M., Townes, T. M., Coward, L., Kirk, M. C., Barnes, S., Alexander, C. B., Rosenfeld, S. S., and Freeman, B. A. (2003). Nitric oxide–dependent generation of reactive species in sickle cell disease. *J. Biol. Chem.* **278**, 4194–4204.

Aulak, K. S., Koeck, T., Crabb, J. W., and Stuehr, D. J. (2004a). Dynamics of protein nitration in cells and mitochondria. *Am. J. Physiol. Heart. Circ. Physiol.* **286**, H30–H38.

Aulak, K. S., Koeck, T., Crabb, J. W., and Stuehr, D. J. (2004b). Proteomic method for identification of tyrosine-nitrated proteins. *Methods Mol. Biol.* **279**, 151–166.

Aulak, K. S., Miyagi, M., Yan, L., West, K. A., Massillon, D., Crabb, J. W., and Stuehr, D. J. (2001). Proteomic method identifies proteins nitrated *in vivo* during inflammatory challenge. *Proc. Natl. Acad. Sci. USA* **98**, 12056–12061.

Beal, M. F. (2002). Oxidatively modified proteins in aging and disease. *Free Radic. Biol. Med.* **32**, 797–803.

Beckman, J. S., Ye, Y. Z., Anderson, P. G., Chen, J., Accavitti, M. A., Tarpey, M. M., and White, C. R. (1994). Extensive nitration of protein tyrosines in human atherosclerosis detected by immunohistochemistry. *Biol. Chem. Hoppe. Seyler* **375,** 81–88.

Castegna, A., Thongboonkerd, V., Klein, J. B., Lynn, B., Markesbery, W. R., and Butterfield, D. A. (2003). Proteomic identification of nitrated proteins in Alzheimer's disease brain. *J. Neurochem.* **85,** 1394–1401.

Elfering, S. L., Haynes, V. L., Traaseth, N. J., Ettl, A., and Giulivi, C. (2004). Aspects, mechanisms, and biological relevance of mitochondrial protein nitration sustained by mitochondrial nitric oxide synthase. *Am. J. Physiol. Heart. Circ. Physiol.* **286,** H22–H29.

Franze, T., Weller, M. G., Niessner, N., and Pöschl, U. (2003). Enzyme immunoassays for the investigation of protein nitration by air pollutants. *Analyst* **128,** 824–831.

Gow, A. J., Farkouth, C. R., Munson, D. A., Posencheg, M. A., and Ischiropoulos, H. (2004). Biological significance of nitric oxide–mediated protein modifications. *Am. J. Physiol. Lung Cell Mol. Physiol.* **287,** L262–L268.

Greenacre, S. A., and Ischiropoulos, H. (2001). Tyrosine nitration: Localization, quantification, consequences for protein function and signal transduction. *Free Radic. Res.* **34,** 541–581.

Han, D. K., Eng, J., Zhou, H., and Aebersold, R. (2001). Quantitative profiling of differentiation-induced microsomal proteins using isotope-coded affinity tags and mass spectrometry. *Nat. Biotechnol.* **19,** 946–951.

Herold, S. (2004). Nitrotyrosine, dityrosine, and nitrotryptophan formation from metmyoglobin, hydrogen peroxide, and nitrite. *Free Radic. Biol. Med.* **36,** 565–579.

Ischiropoulos, H., and Beckman, J. S. (2003). Oxidative stress and nitration in neurodegeneration: Cause, effect, or association? *J. Clin. Invest.* **111,** 163–169.

Kanski, J., Alterman, M. A., and Schöneich, C. (2003). Proteomic identification of age-dependent protein nitration in rat skeletal muscle. *Free Radic. Biol. Med.* **35,** 1229–1239.

Kanski, J., Behring, A., Pelling, J., and Schöneich, C. (2005). Proteomic identification of 3-nitrotyrosine containing cardiac proteins in Fisher 344/BNF1 rats: Effects of biological aging. *Am. J. Physiol. Heart. Circ. Physiol.* **288,** H371–H381.

Koeck, T., Fu, X., Hazen, S. L., Crabb, J. W., Stuehr, D. J., and Aulak, K. S. (2004a). Rapid and selective oxygen-regulated protein tyrosine denitration and nitration in mitochondria. *J. Biol. Chem.* **279,** 27257–27262.

Koeck, T., Levison, B., Hazen, S. L., Crabb, J. W., Stuehr, D. J., and Aulak, K. S. (2004b). Tyrosine nitration impairs mammalian aldolase C activity. *Mol. Cell Prot.* **3,** 548–557.

Lubman, D. M., Kachman, M. T., Wang, H., Gong, S., Yan, F., Hamler, R. L., O'Neil, K. A., Zhu, K., Buchanan, N. S., and Barder, T. J. (2002). Two-dimensional liquid separations-mass mapping of proteins from human cancer cell lysates. *J. Chromatogr. B* **782,** 183–196.

MacMillan-Crow, L. A., and Thompson, J. A. (1999). Immunoprecipitation of nitrotyrosine-containing proteins. *Methods Enzymol.* **301,** 135–145.

Miyagi, M., Skaguchi, H., Darrow, R. M., Yan, L., West, K. A., Aulak, K., Stuehr, D. J., Hollyfield, J. G., Organisciak, D. T., and Crabb, J. W. (2002). Evidence that light modulates protein nitration in rat retina. *Mol. Cell Prot.* **1,** 293–302.

Molloy, M. P. (2000). Two-dimensional electrophoresis of membrane proteins using immobilized pH gradients. *Anal. Biochem.* **280,** 1–10.

Murray, J., Taylor, S. W., Zhang, B., Gosh, S. S., and Capaldi, R. A. (2003). Oxidative damage to mitochondrial complex I due to peroxynitrite; identification of reactive tyrosines by mass spectrometry. *J. Biol. Chem.* **278,** 27223–27230.

Nikov, G., Bhat, V., Wishnok, J. S., and Tannenbaum, S. R. (2003). Analysis of nitrated proteins by tyrosine-specific affinity probes and mass spectrometry. *Anal. Biochem.* **320,** 214–222.

Roepstorff, P., and Fohlman, J. (1984). Proposal for a common nomenclature for sequence ions in mass spectra of peptides. *Biomed. Mass. Spectrom.* **11,** 601.

Rout, M. P., Aitchison, J. D., Suprapto, A., Hjertaas, K., Zhao, Y., and Chait, B. T. (2000). The yeast nuclear pore complex: Composition, architecture, and transport mechanism. *J. Cell Biol.* **148,** 635–651.

Sarver, A., Scheffler, N. K., Shetlar, M. D., and Gibson, B. W. (2001). Analysis of peptides and proteins containing nitrotyrosine by matrix-assisted laser desorption/ionization mass spectrometry. *J. Am. Soc. Mass. Spectrom.* **12,** 439–448.

Schopfer, F. J., Baker, P. R., and Freeman, B. A. (2003). NO-dependent protein nitration: A cell signaling event or an oxidative inflammatory response? *Trends Biochem. Sci.* **28,** 646–654.

Sharov, V., Galeva, N. A., Knyushko, T. V., Bigelow, D. J., Williams, T. D., and Schöneich, C. (2002). Two-dimensional separation of the membrane protein sarcoplasmic reticulum Ca-ATPase for high-performance liquid chromatography-tandem mass spectrometry analysis of posttranslational protein modifications. *Anal. Biochem.* **308,** 328–335.

Turko, I. V., Li, L., Aulak, K. S., Stuehr, D. J., Chang, J.-Y., and Murrad, F. (2003). Protein tyrosine nitration in the mitochondria from diabetic mouse heart. *J. Biol. Chem.* **278,** 33972–33977.

Turko, I. V., and Murrad, F. (2002). Protein nitration in cardiovascular diseases. *Pharmacol. Rev.* **54,** 619–634.

Washburn, M. P., Wolters, D., and Yates, J. R., III. (2001). Large-scale analysis of the yeast proteome by multidimensional protein identification technology. *Nat. Biotech.* **19,** 242–247.

Zhu, M. M., Rempel, D. L., Zhao, J., Giblin, D. E., and Gross, M. L. (2003). Probing Ca2+-induced conformational changes in porcine calmodulin by H/D exchange and ESI-MS: Effect of cations and ionic strength. *Biochemistry* **42,** 15388–15397.

# [17] HPLC-Electrochemical Detection of Tocopherol Products as Indicators of Reactive Nitrogen Intermediates

*By* Kenneth Hensley and Kelly S. Williamson

## Abstract

The nitric oxide (NO) free radical serves diverse functions in mammalian physiology, facilitating processes that range from vasodilation to neurotransmission to host–pathogen defense. Despite the fascinating biochemical utility of this small diatomic gas, NO and its derived oxidation or reduction products [reactive nitrogen species (RNSs) or reactive nitrogen intermediates (RNIs)] can be deleterious to cells and tissues under conditions of pathophysiology. Recent years have witnessed a tremendous and continuing scientific interest in RNI, both as targets for pharmacotherapy and as biomarkers for disease. Accordingly, methods have been

METHODS IN ENZYMOLOGY, VOL. 396
0076-6879/05 $35.00
DOI: 10.1016/S0076-6879(05)96017-5

developed to quantify RNI in real time under controlled experimental conditions. Such methods usually employ either electron paramagnetic resonance (EPR) spin traps (see Chapter 45) or electrochemical sensors (Cserey and Gratzl, 2001). Nonetheless, the transient nature of NO and RNIs often precludes their routine assessment in animal experiments or human clinical studies where tissue must be archived and stored at a later time. To circumvent these limitations, methods have been invented to detect and quantify stably nitrated products of RNI reaction with ambient biomolecules. This chapter describes the theory and methodology for detection of lipid-phase tocopherol nitration products, especially 5-nitro-$\gamma$-tocopherol (5-NO$_2$-$\gamma$T) by high-performance liquid chromatography with electrochemical detection (HPLC-ECD).

## Principles of Electrochemical Array Detection

Electrochemical detection (ECD) was originally coupled with liquid chromatographic separation technologies to measure dopamine metabolites, and detailed reviews of ECD technology have been published (Kissinger *et al.*, 1974; Krstulovic, 1982; Mefford, 1985; Riggin and Kissinger, 1977). ECDs are either amperometric or coulometric. Amperometric detectors consist simply of a working electrode (often glassy carbon) embedded in the wall of a narrow lumen through which analyte is passed. As an analyte flows across the electrode surface, a chemical reaction can occur whereby the analyte loses one or more electrons (*i.e.*, the analyte is oxidized). These electrons are detected as a current flux through the working electrode. The current flowing through the electrochemical cell at any moment is a linear function of the number of molecules being oxidized at the cell surface at that point in time (Mefford, 1985). The intrinsic sensitivity is, thus, dependent on two principal factors: the oxidation potential of the analyte and the equilibrium constant for adsorption onto the cell surface.

Amperometric detectors are inefficient because of mass-transfer limitation. Only a small fraction of total analyte in the flow stream, typically less than 10%, actually diffuses onto the electrode surface and experiences electrochemical conversion (Mefford, 1985). Coulometric detectors are designed to circumvent this limitation somewhat. In the coulometric scheme, the analyte flows through a porous matrix (typically porous carbon). In practice, coulometric detectors are two to five times more sensitive than amperometric detectors, with detection limits in the high femtomole to low picomole range and a dynamic detection range of 2–3 orders of magnitude (Jane *et al.*, 1985; Mefford, 1985). The major disadvantage of electrochemical detection is the necessity for employing mobile phases of high ionic strength, which implies a significant aqueous component. Thus, normal-phase chromatography is seldom employed when ECD is used, and this limits certain separations.

If several coulometric detectors are arranged in series with systematically incremented potentials, a two-dimensional (2D) ECD array chromatogram can be generated. An ECD array chromatogram provides several strategic advantages over a single channel recording. First, the analyst can adjust the proximal cells in the flow stream to low potential to oxidize interfering substances and simplify the chromatogram on latter channels. Second, the polarity of the first several cells in a series can be adjusted to reduce rather than oxidize specific analytes within the flow stream. The reduction products can then be measured oxidatively on latter channels. Sometimes this strategy will increase sensitivity dramatically (Musch et al., 1985; Shigenaga et al., 1997). Last, the multidimensionality of an ECD array can provide a diagnostic plot called a "hydrodynamic voltamogram." When the oxidation potentials of several cells are arranged judiciously and matched to the velocity of the flow stream, an analyte can be made to oxidize across three to four channels. The hydrodynamic voltamogram is generated by plotting the cumulative fraction of total current response on each of these successive channels. This plot is characteristic for each analyte under a specified set of chromatographic parameters (Hensley et al., 1997, 1999, 2000; Rizzo et al., 1991). In particular, the interpolated plot allows determination of a half-maximal oxidation potential. A chromatographic peak that matches a standard peak with respect to both retention time and hydrodynamic voltametry is, therefore, assignable with a high degree of confidence.

ECDs have been coupled with high-performance liquid chromatography (HPLC) separations for the detection of many biomolecules including abnormal DNA bases, unnatural amino acids, and drug metabolites (Hensley et al., 1999). The most practical analytes for HPLC ECD have certain key characteristics. First, they should be oxidizable at intermediate potentials (200–900 mV). Below this potential, the analytes tend to autoxidize, which necessitates rapid sample processing; above this potential, the target analytes are invisible to the ECD. Second, the analytes should be chemically stable in the biological milieu and at acidic pH, because coulometric cells are vulnerable to highly basic solutions. Last, the compounds should be reasonably hydrophobic so they are separable on reverse-phase HPLC columns. Conveniently, phenolic compounds satisfy all these criteria. Thus, nitrated phenolic biomolecules, particularly 3-NO$_2$-tyrosine (3-NO$_2$-Tyr) and 5-NO$_2$-$\gamma$-tocopherol, have emerged as favored species to measure by HPLC-ECD as biomarkers for exposure to RNI (Hensley et al., 1997, 1998, 1999, 2000; Williamson et al., 2002, 2003a,b).

Detailed discussions have been published elsewhere regarding the chemistry of RNI interaction with phenolic substrates leading to the formation of stably nitrated products (Hensley *et al.*, 2000). Therefore, in this chapter we omit these details and concentrate on the technical aspects and advantages of measuring nitrated tocopherol as an indicator of lipid-phase nitration.

### 5-NO$_2$-$\gamma$-Tocopherol as an Indicator for Lipid-Phase Nitration

The earliest application of HPLC-ECD to measurement of nitrated biomolecules included investigations by a number of groups into 3-NO$_2$-Tyr (Hensley *et al.*, 1997, 1998, 1999, 2000; Shigenaga *et al.*, 1997; Williamson *et al.*, 2003b). This analyte was successfully measured in various biological matrices including mammalian cell cultures (Hensley *et al.*, 1997) and human brain tissue, where protein tyrosine nitration was found to correlate with protein-bound 3,3′-dityrosine and with brain regional patterns of neuropathology (Hensley *et al.*, 1998). Unfortunately, a number of limitations have hindered the routine use of HPLC-ECD to measure 3-NO$_2$-Tyr. This analyte is present at very low levels in mammalian protein (typically 0.01–0.1% of total protein tyrosine residues) and, generally, at even lower levels in free tyrosine (personal observations). Liberation of 3-NO$_2$-Tyr from protein necessitates exhaustive protease digestion, which is invariably inefficient and difficult to control with precision (Hensley *et al.*, 1997, 1998; Shigenaga *et al.*, 1997). The oxidation potential of 3-NO$_2$-Tyr is approximately 750 mV which is accessible with HPLC-ECD experiments but with a loss of optimum sensitivity (Hensley *et al.*, 1997, 1998; Shigenaga *et al.*, 1997). Certain pre- or post-column chemical reduction strategies have been developed to convert nitrotyrosine to amino-tyrosine, which partially circumvents this shortcoming (Ohshima *et al.*, 1999; Shigenaga *et al.*, 1997). Perhaps the worst limitation is the complexity of the analyte matrix, which, even given the selectivity of HPLC-ECD, makes for a very difficult assignment of 3-NO$_2$-Tyr peaks, especially when other tyrosine oxidation products such as chlorotyrosine are present in appreciable quantities (Hensley *et al.*, 1999).

For these reasons, we have sought to exploit other analytes for use in monitoring biological RNIs. In considering what possible phenolic substrates might be present in mammalian systems at sufficient quantity to permit accumulation of significant nitrated products, our attention was drawn immediately to $\gamma$-tocopherol ($\gamma$T) (Fig. 1) (Hensley *et al.*, 1999, 2000; Williamson *et al.*, 2002, 2003a). $\gamma$-Tocopherol is a natural component of the diet, found at various levels in most vegetable and animal tissues. In fact, $\gamma$T is often present in plant oils at greater concentrations than the

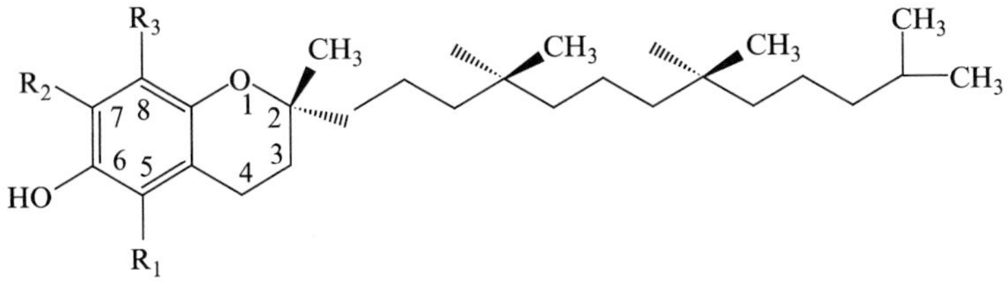

| R$_1$ | R$_2$ | R$_3$ | |
|-------|-------|-------|---|
| CH$_3$ | CH$_3$ | CH$_3$ | $\alpha$-tocopherol |
| CH$_3$ | CH$_3$ | H | $\beta$-tocopherol |
| HC | CH$_3$ | CH$_3$ | $\gamma$-tocopherol |
| H | H | CH$_3$ | $\delta$-tocopherol |
| H | H | H | Tocol |
| NO$_2$ | CH$_3$ | CH$_3$ | 5-NO$_2$-$\gamma$-tocopherol |

FIG. 1. Chemical structures of $\alpha$T, $\gamma$T, its nitration product 5-NO$_2$-$\gamma$T, and other tocopherols discussed in the text.

better-known $\alpha$-tocopherol ($\alpha$T, vitamin E) (Hensley et al., 2004; Lehmann et al., 1986; Traber et al., 1992). The only structural difference in $\alpha$T and $\gamma$T is the presence of a free position within the chroman head group of $\gamma$T where a nitration reaction is likely to occur, yielding 5-NO$_2$-$\gamma$T (Fig. 1). In humans and other mammals, the ratio of plasma $\alpha$T:$\gamma$T is regulated in part by hepatic tocopherol transfer systems that generally accumulate $\alpha$T preferentially into low-density lipoproteins (Traber et al., 1992). As a result, human plasma contains 5–10 times more $\alpha$T than $\gamma$T, although appreciable levels of $\gamma$T circulate in the bloodstream and may accumulate to near-equivalent amounts within tissue (Burton et al., 1998; Hensley et al., 2004). The biology of $\gamma$T versus $\alpha$T and a discussion of possible $\gamma$T relevance to mammalian nutrition are subjects of a 2004 review (Hensley et al., 2004) and are not reiterated extensively in this chapter.

5-NO$_2$-$\gamma$T offers several major features that recommend its investigation as a biomarker of RNI: (1) Its oxidation and hydrophobicity characteristics are very well suited for separation in both the temporal and the electrochemical dimension on HPLC ECD systems; (2) tocopherols can be extracted easily using organic solvents, yielding vastly simplified analytical matrices with very few contaminating peaks from unidentified analytes; and (3) $\gamma$T is likely to be sequestered within lipid microenvironments where NO autoxidation is known to occur hundreds of times faster than in the aqueous compartments (Liu *et al.*, 1998). Accordingly, we have exploited these features and succeeded in measuring elevations of 5-NO$_2$-$\gamma$T within cell culture exposed to NO-inducing inflammogens (Hensley *et al.*, 2000) and in the Alzheimer's disease–afflicted brain (Williamson *et al.*, 2002), where regional patterns of 5-NO$_2$-$\gamma$T correlate very well with protein-bound 3-NO$_2$-Tyr (Hensley *et al.*, 1998). Other researchers have used HPLC-ECD to measure 5-NO$_2$-$\gamma$T in sera from experimentally inflamed rats (Christen *et al.*, 2002). The remainder of the chapter provides detailed methodology suitable to the simultaneous HPLC ECD measurement of 5-NO$_2$-$\gamma$T, along with $\alpha$T and $\gamma$T.

## Materials

Most reagents are commercially available. 5-NO$_2$-$\gamma$T was synthesized from $\gamma$T by published methods (Hensley *et al.*, 2000). The general protocol is as follows. A solution of approximately 10 ml of 10 mg/ml $\gamma$T in hexane is exposed to NO$_2$ (g) or NO (g) without special precautions to exclude air from the reaction. The gas is gently introduced into the head space above the solution, which is swirled gently. The organic solution turns reddish, indicating a reaction has occurred. The reaction mixture contains 5-NO$_2$-$\gamma$T, $\gamma$-tocopheryl quinone, and unreacted $\gamma$T. 5-NO$_2$-$\gamma$T is purified by silica gel chromatography by preparative-scale thin-layer or column format. For thin-layer chromatography, the mobile phase is 95% methylene chloride plus 5% methanol. For column separations, the methanol is increased to 20%. 5-NO$_2$-$\gamma$T elutes as a yellow band near the solvent front; $\gamma$-tocopheryl quinone elutes as a reddish band with $R_f$ approximately 0.6. Solvent is removed under vacuum, and 5-NO$_2$-$\gamma$T redissolved in methanol for use as a standard; concentration is adjusted using an extinction coefficient for 5-NO$_2$-$\gamma$T of $\varepsilon = 6750\ M^{-1}\ cm^{-1}$ at 302 nm (Table I).

TABLE I

MOLAR EXTINCTION COEFFICIENTS AND PLASMA REFERENCE VALUES FOR THE TOCOPHEROL SPECIES DISCUSSED IN THE TEXT[a]

| Analyte | $\varepsilon(M^{-1}\ cm^{-1}, \lambda_{max})$ | Plasma values (mean, median) |
|---|---|---|
| $\alpha$-Tocopherol | 3060, 294 nm | 37.9 $\mu M$, 27.7 $\mu M$ |
| $\gamma$-Tocopherol | 3966, 298 nm | 2.3 $\mu M$, 1.6 $\mu M$ |
| $\delta$-Tocopherol | 3677, 298 nm | 325 n$M$, 277 n$M$ |
| Tocol | 3389, 298 nm | 125 n$M$, 226 n$M$ |
| 5-NO$_2$-$\gamma$-tocopherol | 6750, 302 nm | 19.5 n$M$, 11.0 n$M$ |

[a] Plasma values are from a survey of 81 presumptively healthy individuals in Oklahoma City, Oklahoma, and surrounding areas during 1999–2000.

## Methods

### Tocopherol Extractions

Tocopherols can be extracted quantitatively from blood plasma, other biological fluids, or tissues. For liquid samples such as plasma, the following procedure is employed. Samples are treated, at the time of acquisition, with 10 $\mu$l/ml of a 10-mg/ml ethanolic solution of butylated hydroxytoluene (BHT) to diminish autoxidative processes during prolonged storage. Each 1-ml sample is extracted in 6 ml of hexane, which can be scaled proportionately to smaller sample volumes. Hexane is dried under vacuum or under a stream of inert gas. The residue is reconstituted in methanol. Tissue samples are prepared by homogenization in 10 m$M$ sodium acetate, pH 7.0, containing 0.1% Triton X-100, and adjusted to 5 mg/ml protein concentration. Each 1-ml sample volume is extracted in a manner identical to that described earlier.

### Instrumental Analysis

HPLC is accomplished using a premium reverse-phase C$_{18}$ column such as the Toso-Haas (Montgomeryville, PA) ODS-80T$_M$ (5-$\mu$m particle size, 4.6 × 250 mm) or other column of similar manufacture. The mobile phase is 83% acetonitrile, 12% methanol, 0.2% acetic acid, and 30 m$M$ lithium acetate with a flow rate of 2 ml/min. Lithium is preferred to other salts because the acetate remains quite soluble in mobile phases of high organic solvent composition. Typical injection volume is 60 $\mu$l. Electrochemical array detection is accomplished on a 12-channel coulometric array (ESA Coulochem 5600) or similar instrument. Oxidation potentials for tocopherol measurement under these conditions are 200, 300, 400, 525, 600, 625, 650, 675, 750, 825, and 900 mV. Under these conditions, 5-NO$_2$-$\gamma$T elutes at approximately 26 min retention time, whereas $\alpha$T and $\gamma$T elute at approximately 15 min and 12 min, respectively (Fig. 2) (exact retention times vary somewhat depending on column). A photodiode array detector can be included upstream from the electrochemical array, allowing simultaneous ultraviolet–visible (UV–Vis) analysis of non-electrochemically active analytes including $\alpha$-tocopheryl quinone (Hensley et al., 2000).

Figure 2 illustrates a typical single channel recording extracted from an HPLC ECD analysis of normal human plasma. In a single run, $\alpha$T, $\gamma$T, $\delta$T, and 5-NO$_2$-$\gamma$T were well resolved and measurable within the calibration range of the assay. Low-concentration tocopherol species, $\delta$T, and tocol (Fig. 1, Table I) are also detectable in the plasma extracts (Fig. 2). Detection limits for 5-NO$_2$-$\gamma$T are approximately 5 n$M$ (<1 pmol on column) (Fig. 3). Figure 4 demonstrates hydrodynamic voltamograms of

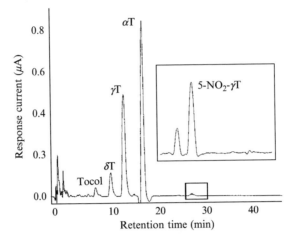

Fig. 2. High-performance liquid chromatography–electrochemical chromatogram of tocopherols extracted from normal human blood plasma. Inset (boxed) peak corresponds to 5-NO$_2$-$\gamma$T. For clarity, only the 600-mV channel is illustrated.

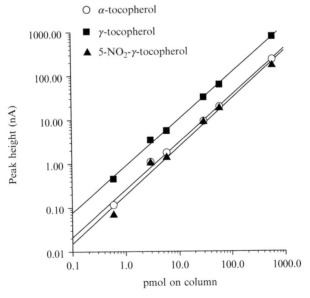

Fig. 3. Typical calibration curves for $\gamma$T and 5-NO$_2$-$\gamma$T as measured by high-performance liquid chromatography–electrochemical detection.

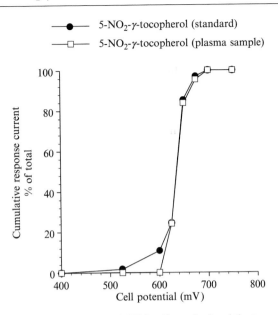

FIG. 4. Hydrodynamic voltamogram of 5-NO₂-γT standard and the temporally co-eluting peak in a human plasma extract.

plasma 5-NO$_2$-$\gamma$T and a standard 5-NO$_2$-$\gamma$T injection, at approximately the same concentration. Under reverse-phase chromatographic conditions, $\gamma$T is not resolvable from its isomer $\beta$-tocopherol. Thus, the analyst should be aware that the $\gamma$T peak is a sum of $\gamma$T plus $\beta$T; however, $\beta$T is a minor component of most biological samples, accounting for perhaps 5–20% of this peak (Handelman *et al.*, 1985).

Discussion

Biological nitration reactions are receiving increased scrutiny in basic and clinical science. In particular, lipid-phase nitration reactions may yield stably nitrated biomolecules, such as 5-NO$_2$-$\gamma$T, which may be valuable biomarkers and ι esearch tools. Accordingly, methods have been developed and validated to measure 5-NO$_2$-$\gamma$T along with precursor $\gamma$T and other salient tocopherols using HPLC with electrochemical array detection. These methods allow the economical and routine quantitation of low-nanomolar 5-NO$_2$-$\gamma$T in normal human plasma while allowing simultaneous quantitation of more prevalent $\gamma$T and $\alpha$T that are present at 100–500 fold higher concentrations. Concentrations of tocopherols including 5-$NO_2$-$\gamma$T that were measured in normal human plasma by HPLC-ECD (Table I) are in fairly close agreement to values measured by liquid chromatography–mass spectrometry (LC-MS) (Leonard *et al.*, 2003). For instance, LC-MS was used to obtain mean values for 5-$NO_2$-$\gamma$T of approximately 4 n$M$ in nonsmokers and 8 n$M$ in smokers, whereas $\alpha$T and $\gamma$T were reported to be 16 $\mu M$ and 1.8 $\mu M$, respectively, in the same study (Leonard *et al.*, 2003). In other work, HPLC ECD has been successfully applied to the analysis of 5-$NO_2$-$\gamma$T in diseased human brain tissue (Williamson *et al.*, 2002) and in sera from acutely inflamed rats (Christen *et al.*, 2002). Further application of this technique may yield valuable insights into the epidemiology of diseases involving RNIs.

## Acknowledgments

This work was supported in part by the National Institutes of Health (NS044154, HL54502), the Oklahoma Center for Advancement of Science and Technology (HR02-149R), the American Heart Association (AHA 0051176Z), and the ALS Association. We thank ESA Inc. (Chelmsford, MA) for providing some equipment used in this work.

## References

Burton, G. W., Traber, M. G., Acuff, R. V., Walters, D. N., Kayden, H., Hughes, L., and Ingold, K. U. (1998). Human plasma and tissue $\alpha$-tocopherol concentrations in response to supplementation with deuterated natural and synthetic vitamin E. *Am. J. Clin. Nutr.* **67,** 669–684.

Christen, S., Jiang, Q., Shigenaga, M. K., and Ames, B. N. (2002). Analysis of plasma tocopherols alpha, gamma, and 5-nitro-gamma in rats with inflammation by HPLC coulometric detection. *J. Lipid Res.* **43,** 1978–1985.

Cserey, A., and Gratzl, M. (2001). Stationary-state oxidized platinum microsensor for selective and on-line monitoring of nitric oxide in biological preparations. *Anal. Chem.* **73,** 3965–3974.

Handelman, G. J., Machliln, L. J., Fitch, K., Weiter, J. J., and Dratz, E. A. (1985). Oral alpha-tocopherol supplements decrease plasma gamma-tocopherol levels in humans. *J. Nutr.* **115,** 807–813.

Hensley, K., Maidt, M. L., Pye, Q. N., Stewart, C. A., Wack, M., Tabatabaie, T., and Floyd, R. A. (1997). Quantitation of protein-bound 3-nitrotyrosine and 3,4-dihydroxyphenylalanine by high-performance liquid chromatography with electrochemical array detection. *Anal. Biochem.* **251,** 187–195.

Hensley, K., Maidt, M. L., Yu, Z. Q., Sang, H., Markesbery, W. R., and Floyd, R. A. (1998). Electrochemical analysis of protein nitrotyrosine and dityrosine in the Alzheimer brain indicates region-specific accumulation. *J. Neurosci.* **18,** 8126–8132.

Hensley, K., Williamson, K. S., Maidt, M. L., Gabbita, S. P., Grammas, P., and Floyd, R. A. (1999). Determination of biological oxidative stress using high performance liquid chromatography with electrochemical detection (HPLC-ECD). *J. High Res. Chrom.* **22,** 429–437.

Hensley, K., Williamson, K. S., and Floyd, R. A. (2000). Measurement of 3-nitrotyrosine and 5-nitro-$\gamma$-tocopherol by HPLC with electrochemical detection. *Free Radic. Biol. Med.* **28,** 520–528.

Hensley, K., Benaksas, R., Bolli, P., Comp, P., Grammas, P., Hamdheydari, L., Mou, S., Pye, Q. N., Stoddard, M. F., Wallis, G., Williamson, K. S., West, M., Wechter, W. J., and Floyd, R. A. (2004). New perspectives on vitamin E: Gamma tocopherol and carboxyethylhydroxychroman derivatives (CEHC) in biology and medicine. *Free Radic. Biol. Med.* **36,** 1–15.

Jane, A., McKinnon, A., and Flanagan, A. (1985). High-performance liquid chromatographic analysis of basic drugs on silica columns using non-aqueous ionic eluents. II. Application of UV, fluorescence and electrochemical oxidation detection. *J. Chrom.* **323,** 191–225.

Kissinger, P. T., Felice, L. J., and Riggin, R. M. (1974). Electrochemical detection of selected organic components in the eluate from high-performance liquid-chromatography. *Clin. Chem.* **20,** 992–997.

Krstulovic, A. M. (1982). Investigations of catecholamine metabolism using high-performance liquid chromatography: Analytical methodology and clinical applications. *J. Chromatogr.* **229,** 1–34.

Lehmann, J., Martin, H. L., Lashley, E. L., Marshall, W., and Judd, J. T. (1986). Vitamin E in foods from high and low linoleic acid diets. *J. Am. Diet. Assoc.* **86,** 1209–1216.

Leonard, S. W., Bruno, E., Paterson, B. C., Schock, J., Atkinson, J., Bray, T. M., Cross, C. E., and Traber, M. G. (2003). 5-Nitro-gamma-tocopherol increases in human plasma exposed to cigarette smoke *in vitro* and *in vivo*. *Free Radic. Biol. Med.* **35,** 1560–1567.

Liu, X., Miller, M. J. S., Joshi, M. S., Thomas, D. D., and Lancaster, J. R., Jr. (1998). Accelerated reaction of nitric oxide with O$_2$ within the hydrophobic interior of biological membranes. *Proc. Natl. Acad. Sci. USA* **95,** 2175–2179.

Mefford, I. N. (1985). Biomedical uses of high-performance liquid chromatography with electrochemical detection. *Meth. Biochem. Anal.* **31,** 221–258.

Musch, G., DeSmet, M., and Massert, D. (1985). Expert system for pharmaceutical analysis. I. Selection of the detection system in high-performance liquid chromatographic analysis: UV versus amperometric detection. *J. Chrom.* **348,** 97–110.

Ohshima, H., Cellan, I., Chazotte, L., Pignatelli, B., and Mower, H. F. (1999). Analysis of 3-nitrotyrosine in biological fluids and protein hydrolysates by high-performance liquid chromatography using a postseparation, on-line reduction column and electrochemical detection: Results with various nitrating agents. *Nitric Oxide* **3,** 132–141.

Riggin, R. M., and Kissinger, P. T. (1977). Determination of catecholamines in urine by reverse-phase liquid chromatography with electrochemical detection. *Anal. Chem.* **49,** 2109–2111.

Rizzo, V., d'Eril, G. M., Achilli, G., and Cellerino, G. P. (1991). Determination of neurochemicals in biological fluids by using an automated high-performance liquid chromatographic system with a coulometric array detector. *J. Chromatogr.* **536,** 229–236.

Shigenaga, M. K., Lee, H. H., Blount, B. C., Christen, S., Shigeno, E. T., Yip, H., and Ames, B. N. (1997). Inflammation and NO(X)-induced nitration: Assay for 3-nitrotyrosine by HPLC with electrochemical detection. *Proc. Natl. Acad. Sci. USA* **94,** 3211–3215.

Traber, M. G., Burton, G. W., Hughes, L., Ingold, K. U., Hidaka, H., Malloy, M., Kane, J., Hyams, J., and Kayden, H. J. (1992). Discrimination between forms of vitamin E by humans with and without genetic abnormalities of lipoprotein metabolism. *J. Lipid Res.* **33,** 1171–1182.

Venkataraman, S., Martin, S. M., and Buettner, G. R. (2002). Electron paramagnetic resonance for quantitation of nitric oxide in aqueous solutions. *Methods Enzymol.* **359,** 3–18.

Williamson, K. S., Gabbita, S. P., Mou, S., West, M., Pye, Q. N., Markesbery, W. R., Cooney, R. V., Grammas, P., Reimann-Phillip, U., Floyd, R. A., and Hensley, K. (2002). The nitration product 5-nitro-gamma-tocopherol is increased in the Alzheimer brain. *Nitric Oxide* **6,** 221–227.

Williamson, K. S., Hensley, K., and Floyd, R. A. (2003a). HPLC-electrochemical and photodiode array analysis of tocopherol oxidation and nitration products in human plasma. *In* "Methods in Biological Oxidative Stress" (K. Hensley and R. A. Floyd, eds.), pp. 167–176. Humana Press, NJ.

Williamson, K. S., Hensley, K., and Floyd, R. A. (2003b). HPLC-ECD analysis of 3-nitrotyrosine in human plasma. *In* "Methods in Biological Oxidative Stress" (K. Hensley and R. A. Floyd, eds.), pp. 151–157. Humana Press, NJ.

## [18]  Hydrophobic Tyrosyl Probes for Monitoring Nitration Reactions in Membranes

*By* Hao Zhang, Joy Joseph, and B. Kalyanaraman

### Abstract

In this chapter, the upregulation of the proteasomal enzymes in response to nitric oxide (NO) in endothelial cells is discussed. The cytoprotective effects of NO are discussed in relation to iron homeostasis, oxidative stress, and the cellular proteasomal function. The role of NO-dependent proteolytic signaling mechanism is important to the overall understanding of the redox signaling in oxidative endothelial cell injury.

### Introduction

Protein nitration has been detected in several inflammatory diseases (Beckmann *et al.*, 1994; Greenacre and Ischiropoulos, 2001; Leeuwenburgh *et al.*, 1997). Nitration of tyrosine or tyrosyl residues associated with peptides and proteins is perceived to be a diagnostic marker of reactive nitrogen species (RNSs) (*e.g.*, peroxynitrite, $OONO^-/ONOOH$, and nitrogen-dioxide radical, $NO_2$) (Eiserich *et al.*, 1998; Leeuwenburgh *et al.*, 1997; Van der Vliet *et al.*, 1995). Nitration of protein tyrosyl group has been postulated to modulate protein metabolism, protein function, and signal transduction pathways (Greenacre and Ischiropoulos, 2001). A number of reports indicate that hydrophobic protein tyrosyl residues are preferentially nitrated by reactive RNSs (Greenacre and Ischiropoulos, 2001; Viner *et al.*, 1999; Yamamoto *et al.*, 2002). Thus, *in vitro* mechanistic studies using "free" tyrosine in the aqueous phase may not adequately reflect the mechanisms of nitration of protein tyrosyl residues *in vivo* (Pfeiffer and Mayer, 1998; Pfeiffer *et al.*, 2000, 2001). In addition, the efficiency of nitration of tyrosine induced by peroxynitrite in the aqueous phase has been reported to be remarkably low (1–4%) (Goss *et al.*, 1999; Ohshima *et al.*, 1999; Souza

METHODS IN ENZYMOLOGY, VOL. 396                    0076-6879/05 $35.00
          DOI: 10.1016/S0076-6879(05)96018-7

*et al.*, 1999). In contrast, nitration of membrane-incorporated $\gamma$-tocopherol (a membrane-associated chromanol) (Goss *et al.*, 1999; Jiang *et al.*, 2000; Zhang *et al.*, 2001) is increased threefold to fivefold. A plausible reason for the increased efficiency of lipid-phase nitration is the greater membrane permeability of RNS coupled with decreased lateral diffusion of membrane-bound tyrosyl free radical, making radical–radical dimerization unlikely (Zhang *et al.*, 2001). Although the chemistry of tyrosine nitration in aqueous solution has been well established, very little information exists regarding the chemistry of tyrosine nitration in the hydrophobic interior of membranes or hydrophobic domains of proteins (Beckmann *et al.*, 1994; Goldstein and Czapski, 1998; Goldstein *et al.*, 2000; Lymar and Hurst, 1995; Lymar *et al.*, 1996; Pfeiffer and Mayer, 1998; Pfeiffer *et al.*, 2000; Uppu *et al.*, 1996; Van der Vliet *et al.*, 1995). Both $ONOO^-/ONOOH$ and $\cdot NO_2$ freely cross the lipid membrane through an ion transport channel or by passive diffusion at rates similar to that of water (Denicola *et al.*, 1998; Groves, 1999; Marla *et al.*, 1997). As many protein tyrosyl residues exist in the hydrophobic regions of membrane proteins or in the interior of proteins in a biological setting, it is essential to understand the physicochemical basis for nitration reactions in hydrophobic membranes.

To understand the chemistry of nitration reactions in a hydrophobic environment, we established *in vitro* models of bilayer membrane with membrane-incorporated tyrosine analogues [*i.e.*, *N*-*t*-BOC L-tyrosine *t*-butyl ester (BTBE) and several transmembrane peptides with tyrosyl residues at different positions]. Results of those studies indicate that $ONOO^-/ONOOH$ at low concentrations preferentially nitrates tyrosyl residues in a hydrophobic environment (Zhang *et al.*, 2001, 2003). Results also indicate that nitration but not oxidation of tyrosyl residues is a major reaction pathway for $OONO^-/ONOOH$ in membranes and show that dityrosine formation is not a significant reaction process for tyrosyl radicals in membranes in which lateral diffusion of tyrosyl radicals is hindered. We also demonstrate that the nitration yields of tyrosyl peptides are dependent on both the hydrophobicity of environment in which they reside and on the nitrating species. Both $ONOO^-$ and $\cdot NO_2$ prefer to nitrate the tyrosyl group in a hydrophobic environment rather than in the aqueous phase. However, $ONOO^-$ induced greater nitration of the tyrosyl group buried deeper in the bilayer membrane. In contrast, $\cdot NO_2$ nitrated the tyrosyl group located closer to the surface of the membrane (Zhang *et al.*, 2003). BTBE was also used successfully to assess the antioxidant and antinitration effects of polyphenolics (Schroeder *et al.*, 2001) and to monitor nitration/oxidation reactions inside the red blood cells caused by $\cdot NO_2$ exposure or $ONOO^-$.

Methods

*Syntheses of Membrane-Incorporated Tyrosine Analogue*

N-t-*BOC* L-*tyrosine* t-*butyl ester*: To incorporate the tyrosyl group into a hydrophobic membrane, we invested the stability of several tyrosine derivatives under our experimental conditions. Initially, we studied the stability of tyrosine methyl ester, ethyl ester, and *t*-butyl ester in 0.1 *M* phosphate buffer, pH 7.4. Incubation of those tyrosine esters in buffer showed that both tyrosine methyl and ethyl esters were not suitable for this purpose because they underwent a slow hydrolysis of the ester bond to form the corresponding tyrosine. Although the tyrosine butyl ester was resistant to hydrolysis, its efficiency of incorporation into liposome was only 53%. To increase the hydrophobicity of tyrosine esters, we synthesized a new compound, *N-t*-BOC L-tyrosine *t*-butyl ester (BTBE) (Fig. 1). Analysis of its stability and incorporation efficiency showed that BTBE was totally incorporated into liposomes without undergoing hydrolysis for 40 h. Thus, BTBE appears to be an ideal hydrophobic tyrosyl probe for investigating nitration and oxidation reactions in membranes.

L-Tyrosine *t*-butyl ester (237 mg, 1 mmol) was dissolved in 10 ml of dioxane and kept in an ice bath. Sodium carbonate (106 mg) was dissolved in 10 ml of water and added to the dioxane solution followed by the addition of a solution of 240 mg *t*-butyl pyrocarbonate in 5 ml of dioxane. The mixture was stirred in an ice bath for 1 h and acidified to pH 2–3 with hydrochloric acid. The mixture was then extracted with 3 × 20 ml ethyl acetate, and the extract was treated with anhydrous $Na_2SO_4$ and evaporated to dryness. The resulting white solid (yield: 220 mg) was recrystallized from a 1:1 mixture of ethyl acetate and hexane. Purity was ascertained by high-performance liquid chromatography (HPLC).

*Peptide synthesis and purification*: Figure 1 shows the sequences of the tyrosyl peptides used in this study. The peptides were acetylated and amidated in order to stabilize their $\alpha$-helical secondary structure. The peptides were chemically synthesized using a standard Fmoc solid-phase peptide synthesis chemistry on an Advanced Chemtech Model 90 synthesizer (Louisville, KY). Rink Amide MBHA resin (loading 0.72 mmol/g) was used as a solid support. Fmoc-protected amino acids were coupled as 1-hydroxybenzotriazole (HOBt) esters. All amino acids were double-coupled using HOBt/diisopropylcarbodiimide (DIC). The following steps

*N-t-*BOC L-tyrosine *t*-butyl ester
(BTBE)

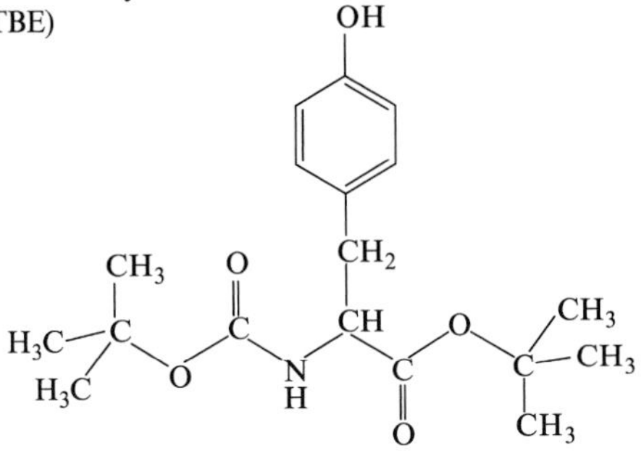

Y-4 transmembrane peptide

Ac-NH-KKAYALALALALALALALALALALALAKK-CONH$_2$

Y-8 transmembrane peptide

Ac-NH-KKALALAYALALALALALALAKK-CONH$_2$

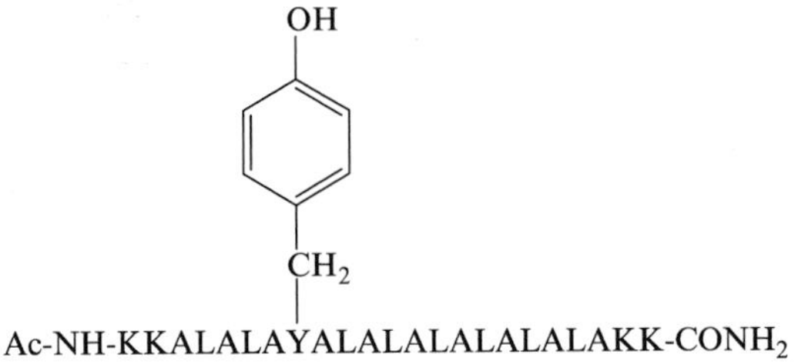

Y-12 transmembrane peptide

Ac-NH-KKALALALALAYALALALALAKK-CONH$_2$

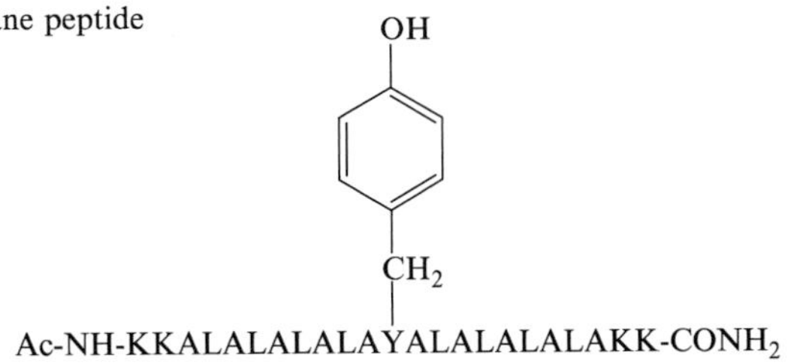

FIG. 1. The structures of hydrophobic *N-t-*BOC L-tyrosine *t*-butyl ester and tyrosyl transmembrane peptides.

were performed in the reaction vessel for each double-coupling: deprotection of the Fmoc group with 20% piperidine in $N$-methylpyrrolidone (NMP) for 30 min (two times), three NMP washes, two dichloromethane (DCM) washes, first coupling for 1 h with fivefold excess of Fmoc amino acid in 0.5 $M$ HOBt and 0.5 $M$ DIC, second coupling using a fresh addition of the same reagent for another 1 h, three NMP washes, and two DCM washes. Final acetylation was performed using an acetic anhydride/HOBt/DIC for 30 min (twice). The resin was washed twice with DCM and thrice with methanol, and then dried under vacuum before cleavage. The peptide was deprotected and cleaved from the resin with 90% trifluoroacetic acid (TFA) containing triisopropylsilane for 3 h at room temperature. The resin was removed by filtration and washed with TFA, and the combined TFA filtrates were evaporated to dryness under a steam of dry $N_2$ gas. The oily residue was washed three times with cold ether to remove scavengers and the dry crude peptide was dissolved in acetonitrile/$H_2O$ (1:1) and lyophilized. The crude peptides were purified by a semipreparative reverse-phase HPLC (RP-HPLC) on a Vydac RP-C8 (10 × 250 mm) column (Hesperia, CA) using a $CH_3CN$/water gradient (20–100% $CH_3CN$ over 50 min) containing 0.1% TFA at a flow rate of 3 ml/min with detection at 220 nm. Elution of the peptides generally occurred between 50% and 70% $CH_3CN$. Purity and homogeneity of the peptides were cross-checked by an analytical RP-HPLC and the purified peptides (~95% pure) characterized by LC-mass spectrometry (MS) on an Agilent 1100 series LC-MS.

## Syntheses of Nitration and Oxidation Products of Membrane-Incorporated Tyrosine Analogues

3-Nitro $N$-$t$-BOC L-tyrosine $t$-butyl ester ($NO_2$-BTBE): Figure 2A and B show the nitration and oxidation products of BTBE. Tetranitromethane (60 mg) was mixed with 50 mg BTBE in 4 ml methanol. The reaction mixture was kept in the dark and stirred overnight at room temperature. The mixture was then mixed with 4 ml phosphate buffer (0.1 $M$, pH 7.4) and extracted with 2 × 8 ml chloroform. The chloroform layer was removed and dried under a stream of nitrogen gas. The crude product was redissolved in methanol and further purified in a C-18 preparative column (Beckman, 10 × 250 mm, 5 $\mu$m) using a methanol gradient, increasing the concentration from 70% to 78% in 100 min. The retention times for BTBE and its nitro derivative (i.e., $NO_2$-BTBE) under these conditions were 35 and 56 min, respectively. Fractions containing $NO_2$-BTBE were pooled, extracted with chloroform, and

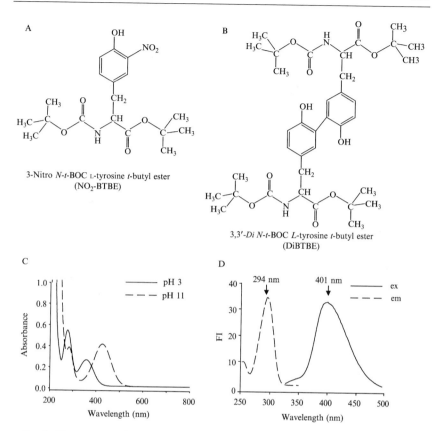

FIG. 2. The spectra and structure of BTBE oxidation/nitration product. (A) The structure of NO$_2$-BTBE; (B) the structure of DiBTBE; (C) the ultraviolet–visible (UV-vis) spectra of NO$_2$-BTBE (0.1 m$M$) taken at different pH values; and (D) fluorescence spectrum of DiBTBE obtained in 70% methanol and 30% phosphate buffer (pH 3, 15 m$M$).

dried under nitrogen gas. The purity of NO$_2$-BTBE was determined in a C-18 analytical HPLC. Isocratic elution at 70% methanol for 30 min showed a single peak (11 min). No additional peak was detected on elution with 100% methanol. The ultraviolet–visible (UV–vis) spectral characteristics of the pure product were similar to those of NO$_2$-Tyr. The UV-vis spectrum of NO$_2$-BTBE taken in 70% methanol containing 15 m$M$ phosphate (pH 3.0) showed an absorbance maximum at 354 nm. This absorbance maximum shifted to 424 nm (extinction coefficient = 4000 $M^{-1}$ cm$^{-1}$) in 70% methanol containing 0.01 $M$ NaOH (Fig. 2C and D).

3,3'-Di *N-t*-BOC L-tyrosine *t*-butyl ester (DiBTBE): BTBE (60 mg) was suspended in 80-ml reaction solution containing 15% methanol and 50 m$M$ acetate buffer, pH 4.7. After stirring for 30 min, 400 $\mu$l hydrogen peroxide (30%) and 200 units horseradish peroxidase were added to BTBE solution and kept stirring at room temperature for 90 min. The reaction mixture was extracted twice with an equal volume of chloroform. The chloroform extracts were combined and dried under a stream of nitrogen gas. Dried extract was redissolved in methanol and separated in a C-18 preparative column (Beckman, 10 × 250 mm, 5 $\mu$m) preequilibrated with 77% methanol in 15 m$M$ phosphate, pH 3.0, and eluted with a linear increase of methanol concentration from 77% to 80% in 70 min. The retention times for BTBE and its dimer (*i.e.*, DiBTBE) were 22 and 60 min, respectively. The fractions containing DiBTBE were pooled, extracted by chloroform, and dried by nitrogen gas. The purity of DiBTBE was verified using HPLC.

Synthesis of nitro peptides (NO$_2$-peptides): 10 mg/ml peptide dissolved in methanol/1% TFA was mixed with nitronium tetrafluoroborate (5 mg/ml). After 20 min, the product was purified by a preparative HPLC (C-18, 250 × 10 mm), preequilibrated with 50% CH$_3$CN in 0.1% TFA. The NO$_2$-peptide was eluted by a linear CH$_3$CN gradient (0.025%/min). Collected nitro peptide peak was dried under a stream of nitrogen and redissolved in methanol for analysis.

## Incorporation of BTBE and Tyrosyl Peptide into DLPC Liposome

Methanolic solution of BTBE or tyrosyl peptides was added to DLPC dissolved in methanol. The mixture was then dried under a stream of nitrogen gas and kept in a vacuum desiccator overnight. Multilamellar liposomes were formed by thoroughly mixing the dried lipid in phosphate buffer (100 mM, pH 7.4) containing DTPA (100 $\mu M$). Unilamellar liposomes were prepared by five cycles of freeze-thawing using liquid nitrogen, followed by five cycles of extrusion through a 0.2-$\mu$m polycarbonate filter (Nucleopore, Pleasanton, CA) in an extrusion apparatus (Lipex Biomembranes, Inc., Vancouver, BC). Prepared liposomes were stored at 4° and used within 2 weeks.

## Preparation of Samples for HPLC Analysis

*Extraction of BTBE*: A 200-$\mu$l reaction mixture containing liposomes was mixed with 200 $\mu$l methanol and vortexed for 1 min to dissolve the liposomes. The resulting solution was mixed with 400 $\mu$l chloroform and 80 $\mu$l NaCl (5 $M$), vortexed for 2 min, centrifuged at 5000 rpm for 10 min, and the top layer (aqueous phase) was removed. The chloroform layer was dried under a stream of nitrogen gas and dissolved in 15 m$M$ phosphate buffer, pH 3.0/70% methanol for HPLC analysis.

*Preparing tyrosyl peptides for HPLC*: A 30 $\mu$l of reaction mixture was diluted with 600 $\mu$l distilled water, centrifuged at 12800 rpm for 120 min, and the pelleted liposome was dried by a speed vac. The dried liposome was dissolved in methanol containing 2.5% TFA for HPLC analysis.

## HPLC Analyses of Nitration/Oxidation Products

*HPLC analysis of BTBE products*: The extract in 70% methanol and 15 m$M$ phosphate buffer (pH 3.0) was injected into the HPLC with a C-18 RP column (Partisil ODS-3, 250 × 4.6 mm, Alltech), eluted by a linear increase of methanol (70–75%) in 15 m$M$ phosphate (pH 3.0) for 30 min. The UV detection at 280 nm was used to monitor BTBE and $NO_2$-BTBE. Fluorescence detection at 294 nm (excitation)/401 nm (emission) was used to monitor DiBTBE. BTBE, its corresponding nitro derivative, and DiBTBE were eluted at 8.3, 11, and 23 min, respectively.

*HPLC analysis of tyrosyl peptides products*: Typically, 2 $\mu$l of sample was injected into a capillary HPLC system (HP1100) with a cap C-18 column (150 × 2 mm) equilibrated with 50% $CH_3CN$ in 0.5% TFA. The peptide and its nitration product were separated by a linear increase of $CH_3CN$ concentration (0.1%/min) at a flow rate of 15 $\mu$l/min. The elution was monitored by a variable UV detector at 280 and 350 nm. The peptide and the $NO_2$-peptide were eluted at 12 and 17.5 min, respectively.

## Results and Discussion

To probe nitration reactions in a hydrophobic environment, we synthesized a series of hydrophobic tyrosine probes that were incorporated into a hydrophobic membrane and established liposome models with those probes. We verified the intramembrane locations of tyrosyl residues using various techniques including fluorescence quenching by membrane-incorporated nitroxide and EPR spin labeling. We also synthesized the authentic nitration and oxidation products of those hydrophobic tyrosyl probes and HPLC methods for analysis of nitration and oxidation yields. Figure 3 summarizes the overall experimental design for investigating nitration reactions of hydrophobic membrane-incorporated tyrosyl probes.

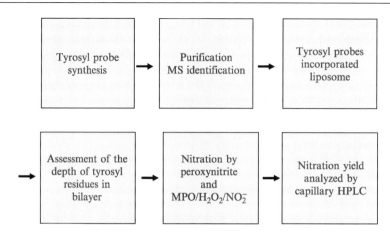

FIG. 3. The experimental design of the *in vitro* model for investigating the tyrosyl nitration/oxidation in a hydrophobic membrane.

## Characterization of Hydrophobic Tyrosyl Analogue and Transmembrane Peptides

BTBE, DiBTBE, $NO_2$-BTBE, tyrosyl transmembrane peptide, and their nitration products ($NO_2$-peptides) were characterized by MS, UV, and fluorescence analyses. Mass spectral analysis of purified compounds showed that all compounds had the molecular weight as expected (Zhang *et al.*, 2001, 2003). Both BTBE and tyrosyl transmembrane peptides had absorption maximum around 280 nm, which indicated the presence of the tyrosyl group in both BTBE and tyrosyl peptides. The UV–vis spectral characteristics of the pure $NO_2$-BTBE and $NO_2$-peptides (Fig. 2C) were also similar to those of $NO_2$-Tyr. The UV–vis spectrum of $NO_2$-BTBE taken in a 70% methanol solution containing 15 m$M$ phosphate (pH 3.0) showed an absorbance maximum at 354 nm. This absorbance maximum shifted to 424 nm (extinction coefficient = 4000 $M^{-1}$ cm$^{-1}$) in 70% methanol containing 0.01 $M$ NaOH. UV–vis also showed the typical nitrotyrosine spectrum. After adding NaOH, the 350-nm absorption peak in MeOH was pH sensitive shifted to 430 nm with an extinction coefficient of 4100 $M^{-1}$ cm$^{-1}$. DiBTBE, a dimer of BTBE through dityrosine formation, showed an absorption maximum at 290 nm, similar to that was reported for dityrosine (DiTyr). The DiBTBE exhibits a fluorescent spectrum (Fig. 2D) similar to that of authentic DiTyr. The UV–vis and fluorescent characteristics of the relevant compounds are summarized in Table I.

TABLE I
THE UV–VIS AND FLUORESCENT CHARACTERISTICS OF THE RELEVANT COMPOUNDS

| Compound | Absorption maximum (nm) | Solvent | Extinction coefficient $(M^{-1} cm^{-1})$ | Fluorescence | |
| | | | | Ex (nm) | Em (nm) |
|---|---|---|---|---|---|
| BTBE | 278 | Methanol | 1200 | | |
| DiBTBE | 290 | Methanol | 5200 | 294 | 401 |
| NO$_2$-BTBE | 424 | 70% methanol 0.1 $M$ NaOH | 4000 | | |
| NO$_2$-peptides | 430 | Methanol 0.05 $M$ NaOH | 4100 | | |

*Location of BTBE and Tyrosyl Peptides in the Bilayer: Fluorescence and Fluorescence Quenching Experiments*

The depths of tyrosyl groups of BTBE and peptides anchored in bilayer membrane were analyzed by fluorescence and fluorescence quenching experiments. Compared to the fluorescence spectrum of BTBE taken in 70% MeOH and 15 m$M$ phosphate buffer, the BTBE fluorescence was significantly enhanced and slightly blue shifted when BTBE was incorporated DLPC liposomes, indicating that its phenolic group was located at the hydrophobic phase of the bilayer (Fig. 4A). This is consistent with the very poor solubility of BTBE in aqueous solvents. To assess the localization of the tyrosine residues in these transmembrane peptides with respect to the membrane bilayer, we examined their intrinsic fluorescence properties. Direct measure of the intrinsic fluorescence of this series of peptides in liposome showed that the fluorescence intensity increased in the order of Y-12 > Y-8 > Y-4, indicating a progressively increasing hydrophobic environment (membrane depth) for the tyrosine residues (Fig. 4B).

To further confirm these results, we incorporated nitroxide-labeled phospholipids into liposome and investigated the quenching effect of nitroxide on tyrosyl fluorescence (Fig. 5). BTBE fluorescence in liposomes was strongly quenched by nitroxide-labeled phospholipids in the order 5-PCSL > 16-PCSL > 12-PCSL (Fig. 4C). That the extent of fluorescence quenching by 16-PCSL falls between 5 and 12 PCSL is consistent with the known propensity of nitroxides located at the 16 position to fold back toward the membrane surface (Feix *et al.*, 1984, 1987). These results show that BTBE is located in the lipid phase of the bilayer and is distributed throughout the bilayer, with the highest concentration present near the glycerol backbone (*i.e.*, at the level of 5-PCSL).

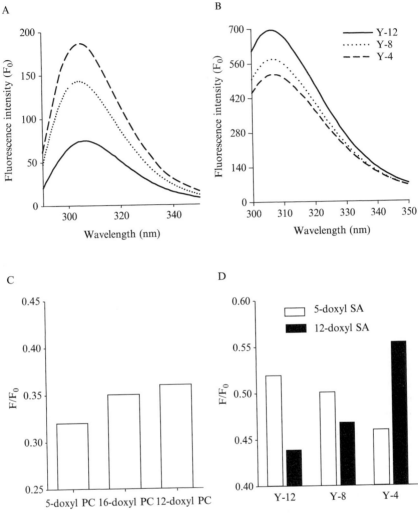

Fig. 4. Fluorescence and fluorescence-quenching experiments. (A) Fluorescence spectra of BTBE (0.3 m$M$) was obtained in methanol (—), cyclohexane (...... ...), or in liposomes (30m$M$) (-----) in phosphate (0.1 $M$) buffer containing diethylenetriaminepentaacetic acid (DTPA) (0.1 m$M$). (B) Fluorescence spectra of tyrosyl peptides (0.3 m$M$) incorporated into liposomes (30 m$M$ DLPC in 0.1 $M$ phosphate buffer containing (0.1 $\mu M$ DTPA) were obtained using an excitation wavelength of 278 nm. (C) The effect of 5-doxyl, 12-doxyl, or 16-doxyl PCSL (3 m$M$) on BTBE (0.3 m$M$) incorporated into DLPC liposomes (30 m$M$). (D) The effect of 5-doxyl and 12-doxyl stearic acid nitroxides (1 m$M$) on the ratio of the fluorescence intensity of tyrosyl peptides in the absence ($F_0$) and in the presence of nitroxide (F).

5-doxyl PC

12-doxyl PC

16-doxyl PC

5-doxyl SA

12-doxyl SA

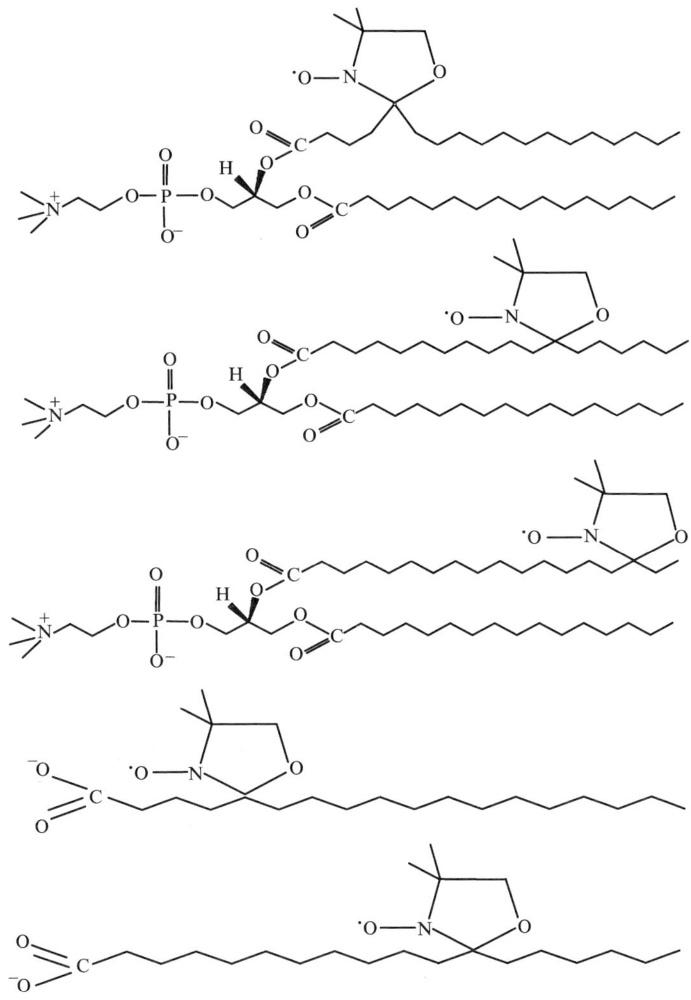

FIG. 5. Structures of membrane anchored nitroxides.

The fluorescence of tyrosyl transmembrane peptides (Y-4, Y8, and Y12) were also quenched by nitroxide labeled fatty acids. As shown in Fig. 4D, a similar order of sensitivity to fluorescence quenching by nitroxides attached at the 5-position of stearic acid (5-doxyl SA) and at the 12-position of stearic acid (12-doxyl SA) was observed. Fluorescence quenching by 5-doxyl SA, which localizes near the glycerol backbone of phospholipid bilayers, was greater for Y-4 than for Y-8 or Y-12 (Fig. 4D),

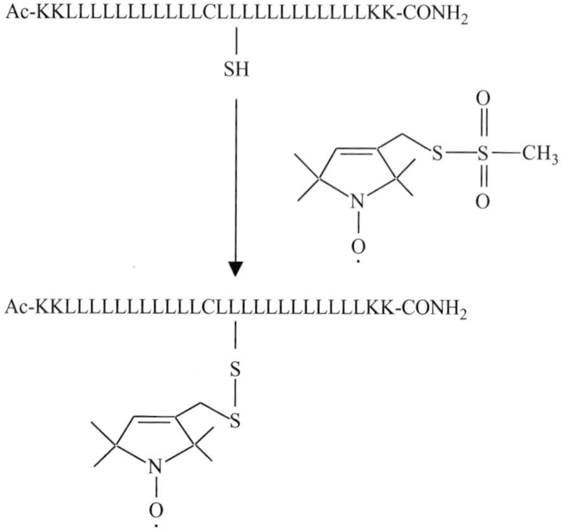
Ac-KKLLLLLLLLLLLCLLLLLLLLLLLKK-CONH2
SH
O
S
S
CH3
O
N
O
Ac-KKLLLLLLLLLLLCLLLLLLLLLLLKK-CONH2
S
S
N
O

FIG. 6. Spin labeling of membrane-anchored cysteinyl probe.

further indicating that the tyrosine residue of Y-4 resides just below the membrane surface. On the other hand, fluorescence quenching by 12-doxyl SA was greater for Y-12 than for Y-4 indicating that Y-8 and Y-12 are located deeper in the bilayer. These data are all consistent with the expected transmembrane orientation of these peptides (de Planque *et al.*, 1999, 2001; Harzer and Bechinger, 2000; Zhang *et al.*, 1995a,b) and indicate that incorporation into a repeating leucine-alanine motif anchored at each end by lysine residues is an effective strategy for positioning tyrosine residues across the lipid bilayer. Alternatively, the hydrophobic sulfhydryl group can be labeled to give a nitroxide probe specifically located in the bilayer membrane (Fig. 6). ESR microwave power saturation data of these nitroxides indicate the approximate location of specific cysteinyl probes in the bilayer.

### *Analysis of Reaction Products by HPLC*

We have extracted BTBE and its oxidation/nitration products (DiBTBE and $NO_2$-BTBE) from the reaction mixture using methanol/chloroform extraction. After the addition of 5 *M* NaCl into the MeOH/aqueous solution mixture (1:5 v/v), BTBE, DiBTBE, and $NO_2$-BTBE were successfully extracted into chloroform phase. The extraction efficiencies for BTBE, $NO_2$-BTBE, and DiBTBE using the procedure we developed were 95.3, 99.4, and 101%, respectively. For analysis of nitration yield of tyrosyl transmembrane peptide, we isolated the peptide containing liposome by centrifugation and dried the pelleted liposome. The dried liposome was readily dissolved in methanol containing 2.5% TFA that is suitable for HPLC analysis. Using this procedure, the recoveries of peptide and nitropeptide were measured to be greater than 95%.

The BTBE extract in 70% methanol and 15 m$M$ phosphate buffer (pH 3.0) were injected into the HPLC with a C-18 RP column (Partisil ODS-3, 250 × 4.6 mm, Alltech), eluted by a linear increase of methanol (70–75%) in 15 m$M$ phosphate (pH 3.0) for 30 min. The UV detection at 280 nm was used to monitor BTBE, and $NO_2$-BTBE and fluorescence detection at 294 (excitation)/401 nm (emission) was used to monitor DiBTBE. BTBE, its corresponding nitro derivative, and dimer product were eluted at 8.3, 11, and 23 min (Fig. 7A and B), respectively. Repeated injections of authentic compounds showed no noticeable loss of these compounds under these conditions for up to 40 h at room temperature.

Peptide nitration yield was analyzed by a capillary HPLC system. A 2-$\mu$l sample was injected into a capillary HPLC system (HP1100) with a capillary C-18 column (150 × 2 mm) equilibrated with 50% $CH_3CN$ in 0.5% TFA. The peptide and its nitration product were separated by a linear increase of $CH_3CN$ concentration (0.1%/min) at a flow rate of 15 $\mu$l/min. The elution was monitored by a variable UV detector at 280 and 350 nm. Peptide and $NO_2$-peptide were eluted at 12 and 17.5 min (Fig. 7C), respectively.

### Studies of Nitration/Oxidation Reaction in Membrane Using BTBE and Tyrosyl Transmembrane Peptides

Although the chemistry of tyrosine nitration in aqueous solution has been well established, very little information exists regarding the chemistry of tyrosine nitration in the hydrophobic interior of membranes or hydrophobic domains of proteins. Previously, we showed that both SIN-1 and peroxynitrite preferentially nitrated membrane-bound $\gamma$-tocopherol as compared to tyrosine in the aqueous solution (Goss *et al.*, 1999). $\gamma$-Tocopherol underwent nitration at the 5-position of the chromanol ring to form 5-$NO_2$-$\gamma$-tocopherol. This finding raised important questions with respect to the mechanisms of phenolic nitration by peroxynitrite in the hydrophobic phase. To understand the nitration reaction in a hydrophobic environment, we compared the nitrating and oxidizing reactions of peroxynitrite using our newly synthesized tyrosine probes.

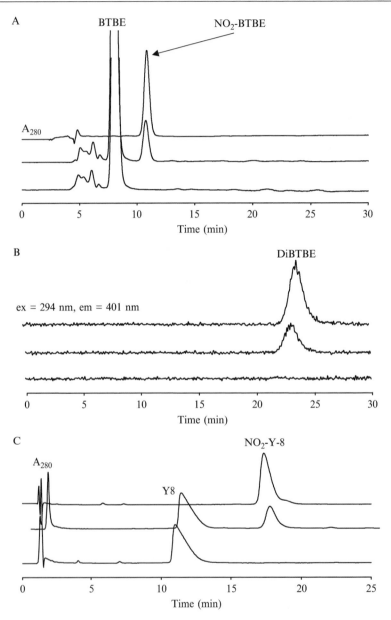

FIG. 7. High-performance liquid chromatography (HPLC) analysis of nitration and oxidation products of membrane-anchored tyrosyl probes. (A) HPLC analysis of $NO_2$-BTBE. (Top) A typical HPLC trace of authentic $NO_2$-BTBE (25 $\mu M$), (middle) the incubation mixture consisted of BTBE (0.3 m$M$) in DLPC (30 m$M$) liposomes and peroxynitrite (50 $\mu M$),

*Bolus addition of peroxynitrite*: A bolus addition of peroxynitrite (50 $\mu M$) to DLPC liposomes containing BTBE (300 $\mu M$) resulted in the formation of a product eluting at 11 min (Fig. 7A, middle trace). This peak was assigned to the nitration product, $NO_2$-BTBE, by comparison with the authentic standard (Fig. 7A, top trace). HPLC fluorescence analysis of the same mixture indicated a product with a retention time of 23 min (Fig. 7B, middle trace), which was attributed to a dimeric oxidation product of BTBE, based on the HPLC profile of authentic DiBTBE (Fig. 7B, top trace). Under identical conditions, bolus addition of peroxynitrite to a phosphate buffer containing tyrosine (300 $\mu M$) also yielded both the nitrated and the oxidized products (i.e., $NO_2$-Tyr and DiTyr) as monitored by HPLC UV and fluorescence detection. The results showed that nitration yield of BTBE in membrane (5.3 $\mu M$) was higher than the nitration yield of tyrosine in aqueous solution (4.2 $\mu M$) (Fig. 8A). On the other hand, the oxidation product of BTBE and tyrosine is significantly lower than nitration reaction (Fig. 8B). To further confirm that the nitration yield of tyrosyl residues could be enhanced by the hydrophobicity of its environment, we analyzed the nitration yields of tyrosyl residues anchored at different depths in DLPC liposomes. HPLC analysis (Fig. 7C) showed that the nitration yields of tyrosyl peptides in membranes were significantly higher than that of tyrosine in aqueous phase. Results showed that the formation of nitrated peptides increased as the depth of tyrosyl residues in the bilayer was increased ($NO_2$-Y-4, 5.8 $\mu M$; $NO_2$-Y-8, 6.8 $\mu M$ and $NO_2$-Y-12, 7.4 $\mu M$) (Fig. 8A). This result indicated that hydrophobicity can enhance the nitration reaction induced by a bolus addition of peroxynitrite.

*Slow infusion of peroxynitrite*: Previous reports suggest that the rate at which peroxynitrite is added to a solution containing tyrosine can drastically alter the nitration yield (Goldstein *et al.*, 2000; Lymar *et al.*, 1996;

---

and (bottom) same as above except without peroxynitrite. HPLC traces were monitored at 280 nm. (B) HPLC analysis of DiBTBE (top) authentic DiBTBE (2 $\mu M$), (middle) and (bottom) traces correspond to incubations identical to (middle) and (bottom) traces shown in (A). HPLC traces were followed by fluorescence analysis, and (C) HPLC traces of nitrated tyrosyl peptides in liposome. Y-8 peptide (0.3 m$M$) in 30 m$M$ DLPC liposome was reacted with peroxynitrite in 0.1 $M$ phosphate buffer containing 100 $\mu M$ DTPA and analyzed by capillary HPLC. HPLC traces were monitored at 280 nm. (top) The authentic 20 $\mu M$ $NO_2$-Y-8 peptide, (middle) the nitrated reaction product from the reaction between peroxynitrite and Y-8 peptide, and (bottom) same as above but in the presence of decomposed peroxynitrite. HPLC traces monitored at 280 nm that showed both original Y-8 peptide (12 min) and nitration product (17.5 min).

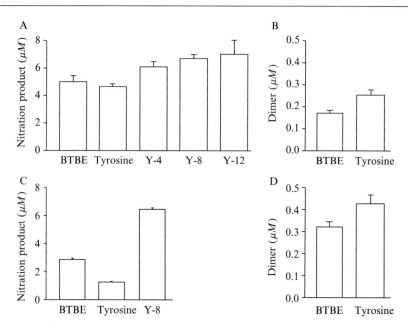

Fig. 8. Nitration and oxidation of tyrosine in aqueous phase and BTBE or tyrosyl peptides in membranes by peroxynitrite. (A) Bolus addition: BTBE (0.3 mM) or tyrosyl peptide (0.3 mM) in DLPC (30 mM) liposomes was incubated with peroxynitrite (50 $\mu M$) in sodium phosphate buffer (0.1 $M$, pH 7.4) and analyzed by HPLC (n = 3 ± SD). The comparable aqueous reaction was performed using tyrosine (0.3 mM). (B) The oxidation products under conditions same as (A). (C) Slow infusion: BTBE (0.3 mM) or Y-8 peptides in DLPC liposomes (30 mM) were infused peroxynitrite at 5 $\mu M$/min (final peroxynitrite concentration, 50 $\mu M$) in phosphate buffer (0.1 $M$, pH 7.4) and analyzed by HPLC. The corresponding aqueous reaction was performed using tyrosine (0.3 mM), and (D) the oxidation products under same condition as (C).

Ohlin *et al.*, 1998; Pfeiffer and Mayer, 1998; Pfeiffer *et al.*, 2000; Uppu *et al.*, 1996). Compared to the bolus addition of peroxynitrite, a slow infusion of the same amount of peroxynitrite into tyrosine solution dramatically decreased the nitration yield while increasing the oxidation yield. The addition of peroxynitrite in a slow infusion to DLPC liposomes containing BTBE or to phosphate buffer containing tyrosine gave, however, a different product profile. As shown in Fig. 8C, when peroxynitrite was infused at 5 $\mu M$/min, the yield of $NO_2$-BTBE was about twofold higher than that of $NO_2$-Tyr. However, in both cases, the yield of the oxidation products (i.e., DiBTBE and DiTyr) was considerably less than that of the nitration products. More importantly, even during slow infusion of peroxynitrite, the yield of $NO_2$-BTBE was nearly 10 times higher than that of DiBTBE,

the oxidation product of BTBE (Fig. 8D). A slow infusion of peroxynitrite
(5 $\mu M$/min; final concentration, 50 $\mu M$) into DLPC liposome containing
Y-8 peptide also induced a fivefold increase in nitration (6.2 $\mu M$) compared
to the nitration of tyrosine (1.2 $\mu M$) in aqueous phase (Fig. 8C). These data
indicate that a slow infusion of peroxynitrite induced nitration of the
tyrosine probe in bilayer membrane that was much more efficient than
the nitration of tyrosine in aqueous solution under same conditions. These
findings indicate that even at low infusion rates of peroxynitrite, nitration
of tyrosine is more efficient than the oxidation reaction.

*SIN-1–induced nitration*: To investigate the effect of peroxynitrite
co-generated *in situ* on nitration/oxidation of BTBE in membranes and
tyrosine in the aqueous solution, DLPC liposomes containing BTBE
(300 $\mu M$) or tyrosine (300 $\mu M$ in phosphate buffer) were incubated with
SIN-1 (1000 $\mu M$). Nitration of tyrosine in aqueous phase by SIN-1, which
generates equal amounts of NO and superoxide, gave very low yields of
nitrotyrosine ($\sim$0.5 $\mu M$). Figure 9A shows that the nitration of BTBE in
the membrane was greatly enhanced compared to the nitration of tyrosine
in the aqueous solution. Again the overall oxidation of tyrosine and BTBE
in the presence of peroxynitrite was still a relatively minor reaction (Fig.
9B). These results strongly suggest that nitration but not oxidation of
tyrosyl residue is a dominant process of peroxynitrite-dependent reactions,
especially in the hydrophobic membrane regions. DLPC liposomes containing tyrosyl peptides (300 $\mu M$) or tyrosine (300 $\mu M$) in phosphate buffer
were incubated with SIN-1 (1 mM) (Fig. 9A). Under similar conditions,
the nitration yield of Y-12 peptide in membrane was about 18 times higher
(9 $\mu M$) than that of tyrosine in aqueous solution (Fig. 9A). Incubation of

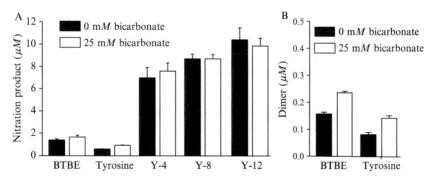

FIG. 9. Nitration of tyrosine in aqueous solution and tyrosyl groups in membranes by SIN-
1. BTBE (0.3 mM) or tyrosyl peptide in DLPC (30 mM) liposomes was incubated with SIN-1
and bicarbonate (0 or 25 mM) in sodium phosphate buffer (0.1 $M$, pH 7.4) and analyzed
by HPLC (n = 3 ± SD). The comparable aqueous reaction was performed using tyrosine
(0.3mM). (A) Nitration and (B) oxidation in membranes.

SIN-1(1 m$M$) with DLPC liposome containing these peptides (Y-4, Y-8, and Y-12) revealed that nitration of those peptides was enhanced with increasing membrane depth for the tyrosine residues (Fig. 9A).

*The effect of bicarbonate*: Several investigators have demonstrated that peroxynitrite-dependent nitration and oxidation product profiles are altered in the presence of $CO_2$ (Denicola, 1996; Lymar and Hurst, 1995; Lymar *et al.*, 1996; Uppu *et al.*, 1996). Peroxynitrite anion reacts with $CO_2$ to form a transient adduct, nitrosoperoxycarbonate ($ONOOCO_2^-$), which decomposes in part to form $CO_3^{\cdot -}$ and $^{\cdot}NO_2$ radicals (Scheme 1). Both species can oxidize tyrosine to the tyrosyl radical, which will rapidly react with $^{\cdot}NO_2$ or another tyrosyl radical to form $NO_2$Tyr and dityrosine. Typically, nitration of tyrosine or tyrosine analogues is enhanced in the aqueous phase in the presence of ONOO and $CO_2$ (Denicola *et al.*, 1996; Gow *et al.*, 1996; Uppu *et al.*, 1996). However, the situation is different in membranes. The transmembrane diffusion of $OONOCO_2^-$ across the bilayer is probably limited due to its increased rate of decay as compared to $ONOO^-$ (Denicola *et al.*, 1996). In addition, the anionic $CO_3^{\cdot -}$ is not sufficiently membrane permeable. This is consistent with the previous findings that bicarbonate inhibits nitration of hydrophobic phenolic probes, $\gamma$-tocopherol (Goss *et al.*, 1999; Jiang *et al.*, 2000; Ohshima *et al.*, 1999). Thus, we compared the effects of $HCO_3^-$ on nitration and oxidation of BTBE in membranes with those of tyrosine in the aqueous solution. As shown in Fig. 8, $HCO_3^-$ significantly enhanced the nitration and oxidation of tyrosine. In marked contrast, $HCO_3^-$ had little or no effect on the nitration and oxidation reactions of BTBE (Fig. 9A and B).

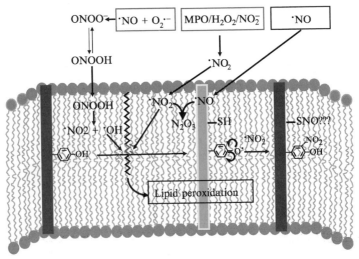

SCHEME. 1. Proposed Mechanism for Transmembrane Tyrosine Nitration and Oxidation. (See color insert.)

Nitration of peptides in DLPC liposomes by a bolus addition of peroxynitrite was inhibited by bicarbonate (data not shown). The inhibitory effect of bicarbonate on the nitration of peptides in membrane increased in the following order: Y-12 (60%) > Y-8 (30%) > Y-4 (20%). Incubation of peptides incorporated in DLPC liposome with SIN-1 showed bicarbonate has little or no effect on the nitration yields induced by $ONOO^-$ co-generated *in situ* (Fig. 9A). Based on these findings, we conclude that bicarbonate does not affect nitration of tyrosyl peptides in membrane exposed to co-generated $O_2^{\cdot-}$ and $\cdot NO$ at the same flux or to slowly infused peroxynitrite.

*Nitration induced by* $MPO/H_2O_2/NO_2^-$ *system:* Two major pathways have been suggested to be responsible for tyrosine nitration *in vivo* (Beckmann *et al.*, 1994; Eiserich *et al.*, 1998; Lymar and Hurst, 1995). These involve either the nitrative chemistry of peroxynitrite or the catalytic action of heme peroxidases using $H_2O_2$ and nitrite anion as substrate. We investigated the tyrosine nitration in membrane by $MPO/H_2O_2/NO_2^-$ system. Peptides (0.3 m$M$) in 30 m$M$ DLPC liposomes were incubated with 0.5 m$M$ $NaNO_2$, 0.2 m$M$ $H_2O_2$, and 50 n$M$ MPO in 0.1 $M$ phosphate buffer containing 100 $\mu M$ DTPA at 37° for 1 h. Samples were extracted and analyzed by HPLC as described. Investigation of the nitration profile of Y-4, Y-8, and Y-12 obtained in $MPO/H_2O_2/NO_2^-$ system produced an interesting yet unexpected result (Fig. 10). In contrast to peroxynitrite-induced nitration profile, incubation of DLPC liposome containing Y-4, Y-8, and Y-12 in the presence of MPO, $H_2O_2$, and $NO_2^-$ caused a decrease in the nitration of peptide with increasing membrane depths (Fig. 10).

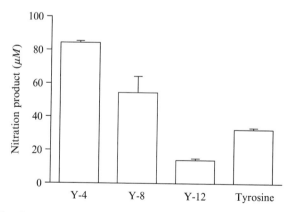

FIG. 10. Nitration of tyrosine in aqueous solution and tyrosyl groups in membranes by MPO system. Tyrosine (0.3 m$M$) or tyrosyl peptide (0.3 m$M$) in DLPC (30 m$M$) liposomes was incubated with $NaNO_2$ (0.5 m$M$), $H_2O_2$ (0.2 m$M$), and myeloperoxidase (50 n$M$) in a phosphate buffer (0.1 $M$, pH 7.4) containing DTPA (100 $\mu M$) at 37° for 1 h. The sample was extracted and analyzed by high-performance liquid chromatography.

Thus, it appears that the differential nitration of tyrosine residues located at different depths could be used to distinguish between $^\bullet NO_2$ and $ONOO^-/ONOOH$–dependent nitration of tyrosyl peptides in membranes.

## Summary

We have synthesized and characterized several membrane-incorporated tyrosine analogues (i.e., N-t-BOC L-tyrosine t-butyl ester and several trans-membrane peptides with tyrosyl residues at different positions). The results of fluorescence and quenching experiments showed that the membrane-anchored tyrosine analogues were incorporated into the hydrophobic bilayer. Using these newly synthesized membrane-anchored tyrosine analogues, we established an in vitro liposome model for analyzing nitration reactions in hydrophobic membranes. Results indicate that slow infusion of $ONOO^-/ONOOH$ or peroxynitrite co-generated in situ at low concentration preferentially nitrated tyrosyl residues in a hydrophobic environment. Results indicate that nitration but not oxidation of tyrosyl residues is a major reaction pathway for peroxynitrite in membranes and show that dityrosine formation is not a significant reaction process for tyrosyl radicals in membranes where lateral diffusion of tyrosyl radicals is hindered. We also demonstrated that the nitration yields of tyrosyl peptides are dependent on both the hydrophobicity of the environment in which they are located and the nitrating species. Even both peroxynitrite and nitrogen dioxide radical prefer to nitrate tyrosyl group in a hydrophobic environment rather than the aqueous phase, peroxynitrite induced more nitration of the tyrosyl groups buried deep in the bilayer membrane, but nitrogen dioxide preferred the tyrosyl group close to the surface of the membrane. We propose a radical mechanism for the tyrosine nitration and oxidation in hydrophobic membrane (Scheme 1). In addition to investigating the chemical reaction in the bilayer membrane, BTBE has been used to assess the antioxidant and antinitration effects of polyphenolics (Schroeder et al., 2001). BTBE was successfully incorporated into red blood cells' membranes to monitor nitration/oxidation reactions inside red blood cells caused by $^\bullet NO_2$ exposure or peroxynitrite.

## References

Beckmann, J. S., Ye, Y. Z., Anderson, P. G., Chen, J., Accavitti, M. A., Tarpey, M. M., and White, C. R. (1994). Extensive nitration of protein tyrosines in human atherosclerosis detected by immunohistochemistry. Biol. Chem. Hoppe-Seyler 375, 81.

Denicola, A., Freeman, B. A., Trujillo, M., and Radi, R. (1996). Peroxynitrite reaction with carbon dioxide/bicarbonate: Kinetics and influence on peroxynitrite-mediated oxidations. *Arch. Biochem. Biophys.* **333**, 49.

Denicola, A., Souza, J. M., and Radi, R. (1998). Diffusion of peroxynitrite across erythrocyte membranes. *Proc. Natl. Acad. Sci. USA* **95**, 3566.

de Planque, M. R., Goormaghtigh, E., Greathouse, D. V., Koeppe, R. E., II, Kruijtzer, J. A., Liskamp, R. M., de Kruijff, B., and Killian, J. A. (2001). Sensitivity of single membrane-spanning alpha-helical peptides to hydrophobic mismatch with a lipid bilayer: Effects on backbone structure, orientation, and extent of membrane incorporation. *Biochemistry* **40**, 5000.

de Planque, M. R., Kruijtzer, J. A., Liskamp, R. M., Marsh, D., Greathouse, D. V., Koeppe, R. E., II, de Kruijff, B., and Killian, J. A. (1999). Different membrane anchoring positions of tryptophan and lysine in synthetic transmembrane alpha-helical peptides. *J. Biol. Chem.* **274**, 20839.

Eiserich, J. P., Hristova, M., Cross, C. E., Jones, A. D., Freeman, B. A., Halliwell, B., and van der Vliet, A. (1998). Formation of nitric oxide-derived inflammatory oxidants by myeloperoxidase in neutrophils. *Nature* **391**, 393.

Feix, J. B., Popp, C. A., Venkatramu, S. D., Park, J. H., and Hyde, J. S. (1984). An electron-electron double-resonance study of interactions between [14N]- and [15N]stearic acid spin-label pairs: Lateral diffusion and vertical fluctuations in dimyristoylphosphatidylcholine. *Biochemistry* **23**, 2293.

Feix, J. B., Yin, J.-J., and Hyde, J. S. (1987). Interactions of 14N:15N stearic acid spin-label pairs: Effects of host lipid alkyl chain length and unsaturation. *Biochemistry* **26**, 3850.

Goldstein, S., and Czapski, G. (1998). Formation of Peroxynitrate from the Reaction of Peroxynitrite with $CO_2$: Evidence for Carbonate Radical Production. *J. Am. Chem. Soc.* **120**, 3458.

Goldstein, S., Czapski, G., Lind, J., and Merenyi, G. (2000). Tyrosine nitration by simultaneous generation of (.)NO and O-(2) under physiological conditions. How the radicals do the job. *J. Biol. Chem.* **275**, 3031.

Goss, S. P. A., Hogg, N., and Kalyanaraman, B. (1999). The effect of alpha-tocopherol on the nitration of gamma-tocopherol by peroxynitrite. *Arch. Biochem. Biophys.* **363**, 333.

Gow, A., Duran, D., Thom, S. R., and Ischiropoulos, H. (1996). Carbon dioxide enhancement of peroxynitrite-mediated protein tyrosine nitration. *Arch. Biochem. Biophys.* **333**, 42.

Greenacre, S. A., and Ischiropoulos, H. (2001). Tyrosine nitration: Localisation, quantification, consequences for protein function and signal transduction. *Free Radic. Res.* **34**, 541.

Groves, J. T. (1999). Peroxynitrite: Reactive, invasive and enigmatic. *Curr. Opin. Chem. Biol.* **3**, 226.

Harzer, U., and Bechinger, B. (2000). Alignment of lysine-anchored membrane peptides under conditions of hydrophobic mismatch: A CD, 15N and 31P solid-state NMR spectroscopy investigation. *Biochemistry* **39**, 13106.

Jiang, Q., Elson-Schwab, I., Courtemanche, C., and Ames, B. N. (2000). gamma-tocopherol and its major metabolite, in contrast to alpha-tocopherol, inhibit cyclooxygenase activity in macrophages and epithelial cells. *Proc. Natl. Acad. Sci. USA* **97**, 11494.

Leeuwenburgh, C., Hardy, M. M., Hazen, S. L., Wagner, P., Ohshishi, S., Steinbrecher, U. P., and Heinecke, J. W. (1997). Reactive nitrogen intermediates promote low density lipoprotein oxidation in human atherosclerotic intima. *J. Biol. Chem.* **272**, 1433.

Lymar, S. V., and Hurst, J. K. (1995). Rapid reaction between peroxynitrite ion and carbon dioxide: Implications for biological activity. *J. Am. Chem. Soc.* **117**, 8867.

Lymar, S. V., Jiang, Q., and Hurst, J. K. (1996). Mechanism of carbon dioxide-catalyzed oxidation of tyrosine by peroxynitrite. *Biochemistry* **35**, 7855.

Marla, S. S., Lee, J., and Groves, J. T. (1997). Peroxynitrite rapidly permeates phospholipid membranes. *Proc. Natl. Acad. Sci. USA* **94,** 14243.

Ohlin, H., Pavlidis, N., and Ohlin, A. K. (1998). Effect of intravenous nitroglycerin on lipid peroxidation after thrombolytic therapy for acute myocardial infarction. *Am. J. Cardiol.* **82,** 1463.

Ohshima, H., Celan, I., Chazotte, L., Pignatelli, B., and Mower, H. F. (1999). Analysis of 3-nitrotyrosine in biological fluids and protein hydrolyzates by high-performance liquid chromatography using a postseparation, on-line reduction column and electrochemical detection: Results with various nitrating agents. *Nitric Oxide* **3,** 132.

Pfeiffer, S., and Mayer, B. (1998). Lack of tyrosine nitration by peroxynitrite generated at physiological pH. *J. Biol. Chem.* **273,** 27280.

Pfeiffer, S., Lass, A., Schmidt, K., and Mayer, B. (2001). Protein tyrosine nitration in mouse peritoneal macrophages activated *in vitro* and *in vivo*: Evidence against an essential role of peroxynitrite. *FASEB J.* **15,** 2355.

Pfeiffer, S., Schmidt, K., and Mayer, B. (2000). Dityrosine formation outcompetes tyrosine nitration at low steady-state concentrations of peroxynitrite. Implications for tyrosine modification by nitric oxide/superoxide *in vivo*. *J. Biol. Chem.* **275,** 6346.

Schroeder, P., Zhang, H., Klotz, L. O., Kalyanaraman, B., and Sies, H. (2001). (−)-Epicatechin inhibits nitration and dimerization of tyrosine in hydrophilic as well as hydrophobic environments. *Biochem. Biophys. Res. Commun.* **289,** 1334.

Souza, J. M., Daikhin, E., Yudkoff, M., Raman, C. S., and Ischiropoulos, H. (1999). Factors determining the selectivity of protein tyrosine nitration. *Arch. Biochem. Biophys.* **371,** 169.

Uppu, R. M., Squadrito, G. L., and Pryor, W. A. (1996). Acceleration of peroxynitrite oxidations by carbon dioxide. *Arch. Biochem. Biophys.* **327,** 335.

Van der Vliet, A., Eiserich, J. P., O'Neill, C. A., Halliwell, B., and Cross, C. E. (1995). Tyrosine modification by reactive nitrogen species: A closer look. *Arch. Biochem. Biophys.* **319,** 341.

Viner, R. I., Ferrington, D. A., Williams, T. D., Bigelow, D. J., and Schoneich, C. (1999). Protein modification during biological aging: Selective tyrosine nitration of the SERCA2a isoform of the sarcoplasmic reticulum Ca2+-ATPase in skeletal muscle. *Biochem. J.* **340,** 657.

Yamamoto, T., Maruyama, W., Kato, Y., Yi, H., Shamoto-Nagai, M., Tanaka, M., Sato, Y., and Naoi, M. (2002). Selective nitration of mitochondrial complex I by peroxynitrite: Involvement in mitochondria dysfunction and cell death of dopaminergic SH-SY5Y cells. *J. Neural. Transm.* **109,** 1.

Zhang, H., Bhargava, K., Keszler, A., Feix, J., Hogg, N., Joseph, J., and Kalyanaraman, B. (2003). Transmembrane nitration of hydrophobic tyrosyl peptides. Localization, characterization, mechanism of nitration, and biological implications. *J. Biol. Chem.* **278,** 8969.

Zhang, H., Joseph, J., Feix, J., Hogg, N., and Kalyanaraman, B. (2001). Nitration and oxidation of a hydrophobic tyrosine probe by peroxynitrite in membranes: Comparison with nitration and oxidation of tyrosine by peroxynitrite in aqueous solution. *Biochemistry* **40,** 7675.

Zhang, Y. P., Lewis, R. N., Henry, G. D., Sykes, B. D., Hodges, R. S., and McElhaney, R. N. (1995a). Peptide models of helical hydrophobic transmembrane segments of membrane proteins. 1. Studies of the conformation, intrabilayer orientation, and amide hydrogen exchangeability of Ac-K2-(LA)12-K2-amide. *Biochemistry* **34,** 2348.

Zhang, Y. P., Lewis, R. N., Hodges, R. S., and McElhaney, R. N. (1995b). Peptide models of helical hydrophobic transmembrane segments of membrane proteins. 2. Differential scanning calorimetric and FTIR spectroscopic studies of the interaction of Ac-K2-(LA)12-K2-amide with phosphatidylcholine bilayers. *Biochemistry* **34,** 2362.

# Section III

# Peroxynitrite

# [19]   Synthesis of Peroxynitrite from Nitrite and Hydrogen Peroxide

*By* Kristine M. Robinson and Joseph S. Beckman

## Abstract

We report a simple method for the synthesis of peroxynitrite from nitrite and hydrogen peroxide that can generate hundreds of milliliters of 180 m$M$ peroxynitrite within 1 h from start to finish. It can be scaled down to make small quantities of isotope-labeled peroxynitrite. The method requires only a syringe pump and tubing connectors and is feasible for any biochemical laboratory. Unreacted hydrogen peroxide is eliminated with manganese dioxide, using an improved preparation compared to commercially available manganese dioxide. A number of contaminants were detected by mass spectrometry in peroxynitrite solutions cleaned with commercially purchased manganese dioxide. Nitrite contamination of the peroxynitrite solution is less than 2% as determined using the Griess method. The residual contaminants are principally 0.28 $M$ sodium chloride and 0.1 $M$ sodium hydroxide, which pose few problems when peroxynitrite is diluted for use in biological experiments.

## Introduction

We describe here a simple synthesis of peroxynitrite (ONOO$^-$) from nitrite and hydrogen peroxide for biochemical studies, which can be safely and rapidly conducted using readily available equipment in any laboratory. Residue nitrite and hydrogen peroxide, the two most troublesome contaminants, are less than 2% of the peroxynitrite concentration. Many syntheses for peroxynitrite have been reported but are potentially hazardous and require expensive specialized equipment. The cleanest method produces solid tetramethylammonium peroxynitrite without contaminants (Bohle *et al.*, 1994). However, this method requires grinding potassium superoxide at $-77°$ in highly toxic liquid ammonia in a potentially explosive reaction. Several fume hoods have been destroyed during the synthesis of the intermediate tetramethylammonium superoxide. Solid peroxynitrite is commercially available but is expensive, even for small amounts. Significant contamination with nitrite can result, even when pure peroxynitrite is dissolved in alkaline solutions unless considerable care is taken (Latal *et al.*, 2004). Alternatively, peroxynitrite may be synthesized by bubbling ozone through alkaline solutions of azide (Pryor *et al.*, 1995). Ozone needs to be generated by an ozone

METHODS IN ENZYMOLOGY, VOL. 396                                    0076-6879/05 $35.00
DOI: 10.1016/S0076-6879(05)96019-9

generator from oxygen, and ozone is an extremely strong pulmonary and eye irritant. Residual sodium azide is toxic and a significant confounding factor in biological experiments using peroxynitrite. If the azide reaction is run to completion, significant contamination with nitrite occurs. It is also possible to synthesize peroxynitrite by crushing solid potassium superoxide with sand in an anaerobic system and then flushing with nitric oxide gas (Koppenol *et al.*, 1995). This method requires substantial plumbing with stainless steel tubing to handle nitric oxide and significant time, and it yields relatively small quantities of peroxynitrite. The peroxynitrite is contaminated with a large amount of hydrogen peroxide formed by the dismutation of unreacted superoxide.

A simpler method of synthesis was first performed in 1902 when Baeyer and Villiger (1901) published a method of mixing nitrite with hydrogen peroxide. In 1968, other authors (Hughes and Nicklin, 1968, 1970a,b; Hughes *et al.*, 1971) reported a synthesis also based on the reaction between nitrite and hydrogen peroxide in which the hydrogen peroxide was "thrown" into a nitrite solution and produced peroxynitrite in yields between 40 and 50%. In 1973, Reed *et al.* (1974) reported a method of reacting nitrite and hydrogen peroxide in a quenched flow reactor. This apparatus was made of a Lucite rod bored for the inlet of reactants. Yields increased to approximately 82%. Beckman *et al.* (1994) also published a method of peroxynitrite synthesis that used a vacuum system to pull reactants through the mixing chamber. This method had variable yields, depending on the time of day. Yields were highest when syntheses were performed at night because the house vacuum had a stronger pull. Saha *et al.* (1998) also searched for optimal conditions with fewer contaminants and published a method in which they constructed a rapid mixer commonly used in stopped flow machines. This required fine machining of two specialized, small rapid mixing chambers.

We herein report a simpler version of the Saha *et al.* method that requires only a syringe pump, four T-connectors, and flexible tubing that optimizes synthesis conditions. The method can produce hundreds of milliliters of approximately 180 m$M$ peroxynitrite in less than an hour. It can be scaled down to make small quantities of isotope-labeled peroxynitrite. An improved procedure for making manganese dioxide is also described to better remove residual hydrogen peroxide. The major contaminant in the peroxynitrite synthesized by this method is nitrate, which generally has little influence on the use of peroxynitrite for most experiments.

## The Chemistry of the Reaction

Peroxynitrite is synthesized by reacting acidified hydrogen peroxide with sodium nitrite. The p$K_a$ of nitrite is approximately 3.4. Under moderately acidic conditions, nitrous acid is an efficient nitrosonium donor. The nucleophilic attack on hydrogen peroxide produces peroxynitrous acid:

$$NO_2^- + H^+ \longrightarrow HO^{\delta-} \text{ - - - } {}^{\delta+}N = O$$

$$\downarrow +H^+$$

$$H_2O$$

Under acidic pH, peroxynitrous acid has a half-life of less than 1 s, but the reaction can be quickly quenched with an excess of base to yield peroxynitrite anion in high concentrations.


Synthesis of Peroxynitrite

Typically, three solutions are freshly prepared in high-quality water: $0.7\ M$ HCl + $0.6\ M$ $H_2O_2$ (14.5 ml of concentrated HCl + 17.0 ml of 30% $H_2O_2$ into a final volume of 250 ml), $0.6\ M$ sodium nitrite (10.4 g into 250 ml of water), and $3\ M$ sodium hydroxide (30 g into 250 ml of $H_2O$). Yields are highest when all solutions are at room temperature. The use of higher concentrations of reactants can result in the hazardous generation of toxic gases that can blow off the tubing. Goggles should be worn during the synthesis in the event that the tubing disconnects. High-quality sodium nitrite should be freshly purchased and reserved for the synthesis. Five 30-ml disposable plastic syringes (Becton Dickinson) are needed. Two are filled with sodium nitrite, two filled with the acidified hydrogen peroxide, and the last syringe is filled with $3\ M$ sodium hydroxide. Tubing should be connected to syringes using plastic Luer lock connectors and not by using steel needles. The tubing is connected using plastic Ts, as shown in Fig. 1. The nitrite and peroxide reactants flow through Tygon tubing, and Teflon tubing was used after the first T junction for the peroxynitrite product flow.

The indicated lengths of tubing have been varied to optimize yields. The tubing is 3 mm inside diameter, except for the tube measuring 5.8 cm, which is approximately 2 mm in diameter. Any metal should be avoided as described in this method, such as the use of metal syringe tips, T connections, and tubing. Trace contamination with transition metals catalyzes the alkaline decomposition of peroxynitrite to nitrite and dioxygen (Pfeiffer *et al.*, 1997). The principal advantage of this arrangement is to have acidified nitrite and hydrogen peroxide mix in the first set of T junctions and then to be further mixed at the second T junction.

The five syringes are loaded side by side onto a syringe pump. We successfully secured five 30-ml syringes on a Harvard Syringe Pump 22 with laboratory tape; however, syringe pumps are commercially available

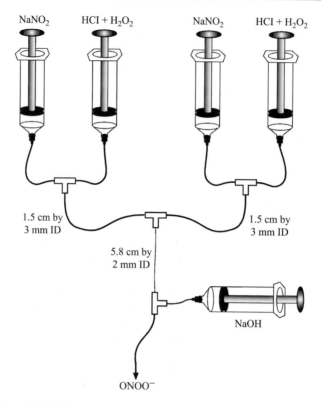

Fig. 1. First, 0.7 $M$ HCl + 0.6 $M$ $H_2O_2$ and 0.6 $M$ NaNO$_2$ are loaded into syringes as in the diagram. The reaction is quenched with 3 $M$ NaOH in the syringe downstream. The lengths of the tubing have been optimized for the highest yields. Metals are avoided in any connections. All five syringes are loaded side by side onto a syringe pump, which is run at a maximum flow of 17 ml/min.

that will hold five syringes. The syringe pump is run at its maximum rate (17 ml/min using 30-ml disposable syringes). The flow is diverted to waste until a yellow solution emerges. The yellow peroxynitrite solution is then collected on ice in a separate flask. The concentration of peroxynitrite ranges from 180 to 190 m$M$. The concentration of peroxynitrite can be easily measured spectrally at 302 nM. For absorption measurements, the stock solution needs to be diluted 2 $\mu$l into 1000 $\mu$l of 100 m$M$ NaOH. It is necessary to repeatedly invert the cuvette to ensure adequate mixing. Repeating the dilution and absorption measurements is recommended for accurately determining the concentration. The first reported extinction coefficient was estimated to be 1670 $M^{-1}$ cm$^{-1}$ and is the most likely used value. However, the extinction coefficient has been more accurately determined using pure tetramethylammonium peroxynitrite to be 1700 $M^{-1}$ cm$^{-1}$ (Bohle et al., 1994).

The synthesis of peroxynitrite gives an approximate 70% yield. The remainder is mostly nitrate. It should be noted that the solution contains approximately 0.28 $M$ NaCl and 0.1 $M$ NaOH. If it is desired to avoid NaCl, 0.7 $M$ nitric acid may be used in place of HCl in the starting hydrogen peroxide solution (White et al., 1997).

Approximately 100 ml of the fresh peroxynitrite solution is transferred to a 250-ml beaker whose inner flat surface is thinly covered in manganese dioxide flakes (this synthesis is described later in this chapter). The beaker is submerged in an ice bath, and the mixture is left to react on ice for 15 min without stirring. This length of time is adequate to consume any unreacted hydrogen peroxide. The peroxynitrite can then be cleaned from manganese dioxide by vacuum filtration using a No. 2 paper filter. Residual $H_2O_2$ is assayed as described later in this chapter. Solutions of peroxynitrite are then frozen in 1-ml aliquots at $-80°$ for up to 1 year. We no longer attempt to freeze-fractionate peroxynitrite because this leads to a large increase in nitrite contamination relative to the increase in peroxynitrite concentration (Beckman et al., 1990).

We were concerned that plasticizers might leach from the tubing into the reactants during the synthesis process. When the synthesized peroxynitrite was analyzed by electrospray mass spectrometry in positive ion mode, a number of unidentifiable peaks appeared after treatment with commercially available MnO$_2$ but not before treatment. These peaks were not present when MnO$_2$ was synthesized, which is easily done as described later in this chapter. This MnO$_2$ was also far more effective at catalyzing hydrogen peroxide decomposition.

## Manganese Dioxide Synthesis

Residual hydrogen peroxide is eliminated by cleaning with manganese dioxide (MnO$_2$), which catalytically decomposes hydrogen peroxide. To prepare MnO$_2$, 8 g of potassium permanganate (caustic) is dissolved in 50 ml water, and 500 ml of 95% ethanol is slowly added. The reaction is stirred overnight. The dark brown precipitate is collected the next day by vacuum filtration and washed with 2–3 liters of water. It is allowed to dry completely for several days. Yields are approximately 97%. The brown product is broken into small, approximately 5-mm flakes. Excessively crushing the MnO$_2$ should be avoided because as the MnO$_2$ becomes increasingly powdery, it becomes difficult to filter the MnO$_2$ out of the peroxynitrite. Once the MnO$_2$ has been filtered out of the peroxynitrite solution, it can be left to dry and used again. However, once the $MnO_2$ becomes too powdery, it is discarded.

Hydrogen peroxide concentrations are determined using the horseradish peroxidase (HRP)–based assay, using 4 m$M$ 3,5-dichloro-2-hydroxybenzene sulfonic acid (sodium salt), 3 m$M$ 4-aminoantipyrine, and 0.01 unit/ml HRP in a 1-ml final volume assay. When performing the HRP assay, one should first decompose peroxynitrite solutions by diluting peroxynitrite into buffer (100 m$M$ potassium phosphate, 40 $\mu M$ DTPA, pH 7.3), and then incubate them at 37° for 15 min.

Peroxynitrite interferes with the peroxidase assay and, therefore, needs to be decomposed first. Caution needs to be taken to ensure that the peroxynitrite is diluted adequately because excess hydrogen peroxide will lead to inaccurate results. For example, before the peroxynitrite is cleaned with $MnO_2$, it is necessary to dilute the peroxynitrite into buffer by 1/2000 using a series of dilutions (first 10 $\mu$l $ONOO^-$ into 100 $\mu$l buffer, then 5 $\mu$l into the 1 ml assay). Once the peroxynitrite has been cleaned with $MnO_2$, a dilution factor of 1/20 is adequate. In this case, 50 $\mu$l of peroxynitrite is added to 1 ml buffer and then the aminoantipyrine and 3,5-dichloro-2-hydroxybenzene sulfonic acid reactants can be added after the peroxynitrite has been decomposed. The reactants are incubated in the dark for 10 minutes, and then the absorbance is checked at 510 n$M$. Concentrations of hydrogen peroxide are calculated using a standard curve prepared from stock $H_2O_2$ solutions.

Before cleaning with $MnO_2$, the fresh peroxynitrite contained approximately 16 m$M$ $H_2O_2$. After cleaning the peroxynitrite for 15 min with manganese dioxide as described earlier in this chapter, hydrogen peroxide levels decreased to approximately 80 $\mu M$. The lower limit of detection in stock solutions was approximately 50 $\mu M$ $H_2O_2$. After synthesis, the peroxynitrite concentration of peroxynitrite ranges from 160 to 180 m$M$ and decreases to 140–160 m$M$ after $MnO_2$ cleaning. Using an excess of 0.7 $M$ $H_2O_2$ in the synthesis can increase yields to 170–190 m$M$ peroxynitrite. However, this increases the $H_2O_2$ contamination to 50 m$M$ before the $MnO_2$ treatment. This excess $H_2O_2$ contamination is only decreased to approximately 10 m$M$ by treating with $MnO_2$ for 15 min. Therefore, it is not advantageous to use the excess $H_2O_2$ in the peroxynitrite synthesis.

Working with Peroxynitrite

It is essential to recognize that several contaminants are present in high concentrations in the final solution of peroxynitrite. These are principally sodium chloride, nitrite, nitrate, oxygen, and sodium hydroxide. By using the Griess method, we find levels of nitrite to be approximately 3.5 m$M$ in 180 m$M$ peroxynitrite solutions. Latal *et al.* (2004) reported that a larger contamination (20 m$M$) of nitrite can result from thawing peroxynitrite at warmer temperatures. We did not observe a similar buildup of nitrite when thawing 1-ml aliquots at 0, 25, or 37°. Possibly, the increase in nitrite contamination from thawing at 37° may be specific to the dissolution of the tetramethylammonium salt or affected by differences in salt or alkali concentrations. However, thawing peroxynitrite solutions on ice is generally preferable because it reduces the decomposition of peroxynitrite that occurs at warmer temperatures.

Nitrite may also accumulate as peroxynitrite is stored on ice and is accompanied by the formation of oxygen bubbles along the tube walls of peroxynitrite solutions as they are thawed. It is important to perform a control with the reverse order of addition to allow peroxynitrite to decay. This control experiment will help reveal effects due to the contaminants rather than the peroxynitrite itself. The present synthesis is a simple means to prepare either very small or large amounts of peroxynitrite, which is suitable for many biological experiments. However, experiments with peroxynitrite should be carefully planned with appropriate controls to reveal any artifacts resulting from residual contaminants. Nitrate and nitrite contaminants are also produced during the decomposition of pure peroxynitrite. Other considerations about working with stock solutions of peroxynitrite have been described (Beckman *et al.*, 1996).

### Acknowledgments

We thank Professor Wim Koppenol (ETH Zurich) for telling us how to make manganese dioxide. This work was supported by the grants P01 AT002034, by R01 NS033291, and by P01 ES00040 funded by the National Institutes of Health. This work was made possible in part through the support from the Oregon State Environmental Health Sciences Center (ES00210) and made use of the mass spectrometry center.

### References

Baeyer, A., and Villiger, V. (1901). Uber die salpetrige Säure. *Ber.* **34,** 755–763.
Beckman, J. S., Beckman, T. W., Chen, J., Marshall, P. M., and Freeman, B. A. (1990). Apparent hydroxyl radical production by peroxynitrite: Implications for endothelial injury from nitric oxide and superoxide. *Proc. Natl. Acad. Sci. USA* **87,** 1620–1624.
Beckman, J. S., Chen, J., Ischiropoulos, H., and Crow, J. P. (1994). Oxidative chemistry of peroxynitrite. *Methods Enzymol.* **233,** 229–240.
Beckman, J. S., Wink, D. A., and Crow, J. P. (1996). *In* "Methods in Nitric Oxide Research" (M. Feelisch and J. S. Stamler, eds.), pp. 61–70. John Wiley & Sons, Chichester.

Bohle, D. S., Hansert, B., Paulson, S. C., and Smith, B. D. (1994). Biomemetic synthesis of the putative cytotoxin peroxynitrite, ONOO$^-$, and its characterization as a tetramethylammonium salt. *J. Am. Chem. Soc.* **116,** 7423–7424.

Hughes, M. N., and Nicklin, H. G. (1968). The chemistry of pernitrites. Part I. Kinetics of decompositions of pernitrous acid. *J. Chem. Soc. A* **1968,** 450–452.

Hughes, M. N., and Nicklin, H. G. (1970a). The chemistry of peroxonitrites. Part II. Copper (II)-catalysed reaction between hydroxylamine and peroxonitrite in alkali. *J. Chem. Soc. A* **1970,** 925–928.

Hughes, M. N., and Nicklin, H. G. (1970b). A possible role for the species peroxonitrite in nitrification. *Biochim. Biophys. Acta* **222,** 660–661.

Hughes, M. N., Nicklin, H. G., and Sackrule, W. A. C. (1971). The chemistry of peroxonitrites. Part III. The reaction of peroxonitrite with nucleophiles in alkali, and other nitrite producing reactions. *J. Chem. Soc. A* 3722–3725.

Koppenol, W. H., Kissner, R., and Beckman, J. S. (1995). *In* "Methods in Enzymology" (L. Packer, ed.), Vol. 269, pp. 296–302. Academic Press, San Diego.

Latal, P., Kissner, R., Bohle, D. S., and Koppenol, W. H. (2004). Preventing nitrite contamination in tetramethylammonium peroxynitrite solutions. *Inorg. Chem.* **43,** 6519–6521.

Pfeiffer, S., Goren, A. C. F., Schmidt, K., Werner, E. R., Hansert, B., Bohle, D., and Mayer, B. (1997). Metabolic fate of peroxynitrite in aqueous solution. Reaction with nitric oxide and pH-dependent decomposition to nitrite and oxygen in a 2:1 stoichiometry. *J. Biol. Chem.* **272,** 3465–3470.

Pryor, W. A., Cueto, R., Jin, X., Koppenol, W. H., Ngu-Schwemlein, M., Squadrito, G. L., Uppu, P. L., and Uppu, R. M. (1995). A practical method for preparing peroxynitrite solutions of low ionic strength and free of hydrogen peroxide. *Free Radic. Biol. Med.* **18,** 75–83.

Reed, J. W., Ho, H. H., and Jolly, W. L. (1974). Chemical syntheses with a quenched flow reactor. Hydroxytrihydroborate and peroxynitrite. *J. Am. Chem. Soc.* **96,** 1248–1249.

Saha, A., Goldstein, S., Cabelli, D., and Czapski, G. (1998). Determination of optimal conditions for synthesis of peroxynitrite by mixing acidified hydrogen peroxide with nitrite. *Free Radic. Biol. Med.* **24,** 653–659.

White, C. R., Spear, N., Thomas, S., Green, I., Crow, J. P., Beckman, J., Patel, R. P., and Darley-Usmar, V. M. (1997). *In* "Methods in Molecular Biology: Nitric Oxide Protocols" (M. A. Titheradge, ed.), Vol. 100, pp. 217–232. Humana Press, Totowa, NJ.

# [20]  Peroxynitrite-Dependent Upregulation of Src Kinases in Red Blood Cells: Strategies to Study the Activation Mechanisms

By Mauro Serafini, Cinzia Mallozzi,
Anna Maria Michela Di Stasi, and
Maurizio Minetti

## Abstract

Several studies have demonstrated that treatment of cells with oxidants, and in particular with peroxynitrite, may cause the upregulation of tyrosine phosphorylation signaling. In erythrocytes, peroxynitrite induces tyrosine phosphorylation of the major intrinsic membrane protein, band 3. A closer look at the enzymes involved revealed that the effect of peroxynitrite was due to the inhibition of phosphotyrosine phosphatases and/or to the activation of *src* kinases. The activity of *src* kinases is modulated not only by phosphatases and other kinases but also through redox modification of cysteine residues: Peroxynitrite can, thus, affect *src* kinase activity by means of direct and indirect mechanisms. In this chapter, we describe the different pathways leading to *src* kinase activation and the experimental procedures that can be performed to reveal the activation mechanism. The aim is to provide a more general strategy adaptable to different cell types and different oxidants.

## Introduction

Early research into the biological activity of nitric oxide radical (NO) tended to consider peroxynitrite a "toxic" or "ugly" aspect of NO chemistry (Beckman and Koppenol, 1996; Trauner, 2003). Peroxynitrite is a nonradical species produced by the fast radical–radical reaction between NO and superoxide anion (Beckman *et al.*, 1990) and is able to oxidize almost all major biological targets including DNA, proteins, lipids, carbohydrates, and low-molecular-weight antioxidants (Ischiropoulos and Al-Mehdi, 1995; King *et al.*, 1992; Radi *et al.*, 1991a,b). Initially, peroxynitrite was considered a powerful hydroxyl radical-generating toxin and, in consideration of the high reactivity of the hydroxyl radical, regulation of peroxynitrite-dependent reactions in biological systems appeared to be a difficult if not impossible task.

Studies from several laboratories have helped to change this concept, at least in part. Studies performed in more relevant biological conditions have

METHODS IN ENZYMOLOGY, VOL. 396
0076-6879/05 $35.00
DOI: 10.1016/S0076-6879(05)96020-5

suggested that in tissues peroxynitrite interacts with relatively few biomolecules (Radi, 1998). This new concept is based on results showing that direct reactions of peroxynitrite with thiols, metalloproteins, and $CO_2$ prevent its decomposition to hydroxyl radical. In fact, the rate at which peroxynitrite decomposes is slower than its reaction with the above biological targets, thus hindering or largely preventing its hydroxyl radical–dependent toxicity. Although the presence of preferred targets does not abolish the cytotoxic potential of peroxynitrite, because secondary oxidants are generated, it opens the hypothesis that this oxidant may be not just an "unwanted rare event" but could be used to trigger specific cellular pathways. Peroxynitrite, like other oxidants, induces a cellular stress response (Klotz et al., 2002), thus the possibility that low, controlled levels of peroxynitrite are continuously generated in nonpathological conditions cannot be completely ruled out (Cadenas, 2004).

Redox reactions have gained importance as major chemical pathways that modulate early events in the process of signal transduction leading to cell growth, differentiation, and death. Because one of the characteristic reactions of peroxynitrite is the nitration of protein tyrosine residues, it has been hypothesized that this oxidant may reduce tyrosine phosphorylation signaling (Kong et al., 1996). However, it has been found that treatment of different cell types with preformed peroxynitrite promotes rather than inhibits tyrosine phosphorylation signaling (Di Stasi et al., 1999; Mallozzi et al., 1997, 1999). Tyrosine phosphorylation, regulated by protein tyrosine phosphatases (PTPs) and kinases (PTKs), is important in signaling pathways and is unusually upregulated in cellular processes underlying tumorigenesis.

The PTK enzymes occur as transmembrane receptors or as non-receptor cytoplasmic proteins, and src kinases are one of the most important groups of non-receptor PTKs. Uncontrolled signaling resulting from the deregulation of tyrosine kinase activity has been implicated in the onset and progression of various human malignancies. The study of the selection of tyrosine kinase substrates, the regulation of their activity, and their biological function is, therefore, a subject of considerable interest.

After briefly reviewing the known mechanisms of regulation of src kinase activity, we outline the experiments that have been performed to elucidate the regulation of peroxynitrite-mediated src activation in red blood cells.

### src Tyrosine Kinase Structure and Regulation

The src family of tyrosine kinases is a closely related group of non-receptor kinases involved in signaling pathways that in response to the activation of cell-surface receptors by growth factors, cytokines, or cell-surface ligands, control the growth and differentiation of cells [for a review of *src* kinases, see Thomas and Brugge (1997)]. The regulation of *src* kinases is a complex and finely tuned phenomenon, involving both direct and indirect processes. There are at least nine members of this family: c-*src*, *fyn, lck, hck, lyn, fgr, blk, yes,* and *yrk*. These kinases range in size from 52 to 62 kDa, share a common architecture, and are organized into a set of modular domains: catalytic or *src* homology 1 (SH1), SH2, SH3, unique, and SH4 (Fig. 1). The catalytic domain confers tyrosine kinase activity and contains the "activation loop," within which is a tyrosine residue (Tyr 416, c-*src* numbering) that is important for the regulation of kinase activity.

### src Kinase Domains

The SH2 and SH3 domains play a dual role in *src* regulation, being required both to keep *src* kinases in an inactive state and to target the enzyme to specific substrates by directing protein-protein interactions (Lee *et al.*, 1995; Liu *et al.*, 1993; Moraefi *et al.*, 1997; Sicheri and Kuriyan, 1997). The fact that many *in vivo* substrates of PTKs contain the SH2 and/or SH3 domain-binding motif supports the notion that specificity arises outside the catalytic domain. SH2 domains contain about 90 amino acids and typically bind to specific peptide motifs that contain phosphorylated tyrosine. The canonical binding motifs for SH3 domains are proline-rich sequences with a

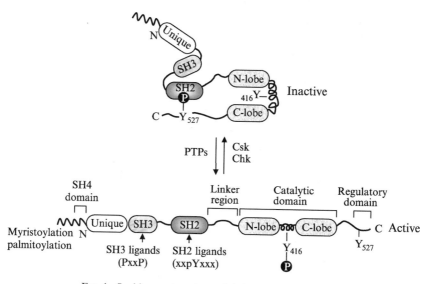

FIG. 1. *Src* kinases domains and their activation pathways.

core consensus sequence of PxxP. The SH4 domain is a short sequence (about 15 amino acids) at the N-terminus of the protein that contains signals for lipid modifications, such as myristoylation and/or palmitoylation, which are important for anchoring the protein to cell membranes. At the C-terminus of the protein is the regulatory domain, a short sequence (15–17 amino acids) containing a tyrosine residue (Tyr 527) that, when phosphorylated, can bind intramolecularly to the SH2 domain, giving rise to an inactive "closed" form of the enzyme (Fig. 1). The inactive conformation of *src* kinases is further stabilized by the binding of the SH3 domain to the linker region between the SH2 and catalytic domains (Williams *et al.*, 1998).

### Regulation of *src* Activity by Phosphorylation/Dephosphorylation of Tyrosine Residues

*Src* kinase is activated by autophosphorylation of Tyr 416, a reaction that has been shown to occur as an intermolecular event triggered by the kinase itself (Cooper and MacAuley, 1988) or by other still unknown tyrosine kinases (crosstalk) (Chiang and Sefton, 2000). The phosphorylation of Tyr 527 is catalyzed by two non-receptor tyrosine kinases, csk (C-terminal *src* kinase) (Okada and Nakagawa, 1989) and chk (csk homologous kinase) (Klages *et al.*, 1994). Moreover, *src* kinases can be activated by dephosphorylation of phosphorylated Tyr 527, which prevents the intramolecular interaction with the SH2 domain. This process is controlled by PTPs through still incompletely defined mechanisms. Data suggest that several candidates, including receptor-like PTPα (Zheng *et al.*, 1992) and PTPλ (Fang *et al.*, 1994), and the cytoplasmic PTP1C/SHP-1 (Somani *et al.*, 1997) and PTP1B (Bjorge *et al.*, 2000b), may dephosphorylate the phospho-Tyr 527. Elevated activity or expression of several of these PTPs correlates with enhanced levels of c-*src* kinase activity in a number of transformed cells (Bjorge *et al.*, 2000a). *Src* tyrosine kinases can also be activated by the binding of extrinsic ligands for the SH2 or SH3 domain, which occludes the intramolecular interactions and thereby promotes the active conformation. Such extrinsic ligands can be found in cellular signaling and scaffolding proteins or can be engineered. It has been demonstrated that a phosphotyrosine-containing peptide identified as binding the *src* SH2 domain with high affinity (pYEEI) stimulates c-*src* tyrosine kinase activity *in vitro* by competing with the phospho-Tyr 527 tail, resulting in the displacement of phospho-Tyr 527 from the SH2 domain (Liu *et al.*, 1993). The HIV-1 Nef protein, which is a high-affinity ligand for the SH3 domain of *hck*, causes a more marked activation of the enzyme than binding of the SH2 domain, indicating the importance of the SH3 domain in regulating the catalytic activity (Lee *et al.*, 1995).

Other modulators of *src* kinase activity bind directly to the catalytic domain; these include small-molecule inhibitors that mimic adenosine triphosphate (ATP) and, thereby, block phosphoryl transfer to substrates, such as PP2 [4-amino-5-(4-chlorophenyl)-7-(*t*-butyl) pyrazolo (3,4-*d*)pyrimidine], as well as intracellular G proteins, which can either activate or inhibit kinase activity.

## Regulation of src Activity by Oxidation/Reduction of Cysteine Residues

Evidence has been accumulated that suggests another regulatory mechanism in which the activation of *src* kinases also occurs independently of tyrosine residues and is triggered by redox-linked protein modifications involving the oxidation of specific cysteines. It has been shown that alkylation of these residues results in a loss of enzymatic activity (Senga *et al.*, 2000; Veillette *et al.*, 1993), while oxidation causes enzyme activation (Akhand *et al.*, 1999). *Src* kinases contain several cysteines (Fig. 2), and Senga *et al.* (2000) reported that of the 10 cysteine residues of v-*src*, four cysteines (Cys 483, Cys 487, Cys 496, and Cys 498) clustered in the carboxylterminal portion of the kinase domain and are important for its activity. These four cysteines are critical for inhibition by the alkylating agents and, therefore, have an allosteric repressor effect on the catalytic activity of *src* in a SH2/phosphoTyr 527–independent manner (Oo *et al.*, 2003).

## Activation of Tyrosine Phosphorylation Signal in Erythrocytes by Peroxynitrite

In erythrocytes, peroxynitrite stimulates a tyrosine-dependent signal transduction pathway that leads to (1) the enhancement of band3 tyrosine phosphorylation, (2) translocation of glycolytic enzymes to the cytoplasm

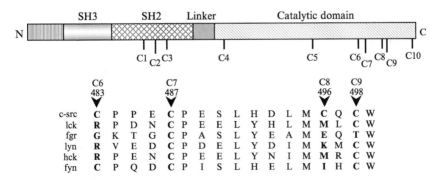

FIG. 2. Position of cysteine residues in some members of *src* kinase family.

and a consequent increase in glucose metabolism, (3) inhibition of PTPs, and (4) activation of *src* kinases (Mallozzi *et al.*, 1997).

### Inhibition of PTPs

The rise in band3 tyrosine phosphorylation appears to be mediated by a reversible inhibition of PTP activity (Mallozzi *et al.*, 1997). One of the effects of peroxynitrite is inhibition of the phosphatase activity that allows kinases to become predominant. This effect has been demonstrated at relatively low micromolar peroxynitrite concentrations in cells and in isolated PTPs (Mallozzi *et al.*, 1997; Takakura *et al.*, 1999). These PTPs contain the conserved active site motif CysXXXXXArg common to all enzymes that use a Cys to form a thiol-phosphate covalent intermediate during their catalytic cycle. Thus, the peculiar vulnerability of PTPs to peroxynitrite is consistent with a peroxynitrite-induced oxidative modification of critical cysteine residues.

### Activation of src Kinases

Peroxynitrite is also able to activate selectively three PTKs of the *src* family expressed in red blood cells, *lyn, hck*, and *fgr*, as demonstrated both in intact cells and in kinase immunocomplexes treated with peroxynitrite (Mallozzi *et al.*, 1999). The upregulation of phosphotyrosine signaling in red blood cells has been observed after a bolus addition of authentic peroxynitrite (10–100 $\mu M$) and after treatment of erythrocytes with 100–300 $\mu M$ SIN-1 (3-morpholinosydnonimine), a drug that slowly and simultaneously releases NO and the superoxide anion. At higher concentrations, signs of cell damage are found, as evidenced by massive hemoglobin oxidation, irreversible protein cross-linking, nitration of tyrosines, and inactivation of *src* kinases (Mallozzi *et al.*, 1997). Treatment of kinase immunoprecipitates with 0.1–1.0 $\mu M$ peroxynitrite also activated *lyn, hck*, and *fgr* kinases. The activation of *src* kinases by peroxynitrite was observed also in brain tissues. In rat brain synaptosomes and in cultured cerebellar granule cells, peroxynitrite up to a 100 $\mu M$ dose dependently stimulated the phosphotyrosine signal upregulating the activity of the *src* tyrosine kinases c-*src, fyn*, and *lyn* (Di Stasi *et al.*, 1999; Mallozzi *et al.*, 1999).

Treatment with peroxynitrite of immunoprecipitates causes *src* activation/inhibition, thus supporting the hypothesis that it can also modulate PTKs through direct mechanisms. The inhibition of *src* kinases by peroxynitrite at high concentration may be due to oxidative modification(s), leading to the loss of catalytic activity, whereas enzyme activation at 0.1–1.0 $\mu M$ peroxynitrite may be linked to oxidative modifications of cysteine residues. It has been shown that the activities of some *src* kinases (both autophosphorylation and phosphorylation of substrates) are increased by a mechanism involving sulfhydryl oxidation (Akhand *èt al.*, 1999; Mallozzi *et al.*, 2001a). Both cysteine and tyrosine residues are preferential targets of peroxynitrite and modification of one or more of these regulatory residues may, therefore, well explain activation/inactivation of the kinases. Moreover, even if cysteines are not modified directly by an oxidant molecule, they may act as a "sink" for other oxidizable residues, including tyrosines (Prutz *et al.*, 1986).

### Regulation through the Nitration of Peptides Mimicking Protein Domains with SH2 Affinity

We have reported that *lyn* upregulation in erythrocytes may occur through the nitration of a tyrosine residue in a model peptide with high affinity for the SH2 domain ($NO_2$-YEEI). Using this peptide, we propose a new mechanism for *src* kinase activation in which the *src*-SH2 domain recognizes a nitrotyrosine peptide with binding affinity for SH2, allowing the displacement of phospho-Tyr 527 and enzyme activation (Fig. 3). More physiologically relevant peptides consisting of the last 15 amino acids of the cytoplasmic band3 tail have also been used to demonstrate *src* activation by nitrated peptides (Mallozzi *et al.*, 2001b) (Fig. 4). We developed a kinase

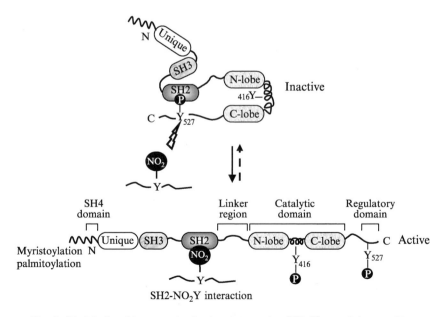

FIG. 3. Model of *src* kinases activation by nitrotyrosine ($NO_2Y$)-containing peptides.

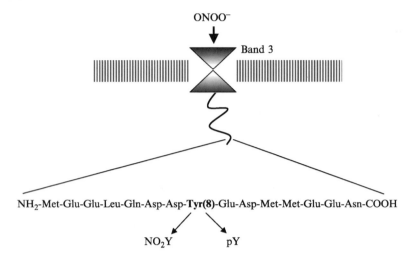

ONOO⁻

Band 3

NH₂-Met-Glu-Glu-Leu-Gln-Asp-Asp-**Tyr(8)**-Glu-Asp-Met-Met-Glu-Glu-Asn-COOH

NO₂Y          pY

FIG. 4. Sequence of C-terminal cytoplasmic tail of band 3.

assay that, when used with a pull-down assay, permits identification of the interaction between the SH2/SH3 domains and nitrotyrosine-containing target proteins.

Conclusions

The activation of tyrosine phosphorylation signaling by oxidants, and in particular by peroxynitrite, may depend on both direct and indirect mechanisms. Direct activation of *src* kinases may occur through oxidation of critical cysteine residues while indirect mechanisms are (1) inhibition of phosphotyrosine phosphatases, (2) crosstalk between *src* and non-*src* kinases, and (3) nitration of specific tyrosine residues with SH2 affinity. The following experimental protocols have been used to clarify the activation mechanism in erythrocytes.

Methods

*Peroxynitrite Preparation*

Different methods can be used to synthesize peroxynitrite (Uppu *et al.*, 1996). One simple method uses nitrite and hydrogen peroxide followed by stabilization at alkaline pH (Radi *et al.*, 1991). At acidic pH, hydrogen peroxide oxidizes nitrite to peroxynitrous acid, and Na(OH) stabilizes the reaction product as the peroxynitrite anion.

*Solution A*: 10 ml of 1.0 $M$ NaNO$_2$ in distilled water.
*Solution B*: 10 ml of 1.0 $M$ H$_2$O$_2$ in 1 $M$ HCl.
*Solution C*: 10 ml of 1.5 $M$ Na(OH). All solutions are ice-cooled. Distilled water used to prepare solutions A–C is treated with Chelex 100 (Bio-Rad, Richmond, CA) to remove traces of transition metals.

*Procedure*

1. Rapid mixing of ice-cold solutions A and B is followed immediately ($\leq$1 s) by the addition of ice-cold solution C. The yellow indicates the formation of the peroxynitrite anion ($\varepsilon_{302} = 1700\ M^{-1}\ cm^{-1}$).
2. Eliminate excess hydrogen peroxide by treatment with MnO$_2$ (6 mg/ml; 30–60 min at 4°).
3. Remove MnO$_2$ by centrifugation (5 min, 15,000$g$ at 4°) and filtration (0.2 $\mu$m).
4. Freeze-fractionation of peroxynitrite solution at $-80°$ and collection of the yellow top layer, which typically contains 200–500 m$M$ peroxynitrite, 50–200% nitrite, $\leq$1–2% hydrogen peroxide and 40–50% nitrate. To modify the concentration of contaminants, change the NaNO$_2$: H$_2$O$_2$ ratio.

*Peroxynitrite Treatment of Erythrocytes*

1. Wash the erythrocytes three times with isotonic phosphate-buffered saline (PBS), pH 7.4.
2. Add peroxynitrite as a bolus and under rapid stirring to cells suspended (2% hematocrit) in 30 m$M$ NaCl, 80 m$M$ phosphate buffer, 0.1 m$M$ diethylenetriaminepentaacetic acid (DTPA), pH 7.2. Incubate at 37° for 5 min. The phosphate buffer is treated extensively with Chelex 100, and all samples contain 0.1 m$M$ DTPA (Sigma Chemical Co.). Decomposed peroxynitrite is obtained by adding peroxynitrite to the phosphate buffer and leaving for 5 min at room temperature before the addition of erythrocytes (reversed order of addition).
3. After peroxynitrite treatment, wash the erythrocytes twice in PBS.

*Determination of PTP Activity in the Membrane Fraction*

*Lysis buffer:* ice-cold 5 m$M$ phosphate buffer (pH 8.0), 0.2 m$M$ phenylmethylsulfonylfluoride (PMSF)
*Reaction buffer:* 25 m$M$ HEPES buffer (pH 7.3), 0.1 m$M$ PMSF, 20 m$M$ MgCl$_2$

*Procedure*

1. Lyse cold pelleted red blood cells with lysis buffer (1:10 vol/vol).

2. Pellet the membrane fraction by centrifugation at 20,000 rpm at 4° for 15 min.

3. Remove the supernatant and wash the pellet extensively with cold lysis buffer until a white hemoglobin-free pellet is obtained.

4. Suspend 20 $\mu$l of pelleted membranes (protein content ∼3 mg/ml) in 100 $\mu$l reaction buffer.

5. Start the reaction by adding 15 $\mu$l substrate (p-nitrophenyl phosphate 0.1 $M$) and incubate at 37° for 30 min.

6. Stop the reaction by adding 865 $\mu$l 0.1 m$M$ NaOH. Centrifuge the samples at 10,000 rpm for 10 min, and read the absorbance of the supernatant at 410 nm in a spectrophotometer (this wavelength reads the yellow of p-nitrophenol released from p-nitrophenyl phosphate).

## Immunoprecipitation Protocol

*Lysis buffer*: 25 m$M$ Tris–HCl (pH 7.5), 150 m$M$ NaCl, 1% (wt/vol) Triton X-100, 1% (vol/vol) Na-deoxycholate, 0.1% (vol/vol) sodium dodecylsulfate (SDS), 10 $\mu$g/ml leupeptin, 10 $\mu$g/ml aprotinin, 1 m$M$ PMSF, 0.1 m$M$ sodium orthovanadate (Na$_3$VO$_4$).

*Antibodies*: rabbit polyclonal anti-*fgr* (200 $\mu$g/ml), anti-*hck* (200 $\mu$g/ml), anti-*lyn* (200 $\mu$g/ml) from Santa Cruz Biotechnology (Santa Cruz, CA).

*Other reagents*: Immunopure Trysacryl immobilized Protein A from Pierce (Rockford, IL); Protein A/G PLUS-agarose from Santa Cruz Biotechnology. [$\gamma^{32}$P]ATP (>3000 Ci/mmol) can be obtained from DuPont NEN (Boston, MA).

*Procedure*

1. Lyse 150 $\mu$l of washed and packed erythrocytes with 150 $\mu$l of 4X ice-cold lysis buffer, and dilute with 300 $\mu$l PBS.

2. After 10 min of incubation in ice, add 400 $\mu$l lysis buffer 1X (final volume 1 ml), and clarify the lysate at 12,000$g$ for 10 min at 4°.

3. Preclearing: This step allows nonspecific binding to be reduced. Add 20 $\mu$l (50% slurry) Protein A/G PLUS-agarose, and mix by means of a rotating wheel for 1 h at 4°. Clarify by centrifugation. The precleared beads may be included in the Western blot analysis, to determine the amount of nonspecific binding in the extracts.

4. Preadsorption of the antibody: This step reduces the time necessary for the immunoprecipitation procedure. Sufficient antibody preadsorbed

on Protein A-Trysacryl can be prepared in a single tube for all the samples [1 $\mu$g antibody + 20 $\mu$l Protein A-Trysacryl (50% slurry) for each sample] by incubation for 3 h at 4° in a rotating wheel. Wash twice in PBS, and suspend the beads in lysis buffer again. Alternatively, the antibody (1 $\mu$g/ 100 $\mu$g protein) can be added to the lysate and incubated overnight at 4° in a rotating wheel, and the immune complex can be precipitated by Protein A-Trysacryl 40 $\mu$l (50% slurry).

5. Incubate lysate with the different preadsorbed antibodies (20 $\mu$l) overnight at 4° in a rotating wheel.

6. Pellet the precipitated immune complex beads, saving the supernatant (flow-through control).

7. Wash the beads twice with 1 ml lysis buffer and twice with 1 ml 50 m$M$ Tris, 150 m$M$ NaCl, pH 7.5 (TBS).

8. Perform kinase assay.

*Treatment of the Immune Complex with Peroxynitrite*

1. Suspend the beads in 500 $\mu$l of TBS, add peroxynitrite and mix vigorously.
2. After 5 min at room temperature, spin at 5000 rpm for 1 min at 4°.
3. Wash twice with 1 ml TSB.
4. Perform kinase assay.

*Kinase Assay*

*Reaction buffer:* 20 m$M$ Tris (pH 7.4), 10 m$M$ MnCl$_2$, 0.1 m$M$ Na$_3$VO$_4$, [$\gamma^{32}$P] ATP (>3000 Ci/mmol) (10 $\mu$Ci/100 $\mu$l reaction buffer). Sufficient reaction buffer can be prepared in a single tube for all samples.

*Procedure*

1. Wash beads once with reaction buffer without [$\gamma^{32}$P] ATP.
2. Add 20 $\mu$l of reaction buffer with [$\gamma^{32}$P] ATP to each sample.
3. Perform the kinase assay for 10 min at 22°.
4. Stop the reaction by adding 10 $\mu$l 4X Laemmli sample buffer. The mixture is boiled for 5 min and subjected to 10% SDS–polyacrylamide gel electrophoresis (PAGE). The gels are then dried and exposed to X-ray film for autoradiography. The [$^{32}$P]-labeled proteins can be quantified by a Phosphor Imager Instrument (Packard, Camberra, CO).

## Determination of Lyn Activation Induced by Peptides

*Peptide sequences*: NH$_2$-Thr-Glu-Pro-Gln-Tyr-Glu-Glu-Ile-Pro-Ile-COOH in which the Tyr is unphosphorylated (Y), phosphorylated (pY), or nitrated at the 3 position (NO$_2$Y). The sequences of scrambled peptides in which the Tyr is phosphorylated or substituted with 3-nitrotyrosine are: NH$_2$-Glu(Ac)-Gln-Glu-Pro-Tyr-Ile-Pro-Ile-Glu-COOH (Sc1), and NH$_2$-Tyr-Pro-Glu-Pro-Glu-Ile-Gln-Ile-Glu-COOH (Sc2). The sequence of band3–derived peptides in which the Tyr is unphosphorylated (band 3-Y), phosphorylated (band 3-pY), or substituted with 3-nitrotyrosine (band 3-NO$_2$Y) is NH$_2$-Met-Glu-Glu-Leu-Gln-Asp-Asp-Tyr-Glu-Asp-Met-Met-Glu-Glu-Asn-COOH (Fig. 4). All peptides can be obtained from Neosystem (Strasbourg, France).

### Procedure

1. Immunoprecipitate *lyn* from erythrocyte lysate using agarose-conjugated anti-*lyn* polyclonal antibody (Santa Cruz, CA).
2. Suspend the beads in 7.5 $\mu$l TBS and add 2.5 $\mu$l peptides (dissolve all peptides in distilled water at 4 m$M$). Peptide final concentrations of 1 m$M$.
3. Incubate for 2 h at 4° and perform kinase assay by adding 10 $\mu$l reaction buffer with [$\gamma^{32}$P] ATP.

## Solid-Phase Binding Assay

Solid-phase binding assay detects the binding of biotin-labeled band3–derived peptides with immunoprecipitated *lyn* kinase or purified GST-fusion proteins containing *lyn* SH2 or SH3 domains. The biotin-labeled peptides are synthesized by Neosystem (Strasbourg, France).

*Materials.* Recombinant GST-fusion proteins: *lyn* 131–243 (SH2), *lyn* 27–131 (SH3), and *lyn* 1–243 (SH2/SH3). Biotin-labeled peptides constructed on band 3: (band 3-Y), (band 3-pY), and (band 3-NO$_2$Y). Glutathione-Sepharose 4B from Amersham Pharmacia Biotech (Uppsala, Sweden).

### Procedure

1. Immunoprecipitate *lyn* from erythrocyte lysate using agarose-conjugated anti-*lyn* polyclonal antibody.
2. Incubate the immune complex with 1 m$M$ biotinylated peptides for 2 h at 4° (final volume of 10 $\mu$l).
3. Wash the beads extensively in TBS and suspend them in 10 $\mu$l TBS/ 1% SDS, spin 1 min at 4°, 5000 rpm.

4. Spot 5 $\mu$l of supernatant on dry nitrocellulose paper.

5. Alternatively, 5 $\mu$g of GST-*lyn* SH2, GST-*lyn* SH3, or GST alone is incubated for 2 h at 4° with 1 m$M$ biotinylated peptides (final volume of 20 $\mu$l). The complex is pulled out by adding 10 $\mu$l of glutathione-Sepharose beads (1 h at 4°). Wash the beads extensively with TBS and dissociate the bound proteins from the beads by 10 $\mu$l of 1% SDS. Remove the beads by centrifugation and spot 5 $\mu$l of supernatant on nitrocellulose.

6. Block the membrane with 3% bovine serum albumin/TBS–Tween 20 for 2 h at room temperature.

7. Incubate the nitrocellulose for 1 h with biotin–avidin–peroxidase complex (Vectastain ABC Kit, Vector Laboratories, Burlingame, CA).

8. Wash in TBS–Tween 20 for 30 min.

9. Detection by enhanced chemiluminescence (ECL Kit, Pierce). The filters are also subjected to immunoblot analysis using polyclonal anti-GST antibodies or polyclonal anti-*lyn* antibodies.

## Acknowledgment

This work was supported in part by Istituto Superiore di Sanità, Roma, Italy, research project C3AD, cap. 524.

## References

Akhand, A. A., Pu, M., Senga, T., Kato, M., Suzuki, H., Miyata, T., Hamaguchi, M., and Nakashima, I. (1999). Nitric oxide controls *src* kinase activity through a sulfhydryl group modification-mediated Tyr-527–independent and Tyr-416–linked mechanism. *J. Biol. Chem.* **274**, 25821–25826.

Beckman, J. S., Beckman, T. W., Chen, J., and Marshall, P. A. (1990). Apparent hydroxyl radical production by peroxynitrite: Implications for endothelial injury from nitric oxide and superoxide. *Proc. Natl. Acad. Sci. USA* **87**, 1620–1624.

Beckman, J. S., and Koppenol, W. H. (1996). Nitric oxide, superoxide, and peroxynitrite: The good, the bad, and the ugly. *Am. J. Physiol.* **271**, C1424–C1437.

Bjorge, J. D., Jakymiw, A., and Fujita, D. J. (2000a). Selected glimpses into the activation and function of *src* kinase. *Oncogene* **19**, 5620–5635.

Bjorge, J. D., Pang, A., and Fujita, D. J. (2000b). Identification of protein-tyrosine phosphatase 1B as the major tyrosine phosphatase activity capable of dephosphorylating and activating c-Src in several human breast cancer cell lines. *J. Biol. Chem.* **275**, 41439–41446.

Cadenas, E. (2004). Mitochondrial free radical production and cell signaling. *Mol. Aspects Med.* **25**, 17–26.

Chiang, G. G., and Sefton, B. M. (2000). Phosphorylation of a src kinase at the autophosphorylation site in the absence of *src* kinase activity. *J. Biol. Chem.* **275**, 6055–6058.

Cooper, J. A., and MacAuley, A. (1988). Potential positive and negative autoregulation of p60c-*src* by intermolecular autophosphorylation. *Proc. Natl. Acad. Sci. USA* **85**, 4232–4236.

Di Stasi, A. M. M., Mallozzi, C., Macchia, G., Petrucci, T. C., and Minetti, M. (1999). Peroxynitrite induces tyrosine nitration and modulates tyrosine phosphorylation of synaptic proteins. *J. Neurochem.* **73,** 727–735.

Fang, K. S., Sabe, H., Saito, H., and Hanafusa, H. (1994). Comparative study of three protein-tyrosine phosphatases. Chicken protein-tyrosine phosphatase lambda dephosphorylates c-Src tyrosine 527. *J. Biol. Chem.* **269,** 20194–20200.

Ischiropoulos, H., and Al-Mehdi, A. B. (1995). Peroxynitrite-mediated oxidative protein modifications. *FEBS Lett.* **364,** 279–282.

King, P. A., Anderson, V. E., Edwards, J. O., Gustafson, G., Plumb, R. C., and Suggs, J. W. (1992). A stable solid that generates hydroxyl radical upon dissolution in aqueous solution: reaction with proteins and nucleic acid. *J. Am. Chem. Soc.* **14,** 5430–5432.

Klages, S., Adam, D., Class, K., Fargnoli, J., Bolen, J. B., and Penhallow, R. C. (1994). Ctk: A protein-tyrosine kinase related to Csk that defines an enzyme family. *Proc. Natl. Acad. Sci. USA* **91,** 2597–2601.

Klotz, L., Schroeder, P., and Sies, H. (2002). Peroxynitrite signaling: Receptor tyrosine kinases and activation of stress-responsive pathways. *Free Radic. Biol. Med.* **33,** 737–743.

Kong, S.-K., Yim, M., Stadman, E. R., and Chock, P. B. (1996). Peroxynitrite disables the tyrosine phosphorylation regulatory mechanism: Lymphocyte-specific tyrosine kinase fails to phosphorylate nitrated cdc2(6-20)NH2 peptide. *Proc. Natl. Acad. Sci. USA* **93,** 3377–3382.

Lee, C. H., Leung, B., Lemmon, M. A., Zheng, J., Cowburn, D., Kuriyan, J., and Saksela, K. (1995). A single amino acid in the SH3 domain of Hck determines its high affinity and specificity in binding to HIV-1 Nef protein. *EMBO J.* **14,** 5006–5015.

Liu, X., Bordeur, S. R., Gish, G., Songyang, Z., Cantley, L. C., Laudano, A. P., and Pawson, T. (1993). Regulation of c-src tyrosine kinase activity by the *src* SH2 domain. *Oncogene* **8,** 1119–1126.

Mallozzi, C., Di Stasi, A. M. M., and Minetti, M. (1999). Activation of src tyrosine kinases by peroxynitrite. *FEBS Lett.* **456,** 201–206.

Mallozzi, C., Di Stasi, A. M. M., and Minetti, M. (2001a). Peroxynitrite-dependent activation of *src* tyrosine kinases *lyn* and *hck* in erythrocytes is under mechanistically different pathways of redox control. *Free Radic. Biol. Med.* **30,** 1108–1117.

Mallozzi, C., Di Stasi, A. M. M., and Minetti, M. (2001b). Nitrotyrosine mimics phosphotyrosine binding to the SH2 domain of the *src* family tyrosine kinase *lyn*. *FEBS Lett.* **503,** 189–195.

Mallozzi, C., Di Stasi, A. M. M., and Minetti, M. (1997). Peroxynitrite modulates tyrosine-dependent signal transduction pathway of human erythrocyte band 3. *FASEB J.* **11,** 1281–1290.

Moraefi, I., LaFevre-Bernt, M., Sicheri, F., Huse, M., Lee, C. H., Kuriyan, J., and Miller, W. T. (1997). Activation of the Src-family tyrosine kinase Hck by SH3 domain displacement. *Nature* **385,** 650–653.

Okada, M., and Nakagawa, H. (1989). A protein tyrosine kinase involved in regulation of pp60c-src function. *J. Biol. Chem.* **264,** 20886–20893.

Oo, M. L., Senga, T., Thant, A. A., Amin, A. R., Huang, P., Mon, N. N., and Hamaguchi, M. (2003). Cysteine residues in the C-terminal lobe of Src: Their role in the suppression of the Src kinase. *Oncogene* **22,** 1411–1417.

Prutz, W. A., Butler, J., Land, E. J., and Swallow, A. J. (1986). Unpaired electron migration between aromatic and sulfur peptide units. *Free Radic. Res. Commun.* **2,** 69–75.

Radi, R. (1998). Peroxynitrite reactions and diffusion in biology. *Chem. Res. Toxicol.* **11,** 720–721.

Radi, R., Beckman, J. S., Bush, K. M., and Freeman, B. A. (1991a). Peroxynitrite oxidation of sulfhydryls: Cytotoxic potential of superoxide and nitric oxide. *J. Biol. Chem.* **266,** 4244–4250.

Radi, R., Beckman, J. S., Bush, K. M., and Freeman, B. A. (1991b). Peroxynitrite-induced membrane lipid peroxidation: The cytotoxic potential of superoxide and nitric oxide. *Arch. Biochem. Biophys.* **288,** 481–487.

Senga, T., Miyazaki, K., Machida, K., Iwata, H., Matsuda, S., Nakashima, I., and Hamaguchi, M. (2000). Clustered cysteine residues in the kinase domain of v-Src: Critical role for protein stability, cell transformation and sensitivity to herbimycin A. *Oncogene* **19,** 273–279.

Sicheri, F., and Kuriyan, J. (1997). Structures of src-family tyrosine kinases. *Curr. Opin. Struc. Biol.* **7,** 777–785.

Somani, A. K., Bignon, J. S., Mills, G. B., Siminovitch, K. A., and Branch, D. R. (1997). Src kinase activity is regulated by the SHP-1 protein-tyrosine phosphatase. *J. Biol. Chem.* **272,** 21113–21119.

Takakura, K., Beckman, J. S., MacMillan-Crow, L. A., and Crow, J. P. (1999). Rapid and irreversible inactivation of protein tyrosine phosphatases PTP1B, CD45, and LAR by peroxynitrite. *Arch. Biochem. Biophys.* **369,** 197–207.

Thomas, S. M., and Brugge, J. S. (1997). Cellular functions regulated by src family kinases. *Annu. Rev. Cell. Dev. Biol.* **13,** 513–609.

Trauner, M. (2003). When bile ducts say NO: The good, the bad, and the ugly. *Gastroenterology* **124,** 847–851.

Uppu, R. M., Squadrito, G. L., Cueto, R., and Pryor, W. A. (1996). Selecting the most appropriate synthesis of peroxynitrite. *Methods Enzymol.* **269,** 285–296.

Veillette, A., Dumont, S., and Fournel, M. (1993). Conserved cysteine residues are critical for the enzymatic function of the lymphocyte-specific tyrosine protein kinase p56[lck]. *J. Biol. Chem.* **268,** 17547–17553.

Williams, J. C., Wierenga, R. K., and Saraste, M. (1998). Insights into *src* kinase functions: Structural comparison. *TIBS* **23,** 179–184.

Zheng, X. M., Wang, Y., and Pallen, C. J. (1992). Cell transformation and activation of pp60c-*src* by overexpression of a protein tyrosine phosphatase. *Nature* **359,** 336–339.

# [21] Hemoglobin and Red Blood Cells as Tools for Studying Peroxynitrite Biochemistry

By NATALIA ROMERO and RAFAEL RADI

## Abstract

Oxyhemoglobin represents a relevant intravascular sink of peroxynitrite. Indeed, peroxynitrite undergoes a fast isomerization (k = $1.7 \times 10^4$ $M^{-1}s^{-1}$) to nitrate in the presence of oxyhemoglobin; the reaction mechanism is complex and leads to methemoglobin and superoxide radical as additional products and a small amount ($\sim$10%) of transient species, including ferrylhemoglobin, nitrogen dioxide, and globin-derived radicals.

METHODS IN ENZYMOLOGY, VOL. 396
0076-6879/05 $35.00
DOI: 10.1016/S0076-6879(05)96021-7

The mechanism of the reaction could be solved only after extensive quantitative analysis of reactants, intermediates, and products and setting up experimental conditions that favor direct reactions of peroxynitrite with hemoglobin versus peroxynitrite decay through proton- or carbon dioxide-catalyzed homolysis. Additionally, oxyhemoglobin has been used as a "reporter" molecule of peroxynitrite diffusion from extracellular to intracellular compartments, using red blood cells (RBCs) as a model system. In RBCs, peroxynitrite diffusion across the membrane is favored by the large abundance of anion channels, and average transit distances can vary as a function of cell density. Indeed, we have developed a mathematical model that incorporates competition between the extracellular consumption of peroxynitrite and the permeation to the erythrocytes as a function of the average diffusion distances. The RBC model presented herein serves to estimate biological diffusion distances of peroxynitrite in the presence of relevant molecular targets, and the theoretical approach can be successfully applied to study the diffusion of peroxynitrite in other cellular/tissue systems.

## Introduction

Peroxynitrite is a strong oxidant, and cytotoxic species formed *in vivo* as a result of the diffusion-controlled reaction between nitric oxide ($^\bullet$NO) and superoxide anion ($O_2^{\bullet-}$) radicals (Radi, 2004). Peroxynitrite formed in biological systems has different possible fates including (a) direct reactions (one- or two-electron oxidations) with several targets and (b) $H^+$- and $CO_2$-mediated homolysis to hydroxyl radical ($^\bullet$OH) and nitrogen dioxide ($^\bullet$NO$_2$) or carbonate radical ($CO_3^{\bullet-}$) and $^\bullet$NO$_2$, respectively; peroxynitrite-derived radicals participate in one-electron oxidation and nitration reactions. The reaction chemistry of peroxynitrite with transition metal-containing proteins such as hemoglobin can be of mixed mechanism, with an initial fast direct reaction with the metal followed by secondary radical pathways. The kinetics of peroxynitrite decomposition and secondary reactions are largely influenced by target concentration (Radi, 1996; Radi *et al.*, 2000).

In spite of its high reactivity in biological systems, peroxynitrite is able to diffuse distances on the order of 5–20 $\mu$m and undergo target molecule reactions in compartments different than where it was originally formed (Denicola *et al.*, 1998; Macfadyen *et al.*, 1999; Marla *et al.*, 1997). The intravascular formation of peroxynitrite by endothelial and/or inflammatory cells contributes to disease conditions such as hypertension, atherogenesis, and diabetic endotheliopathy (Radi, 2004; Sowers, 2002). We have previously postulated the relevance of the oxyhemoglobin [oxyHb, Hb(Fe$^{II}$)O$_2$] reaction with peroxynitrite *in vivo* (Denicola *et al.*, 1998) and the role of oxyHb in red blood cells (RBCs) as a relevant intravascular sink of peroxynitrite. The aim of this work is to describe methodology and results obtained from the study of peroxynitrite reaction with hemoglobin and RBCs and to discuss their usefulness for the study of peroxynitrite reactivity and diffusivity.

## Oxyhemoglobin and Peroxynitrite

Peroxynitrite can oxidize transition metal centers, especially metal-porphyrin systems as heme proteins, with high kinetics constants (Radi *et al.*, 2000). These oxidations can be one- or two-electron processes depending on the particular metal center and its starting oxidation state. One-electron reduction of peroxynitrite yields nitrogen dioxide ($^\bullet NO_2$) [Eq. (1)], whereas two-electron reduction yields nitrite ($NO_2^-$) [Eq. (2)]; it has also been established that some metal complexes catalyze peroxynitrite isomerization to nitrate ($NO_3^-$) [Eq. (3)] (Bourassa *et al.*, 2001; Herold and Shivashankar, 2003; Jensen and Riley, 2002; Romero *et al.*, 2003).

Both one-electron oxidation or isomerization reactions go through the formation of intermediate complexes where the metal ion acts as a Lewis acid and promotes peroxynitrite decomposition via O-O bond homolysis, to yield $^\bullet OH$ and $^\bullet NO_2$ inside a solvent cage. The rate of radical diffusion out of the solvent cage could in some cases be slow, favor the recombination of caged radicals to $NO_3^-$, and result in low $^\bullet NO_2$ yields, whereas in other cases, the caged radicals recombination is the slow step, and the free radical yield is close to 100% (Ferrer-Sueta *et al.*, 2002).

$$\text{One-electron oxidation: } X\text{-}Me^n + ONOO^- \rightarrow$$
$$X\text{-}Me^{n+1}{=}O + {}^\bullet NO_2 \tag{1}$$

$$\text{Two-electron oxidation: } X\text{-}Me^n + ONOO^- \rightarrow$$
$$X\text{-}Me^{n+2}{=}O + NO_2^- \tag{2}$$

$$\text{Isomerization: } X\text{-}Me^n + ONOO^- \rightarrow X\text{-}Me^n + NO_3^- \tag{3}$$

Typically, reactions of peroxynitrite with transition metal centers concomitantly produce, in addition to $^\bullet NO_2$, oxo-metal complexes, which are strong oxidants and promote secondary redox transitions such as one-electron oxidation of tyrosine. Formation of tyrosyl radical, concomitantly with $^\bullet NO_2$, promotes nitration reactions (Radi, 2004).

Because of its high concentration in the intravascular compartment and the abundance of anion channels in the RBC membrane, oxyHb has been proposed as one of the most important "sinks" for intravascularly formed peroxynitrite (Minetti *et al.*, 2000; Romero *et al.*, 1999, 2003). Although different reports consistently demonstrated methemoglobin [metHb, $Hb(Fe^{III})$] as the main final product of the reaction (Alayash et al., 1998; Denicola et al., 1998; Exner and Herold, 2000; Romero et al., 2003), the mechanism of metHb formation was far from obvious. Initial attempts to define the mechanism involved the formation of a high oxidation state of hemoglobin, namely ferrylhemoglobin [ferrylHb, $Hb(Fe^{IV})$], as a transient intermediate that would be reduced back to metHb by peroxynitrite (Exner and Herold, 2000) or globin residues adjacent to the heme moiety (Minetti et al., 2000). However, peroxynitrite oxidation to oxyHb by a two-electron oxidation process should also result in stoichiometric formation of nitrite and molecular oxygen in addition to ferrylHb [Eq. (4)]:

$$Hb(Fe^{II})O_2 + ONOOH \rightarrow Hb(Fe^{IV})=O + O_2 + NO_2^- \qquad (4)$$

Surprisingly, when Romero et al. (2003) performed product analysis, ferrylHb and molecular oxygen were obtained at much lower yields than expected and no $NO_2^-$ formation was detected. Thus, to establish the reaction mechanism, it is critical to perform a quantitative analysis of all the expected, final products and evaluate possible intermediates.

Another important consideration for mechanistic studies on the *direct* reactions of peroxynitrite with targets is that appropriate peroxynitrite/ target ratios must be used (i.e., large excess of target) to minimize the fraction of peroxynitrite that undergoes homolysis to $^\bullet OH$ and $^\bullet NO_2$ ($k = 0.26 \text{ s}^{-1}$ at $25°$ and pH 7.4 and $k = 0.9 \text{ s}^{-1}$ at $37°$ and pH 7.4). Under appropriate conditions, most peroxynitrite will react directly with the target, and secondary radical reactions observed would depend on intermediate products arising from the direct reaction and not from the radicals arising from peroxynitrite homolysis.

Observations from our group indicate that the reaction of oxyHb with peroxynitrite comprises a complex mechanism that results in the isomerization of peroxynitrite to $NO_3^-$ and the one-electron oxidation of oxyHb to metHb at the expense of the one-electron reduction of the ferrous heme-bound molecular oxygen to yield superoxide radical anion ($O_2^{\bullet-}$) (Romero et al., 2003). In this mechanism (Fig. 1), a metHb-peroxynitrite intermediate complex is formed, followed by a rapid ferric heme-catalyzed peroxynitrite decomposition to "caged" ferrylHb and $^\bullet NO_2$. Recombination of the $^\bullet NO_2$ radical with ferrylHb within the heme cavity would mainly yield metHb and $NO_3^-$, whereas a small fraction of caged intermediates diffuse apart and yield free ferrylHb and $^\bullet NO_2$. These two latter species are responsible for secondary oxidation reactions such as oxidation of the globin moiety to yield amino acid–derived radicals (e.g., tyrosyl and cysteinyl radicals) that promote intermolecular cross-linking. The

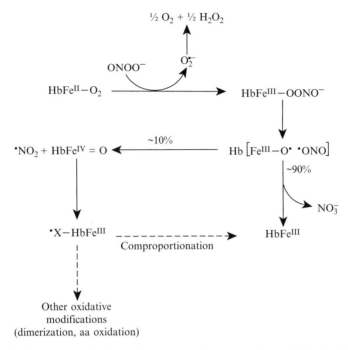

FIG. 1. Proposed mechanisms for the reaction of oxyhemoglobin (oxyHb) with peroxynitrite. Peroxynitrite reacts with oxyHb, displacing $O_2^{\bullet-}$ and forming a methemoglobin (metHb)–peroxynitrite complex. Ferric heme catalyzes peroxynitrite homolysis to ferrylhemoglobin (ferrylHb) and $^{\bullet}NO_2$, which mainly (~90%) recombine within the heme cavity yielding metHb and $NO_3^-$ while a minor fraction (~10%) of $^{\bullet}NO_2$ diffuses outside the heme. FerrylHb is reduced back to metHb by a proximal amino acid residue (−X), yielding globin-centered radicals (e.g., tyrosyl and cysteinyl radicals) that finally result in one-electron oxidation of another oxyHb molecule to yield metHb (comproportionation reaction), interchain cross-linking or globin-oxidized products. In addition, $O_2^{\bullet-}$ released in the first step dismutates spontaneously to yield $O_2$ and $H_2O_2$. Adapted from Romero *et al.* (2003). Alternative mechanisms for the isomerization reaction are under investigation (Boccini and Herold, 2004).

proposed isomerization mechanism has some analogy to the one proposed for the reaction of oxymyoglobin with $^{\bullet}NO$ (Wade and Castro, 1996).

## Experimental Approach and Results

### Purification of Human Oxyhemoglobin and Spectroscopic Analysis

Purified human oxyHb is prepared from fresh human erythrocytes by ion-exchange chromatography as described by Winterbourn (1990). The different oxidation states of hemoglobin could be distinguished by differences in their absorption spectra in the visible region. OxyHb presents two peaks in the 500–700 nm region, with a maximum at 545 and 577 nm ($\varepsilon_{577\,nm} = 15.0\ mM^{-1}\ cm^{-1}$). One-electron oxidation of the ferrous heme yields metHb, characterized by the disappearance of the peaks around 550 nm and a characteristic shoulder at 630 nm ($\varepsilon_{630\,nm} = 3.63\ mM^{-1}\ cm^{-1}$). The high oxidation state of hemoglobin, ferrylHb, is characterized by a broad absorption band around 550 nm (with a maximum at 548 and 582 nm) and the absence of the 630 nm shoulder. Because this species is highly reactive ($E^{o'} = 0.99$ V) (Koppenol and Liebman, 1983), its detection in biological systems is usually performed in the presence of sodium sulfide, which yields sulfhemoglobin (after nucleophilic addition of sulfide to the oxoferryl heme) with a characteristic peak at 617 nm ($\varepsilon_{617\,nm} = 24.0\ mM^{-1}\ cm^{-1}$) (Berzofsky et al., 1971). If only oxyHb and metHb are present, concentration of both species can be calculated monitoring absorbances at 577 and 630 nm as described by Winterbourn (1990): oxyHb (concentration of heme, $\mu M$) = 66 $A_{577}$ − 80 $A_{630}$; metHb (concentration of heme, $\mu M$) = 279 $A_{630}$ − 3 $A_{577}$.

### Kinetic Studies

Kinetic studies of peroxynitrite reaction with oxyHb cannot be performed following peroxynitrite decay at 302 nm as usually chosen (Radi, 1996), because of (1) changes in hemoglobin absorbance during its oxidation and (2) the high concentrations of protein required to be under conditions of hemoglobin excess, which interfere with peroxynitrite measurements. Instead, initial velocities studies were performed following disappearance of oxyHb at 577 nm ($\Delta\varepsilon = 11.000\ M^{-1}\ cm^{-1}$) using a stopped-flow spectrophotometer as described previously (Denicola et al., 1998; Radi, 1996). OxyHb (25 $\mu M$) in 200 mM potassium phosphate buffer, 0.2 mM diethylenetriaminepentaacetic acid (DTPA), pH 7.0, was mixed with peroxynitrite (50–300 $\mu M$) in NaOH (20 mM) at 37° and pH 7.4 (measured at the outlet of the stopped flow); the first 0.02–0.05 s portion of the reaction was fit to a linear plot to obtain the initial rates at different peroxynitrite concentrations. Initial rates of oxyHb oxidation linearly increased with initial peroxynitrite concentration, and from the slope of this plot, an apparent second-order kinetic constant of $1.7 \times 10^4\ M^{-1}\ s^{-1}$ was obtained.

### Detection of Transient Intermediates and Final Products

DETERMINATION OF NITRITE AND NITRATE. To study the mechanism of peroxynitrite reactions, it is important to determine whether peroxynitrite is isomerized to $NO_3^-$, reduced by one electron to $^\bullet NO_2$ (that will finally

TABLE I

EXPECTED AND EXPERIMENTALLY OBTAINED YIELDS OF PRODUCTS OF THE REACTION OF OXYHB
WITH A SUBSTOICHIOMETRIC AMOUNT OF PEROXYNITRITE, CONSIDERING POSSIBLE DIFFERENT
REACTION MECHANISMS (ONE- OR TWO-ELECTRON OXIDATION OR ISOMERIZATION)

| | Yield[a] | | | | | |
| --- | --- | --- | --- | --- | --- | --- |
| | metHb | ferrylHb | $O_2$ | $H_2O_2$ | $NO_3^-$ | $NO_2^-$ |
| One-electron oxidation | 1 | – | 1.00 | – | 0.5 | 0.5 |
| Two-electron oxidation | – | 1.0 | 1.00 | – | – | 1.0 |
| Isomerization[b] | 1 | – | 0.50 | 0.5 | 1.0 | – |
| Experimental values | 1 (final) | ~0.1 (transient) | 0.35 | ~0.5 | ~1.0 | n.d.[c] |

[a] Yields are expressed as fractional values and are referred to with respect to added peroxynitrite.
[b] Assuming proposed isomerization mechanism (Romero et al., 2003).
[c] n.d., not detected.

yield ½ $NO_2^-$ and ½ $NO_3^-$ per •$NO_2$) or by two electrons to $NO_2^-$ (Table I), so product analysis of $NO_x^-$ is important for elucidation of reaction mechanisms.

Determination of $NO_2^-$ is carried out by the Griess method (Schmidt and Kelm, 1996), and to determine total $NO_2^-$ plus $NO_3^-$, a solution of vanadium (III) chloride (25 mM final concentration) is added to the Griess reagents to reduce $NO_2^-$ to $NO_3^-$, as described by Miranda et al. (2001). Because $NO_2^-$ is not an inert product of the reaction and slowly reacts with oxyHb (Kosaka et al., 1981; Pietraforte et al., 2004), it is important that immediately after each treatment, the samples are passed through a desalting column (such as a Hi-Trap Desalting column, Amersham Bioscience) to separate $NO_x^-$ from oxyHb. $NO_2^-$ and $NO_3^-$ analysis could also be carried out by anion exchange chromatography with conductivity detection as described by Herold (1999). Analysis of nitrogen oxides after oxyHb reaction with a limiting amount of peroxynitrite resulted in a stoichiometric isomerization to nitrate (Table I).

DETECTION OF A FERRYLHB INTERMEDIATE. Studies of the spectral changes occurring during peroxynitrite reaction with oxyHb using a fast response photodiode array accessory of the stopped-flow spectrophotometer (SX17MV; Applied Photophysics, Leatherhead, England) showed that the detected ferrylHb species has a very short half-life (i.e., milliseconds), and can only be clearly observed when trapped with sodium sulfide, and even the sulfhemoglobin yields obtained were low (10–15% of metHb yield). The lack of stability of this ferrylHb contrasts with the relatively stable intermediate obtained from the reaction of oxyHb and metHb under *excess* hydrogen peroxide ($H_2O_2$), which can be observed *directly* by spectrophotometric techniques (D'Agnillo and Alayash, 2000; Giulivi and Davies, 1990). We hypothesize that under *excess* oxidant, metHb is formed and subsequently oxidized by two electrons, yielding the stable oxo-iron (IV) complex plus a globin-derived radical, analogue to peroxidases compound I:

$$Hb(Fe^{II})O_2 \xrightarrow{H_2O_2} \dashrightarrow Hb(Fe^{III}) \xrightarrow{2e^-} \underset{(relatively\ stable\ species)}{{}^\bullet X\text{-}Hb(Fe^{IV}) = O} \tag{5}$$

In contrast, the two-electron oxidation product of oxyHb with a *limiting* amount of oxidant (peroxynitrite or hydrogen peroxide) yields an unstable ferrylHb that rapidly undergoes a one-electron reductive step to metHb, probably at the expense of the globin moiety and lastly yielding an amino acid–derived radical [Eq. (6)]:

$$Hb(Fe^{II})O_2 \xrightarrow{2e^-} \underset{(unstable\ species)}{Hb(Fe^{IV}) = O} \longrightarrow {}^\bullet X\text{-}Hb(Fe^{III}) \tag{6}$$

OXYGEN AND HYDROGEN PEROXIDE YIELDS. One important difference between oxyHb and most other transition metal-containing proteins is that oxyHb has a diatomic oxygen molecule bound to the ferrous penta-coordinated heme that cannot bind to the hexa-coordinated ferric heme. Consequently, additional product analysis should be performed concerning the fate of the oxygen molecule. Oximetry studies with a Clark-type electrode reveal oxygen evolution from the reaction of peroxynitrite with oxyHb, but yields are only near 30% with respect to metHb formed (Table I). On the other hand, the known two-electron oxidation of oxyHb by $H_2O_2$ or the one-electron oxidation by $Fe(CN)_6^{3-}$ results in oxygen yields of approximately 70% and 100%, respectively. We, therefore, evaluated whether oxygen molecules bound to the ferrous heme were reduced by one or two electrons to $O_2^{\bullet-}$ and/or $H_2O_2$, respectively, during the reaction of oxyHb with peroxynitrite.

$H_2O_2$ was determined by chemiluminescence techniques based on the pseudo-peroxidase activity of hemoglobin; this methodology overcomes the loss of $H_2O_2$ from reactions with oxyHb and metHb that would occur during sample processing that other techniques require. Briefly, peroxynitrite (250 and 500 $\mu M$) is added to oxyHb (1.0 m$M$) in 100 m$M$ phosphate buffer, 0.1 m$M$ DTPA, pH 7.4, and immediately after, 100 $\mu$l of the sample is placed in a tube containing 3.0 ml of luminol (10 $\mu M$) in 200 m$M$ phosphate buffer, 0.3 m$M$ DTPA, pH 7.8. A luminescence signal, partially inhibitable with catalase, was obtained, indicative of $H_2O_2$ formation at 50% yield with respect to peroxynitrite (Table I).

ANALYSIS OF PROTEIN RADICAL FORMATION

• *EPR spin trapping*: Although some amino acid–derived radicals give relatively stable EPR signals that could be analyzed by direct EPR such as tyrosyl-derived radical from ribonucleotide reductase (Barlow *et al.*, 1983), the use of spin-trapping techniques allows detection of short-life radicals because of the stability of the adduct formed between the protein radical and the spin-trap molecule, and it provides information about the identity of the amino acid–derived radical.

Peroxynitrite (1 m$M$) reaction with oxyHb (1 m$M$) in the presence of 2-methyl-2-nitroso propane (MNP) (20 m$M$) results in an immobilized EPR signal characteristic of MNP and amino acid–derived radical adduct. When the spin adduct was subjected to nonspecific proteolysis with pronase (20 mg/ml final concentration), an isotropic three-line spectrum with a hyperfine coupling constant of 15.4 G was obtained, characteristic of a radical centered on a tertiary carbon atom. The spectrum obtained using a smaller modulation amplitude and MNP-d$^9$ as the spin trap provides a super-hyperfine structure that strongly supports the idea that a tyrosyl radical is formed during the reaction.

When reaction was performed in the presence of DMPO (100 m$M$), a four-line spectrum was obtained with a $\beta$-hydrogen splitting constant value of 15.4 G, consistent with those previously assigned to DMPO/protein-thiyl radical adducts. No signal was obtained when reactive Cys-93 of hemoglobin was previously blocked with $N$-ethylmaleimide (NEM).

• *Immuno-spin trapping*: Reaction of the protein-centered radical with the spin trap DMPO yields a nitroxide radical adduct that decays on the order of 1 min as a result of the oxidation to the corresponding globin radical–derived nitrone. The nitrone formed, in spite of being an EPR silent species, is a stable protein modification that can be detected by immuno-spin trapping with the use of a novel antibody developed against DMPO-nitrone adduct (Detweiler *et al.*, 2002). Using this antibody with the oxyHb samples treated with peroxynitrite at 2:1 ratios in the presence of DMPO (50 m$M$), formation of a DMPO nitrone-globin adduct is detected. Interestingly, when oxyHb is pretreated with NEM, more adducts are detected. This indicates that, although under our experimental conditions only DMPO-thiyl radicals were detected by EPR, DMPO can also add to amino acid–derived radicals different from cysteinyl such as tyrosyl radical (as detected with MNP) to yield an unstable nitroxide that rapidly evolves to the stable nitrone.

Proteolysis techniques followed by liquid chromatography/mass-spectrometry analysis of the DMPO-Hb adduct obtained by reaction of oxyHb with another oxidant, $H_2O_2$, have identified sites of hemoglobin-radical formation at Cys-93 of the $\beta$-chain, and Tyr-42, Tyr-24, and His-20 of the $\alpha$-chain (Deterding et al., 2004).

DETECTION OF NITROTYROSINE. As the main reaction of peroxynitrite with oxyHb is the isomerization to $NO_3^-$, and $^\bullet NO_2$ is only marginally formed, nitration catalysis is not expected to occur. To confirm that, oxyHb (1.0, 2.0, or 4.0 m$M$) was exposed to increasing peroxynitrite concentration (0.25–2.0 m$M$), and tyrosine nitration was assessed by Western blot using a highly specific anti-nitrotyrosine antibody produced in our laboratory (Brito et al., 1999). No 3-nitrotyrosine in hemoglobin was detected under conditions in which peroxynitrite homolysis is not relevant (i.e., under a high excess of hemoglobin over peroxynitrite), which supports the concept that yields of ferrylHb and $^\bullet NO_2$ are modest under biologically relevant conditions and further supports the idea of oxyHb serving an inhibitory role on peroxynitrite-dependent nitrations (Herold et al., 2002).

## Red Blood Cells and Peroxynitrite

Because of its reactivity with various targets, the biological half-life of peroxynitrite is short, on the order of 10–20 ms (Denicola et al., 1998; Radi et al., 2000; Romero et al., 1999). This poses the question of whether peroxynitrite formed in one cell or compartment can diffuse sufficient distances to exert toxic effects on other target cells or compartments. An additional limitation arises because the predominant form of peroxynitrite at physiological pH is negatively charged and, therefore, may diffuse poorly through biological membranes that are not abundant in anion channels. We have developed and characterized an RBC model to define peroxynitrite diffusion distances through biological milieu (Denicola et al., 1998; Romero et al., 1999). With this model, we have found that peroxynitrite is capable of crossing the erythrocyte membrane and exerting intracellular effects, even in the presence of physiological concentrations of extracellular targets (Denicola et al., 1998; Romero et al., 1999). Transmembrane diffusion occurs via anion channel–dependent and independent mechanisms. Indeed, although peroxynitrite anion permeates through the bicarbonate-chloride exchanger or band 3 of the erythrocyte, peroxynitrous acid crosses the membrane by passive diffusion.

In the intravascular compartment, the diffusion of peroxynitrite to RBC competes with other routes of peroxynitrite consumption and results in a rapid and almost complete intraerythrocytic isomerization of peroxynitrite to nitrate (Romero et al., 1999, 2003).

*Experimental Approach and Results*

*Oxidation of Intracellular oxyHb by Peroxynitrite in the Presence of Extracellular Targets*

RBCs are obtained from freshly heparinized human blood. The concentration of hemoglobin used in intact RBC experiments is expressed as the concentration of oxyHb released after lysis, which is achieved by pelleting the cells and resuspending them in the same volume of distilled water. The RBC percentages are calculated assuming that a hemoglobin (tetramer) concentration of 5 m$M$ (20 m$M$ heme) corresponds to a 100% of RBC. Normal hematocrit corresponds to 45% RBCs in blood, which represents $4.5 \times 10^9$ RBC/ml.

Suspensions of RBC (2–40%) in isotonic buffer (80 m$M$ potassium phosphate buffer/40 m$M$ NaCl/ 10 m$M$ KCl, pH 7.3) in the absence or presence of the extracellular target (e.g., plasma, albumin, and carbon dioxide; see later discussion) are incubated for 10 min at room temperature with nigericin (10 $\mu$g/ml) in order to equilibrate intracellular and extracellular pH. Peroxynitrite is added under vigorous vortexing, and concentrations used should be near fourfold smaller than the corresponding heme concentration to always ensure an excess of oxyHb. Immediately after peroxynitrite addition, the RBC suspension is centrifuged (8000$g$, 10 s), the supernatant set aside, and the pellet resuspended in the same volume with distilled water to lysis. The whole procedure after peroxynitrite addition should be completed in less than 2 min to avoid reaction of RBC with nitrite present as contaminant of peroxynitrite solutions. After appropriate dilutions, the absorbances at 577 and 630 nm of the lysates are measured, and yields of intracellular oxyHb oxidation are calculated as the ratio between metHb and total Hb concentrations. The percentage of inhibition of oxyHb oxidation by extracellular targets is calculated as follows:

$$\% \text{ inhibition} = \left(1 - \frac{\% \text{metHb}_{(+\text{Target})}}{\% \text{metHb}_{(-\text{Target})}}\right) \times 100, \qquad (7)$$

where $\% \text{metHb}_{(+\text{Target})}$ and $\% \text{metHb}_{(-\text{Target})}$ represent the yield of oxyHb oxidation in the presence and absence of the extracellular target, respectively.

Because diffusion of the anionic and the protonated form of peroxynitrite occurs through different mechanisms, diffusion can be modulated by (1) changing extracellular pH (i.e., lowering pH buffer will favor the passive diffusion through the membrane over the anion channel–dependent mechanism and vice versa) and (2) blocking the anionic channel of RBCs with the stilbene derivatives 4,4′-diisothiocyanatostilbene-2,

2′-disulfonic acid (DIDS), 4-acetamido 4′-isothiocyanatostilbene 2,2′-disulfonic acid (SITS), and phenyl isothiocyanate (PITS) (Denicola *et al.*, 1998; Macfadyen *et al.*, 1999; Romero *et al.*, 1999). For example, a stock of RBC (50% v/v) is incubated with DIDS (10 m$M$) for 30 min at room temperature. After that, cells are washed to remove excess of DIDS, centrifuged, and resuspended in isotonic phosphate buffer, and the oxidation experiments are performed with the same procedure as described earlier.

### Estimation of Peroxynitrite Diffusion Distances

In the RBC model described earlier in this chapter, two competing processes occur: the reaction of peroxynitrite with the extracellular target and the diffusion of peroxynitrite to the RBCs. In biological systems, $CO_2$ represents probably the most important extracellular target that limits peroxynitrite diffusion because of its high concentration (1–2 m$M$) and the kinetic constant ($k_2 = 4.6 \times 10^4\ M^{-1}\ s^{-1}$ at 37° and pH 7.4) (Denicola *et al.*, 1996). This reaction leads to a fractional amount of the secondary species $CO_3^{\bullet-}$ or $^{\bullet}NO_2$, which have a half-life even smaller than that of peroxynitrite, and in the case of $CO_3^{\bullet-}$, its diffusivity across biomembranes is likely to be limited. As peroxynitrite reacts with extracellular $CO_2$, the amount that reaches intraerythrocytic oxyHb will decrease as the diffusion distances increase. We have developed a theoretical model to estimate peroxynitrite diffusion distances as a function of cell density and the percentage of peroxynitrite that would reach a target located at a certain distance from its site of production. According to this model (Romero *et al.*, 1999), the average diffusion distance ($\Delta x$) of extracellularly added peroxynitrite in cell suspensions can be calculated as follows:

$$\Delta x = r \cdot \sqrt{\frac{3}{4}\left(\sqrt[3]{\frac{4n\pi}{3}}\right)^2 - 1}, \tag{8}$$

where $n$ represents the ratio between the total volume and the cell volume ($V_t/V_{cell}$) of the suspension and $r$ the cell radius (calculated from the known cell volume and assuming a spherical cell). The percentage of added peroxynitrite that effectively reaches the target cell is determined from the following equation:

$$\ln\frac{[ONOO^-]_x}{[ONOO^-]_o} = \frac{-\ln2 \cdot \Delta \cdot x^2}{2D_{ONOO^-} \cdot t_{1/2}}, \tag{9}$$

where $t_{1/2}$ represents the half-life of peroxynitrite in the extracellular medium and $D_{ONOO^-}$ the diffusion coefficient of peroxynitrite considered to be similar to that of $NO_3^-$, 1500 $\mu m^2\ s^{-1}$ (Lide, 1990). Assuming a model in which the only cause of decay is the proton-catalyzed decomposition of peroxynitrite, the half-life at pH 7.4 is 2.7 s and 0.7 s at 25° and 37°, respectively. In the presence of 1.3 m$M$ $CO_2$ in the extracellular medium, the half-life of peroxynitrite is reduced to 24 ms (at 25°) or 12.3 ms (at 37°). From this theoretical model, we can calculate that in the presence of physiological concentrations of $CO_2$, peroxynitrite is able to diffuse distances between 5 and 10 $\mu$m, which represents one or two cellular diameters. The theoretical model was corroborated experimentally by performing diffusion experiments with different RBC densities in the presence of 25 m$M$ sodium bicarbonate (in equilibrium with 1.3 m$M$ $CO_2$) in the extracellular medium as shown in Fig. 2. Although the "diffusion model" was developed in RBCs, it has been successfully applied to other cellular models, as in studies of peroxynitrite toxicity toward *Trypanosoma cruzi* at different cell densities (Alvarez *et al.*, 2004).

Related approaches include the preparation of erythrocyte ghosts containing variable amounts of oxyHb and serving to define the role that the intracellular consumption of peroxynitrite can play in modulating oxidative reactions extracellularly and in the membrane (Minetti *et al.*, 2000).

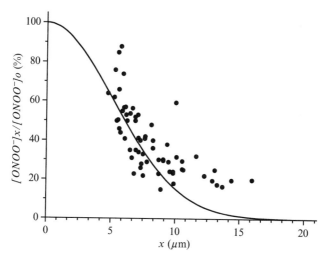

FIG. 2. Peroxynitrite diffusion distances in the presence of $CO_2$. The ratio of extracellular peroxynitrite that effectively reaches an intracellular target at a distance $x$ of the site of production respect to initial peroxynitrite concentration was determined from Eq. (8) and considering the half-life of peroxynitrite in the extracellular medium as 12.3 ms (solid line). Experimental data (●) was extracted from Romero *et al.* (1999).

## Concluding Remarks

a. The reaction of oxyHb with peroxynitrite is unusual because it not only promotes the isomerization of peroxynitrite to $NO_3^-$ but also results in a redox change of the metal center to metHb. One-electron oxidation of the heme is due to the concomitant formation of $O_2^{\bullet-}$, the third reaction product that arises from the reduction of ferrous heme-bound oxygen. Secondary and low-yield intermediates are an unstable ferrylHb, $^{\bullet}NO_2$, and globin-derived radicals.

b. Because of its large abundance in RBCs and fast reaction, oxyHb represents a key intravascular target of peroxynitrite and can be easily used as a reporter molecule, following spectrophotometrically its oxidation to metHb.

c. The RBC model is advantageous for studying the diffusion properties of peroxynitrite because intercellular distances vary according to erythrocyte density, and it permits the determination of the impact of extracellular reactions on peroxynitrite diffusion to the cell compartment.

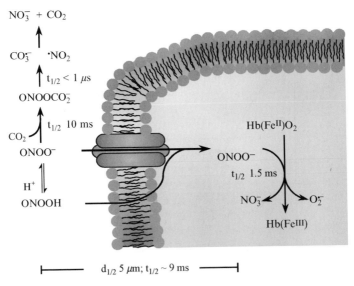

Fig. 3. Scheme representing competition between peroxynitrite diffusion to red blood cells (RBCs) and reaction with extracellular $CO_2$. Peroxynitrite can undergo extracellular reactions or diffuse to the erythrocyte through anion channels ($ONOO^-$) or passive diffusion ($ONOOH$). Once inside RBCs, peroxynitrite is mainly isomerized to $NO_3^-$ by oxyhemoglobin (oxyHb). The half-time of diffusion of 9 ms corresponds to a mean diffusion distance of $5\,\mu m$ at 45% hematocrit. Adapted from Romero *et al.* (1999).

Further, the RBC membrane is rich in anion channels, which favor the permeation of the anionic form ($\sim$80% of peroxynitrite at pH 7.4).

d. The model studies reported herein serve to emphasize the role of oxyHb in RBCs as a relevant intravascular sink of peroxynitrite (Fig. 3).

## Acknowledgments

This work was supported by grants from the Howard Hughes Medical Institute, Fogarty-National Institutes of Health, and the Guggenheim Foundation. N. R. was partially funded by Programa de Desarrollo de Ciencias Básicas (PEDECIBA, Uruguay) and R. R. is a Howard Hughes International Research Scholar.

## References

Alayash, A. I., Ryan, B. A., and Cashon, R. E. (1998). Peroxynitrite-mediated heme oxidation and protein modification of native and chemically modified hemoglobins. *Arch. Biochem. Biophys.* **349,** 65–73.

Alvarez, M. N., Piacenza, L., Irigoin, F., Peluffo, G., and Radi, R. (2004). Macrophage-derived peroxynitrite diffusion and toxicity to *Trypanosoma cruzi. Arch. Biochem. Biophys.* **432,** 222–232.

Barlow, T., Eliasson, R., Platz, A., Reichard, P., and Sjoberg, B. M. (1983). Enzymic modification of a tyrosine residue to a stable free radical in ribonucleotide reductase. *Proc. Natl. Acad. Sci. USA* **80,** 1492–1495.

Berzofsky, J. A., Peisach, J., and Blumberg, W. E. (1971). Sulfheme proteins. I. Optical and magnetic properties of sulfmyoglobin and its derivatives. *J. Biol. Chem.* **246,** 3367–3377.

Boccini, F., and Herold, S. (2004). Mechanistic studies of the oxidation of oxyhemoglobin by peroxynitrite. *Biochemistry* **43,** 16393–16404.

Bourassa, J. L., Ives, E. P., Marqueling, A. L., Shimanovich, R., and Groves, J. T. (2001). Myoglobin catalyzes its own nitration. *J. Am. Chem. Soc.* **123,** 5142–5143.

Brito, C., Naviliat, M., Tiscornia, A. C., Vuillier, F., Gualco, G., Dighiero, G., Radi, R., and Cayota, A. M. (1999). Peroxynitrite inhibits T lymphocyte activation and proliferation by promoting impairment of tyrosine phosphorylation and peroxynitrite-driven apoptotic death. *J. Immunol.* **162,** 3356–3366.

D'Agnillo, F., and Alayash, A. I. (2000). Interactions of hemoglobin with hydrogen peroxide alters thiol levels and course of endothelial cell death. *Am. J. Physiol. Heart Circ. Physiol.* **279,** H1880–H1889.

Denicola, A., Freeman, B. A., Trujillo, M., and Radi, R. (1996). Peroxynitrite reaction with carbon dioxide/bicarbonate: Kinetics and influence on peroxynitrite-mediated oxidations. *Arch. Biochem. Biophys.* **333,** 49–58.

Denicola, A., Souza, J. M., and Radi, R. (1998). Diffusion of peroxynitrite across erythrocyte membranes. *Proc. Natl. Acad. Sci. USA* **95,** 3566–3571.

Deterding, L. J., Ramirez, D. C., Dubin, J. R., Mason, R. P., and Tomer, K. B. (2004). Identification of free radicals on hemoglobin from its self-peroxidation using mass spectrometry and immuno-spin trapping: observation of a histidinyl radical. *J. Biol. Chem.* **279,** 11600–11607.

Detweiler, C. D., Deterding, L. J., Tomer, K. B., Chignell, C. F., Germolec, D., and Mason, R. P. (2002). Immunological identification of the heart myoglobin radical formed by hydrogen peroxide. *Free Radic. Biol. Med.* **33**, 364–369.

Exner, M., and Herold, S. (2000). Kinetic and mechanistic studies of the peroxynitrite-mediated oxidation of oxymyoglobin and oxyhemoglobin. *Chem. Res. Toxicol.* **13**, 287–293.

Ferrer-Sueta, G., Quijano, C., Alvarez, B., and Radi, R. (2002). Reactions of manganese porphyrins and manganese-superoxide dismutase with peroxynitrite. *Methods Enzymol.* **349**, 23–37.

Giulivi, C., and Davies, K. J. (1990). A novel antioxidant role for hemoglobin. The comproportionation of ferrylhemoglobin with oxyhemoglobin. *J. Biol. Chem.* **265**, 19453–19460.

Herold, S. (1999). Mechanistic studies of the oxidation of pyridoxalated hemoglobin polyoxyethylene conjugate by nitrogen monoxide. *Arch. Biochem. Biophys.* **372**, 393–398.

Herold, S., and Shivashankar, K. (2003). Metmyoglobin and methemoglobin catalyze the isomerization of peroxynitrite to nitrate. *Biochemistry* **42**, 14036–14046.

Herold, S., Shivashankar, K., and Mehl, M. (2002). Myoglobin scavenges peroxynitrite without being significantly nitrated. *Biochemistry* **41**, 13460–13472.

Jensen, M. P., and Riley, D. P. (2002). Peroxynitrite decomposition activity of iron porphyrin complexes. *Inorg. Chem.* **41**, 4788–4797.

Koppenol, W. H., and Liebman, J. F. (1983). The oxidizing nature of the hydroxyl radical: A comparison with the ferryl ion ($FeO^{2+}$). *J. Phys. Chem.* **88**, 99–101.

Kosaka, H., Imaizumi, K., and Tyuma, I. (1981). Mechanism of autocatalytic oxidation of oxyhemoglobin by nitrite. An intermediate detected by electron spin resonance. *Biochim. Biophys. Acta* **702**, 237–241.

Lide, D. R. (1990). "Handbook of Chemistry and Physics." CRC, Boca Ratón, FL.

Macfadyen, A. J., Reiter, C., Zhuang, Y., and Beckman, J. S. (1999). A novel superoxide dismutase-based trap for peroxynitrite used to detect entry of peroxynitrite into erythrocyte ghosts. *Chem. Res. Toxicol.* **12**, 223–229.

Marla, S. S., Lee, J., and Groves, J. T. (1997). Peroxynitrite rapidly permeates phospholipid membranes. *Proc. Natl. Acad. Sci. USA* **94**, 14243–14248.

Minetti, M., Pietraforte, D., Carbone, V., Salzano, A. M., Scorza, G., and Marino, G. (2000). Scavenging of peroxynitrite by oxyhemoglobin and identification of modified globin residues. *Biochemistry* **39**, 6689–6697.

Miranda, K. M., Espey, M. G., and Wink, D. A. (2001). A rapid, simple spectrophotometric method for simultaneous detection of nitrate and nitrite. *Nitric Oxide* **5**, 62–71.

Pietraforte, D., Salzano, A. M., Scorza, G., and Minetti, M. (2004). Scavenging of reactive nitrogen species by oxygenated hemoglobin: Globin radicals and nitrotyrosines distinguish nitrite from nitric oxide reaction. *Free Radic. Biol. Med.* **37**, 1244–1255.

Radi, R. (1996). Kinetic analysis of reactivity of peroxynitrite with biomolecules. *Methods Enzymol.* **269**, 354–366.

Radi, R. (2004). Nitric oxide, oxidants, and protein tyrosine nitration. *Proc. Natl. Acad. Sci. USA* **101**, 4003–4008.

Radi, R., Denicola, A., Alvarez, B., Ferrer-Sueta, G., and Rubbo, H. (2000). The biological chemistry of peroxynitrite. *In* "Nitric Oxide: Biology and Pathobiology" (L. J. Ignarro, ed.), pp. 57–82. Academic Press, NY.

Romero, N., Denicola, A., Souza, J. M., and Radi, R. (1999). Diffusion of peroxynitrite in the presence of carbon dioxide. *Arch. Biochem. Biophys.* **368**, 23–30.

Romero, N., Radi, R., Linares, E., Augusto, O., Detweiler, C. D., Mason, R. P., and Denicola, A. (2003). Reaction of human hemoglobin with peroxynitrite. Isomerization to nitrate and secondary formation of protein radicals. *J. Biol. Chem.* **278**, 44049–44057.

Schmidt, H. H. W., and Kelm, M. (1996). Determination of nitrite and nitrate by the Griess reaction. *In* "Methods in Nitric Oxide Research" (M. Feelisch and J. Stamler, eds.), pp. 491–498. Wiley, NY.

Sowers, J. R. (2002). Hypertension, angiotensin II, and oxidative stress. *N. Engl. J. Med.* **346,** 1999–2001.

Wade, R. S., and Castro, C. E. (1996). Reactions of oxymyoglobin with NO, $NO_2$, and $NO_2^-$ under argon and in air. *Chem. Res. Toxicol.* **9,** 1382–1390.

Winterbourn, C. C. (1990). Oxidative reactions of hemoglobin. *Methods Enzymol.* **186,** 265–272.

# [22]   Quantification of 3-Nitrotyrosine Levels Using a Benchtop Ion Trap Mass Spectrometry Method

*By* Stephen J. Nicholls, Zhongzhou Shen, Xiaoming Fu, Bruce S. Levison, and Stanley L. Hazen

## Abstract

Oxidative damage by reactive nitrogen species is linked to the pathogenesis of numerous inflammatory disorders, including atherosclerosis. 3-Nitrotyrosine ($NO_2Tyr$), a posttranslational modification of proteins generated by reactive nitrogen species, serves as a "molecular fingerprint" for protein modification by nitric oxide (NO)–derived oxidants. Studies demonstrate that systemic levels of protein-bound $NO_2Tyr$ serve as an independent predictor of cardiovascular risks and are modulated by statin therapy. Measurement of $NO_2Tyr$ in biological matrices may thus serve both as a quantitative index of nitrative stress *in vivo* and an important new prognostic marker of clinical relevance. Analytical methods for the accurate detection and quantification of trace levels of $NO_2Tyr$ in biological tissues and fluids are, thus, of considerable interest. Here, we describe a rapid, sensitive, and specific method for the quantification of $NO_2Tyr$ in biological matrices using readily available benchtop ion-trap mass spectrometry instrumentation (e.g., LCQDeca) combined with high-performance liquid chromatography (HPLC) interface. Through judicious use of stable isotopically labeled precursors as synthetic internal standards, the tandem mass spectrometric method described simultaneously adjusts for potential intrapreparative sample losses and monitors potential artifactual generation of $NO_2Tyr$ during processing. The described method permits rapid and reproducible quantification of $NO_2Tyr$ in biological and clinical specimens at the 100 fmol on column detection limit and should

METHODS IN ENZYMOLOGY, VOL. 396
0076-6879/05 $35.00
DOI: 10.1016/S0076-6879(05)96022-9

prove useful for studies defining the impact of reactive nitrogen species in cardiovascular disease and other inflammatory disorders.

Introduction

Oxidative modifications of proteins is linked to both physiological aging (Beal, 2002; Drew and Leeuwenburgh, 2002) and numerous pathological states, including atherosclerosis (Moriel and Abdalla, 1997), asthma (Andreadis *et al.*, 2003), and neurodegenerative disorders (Smith *et al.*, 1997; Tanaka *et al.*, 1997). Modification of biological targets by reactive nitrogen species is a key event in these processes. 3-Nitrotyrosine ($NO_2Tyr$) is formed by the reaction of tyrosine with reactive nitrogen species such as peroxynitrite, a potent oxidant formed by the reaction of nitric oxide (NO) and the superoxide anion (Beckman *et al.*, 1990). Alternative pathways for generation of NO-derived oxidants and formation of $NO_2Tyr$ involve the recruitment and activation of the leukocyte-derived enzymes, myeloperoxidase and eosinophil peroxidase, in inflammatory conditions (Brennan *et al.*, 2002). Both peroxynitrite- and peroxidase-generated reactive nitrogen species have been linked to the initiation of lipid peroxidation, protein nitration, and oxidative conversion of low-density lipoprotein (LDL) into an atherogenic form (Graham *et al.*, 1993; Podrez *et al.*, 1999). Clinical studies demonstrate that serum levels of protein-bound nitrotyrosine independently predict atherosclerotic risk and burden, demonstrating stronger correlations with disease presence than currently employed clinical markers for cardiac risks, such as LDL cholesterol and C-reactive protein (CRP) (Shishehbor *et al.*, 2003a). Further, plasma levels of protein-bound $NO_2Tyr$ are substantially reduced following statin therapy independent of changes in LDL cholesterol, suggesting that alterations in $NO_2Tyr$ levels may serve as a method for monitoring the antioxidant and antiinflammatory actions of medications such as statins (Shishehbor *et al.*, 2003a,b).

Because of its potential significance, a number of methods for the separation, detection and quantitation of $NO_2Tyr$ in biological samples have been developed. The most widely used methods include immunochemical techniques employing antibodies specific for $NO_2Tyr$ (Abe *et al.*, 1995; Giasson *et al.*, 2000; Haddad *et al.*, 1994; Szabo *et al.*, 1995; Viera *et al.*, 1999). Immunological approaches have been used to identify cellular sources of reactive nitrogen species and specific protein targets for modification (Fries *et al.*, 2003; Vadseth *et al.*, 2004; Zheng *et al.*, 2004). Thus far, however, antibody-based methods that provide quantitatively comparable results to those observed with more rigorous analytical approaches, such as mass spectrometry (MS), have not yet been reported. Methods incorporating chromatographic separation [high-performance liquid chromatography

(HPLC)] of $NO_2$Tyr in combination with various detection systems, such as ultraviolet absorption (Crow and Ischiropoulos, 1996), electrochemical detector (ECD) (Shigenaga et al., 1997), and fluorescence detectors (Kamisaki et al., 1996), have also been reported. These methods are typically complicated by the need for chemical derivatization to permit adequate sensitivity to detect $NO_2$Tyr within biological specimens. Moreover, they do not permit simultaneous monitoring for intrapreparative generation of artifactual $NO_2$Tyr formation during sample handling. Gas chromatography MS (GC-MS) assays have been reported for $NO_2$Tyr and offer high sensitivity and specificity for the quantitation of both free and protein-bound forms of $NO_2$Tyr (Crowley et al., 1998; Leeuwenburgh et al., 1997; Schwedhelm et al., 1999). These methods, however, require derivatization, enhancing the complexity of the assay.

Liquid chromatography MS (LC-MS), which is reported to provide increased sensitivity and specificity, is becoming increasingly used. Most quantification using LC-MS has been performed by triple quadrupole MS (Althaus et al., 2000; Frost et al., 2000; Yi et al., 2000). With the introduction of lower cost benchtop LC-MS systems based on ion trap technology, there exists the opportunity for the development of a more widely available LC/electrospray ionization (ESI)/MS/MS–based method than those employing the more expensive triple quadrupole MS instruments. Here, we describe a novel, sensitive, and specific method for $NO_2$Tyr quantification in biological tissues and fluids using HPLC with online ESI tandem MS on a benchtop ion trap MS. The method is applicable for the quantification of trace levels of protein-bound and free $NO_2$Tyr within biological matrices. A key feature of the stable isotope dilution method is the simultaneous monitoring for artifactual generation of $NO_2$Tyr during sample processing. LC/ESI/MS/MS determination of $NO_2$Tyr on a benchtop ion trap instrument, such as the Finnegan LCQDeca, has the potential to serve as a highly sensitive and specific clinical diagnostic method for defining cardiovascular risk and responses to therapy and for exploring the role of protein oxidation by reactive nitrogen species in vivo.

## Experimental Procedures

### Materials

Organic solvents (HPLC-grade), $H_2O_2$ (30%; ACS grade), $H_3PO_4$, $NaH_2PO_4$, and $Na_2HPO_4$ were obtained from Fisher Chemical Company (Pittsburgh, PA). Chelex 100 resin (200–400 mesh, sodium form) was obtained from Bio-Rad (Hercules, CA). Methane sulfonic acid were purchased from Fluka Chemical Company (Ronkonkoma, NY). Heavy isotope-labeled tyrosine ([$^{13}C_6$]-tyrosine; [$^{13}C_9{}^{15}N_1$]-tyrosine) was obtained from Cambridge Isotope Laboratories (Andover, MA). Solid-phase extraction columns (Supelclean LC-C-18 SPE minicolumns) were obtained from Supelco (Bellefonte, PA). All other reagents were purchased from Sigma Chemical Company (St. Louis, MO) unless otherwise indicated.

*Samples*

Animal studies described were performed using approved protocols from the Animal Research Committee of the Cleveland Clinic Foundation. Bronchoalveolar lavage samples were obtained from age- and sex-matched C57BL/6 J EPO knockout (KO) and wild-type mice before and after ovalbumin (OVA) challenge. Mice were sensitized or sham treated with either saline or 20 $\mu$g of OVA (grade IV, Sigma) and 2.25 mg of Imject Alum (Pierce) on days 0 and 14. On days 24, 25, and 26, animals were challenged with 20-min inhalations of a 1% ovalbumin or saline aerosol. On day 28, bronchoalveolar lavage fluid was obtained by saline lavage. Bronchoalveolar lavage and plasma specimens were collected and aliquoted into 1.5-ml centrifuge tubes with the addition of 100 $\mu M$ DTPA and 100 $\mu M$ butylated hydroxytoluene (BHT). They were subsequently stored frozen under $N_2$ at $-80°$ until they were ready for use. All human clinical specimens were obtained from subjects who gave written, informed consent for a study protocol approved by the Institutional Review Board of the Cleveland Clinic Foundation.

*Methods*

*Preparation of Ring-Labeled Nitrotyrosine Standards*

Ring-labeled standards may be prepared by the reaction of stable isotope-labeled tyrosine (Tyr) species with tetranitromethane (Aldrich) (Riordan *et al.*, 1966). Briefly, 10.2 mg [$^{13}C_6$]-labeled Tyr (Cambridge Isotope Labs) was suspended in 1 ml Chelex-treated water and combined with 50 $\mu$l of 2.5 $M$ sodium hydroxide, and 7 $\mu$l neat tetranitromethane was added using a metal-free plastic Pipetman tip. The solution was vortexed, resulting in an immediate color change to yellow and then subsequently brown. The solution was left at room temperature overnight for 18 h. $NO_2$Tyr within the mixtures were purified by first using a solid-phase extraction column (10 $\times$ 3 ml Supelclean LC-C-18 SPE minicolumn; Supelco Bellefonte, PA), which had been prepared by sequential rinses of methanol (2 $\times$ 2 ml) and Chelex-treated water (3 $\times$ 2 ml). Aliquots ($\sim$100 $\mu$l) of the reaction mixture were diluted with 1 ml of Chelex-treated water and were drawn into the resin under vacuum at 7 psi. The columns were rinsed with Chelex-treated water (4 $\times$ 2 ml) until the eluate ran clear. $NO_2$Tyr was then eluted with

30% methanol in Chelex-treated water (v/v, $2 \times 2$ ml). The methanol eluate was concentrated to dryness using vacuum centrifugation at $37°$.

NO$_2$Tyr was then isolated to homogeneity by HPLC analysis. Runs were performed using 1-ml injections of an aqueous solution containing 1–2 mg/ml (based on starting material) in 6% methanol in water at 3 ml/min on a Beckman Ultrasphere ODS 5 $\mu$m particle diameter, 10 mm inside diameter (I.D.) $\times$ 250 mm length column monitored at 254 and 276 nm. Nonlabeled NO$_2$Tyr, which is readily commercially available (e.g., Aldrich), is used as on external standard to identify retention time and confirm chromatographic behavior of NO$_2$Tyr before injection and purification of the isotopically labeled NO$_2$Tyr synthetic species. HPLC analyses using this system give baseline resolved peaks eluting at approximately 6.1 (Tyr) and 13.0 (NO$_2$-Tyr) min. Successful production of NO$_2$Tyr is indicated by an increase in its chromatographic peak area accompanied by a reduction in Tyr. An alternative synthetic route to production of isotopically labeled NO$_2$Tyr involves reaction with a molar equivalent of peroxynitrite. Ring-labeled [$^{13}$C$_6$]-NO$_2$Tyr or universal-labeled NO$_2$Tyr ([$^{13}$C$_9$$^{15}$N$_1$]-NO$_2$Tyr) may thus also be synthesized by the reaction of [$^{13}$C$_6$]-Tyr or [$^{13}$C$_9$$^{15}$N$_1$]-Tyr, respectively, with an equimolar amount of peroxynitrite in phosphate buffer (pH 7.0), and purified using a solid-phase extraction column (Supelclean LC-C-18 SPE minicolumn; Supelco Bellefonte, PA), followed by HPLC, as described earlier. Briefly, [$^{13}$C$_6$]-NO$_2$Tyr and [$^{13}$C$_9$$^{15}$N$_1$]-NO$_2$Tyr are prepared by mixing a solution of [$^{13}$C$_6$]-Tyr (250 $\mu$l, 2 m$M$) or [$^{13}$C$_9$N$_1$]-Tyr (250 $\mu$l, 2 m$M$) in 150 m$M$ sodium phosphate buffer (pH 7.0) with peroxynitrite (2.5 $\mu$l, 200 m$M$) for 5 min. Then, 50 $\mu$l of this mixed solution may be diluted to 150 $\mu$l with H$_2$O and then loaded onto a solid-phase extraction column (3 ml, C-18 SPE tube; Supelco) that is preconditioned with methanol ($3 \times 2$ ml) and then H$_2$O ($3 \times 2$ ml). After loading, the minicolumn is washed with aqueous trifluoroacetic acid (TFA) (0.1%, $3 \times 2$ ml), and then [$^{13}$C$_6$]-NO$_2$Tyr or [$^{13}$C$_9$$^{15}$N$_1$]-NO$_2$Tyr subsequently is eluted from the column using 30% methanol in 0.1% TFA water (2 ml), and dried under vacuum. Dried products are reconstituted with water, and then structural identity (and isotopic purity) is confirmed by LC/ESI/MS/MS. HPLC analysis is used to quantify the concentration of synthetic-labeled NO$_2$Tyr species using an external calibration curve based on solutions of authentic NO$_2$Tyr. It should be noted that universal-labeled NO$_2$Tyr ([$^{13}$C$_9$$^{15}$N$_1$]-NO$_2$Tyr) is not used as an internal standard during the assay. Rather, its use is limited to the identification of parent–daughter transitions and standard curves for quantification that are subsequently used to monitor for potential artificial formation of universal-labeled NO$_2$Tyr from the universal-labeled precursor amino acid [$^{13}$C$_9$$^{15}$N$_1$]-Tyr.

### Quantification of Free 3-Nitrotyrosine and Tyrosine in Plasma

For analysis of the free amino acids $NO_2Tyr$ and Tyr, plasma samples are first supplemented with internal standards (universal-labeled $[^{13}C_9{}^{15}N_1]$-Tyr and ring-labeled $[^{13}C_6]$-$NO_2Tyr$), and then the amino acids are extracted by passage over a Supelclean LC-$C_{18}$ SPE minicolumn (3 ml) as described later. Briefly, the plasma samples and the internal standards were loaded onto minicolumns that had been washed and then preequilibrated with 0.1% (TFA). The minicolumns were then washed with 0.1% TFA (2 × 2 ml). $NO_2Tyr$ and small amounts of Tyr were then eluted off of the column with 2 ml of 30% (v/v) methanol plus 0.1% (v/v) TFA. This eluate was dried and reconstituted in 200 $\mu$l of $H_2O$ and then subjected to LC/ESI/MS analysis. This method, by design, results in only partial recovery of the free Tyr (and its corresponding internal standard, $[^{13}C_9{}^{15}N_1]$-Tyr, thus permitting accurate quantification of the Tyr despite the partial recovery), because the dynamic range of the mass detector for simultaneous quantification of $NO_2Tyr$ and Tyr in the same injection (within the eluate) is difficult unless only partial Tyr recovery occurs.

### Protein Hydrolysis of Samples

To minimize artifactual nitration of proteins during acid hydrolysis, free nitrite/nitrate in specimens is first removed during protein precipitation in a combined delipidation and desalting step, using two sequential extractions with a single-phase mixture composed of $H_2O$/methanol/$H_2O$-saturated diethyl ether (1:3:8, v/v/v). For each extraction, mixtures were vortexed and incubated in ice/water bath for several minutes to facilitate protein precipitation, and the protein precipitate was isolated by centrifugation at 4000$g$ at 0–10° (the cold temperature facilitates protein precipitation). Universal labeled Tyr (2 nmol $[^{13}C_9{}^{15}N_1]$-Tyr) is added to the protein precipitate to both quantify endogenous Tyr levels and to monitor for potential artifactual generation of $NO_2Tyr$ by monitoring the precursor-> product ion transitions appropriate for the $[^{13}C_9{}^{15}N_1]$-$NO_2Tyr$ isotopomer (Table I). $^{13}C$-labeled $NO_2Tyr$ (2 pmol $[^{13}C_6]$-$NO_2Tyr$) is also added to the protein pellet and serves as an internal standard to quantify endogenous $NO_2Tyr$ levels in the sample. The level of the $[^{13}C_9{}^{15}N_1]$-Tyr label added to specimens (2 pmol) is selected to approximate (within twofold to fivefold) the endogenous level of Tyr (natural abundance $^{12}C$ isotopomer) present in the sample analyzed. Similarly (but less critically), the level of $[^{13}C_6]$-$NO_2Tyr$ internal standard added to a specimen is ideally roughly comparable (within 1–2 orders of magnitude) to that anticipated to be present in the sample. The masses of standards given above are reasonable for analysis of a 5–10 $\mu$l aliquot of plasma. Before initiating acid hydrolysis, acid mixtures

TABLE I

IONS MONITORED AND ACQUISITION PARAMETERS USED FOR THE DETECTION OF NATIVE AND LABELED FORMS OF NITROTYROSINE (NO$_2$TYR) AND TYROSINE (TYR)

| Analyte | Precursor Ions [M + H]$^+$ | MS/MS scan range | Product Ions monitored[a] | Collision energy (%) |
|---|---|---|---|---|
| [$^{12}$C$_6$]-NO$_2$Tyr | 227 | 120–250 | 181, 210 | 25 |
| [$^{13}$C$_6$]-NO$_2$Tyr | 233 | 120–250 | 187, 216 | 25 |
| [$^{13}$C$_9$$^{15}$N$_1$]-NO$_2$Tyr | 237 | 120–250 | 190, 219 | 25 |
| [$^{12}$C$_6$]-Tyr | 182 | 120–200 | 136, 165 | 24 |
| [$^{13}$C$_9$$^{15}$N$_1$]-Tyr | 192 | 120–200 | 145, 174 | 24 |

[a] The sum of the responses of the ions was used for quantification.

were degassed under vacuum and then sealed under a blanket of argon in mininert (Pierce) capped vials. Proteins then were hydrolyzed by 6 N HCl (0.5 ml) supplemented with 1% phenol (to serve as scavenger of potential oxidants that might modify Tyr) for 16–24 h at 110°.

### C$_{18}$ Solid-Phase Extraction of 3-Nitrotyrosine and Tyrosine

Protein hydrolysates were diluted with 500 $\mu$l Chelex-treated H$_2$O. The hydrolysates were desalted and purified over a Supelclean LC-C$_{18}$ SPE minicolumn (3 ml) in order to prolong HPLC column life and remove polymers, polyphenols, and potential particulate debris generated during sample hydrolysis before LC/ESI/MS/MS. Briefly, the hydrolysates were loaded onto minicolumns, which had been washed with methanol (3 × 2 ml) and preequilibrated with 0.1% TFA. Salts and most of the Tyr were eluted from the column through sequential washes (2 × 2 ml) of 0.1% TFA. NO$_2$Tyr and a small amount of Tyr (and their corresponding isotopically labeled internal standards) are then eluted off the column with 2 ml of 30% (v/v) methanol in 0.1% (v/v) TFA in H$_2$O. Fractions are dried under vacuum, reconstituted with either 100 $\mu$l of H$_2$O or initial mobile phase (H$_2$O with 0.1% formic acid), and either centrifuged or passed through a 0.22-$\mu$m filter (Millipore Co., Bedford, MA) before LC/MS/MS analysis via autosample injection.

### LC/ESI/MS Quantification of Tyrosine and 3-Nitrotyrosine

MS was performed using ESI and detection with an ion trap mass spectrometer LCQDeca (ThermoFinnegan, San Jose, CA) interfaced with a Thermo P4000 HPLC Pump and an AS3000 autosampler (Thermo Finnegan, San Jose, CA). Samples were injected onto a Prodigy ODS(2) C-18 column (Phenomenex, 5 $\mu$m particle, 2.0 × 150 mm) at a flow rate of

200 $\mu$l/min solvent [H$_2$O with 0.1% (v/v) formic acid]. The separation was generated using a linear gradient with a second mobile phase [acetonitrile with 0.1% (v/v) formic acid] over the course of 30 min as follows: 0% for 2 min, 0–40% over 14 min, and increase to 95% for 2 min, then immediately back to initial conditions (0% acetonitrile) for the remainder. Analytes are monitored using ESI MS (ESI/MS/MS) in positive ion (MS/MS) mode with selected reaction monitoring (SRM) of two scan events (182, 192 Tyr) for the first 9 min and three scan events (227, 233, 237 NO$_2$Tyr) for the remainder. The response was optimized with a spray voltage setting of 5 kV and a spray current of 80 $\mu$A. The heated capillary voltage was set at 10 V, and the temperature to 350°. Nitrogen was used as both sheath and auxiliary gas at flow rates of 70 and 30 (arbitrary units), respectively. The precursor ion isolation width was 1.0 and 3.0 amu for 3-nitrotyrosine and tyrosine, respectively. The analyte abundance was evaluated by measuring the chromatographic peak areas of selected product ions extracted from the full scan total ion chromatograms, according to the corresponding ion trap product ion spectra (Figs. 1 and 2). The ions monitored and acquisition parameters for each analyte are summarized in Table I and included 3-NO$_2$[$^{12}$C$_6$]Tyr [mass-to-charge ($m/z$) 227 $\rightarrow$ 181, and 210]; 3-NO$_2$[$^{13}$C$_6$]Tyr ($m/z$ 233 $\rightarrow$ 187, and 216); 3-NO$_2$[$^{13}$C$_9$$^{15}$N$_1$]Tyr ($m/z$ 237 $\rightarrow$ 190, and 219); [$^{12}$C$_6$]Tyr ($m/z$ 182 $\rightarrow$ 136, and 165); and [$^{13}$C$_9$$^{15}$N$_1$]Tyr ($m/z$ 192 $\rightarrow$ 145, and 174). The maximum ion injection time was 200 ms; each scan was composed of one microscan, resulting in a minimum sampling rate of about 9–10 points per chromatographic peak. Using the HPLC conditions employed, Tyr and NO$_2$Tyr demonstrate base line separation. Therefore, the LCQDeca was programmed for analysis over 0–9 min for detection of Tyr isotopomers and from 9 min to the end of the gradient for detection of NO$_2$Tyr isotopomers. For increased sample throughput, we routinely use two HPLC systems and columns, with column switching (i.e., a multiplex arrangement), and data acquisition timed to the appropriate retention time period for each analytical run.

### Adjusting for Artificial Nitrotyrosine Generation

A major issue with NO$_2$Tyr detection is the potential for artifactual nitration in samples acidified without total removal of residual nitrite. A major advance made possible by MS is the addition of an isotopically labeled internal standard of the precursor (Tyr) with simultaneous monitoring for the isotopically labeled NO$_2$Tyr to determine whether artifactual nitration occurred during sample processing. The methods described are optimized to the point that artifactual nitration is rarely observed. Nonetheless, we

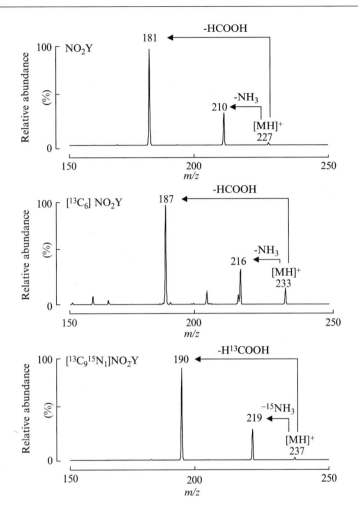

FIG. 1. Mass spectra and fragmentation patterns of native (top), heavy isotope (ring) labeled (middle), and universally labeled (bottom) forms of 3-nitrotyrosine detected by LCQDeca. In addition to the molecular ion ($m/z$ 227, 233, and 237 for ([$^{12}$C]-nitrotyrosine, [$^{13}$C$_6$]-nitrotyrosine, and [$^{13}$C$_9$$^{15}$N$_1$]-nitrotyrosine, respectively), the other fragments represent loss of NH$_3$ [(M-17), $m/z$ 210, 216, and 219] and loss of HCOOH [(M-46), $m/z$ 181, 187, and 190], respectively.

believe it is important to incorporate this methodology into the assay, because occasional surprises do happen.

The level of native NO$_2$Tyr is determined offline after adjusting for the level of artifactual generation (which should be minimal, if at all observed, if the desalting step is complete). The key to this modification relies on the

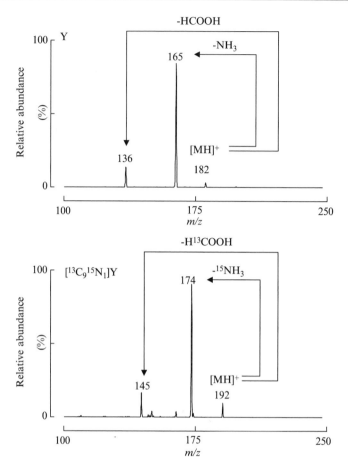

FIG. 2. Mass spectra and fragmentation patterns of native (top) and heavy isotope (universal) labeled (bottom) forms of tyrosine detected by LCQDeca. In addition to the molecular ion (m/z 182 and 192 for ($[^{12}C]$-tyrosine and $[^{13}C_9{}^{15}N_1]$-tyrosine, respectively), the other fragment ions represent loss of $NH_3$ [(M-17), m/z 165, 174] and loss of HCOOH [(M-46), m/z 136 and 145], respectively, from the $[^{12}C]$ and $[^{13}C_9{}^{15}N_1]$ isotopomers.

assumption that Tyr and its universal labeled form theoretically undergo nitration at a similar rate by the following reactions:

$$Tyr \rightarrow NO_2Tyr \qquad (1)$$

$$u\text{-}Tyr \rightarrow u\text{-}NO_2Tyr \qquad (2)$$

The molar ratio of the resulting nitrated species, if formed intrapreparatively, should equal the molar ratio of Tyr and universal (u) labeled Tyr present in the specimen, as depicted here:

$$\text{molar ratio Tyr: u-Tyr} = \text{molar ratio NO}_2\text{Tyr: u-NO}_2\text{Tyr} \qquad (3)$$

When the chromatographic peak area is plotted versus the molar amount of both Tyr and $NO_2$Tyr, the relationship approaches linearity with a slope of 1 and a minimal "y" intercept. As a result, the peak area and molar amount are essentially equivalent. The $NO_2$Tyr signal that is detected comprises both the native form and that formed artificially. In contrast, because no universal labeled $NO_2$Tyr is used during this method, any detected must arise artificially. As the molar amount of native and universal labeled Tyr are known, the amount of artificial $NO_2$Tyr present can be calculated by the following equation:

$$\text{Artificial NO}_2\text{Tyr} = (\text{molar ratio Tyr: u-Tyr}) *$$
$$(\text{molar amount u-NO}_2\text{Tyr}) \qquad (4)$$

This normalized level of artificial $NO_2$Tyr can then be subtracted from the amount of $NO_2$Tyr detected to reveal the amount of native $NO_2$Tyr originally present in the specimen. If artifactual nitration is detected, its level should represent less than 10% of the natural abundance $NO_2$Tyr level, and if the level is greater, then the sample is reanalyzed, incorporating an extra desalting/delipidation precipitation step, as outlined earlier, to remove more residual nitrite in the sample before acidification. Though not described herein, base hydrolysis of protein precipitates within plastic vials using NaOH is also feasible. The pH of the base hydrolysates, however, must then be neutralized before $NO_2$Tyr and Tyr bind adequately to C-18 minicolumns for subsequent minicolumn extraction and analysis.

### Monitoring 3-Nitrotyrosine in Solid Tissue

The method described earlier in this chapter for $NO_2$Tyr quantification can also be applied to the monitoring of nitrative stress in solid tissue samples or even deparaffinized tissue block [e.g., several 50-$\mu$m shavings (Samoszuk et al., 2002)] as described in the demonstration of the association between nitrotyrosine levels and microvascular density in human breast cancer.

For exemplary purposes in the present method, we provide data quantifying $NO_2$Tyr levels in lung and bronchoalveolar lavage fluid after OVA challenge (Brennan et al., 2002), an asthma model. Lung tissue samples from OVA-sensitized and -challenged mice were immediately rinsed in ice-cold buffer (65 m$M$ sodium phosphate, pH 7.4) with 100 $\mu M$ DTPA,

100 $\mu M$ BHT, and 10 m$M$ aminotriazole and were subsequently stored frozen under argon at $-80°$ until they were ready for analysis. Frozen tissues were pulverized in a stainless-steel mortar and pestle at liquid nitrogen temperature. Tissue powder was then transferred to a 13 × 100 mm heavy-walled, threaded glass centrifuge tube and suspended in water supplemented with 100 $\mu M$ DTPA and 10 m$M$ aminotriazole. Aliquots of samples were then processed as described earlier for NO$_2$Tyr analysis.

## Results

### Spectral Characterization of Tyrosine and 3-Nitrotyrosine

Chromatographic spectra, detected by ESI/MS/MS in the positive ion mode, using SRM to monitor NO$_2$Tyr, Tyr, and their corresponding heavy isotope labeled forms (both [$^{13}C_6$] ring-labeled and [$^{13}C_9{}^{15}N_1$] universally labeled isotopomers) are presented in Figs. 1 and 2. Table I lists precursor ions and characteristic product ions monitored for each isotopomer in MS quantitative SRM studies. Both NO$_2$Tyr and Tyr isotopomers are easily fragmented from their precursor to product species with the neutral loss of NH$_3$ and HCOOH. To monitor for the potential generation of artificial NO$_2$Tyr during the protein hydrolysis stage, universally labeled [$^{13}C_9{}^{15}N_1$] Tyr was added before acid hydrolysis, and an SRM channel was selected to detect [$^{13}C_9{}^{15}N_1$]NO$_2$Tyr. The bottom panel of Fig. 1 shows the spectrum of [$^{13}C_9{}^{15}N_1$]NO$_2$Tyr, revealing fragment ions consistent with cleavage of NH$_3$ and HCOOH, generating $m/z$ 219 and 190, respectively.

### LC/MS Detection of Fentomole Levels of 3-Nitrotyrosine

There was a near linear correlation between the chromatographic peak area of NO$_2$Tyr and the concentration analyzed (Fig. 3). This relationship was demonstrated over a wide concentration range. Furthermore, detection of NO$_2$Tyr with this method demonstrated a high sensitivity. The limit of detection of NO$_2$Tyr (signal/noise > 5) is reproducibly ≤100 fmol of mass applied to the column.

### Standard Curves of 3-Nitrotyrosine and Tyrosine

Standard curves were generated by analysis of a fixed amount of isotopically labeled internal standard with varying concentrations of the indicated isotopomer of analyte and then plotting the mole ratio of analyte to internal standard against the peak area ratio for each monitored isotopomer from its characteristic precursor → product ion transition (see Table I for transitions monitored for each analyte). Both NO$_2$Tyr ($r^2 = 0.998$) and Tyr ($r^2 = 0.999$) demonstrate good linearity, as illustrated in Figs. 4 and 5,

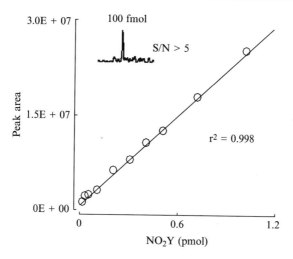

FIG. 3. Limit of detection for nitrotyrosine by LCQDeca. The limit of detection, depicted by the spectral pattern in the inset, was 100 fmol with a signal/noise ratio greater than 5. A linear relationship is demonstrated between the molar amount of nitrotyrosine present and the chromatographic peak area recorded.

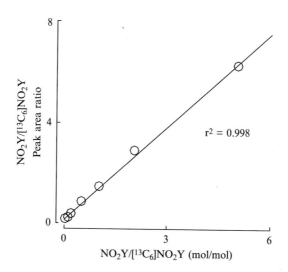

FIG. 4. Standard calibration curve for the detection of nitrotyrosine by LCQDeca. Increasing amounts of native nitrotyrosine were added to 2 pmol of ring-labeled nitrotyrosine, which was subsequently subjected to protein hydrolysis before liquid chromatography mass spectrometry analysis. A linear relationship is demonstrated between the peak area ratio of native to ring-labeled forms and the relative amounts present in the sample assayed.

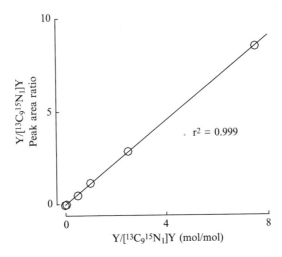

F$_{IG}$. 5. Standard calibration curve for the detection of tyrosine by LCQDeca. Increasing amounts of native tyrosine were added to 5 nmol of universal labeled [$^{13}$C9$^{15}$N1]-tyrosine, which was subsequently subjected to protein hydrolysis conditions before liquid chromatography mass spectrometry analysis. A linear relationship is demonstrated between the peak area ratio of native to universal labeled forms and the relative amounts present in the sample assayed.

respectively. Analysis of multiple independent and replicate samples demonstrated that the LCQDeca assay was highly reproducible, with inter-sample and intrasample coefficients of variance both less than 10%.

### Example: LC/MS Quantification of 3-Nitrotyrosine in EPO Knockout Mice following Ovalbumin Challenge Model of Asthma

For illustrative purposes, protein-bound NO$_2$Tyr levels within lung lavages recovered from wild type and EPO KO mice were determined using the stable isotope dilution tandem MS method described. Animals were presensitized to ovalbumin and then challenged with either nebulized normal saline or OVA. Figure 6 illustrates typical results for a specimen monitored for both NO$_2$Tyr (channels: natural abundance isotopomer [$^{12}$C]; the ring-labeled internal standard isotopomer for quantification [$^{13}$C$_6$]; and the universal labeled isotopomer [$^{13}$C$_9$$^{15}$N$_1$] for detection of any potential artifactual generation of NO$_2$Tyr) and Tyr (channels: natural abundance isotopomer [$^{12}$C]; and the universal labeled isotopomer [$^{13}$C$_9$$^{15}$N$_1$] served both as internal standard for quantification of natural abundance Tyr, as well as the precursor for detection of any potential artifactual generation of NO$_2$Tyr). Quantification of protein-bound

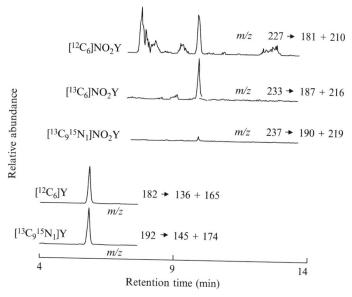

FIG. 6. Typical selected reaction monitoring (SRM) chromatographic spectra demonstrated in the determination of nitrotyrosine and tyrosine as detected by LCQDeca. Spectra are depicted according to their relative abundance and retention times. In this unusual example, selected for illustrative purposes, a small amount of artifactual [$^{13}C_9 {}^{15}N_1$]-nitrotyrosine is observed. Typically, the amount of artifactual nitrotyrosine present should be less than 5% of the total. If this is not the case, the preparative method is repeated, with additional desalting steps if needed.

$NO_2Tyr$ was readily achieved in protein hydrolysates from lungs of sensitized wild-type and eosinophil peroxidase KO mice after saline versus OVA challenge (Fig. 7).

For illustrative purposes, the chromatogram shown in Fig. 6 demonstrates a small spectral signal of artifactually generated [$^{13}C_9 {}^{15}N_1$]$NO_2Tyr$ (this is typically not seen under the conditions employed in the assay). Calculations demonstrated that endogenous $NO_2Tyr$ levels ($^{12}C$ isotopomer) are in vast excess to that of the [$^{13}C_9 {}^{15}N_1$]$NO_2Tyr$ isotopomer (whose level accounted for <5% of the native $NO_2Tyr$ after adjustments for differences in [$^{13}C_9 {}^{15}N_1$]-labeled Tyr versus [$^{12}C$]-labeled Tyr).

In the example shown, levels of $NO_2Tyr$ were significantly higher in wild-type compared to eosinophil peroxidase KO mice following allergen challenge (Fig. 7). These results are consistent with our previous finding of eosinophils as a major source of $NO_2Tyr$ in allergen-challenged lung tissues (Brennan et al., 2002; MacPherson et al., 2001). The increase in $NO_2Tyr$

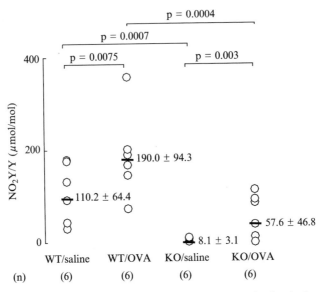

FIG. 7. Levels of nitrotyrosine, normalized for tyrosine, in bronchoalveolar lavage samples from allergen (ovalbumin)-sensitized wild-type (WT) and eosinophil peroxidase (EPO) knockout (KO) mice subjected to either saline or ovalbumin (OVA) aerochallenge. WT mice demonstrated higher levels, both at baseline and in the stimulated state. The increase in level seen in KO mice after stimulation with OVA suggests that there are EPO-independent pathways in nitrotyrosine generation.

formed following OVA challenge in eosinophil KO mice suggests that there are eosinophil peroxidase–independent pathways for generating $NO_2Tyr$ in airways following allergen challenge.

## LC/MS Analysis of Free 3-Nitrotyrosine in Plasma

Using the assay described herein, where all vestiges of artifactual nitration are removed as monitored using isotopically labeled tyrosine precursors, we note that free $NO_2Tyr$ levels in biological tissues and fluids (plasma, serum, tissue homogenates, bronchoalveolar lavage fluid, cerebrospinal fluid, and joint aspirates) are extremely low ($<10$ n$M$). Figure 8 illustrates the detection of endogenous free $NO_2Tyr$ in plasma of an individual with cardiovascular disease (a condition that has increased protein-bound $NO_2Tyr$). The chromatograms depicted in the figure inset clearly demonstrate detectable endogenous free $NO_2Tyr$ within the plasma sample. The concentration of $NO_2Tyr$ calculated from stable isotope standard curves is

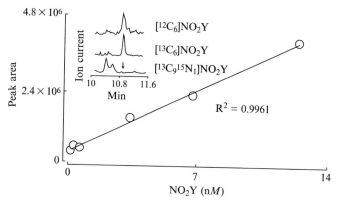

FIG. 8. Detection of free nitrotyrosine in human plasma by LCQDeca. The example is from a subject with coronary artery disease and a higher than usual level of free nitrotyrosine (plasma concentrations are typically <500 pM). Whereas the free level is usually low *in vivo*, the curve demonstrates that spiking the plasma sample with aliquots of increased concentrations of 3-nitrotyrosine results in a linear increase in the chromatographic peak area detected.

approximately 1 n$M$ in this sample. To confirm this low level and to demonstrate the validity of the assay, sequentially increased amounts of free NO$_2$Tyr were added to the plasma sample, and the amount of NO$_2$Tyr detected is shown. Using this method of standard addition, we demonstrate a linear increase in the chromatographic peak area detected and confirm that endogenous levels of free NO$_2$Tyr are exceedingly low ($\sim$1 n$M$ in this specimen). Analysis of multiple independent samples with replicates also reveals that this LC/MS/MS technique was highly reproducible, with inter-sample and intrasample coefficients of variance both less than 15% when monitoring trace levels of free NO$_2$Tyr in biological specimens.

## General Comments

The emergence of the concept that reactive nitrogen species play a critical role in the pathogenesis of various inflammatory processes has highlighted the need to develop accurate analytical methods to quantify nitrative stress *in vivo*. For example, nitrative stress contributes to the inflammatory cascade that promotes atherosclerotic plaque development (Podrez *et al.*, 1999). Serum levels of protein-bound NO$_2$Tyr correlate with the prevalence of coronary artery disease, and systemic levels of protein-bound NO$_2$Tyr decline in response to treatment with atorvastatin (Shishehbor *et al.*, 2003a). For use as a biomarker of clinical risk, such analytical methods need to be performed in a rapid, simple, and reproducible fashion. The benchtop ion trap LC/MS/MS method described herein for the measurement of both free and protein-bound forms of $NO_2Tyr$ fulfills these criteria.

A critical component of this method involves its use of heavy isotope–labeled internal standards. Their inclusion assists in the detection of artifactual $NO_2Tyr$ generated during the method. This follows our report on the use of isotopomeric standards in the detection of bromotyrosine in patients with asthma (Wu *et al.*, 1999, 2000) and their use in the detection of free and protein-bound $NO_2Tyr$ by others (Frost *et al.*, 2000; Yi *et al.*, 2000). As discussed, the key to accurately determining the level of artifactual-generated $NO_2Tyr$ depends on adjusting the level detected in response to the ratio of universal-labeled to natural forms of Tyr. We have found that the amount of artifactual generation is usually minimal under the sample processing conditions described. In the rare circumstance in which the artifactual species represents more than 5% of the total $NO_2Tyr$ signal, the sample preparation is repeated with a more exhaustive desalting step. The aim should ideally be to achieve a molar ratio of universal to natural tyrosine that approximates 1:1. Modifications to this protocol are required in certain circumstances. For example, samples from patients with acute coronary syndromes receiving nitrate therapy require an extra desalting step to remove excess nitrite observed in the pharmacologically enhanced plasma.

The present method demonstrates high sensitivity and specificity for the simultaneous detection of both Tyr and $NO_2Tyr$. The detection limit for standard samples was low (100 fmol), consistent with previous reports using triple quadrupole LC/MS/MS systems (Shishehbor *et al.*, 2003a). Similarly, the present method demonstrates a low endogenous level of free $NO_2Tyr$, with a limit of detection well under 1 n$M$ (100 fmol/100 $\mu$l plasma). This compares favorably with previous reports.

The precursor-product transitions monitored are another factor of major importance. The sensitivity and specificity of the assay can be further enhanced by changing the precursor-product transitions monitored (Table I). Taking a precursor ion to a single product ion increases specificity. In contrast, the sum of two product ions increases sensitivity. We initially performed all studies using a precursor to facilitate single product ion transition. However, comparisons with data obtained using the sum of two characteristic product ions as described in Table I demonstrate negligible alterations in results when analyzing human plasma or serum specimens. Changing the precursor-product transition necessitates the generation of  new standard curves when calculating levels for measured samples. Similarly, when applying the assay to different biological matrices, the transitions should be reexamined to ensure they yield comparable results. The chromatographic peaks monitored for all isotopomers should demonstrate baseline resolution from their corresponding unmodified amino acids to facilitate integration of peaks.

Several caveats should be highlighted with regard to the use of this method. The desalting procedure requires additional extraction steps in conditions in which significant endogenous nitrite/nitrate levels preside, such as in salivary fluid or endobronchial aspirates. Furthermore, the protein hydrolysis stage can also be revised. The method described incorporates acid hydrolysis using HCl. This, however, cannot be employed for simultaneous determination of chlorotyrosine. Therefore, in situations in which it is desirable to monitor chlorotyrosine in addition to $NO_2Tyr$, hydrolysis using HBr is preferable. Alternatively, methane sulfonic acid can be used in the determination of a wide range of analytes (we use this acid in when simultaneous quantification of $NO_2Tyr$, chlorotyrosine, and bromotyrosine is desired).

In summary, we have described a robust, stable isotope dilution method for the rapid and accurate detection of free and protein-bound $NO_2Tyr$ in biological matrices using a benchtop ion trap LC/MS/MS system. The use of internal isotopomeric standards allows for the adjusting for intrapreparative losses during sample handling, and simultaneous quantification of any potential artifactual generation of $NO_2Tyr$ during the sample analysis. The described method demonstrates high sensitivity, specificity, and reproducibility, and it can be performed on multiple samples simultaneously. Given the increasing availability of benchtop ion trap MS systems, the present method provides an analytical tool for monitoring nitrative stress in a range of biological matrices. Quantification of $NO_2Tyr$ will both help elucidate the role of nitrative stress in the pathogenesis of various disease processes and potentially serve as a biomarker to predicting cardiovascular risks and responses to therapy.

## Acknowledgments

This work was supported by National Institutes of Health grants P01 HL076491, HL62526, HL70621, HL51469, HL077692, and HL61878. S. J. N. is supported by a Ralph Reader overseas research fellowship from the National Heart Foundation of Australia. Support was also provided by the General Clinical Research Center at the Cleveland Clinic Foundation (MO1RR018390).

## References

Abe, K., Pan, U. H., Watanabe, M., Kato, T., and Itoyama, Y. (1995). Induction of nitrotyrosine-like immunoreactivity in the lower motor neuron of amyotrophic lateral sclerosis. *Neurosci. Lett.* **199,** 152–154.

Althaus, J. S., Schmidt, K. R., Fountain, S. T., Tseng, M. T., Carroll, R. T., Galatsis, P., and Hall, E. D. (2000). LC-MS/MS detection of peroxynitrite-derived 3-nitrotyrosine in rat microvessels. *Free Radic. Biol. Med.* **29,** 1085–1095.

Andreadis, A. A., Hazen, S. L., Comhair, S. A., and Erzurum, S. C. (2003). Oxidative and nitrosative events in asthma. *Free Radic. Biol. Med.* **35,** 213–225.

Beal, M. F. (2002). Oxidatively modified proteins in aging and disease. *Free Radic. Biol. Med.* **32,** 797–803.

Beckman, J. S., Beckman, T. W., Chen, J., Marshall, P. A., and Freeman, B. A. (1990). Apparent hydroxyl radical production by peroxynitrite: Implications for endothelial injury from nitric oxide and superoxide. *Proc. Natl. Acad. Sci. USA* **87,** 1620–1624.

Brennan, M. L., Wu, W., Fu, X., Shen, Z., Song, W., Frost, H., Vadseth, C., Narine, L., Lenkiewicz, E., Borchers, M. T., Lusis, A. J., Lee, J. J., Lee, N. A., Abu-Soud, H. M., Ischiropoulos, H., and Hazen, S. L. (2002). A tale of two controversies: defining both the role of peroxidases in nitrotyrosine formation *in vivo* using eosinophil peroxidase and myeloperoxidase-deficient mice, and the nature of peroxidase-generated reactive nitrogen species. *J. Biol. Chem.* **277,** 17415–17427.

Crow, J. P., and Ischiropoulos, H. (1996). Detection and quantitation of nitrotyrosine residues in proteins: *In vivo* marker of peroxynitrite. *Methods Enzymol.* **269,** 185–194.

Crowley, J. R., Yarasheski, K., Leeuwenburgh, C., Turk, J., and Heinecke, J. W. (1998). Isotope dilution mass spectrometric quantification of 3-nitrotyrosine in proteins and tissues is facilitated by reduction to 3-aminotyrosine. *Anal. Biochem.* **259,** 127–135.

Drew, B., and Leeuwenburgh, C. (2002). Aging and the role of reactive nitrogen species. *Ann. NY Acad. Sci.* **959,** 66–81.

Fries, D. M., Paxinou, E., Themistocleous, M., Swanberg, E., Griendling, K. K., Salvemini, D., Slot, J. W., Heijnen, H. F., Hazen, S. L., and Ishciropoulos, H. (2003). Expression of inducible nitric-oxide synthase and intracellular protein tyrosine nitration in vascular smooth muscle cells: Role of reactive oxygen species. *J. Biol. Chem.* **278,** 22901–22907.

Frost, M. T., Halliwell, B., and Moore, K. P. (2000). Analysis of free and protein-bound nitrotyrosine in human plasma by a gas chromatography/mass spectrometry method that avoids nitration artifacts. *Biochem. J.* **345,** 453–458.

Giasson, B. I., Duda, J. E., Murray, I. V., Chen, Q., Souza, J. M., Hurting, H. I., Ischiropoulos, H., Trojanowski, J. Q., and Lee, V. M. (2000). Oxidative damage linked to neurodegeneration by selective alpha-synuclein nitration in synucleinopathy lesions. *Science* **290,** 985–989.

Graham, A., Hogg, N., Kalyanaraman, B., O'Leary, V., Darley-Usmar, V., and Moncada, S. (1993). Peroxynitrite modification of low-density lipoprotein leads to recognition by the macrophage scavenger receptor. *FEBS Lett.* **330,** 181–185.

Haddad, I. Y., Pataki, G., Hu, P., Galliani, C., Beckman, J. S., and Matalon, S. (1994). Quantitation of nitrotyrosine levels in lung sections of patients and animals with acute lung injury. *J. Clin. Invest.* **94,** 2407–2413.

Kamisaki, Y., Wada, K., Nakamoto, K., Kishimoto, Y., Kitano, M., and Itoh, T. (1996). Sensitive determination of nitrotyrosine in human plasma by isocratic high-performance liquid chromatography. *J. Chromatogr. B Biomed. Appl.* **685,** 343–347.

Leeuwenburgh, C., Hardy, M. M., Hazen, S. L., Wagner, P., Oh-ishi,, S., Steinbrecher, U. P., and Heinecke, J. W. (1997). Reactive nitrogen intermediates promote low density lipoprotein oxidation in human atherosclerotic intima. *J. Biol. Chem.* **272,** 1433–1436.

MacPherson, J. C., Comhair, S. A., Erzurum, S. C., Klein, D. F., Lipscomb, M. F., Kavuru, M. S., Samoszuk, M. K., and Hazen, S. L. (2001). Eosinophils are a major source of nitric oxide-derived oxidants in severe asthma: Characterization of pathways available to eosinophils for generating reactive nitrogen species. *J. Immunol.* **166,** 5763–5772.

Moriel, P., and Abdalla, D. S. (1997). Nitrotyrosine bound to beta-VLDL-apoproteins: A biomarker of peroxynitrite formation in experimental atherosclerosis. *Biochem. Biophys. Res. Commun.* **232,** 332–335.

Podrez, E. A., Schmitt, D., Hoff, H. F., and Hazen, S. L. (1999). Myeloperoxidase-generated reactive nitrogen species convert low density lipoprotein into an atherogenic form *in vitro. J. Clin. Invest.* **103,** 1547–1560.

Riordan, J. F., Sokolovsky, M., and Vallee, B. L. (1966). Tetranitromethane. A reagent for the nitration of tyrosine and tyrosyl residues of proteins. *J. Am. Chem. Soc.* **88,** 4104–4105.

Samoszuk, M., Brennan, M. L., To, V., Leonor, L., Zheng, L., Fu, X., and Hazen, S. L. (2002). Association between nitrotyrosine levels and microvascular density in human breast cancer. *Breast Cancer Res. Treat.* **74,** 271–278.

Schwedhelm, E., Tsikas, D., Gutzki, F. M., and Frolich, J. C. (1999). Gas chromatographic-tandem mass spectrometric quantification of free 3-nitrotyrosine in human plasma at the basal state. *Anal. Biochem.* **276,** 195–203.

Shigenaga, M. K., Lee, H. H., Blount, B. C., Christen, S., Shigeno, E. T., Yip, H., and Ames, B. N. (1997). Inflammation and NO(X)-induced nitration: Assay for 3-nitrotyrosine by HPLC with electrochemical detection. *Proc. Natl. Acad. Sci. USA* **94,** 3211–3216.

Shishehbor, M. H., Aviles, R. J., Brennan, M. L., Fu, X., Goormastic, M., Pearce, G. L., Gokce, N., Keaney, J. F., Jr., Penn, M. S., Sprecher, D. L., Vita, J. A., and Hazen, S. L. (2003a). Association of nitrotyrosine levels with cardiovascular disease and modulation by statin therapy. *JAMA* **289,** 1675–1680.

Shishehbor, M. H., Brennan, M. L., Aviles, R. J., Fu, X., Penn, M. S., Sprecher, D. L., and Hazen, S. L. (2003b). Statins promote potent systemic antioxidant effects through specific inflammatory pathways. *Circulation* **108,** 426–431.

Smith, M. A., Richey Harris, P. L., Sayre, L. M., Beckman, J. S., and Perry, G. (1997). Widespread peroxynitrite-mediated damage in Alzheimer's disease. *J. Neurosci.* **17,** 2653–2657.

Szabo, C., Salzman, A. L., and Ischiropoulos, H. (1995). Endotoxin triggers the expression of an inducible isoform of nitric oxide synthase and the formation of peroxynitrite in the rat aorta *in vivo. FEBS Lett.* **363,** 235–238.

Tanaka, K., Shirai, T., Nagata, E., Dembo, T., and Fukuuchi, Y. (1997). Immunohistochemical detection of nitrotyrosine in postischemic cerebral cortex in gerbil. *Neurosci. Lett.* **235,** 85–88.

Vadseth, C., Souza, J. M., Thomson, L., Seagraves, A., Nagaswami, C., Scheiner, T., Torbet, J., Vilaire, G., Bennett, J. S., Murciano, J. C., Muzykantov, V., Penn, M. S., Hazen, S. L., Weisel, J. W., and Ischiropoulos, H. (2004). Pro-thrombotic state induced by post-translational modification of fibrinogen by reactive nitrogen species. *J. Biol. Chem.* **279,** 8820–8826.

Viera, L., Ye, Y. Z., Estevez, A. G., and Beckman, J. S. (1999). Immunohistochemical methods to detect nitrotyrosine. *Methods Enzymol.* **201,** 373–381.

Wu, W., Chen, Y., d'Avignon, A., and Hazen, S. L. (1999). 3-Bromotyrosine and 3,5--dibromotyrosine are major products of protein oxidation by eosinophil peroxidase: Potential markers for eosinophil-dependent tissue injury *in vivo. Biochemistry* **38,** 3538–3548.

Wu, W., Samoszuk, M. K., Comhair, S. A., Thomassen, M. J., Farver, C. F., Dweik, R. A., Kavuru, M. S., Erzurum, S. C., and Hazen, S. L. (2000). Eosinophils generate brominating oxidants in allergen-induced asthma. *J. Clin. Invest.* **105,** 1455–1463.

Yi, D., Ingelse, B. A., Duncan, M. W., and Smythe, G. A. (2000). Quantification of 3-nitrotyrosine in biological tissues and fluids: Generating valid results by eliminating artifactual formation. *J. Am. Soc. Mass Spectrom.* **11,** 578–586.

Zheng, L., Nukuna, B., Brennan, M. L., Sun, M., Goormastic, M., Settle, M., Schmitt, D., Fu, X., Thomson, L, Fox, P. L., Ischiropoulos, H., Smith, J. D., Kinter, M., and Hazen, S. L. (2004). Apolipoprotein A-I is a selective target for myeloperoxidase-catalyzed oxidation and functional impairment in subjects with cardiovascular disease. *J. Clin. Invest.* **114,** 529–541.

# [23] Mapping Sites of Tyrosine Nitration by Matrix-Assisted Laser Desorption/Ionization Mass Spectrometry

By ILLARION V. TURKO and FERID MURAD

## Abstract

Protein tyrosine nitration is an important part of nitric oxide biology. This posttranslational modification occurs under normal physiological conditions and is substantially enhanced under various pathological conditions. Studies reveal that protein tyrosine nitration is a dynamic and selective process that influences protein function and turnover and can be considered a diagnostic biomarker of pathology. The identification of nitrated tyrosine residues directly within any given nitrated protein is important for studies on *in vivo* mechanisms of nitration and for the explanation of functional consequences of nitration. Specific nitrated tyrosines in given proteins may be also more informative as oxidative biomarkers than overall nitrotyrosine levels. However, localization of the sites of nitration remains a methodological challenge. Mass spectrometry (MS) is an ideal method for identifying nitrated tyrosines in proteins because of its sensitivity and specificity. This chapter is not intended to thoroughly discuss the various MS-based approaches for nitrotyrosine identification and merely focuses on the analysis of peptides containing nitrotyrosine by matrix-assisted laser desorption ionization MS (MALDI-MS). The data summarized show that the MALDI-MS pattern of a tyrosine-nitrated peptide includes the unique combination of ions that provides unequivocal evidence for the presence of nitrotyrosine in a given peptide and could be used for mapping sites of tyrosine nitration in proteins.

METHODS IN ENZYMOLOGY, VOL. 396
Copyright 2005, Elsevier Inc. All rights reserved.

0076-6879/05 $35.00
DOI: 10.1016/S0076-6879(05)96023-0

## Introduction

Increased protein tyrosine nitration has been reported in virtually all tissues that experience oxidative stress as a part of the normal cellular metabolism or as a part of a specific disease state (Greenacre and Ischiropoulos, 2001; Turko and Murad, 2002). Tyrosine nitration may affect protein structure and function. A gain or inhibition of function, as well as no effect on function, has been reported for some nitrated proteins (Balafanova *et al.*, 2002; Gole *et al.*, 2000; Greenacre and Ischiropoulos, 2001; Turko and Murad, 2002). However, the inhibition of function is a more common consequence of protein nitration (Greenacre and Ischiropoulos, 2001; Turko and Murad, 2002). It has also been shown that nitration of a tyrosine residue may prevent the subsequent phosphorylation of that residue (Gow *et al.*, 1996; Kong *et al.*, 1996). Alternatively, nitration of tyrosine residues may simulate phosphorylation (MacMillan-Crow *et al.*, 2000; Mallozzi *et al.*, 2001) and result in the constitutively active proteins. Furthermore, tyrosine nitration may change the rate of proteolytic degradation of nitrated proteins and favor either its faster clearance or the accumulation of nitrated proteins in cells. Cumulatively, this suggests that protein nitration is a dynamic process that affects protein properties and turnover and may be involved in a variety of functions, possibly including disease initiation and progression. The current field is immature, and many studies are yet to be done.

It was believed that nitrotyrosine *in vivo* was formed almost exclusively by the reaction of tyrosine with peroxynitrite ($ONOO^-$), which is a powerful oxidant produced by the rapid almost diffusion-limited reaction between nitric oxide ($^\bullet NO$) and superoxide ($O_2^{\bullet -}$) (Beckman *et al.*, 1992; Ischiropoulos *et al.*, 1992). In addition to the uncatalyzed nitration by $ONOO^-$, secondary reactions of $ONOO^-$ with $CO_2$, transition metals, and myeloperoxidase (MPO) have been found to catalyze the nitration of tyrosine residues (Beckman *et al.*, 1992; Ischiropoulos *et al.*, 1992; Lymar *et al.*, 1996; Van der Vliet *et al.*, 1997). These reactions increase the rate of tyrosine nitration and may explain the ability of $ONOO^-$ to nitrate proteins despite the presence of potentially $ONOO^-$-specific scavengers in biological samples. In addition to $ONOO^-$, nitrite-dependent heme peroxidase reactions may also give rise to protein tyrosine nitration (Eiserich *et al.*, 1998; Pfeiffer *et al.*, 2001). It has been shown that heme peroxidase enzymes (myeloperoxidase, eosinophil peroxidase, and horse radish peroxidases) in the presence of nitrite ($NO_2^-$) and hydrogen peroxide ($H_2O_2$) can nitrate proteins in various biological samples (Eiserich *et al.*, 1998; Sampson *et al.*, 1998; Wu *et al.*, 1999). Because of the heme moiety, many hemoproteins also possess pseudo-peroxidase activity and may contribute to tyrosine nitration *in vivo* (Grzelak *et al.*, 2001; Kilinc *et al.*, 2001; Ogino *et al.*, 2001). Heme itself is a complex of protoporphyrin IX with iron. Hypervalent states of the heme iron may catalyze generation of the free radical species. We (Bian *et al.*, 2003), and others (Thomas *et al.*, 2002) have reported that free iron/heme can cause protein nitration *in vitro* and *in vivo*. Currently, the relative contribution and importance of multiple mechanisms of protein nitration *in vivo* remain poorly studied.

Apart from the mechanism of tyrosine nitration, its biological significance is also a subject of great interest. Once formed, nitrotyrosine is a relatively stable product and may be used as a "footprint" of a specific disease state. Various methods for measuring nitrotyrosine have been developed, including immunochemical methods (Ye *et al.*, 1996) and chromatographic methods, using either high-performance liquid chromatography (HPLC) separation combined with electrochemical detection (Crow, 1999; Shigenaga, 1999) or gas chromatography–mass spectrometry techniques (Crowley *et al.*, 1998; Jiang and Balazy, 1998). The detailed analysis of quantitative approaches for measuring nitrotyrosine is beyond the scope of this chapter but can be found elsewhere (Balazy, 2002; Herce-Pagliai *et al.*, 1998).

A much more challenging analytical task is to identify this modification directly within any given protein. Site-specific information will be important for studies on *in vivo* mechanisms of nitration and for the explanation of functional consequences of nitration. Specific nitrated tyrosines in given proteins may also be more informative as oxidative biomarkers than overall nitrotyrosine levels.

MS is an ideal method for identifying nitrated tyrosines in proteins because of its sensitivity and specificity. Several studies have employed electrospray ionization MS to identify nitrated tyrosines in proteins and peptides (Kanski *et al.*, 2003; MacMillan-Crow *et al.*, 1998; Murray *et al.*, 2003). These studies demonstrated a typical addition of 45 Da to the molecular ion of the nitrotyrosine-containing peptide. This chapter is not intended to thoroughly discuss the various MS-based approaches for nitrotyrosine identification, but we focus on the analysis of peptides containing nitrotyrosine by matrix-assisted laser desorption/ionization MS (MALDI-MS).

Chemicals

Human synthetic angiotensin II (Asp-Arg-Val-Tyr-Ile-His-Pro-Phe, $MH^+$ $m/z$ 1047.2) was purchased from Sigma (St. Louis, MO). $ONOO^-$ was obtained from Upstate Biotechnology (Lake Placid, NY). Human polymorphonuclear leukocytes MPO came from Calbiochem-Novabiochem Corporation (San Diego, CA).

### Preparation of Nitrated Angiotensin II

First, 0.8 mM solution of angiotensin II in 50 mM $NH_4HCO_3$ (pH 7.8) was treated with either (1) single bolus addition of 1.0 mM $ONOO^-$ or (2) 0.5 mM $NaNO_2$, 200 $\mu$g/ml glucose with 40 ng/ml glucose oxidase (to generate $H_2O_2$), and 100 nM MPO for 30 min at room temperature. The samples were then dried using Vacufuge from Eppendorf AG (Hamburg, Germany) and kept at $-80°$.

### MALDI-MS Analysis of Nitrated Angiotensin II

Dry nitrated angiotensin II samples were dissolved in 50% acetonitrile/5% trifluoroacetic acid. We then mixed 0.5-$\mu$l aliquot of samples with 0.5 $\mu$l aliquot of matrix (10 mg/ml $\alpha$-cyano-4-hydroxycinnamic acid in 50% acetonitrile/0.1% trifluoroacetic acid) on a stainless-steel MALDI plate and air-dried the mixture. Spectra were acquired on an Applied Biosystems Voyager-DE STR MALDI-TOF MS equipped with a nitrogen laser (337 nm) and operated in the delayed extraction and reflector mode.

Figure 1 shows a pattern of peaks generated for angiotensin II nitrated in $ONOO^-$- and MPO-dependent reactions. These patterns are similar and include ions of original nonnitrated angiotensin II at $m/z$ 1047.2 ($MH^+$) and nitrated angiotensin II at $m/z$ 1092.2 ($MH^+ + 45$). In addition, three more ions are seen at $m/z$ 1060.2 ($MH^+ + 13$), $m/z$ 1062.2 ($MH^+ + 15$), and $m/z$ 1076.2 ($MH^+ + 29$). The source of these ions was explained in original papers by Sarver et al. (2001) and Petersson et al. (2001). It appears that these ions represent various photodecomposition and reductive products of nitrotyrosine (Fig. 2). Specifically, ($MH^+ + 29$) ion represents the loss of a single oxygen from the aromatic nitro group to form the nitroso derivative of tyrosine. ($MH^+ + 15$) ion represents the reduction of nitro group to amino group. ($MH^+ + 13$) ion is likely a nitrene derivative of tyrosine (Fig. 2). Phenyl nitrene is relatively stable or may be rearranged in a seven atom ring, dehydroazepine-type isomer (not shown in Fig. 2). Whatever the nature of ($MH^+ + 13$) ion, taken all together, these ions create a unique pattern of MALDI-MS peaks corresponding to any nitrotyrosine-containing peptide.

To examine the effect of different matrices on the formation of these photodecomposition fragments, Petersson et al. (2001) used dihydrobenzoic acid, 2,4,6-trihydroxyacetophenone/nitrocellulose, and $\alpha$-cyano-4-

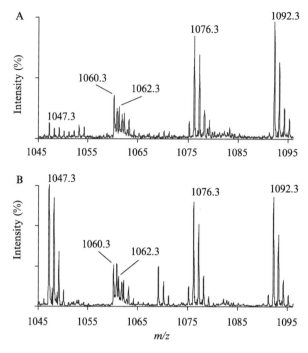

FIG. 1. Matrix-assisted laser desorption ionization mass spectrometry (MALDI-MS) pattern for tyrosine nitrated angiotensin II. Angiotensin II (MH$^+$ $m/z$ 1047.3) was nitrated with ONOO$^-$ (A) or with NO$_2^-$ /H$_2$O$_2$/MPO (B).

hydroxycinnamic acid. Some differences in the abundance of the ions were observed with these matrices; however, the pattern remains the same. Other variable conditions of acquiring a MALDI spectrum, including number of laser shots, laser power, and peptide concentration, were studied by Sarver *et al.* (2001). They found that intensity and number of laser pulses at the same spot apparently have no effect on the relative abundance of the ions creating the pattern. A noticeable difference was observed when the amount of spotted peptide was varied. Lower concentrations of peptide favor abundance of (MH$^+$ + 15) and (MH$^+$ + 13) ions; however, the pattern itself remains the same.

Using strong oxidizing and nitrating compound, tetranitromethane, tyrosine could be converted in double-nitrated derivative, namely 3,5-dinitrotyrosine. However, under physiologically relevant conditions of nitration used in this study, no evidence was found for the presence of a doubly nitrated tyrosine derivative of angiotensin II.

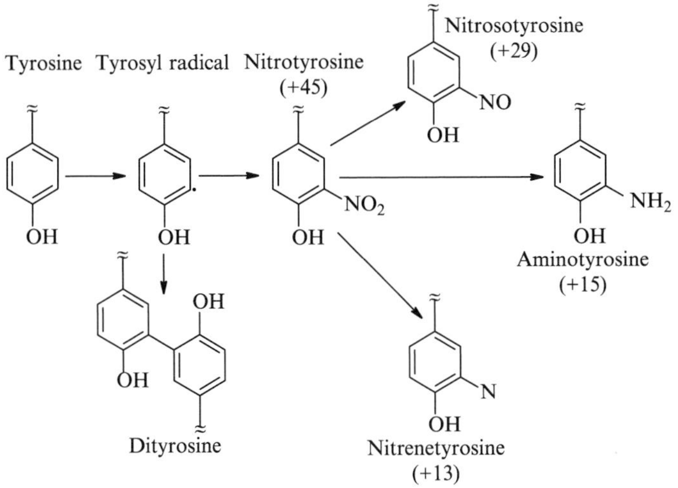

FIG. 2. Generation of dityrosine and nitrotyrosine and likely products of nitrotyrosine photochemical decomposition.

An additional interest may have a peak in nitrated angiotensin II found at $m/z$ 2091.2 (Fig. 3) that corresponds to $[(2M + H)^+ - 2]$ and is likely a dimer of angiotensin II cross-linked through dityrosine (Fig. 2). Figure 2 shows that formation of nitrotyrosine begins with the generation of tyrosyl radical, which could be then nitrated. Another possibility is a radical isomerization followed by diradical reaction that results in formation of dityrosine (Giulivi et al., 2003). Presumably, formation of dityrosine may outcompete tyrosine nitration under certain conditions. Indeed, dityrosine was found in proteins in vivo under conditions of oxidative stress (Giulivi et al., 2003; Malencik and Anderson, 2003). Our in vitro data (Figs. 1 and 3) demonstrate that both nitrating reactions, $ONOO^-$- and MPO-dependent, generate nitrotyrosine and dityrosine simultaneously. Because intensity of MALDI signals cannot be interpreted unambiguously, we do not speculate which nitrating reaction may favor nitrotyrosine over dityrosine formation and vice versa.

## MALDI-MS Pattern for Nitrotyrosine in Proteins

MALDI-MS pattern for nitrotyrosine-containing peptides has been used for identification of specific tyrosine residues nitrated in proteins (Aslan et al., 2003; Borger et al., 2003; Kuhn et al., 2002; Turko et al.,

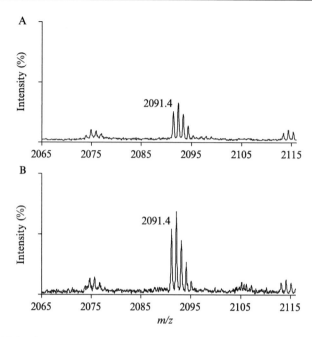

Fɪɢ. 3. Molecular ion region of matrix-assisted laser desorption ionization (MALDI) spectrum of nitrated angiotensin II that likely corresponds to the dimer of angiotensin II cross-linked through dityrosine (MH$^+$ $m/z$ 2091.4). Angiotensin II (MH$^+$ $m/z$ 1047.3) was nitrated with ONOO$^-$ (A) or with NO$_2^-$/H$_2$O$_2$/MPO (B).

2003). Aslan *et al.* (2003) immunoprecipitated nitrated actin from liver and kidney extracts of knockout-transgenic sickle cell disease mice and used the MALDI-MS pattern for identification of tyrosine-nitrated peptides. Specific nitrated tyrosine residues were further identified by MS/MS spectrum of the ion at $m/z$ (MH$^+$ + 45). Another example is identification of three nitrated tyrosine residues in ONOO$^-$-treated tyrosine hydroxylase (Borger *et al.*, 2003; Kuhn *et al.*, 2002). We also observed the MALDI-MS nitrotyrosine pattern in several mitochondrial proteins after treatment of purified mouse heart mitochondria with ONOO$^-$ or with NO$_2^-$/H$_2$O$_2$/ MPO (Turko *et al.*, 2003).

Concluding Remarks

Studies on the biological role of protein tyrosine nitration need a highly selective method for the analysis and characterization of peptides containing nitrotyrosine. This chapter reviewed the MALDI-MS pattern of photochemical reaction products that has been observed for tyrosine nitrated peptides (Petersson *et al.*, 2001; Sarver *et al.*, 2001). The structures for some of these products are not known with certainty, and relative abundance of ions in the pattern may slightly vary depending on conditions of spectrum acquiring; however, the pattern itself always remains the same. The pattern includes original nonnitrated peptide at *m/z* MH$^+$ and four additional ions at *m/z* (MH$^+$ + 13), (MH$^+$ + 15), (MH$^+$ + 29), and (MH$^+$ + 45). This unique combination of ions provides unequivocal evidence for the presence of nitrotyrosine in a given peptide. One may argue that *in vivo* appearance of nitrotyrosine is lower than that used in various *in vitro* experiments, and it might be difficult to find the pattern in biological samples. Immunoprecipitation of *in vivo*–nitrated proteins with anti-nitrotyrosine antibody can overcome this problem (Aslan *et al.*, 2003). A paper by Aslan *et al.* (2003) is a good example of enrichment of *in vivo*–derived samples, clear identification of the nitrated peptides based on MALDI-MS pattern, and confirmation of nitration by electrospray ionization MS/MS. All current data suggest that the MALDI-MS pattern for tyrosine-nitrated peptides could be used for mapping sites of tyrosine nitration in proteins exposed to oxidative stress or disease.

## Acknowledgments

This work was supported by the National Institutes of Health (grant No. GM61731), the John S. Dunn Foundation, the G. Harold and Leila Y. Mathers Charitable Foundation, the Welch Foundation, the United States Army, and the University of Texas.

## References

Aslan, M., Ryan, T. M., Townes, T. M., Coward, L., Kirk, M. C., Barnes, S., Alexander, C. B., Rosenfeld, S. S., and Freeman, B. A. (2003). Nitric oxide–dependent generation of reactive species in sickle cell disease. Actin tyrosine nitration induces defective cytoskeletal polymerization. *J. Biol. Chem.* **278**, 4194–4204.

Balafanova, Z., Bolli, R., Zhang, J., Zheng, Y., Pass, J. M., Bhatnagar, A., Tang, X. L., Wang, O., Cardwell, E., and Ping, P. (2002). Nitric oxide induces nitration of PKCepsilon, facilitating PKCepsilon translocation via enhanced PKCepsilon-RACK2 interactions: A novel mechanism of NO-triggered activation of PKCepsilon. *J. Biol. Chem.* **277**, 15021–15027.

Balazy, M. (2002). Gas chromatography/mass spectrometry assay for 3-nitrotyrosine. *Methods Enzymol.* **359**, 390–399.

Beckman, J. S., Ischiropoulos, H., Zhu, L., van der Woerd, M., Smith, C. D., Harrison, J., Martin, J. C., and Tsai, J.-H. M. (1992). Kinetics of superoxide dismutase and iron-catalyzed nitration of phenolics by peroxynitrite. *Arch. Biochem. Biophys.* **298**, 438–445.

Bian, K., Gao, Z., Weisbrodt, N., and Murad, F. (2003). The nature of heme/iron-induced protein tyrosine nitration. *Proc. Natl. Acad. Sci. USA* **100,** 5712–5717.

Borger, C. R., Kuhn, D. M., and Watson, J. T. (2003). Mass mapping sites of nitration in tyrosine hydroxylase: Random vs selective nitration of three tyrosine residues. *Chem. Res. Toxicol.* **16,** 536–540.

Crow, J. P. (1999). Measurement and significance of free and protein-bound 3-nitrotyrosine, 3-chlorotyrosine, and free 3-nitro-4-hydroxyphenylacetic acid in biological samples: A high-performance liquid chromatography method using electrochemical detection. *Methods Enzymol.* **301,** 151–160.

Crowley, J. R., Yarasheski, K., Leeuwenburgh, C., Turk, J., and Heinecke, J. W. (1998). Isotope dilution mass spectrometric quantification of 3-nitrotyrosine in proteins and tissues is facilitated by reduction to 3-aminotyrosine. *Anal. Biochem.* **259,** 127–135.

Eiserich, J. P., Hristova, M., Cross, C. E., Jones, A. D., Freeman, B. A., Halliwell, B., and Van der Vliet, A. (1998). Formation of nitric oxide–derived inflammatory oxidants by myeloperoxidase in neutrophils. *Nature (London)* **391,** 393–397.

Giulivi, C., Traaseth, N. J., and Davies, K. J. A. (2003). Tyrosine oxidation products: Analysis and biological relevance. *Amino Acids* **25,** 227–232.

Gole, M. D., Souza, J. M., Choi, I., Hertkorn, C., Malcolm, S., Foust, R. F., III, Finkel, B., Lanken, P. N., and Ischiropoulos, H. (2000). Plasma proteins modified by tyrosine nitration in acute respiratory distress syndrome. *Am. J. Physiol. Lung Cell Mol. Physiol.* **278,** L961–L967.

Gow, A. J., Duran, D., Malcolm, S., and Ischiropoulos, H. (1996). Effects of peroxynitrite-induced protein modifications on tyrosine phosphorylation and degradation. *FEBS Lett.* **385,** 63–66.

Greenacre, S. A. B., and Ischiropoulos, H. (2001). Tyrosine nitration: Localization, quantification, consequences for protein function and signal transduction. *Free Radic. Res.* **34,** 541–581.

Grzelak, A., Balcerczyk, A., Mateja, A., and Bartosz, G. (2001). Hemoglobin can nitrate itself and other proteins. *Biochim. Biophys. Acta* **1528,** 97–100.

Herce-Pagliai, C., Kotecha, S., and Shuker, D. E. G. (1998). Analytical methods for 3-nitrotyrosine as a marker of exposure to reactive nitrogen species: A review. *Nitric Oxide* **2,** 324–336.

Ischiropoulos, H., Zhu, L., Chen, J., Tsai, J.-H. M., Martin, J. C., Smith, C. D., and Beckman, J. S. (1992). Peroxynitrite-mediated tyrosine nitration catalyzed by superoxide dismutase. *Arch. Biochem. Biophys.* **298,** 431–437.

Jiang, H., and Balazy, M. (1998). Detection of 3-nitrotyrosine in human platelets exposed to peroxynitrite by a new gas chromatography/mass spectrometry assay. *Nitric Oxide* **2,** 350–359.

Kanski, J., Alterman, M. A., and Schoneich, C. (2003). Proteomic identification of age-dependent protein nitration in rat skeletal muscle. *Free Rad. Biol. Med.* **35,** 1229–1239.

Kilinc, K., Kilinc, A., Wolf, R. E., and Grisham, M. B. (2001). Myoglobin-catalyzed tyrosine nitration: No need for peroxynitrite. *Biochem. Biophys. Res. Commun.* **285,** 273–276.

Kong, S.-K., Yim, M. B., Stadtman, E. R., and Chock, P. B. (1996). Peroxynitrite disables the tyrosine phosphorylation regulatory mechanism: Lymphocyte-specific tyrosine kinase fails to phosphorylate nitrated cdc2(6-20)$NH_2$ peptide. *Proc. Natl. Acad. Sci. USA* **93,** 3377–3382.

Kuhn, D. M., Sadidi, M., Liu, X., Kreipke, C., Geddes, T., Borges, C., and Watson, J. T. (2002). Peroxynitrite-induced nitration of tyrosine hydroxylase. Identification of tyrosines 423, 428, and 432 as sites of modification by matrix-assisted laser desorption ionization time-of-flight mass spectrometry and tyrosine-scanning mutagenesis. *J. Biol. Chem.* **277,** 14336–14342.

Lymar, S. V., Jiang, Q., and Hurst, K. (1996). Mechanism of carbon dioxide–catalyzed oxidation of tyrosine by peroxynitrite. *Biochemistry* **35,** 7855–7861.

MacMillan-Crow, L. A., Crow, J. P., and Thompson, J. A. (1998). Peroxynitrite-mediated inactivation of manganese superoxide dismutase involves nitration and oxidation of critical tyrosine residues. *Biochemistry* **37,** 1613–1622.

MacMillan-Crow, L. A., Greendorfer, J. S., Vickers, S. M., and Thompson, J. A. (2000). Tyrosine nitration of c-SRC tyrosine kinase in human pancreatic ductal adenocarcinoma. *Arch. Biochem. Biophys.* **377,** 350–356.

Malencik, D. A., and Anderson, S. R. (2003). Dityrosine as a product of oxidative stress and fluorescent probe. *Amino Acids* **25,** 233–247.

Mallozzi, C., Di Stasi, A. M., and Minetti, M. (2001). Nitrotyrosine mimics phosphotyrosine binding to the SH2 domain of the src family tyrosine kinase lyn. *FEBS Lett.* **503,** 189–195.

Murray, J., Taylor, S. W., Zhang, B., Ghosh, S. S., and Capaldi, R. A. (2003). Oxidative damage to mitochondrial complex I due to peroxynitrite. Identification of reactive tyrosines by mass spectrometry. *J. Biol. Chem.* **278,** 37223–37230.

Ogino, K., Kodama, N., Nakajima, M., Yamada, A., Nakamura, H., Nagase, H., Sadamits, Y. D., and Maekawa, T. (2001). Catalase catalyzes nitrotyrosine formation from sodium azide and hydrogen peroxide. *Free Radic. Res.* **35,** 735–747.

Petersson, A. S., Steen, H., Kalume, D. E., Caidahl, K., and Roepstorff, P. (2001). Investigation of tyrosine nitration in proteins by mass spectrometry. *J. Mass Spectrom.* **36,** 616–625.

Pfeiffer, S., Lass, A., Schmidt, K., and Mayer, B. (2001). Protein tyrosine nitration in mouse peritoneal macrophages activated *in vitro* and *in vivo*: Evidence against an essential role of peroxynitrite. *FASEB J.* **15,** 2355–2364.

Sampson, J. B., Ye, Y. Z., Rosen, H., and Beckman, J. S. (1998). Myeloperoxidase and horseradish peroxidase catalyze tyrosine nitration in proteins from nitrite and hydrogen peroxide. *Arch. Biochem. Biophys.* **356,** 207–213.

Sarver, A., Scheffler, K., Shetlar, M., and Gibson, B. W. (2001). Analysis of peptides and proteins containing nitrotyrosine by matrix-assisted laser desorption/ionization mass spectrometry. *J. Am. Soc. Mass Spectrom.* **12,** 439–448.

Shigenaga, M. K. (1999). Quantification of protein-bound 3-nitrotyrosine by high-performance liquid chromatography with electrochemical detection. *Methods Enzymol.* **301,** 27–40.

Thomas, D. D., Espey, M. G., Vitek, M. P., Miranda, K. M., and Wink, D. A. (2002). Protein nitration is mediated by heme and free metals through Fenton-type chemistry: An alternative to the $NO/O_2^-$ reaction. *Proc. Natl. Acad. Sci. USA* **99,** 12691–12696.

Turko, I. V., Li, L., Aulak, K. S., Stuehr, D. J., Chang, J. Y., and Murad, F. (2003). Protein tyrosine nitration in the mitochondria from diabetic mouse heart. Implications to dysfunctional mitochondria in diabetes. *J. Biol. Chem.* **278,** 33972–33977.

Turko, I. V., and Murad, F. (2002). Protein nitration in cardiovascular diseases. *Pharmacol. Rev.* **54,** 619–634.

Van der Vliet, A., Eiserich, J. P., Hallowell, B., and Cross, C. E. (1997). Formation of reactive nitrogen species during peroxidase-catalyzed oxidation of nitrite. A potential additional mechanism of nitric oxide–dependent toxicity. *J. Biol. Chem.* **272,** 7617–7625.

Wu, W., Chen, Y., and Hazen, S. L. (1999). Eosinophil peroxidase nitrates protein tyrosyl residues. Implications for oxidative damage by nitrating intermediates in eosinophilic inflammatory disorders. *J. Biol. Chem.* **274,** 25933–25944.

Ye, Y. Z., Strong, M., Huang, Z. Q., and Beckman, J. S. (1996). Antibodies that recognize nitrotyrosine. *Methods Enzymol.* **269,** 201–209.

# [24]   Peroxynitrite in the Pathogenesis of Parkinson's Disease and the Neuroprotective Role of Metallothioneins

By MANUCHAIR EBADI, SUSHIL K. SHARMA, PEDRAM GHAFOURIFAR, HOLLY BROWN-BORG, and H. EL REFAEY

## Abstract

Parkinson's disease (PD) is characterized by a progressive loss of dopaminergic neurons in the substantia nigra zona compacta and in other subcortical nuclei associated with a widespread occurrence of *Lewy bodies.* The causes of cell death in Parkinson's disease are still poorly understood, but a defect in mitochondrial oxidative phosphorylation and enhanced oxidative stress has been proposed. We have examined 3-morpholinosydnonimine (SIN-1)–induced apoptosis in control and metallothionein-overexpressing dopaminergic neurons, with a primary objective to determine the neuroprotective potential of metallothionein (MT) against peroxynitrite-induced neurodegeneration in PD. SIN-1 induced lipid peroxidation and triggered plasma membrane blebbing. In addition, it caused DNA fragmentation, $\alpha$-synuclein induction, and intramitochondrial accumulation of metal ions (copper, iron, zinc, and calcium), and it enhanced the synthesis of 8-hydroxy-2-deoxyguanosine. Furthermore, it downregulated the expression of Bcl-2 and poly(adenosine diphosphate-ribose) polymerase, but upregulated the expression of caspase-3 and Bax in dopaminergic (SK-N-SH) neurons. SIN-1 induced apoptosis in aging mitochondrial genome knockout cells, $\alpha$-synuclein–transfected cells, metallothionein double-knockout cells, and caspase-3–overexpressed dopaminergic neurons. SIN-1–induced changes were attenuated with selegiline or in metallothionein-transgenic striatal fetal stem cells. SIN-1–induced oxidation of dopamine (DA) to dihydroxyphenylacetaldehyde (DopaL) was attenuated in metallothionein-transgenic fetal stem cells and in cells transfected with a mitochondrial genome, and was enhanced in aging mitochondrial genome knockout cells, in metallothionein double-knockout cells, and caspase-3 gene–overexpressing dopaminergic neurons. *Selegiline, melatonin, ubiquinone,* and *metallothionein* suppressed SIN-1–induced downregulation of a mitochondrial genome and upregulation of caspase-3 as determined by reverse transcription polymerase chain reaction. These studies provide evidence that *nitric oxide synthase* activation and *peroxynitrite ion* overproduction may be involved in the etiopathogenesis

METHODS IN ENZYMOLOGY, VOL. 396                          0076-6879/05 $35.00
Copyright 2005, Elsevier Inc. All rights reserved.              DOI: 10.1016/S0076-6879(05)96024-2

of PD, and that metallothionein gene induction may provide *neuroprotection.*

Introduction

Parkinson's disease (PD) is characterized by a progressive loss of dopaminergic neurons in the substantia nigra zona compacta and in other subcortical nuclei associated with a widespread occurrence of Lewy bodies. The causes of cell death in PD are still poorly understood, but oxidative stress, resulting either from excess generation or from reduced scavenging of free radicals (Ebadi *et al.,* 1991); defect in mitochondrial oxidative phosphorylation and nitrative stress, resulting from induction of nitric oxide synthase (NOS) have been proposed (Ebadi and Sharma, 2003; Ghafourifar and Colton, 2003). Mitochondria release binds zinc when exposed to nitric oxide (NO). At inflammatory sites, both MTs and NOS are induced, and the zinc released from MTs by NO suppresses NOS induction. NO- and peroxynitrite (ONOO$^-$)-mediated zinc release from MTs is suppressed by reduced glutathione (GSH), but not by oxidized glutathione (Ebadi *et al.,* 1999). Contrary to the peroxynitrite-induced activation of guanylyl cyclase, where GSH is needed, the zinc released from MTs by peroxynitrite is suppressed by reduced glutathione (Ebadi *et al.,* 1999; Khatai *et al.,* 2004). In addition, zinc, the major natural metal ligand in MTs and suppressor of inducible NOS (iNOS), is released more readily under the influence of NO.

Studies have shown that MTs may react directly with peroxynitrite to prevent DNA and lipoprotein damage induced by reactive nitrogen species (Cai *et al.,* 2000). We examined 3-morpholinosydnonimine (SIN-1) (a potent peroxynitrite ion generator)–induced apoptosis in control and MTs-overexpressing dopaminergic neurons, with a primary objective to determine the neuroprotective potential of MTs against peroxynitrite-induced neurodegeneration in PD (Sharma and Ebadi, 2003). SIN-1 produced both oxidative and nitrative stress to cause apoptosis in human dopaminergic (SK-N-SH) neurons. SIN-1 induced lipid peroxidation and triggered plasma membrane blebbing. In addition, it caused DNA fragmentation, $\alpha$-synuclein nitration, intramitochondrial accumulation of metal ions (copper, iron, zinc, and calcium), and enhanced the synthesis of 8-hydroxy-2-deoxyguanosine (8-OH, 2dG). Furthermore, it downregulated the expression of Bcl-2 and poly[adenosine diphosphate (ADP)-ribose] polymerase but upregulated the expression of caspase-3 and Bax in dopaminergic (SK-N-SH) neurons. SIN-1 induced apoptosis in aging mitochondrial genome knockout (RhO$_{mgko}$) cells, $\alpha$-Syn–transfected cells, MT-double-knockout (MT$_{dko}$) cells, and caspase-3–overexpressing dopaminergic neurons. SIN-1–induced changes were attenuated with selegiline or in MT-transgenic striatal fetal stem cells. SIN-1–induced oxidation of DA to dihydroxyphenylacetaldehyde (DOPAL) was attenuated in $MT_{trans}$ fetal stem cells and in cells transfected with a mitochondrial genome, and it was enhanced in $RhO_{mgko}$ cells, in $MT_{dko}$ cells, and caspase-3 gene–overexpressing dopaminergic neurons. Selegiline, melatonin, ubiquinone, and MTs suppressed SIN-1–induced downregulation of mitochondrial genomes and upregulation of caspase-3 as determined by reverse transcription polymerase chain reaction (RT-PCR).

The synthesis of mitochondrial 8-hydroxy-2-deoxyguanosine and apoptosis-inducing factors were increased, following exposure to 1-methyl-4-phenylpyridinium ($MPP^+$) ion or rotenone. Pretreatment with selegiline or MTs suppressed $MPP^+$, 6-OHDA, and rotenone-induced increases in mitochondrial 8-OH, 2dG accumulation. Transfection of $RhO_{mgko}$ neurons with mitochondrial genome–encoding complex-1 attenuated SIN-1–induced increase in lipid peroxidation. SIN-1 induced the expression of $\alpha$-Syn, caspase-3, and 8-OH, 2dG, and augmented $\alpha$-Syn nitration. These effects were attenuated by MTs gene overexpression, indicating that NOS activation and peroxynitrite ion overproduction might be involved in the etiopathogenesis of PD, and that MT gene induction may provide neuroprotection.

We have discovered that MTs inhibit peroxynitrite-induced $\alpha$-Syn nitration and apoptosis to provide dopaminergic neuroprotection in SK-N-SH neurons and in $MT_{trans}$ mice (Sharma and Ebadi, 2003). Because MTs attenuate peroxynitrite apoptosis, $MT_{trans}$ fetal stem cells may be used to determine the graft outcome in homozygous weaver mutant ($WM_{homo}$) mice exhibiting significantly reduced striatal $^{18}$F-DOPA uptake, progressive dopaminergic degeneration, and parkinsonism. Furthermore, interventions to induce brain regional MTs would have therapeutic potential in PD and other neurodegenerative disorders of unknown etiopathogenesis. Here, we describe briefly the neuroprotective role of MTs in peroxynitrite ion-induced oxidative and nitrative stress in cellular and genetic models of PD.

## Materials and Methods

### Animals

Experimental animals were housed in temperature- and humidity-controlled rooms with 12-h day and 12-h night cycles and were provided with commercially prepared chow and water *ad libitum*. The animals were acclimated to laboratory conditions for at least 4 days before experimentation.

Care was taken to avoid any distress to animals during the experiment. Breeder pairs of control wild-type (control$_{wt}$) C57BJ6, metallothionein double-knockout (MT$_{dko}$), metallothionein-transgenic (MT$_{trans}$), $\alpha$-Syn knockout ($\alpha$-Syn$_{ko}$), and homozygous weaver mutant (WM$_{homo}$) mice were purchased from Jackson Labs (Minneapolis, MN). Detailed information regarding these genotypes is available from Jackson Labs' web page (www. jax.org/jaxmice). $\alpha$-Syn–metallothionein triple knockout ($\alpha$-Syn-MT$_{tko}$) mice were prepared by cross-breeding MT$_{dko}$ female mice with $\alpha$-Syn$_{ko}$ males as described in one of our publications (Sharma and Ebadi, 2004). The breeder colony was maintained in an air-conditioned animal facility in hepa-filtered cages with free access to water and lab chow. The zinc, copper, and iron content of the lab chow were monitored by atomic absorption spectrometery to maintain their adequate supply. PCR analysis of the tail DNA was done for genotyping. Nonresponders were excluded from the study. The number of animals used in each experimental group are presented in figure legends. The heterozygous weaver mutant mice (WM$_{hetro}$) were excluded from this study, because they do not exhibit progressive neurodegenerative changes as observed in the WM$_{homo}$ mice and because of the cost involved in their breeding.

*Equipment*

The RDS-111 cyclotron was purchased from CTI Molecular Imaging (Knoxville, TN). BBS2V for $^{18}$F2 delivery and Manuela Hot cell for dose calibration and dose fractionation were purchased from Comcer (Castelbolognese, Italy). A GINA Star $^{18}$F-DOPA synthesis module was purchased from Raytest (Isotopenmessgerate, GmBH, Germany). A high-performance liquid chromatography (HPLC) for $^{18}$F-DOPA quality control was purchased from Agilent Technologies (Germany). A high-resolution micro–positron emission tomography (microPET) scanner (R4) was purchased from Concorde Microsystems (Knoxville, TN).

*Chemicals*

Sodium dihydrogen phosphate (NaH$_2$PO$_4$), ethanol, diammonium hydrogen phosphate (NH$_4$)$_2$HPO$_4$, ammonium dihydrogen phosphate (NH$_4$H$_2$PO$_4$), ammonium hydroxide (NH$_4$OH), hydrobromic acid (HBr), trichlorofluoromethane (Freon), and ascorbic acid were purchased from Aldrich Chemicals (St. Louis, MO). Tri-boc, trimethyl stannalyl L-dihydroxyl phenylalanine (Precursor), and cold F-DOPA were purchased from Advanced Biochemical Compounds (ABX, GmBH, Radeberg, Germany). $^{18}$O gas was purchased from Isotec (Miamisburg, OH).

## Methods

### Digital Fluorescence Microscopy

A detailed procedure to grow cultured dopaminergic (SK-N-SH) neurons, prepare mitochondrial genome knockout ($RhO_{mgko}$), and cell transfection is described in Sharma and Ebadi (2003). Briefly, the SK-N-SH neurons were grown in polylysine–coated multichambered slides using complete Dulbecco's Modified Eagle Medium (DMEM), supplemented with 10% fetal bovine serum, 3.7 g/liter sodium bicarbonate, and high glucose and glutamine. After 48 h of incubation at 37° in a 5% $CO_2$ incubator, they were washed three times with Dulbecco's phosphate-buffered saline (PBS), were exposed to 3% goat serum for 2 h at room temperature to block nonspecific binding, and were washed three times in PBS. The sections were incubated for 1 h in primary antibody as per manufacturer's recommendations, and were washed three times with PBS. They were then incubated in fluorescein-isothiocyanate (FITC)–conjugated immunoglobulin G (IgG) (1:10,000) secondary antibody for 2 h at room temperature, washed three times, counterstained with 4′,6′,-diamidi-no-2-phenylindole dihydrochloride (DAPI) and ethidium bromide (5 nm) for 30 s, and again washed three times in PBS. The slides were mounted in Fluor-Mounting Medium (Trevigen) and allowed to dry at room temperature in a dark chamber. The slides were examined under digital fluorescent microscope (Leeds Instruments, Minneapolis, MN) set at three wavelengths (blue for DAPI, green for FITC, and red for ethidium bromide). (DAPI stains preferentially structurally intact nuclear DNA, whereas ethidium bromide stains fragmented DNA.) The fluorescent images were captured using a SpotLite digital camera and analyzed by ImagePro software. Target accentuation and background inhibition software were employed to improve the quality of fluorescent images.

### Immunoblotting

The cell lysates were prepared in Laemmli buffer with protease inhibitor cocktail, as described in Sharma and Ebadi (2003). In brief, lysates containing 15 $\mu$g of protein were subjected to 10% sodium dodecylsulfate (SDS) polyacrylamide gel electrophoresis. Proteins were transferred on to nitrocellulose paper by electroblotting using 50 mA current strength for 1.5 h, and transfer efficiency was checked by Ponceau red stain. The blots were incubated overnight in 5% nonfat milk for nonspecific binding, washed three times in PBS (pH 7.4), and subjected to recommended dilution of primary antibodies overnight, washed three times with PBS (pH 7.4), and exposed to 1:2000 secondary antibody (HRP-labeled anti-mouse IgG) for 2 h and washed three times with PBS (pH 7.4) containing 0.1% Tween 20. Amersham chemiluminescent kit was used for developing the autoradiograms. Immunoblots were scanned with precalibrated Bio-Rad GS-810 high-resolution densitometric scanner and analyzed by Utility-1 software.

## $^{18}$F-DOPA Synthesis

$^{18}$F-DOPA synthesis required a single reaction vessel and involved essentially two synthesis steps: (1) electrophilic substitution and (2) fluorodestannlyzation. Briefly, all the $^{18}$F-DOPA synthesis steps were computer controlled and were performed remotely from the cyclotron and the hot cells. The reagent vessels were cleaned (X3) with Millipore water and (X3) with acetonitrile. Acetonitrile was evaporated before initiating $^{18}$F-DOPA synthesis. The reagent vessels were filled with the following chemicals in a sequence: [Vessel 1: 3.5 ml of precursor (30 mg triboc, trimethyl stanalyl L-dihydroxyl phenyl alanine in Freon); vessel 2: 48% of 1.8 ml hydrobromic acid (HBr); vessel 3: 1.2 ml buffer (Composition: $[NH_4]_2HPO_4$ [1 $M$] + $[NH_4H_2PO_4$ (1 $M$) (50/50)]); and vessel 4: 1.1 ml of 25% $NH_4OH$.] The Dewar flask was filled with liquid nitrogen to trap fluorine from Freon in the reaction vessel. Cold runs of F-DOPA were performed in the synthesis module to monitor the position of the $^{18}$F-DOPA during synthesis before running the cyclotron beam. The RDS-111 cyclotron was initialized for the first 20 min, employing Supervisory Control and Documentation Analysis (SCADA) computer commands via Virtual Memory EuroBus (VME) Crate. The radiofrequency (RF) synthesizer and the magnet were conditioned before running the proton beam on the Faraday cup (20 $\mu A$, 10 min) for initial warm up. The cyclotron was run using the following parameters: Dee voltage 36.83 KV, RF 72.24 MHz, main magnet current 231.83 A, high vacuum $5.14 \times 10^{-6}$ Torr, ion source power supply (ISPS) current 0.09 A, ISPS voltage 0.6 KV, bias voltage 13.93 kV, bias current 3.84 mA, ion source gas flow 5.71 sccm, and 75% foil transmission efficiency. At least 3–5 $F_2$ passivations of 30 $\mu A$ for 20 min were performed to condition the delivery lines for $^{18}F_2$ production. The precursor was cooled in Freon down to $-20°$ within 10 min to trap the entire $^{18}F_2$. The reaction vessel was then heated to $30°$ for 4 min, and Freon was evaporated under vacuum at $50°$, after which HBr was added in the residue. The reaction mixture was then heated to $130°$ for 10 min and then cooled to $70°$ for 1 min and was neutralized by adding $NH_4OH$. The reaction vessel was further cooled down to $40°$, and the buffer was added to adjust the pH to 4. The resulting solution was trapped in the HPLC injection loop (4.5 ml), automatically injected, and separated by an HPLC reverses phase column

(Nucleosil 100 $C_{18}$, 7 $\mu M$, 16 × 250 mm), using 2% ethanol in 0.5 $M$ $NaH_2PO_4$ (pH 4–5), at a flow rate of 8 ml/min. The eluant was monitored for ultraviolet (UV) absorbance (282 nm) and radioactivity. The peak of interest in HPLC was obtained between 5.8 and 7 min, depending on the amount of ethanol added in the mobile phase. We used 2% ethanol, which provided [18]F-DOPA peak at 5.8 min. We eluted the peak of interest in the collection vial containing 100 $\mu l$ of 150 mg/liter ascorbic acid in 1.2 ml of 0.5 $M$ NaOH, pH 4–5 as a stabilizer.

## [18]F-DOPA Uptake

The animals were anesthetized with 1.6 mg/g of body weight tribromoethane intraperitoneally. [18]F-DOPA uptake was studied by injecting 250 $\mu Ci$ of [18]F-DOPA intravenously through the caudal vein and high-resolution microPET quick scanning after 60 min. The uptake of [18]F-DOPA in the striatal region was collected in list mode by microPET-Manager and quantitated by ASIPro computer software. [18]F-DOPA uptake in the striatum was also estimated by dissolving 10 mg of tissue in 1 N NaOH at 80° and counting the radioactivity in a well-type gamma scintillation counter above background to correlate and confirm the microPET imaging data.

### Statistical Analysis

The data were analyzed with Sigma–Stat (version 3.02), employing repeated measures of analysis of variance (ANOVA). $p$ Values less than .05 were taken as statistically significant.

## Observations and Results

### SIN-1 Accentuates 6-OHDA Hemiparkinsonism

Local microinjection of 6-OHDA in the substantia nigra of mice induced hemiparkinsonism. 6-OHDA–induced hemiparkinsonism is presented in Figs. 1A–D, whereas Figs. 1E and F demonstrate 6-OHDA– and SIN-1–induced hemiparkinsonism. Co-administration of SIN-1 accentuated 6-OHDA–induced hemiparkinsonism, as illustrated in Fig. 1.

### MPTP-Induced Striatal Release of Zinc Inhibits Peroxynitrite

Because MPTP is known to induce iNOS and participate in peroxynitrite synthesis, and zinc released from MTs is involved in iNOS inhibition, we have examined MPTP-induced zinc and oxidized iron release from $MT_{trans}$ and $MT_{dko}$ mice striatum employing microdialysis. The microdialysates

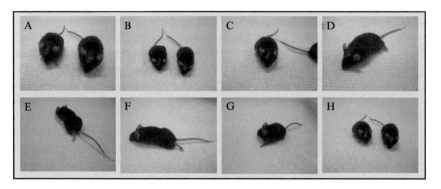

Fig. 1. (A) Postural irregularity. (B) Drooping body posture. (C) Body tremors. (D) Neck muscle rigidity. (E) Unilateral hind-limb extension. (F) Walking difficulty. (G) Stiff tail. (H) Reduced mobility. SIN-1 (100 ng) accentuated 6-OHDA hemiparkinsonism as illustrated in (E–H). Note that pictures were taken 6 h after intranigral 6-OH-DA and/or SIN-1 microinjection. SN-pc 6-OH-DA: 500 ng + SIN-1 (100 ng)/500 nl in PBS pH 7.5 m 0.15 $M$, 0.01% ascorbic acid. (See color insert.)

were analyzed with HPLC with electrochemical detection (ECD). MPTP-induced striatal release of zinc was significantly increased in $MT_{trans}$ mice as compared to $MT_{dko}$ mice, whereas MPTP-induced release of oxidized iron was significantly increased in $MT_{dko}$ and reduced in $MT_{trans}$ mice. A kinetic analysis of MPTP-induced striatal release of zinc and oxidized iron is presented in Fig. 2.

*SIN-1–Induced Apoptosis in SK-N-SH Neurons*

SIN-1 induced concentration and time-dependent phosphatidyl serine externalization, lipid peroxidation, caspase-3 activation, Bcl-2 downregulation, and Bax upregulation in human dopaminergic (SK-N-SH) neurons (Sharma and Ebadi, 2003). SIN-1–induced apoptotic changes were attenuated by transfecting the SK-N-SH neurons with $MT-1_{sense}$ oligonucleotides and were accentuated by transfecting with $MT-1_{antisense}$ oligonucleotides. Transfection with scrambled sequences did not produce any significant change. In mitochondrial genome knockout ($RhO_{mgko}$) neurons, SIN-1 produced severe apoptotic changes. Transfection of $RhO_{mgko}$ neurons with mitochondrial genome encoding complex-1 attenuated SIN-1–induced apoptosis (Sharma *et al.*, 2004).

*MTs Prevent $MPP^+$ Apoptosis in SK-N-SH Neurons*

We have shown that MPTP-induced protein nitration is attenuated by MT overexpression in striatal dopaminergic neurons of mice (Ebadi and Sharma, 2003). In view of the aforementioned findings, we questioned

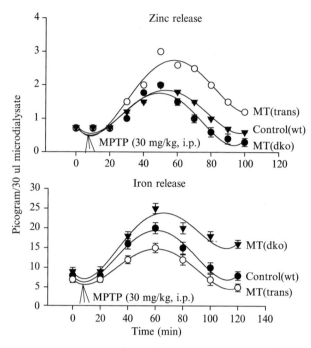

Fig. 2. Quantitative estimation of zinc ($Zn^{2+}$) and oxidized iron ($Fe^{3+}$) released from the striatal microdialysates of control$_{wt}$, MT$_{dko}$, and MT$_{trans}$ mice. Zinc and iron were estimated from 10 $\mu$l microdialysates using Varian Atomic Absorption Spectrometer as described in Sharma and Ebadi (2003).

whether MTs might prevent MPP$^+$ apoptosis in SK-N-SH neurons. Overnight treatment of MPP$^+$ induced apoptosis in control$_{wt}$ neurons. MPP$^+$ apoptosis was attenuated in MT-1$_{sense}$-oligonucleotide–transfected neurons and accentuated in MT-1$_{antisense}$ oligonucleotide–transfected neurons. Transfection of RhO$_{mgko}$ neurons with MT-1$_{sense}$ oligonucleotides also suppressed MPP$^+$ apoptosis, whereas transfection of RhO$_{mgko}$ neurons with MT-1$_{antisense}$ oligonucleotides accentuated MPP$^+$ apoptosis, indicating the primary involvement of peroxynitrite in dopaminergic degeneration and the neuroprotective potential of MTs.

### Attenuation of Peroxynitrite Apoptosis and α-Syn Nitration by MTs

To provide direct evidence of the neuroprotective role of MTs against peroxynitrite apoptosis, we exposed SK-N-SH neurons to SIN-1. Overnight exposure to SIN-1 (10 $\mu M$) induced apoptosis in SK-N-SH neurons. SIN-1 apoptosis was attenuated in MT-1$_{sense}$ oligonucleotides transfected

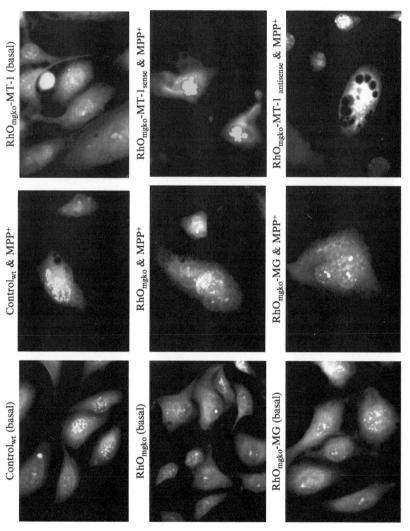

Fig. 3. *(continued)*

SK-N-SH neurons and accentuated in MT-$1_{antisense}$ oligonucleotide-transfected SK-N-SH neurons.

## Induction of NF-κB and Inhibition of Complex-1 in WM$_{homo}$ Mice

To provide evidence of oxidative and nitrative stress and induction of proinflammatory cytokines in the etiopathogenesis of PD, we have used WM$_{homo}$ mice. We have discovered that proinflammatory transcription factor nuclear factor-κB (NF-κB) is induced, whereas mitochondrial complex-1 gene is downregulated in WM$_{homo}$ mice exhibiting progressive dopaminergic degeneration and severe parkinsonism, characterized by body tremors, stiff neck, postural irregularities, and walking difficulties. Chronic treatment of rotenone to human dopaminergic (SK-N-SH) neurons also induced NF-κB and inhibited mitochondrial complex-1 as determined by immunoblotting and RT-PCR (Ebadi *et al.*, 2004).

## $^{18}$F-DOPA Uptake

$^{18}$F-DOPA uptake was significantly reduced in the central nervous system (CNS) of weaver mutant mice. $^{18}$F-DOPA was de-localized in the kidneys, indicating downregulation of synaptosomal dopamine transporter (sDAT) and DOPA decarboxylase in these animals.

## Discussion

### SIN-1 Accentuates 6-OHDA Hemiparkinsonism

We have demonstrated that local microinjection of 6-OHDA in the substantia nigra of mice induces hemiparkinsonism, which is accentuated by SIN-1 co-administration. SIN-1 is a potent peroxynitrite ion generator and induces oxidative and nitrative stress. MTs suppress SIN-1–induced peroxynitrite synthesis by donating zinc ions. Zinc released from MTs

---

FIG. 3. Multiple fluorochrome analysis of MPP$^+$ (100 $\mu M$) apoptosis in SK-N-SH neurons. Overnight treatment of MPP$^+$ induced DNA condensation in control$_{wt}$ neurons and nuclear DNA fragmentation in RhO$_{mgko}$ neurons. Transfection of RhO$_{mgko}$ neurons with either complex-1 gene or MT-$1_{sense}$ oligonucleotides attenuated MPP$^+$ apoptosis, whereas transfection with MT-$1_{antisense}$ oligonucleotides augmented blebbing, plasma membrane perforations, DNA fragmentation, and condensation. Digital fluorescence images were captured by SpotLite digital camera and analyzed by Image Pro software. (Fluorochromes: Blue: DAPI nuclear DNA stain; Red: JC-1 mitochondrial $\Delta\psi$ marker, Green: Fluorescein isothiocyanate) (RNA, and protein stain). (See color insert.)

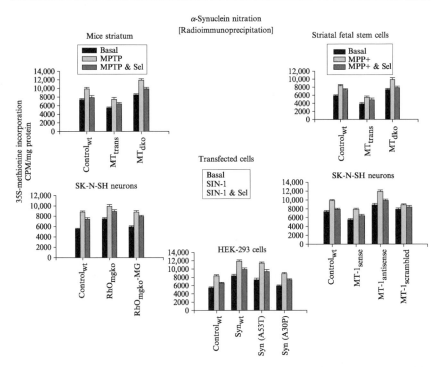

FIG. 4. Aging RhO$_{mgko}$ neurons exhibited enhanced α-Syn nitration upon overnight exposure to SIN-1 (10 $\mu M$). Transfection of aging RhO$_{mgko}$ neurons with complex-1 attenuated SIN-1–induced α-Syn nitration (lower left panel). SIN-1–induced α-Syn nitration was suppressed in MT-1$_{sense}$, enhanced in MT-1$_{antisense}$, and did not produce significant change in MT-1$_{scrambled}$ oligonucleotides transfected neurons (lower right panel). Selegiline pretreatment attenuated SIN-1–induced α-Syn nitration in MT-1$_{sense}$, MT-1$_{antisense}$, and MT-1$_{scrambled}$ oligonucleotide-transfected neurons (lower right panel). SIN-1–induced nitration of α-Syn was also enhanced in α-Syn$_{wt}$, and A53T α-Syn overexpressed HEK cells (lower middle panel). A30P α-Syn mutants did not exhibit significant induction in α-Syn nitration upon SIN-1 exposure. Selegiline pretreatment attenuated SIN-1–induced α-Syn nitration in control$_{wt}$, α-Syn$_{wt}$, A53T, and A30P α-Syn overexpressing HEK cells, suggesting that induction of wild type or A53T mutant α-Syn can enhance α-Syn nitration, and hence, its aggregation in order to induce Lewy body pathology during the progression of sporadic or familial type Parkinson's disease.

participates in NOS inhibition and prevents peroxynitrite synthesis, whereas 6-OHDA, by donating hydroxyl radicals, may react readily with NO to synthesize peroxynitrite. Indeed MT$_{trans}$ mice are genetically resistant to MPTP as compared to MT$_{dko}$ mice, suggesting mitochondrial neuroprotection in PD (Sharma and Ebadi, 2003). Because MTs regulate the intracellular redox potential, we explored whether 6-OHDA could alter the level of zinc and MTs. 6-OHDA reduced the level of zinc and MTs in the

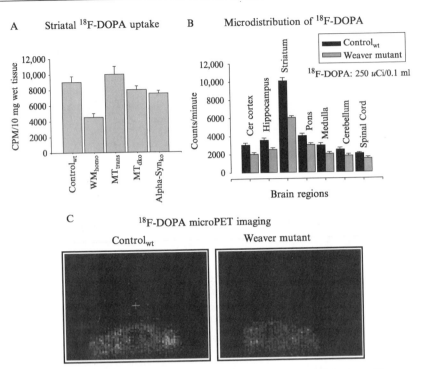

Fig. 5. (A) Histogram demonstrating significantly ($p < .05$) reduced striatal [18]F-DOPA uptake in WM$_{homo}$ mice as compared to control$_{wt}$, MT$_{trans}$, MT$_{dko}$, and $\alpha$-Syn$_{ko}$ mice. Data are mean $\pm$SD of five determinations in each group. (B) Histogram demonstrating [18]F-DOPA microdistribution in the central nervous system (CNS) of control$_{wt}$ and weaver mutant mice. The data mean $\pm$ SD of eight determinations in each experimental group. (C) High-resolution microPET imaging illustrating significantly increased [18]F-DOPA uptake in the CNS of control$_{wt}$ mouse (left panel) as compared to WM$_{homo}$ mice (right panel). The radioactivity is de-delocalized in the kidneys of WM$_{homo}$ mouse. (See color insert.)

striatum but not other brain regions. The effect of DA in stimulating the synthesis of MTs was similar to that of zinc, known to generate the synthesis of MTs; and to that of $H_2O_2$ and $FeSO_4$, known to generate free radicals, indicating that zinc or zinc MTs are altered in conditions where oxidative stress has taken place (Shiraga et al., 1993).

### MTs Inhibit Peroxynitrite Neurotoxicity by Donating Zinc

Halasz et al. (2004) have shown that chronic MPTP enhances NOS activity in the mice striatum. NO reacts readily with superoxide ion to synthesize peroxynitrite ions. Zinc inhibits NOS activity and protects dopaminergic neurons from peroxynitrite ions during MPTP or rotenone-induced oxidative and nitrative stress. MPTP- and rotenone-induced zinc release was significantly increased in $MT_{trans}$ mice as compared to $MT_{dko}$ mice, indicating that MTs may inhibit peroxynitrite by donating zinc, whereas the oxidized form of iron may participate in the Fenton reaction to generate OH radicals in $MT_{dko}$ mice. Furthermore, OH radicals may react with NO to synthesize peroxynitrite ions. Significantly reduced release of zinc in $MT_{dko}$ mice might be responsible for the enhanced genetic susceptibility to MPTP parkinsonism, whereas in $MT_{trans}$ mice, these events might be attenuated by augmented release of zinc. The concentration of zinc is altered in various neurological disorders, including alcoholism, Alzheimer-type dementia, amyotrophic lateral sclerosis (ALS), Down's syndrome, epilepsy, Friedreich's ataxia, Guillain-Barré syndrome, hepatic encephalopathy, multiple sclerosis, PD, Pick's disease, retinitis pigmentosa, retinal dystrophy, schizophrenia, and Wernicke-Korsakoff syndrome. As peroxynitrite has been implicated in the etiopathogenesis of various neuroinflammatory diseases, we investigated the neuroprotective role of MTs (Ebadi and Sharma, 2003; Ebadi et al., 2004; Sharma and Ebadi, 2003) in various cellular and genetic models of PD. MTs are expressed in neurons that sequester zinc in their synaptic vesicles, and the regulation of the expression of MTs is extremely important in terms of maintaining the steady-state level of zinc and controlling redox potentials.

Because several of these disorders, such as PD, are associated with oxidative stress, and because MT is able to prevent the formation of free radicals and peroxynitrite ions, it is believed that cytokine-induced induction of MT provides a long-lasting protection to avert oxidative and nitrative damage (Ebadi et al., 1995).

## MTs-Glutathione Interaction in Peroxynitrite Neurotoxicity

The mobilization of zinc from MT suggests a possible function of MT as a physiological zinc donor. A shift of the glutathione redox balance under conditions of oxidative stress accelerates zinc release from MT. Such a disturbance of zinc metabolism has important consequences for the progression of diseases such as PD, where oxidative stress occurs in affected brain tissue (Maret, 1994, 1995).

## Inhibition of MPTP- and THIQ-Induced Peroxynitrite by MTs

The finding that MPTP elicits parkinsonism in humans suggests that endogenous or xenobiotic neurotoxic compounds may be involved in the etiology of PD. The endogenous neurotoxin, 1-methyl-6, 7-dihydroxy-1, 2,3,4-tetrahydroisoquinoline (salsolinol), has been considered a potential neurotoxin in the etiology of PD. Salsolinol and N-methyl(R)-salsolinol were identified in the brains and cerebrospinal fluid of patients with PD. We have conducted studies to understand the role of salsolinol in oxidative-mediated neuronal toxicity in dopaminergic SH-SY5Y cells, as well as the neuroprotective effects of MTs against salsolinol toxicity in $MT_{trans}$ fetal mesencephalic cells. Salsolinol increased the production of reactive oxygen species (ROS) and decreased glutathione (GSH) levels and cell viability in SH-SY5Y cells.

Salsolinol also decreased intracellular adenosine triphosphate (ATP) levels and induced nuclear condensation in these cells. Salsolinol-induced depletion in cell viability was completely prevented by N-acetylcysteine in SH-SY5Y cells and was prevented by MT in $MT_{trans}$ fetal mesencephalic cells compared to $control_{wt}$ cells. The extent of nuclear condensation and caspase activation was also less in $MT_{trans}$ cells than $control_{wt}$ cells, suggesting that salsolinol causes oxidative and nitrative stress by decreasing the levels of GSH and by increasing ROS production, and these events may lead to the death of dopaminergic neurons, whereas MT overexpression may protect dopaminergic neurons against salsolinol-induced neurotoxicity, by the inhibition of oxidative stress and apoptotic pathways, including caspase-3 activation and $\alpha$-Syn nitration (Wanpen et al., 2004).

*Growth Inhibitory Factor (GIF:MT-III) and PD*

*Growth inhibitory factor* (GIF), a brain-specific member of the MT family (MT-III), has been characterized as an inhibitory substance for neurotrophic factors in Alzheimer's–diseased brains. The exact function of GIF (MT-III), other than the inhibition of neurotrophic factors in PD, remains unknown. GIF prevents neurite extension of neurons in the early period of differentiation and supports the survival of differentiated neurons by scavenging hydroxyl radicals (Uchida et al., 2002). GIF is induced in reactive astrocytes in the cerebral cortex in cases of meningitis, in Creutzfeldt-Jakob disease, and in reactive astrocytes surrounding old cerebral infarcts. On the other hand, GIF is reduced in the subset of reactive astrocytes in lesioned areas of degenerative diseases such as Alzheimer's disease, multiple system atrophy, PD, progressive supranuclear palsy, and ALS. Reduction of GIF is correlated with neuronal loss.

Thus, perturbation in normal neuroglial interaction in degenerative diseases may lead to a reduction of GIF in reactive astrocytes (Uchida, 1994).

*cDNA Microarray Analysis of Oxidative and Nitrative Stress*

Microarray analysis of 5000 genes has shown that heat shock protein (HSP70) is induced 100 times within 30 min of brief spinal ischemia, followed by a 12-fold increase in MT expression within 6–12 h. Immediate early genes [B-cell translocation gene-2 (BTG2), the transcription factors early growth response 1 [erg-1], nerve growth factor–inducible B [NGF1-B], mitogen-activated protein kinase phosphatase, and ptpn16, are important cell signaling regulators (Carmel et al., 2004). Swiss 3T3 cells exposed to aqueous extracts of cigarette smoke, on bubbled PBS expressed mainly antioxidant response genes coding for, for example, heme oxygenase-1 (HO-1), MTs 1/2 (MT-1/2), and HSPs; genes coding for transcription factors, for example, JunB and CAAT/enhancer binding protein (C/EBP); cell cycle–related genes (e.g., gadd34 and gadd45); and genes described as mediators of an inflammatory/immune-regulatory response (e.g., st2, kc, and id3). Among the 10 most upregulated genes, five are known to counteract stress induced by peroxynitrite (Bosio et al., 2002). cDNA microarray analysis of MPTP-intoxicated rats also showed that they exhibited similar induction of redox-sensitive genes (Youdim, 2003).

*Peroxynitrite Induction of Proinflammatory Genes in PD*

Although the exact role of inflammatory cytokines in PD remains unknown, studies have indicated that the inflammatory process in PD is characterized by activation of microglia without reactive astrocytosis, suggesting that the progressive loss of dopaminergic neurons in PD is an ongoing neurodegenerative process with minimal involvement of the surrounding nervous tissue. The absence of reactive astrocytosis in PD may indicate that the inflammatory process in PD is a unique phenomenon in which MTs are induced in reactive astrocytes to exert their antioxidative effects (Mirza et al., 2000).

*MT Neuroprotection in Peroxynitrite Neuroinflammations*

In kainic acid–injected interleukin-6 (IL-6) null mice, reactive astrogliosis and microgliosis were reduced, whereas morphological hippocampal damage, oxidative stress, and apoptotic neuronal death were increased. Because MT I-II levels were lower and those of iNOS higher, these concomitant changes are likely to contribute to the observed increased oxidative stress and neuronal death in the IL-6 null mice, demonstrating that IL-6 deficiency increases neuronal injury and impairs the inflammatory response after kainic acid–induced seizures (Penkova et al., 2001). The loss of zinc from wild-type superoxide dismutase (SOD) approximately doubled its efficiency for catalyzing peroxynitrite-mediated tyrosine nitration, suggesting that one gained function by SOD in ALS may be an indirect consequence of zinc loss. Nitration of protein-bound tyrosines is a permanent modification that can adversely affect protein function. Thus, the toxicity of ALS-associated SOD mutants may be related to enhanced catalysis of protein nitration subsequent to zinc loss. By acting as a high-capacity zinc sink, Neurofilament-L (NF-L) could foster the formation of zinc-deficient SOD within motor neurons (Crow *et al.*, 1997).

We have discovered that selegiline, by preventing the generation of free radicals; MTs, by scavenging free radicals; and neurotrophins, by rescuing dopaminergic neurons all attenuate oxidative stress and provide neuroprotection in PD (Ebadi *et al.*, 1998). Furthermore, MTs provide ubiquinone-mediated neuroprotection in PD (Ebadi *et al.*, 2002). To authenticate this hypothesis, we used control$_{wt}$, MT$_{dko}$, MT$_{trans}$, $\alpha$-Syn$_{ko}$, $\alpha$-Syn-MT$_{tko}$, and weaver mutant mice. Genetically susceptible MT$_{dko}$, $\alpha$-Syn-MT$_{tko}$, and weaver mutant mice possessed significantly reduced striatal coenzyme Q$_{10}$ as compared to genetically resistant MT$_{trans}$ and $\alpha$-Syn$_{ko}$ mice (Ebadi *et al.*, 2002). Chronic intoxication of rotenone also reduced complex-1 expression and increased NF-$\kappa$B expression as determined by immunoblotting and RT-PCR. NF-$\kappa$B is induced in peroxynitrite-induced oxidative and nitrative stress. Indeed, MPTP-, rotenone-, salsolinol-, 6-OHDA–, and SIN-1–induced $\alpha$-Syn nitration was significantly suppressed in SK-N-SH neurons by MT induction (Ebadi and Sharma, 2003). Furthermore, transfection of RhO$_{mgko}$ neurons with MT-1$_{sense}$ oligonucleotides significantly increased coenzyme Q$_{10}$ and enhanced neuritogenesis. These findings suggest that MTs provide coenzyme Q$_{10}$–mediated neuroprotection to avert peroxynitrite damage.

## MT Neuroprotection in Peroxynitrite Apoptosis

Oxidative stress, resulting from either excess generation or reduced scavenging of free radicals and peroxynitrite, has been proposed to play a role in damaging striatal neurons in PD. Because MTs are able to regulate the intracellular redox potential, we conducted experiments to learn whether 6-OHDA, which generates free radicals and is toxic to dopaminergic neurons, could alter the levels of zinc and MTs in the brain. The lesioning of the rat striatum with 6-OHDA (8.0 $\mu$g in 4 $\mu$l 0.02% ascorbic acid) resulted in a reduction in the levels of zinc and MTs in the striatum but not other brain regions. However, the intracerebroventricular administration of 6-OHDA, in a dosage regimen that does not lesion catecholaminergic pathways but causes oxidative stress, enhanced dramatically the level of MTs I mRNA in some brain areas such as hippocampus, arcuate nucleus, choroid plexus, and granular layer of cerebellum, but not in the striatum. The results of these studies suggest that zinc or MTs are altered in conditions where oxidative stress has taken place. Moreover, it is proposed that areas of brain containing high concentrations of iron, such as the striatum, but low levels of inducible MTs are particularly vulnerable to oxidative stress (Rojas *et al.*, 1996). Aschner (1997) has also described the functions of brain MTs and has reported on their propensity to attenuate methyl mercury–induced cytotoxicity in astrocytes.

## MTs Donate Zinc to Inhibit NOS and Peroxynitrite

MTs may serve as the source of zinc for incorporation into proteins, including a number of DNA transcription factors. However, zinc is readily released from MTs by disulfides, increasing concentrations, which are formed under oxidative stress. MTs are very good scavengers of free radicals, and zinc itself can reduce oxidative stress by binding to thiol groups, decreasing their oxidation. Zinc is also a potent inhibitor of NOS. Increased levels of chelatable zinc have been shown to be present in cell cultures of immune cells undergoing apoptosis. Pharmacological doses of zinc cause neuronal death, and some estimates indicate that extracellular concentrations of zinc could reach neurotoxic levels under pathological conditions. Zinc is released in high concentrations from the hippocampus during seizures. Unfortunately, there are contrasting observations about whether this zinc serves to potentiate or decrease seizure activity. Zinc may have an additional role in causing death in at least some neurons damaged by seizure activity and may be involved in the sprouting phenomenon, which may cause recurrent seizure propagation in the hippocampus.

In Alzheimer's disease, zinc has been shown to aggregate $\beta$-amyloid, a form that is potentially neurotoxic. The zinc-dependent transcription factors NF-$\kappa$B and Sp1 bind to the promoter region of the amyloid precursor protein (APP) gene. Zinc also inhibits enzymes that degrade APP to nonamyloidogenic peptides, which degrade the soluble form of $\beta$-amyloid. The changes in zinc metabolism that occur during oxidative stress may be important in neurological diseases where oxidative stress is implicated, such as Alzheimer's disease, PD, and ALS. Zinc is a structural component of SOD-1, mutations of which give rise to one form of familial ALS. After human immunodeficiency virus (HIV) infection, zinc deficiency is found, which may be secondary to immune-induced cytokine synthesis. Zinc is involved in the replication of HIV at a number of sites. These observations may stimulate further research into the role of zinc and MTs in averting peroxynitrite neuropathology in PD (Cuajungco and Lees, 1997).

## Peroxynitrite and MT$_{trans}$ Stem Cell Therapy

Genetic transfer approaches have received consideration as potential treatment modalities for human CNS and peripheral nervous system neurodegenerative disorders, including PD, Alzheimer's disease, and ALS. Transplantation of genetically modified cells into the brain represents a promising strategy for the delivery and expression of specific neurotrophic factors, neurotransmitter-synthesizing enzymes, and cellular regulatory proteins for intervention in neurodegenerative diseases. The use of specific promoters may also provide potential control of gene expression required for dose-specific or time-specific therapeutic strategies. Neural transplantation of genetically modified cells has been successfully employed to reverse functional deficits in animal models of neurodegenerative disorders, including PD. While implanted PC12 cells secrete DA *in vivo* and can ameliorate DA deficiency in parkinsonian rat model systems, these cells either degenerate within 2–3 weeks after implantation (because of peroxynitrite synthesis and lack of neural trophic factor support at the site of implantation) or, in some cases, form a tumor mass leading to the death of the host animal. To address these limitations, genetically modified PC12 cell lines that could synthesize NGF under the control of a zinc-inducible MTs promoter were developed. When implanted in the rat striatum and under *in vivo* zinc stimulation, these cells differentiated, expressed tyrosine hydroxylase, and survived through potential autocrine trophic support (Rohrer *et al.*, 1996).

MT overexpression in pancreatic $\beta$ cells provided resistance to oxidative stress by scavenging ROS, including $H_2O_2$, peroxynitrite released from streptozotocin, SIN-1, and superoxide radical produced by xanthine/xanthine oxidase. MT reduced NO-induced $\beta$-cell death and ROS production, and improved islet cell survival. MT islets synthesized increased insulin than controlled and extended the duration of euglycemia (Li *et al.*, 2003). The benefit of MT was due to protection from ROS because nitrotyrosine synthesis, an indicator of peroxynitrite synthesis, was much lower in MT grafts (Ebadi *et al.*, 2004a).

We have reported that $MT_{trans}$ fetal stem cells are genetically resistant to DOPAL apoptosis as compared to $control_{wt}$ cells, indicating their therapeutic potential in neuronal replacement therapy of PD (Sharma and Ebadi, 2003). Because $WM_{homo}$ mice exhibit progressive dopaminergic degeneration and $MT_{trans}$ striatal fetal stem cells are genetically resistant to peroxynitrite ion-induced apoptosis, $MT_{trans}$ fetal stem cells can be used for successful transplantation in PD. We are now evaluating the graft outcome by longitudinal *in vivo* molecular imaging of dopamine transporter and mitochondrial complex-1 with $^{18}$F-DOPA and $^{18}$F-rotenone, respectively, employing high-resolution microPET imaging, as illustrated in Fig. 6. We have discovered that $^{18}$F-DOPA uptake is significantly reduced in CNS of weaver mutant mice. $^{18}$F-DOPA was de-localized in the kidneys, indicating downregulation of synaptosomal DAT (sDAT) and DOPA decarboxylase in these animals. Indeed, sDAT and tyrosine hydroxylase activities were reduced, whereas lipid peroxidation and $\alpha$-Syn nitration

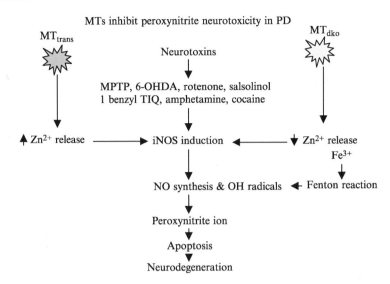

FIG. 6. A proposed model of MT-mediated inhibition of peroxynitrite neurotoxicity. MTs release zinc ions during MPTP or other THIQ-induced oxidative and nitrative stress. Zinc inhibits inducible nitric oxide synthase (iNOS) and thus prevents the synthesis of peroxynitrite ions in $MT_{trans}$ mice. In $MT_{dko}$ mice, due to significantly reduced zinc in the cell, iNOS is induced and NO is synthesized, which participates in peroxynitrite ion synthesis by reacting with OH radicals generated from $Fe^{3+}$-mediated Fenton reaction. Furthermore, zinc deficiency may enhance $Fe^{3+}$-mediated neurotoxicity in $MT_{dko}$ mice. (See color insert.)

were enhanced in weaver mutant mice, confirming oxidative and nitrative stress in the etiopathogenesis of PD (Ebadi *et al.*, 2004a).

Conclusion

As a result of oxidative and nitrative stress, peroxynitrite ions are generated, which induce $\alpha$-Syn nitration, apoptosis, and eventually degeneration of dopaminergic neurons in PD. MT provides neuroprotection as an antioxidant through SH moieties on the cysteine residues and by augmenting glutathione function. MT overexpression inhibits peroxynitrite-induced protein nitration and apoptosis to provide neuroprotection. MTs may also provide neuroprotection by donating, buffering, and scavenging zinc involved in the transcriptional regulation of several redox-sensitive genes and by inhibiting iNOS in PD. Translocation of MTs in the nucleus might regulate the DNA cell cycle through zinc-mediated transcriptional activation of genes involved in cell proliferation and differentiation. Furthermore, peroxynitrite induction of proinflammatory cytokines such as NF-$\kappa$B can

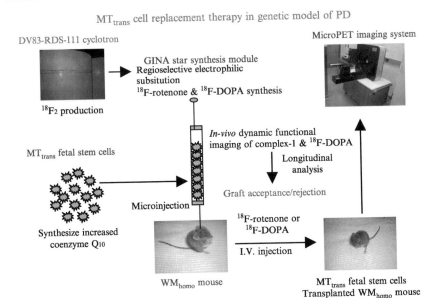

MT$_{trans}$ cell replacement therapy in genetic model of PD

FIG. 7. MT$_{trans}$ fetal stem cells are genetically resistant to dihydroxyphenylacetaldehyde (DOPAL) apoptosis, so they can be transplanted in the striatal region of WM$_{homo}$ mice exhibiting progressive dopaminergic degeneration and parkinsonism. The outcome of the grafts is evaluated by $^{18}$F-DOPA and $^{18}$F-rotenone imaging with high-resolution micro–positron emission tomography (PET) scanning. The procedure to prepare fetal stem cells is described in detail in Sharma and Ebadi (2003). (See color insert.)

be suppressed by MT-mediated coenzyme Q$_{10}$ synthesis. Indeed, MTs enhanced coenzyme Q$_{10}$ synthesis to avert iNOS activity and provided neuroprotection by suppressing peroxynitrite synthesis. Because peroxynitrite is the main culprit in graft rejection, as it enhances NF-$\kappa$B–mediated tumor necrosis factor-$\alpha$ (TNF-$\alpha$) synthesis and MTs attenuate peroxynitrite apoptosis, MT$_{trans}$ fetal stem cells can be implanted in WM$_{homo}$ mice exhibiting dopaminergic degeneration and parkinsonism. Studies in this direction will determine the basic molecular mechanism of graft acceptance/rejection in PD and the exact role of peroxynitrite in the pathogenesis of PD, which may go a long way in the clinical management of PD.

## Acknowledgments

This research was supported by a grant from the Counter Drug Technology Assessment Center, the Office of National Drug Control Policy (No. DATMO5-02C-1252) (M. E.). The authors express their sincere thanks and gratitude to CTI Corporation, Comecer, and Raytest employees for the success of this project. Secretarial assistance of Ms. Dani Stramer is gratefully acknowledged.

## References

Aschner, M. (1997). Astrocyte metallothioneins and their Neuroprotective role. *Ann. NY Acad. Sci.* **825,** 334.

Bosio, A., Knorr, C., Janssen, U., Gerbel, S., Haussmann, H., and Muller, T. (2002). Kinetics of gene expression profiling in Swiss 3T3 cells exposed to aqueous extracts of cigarette smoke. *Carcinogenesis* **23,** 741.

Cai, L., Klein, J., and Kang, Y. (2000). MTs inhibits peroxynitrite-induced DNA and lipoprotein damage. *J. Biol. Chem.* **275,** 38957.

Carmel, J., Kakinohama, O., Mestri, R., Young, W., Marsala, M., and Hart, R. (2004). Mediators of ischemic preconditioning identified by microarray analysis of rat spinal cord. *Expt. Neurol.* **185,** 81.

Crow, J., Sampson, J., Zhuang, Y., Thompson, J., and Beckman, J. (1997). Decreased zinc affinity of amyotrophic lateral sclerosis-associated superoxide dismutase mutants leads to enhanced catalysis of tyrosine nitration by peroxynitrite. *J. Neurochem.* **69,** 1936.

Cuajungco, and Lees, G. (1997). Zinc metabolism in the brain: Relevance to human neurodegenerative disorders. *Neurobiol. Dis.* **4,** 137.

Ebadi, M., Hiramatsu, M., Ramana Kumari, M., Hao, R., and Pfeiffer, R. (1999). Metallothionein in oxidative stress of Parkinson's disease. *In* "Metallothionein" IV (C. Klassen, ed.), p. 341. Kansas City, KS.

Ebadi, M., Iversen, P., Hao, R., Cerutis, D., Rojas, P., Happe, H., Murrin, C., and Pfeiffer, R. (1995). Expression and regulation of brain MTs. *Neurochem. Int.* **27,** 111.

Ebadi, M., Pfeiffer, R., Murrin, L., and Shiraga, H. (1991). MTs and oxidation reactions in PD. *Proc. West. Pharmacol. Soc.* **34,** 285.

Ebadi, M., Ramana Kumari, M., Hiramtsu, M., Hao, R., Pfeifer, R., and Rojas, P. (1998). MTs, neurotrophins and selegiline in providing neuroprotection in PD. *Restor. Neurol. Neurosci.* **12,** 103.

Ebadi, M., and Sharma, S. (2003). Peroxynitrite and mitochondrial dysfunction in the pathogenesis of Parkinson's disease. *Antioxidants Redox Sig.* **5,** 319.

Ebadi, M., Sharma, S., Ajjimaporn, A., and Maanum, S. (2004a). Weaver mutant mouse in progression of neurodegeneration in Parkinson's disease. *In* "Parkinson's Disease," p. 537. CRC Press, Boca Rota, FL.

Ebadi, M., Sharma, S., Muralikrishnan, D., Shavali, S., Eken, J., Sangchot, P., Chetsawang, B., and Brekke, L. (2002). Metallothionein provides ubiquinone-mediated neuroprotection. *Proc. West. Pharmacol. Soc.* **45,** 36.

Ebadi, M., Sharma, S., Wanpen, S., and Amornpan, A. (2004). Coenzyme Q10 inhibits mitochondrial complex-1 down-regulation and nuclear factor-kappa B activation. *J. Cell Mol. Med.* **8,** 213.

Ghafourifar, P., and Colton, C. (2003). Mitochondria and nitric oxide. *Antioxidants Redox Sig.* **5,** 249.

Halasz, A., Palfi, M., Tabi, T., Magyar, K., and Szoko, E. (2004). Altered nitric oxide production in mouse brain after administration of 1-methyl-4-phenyl-1,2,3,6-tetrahydro-pyridin or methamphetamine. *Neurochem. Int.* **44,** 641.

Khatai, L., Goessler, W., Lorencova, H., and Zangger, K. (2004). Modulation of nitric oxide-mediated metal release from MTs by the redox state of glutathione *in vitro. Eur. J. Biochem.* **271,** 2408.

Li, X., Chen, H., and Epstein, P. (2003). Metallothionein protects islets from hypoxia and extends islet graft survival by scavenging most kinds of reactive oxygen species. *J. Biol. Chem.* **279,** 765.

Maret, W. (1994). Oxidative metal release from MTs via zinc-thiol/disulfide interchange. *Proc. Natl. Acad. Sci. USA* **91,** 237.

Maret, W. (1995). MTs/disulfide interactions, oxidative stress, and the mobilization of cellular zinc. *Neurochem. Int.* **27,** 111.

Mirza, B., Hadberg, H., Thompson, P., and Moos, T. (2000). The absence of reactive astrocytosis is indicative of a unique inflammatory process in PD. *Neuroscience* **95,** 425.

Penkova, M., Molinero, A., Carrasco, J., and Hidalgo, J. (2001). Interleukiin-6 deficiency reduces the brain inflammatory response and increases oxidative stress and neurodegeneration after kainic acid-induced seizures. *Neuroscience* **102,** 805.

Rohrer, D., Nilaver, G., Nipper, V., and Machida, C. (1996). Genetically modified PC12 brain grafts: Survivability and inducible nerve growth factor expression. *Cell Transplant.* **5,** 57.

Rojas, P., Cerutis, D., Happe, H., Murrin, L., Hao, R., Pfeiffer, R., and Ebadi, M. (1996). 6-HydroxyDA-mediated induction of rat brain MTs I mRNA. *Neurotoxicology* **17,** 323.

Sharma, S., and Ebadi, M. (2003). Metallothionein attenuates 3-morpholinosydnonimine (SIN-1)-induced oxidative stress in DA ergic neurons. *Antioxidants Redo. Sig.* **5,** 251.

Sharma, S., and Ebadi, M. (2004). An improved method for analyzing coenzyme Q homologues and multiple detection of rare biological samples. *J. Neurosci. Methods* **137,** 1.

Sharma, S., Kheradpezhou, M., Shavali, S., El Refaey, H., Eken, J., Hagen, C., and Ebadi, M. (2004). Neuroprotective actions of coenzyme Q10 in Parkinson's disease. *Methods Enzymol.* **382,** 488.

Shiraga, H., Pfeiffer, R., and Ebadi, M. (1993). The effects of 6-hydroxyDA and oxidative stress on the level of brain MTs. *Neurochem. Int.* **23,** 561.

Uchida, Y. (1994). Growth inhibitory factor prevents neurite extension and the death of cortical neurons caused by high oxygen exposure through hydroxyl radical scavenging. *Biol. Sig.* **3,** 211.

Uchida, Y., Gomi, F., Masumizu, T., and Miura, Y. (2002). Growth-inhibitory factor, metallothionein-like protein, and neurodegenerative disease. *J. Biol. Chem.* **277,** 32353.

Wanpen, S., Govitropong, P., Shavali, S., Sangchot, P., and Ebadi, M. (2004). Salsolinol, a DA-derived tetrahydroisoquinoline, induces cell death by causing oxidative stress in DAergic SH-SY5Y cells, and the said effect is attenuated by MTs. *Brain Res.* **1005,** 67.

Youdim, M. (2003). What have we learnt from CDNA microarray gene expression studies about the role of iron in MPTP induced neurodegeneration and Parkinson's disease? *J. Neural. Transm.* **110,** 1413.

# Section IV

# Signaling and Gene Expression

## [25]  Yeast Model Systems for Examining Nitrogen Oxide Biochemistry/Signaling

By Masaru Shinyashiki, Brenda E. Lopez,
Chester E. Rodriguez, and Jon M. Fukuto

### Abstract

The yeast *Saccharomyces cerevisiae* is an ideal model system for examining fundamental nitrogen oxide biochemistry. The utility of this model system lies in both the similarities and the differences between yeast and mammalian cells. The similarities between the two systems, with regards to many of the fundamental biochemical processes, allow studies in yeast to be extrapolated to mammalian systems. On the other hand, yeast has distinct differences that allow, for example, the facile examination of $O_2$, pH, and genetic dependencies on a number of nitrogen oxide–mediated processes. Thus, the yeast system is amenable to experimentation that is otherwise problematic or impossible in mammalian systems. Herein, we present several examples of the utility of the yeast model system for studying the intimate details of basic nitrogen oxide biochemistry.

### Introduction

The budding yeast *Saccharomyces cerevisiae* is a unicellular eukaryotic organism widely used in research laboratories and by bakers and brewers. Because of the similarities between *S. cerevisiae* and mammalian cells in terms of subcellular structure and metabolic systems, yeast is considered a good model for a multitude of mammalian cellular processes and functions (Botstein, 1991). For example, the similarities between the yeast and mammalian antioxidant and metal metabolism systems (Jamieson, 1998; Nelson, 1999) have allowed the use of *S. cerevisiae* as a model for the study of the biochemistry and genetics of human diseases such as Friedreich's ataxia, amyotrophic lateral sclerosis (ALS), and Menkes/Wilson diseases (Askwith and Kaplan, 1998; Roe *et al.*, 2002).

Clearly, the homology between *S. cerevisiae* and mammalian cells is an important factor in the utility of yeast as an experimental model for mammalian cells. However, yeast cells also possess various notable differences, which are also important to their utility in examining mammalian cell biochemistry. For example, unlike mammalian cells, *S. cerevisiae* exhibit fundamentally different cell biochemistry depending on their

METHODS IN ENZYMOLOGY, VOL. 396
0076-6879/05 $35.00
DOI: 10.1016/S0076-6879(05)96025-4

growth phase (Herman, 2002). In the presence of glucose, *S. cerevisiae* use glycolysis as their energy source and grow exponentially (log phase) undergoing fermentation. In this stage of growth, oxygen is not required, and yeast grows easily under anaerobic conditions. As glucose becomes depleted, yeast then starts developing mitochondrial respiratory and oxidative stress defense systems (diauxic shift). The growth rate decreases after the diauxic shift, and it ceases proliferating as it enters the stationary phase. At this point, adenosine triphosphate (ATP) is produced solely by respiration, typically using the ethanol produced during fermentation as the electron source. The consumption of glucose also leads to the derepression of various genes, including those responsible for mitochondrial activity and oxidant defense. Thus, during the stationary phase, their stress defense systems (i.e., antioxidant systems) are highly induced, and their mitochondria are fully developed and active. Stationary phase cells possess mammalian-like cell characteristics and are, therefore, often used to study aging and oxidative stress responses (Longo *et al.*, 1996; MacLean *et al.*, 2001; Stephen *et al.*, 1995). Thus, by examining yeast at various growth stages, we can also specifically evaluate the effects of cellular stressors/biochemical agents on, for example, glycolytic pathways or respiratory function. Moreover, the ability of yeast cells to grow/survive under anaerobic (glycolytic) and aerobic (respiratory) conditions makes them ideal for evaluating the importance of oxygen on nitrogen oxide–dependent processes. Indeed, we have exploited this aspect of yeast extensively to determine the fundamental biochemistry associated with nitrogen oxide–mediated interactions with several yeast proteins (*vide infra*). *S. cerevisiae* can survive and grow under other conditions that would otherwise be lethal to mammalian cells in culture. For example, yeast tolerates a wide range of pH levels (Shinyashiki *et al.*, 2001) and osmolarity (Klis *et al.*, 2002), two factors that are also important attributes for the study of nitrogen oxide biochemistry (*vide infra*).

Another important and potentially useful difference between yeast and mammalian cells is that *S. cerevisiae* have no polyunsaturated fatty acids (PUFAs) in their membranes (Gunstone *et al.*, 1994). For this reason, yeast cells are highly resistant to membrane lipid peroxidation. However, yeast membranes can be made mammalian-like. Supplementation of yeast media with exogenous PUFAs leads to their incorporation into the yeast cell membranes (Kohlwein and Paltauf, 1984). Significantly, the PUFA-incorporated yeast is more sensitive to oxidative stress mediated by heavy metals (Avery, 2001; Avery *et al.*, 1996) and other agents mediating oxidative stress (Steels *et al.*, 1994). Do *et al.* (1996) showed that coenzyme Q (CoQ) is an important membrane antioxidant because CoQ-delete yeast mutants were found to be particularly sensitive to agents producing cellular oxidative stress. Although this aspect of yeast biology has yet to be exploited to examine the membrane oxidant and antioxidant functions of the nitrogen oxides, this clearly represents an ideal system to determine the role of NO, and related species, on membrane integrity. Moreover, yeast represents an ideal system for evaluating the importance of the membrane as a target for the actions/toxicity of oxidative stressors.

The whole genome sequence of *S. cerevisiae* was completed in 1996 by a worldwide collaboration (Goffeau *et al.*, 1996). The gene sequence information and descriptions of the known biological functions of the gene products are available in databases such as Saccharomyces Genome Database (SGD) and Proteome's fungal database for *S. cerevisiae* (YPD) (Guthrie and Fink, 2002a). Using standard molecular biology techniques, *S. cerevisiae* are amenable to facile gene manipulation. The overexpression of genes in *S. cerevisiae* is easily achieved by the use of high- or low-copy plasmid expression vectors harboring different types of promoters (Mumberg *et al.*, 1995). Gene deletion is easily accomplished as well. However, the Saccharomyces Genome Deletion project (Stanford Genome Technology Center, Palo Alto, CA) has deleted 95% of the approximately 6200 open reading frames, and more than 24,000 mutant strains with different mating type backgrounds are available from several providers, including American Type Culture Collection (Manassas, VA), Invitrogen Corporation (Carlsbad, CA), Open Biosystems (Huntsville, AL), and Euroscarf (Frankfurt, Germany).

The availability of yeast mutants, either commercially or via facile experimental manipulation, makes *S. cerevisiae* a valuable research tool for examining the role and function of specific genes. Yeast also lends itself to the study of protein–protein interactions via the yeast two-hybrid assay conferring the utility of yeast to examine specific biochemical/biophysical events in a whole cell system. Moreover, the ability to express heterologous mammalian genes in yeast allows the study of mammalian gene function and can be a source of mammalian protein. The details of these techniques are available from various sources (Ausubel *et al.*, 2004; Guthrie and Fink, 1991, 2002a,b; Kaiser *et al.*, 1994).

Thus, the yeast *S. cerevisiae* is an ideal model system for examining fundamental nitrogen oxide biochemistry and the associated biological targets. The following is a summary of some of the most important aspects of yeast biology that allow it to be such an important and valuable tool in this regard:

1. Yeast cells grow/survive under either aerobic or anaerobic conditions. Thus, the effect of $O_2$ levels on nitrogen oxide–mediated events can be easily determined. This aspect of yeast biology is important to all the

studies described herein because the presence of $O_2$ and $O_2$-derived species will change the nature of the possible chemistry of all interactions examined. Studies of this nature are not as feasible using mammalian cells.

2. *S. cerevisiae* genetics are fully characterized, allowing facile genetic manipulation as a means of altering the intracellular environment. This aspect of the yeast model system allows nitrogen oxide chemistry to be examined under specific and varied conditions in a whole cell system. For example, antioxidant or metal–sequestering proteins can be deleted or overexpressed in order to predictably change the intracellular environment.

3. *S. cerevisiae* is an established, useful, and highly used model for mammalian cell biochemistry (*i.e.*, cell cycle, DNA repair, cytotoxicity, etc.). It is expected that the basic knowledge gained from studies in yeast can be extrapolated to mammalian systems. That is, yeast studies will serve as important examples of the possible nitrogen oxide chemistry and define the cellular conditions that promote these specific chemical processes.

4. Yeast membranes, unlike mammalian membranes, contain only saturated and monounsaturated fatty acids. Therefore, yeast is not as susceptible to lipid peroxidation chemistry as mammalian cells. However, the composition and nature of yeast membranes can be altered by simply exposing them to exogenous fatty acids in the media. Since it is proposed that critical biological targets for the nitrogen oxides in mammalian systems are membranes, this aspect of yeast allows an examination of the nature and importance of the interaction of nitrogen oxides with membranes. Although this aspect of yeast biology has not yet been exploited for determining the importance of specific interactions between the nitrogen oxides and membrane fatty acids, the utility of yeast in this regard is clear.

5. Finally, compared to mammalian cells, yeast is an inexpensive and technically simple biological system to manipulate and utilize. Because they do not require serum and can be manipulated on lab benchtops (a laminar flow hood is not needed), the expenses of growing and manipulating yeast cells are considerably lower than mammalian cells.

There have been several studies using yeast, in which the pathophysiological and biochemical aspects of nitric oxide (NO) have been examined. For example, the effects of NO on zinc finger transcription factors (Kroncke *et al.*, 1994), metal-thiolate proteins (Chiang *et al.*, 2000a; Cook *et al.*, 2003; Hartmann and Weser, 2000; Kroncke *et al.*, 1994; Shinyashiki *et al.*, 2000, 2001), transmembrane and cytosolic heme proteins (Liu *et al.*, 2000; Shinyashiki *et al.*, 2004), and oxidative stress (Chiang *et al.*, 2000b; Jakubowski *et al.*, 1999) have been examined in yeast. The purpose of this chapter is to illustrate the utility of the yeast model system in examining nitrogen oxide biochemistry in a living cell. To do this, we present parts of several studies that serve to illustrate this point.

## Methods

### General Methods of Yeast Culture

Like *Escherichia coli*, *S. cerevisiae* can be grown on either liquid or solid (2% agarose) media, depending on the experimental requirements. Rich YP medium contains 2% peptone, 1% yeast extract, and appropriate carbon source(s) and is used for general culturing. Synthetic complete (SC) medium contains the minimum nutritional requirements, including a nitrogen source, vitamins, trace elements, amino acids, nucleic acids, salt, and an appropriate carbon source. Typically, SC medium is used for experiments in which deletion or overexpressing strains are being developed and that require a selective medium. Glucose is the most efficient carbon source and is widely used for most experiments unless a non-glucose carbon source is required (a nonfermenting carbon source, a nonrepressing carbon source, etc.). Other possible carbon sources include a hexose (such as galactose), glycerol, ethanol, lactate, and acetate.

Laboratory strains of *S. cerevisiae* are well characterized and are available commercially or from other laboratories. Strains used in this chapter were gifts from Dr. Edith Gralla (UCLA Department of Chemistry and Biochemistry). Cells can be stored for years at $-80°$ in 15% glycerol, for months at $4°$ on agarose plates, and weeks at $4°$ in liquid media. *S. cerevisiae* cultures are usually started from a $-80°$ glycerol stock. To eliminate respiratory-deficient mutants, made by spontaneous mutations, the glycerol stock is streaked on a plate containing glucose deplete medium (YP with 0.1% glucose and 3% glycerol, YPDG). Mitochondrial–deficient mutants can be distinguished by the smaller size of the colony (and thus are called "petite" mutants), because those mutants grow more slowly than wild-type yeast on a glucose deplete medium (Guthrie and Fink, 1991). A single colony obtained from the YPDG plate is then inoculated in the appropriate medium and cultured at $30°$ overnight with shaking at 275 rpm. This culture can be stored at $4°$ and used as a stock culture for at least a week. Before an experiment (usually the day before), the medium in an Erlenmeyer flask or polypropylene centrifuge tube is inoculated with the $4°$ stock culture and grown to the preferred growth phase. The cell growth is typically monitored by optical density at 600 nm ($OD_{600}$). However, the relationship between $OD_{600}$ and the growth phase of yeast cultures varies, depending on many factors such as strain, medium, and spectrophotometer (due to differences in the distance from the cuvette to the detector). For example, the wild-type strain EG217 has an $OD_{600} < 1$ in early log, $> 3$ in late log, and $> 5$ in stationary phase (using a Milton-Roy Spectronic 1001 spectrophotometer).

## Anaerobic and Aerobic Experiments

Yeast cultures (5–10 ml) are transferred into either 35-ml glass centrifuge tubes or 50-ml Erlenmeyer flasks, which are sealed with gas-tight silicon rubber septa. Deoxygenation is then accomplished by passing humidified nitrogen or argon gas through the head space of the culture through inlet and outlet needles. To decrease oxygen to 1–2% (of air saturated) in 10-ml of culture medium, 30–40 min of purging at approximately a 50-ml/min flow rate is required (as determined in control experiments using a Clarke electrode, YSI 5300 Oxygen Monitor, Yellow springs, OH). For aerobic conditions, the head-space gases are allowed to freely exchange with room air.

## Viability Assay

After treatment of cells with desired compounds (*i.e.*, NO-donor compounds, cellular stressors, etc.), cells are washed twice with double-distilled water (DDW). That is, cells are centrifuged at $1000g$ for 5 min (in 50-ml centrifuge tubes) or $15,000g$ for 1 min (in 1.5-ml Microfuge tubes) and the pellet resuspended in DDW. The suspensions are then diluted in DDW and spread on YPD plates (at a density of ~100–200 colonies/10-cm diameter plate for the control). Plates are incubated at 30° for 2–3 days, or until the colonies appear, and are then counted. Viability is represented as the number of colonies on the sample plate normalized to the control plate (% control).

## β-Galactosidase Reporter System

Some of the studies described herein involve the examination of the effects of nitrogen oxides on the activity of the copper-responsive transcription factor Ace1. To monitor Ace1 activity, the wild-type strain EG217 (W303A, *MAT*a *leu2-3,112 trp1-1 his3-11,15 ura3-1 ade2-1*) was transformed with a *CUP1* promoter-lacZ fusion reporter plasmid (YEp-CUP-HSE-M-lacZ) (Santoro *et al.*, 1998) by the method of Kaiser *et al.* (1994). Because the plasmid uses the *URA3* gene as a selection marker, the transformant (217CUP) was cultured in a synthetic dextrose medium without uracil (SD-ura). Ace1 is a copper-responsive transcription factor responsible for the transcription of a number of genes that, in part, protect yeast from high copper levels. Thus, Ace1 is activated by incubation with

100 $\mu M$ CuSO$_4$ in SD-ura media for 60 min. After the cells are washed with DDW and resuspended in buffer, they are lysed by vortexing in the presence of glass beads (0.5 mm diameter, Biospec Products, Inc., Bartlesville, OK) in a microcentrifuge tube. The homogenate is then clarified by centrifugation at 15,000g and 4° for 5 min. $\beta$-Galactosidase activity in the supernatant is measured using o-nitrophenyl $\beta$-D-galactopyranoside as the substrate (Thorvaldsen et al., 1993). The activity is normalized to protein concentration as measured by the Bradford assay (Bradford, 1976).

## Applications

### Effects of NO and HNO on the Transcription Factor Ace1

Copper is an essential nutrient (Mercer, 2001) but can be toxic if it is not carefully regulated (Bertinato and L'Abbe, 2004; Puig and Thiele, 2002). Yeast has evolved a series of metal-responsive transcription factors, which are important in the regulation of metals such as copper (Eide, 1998). Ace1 is a yeast transcription factor involved in copper regulation/detoxification. Under high copper conditions, Ace1 is activated by binding to copper through multiple cysteine thiols (Brown et al., 2002), which results in the transcription of genes encoding proteins involved in the sequestering of copper including CUP1, CRS5, and SOD1 (Eide, 1998). Experimentally, copper-mediated activation of Ace1 can be manipulated by adding to the culture media Cu$^{2+}$ (leading to Ace1 activation) or the copper-specific chelator bathocuproinesulfonic acid (BCS) (leading to Ace-1 inactivation).

We have examined the effects of nitrogen oxides on copper-mediated activation of Ace1 as a means of determining the intimate details of the chemistry between various nitrogen oxides and metal-thiolate proteins. More specifically, using the yeast model system, we have been able to determine the cellular conditions, which either promote or inhibit the disruption of metal-thiolate proteins (as represented by Ace1) by the nitrogen oxides NO and HNO. The activity of Ace1 has been monitored by utilizing the reporter gene system (described earlier) and by directly examining Ace1-DNA binding via an electrophoretic mobility shift assay (EMSA) (Shinyashiki, 2000).

Late-log phase cultures of 217CUP (OD$_{600}$ = 3.0–3.5, 10 ml) in SD-ura with 0.1 m$M$ BCS were prepared under both aerobic and anaerobic conditions. Different levels of exposure of yeast to the nitrogen oxides were accomplished via administration of the NO donor diethylammonium-1-($N$, $N$-diethylamino) diazen-1-ium-1,2-diolate (DEA/NO) (t$_{1/2}$ = 2.5 min at 37°) or the HNO donor Angeli's salt (AS) (t$_{1/2}$ = 2.5 min at 37°) to the yeast cultures. Stock solutions of the donors, made up in basic solutions, were

injected into the cultures using a gas-tight syringe. After incubation at 30°
(30 min for DEA/NO and 45 min for AS), $N_2$ or Ar gas was passed through
the head space to remove remaining volatile nitrogen oxide species. Cells
were washed and $\beta$-galactosidase activity measured as described earlier.

As shown in Fig. 1, NO and HNO inhibited Ace1 activity dose dependently. NO inhibition was found to be oxygen-dependent, whereas HNO
inhibition was oxygen-independent. These results are consistent with
the chemistry of thiol reactions with NO (Goldstein and Czapski, 1996;
Keshive *et al.*, 1996) and HNO (Doyle *et al.*, 1988; Wong *et al.*, 1998). Thus,
we have proposed that Ace thiol residues can be modified by NO in an
oxygen-dependent fashion via the generation of nitrosating species [Eqs.
(1–3)], whereas HNO is capable of directly reacting with thiols in an
oxygen-independent fashion [Eq. (4)].

$$2NO + O_2 \rightarrow 2NO_2 \tag{1}$$

$$NO_2 + NO \rightarrow N_2O_3 \tag{2}$$

$$Ace1\text{-}S^- + N_2O_3 \rightarrow Ace1\text{-}SNO + NO_2^- \tag{3}$$

$$Ace1\text{-}S^- + HNO + H^+ \rightarrow Ace1\text{-}S - NHOH \rightarrow etc. \tag{4}$$

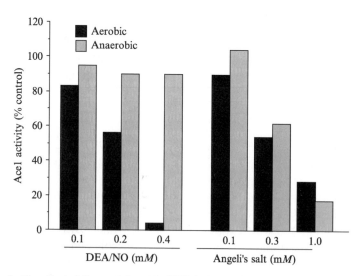

Fig. 1. The effect of $O_2$ on nitric oxide (NO) (DEA/NO)– and nitroxyl (HNO) (Angeli's
salt)–mediated inhibition of copper-dependent Ace1 activation.

The ability of yeast to grow/survive under either aerobic or anaerobic conditions allowed us to determine the oxygen dependency of these processes. To be sure, similar experiments using mammalian cell cultures are problematic or impossible.

Further validation of the mechanism of the proposed NO-mediated chemical modification of Ace1 thiols/thiolates by NO was obtained with several other studies, whereby manipulation of the experimental conditions predictably promoted or inhibited the degree of Ace1 inactivation. For example, it was determined that nitrite, $NO_2^-$, could inhibit copper-mediated Ace1 activity only under acidic conditions (pH 4) and at high concentrations (Shinyashiki et al., 2001). Because these conditions favor formation of dinitrogen trioxide, $N_2O_3$ [Eq. (5)], this result supports the idea that nitrosation of thiols via products of the $NO/O_2$ reactions, Eqs. (1) and (2) (i.e., $N_2O_3$), can serve to inhibit Ace1 activity.

$$2NO_2^- + 2H^+ \rightarrow N_2O_3 + H_2O \qquad (5)$$

Because yeast cells easily survive in a wide pH range while mammalian cells survive only in a very narrow range near physiological pH, this study would not be possible in any mammalian cell system.

Because $NO_2$ is generated from the reaction of NO and $O_2$ [Eq. (1)], there was the possibility that $NO_2$ was the nitrogen oxide species responsible for the disruption of Ace1 activity. We were able to address this by simply incubating yeast cultures anaerobically in the presence of either or both authentic $NO_2$ and NO gas (Shinyashiki et al., 2001). In these experiments, inhibition of Ace1 activity was maximal only in the presence of both $NO_2$ and NO. Again, this result is consistent with the idea that generation of $N_2O_3$ [Eq. (2)] potently inhibits Ace1. Again, the ability to manipulate yeast under anaerobic conditions made these experiments feasible.

## Toxicological Effects of HNO on Different Growth Phases of Yeast

As mentioned earlier, S. cerevisiae demonstrate different characteristics depending on their growth phase. Early log phase cells produce ATP mostly by glycolysis and are less resistant to oxidative stress because of the repression of stress response genes by glucose (Estruch, 2000). It was reported that HNO can be toxic to mammalian cells (Wink et al., 1999). However, the mechanism of toxicity is not established. To begin to address the mechanisms of HNO-mediated cytotoxicity, we used the yeast model system to determine the intracellular environment (i.e., metabolic status) most conducive to this toxicity. Thus, the toxicity of HNO as a function of different yeast growth phases was determined.

Wild-type EG104 (DBY747, *MATa leu2-3,11 his3Δ1 trp1-289$_a$ ura3-53 can$^R$ gal2*) was grown until early log (OD$_{600}$ = 0.3) and late-log phase (OD$_{600}$ = 3). The cells were exposed to various concentrations of the HNO donor AS for 45 min, and the viability assay was performed. HNO exhibited dose-dependent toxicity in both early and late-log phase (Fig. 2A). However, early log phase cells showed a greater susceptibility to HNO-mediated toxicity compared with late-log phase cells, suggesting that cells surviving via glycolysis are more sensitive to HNO toxicity than cells relying on mitochondrial respiration (alternatively, the late-log phase yeast cells, which are respiring, also have well-developed antioxidant systems that may protect against HNO toxicity). Because it is established that HNO is very thiophilic (Bartberger *et al.*, 2001; Wong *et al.*, 1998), we surmised that glycolytic enzymes, which contain crucial thiolates, may be targets for the actions of HNO in early log phase growth. Indeed, De-Master *et al.* (1998) have reported inhibition of the glycolytic and thiol-containing enzyme glyceraldehyde phosphate dehydrogenase (GAPDH) by HNO, and we have confirmed this finding in a whole cell yeast system (Fig. 2B). Thus, we propose that HNO can potently inhibit thiolate proteins (*i.e.*, GAPDH), resulting in their loss of activity and, in the case of yeast, leading to cessation of growth in early log phase because of an inhibition of glycolysis. Clearly, this idea is speculative and requires further examination. However, these studies clearly illustrate the utility of the yeast model for distinguishing between possible targets (*i.e.*, glycolytic pathways *vs* respiratory systems) for nitrogen oxide–mediated toxicity. Interestingly, mammalian cancer cells have been reported to rely more on glycolysis compared to normal cells (Mathupala *et al.*, 1997), so yeast in early log phase can be a good model to study the pathophysiology of nitrogen oxide species against tumor cells.

### NO as an Antioxidant Against Paraquat Toxicity

Many of the proposed biological effects of NO are highly dependent on the cellular environment. For example, NO has been suggested to be both an antioxidant (Gewaltig and Kojda, 2002; Violi *et al.*, 1999) and a prooxidant (Pryor and Squadrito, 1995), depending on its concentration and cellular conditions. The antioxidant properties of NO are, in part, due to the fact that it is a stable radical species capable of intercepting and "quenching" reactive free radicals, which may otherwise react with other biological targets [Eq. (6)]. An antioxidant function of NO is clearly illustrated by its ability to inhibit lipid peroxidation (Rubbo *et al.*, 1994; Wink

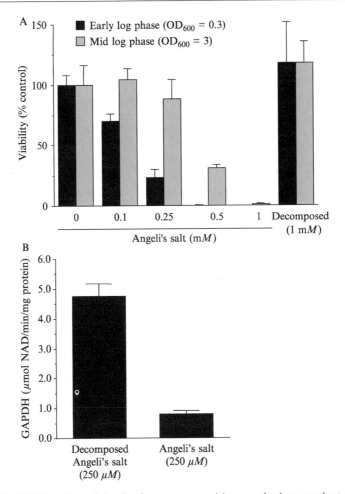

Fig. 2. (A) The effect of the *Saccharomyces cerevisiae* growth phase on the toxicity of nitroxyl (HNO). (B) Inhibition of glyceraldehyde phosphate dehydrogenase (GAPDH) by HNO in whole-cell yeast.

*et al.*, 1993). The prooxidant properties of NO have been attributed, in part, to its ability to react rapidly with superoxide ($O_2^-$) [Eq. (7)], generating a putative oxidant, peroxynitrite ($ONOO^-$) (Pryor and Squadrito, 1995).

$$NO + R^\bullet \rightarrow R - NO \ (R^\bullet = \text{reactive radical species}) \qquad (6)$$

$$NO + O_2^- \rightarrow ONOO^- \rightarrow \text{oxidation chemistry} \qquad (7)$$

Because $O_2^-$ (and the protonated form $HOO^\bullet$) has also been implicated in biological oxidation processes (Halliwell and Gutteridge, 1999), the "quenching" of $O_2^-$ by NO may actually represent an antioxidant function of NO because $ONOO^-$ can also degrade to give the innocuous nitrogen oxide $NO_3^-$ [Eq. (8)]. That is, it can be proposed that NO protects oxidation-sensitive biological targets from $O_2^-$-mediated oxidation processes because it converts $O_2^-$ to the unreactive nitrogen oxide species $NO_3^-$.

$$ONOO^- \rightarrow NO_3^- \tag{8}$$

Thus, the reaction of NO with $O_2^-$ can be envisioned to be either a prooxidant or an antioxidant process.

*S. cerevisiae* is an ideal system to examine and delineate the oxidant and antioxidant properties of NO because cellular environments are easily manipulated genetically. Examples of the utility of yeast in this regard are studies whereby the effect of deletion of Cu,Zn-superoxide dismutase (CuZnSOD) on NO and/or $O_2^-$-mediated toxicity was determined. Like mammalian cells, yeast possesses CuZnSOD, and it has been demonstrated that this enzyme is important for the protection from the toxicity of $O_2^-$ (Gralla and Valentine, 1991). Thus, using paraquat to generate $O_2^-$ in yeast, the effect of NO on the toxicity of paraquat in CuZnSOD-delete (*sod1Δ*) and wild-type yeast was examined.

Stationary phase cultures (5 ml) of wild-type EG103 (DBY746, *MATa leu2-3,112 his3Δ1 trp1-289a ura3-52*) and *sod1Δ* mutant EG118 (EG103 with *sod1a::URA3*) were exposed to paraquat (0–50 m*M*) in the presence or absence

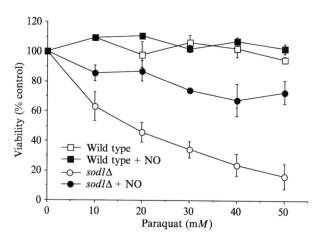

FIG. 3. The effect of nitric oxide (NO) on the toxicity of paraquat in wild-type and CuZnSOD-delete (*sod1Δ*) yeast.

of the NO donor 1,3-propanediamine, $N$-[4-[1-(3-aminopropyl)-2-hydroxy-2-nitrosohydrazino]butyl]-1, 3-propanediamine (SPER/NO) ($t_{1/2}$ = 40 min at 37°) for 4 h at 30°. Viability was assayed as described earlier. As shown in Fig. 3, paraquat exhibited a dose-dependent toxicity on the $sod1\Delta$ strain, whereas the wild-type yeast was not affected by paraquat at the concentrations up to 50 m$M$. SPER/NO (at 1 m$M$, which did not have an effect on the wild-type yeast) protected the $sod1\Delta$ mutant against paraquat-mediated cytotoxicity. These results are consistent with the idea that $O_2^-$ can be toxic to yeast and that the reaction of NO with $O_2^-$ can represent an antioxidant function of NO, rather than a prooxidant function. Again, the versatility of the yeast model system is demonstrated, and the similarity of yeast to mammalian cells allows these results to be extrapolated to mammalian cell systems.

## Summary

The inherent complexities of nitrogen oxide chemistry have made the elucidation of biochemical mechanisms of the actions of nitrogen oxides in mammalian systems difficult. However, the versatility and flexibility of the yeast system allow for greater experimental breadth compared to mammalian cells and the marked similarities of the two systems allow yeast to be a useful model for numerous fundamental biochemical processes in mammalian cells. These few representative studies serve to illustrate the utility of the yeast model system in addressing specific biochemical questions that may otherwise be difficult (or impossible) to address in a mammalian system. Many other examples can be given and, to be sure, the usefulness of yeast goes far beyond the elucidation of the fundamental biochemistry of nitrogen oxide–mediated processes. However, the similarities and differences between yeast and mammalian cells appear ideal for the exploitation of yeast to study the fundamentals of nitrogen oxide biochemistry/toxicology.

## References

Askwith, C., and Kaplan, J. (1998). Iron and copper transport in yeast and its relevance to human disease. *Trends Biochem. Sci.* **23**, 135–138.

Ausubel, F. M., Brent, R., Kingston, R. E., David, D., Moore, D. D., Seidman, J. G., Smith, J. A., and Struhl, K. (2004). "Current Protocols in Molecular Biology." Unlimited Learning Resources, LLC, Winston, Salem.

Avery, S. V. (2001). Metal toxicity in yeast and the role of oxidative stress. *Adv. Appl. Microbiol.* **49**, 111–142.

Avery, S. V., Howlett, N. G., and Radice, S. (1996). Copper toxicity towards *Saccharomyces cerevisiae*: Dependence on plasma membrane fatty acid composition. *Appl. Environ. Microbiol.* **62**, 3960–3966.

Bartberger, M. D., Fukuto, J. M., and Houk, K. N. (2001). On the acidity and reactivity of HNO in aqueous solution and biological systems. *Proc. Natl. Acad. Sci. USA* **98,** 2194–2198.

Bertinato, J., and L'Abbe, M. R. (2004). Maintaining copper homeostasis: Regulation of copper-trafficking proteins in response to copper deficiency or overload. *J. Nutr. Biochem.* **15,** 316–322.

Botstein, D. (1991). Why yeast? *Hospital Practice* **26,** 157–161.

Bradford, M. M. (1976). A rapid and sensitive method for the quantitation of microgram quantities of protein utilizing the principle of protein-dye binding. *Anal. Biochem.* **72,** 248–254.

Brown, K. R., Keller, G. L., Pickering, I. J., Harris, H. H., George, G. N., and Winge, D. R. (2002). Structures of the cuprous-thiolate clusters of the Mac1 and Ace1 transcriptional activators. *Biochemistry* **41,** 6469–6476.

Chiang, K. T., Shinyashiki, M., Switzer, C. H., Valentine, J. S., Gralla, E. B., Thiele, D. J., and Fukuto, J. M. (2000a). Effects of nitric oxide on the copper-responsive transcription factor Ace1 in Saccharomyces cerevisiae: Cytotoxic and cytoprotective actions of nitric oxide. *Arch. Biochem. Biophys.* **377,** 296–303.

Chiang, K. T., Switzer, C. H., Akali, K. O., and Fukuto, J. M. (2000b). The role of oxygen and reduced oxygen species in nitric oxide–mediated cytotoxicity: Studies in the yeast Saccharomyces cerevisiae model system. *Toxicol. Applied Pharmacol.* **167,** 30–36.

Cook, N. M., Shinyashiki, M., Jackson, M. I., Leal, F. A., and Fukuto, J. M. (2003). Nitroxyl-mediated disruption of thiol proteins: Inhibition of the yeast transcription factor Ace1. *Arch. Biochem. Biophys.* **410,** 89–95.

DeMaster, E. G., Redfern, B., and Nagasawa, H. T. (1998). Mechanisms of inhibition of aldehyde dehydrogenase by nitroxyl, the active metabolite of the alcohol deterrent agent cyanamide. *Biochem. Pharmacol.* **55,** 2007–2015.

Do, T. Q., Schultz, J. R., and Clarke, C. F. (1996). Enhanced sensitivity of ubiquinone-deficient mutants of *Saccharomyces cerevisiae* to products of autoxidized polyunsaturated fatty acids. *Proc. Natl. Acad. Sci. USA* **93,** 7534–7539.

Doyle, M. P., Mahapatro, S. N., and Broene, R. D. (1988). Oxidation and reduction of hemoproteins by trioxodinitrate (II). The role of nitrosyl hydride and nitrite. *J. Am. Chem. Soc.* **110,** 593–599.

Eide, D. J. (1998). The molecular biology of metal ion transport in *Saccharomyces cerevisiae.* *Annu. Rev. Nutr.* **18,** 441–469.

Estruch, F. (2000). Stress-controlled transcription factors, stress-induced genes and stress tolerance in budding yeast. *FEMS Microbiol. Rev.* **24,** 469–486.

Gewaltig, M. T., and Kojda, G. (2002). Vasoprotection by nitric oxide: Mechanisms and therapeutic potential. *Cardiovasc. Res.* **55,** 250–260.

Goffeau, A., Barrell, B. G., Bussey, H., Davis, R. W., Dujon, B., Feldmann, H., Galibert, F., Hoheisel, J. D., Jacq, C., Johnston, M., Louis, E. J., Mewes, H. W., Murakami, Y., Philippsen, P., Tettelin, H., and Oliver, S. G. (1996). Life with 6000 Genes. *Science* **274,** 546–567.

Goldstein, S., and Czapski, G. (1996). Mechanism of the nitrosation of thiols and amines by oxygenated •NO Solutions: The nature of the nitrosating intermediates. *J. Am. Chem. Soc.* **118,** 3419–3425.

Gralla, E. B., and Valentine, J. S. (1991). Null mutants of *Saccharomyces cerevisiae* Cu, Zn superoxide dismutase: Characterization and spontaneous mutation rates. *J. Bacteriol.* **173,** 5918–5920.

Gunstone, F. D., Harwood, J. L., and Padley, F. B. (1994). "The Lipid Handbook," 2nd Ed. Chapman & Hall, London.

Guthrie, C., and Fink, G. R. (1991). "Methods in Enzymology," Vol. 194, "Guide to Yeast Genetics and Molecular Biology." Academic Press, San Diego.

Guthrie, C., and Fink, G. R. (2002a). "Methods in Enzymology," Vol. 350, "Guide to Yeast Genetics and Molecular and Cell Biology, Part B." Academic Press, San Diego.

Guthrie, C., and Fink, G. R. (2002b). "Methods in Enzymology," Vol. 351, "Guide to Yeast Genetics and Molecular and Cell Biology, Part C." Academic Press, San Diego.

Halliwell, B., and Gutteridge, J. M. C. (1999). "Free Radicals in Biology and Medicine," 3rd Ed. Oxford University Press, Oxford.

Hartmann, H.-J., and Weser, U. (2000). Copper-release from yeast Cu(I)-metallothionein by nitric oxide (NO). *BioMetals* **13,** 153–156.

Herman, P. K. (2002). Stationary phase in yeast. *Curr. Opin. Microbiol.* **5,** 602–607.

Jakubowski, W., Bilinski, T., and Bartosz, G. (1999). Sensitivity of antioxidant-deficient yeast Saccharomyces cerevisiae to peroxynitrite and nitric oxide. *Biochim. Biophys. Acta* **1472,** 395–398.

Jamieson, D. J. (1998). Oxidative stress responses of the yeast *Saccharomyces cerevisiae. Yeast* **14,** 1511–1527.

Kaiser, C., Michaelis, S., and Mitchell, A. (1994). "Methods in Yeast Genetics." Cold Spring Harbor Laboratory Press, Plainview.

Keshive, M., Singh, S., Wishnok, J. S., Tannenbaum, S. R., and Deen, W. D. (1996). Kinetics of S-nitrosation of thiols in nitric oxide solutions. *Chem. Res. Toxicol.* **9,** 988–993.

Klis, F. M., Pieternella Mol, P., Hellingwerf, K., and Stanley Brul, S. (2002). Dynamics of cell wall structure in *Saccharomyces cerevisiae. FEMS Microbiol. Rev.* **26,** 239–256.

Kohlwein, S. D., and Paltauf, F. (1984). Uptake of fatty acids by yeast, *Saccharomyces uvarum* and *Saccharomycopsis lipolytica. Biochim. Biophys. Acta* **792,** 310–317.

Kroncke, K.-D., Fchsel, K., Schmidt, T., Zenke, F. T., Dasting, I., Wesener, J. R., Bettermann, H., Breunig, K. D., and Kolb-Bachofen, V. (1994). Nitric oxide destroys zinc-sulfur clusters inducing zinc release from metallothionein and inhibition of the zinc finger-type yeast transcription activator LAC9. *Biochem. Biophys. Res. Commun.* **200,** 1105–1110.

Liu, L., Zeng, M., Hausladen, A., Heitman, J., and Stamler, J. S. (2000). Protection from nitrosative stress by yeast flavohemoglobin. *Proc. Natl. Acad. Sci. USA* **97,** 4672–4676.

Longo, V. D., Gralla, E. B., and Valentine, J. S. (1996). Superoxide dismutase activity is essential for stationary phase survival in *Saccharomyces cerevisiae. J. Biol. Chem.* **271,** 12275–12280.

MacLean, M., Harris, N., and Piper, P. W. (2001). Chronological lifespan of stationary phase yeast cells; a model for investigating the factors that might influence the ageing of postmitotic tissues in higher organisms. *Yeast* **18,** 499–509.

Mathupala, S. P., Rempel, A., and Pedersen, P. L. (1997). Aberrant glycolytic metabolism of cancer cells: A remarkable coordination of genetic, transcriptional, post-translational, and mutational events that lead to a critical role for type II hexokinase. *J. Bioenergetics Biomembranes* **29,** 339–343.

Mercer, J. F. B. (2001). The molecular basis of copper-transport diseases. *Trends Mol. Med.* **7,** 64–69.

Mumberg, D., Muller, R., and Funk, M. (1995). Yeast vectors for the controlled expression of heterologous proteins in different genetic backgrounds. *Gene* **14,** 119–122.

Nelson, N. (1999). Metal ion transporters and homeostasis. *EMBO J.* **18,** 4361–4371.

Poon, W. W., Do, T. Q., Marbois, B. N., and Clarke, C. F. (1997). Sensitivity to treatment with polyunsaturated fatty acids is a general characteristic of the ubiquinone-deficient yeast coq mutants. *Mol. Aspects Med.* **18**(Suppl.), S121–S127.

Pryor, W. A., and Squadrito, G. L. (1995). The chemistry of peroxynitrite: A product from the reaction of nitric oxide with superoxide. *Am. J. Physiol.* **268** (*Lung Cell. Mol. Physiol. 12*), L699–L722.

Puig, S., and Thiele, D. J. (2002). Molecular mechanisms of copper uptake and distribution. *Curr. Opinion Chem. Biol.* **6,** 171–180.

Roe, J. A., Wiedau-Pazos, M., Moy, V. N., Goto, J. J., Gralla, E. B., and Valentine, J. S. (2002). *In vivo* peroxidative activity of FALS-mutant human CuZnSODs expressed in yeast. *Free Radic. Biol. Med.* **32,** 169–174.

Rubbo, H., Radi, R., Trujillo, M., Telleri, R., Kalyanaraman, B., Barnes, S., Kirk, M., and Freeman, B. A. (1994). Nitric oxide regulation of superoxide and peroxynitrite-dependent lipid peroxidation. *J. Biol. Chem.* **269,** 26066–26075.

Santoro, N., Johansson, N., and Thiele, D. J. (1998). Heat shock element architecture is an important determinant in the temperature and transactivation domain requirements for heat shock transcription factor. *Mol. Cell. Biol.* **18,** 6340–6352.

Shinyashiki, M., Chiang, K. T., Switzer, C. H., Gralla, E. B., Valentine, J. S., Thiele, D. J., and Fukuto, J. M. (2000). The interaction of nitric oxide (NO) with the yeast transcription factor Ace1: A model system for NO-protein thiol interactions with implications to metal metabolism. *Proc. Natl. Acad. Sci. USA* **97,** 2491–2496.

Shinyashiki, M., Pan, C.-J. G., Switzer, C. H., and Fukuto, J. M. (2001). Mechanisms of nitrogen oxide–mediated disruption of metalloprotein function: An examination of the copper-responsive yeast transcription factor Ace1. *Chem. Res. Toxicol.* **14,** 1584–1589.

Shinyashiki, M., Pan, C.-J., Lopez, B. E., and Fukuto, J. M. (2004). Inhibition of the yeast metal reductase heme protein Fre1 by nitric oxide (NO): A model for inhibition of NADPH oxidase by NO. *Free Radic. Biol. Med.* **37,** 713–723.

Steels, E. L., Learmonth, R. P., and Watson, K. (1994). Stress tolerance and membrane lipid unsaturation in *Saccharomyces cerevisiae* grown aerobically or anaerobically. *Microbiology* **140,** 569–576.

Stephen, D. W. S., Rivers, S. L., and Jamieson, D. J. (1995). The role of the YAP1 and YAP2 genes in the regulation of the adaptive oxidative stress responses of *Saccharomyces cerevisiae. Mol. Microbiol.* **16,** 415–423.

Thorvaldsen, J. L., Sewell, A. K., McCowen, C. L., and Winge, D. R. (1993). Regulation of metallothionein genes by the ACE1 and AMT1 transcription factors. *J. Biol. Chem.* **268,** 12512–12518.

Violi, F., Marino, R., Milite, M. T., and Loffredo, L. (1999). Nitric oxide and its role in lipid peroxidation. *Diabetes Metab. Res. Rev.* **15,** 283–288.

Wink, D. A., Hanbauer, I., Krishna, M. C., DeGraff, W., Gamson, J., and Mitchell, J. B. (1993). Nitric oxide protects against cellular damage and cytotoxicity from reactive oxygen species. *Proc. Natl. Acad. Sci. USA* **90,** 9813–9817.

Wink, D. A., Feelisch, M., Fukuto, J., Christodoulou, D., Jourd'heuil, D., Grisham, M., Vodovotz, V., Cook, J. A., Krishna, M., DeGraff, W., Kim, S., Gamson, J., and Mitchell, J. B. (1999). The cytotoxicity of nitroxyl: Possible implications for the pathophysiological role of NO. *Arch. Biochem. Biophys.* **351,** 66–74.

Wong, P. S.-Y., Hyun, J., Fukuto, J. M., Shirota, F. N., DeMaster, E. G., Shoeman, D. W., and Nagasawa, H. T. (1998). Reaction between S-nitrosothiols and thiols: Generation of nitroxyl (HNO) and subsequent chemistry. *Biochemistry* **37,** 5362–5371.

## [26] Fluorescence Resonance Energy Transfer–Based Assays for the Real-Time Detection of Nitric Oxide Signaling

*By* CLAUDETTE M. ST. CROIX, MOLLY S. STITT,
SIMON C. WATKINS, and BRUCE R. PITT

### Abstract

Low-molecular-weight *S*-nitrosothiols are found in many tissues and affect an array of signaling pathways via decomposition to •NO or exchange of their —NO function with thiol-containing proteins (transnitrosation). We used spectral laser scanning confocal imaging to visualize the effects of the membrane permeant *S*-nitrosothiol, *S*-nitrosocysteine ethyl ester (SNCEE), on a fluorescence resonance energy transfer (FRET) reporter based on the cysteine-rich heavy metal binding protein, metallothionein (FRET-MT) flanked by enhanced cyan and yellow fluorescent proteins (ECFP and EYFP, respectively). We previously showed that FRET can be used to follow metal binding and release by this construct. SNCEE ($50 \mu M$) induced a decrease in energy transfer, as shown by an increase in the peak emission intensity of the donor fluorophore (ECFP) and a decrease in that of the acceptor (EYFP). These changes in intramolecular FRET were reversed by $50 \mu M$ dithiothreitol (DTT), suggesting nitrosothiol-mediated modification of a cysteine residue in MT. Furthermore, the effects of SNCEE on the FRET-MT reporter were not affected by $HbO_2$, which would be expected to block any process involving •NO liberated by decomposition of nitrosothiol but would not necessarily affect transnitrosation. In further support of SNCEE-induced conformational changes in MT, we used live cell imaging of the zinc-sensitive fluorescent indicator FluoZin-3 to show that SNCEE also caused increases in labile $Zn^{2+}$.

### Introduction

Detection of interactions between target proteins and nitric oxide (NO) species *in vivo* typically requires disruptive biochemical techniques that preclude or limit temporospatial information. Although significant technological advancements have promoted an increased reliance on the use of high-performance multimodal optical imaging tools to visualize cellular processes in biological systems, the imaging of NO-based signaling events

METHODS IN ENZYMOLOGY, VOL. 396        0076-6879/05 $35.00
DOI: 10.1016/S0076-6879(05)96026-6

remains especially challenging. Fluorescence resonance energy transfer (FRET) is a nondestructive spectrofluorometric technique. In combination with scanning laser confocal microscopy, FRET is capable of detecting changes in the conformational state of proteins in live cells. The validity of FRET-based approaches to study intracellular signaling pathways in live cells has been demonstrated using fluorescent reporter molecules for calcium (Miyawaki *et al.*, 1997), guanosine 3',5'-cyclic monophosphate (cGMP) (Honda, 2001), 3',5'-cyclic-adenosine monophosphate (cAMP) (Zaccolo *et al.*, 2000), and tyrosine kinase activity (Ting *et al.*, 2001), among others. Our efforts using this methodology to elucidate the role of metallothioneins (MTs) in NO signaling (Pearce *et al.*, 2000; St. Croix *et al.*, 2002, 2004) suggest that FRET is suitable for detection of posttranslational protein modifications caused by NO-related species.

FRET can occur between two fluorophores (i.e., donor and acceptor pair) that have appropriate spectral properties and become closely apposed (<10 nm). The phenomenon of FRET occurs when an overlap in donor emission spectra and acceptor absorption spectra allows the transfer of excitation energy. The FRET effect decreases with the sixth power of the distance between donor and acceptor fluorophores. This relationship allows the use of FRET to measure physical interactions among proteins or among domains within a single protein.

Established methods for detecting FRET between a donor and an acceptor with overlapping excitation and emission spectra require the use of narrow detection bands and automatic switching of optical filters to differentiate between emissions, along with complex mathematical corrections to account for crosstalk between channels. Advances in detector technology, however, now enable the resolution of fluorescent images, providing full spectral information for each voxel of the image without switching of optical filters. Furthermore, using calibration spectra, one can unambiguously separate the crosstalk between overlapping cyan and yellow emissions. The use of this method allows the detection of small, but potentially biologically meaningful, changes in FRET that are common with genetically encoded reporters and are extremely difficult to resolve reliably using more traditional methods relying on bandpass filters. Furthermore, it is quite reasonable to spectrally separate multiple concurrent fluorophores with similar emission spectra.

## Methods

### Cultured Sheep Pulmonary Artery Endothelial Cells.

Sheep pulmonary artery endothelial cells (SPAECs) were cultured from sheep pulmonary arteries obtained from a nearby slaughterhouse (Hoyt *et al.*, 1995). The SPAECs were grown in OptiMEM(Gibco) supplemented with 10% fetal bovine serum, 100 U/ml penicillin, and 100 $\mu$g/ml streptomycin at 37° in an atmosphere with 5% $CO_2$.

## FRET Reporter (FRET-MT)

MT, a cysteine-rich (30 mol%) heavy metal binding protein, is critical to intracellular $Zn^{2+}$ homeostasis with the ability to bind up to seven zinc atoms per mole of MT. *In vitro*, NO is capable of interacting with MT, forming either dinitrosyl iron sulfur complexes (Kennedy *et al.*, 1993; Schwartz *et al.*, 1995) or *S*-nitrosation products, resulting in NO-mediated release of cadmium (Misra *et al.*, 1996), copper (Kawai *et al.*, 2000), or zinc (Kroncke *et al.*, 1994). We followed the example of the $Ca^{2+}$ indicator cameleon-1 (Miyawaki *et al.*, 1997) and constructed a chimera in which a yellow-green fluorescent protein (GFP) variant (EYFP) and a cyan GFP variant (ECFP) were fused to the COOH and $NH_2$ termini, respectively, of human MT-IIA (Pearce *et al.*, 2000) (FRET-MT) (Fig. 1). An E1- and E3-deleted replication-deficient adenoviral vector expressing this chimera was constructed, as previously described (St. Croix *et al.*, 2002), and cells were infected at a multiplicity of infection (MOI) of 50:1 for 24 h before imaging.

## FRET Detection

Cells were bathed in Hanks balanced salt solution (Gibco) containing 100 $\mu M$ EDTA to chelate trace amounts of metal ions in the buffer and prevent extracellular decomposition of *S*-nitrosothiol, and they were

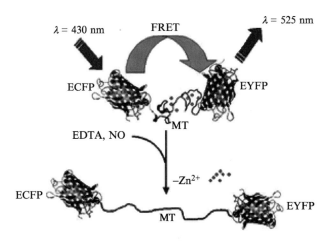

FIG. 1. Scheme of the fluorescence resonance energy transfer–metallothionein (FRET-MT) reporter construct of human type IIa MT, flanked by enhanced cyan fluorescent protein (ECFP) and enhanced yellow fluorescent protein (EYFP) showing unfolding, metal release, and consequent changes in FRET induced by metal chelators or nitric oxide (NO).

imaged at 37° using a thermocontrolled stage insert (Harvard Apparatus, Inc., Holliston, MA). Images were obtained with a 40× oil immersion optic at 512 × 512 pixels using the confocal-based Zeiss spectral imaging system (LSM510 META; Carl Zeiss, Jena, Germany). Cyan was excited at 458 nm (HFT 458). Resolved fluorescence spectra at each pixel were detected by an array of eight spectrally separate photomultiplier tube elements within the META detection head and recorded on a voxel-by-voxel basis during scanning to generate a set of images, each corresponding to the fluorescence wavelength resolved at 10-nm intervals. Color separation of cyan and yellow emission spectra was determined from the resolved image using a linear unmixing algorithm based on reference spectra obtained in cells expressing only cyan or yellow protein.

Changes in the emissions ratio of the acceptor (EYFP, ∼525 nm) to the donor (ECFP, ∼480 nm) were monitored after addition of S-nitrosocysteine ethyl ester. A relative decrease in the ratio was suggestive of conformational changes in the FRET reporter when MT was modified in such a way to lose metal.

### Acceptor Photobleaching

Selective acceptor photobleaching was performed to confirm the existence of FRET under baseline conditions. This technique is based on the principle that the constant or repetitive illumination of the acceptor (in this case, EYFP) leads to the irreversible photochemical destruction of the fluorophore. Accordingly, if FRET was occurring, the donor would become unquenched after bleaching of the acceptor, reflected by an increase in donor emission. A single cell within a field was chosen, and EYFP was photobleached using the 514-nm line (at full power) of the argon laser. FRET transfer efficiency was calculated as a percentage using the formula $E_{FRET} = (I_1 - I_0) \times 100/I_1$, where $I_0$ and $I_1$ represent ECFP intensity at baseline and after photobleaching, respectively (Karpova $et\ al.$, 2003). The same calculation was performed on nonbleached cells within the same field.

### Zinc Detection

Cells were incubated with 5 $\mu M$ FluoZin-3 (Molecular Probes, Eugene, OR) for 20 min at 37°. Temperature was maintained at 37° during imaging using a thermocontrolled stage insert (Bioptechs, Inc., Butler, PA). Images were obtained using a Nikon TE2000E equipped with a 40× oil-immersion objective, Lambda DG4 wavelength switcher, and xenon light source (Sutter Instrument, Novato, CA), charge-couple device camera (Cool-SNAP HQ, Photometrics, Tucson, AZ), and Metamorph software

(Universal Imaging Corp., Downingtown, PA). FluoZin-3 was illuminated at 480 nm, and emitted fluorescence was filtered through an HQ510LP emission filter (Chroma Technology Corp., Rockingham, VT). Cells were exposed to L-SNCEE, and time-dependent changes in fluorescence were monitored as an index of labile $Zn^{2+}$. Cells were then exposed to the zinc-specific chelator $N,N,N',N'$-tetrakis-(2-pyridylmethyl)ethylenediamine (TPEN; Sigma).

## Preparation of Reagents

The hydrochloride of $S$-nitroso-L-cysteine ethyl ester (SNCEE·HCl) was prepared via direct $S$-nitrosation of the hydrochloride of L-cysteine ethyl ester with ethyl nitrite, as previously described (Clancy *et al.*, 2001).

Human hemoglobin (Sigma-Aldrich) was dissolved in Hanks balanced salt solution (Gibco/BRL) and reduced using a 10-fold excess of sodium ascorbate (Sigma-Aldrich). Oxyhemoglobin ($HbO_2$) was purified by passing the solution over a Sephadex G-25 column, and the concentration of $HbO_2$ was measured at an absorbance of 415 nm.

Final concentrations of 100 $\mu M$ $HbO_2$ were used in the described experiments.

## Results and Discussion

Selective acceptor photobleaching was performed as a control, confirming that FRET occurs under baseline conditions with the FRET-MT reporter construct. The images of two endothelial cells expressing FRET-MT (shown in Fig. 1) have undergone linear unmixing using calibration spectra for ECFP and EYFP and, therefore, represent the separate contributions of the cyan and yellow emissions (Fig. 2). The spectral report shows an increase in the peak emission intensity of the donor (ECFP, $\sim$485 nm) and a decrease in that of the acceptor (EYFP, $\sim$525 nm) in the single cell selected for photobleaching. The estimates for FRET efficiency were 24% in the bleached cell and 0.1% in the control cell.

We have previously shown that the FRET-MT construct is sensitive to various NO donors and endothelial NO synthase (eNOS)–derived NO (Pearce *et al.*, 2000; St. Croix *et al.*, 2002, 2004). We have now extended these findings using the spectral imaging approach for FRET detection, in combination with a membrane permeant $S$-nitrosothiol, $S$-nitrosocysteine ethyl ester (SNCEE). The spectral report shown in Fig. 3 demonstrates that SNCEE induced changes in the FRET-MT reporter, as evidenced by an increase in the peak emission intensity of the donor (cyan, $\sim$485 nm) and a decrease in that of the acceptor (yellow, $\sim$525 nm). The bioactivities of

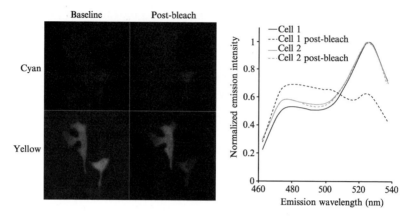

Fig. 2. Confirmation of energy transfer using acceptor photobleaching of the fluorescence resonance energy transfer–metallothionein (FRET-MT) construct. Lung endothelial cells were infected with an adenoviral vector encoding the fluorescent FRET-MT reporter molecule and were imaged 24 h later. FRET was detected in real time, using full spectral confocal imaging. The images show the separation of the two emitted signals (cyan and yellow) following spectral unmixing based on individual calibration spectra for each protein. The graph shows the spectral report provided by the Zeiss software. After selective photobleaching of cell 1, the donor (cyan) was unquenched, resulting in a pronounced increase in the peak emission intensity ($\sim$485 nm) indicative of positive FRET. In contrast, there were no changes in the emission intensity of the unbleached cell 2. (See color insert.)

low-molecular-weight $S$-nitrosothiols may be mediated either by their decomposition to $\cdot$NO or by direct exchange of their $-$NO function with thiol-containing proteins (transnitrosation) (Stamler *et al.*, 1992). The SNCEE-induced changes in intramolecular FRET were reversed by 50 $\mu M$ DTT. Such reversibility is indicative of nitrosothiol-mediated modification of a cysteine residue in MT. Furthermore, the effects of SNCEE on the FRET-MT reporter were not affected by $HbO_2$, which would be expected to block any process involving $^{\bullet}$NO liberated by decomposition of nitrosothiol but would not necessarily affect transnitrosation. In a separate report (St. Croix *et al.*, 2004), we presented additional data showing that in contrast to the effects of the L-isoform of SNCEE on FRET-MT, there was no significant change in FRET in response to equimolar concentrations of D-SNCEE. The preferential effects of $S$-nitroso-L-cysteine on FRET-MT might suggest that the structure of the L-stereoisomer of SNCEE confers access to critical cysteine residues in the MT protein and that the observed changes in energy transfer are via direct transnitrosation of MT.

Consistent with the FRET-based observations suggestive of conformational changes in MT, we used live cell imaging of the $Zn^{2+}$-sensitive

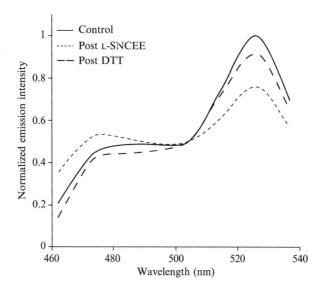

FIG. 3. The spectral report for a single endothelial cell showing L-SNCEE–induced conformational changes in the fluorescence resonance energy transfer–metallothionein (FRET-MT) construct as shown by a decrease in energy transfer with an increase in the peak emission intensity of the donor (cyan ~485 nm) and a decrease in that of the acceptor (yellow ~525 nm). These effects were found to be reversible by dithiothreitol (DTT).

fluorophore, FluoZin-3 (Fig. 4) to show that SNCEE also induced time-dependent increases in intracellular zinc that were reversed by the $Zn^{2+}$-specific chelator, $N,N,N',N'$-tetrakis-(2-pyridylmethyl)ethylenediamine) (TPEN). NO has been shown to increase labile zinc in a number of cell types (Berenjii et al., 1997; Cuajungco and Lee, 1997; Kroncke et al., 1999; Spahl et al., 2003; St. Croix et al., 2002). The absence of an NO-mediated increase in intracellular $Zn^{2+}$ in MT-null cells (Spahl et al., 2003; St. Croix et al., 2002), and the restoration of this response following MT complementation by adenoviral gene transfer, implies a critical role for MT in $Zn^{2+}$ homeostasis (St. Croix et al., 2002).

We showed (St. Croix et al., 2004) that a second FRET reporter cygnet-2 (kindly donated by Roger Y. Tsien, UC San Diego), which detects changes in the activation of guanylyl cyclase indirectly via structural changes in cGMP-dependent PKG Iα (Ting, 2001), was sensitive to both isoforms of SNCEE, likely via the formation of an Fe-nitrosyl heme complex. Furthermore, in contrast to FRET-MT, the scavenging of extracellular •NO by $HbO_2$ blocked the L-SNCEE–mediated effects on the cygnet-2 reporter. The different effects of SNCEE on the two FRET reporters could potentially distinguish between signaling events that are

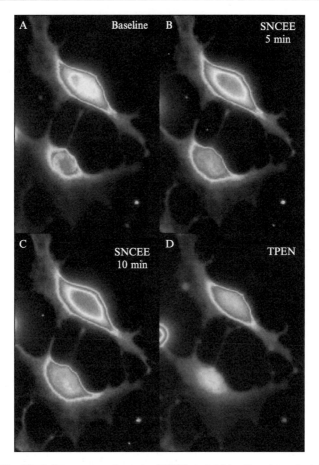

FIG. 4. FluoZin-3 fluorescence showed a TPEN-chelatable, time-dependent increase in response to application of L-SNCEE (50 $\mu M$), reaching a plateau at 10 min with a final mean increase of 88 ± 16% (n = 3). (See color insert.)

mediated by the free radical $^{\bullet}NO$, as is best exemplified by the activation of soluble guanylyl cyclase (sGC), and direct transnitrosation of cysteine thiols.

## Concluding Remarks

FRET between GFP mutants offers a general mechanism to build genetically encoded indicators and to monitor dynamic molecular interactions in living systems. This chapter describes the use of spectral confocal–based imaging for FRET detection, which is capable of providing full spectral information for each voxel of the fluorescent image, thereby obviating problems of signal bleed-through and channel crosstalk, and permitting detection of small but potentially biologically meaningful changes in FRET that are extremely difficult to resolve reliably using the more traditional methods relying on band-pass filters. We found that FRET detection using this spectral imaging approach is a highly effective technique for detecting posttranslation protein modifications induced by NO-related species in live cells.

## References

Berendji, D., Kolb-Bachofen, V., Meyer, K. L., Grapenthin, O., Weber, H., Wahn, V., and Kroncke, K. D. (1997). Nitric oxide mediates intracytoplasmic and intranuclear zinc release. *FEBS Lett.* **405**, 37–41.

Clancy, R., Cederbaum, A. I., and Stoyanovsky, D. A. (2001). Preparation and properties of *S*-nitroso-L-cysteine ethyl ester, an intracellular nitrosating agent. *J. Med. Chem.* **44**, 2035–2038.

Cuajungco, M. P., and Lees, G. J. (1997). Zinc metabolism in the brain: Relevance to human neurodegenerative disorders. *Neurobiol. Dis.* **4**, 137–169.

Honda, A., Adams, S. R., Sawyer, C. L., Lev-Ram, V., Tsien, R. Y., and Dostmann, W. R. (2001). Spatiotemporal dynamics of guanosine $3',5'$-cyclic monophosphate revealed by a genetically encoded, fluorescent indicator. *Proc. Natl. Acad. Sci. USA* **98**, 2437–2442.

Hoyt, D. G., Mannix, R. J., Rusnak, J. M., Pitt, B. R., and Lazo, J. S. (1995). Collagen is a survival factor against LPS-induced apoptosis in cultured sheep pulmonary artery endothelial cells. *Am. J. Physiol.* **269**, L171–L177.

Karpova, T. S., Baumann, C. T., He, L., Wu, X., Grammer, A., Lipsky, P., Hager, G. L., and McNally, J. G. (2003). Fluorescence resonance energy transfer from cyan to yellow fluorescent protein detected by acceptor photobleaching using confocal microscopy and a single laser. *J. Microsc.* **209**, 56–70.

Kawai, K., Liu, S. X., Tyurin, V. A., Tyurina, Y. Y., Borisenko, G. G., Jiang, J. F., St. Croix, C. M., Fabisiak, J. P., Pitt, B. R., and Kagan, V. E. (2000). Antioxidant and antiapoptotic function of metallothioneins in HL-60 cells challenged with copper nitrilotriacetate. *Chem. Res. Toxicol.* **13**, 1275–1286.

Kennedy, M. C., Gan, T., Antholine, W. E., and Petering, D. H. (1993). Metallothionein reacts with $Fe^{2+}$ and NO to form products with a $g = 2.039$ ESR signal. *Biochem. Biophys. Res. Comm.* **196**, 632–636.

Kroncke, K. D., Fchsel, K., Schmidt, T., Zenke, F. T., Dasting, I., Wesener, J. R., Bettermann, H., Breunig, K. D., and Kolb-Bachofen, V. (1994). Nitric oxide destroys zinc-sulfur clusters inducing zinc release from metallothionein and inhibition of the zinc finger-type yeast transcription activator LAC9. *Biochem. Biophys. Res. Comm.* **200**, 1105–1110.

Kroncke, K.-D., and Kolb-Bachofen, V. (1999). Measurement of nitric oxide-mediated effects on zinc homeostasis and zinc finger transcription factors. *Methods Enzymol.* **301**, 26–135.

Misra, R. R., Hochadel, J. F., Smith, G. T., Cook, J. C., Waalkes, M. P., and Wink, D. A. (1996). Evidence that nitric oxide enhances cadmium toxicity by displacing the metal from metallothionein. *Chem. Res. Toxicol.* **9**, 326–332.

Miyawaki, A., Llopis, J., Heim, R., McCaffery, J. M., Adams, J. A., Ikura, M., and Tsien, R. Y. (1997). Fluorescent indicators for $Ca^{2+}$ based on green fluorescent proteins and calmodulin. *Nature* **388**, 882–887.

Pearce, L. L., Gandley, R. E., Han, W., Wasserloos, K., Stitt, M., Kanai, A. J., McLaughlin, M. K., Pitt, B. R., and Levitan, E. S. (2000). Role of metallothionein in nitric oxide signaling as revealed by a green fluorescent fusion protein. *Proc. Natl. Acad. Sci. USA* **97**, 477–482.

Schwarz, M. A., Lazo, J. S., Yalowich, J. C., Allen, W. P., Whitmore, M., Bergonia, H. A., Tzeng, E., Billiar, T. R., Robbins, P. D., Lancaster, J. R., Jr., and Pitt, B. R. (1995). Cytoplasmic metallothionein overexpression protects NIH 3T3 cells from tert-butyl hydroperoxide toxicity. *Proc. Natl. Acad. Sci. USA* **92**, 4452–4456.

Spahl, D. U., Berendji-Grun, D., Suschek, C. V., Kolb-Bachofen, V., and Kroncke, K. D. (2003). Regulation of zinc homeostasis by inducible NO synthase-derived NO: Nuclear metallothionein translocation and intranuclear $Zn^{2+}$ release. *Proc. Natl. Acad. Sci. USA* **100**, 13952–13957.

St. Croix, C. M., Wasserloos, K. J., Dineley, K. E., Reynolds, I. J., Levitan, E. S., and Pitt, B. R. (2002). Nitric oxide–induced changes in intracellular zinc homeostasis are modulated by metallothionein/thionein. *Am. J. Physiol. Lung Cell Mol. Physiol.* **282**, L185–L192.

St. Croix, C. M., Stitt, M. S., Leelavanichkul, K., Wasserloos, K. J., Pitt, B. R., and Watkins, S. C. (2004). Nitric oxide mediated signaling in endothelial cells as determined by spectral fluorescence resonance energy transfer. *Free Radic. Biol. Med.* **37**, 785–792.

Stamler, J. S., Singel, D. J., and Loscalzo, J. (1992). Biochemistry of nitric oxide and its redox-activated forms. *Science* **258**, 1898–1902.

Ting, A. Y., Kain, K. H., Klemke, R. L., and Tsien, R. Y. (2001). Genetically encoded fluorescent reporters of protein tyrosine kinase activities in living cells. *Proc. Natl. Acad. Sci. USA* **98**, 15003–15008.

Zaccolo, M., De Giorgi, F., Cho, C. Y., Feng, L., Knapp, T., Negulescu, P. A., Taylor, S. S., Tsien, R. Y., Pozzan, T. (2000). A genetically encoded, fluorescent indicator for cyclic AMP in living cells. *Nat. Cell Biol.* **2**, 25–29.

# [27] Nitric Oxide Is a Signaling Molecule that Regulates Gene Expression

*By* Lorne J. Hofseth, Ana I. Robles,
Michael G. Espey, and Curtis C. Harris

## Abstract

Nitric oxide (NO) is a dynamic and bioreactive molecule that can both participate in and inhibit the genesis of disease. Its ability to have an impact on a wide range of physiological events stems from its capacity to reversibly alter the expression of specific genes and the activities of a wide range of proteins and signaling pathways. Yet, NO• remains an enigmatic molecule. Recently developed technologies, including gene-chips, two-dimensional electrophoresis, RNA interference, matrix-assisted laser desorption ionization (MALDI)-TOF (time-of-flight) mass spectrometry, and protein arrays will allow us to better understand how NO• and associated reactive

METHODS IN ENZYMOLOGY, VOL. 396
0076-6879/05 $35.00
DOI: 10.1016/S0076-6879(05)96027-8

nitrogen species (RNS) regulate both physiology and disease states, toward the development of treatments using NO• synthase inhibitors or NO• donors.

## Introduction

In 1998, Drs. Ignarro, Furchgott, and Murad received the Nobel Prize in physiology or medicine for their discoveries concerning NO• as a signaling molecule in the cardiovascular system. An explosion of studies followed that have revealed insight into the many roles of NO• in normal and disease processes. A characteristic of NO• is that it can act as both a "friend" and a "foe." NO• has been considered in the treatment of cardiovascular and respiratory diseases, cancer, and erectile dysfunction. However, nitrosative stress can also play a role in the etiology and progression of many diseases, including cardiovascular disease, cancer, asthma, neurodegeneration, obesity, and diabetes. NO• has these many faces because the specific effects of NO• exposure change with NO• levels, the microenvironment in which it acts (including the presence of other free radicals), and the genetic background of the host/tissue (Espey et al., 2000a; Hofseth et al., 2003a; Thomas et al., 2004; Wink and Mitchell, 1998). Each of these factors will dictate the impact of NO• on cell signaling, gene expression, and disease outcome.

## Nitric Oxide Biochemistry

NO• is endogenously formed by a family of enzymes called NO• synthases (NOSs), which use L-arginine as a substrate and molecular oxygen and NADPH as cofactors (Alderton et al., 2001; Marletta, 1994). Because NOS enzymes are responsible for triggering NO signaling cascades, an elaborate network of control mechanisms is available to regulate NOS levels and activity. NOSs are controlled at the transcriptional, posttranscriptional, and translational levels, as well as through posttranslational modifications, protein–protein interactions, and subcellular localization.

NOSs are active as homodimers, and there are four isoforms. Two are $Ca^{2+}$-dependent [NOS1 (neuronal NOS) and NOS3 (endothelial NOS)] and constitutively expressed, whereas the $Ca^{2+}$-independent isoform (NOS2 or iNOS) requires induction. There are, however, some exceptions to this general scheme, as NOS1 and 3 have been shown to be inducible (Forstermann et al., 1995), and iNOS is expressed constitutively in some tissues [e.g., bronchus and ileum (Guo et al., 1995; Hoffman et al., 1997)]. $Ca^{2+}$-dependent neuronal (NOS1) and endothelial isoforms (NOS3) produce low levels of NO• that range from picomolar to nanomolar concentrations. In contrast, iNOS produces a sustained NO• concentration in the micromolar range (Beckman *et al.*, 1990; Espey *et al.*, 2000b; Malinski *et al.*, 1993). The fourth isoform, mitochondrial NOS (mtNOS), has only recently been defined (Elfering *et al.*, 2002; Haynes *et al.*, 2003; Lacza, 2003; Nisoli, 2003). It appears that mtNOS could be the alpha form of NOS1, but with posttranslational modifications (Elfering *et al.*, 2002). Its roles in physiology and pathology have been reviewed (Haynes *et al.*, 2003).

## Nitric Oxide Signaling

NO• is an efficient signaling molecule because (1) it can affect a wide range of cells in its vicinity; (2) it can react with and reversibly generate other free radicals [collectively known as *reactive nitrogen species* (RNS)] that have their own set of bioreactive properties; (3) it, along with its byproducts (RNS), can reversibly and irreversibly posttranslationally modify proteins and lipids; and as evidence suggests here, (4) it has the ability to alter gene transcription in a discriminatory fashion.

Because of their ubiquitous nature, RNS are also unlikely to interact with a single signaling pathway. Although this latter observation means that RNS can influence the expression of many genes (detailed later in this chapter), the pattern of changes is very complicated and often depends on the levels, the microenvironment, and the genetic background of the cell, tissue, or host. The outcome, then, comprises a wide range of physiological and pathological effects observed for RNS that differ, depending on the experimental system (Espey *et al.*, 2001, 2002b; Thomas *et al.*, 2002; Wink and Mitchell, 1998).

RNS signaling can broadly be divided into cyclic guanosine monophosphate (cGMP)–dependent and –independent pathways (Fig. 1). In cGMP-dependent pathways, small fluxes of RNS can bind to the heme iron of soluble guanylate cyclase (sGC) and activate it, which stimulates cGMP formation (Hobbs and Ignarro, 1996; Ignarro, 1992; Stamler, 1994). cGMP then targets a series of proteins ("effector proteins"), including cGMP-dependent protein kinases (PKG) I and II (resulting in substrate phosphorylation), cyclic nucleotide–regulated ion channels (resulting in increased ion flux), and phosphodiesterases [resulting in cGMP and/or cyclic adenosine monophosphate (cAMP) hydrolysis]. The regulation of gene expression by cGMP has been detailed elsewhere (Pilz and Casteel, 2003) and, therefore, will not be covered here.

cGMP-independent pathways are generally mediated by larger fluxes of RNS (Wink and Mitchell, 1998), which may damage DNA directly and reversibly or irreversibly damage posttranslationally modified proteins (Beckman, 1996) and lipids (Baker *et al.*, 2004), leading to a dynamic

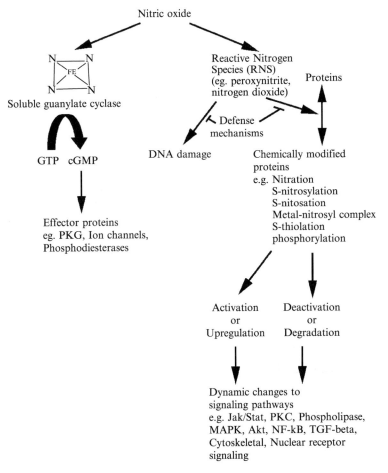

FIG. 1. Nitric oxide signaling pathways.

change in the activity of signaling pathways. Depending on the conditions used, different RNSs (e.g., nitrosoperoxycarbonate, peroxynitrite, and nitrogen dioxide) may be formed. For example, peroxynitrite ($ONOO^-$) may form via the reaction between $NO^\bullet$ and superoxide (Beckman and Koppenol, 1996). $ONOO^-$ can oxidize protein sulfhydryl groups (DeMaster *et al.*, 1995; Kuhn and Geddes, 1999; Radi *et al.*, 1991) and nitrate peptides at tyrosine residues (Beckman, 1996). The presence of nitration is not necessarily indicative of the intermediacy of $ONOO^-$. Nitration can also occur through the oxidation of nitrite, which forms nitrogen dioxide ($NO_2$). A comparison between $ONOO^-$ and $NO_2$ showed the latter was a much more efficient effector species (Espey *et al.*, 2002a). $ONOO^-$ formation requires maintenance of a precise balance between $NO^\bullet$ and superoxide ($O_2^-$), whereas nitrite oxidation can occur through various pathways without such spatial and temporal limitations. Because nitration has been shown to influence protein activity (Kuhn and Geddes, 1999; Schopfer *et al.*, 2003), the nitrotyrosine antibodies have been useful as a biomarker of nitrosative stress. Nitrosative stress has been shown to occur in many conditions such as chronic inflammation, cancer, and cardiovascular diseases (Maeda and Akaike, 1998; Turko and Murad, 2002).

In addition to nitration, a potential key signaling mechanism is the formation of $NO^\bullet$ adducts on protein nucleophiles (such as thiols, amines, and alcohols). This may occur via nitrosation by the RNS dinitrogen trioxide ($N_2O_3$) formed during the autoxidation of $NO^\bullet$ (Espey *et al.*, 2001). Alternatively, $NO^\bullet$ adducts may form through oxidative nitrosylation (Espey *et al.*, 2002b). Improvements in the analysis of nitrosated residues suggest that they may play a more ubiquitous role in physiology than previously thought (Bryan *et al.*, 2004).

A key quality of RNS in cell signaling is their ability to both irreversibly and reversibly modify the structure and activity of key signaling proteins (see abbreviated list in Table I). Sometimes this modification results in protein activation, and sometimes it results in protein deactivation. For example, RNS have been shown to both inhibit (Sommer *et al.*, 2002) and activate phosphatases, kinases, and other signaling proteins, including transcription factors and repair proteins (Chen and Wang, 2004; Hofseth *et al.*, 2003b; Klatt *et al.*, 1999; Kibbe *et al.*, 2000; Laval and Wink, 1994; Lin *et al.*, 2003; Mateo *et al.*, 2003; Namkung-Matthai *et al.*, 2000; Reynaert *et al.*, 2004; Sandau *et al.*, 2001; So *et al.*, 1998; Wink and Laval, 1994; Zhou *et al.*, 2000). Although it is unclear why a particular protein is activated while another is deactivated after posttranslational modification, the types of modifications and the mechanisms of these changes are beginning to be understood. For example, $NO^\bullet$ can donate electrons and react with iron, copper, and zinc, leading to the formation of metal-nitrosyl complexes (Ford and Lorkovic, 2002). Because many enzymes, transcription factors, and other proteins (e.g., iron of the heme moiety of sGC and hemoglobin; enzyme sulfide clusters; zinc-finger proteins) have a transition metal component, the interaction of $NO^\bullet$ with transition metals is fundamental to $NO^\bullet$ signaling. Some messenger RNA (mRNA) binding proteins also regulated by transition metals can be functionally modulated by RNS.

TABLE I
SELECTED MOLECULES THAT INTERACT WITH NITRIC OXIDE IN CANCER, CARDIOVASCULAR
DISEASE, AND NEURONAL PATHOPHYSIOLOGY

| Cancer | Cardiovascular disease | Neurotransmission/ neurodegeneration |
|---|---|---|
| Heme oxygenase-1 | Heme oxygenase-1 | Heme oxygenase-1 |
| HIF-1$\alpha$ | HIF-1$\alpha$ | Parkin |
| Estrogen receptor | Estrogen receptor | Estrogen receptor |
| VEGF | Beta adrenergic receptors | Antioxidant enzymes |
| Kinases | PPAR$\gamma$ | $\beta$-amyloid |
| Phosphatases | G-protein–coupled receptors | Tau protein |
| Bax | Kinases | Synaptophysin |
| p53 | Phosphatases | GADPH |
| Caspases | Actin | Acetylcholine |
| PPAR$\gamma$ | Light chain myosin | Dopamine |
| Microtubules | Fibrinogen | Myelin |
| Histones | Kinases | Presenilin |
| Antioxidant enzymes | Antioxidant enzymes | Huntington disease protein $\alpha$-synuclein |

Other modifications resulting in either loss or gain of protein function include nitration of tryptophan and tyrosine side-chains (by ONOO$^-$ or NO$_2$, described earlier). Key proteins shown to be nitrated include albumin, Cu,Zn superoxide dismutase (SOD), cytochrome P450, iNOS, histone, $\alpha$-tubulin, and actin (Schopfer *et al.*, 2003). One must take caution, however, in evaluating the significance of RNS adducts formed when proteins are exposed outside the context of the intact cell or tissue. Using two-dimensional electrophoresis and MALDI-TOF mass spectrometry, Aulak *et al.* (2001) showed approximately 31 proteins nitrated in lipopolysaccharide (LPS)-treated rats, including many proteins not previously recognized as heavily nitrated by RNS, providing evidence for the presence of extensive protein nitration *in vivo*.

S-nitrosylation (incorporation of a NO$^\bullet$ group in cysteine thiols), S-nitrosation, methionine sulfoxidation (oxidation of methionine residues), carbonyl formation at lysine and arginine residues, dityrosine formation, formation of disulfide bonds, and S-thiolation (S-glutathionylation; protein mixed disulfide formation between cysteine and GSH) are other posttranslational modifications that play a central role in cell signaling (Klatt and Lamas, 2000). Because of their ability to activate and deactivate both kinases and phosphatases, RNS can influence the phosphorylation status of key phosphoproteins such as p53 (Hofseth et al., 2003b; Schneiderhan et al., 2003; Thomas et al., 2004) and pRb (Hofseth et al., unpublished observation; Radisavljevic, 2004). pRb and p53 are cancer suppressor proteins, so this observation is critical to the understanding of RNS in carcinogenesis.

Until recently, we have only been able to study the influence of RNS on a protein-by-protein basis and test a limited amount of that protein's function. With the development of high-throughput technologies such as RNA interference (RNAi), DNA chips, two-dimensional electrophoresis, MALDI-TOF mass spectrometry, protein arrays, and other techniques, we now have the ability to study the expression, modifications, and binding properties of thousands of genes, or the proteins they encode, following RNS exposure.

## Nitric Oxide and Gene Expression

When released from inflammatory cells or produced endogenously by epithelial, endothelial, or neuronal cells, NO$^{\bullet}$ targets a large protein pool. Until the development and use of DNA microarrays, it had been difficult to broadly study the gene expression changes associated with RNS exposure. This technological breakthrough, combined with the more recent development of protein chips, MALDI-TOF, and RNAi, offers a unique opportunity to uncover pathways altered by RNS that may affect cellular phenotype. At present, relatively few studies in the NO$^{\bullet}$ field have reported use of these technologies.

A survey of the peer-reviewed literature yields four studies looking at the effects of RNS on global gene expression in bacteria or plant systems (Firoved et al., 2004; Huang et al., 2002; Ohno et al., 2003; Polverari et al., 2003), three* studies using mammalian cells exposed to RNS (Hemish et al., 2003; Li et al., 2004; Zamora et al., 2002), and two studies that examine the effects of RNS on gene expression in intact rats (Wang et al., 2002) or mice (Okamoto et al., 2004). Although it has long been known that there is an extensive list of individual proteins affected by RNS (abbreviated list in Table I), as will be shown, a key observation arising from microarray studies is that RNS have discriminative properties.

For this review, we wanted to tease out trends in gene expression profile arising from the three published data sets using cultured mammalian cells. Although all three studies used different microarray platforms for analysis, we can generalize a few interesting observations. First, there is a consistently

---

* After this paper went to press, two additional references appeared in the literature examining the influence of nitric oxide on gene expression in mammalian cells (Li et al., 2004; Turpaev et al., 2005).

low percentage of genes that changed significantly in expression after exposure to RNS. Hemish *et al.* (2003) found 560 of approximately 10,000 genes (5%) to be significantly changed (upregulated or downregulated) after exposure of mouse NIH-3T3 cells to NO$^\bullet$ (250 $\mu M$ of the NO$^\bullet$ donor, SNAP). Zamora *et al.* (2002) found 200 of approximately 6500 genes (3%) significantly changed (upregulated or downregulated) after exposure of mouse hepatocytes to NO$^\bullet$ (infection with adenovirus expressing human iNOS). We have characterized p53-dependent apoptotic signaling associated with NO$^\bullet$ exposure (Li *et al.*, 2004). Overall, we found 358 (3.9%) of 9180 genes significantly changed (upregulated or downregulated) in TK6 (p53 wild-type) human lymphoblastoid cells exposed to NO$^\bullet$ (390 $\mu$mol pure NO$^\bullet$ gas).

Second, even though at first glance there is very limited overlap in individual gene expression changes, a more detailed analysis reveals that RNS exposure modulates similar signal transduction pathways in all three experimental systems. Some of the conserved changes include alteration of expression of kinase and phosphatases, heat shock proteins, cyclins, zinc-finger proteins, transcription factors, cell energy (ATP)–related proteins, apoptosis, serine proteinase inhibitors, and members of the ubiquitin family (data not shown).

Third, more extensive overlap is found within the same species and type of exposure. Direct comparison of genes significantly changed by RNS exposure in the two studies examining murine cells (Hemish *et al.*, 2003; Zamora *et al.*, 2002) yields 12 genes in common (Table II). Upon methodically searching for mouse genes with homology to human genes significantly changed in our study (Li *et al.*, 2004) using HomoloGene (*http://www. ncbi.nlm.nih.gov/entrez/query.fcgi*), only nine genes were found to overlap with those of Hemish *et al.* (2003), even though both used exogenous NO$^\bullet$ as the exposure route (Table III). Again, some of the conserved changes include alteration of kinase and phosphatase expression, heat shock proteins, cyclins, zinc-finger proteins, transcription factors, cell energy (ATP)–related proteins, apoptosis, serine proteinase inhibitors, and members of the ubiquitin family. Additionally, there was commonality in extracellular matrix proteins (integrins) that were affected by RNS exposure in both mouse and human cells.

Perhaps not surprisingly, the least amount of overlap was found between our study (Li *et al.*, 2004) and the published data set from Zamora *et al.* (2002), which differ in both species and route of exposure. There were only three genes in common: ribonucleotide reductase M1 polypeptide, proliferating cell nuclear antigen (PCNA), and heme oxygenase-1 (HO-1).

As shown in Tables II and III, the only two proteins with significantly changed expression in all three studies using mammalian cells exposed to

TABLE II
GENES SIGNIFICANTLY CHANGED IN BOTH PUBLISHED STUDIES EXAMINING THE
EFFECTS OF RNS ON MURINE CELL GENE EXPRESSION[a]

| Description | Hemish et al. | Zamora et al. |
|---|---|---|
| Proliferating cell nuclear antigen | Increase (1.3-fold at 12 h) | Increase (5.9-fold) |
| Heme oxygenase-1 | Increase (2.4-fold at 12 h) | Increase (6.2-fold) |
| Macrophage migration inhibitory factor | Increase (1.9-fold at 16 h) | Increase (2-fold) |
| Thymidylate synthase | Increase (max 1.3-fold at 0.5 h) and decrease (max 2.6-fold at 0.75 h) | Increase (4.2-fold) |
| CD81 antigen | Increase (1.5-fold at 16 h)[b] | Decrease (40-fold) |
| Adrenomedullin | Increase (1.4-fold at 8 h) | Increase (3.1-fold) |
| Cystein and glycine-rich protein 1 | Increase (1.5-fold at 4 h) | Increase (5.2-fold) |
| Eukaryotic translation initiation factor 3 | Decrease (1.5-fold at 48 h) | Decrease (6.5-fold) |
| Glutamate-cysteine ligase, catalytic subunit | Increase (1.8-fold at 12 h) and decrease (1.2-fold at 48 h) | Decrease (6-fold) |
| Keratin 19 | Increase (1.4-fold at 4 h) | Increase (4.2-fold) |
| Ferritin light chain 1 | Increase (1.5-fold at 16 h) | Decrease (2.8-fold) |
| Adaptor protein complex AP-1, $\beta_1$-subunit | Decrease (1.14-fold at 12 h) | Increase (2-fold) |

[a] Italicized text indicates that the gene was significantly changed in human TK6 cells exposed to NO• gas (Li et al., 2004).
[b] Initial slight decrease (1.2-fold) at 0.75 h.

NO• were PCNA and HO-1. Both of these genes were increased after exposure in every case. PCNA was first recognized in 1981 as a nuclear antigen associated with cell proliferation and blast transformation in the sera of patients with systemic lupus erythematosus (Takasaki et al., 1981). First used as a proliferation marker, it has now been shown that PCNA functions as a sliding clamp protein of the DNA polymerase complex, is intimately involved in DNA repair, and is an executive molecule controlling critical cellular decision pathways (Matunis, 2002; Paunesku et al., 2001).

TABLE III

GENES SIGNIFICANTLY CHANGED IN BOTH PUBLISHED STUDIES EXAMINING THE EFFECTS OF
EXOGENOUS RNS ON MAMMALIAN CELL GENE EXPRESSION[a]

| Description | Hemish et al. | Li et al. |
|---|---|---|
| Heme oxygenase-1 | Increase (2.4-fold at 12 h) | Increase (2.1-fold at 24 h) |
| Proliferating cell nuclear antigen | Increase (1.3-fold at 12 h) | Increase (2.3-fold at 24 h) |
| Hypoxia inducible factor 1-alpha | Increase (1.3-fold at 8 h) | Increase (1.9-fold at 12 h) |
| Cyclin E2 | Increase (1.6-fold at 12 h) | Increase (2.2-fold at 24 h) |
| Solute carrier family 3 (activators of dibasic and neutral amino acid transport) member 2 | Increase (1.6-fold at 12 h) | Increase (1.8-fold at 24 h) |
| [b]GADD45-$\alpha$ | Increase (8-fold at 16 h) | Increase (3.2-fold at 12 h) |
| [c]p21$^{Cip1/Waf1}$ | Increase (11-fold at 16 h) | Increase (6-fold at 24 h) |
| [d]MDM2 | Increase (8-fold at 16 h) | Increase (3.6-fold at 24 h) |
| Tumor necrosis factor receptor superfamily, member 6 | Decrease (2.2-fold at 8 h) | Increase (3.3-fold at 24 h) |

[a] Italicized text indicates that the gene was significantly changed in mouse hepatocytes overexpressing human inducible nitric oxide synthase (Zamora et al., 2002).
[b] Induced also in a study by Li et al. (2004).
[c] Induced also in studies by Li et al. (2004) and Turpaev et al. (2005).
[d] Induded also in a study by Li et al. (2004).

HO-1 has previously been found to be induced by oxidative or nitrosative stress, cytokines, and other mediators produced during inflammatory processes (Bouton and Demple, 2000; Terry et al., 1999; Vile et al., 1994). HO-1 is involved in the regulatory defense system that leads to the production of mediators that modulate the inflammatory response. For example, it controls intracellular levels of "free" heme (a prooxidant), produces biliverdin (an antioxidant), improves nutritive perfusion via CO release, and fosters the synthesis of the Fe-binding protein ferritin (Bauer and Bauer, 2002). HO-1 activity results in the inhibition of oxidative damage and apoptosis, with significant reductions in inflammatory events (Alcaraz et al., 2003).

An obvious reason for the low overlap in individual gene expression affected by RNS from study to study is the use of different microarray platforms, which can lead to gene representation bias. However, more relevant physiological reasons for the lack of consistency can reflect previous observations that the response to RNS depends on the microenvironment, the genetic background, and the type and level of RNS exposure (Hofseth et al., 2003a).

Such conditions were also different in the three published studies we examined here. It would be interesting to compare the response of the same cell type to different types of RNS and exposure methods or different mammalian cell types to the same RNS and exposure methods in carefully controlled experiments. Such studies have yet to be published, but we are continuing our efforts to delineate the response to RNS in human cells.

Two studies have examined the influence of $NO^{\bullet}$ on global gene expression in animals. Wang *et al.* (2002) examined cells from the thoracic aorta in rats receiving RNS through nitroglycerin infusion. They found that 447 of approximately (11.6%) 3531 genes were significantly changed (upregulated or downregulated). Okamoto *et al.* (2004) observed that 106 of (0.9%) 12,451 genes were significantly changed in lung tissues from $iNOS^{+/+}$ versus $iNOS^{-/-}$ mice exposed to LPS. Only one gene (aldehyde dehydrogenase) was consistently changed in association with RNS in both studies, which again underscores the observation that RNS affects physiology differently, based on the levels used, the surrounding microenvironment, and the genetic background of the tissue/host.

Conclusion

RNS are a ubiquitous and complicated group of free radicals that reversibly and irreversibly posttranslationally modify proteins and dynamically alter the activity of cell signaling pathways. With advances in genomics, and more recently in proteomics, we have the tools to better understand key genes expressed and pathways changed that lead to the plethora of phenotypes associated with RNS exposure. Choosing the appropriate system is critical to proper interpretation of study results. Careful attention has to be paid to the RNS used (e.g., an $NO^{\bullet}$ donor versus pure $NO^{\bullet}$ gas), the environment in which it is used (e.g., media properties or degree of hypoxia in the tissue examined), the levels at which it is used (e.g., low levels can inhibit apoptosis, whereas high levels can stimulate apoptosis), and the genetic background of the host (e.g., cell type and p53 status). Only when these systems are controlled can we compare genomic and proteomic results and perhaps better evaluate the usefulness of RNS or their inhibition in the treatment of cancer, cardiovascular and respiratory diseases, neurodegeneration, and diabetes, which are all among the major causes of death in humans.

References

Alderton, W. K., Cooper, C. E., and Knowles, R. G. (2001). Nitric oxide synthases: Structure, function and inhibition. *Biochem. J.* **357,** 593–615.

Alcaraz, M. J., Fernandez, P., and Guillen, M. I. (2003). Anti-inflammatory actions of the heme oxygenase-1 pathway. *Curr. Pharm. Des.* **9,** 2541–2551.

Aulak, K. S., Miyagi, M., Yan, L., West, K. A., Massillon, D., Crabb, J. W., and Stuehr, D. J. (2001). Proteomic method identifies proteins nitrated *in vivo* during inflammatory challenge. *Proc. Natl. Acad. Sci. USA* **98,** 12056–12061.

Baker, P. R., Schopfer, F. J., Sweeney, S., and Freeman, B. A. (2004). Red cell membrane and plasma linoleic acid nitration products: Synthesis, clinical identification, and quantitation. *Proc. Natl. Acad. Sci. USA* **101,** 11577–11582.

Bauer, M., and Bauer, I. (2002). Heme oxygenase-1: Redox regulation and role in the hepatic response to oxidative stress. *Antioxid. Redox Sig.* **4,** 749–758.

Beckman, J. S., Beckman, T. W., Chen, J., Marshall, P. A., and Freeman, B. A. (1990). Apparent hydroxyl radical production by peroxynitrite: Implications for endothelial injury from nitric oxide and superoxide. *Proc. Natl. Acad. Sci. USA* **87,** 1620–1624.

Beckman, J. S. (1996). Oxidative damage and tyrosine nitration from peroxynitrite. *Chem. Res. Toxicol.* **9,** 836–844.

Beckman, J. S., and Koppenol, W. H. (1996). Nitric oxide, superoxide, and peroxynitrite: The good, the bad, and ugly. *Am. J. Physiol.* **271,** C1424–C1437.

Bouton, C., and Demple, B. (2000). Nitric oxide–inducible expression of heme oxygenase-1 in human cells. Translation-independent stabilization of the mRNA and evidence for direct action of nitric oxide. *J. Biol. Chem.* **275,** 32688–32693.

Bryan, N. S., Rassaf, T., Maloney, R. E., Rodriguez, C. M., Saijo, F., Rodriguez, J. R., and Feelisch, M. (2004). Cellular targets and mechanisms of nitros(yl)ation: An insight into their nature and kinetics *in vivo*. *Proc. Natl. Acad. Sci. USA* **101,** 4308–4313.

Chen, H. H., and Wang, D. L. (2004). Nitric oxide inhibits matrix metalloproteinase-2 expression via the induction of activating transcription factor 3 in endothelial cells. *Mol. Pharmacol.* **65,** 1130–1140.

DeMaster, E. G., Quast, B. J., Redfern, B., and Nagasawa, H. T. (1995). Reaction of nitric oxide with the free sulfhydryl group of human serum albumin yields a sulfenic acid and nitrous oxide. *Biochemistry* **34,** 11494–11499.

Elfering, S. L., Sarkela, T. M., and Giulivi, C. (2002). Biochemistry of mitochondrial nitric-oxide synthase. *J. Biol. Chem.* **277,** 38079–38086.

Espey, M. G., Miranda, K. M., Feelisch, M., Fukuto, J., Grisham, M. B., Vitek, M. P., and Wink, D. A. (2000a). Mechanisms of cell death governed by the balance between nitrosative and oxidative stress. *Ann. NY Acad. Sci.* **899,** 209–221.

Espey, M. G., Miranda, K. M., Pluta, R. M., and Wink, D. A. (2000b). Nitrosative capacity of macrophages is dependent on nitric-oxide synthase induction signals. *J. Biol. Chem.* **14,** 11341–11347.

Espey, M. G., Miranda, K. M., Thomas, D. D., and Wink, D. A. (2001). Distinction between nitrosating mechanisms within human cells and aqueous solution. *J. Biol. Chem.* **276,** 30085–30091.

Espey, M. G., Xavier, S., Thomas, D. D., Miranda, K. M., and Wink, D. A. (2002a). Direct real-time evaluation of nitration with green fluorescent protein in solution and within human cells reveals the impact of nitrogen dioxide vs. peroxynitrite mechanisms. *Proc. Natl. Acad. Sci. USA* **99,** 3481–3486.

Espey, M. G., Thomas, D. D., Miranda, K. M., and Wink, D. A. (2002b). Focusing of nitric oxide mediated nitrosation and oxidative nitrosylation as a consequence of reaction with superoxide. *Proc. Natl. Acad. Sci. USA* **99,** 11127–11132.

Firoved, A. M., Wood, S. R., Ornatowski, W., Deretic, V., and Timmins, G. S. (2004). Microarray analysis and functional characterization of the nitrosative stress response in nonmucoid and mucoid *Pseudomonas aeruginosa*. *J. Bacteriol.* **186,** 4046–4050.

Ford, P. C., and Lorkovic, I. M. (2002). Mechanistic aspects of the reactions of nitric oxide with transition-metal complexes. *Chem. Rev.* **102,** 993–1018.

Forstermann, U., and Kleinert, H. (1995). Nitric oxide synthase: Expression and expressional control of the three isoforms. *Naunyn. Schmiedebergs. Arch. Pharmacol.* **352,** 351–364.

Guo, F. H., De Raeve, H. R., Rice, T. W., Stuehr, D. J., Thunnissen, F. B., and Erzurum, S. C. (1995). Continuous nitric oxide synthesis by inducible nitric oxide synthase in normal human airway epithelium *in vivo. Proc. Natl. Acad. Sci. USA* **92,** 7809–7813.

Haynes, V., Elfering, S. L., Squires, R. J., Traaseth, N., Solien, J., Ettl, A., and Giulivi, C. (2003). Mitochondrial nitric-oxide synthase: Role in pathophysiology. *IUBMB Life* **55,** 599–603.

Hemish, J., Nakaya, N., Mittal, V., and Enikolopov, G. (2003). Nitric oxide activates diverse signaling pathways to regulate gene expression. *J. Biol. Chem.* **278,** 42321–42329.

Hobbs, A. J., and Ignarro, L. J. (1996). Nitric oxide–cyclic GMP signal transduction system. *Methods Enzymol.* **269,** 134–148.

Hoffman, R. A., Zhang, G., Nussler, N. C., Gleixner, S. L., Ford., H. R., and Simmons, R. L. (1997). Constitutive expression of inducible nitric oxide synthase in the mouse ileal mucosa. *Am. J. Physiol.* **272,** G383–G392.

Hofseth, L. J., Hussain, S. P., Wogan, G. N., and Harris, C. C. (2003a). Nitric oxide in cancer and chemoprevention. *Free Radic. Biol. Med.* **34,** 955–968.

Hofseth, L. J., Saito, S., Hussain, S. P., Espey, M. G., Miranda, K. M., Araki, Y., Jhappan, C., Higashimoto, Y., He, P., Linke, S. P., Quezado, M. M., Zurer, I., Rotter, V., Wink, D. A., Appella, E., and Harris, C. C. (2003b). Nitric oxide–induced cellular stress and p53 activation in chronic inflammation. *Proc. Natl. Acad. Sci. USA* **100,** 143–148.

Huang, X., von Rad, U., and Durner, J. (2002). Nitric oxide induces transcriptional activation of the nitric oxide–tolerant alternative oxidase in *Arabidopsis* suspension cells. *Planta* **215,** 914–923.

Ignarro, L. J. (1992). Haem-dependent activation of cytosolic guanylate cyclase by nitric oxide: A widespread signal transduction mechanism. *Biochem. Soc. Trans.* **20,** 465–469.

Kibbe, M. R., Li, J., Nie, S., Watkins, S. C., Lizonova, A., Kovesdi, I., Simmons, R. L., Billiar, T. R., and Tzeng, E. (2000). Inducible nitric oxide synthase (iNOS) expression upregulates p21 and inhibits vascular smooth muscle cell proliferation through p42/44 mitogen-activated protein kinase activation and independent of p53 and cyclic guanosine monophosphate. *J. Vasc. Surg.* **31,** 1214–1228.

Klatt, P., Molina, E. P., and Lamas, S. (1999). Nitric oxide inhibits c-Jun DNA binding by specifically targeted *S*-glutathionylation. *J. Biol. Chem.* **274,** 15857–15864.

Klatt, P., and Lamas, S. (2000). Regulation of protein function by *S*-glutathiolation in response to oxidative and nitrosative stress. *Eur. J. Biochem.* **26,** 74928–74944.

Kuhn, D. M., and Geddes, T. J. (1999). Peroxynitrite inactivates tryptophan hydroxylase via sulfhydryl oxidation. Coincident nitration of enzyme tyrosyl residues has minimal impact on catalytic activity. *J. Biol. Chem.* **274,** 29726–29732.

Lacza, Z. (2003). Mitochondrial nitric oxide synthase is not eNO•S, nNO•S or iNO•S. *Free Radic. Biol. Med.* **35,** 1217–1228.

Laval, F., and Wink, D. A. (1994). Inhibition by nitric oxide of the repair protein, O6-methylguanine-DNA-methyltransferase. *Carcinogenesis* **15,** 443–447.

Li, C. Q., Robles, A. I., Hanigan, C. L., Hofseth, L. J., Trudel, L. J., Harris, C. C., and Wogan, G. N. (2004). Apoptotic signaling pathways induced by nitric oxide in human lymphoblastoid cells expressing wild-type or mutant p53. *Cancer Res.* **64,** 3022–3029.

Li, L., Zhang, J., Block, E. R., and Patel, J. M. (2004). Nitric oxide-modulated marker gene expression of signal transduction pathways in lung endothelial cells. *Nitric Oxide* **11,** 290–297.

Lin, Y., Ceacareanu, A. C., and Hassid, A. (2003). Nitric oxide–induced inhibition of aortic smooth muscle cell motility: Role of PTP-PEST and adaptor proteins p130cas and Crk. *Am. J. Physiol. Heart Circ. Physiol.* **285,** 710–721.

Maeda, H., and Akaike, T. (1998). Nitric oxide and oxygen radicals in infection, inflammation, and cancer. *Biochemistry (Mosc)* **63,** 854–865.

Malinski, T., Taha, Z., Grunfeld, S., Patton, S., Kapturczak, M., and Tomboulian, P. (1993). Diffusion of nitric oxide in the aorta wall monitored in situ by porphyrinic microsensors. *Biochem. Biophys. Res. Commun.* **193,** 1076–1082.

Marletta, M. A. (1994). Nitric oxide synthase: Aspects concerning structure and catalysis. *Cell* **7892,** 7–78930.

Mateo, J., Garcia-Lecea, M., Cadenas, S., Hernandez, C., and Moncada, S. (2003). Regulation of hypoxia-inducible factor-1alpha by nitric oxide through mitochondria-dependent and -independent pathways. *Biochem. J.* **376,** 537–544.

Matunis, M. J. (2002). On the road to repair: PCNA encounters SUMO and ubiquitin modifications. *Mol. Cell* **10,** 441–442.

Namkung-Matthai, H., Diwan, A., Mason, R. S., Murrell, G. A., and Diamond, T. (2000). Nitric oxide regulates alkaline phosphatase activity in rat fracture callus explant cultures. *Redox Rep.* **5,** 126–127.

Nisoli, E. (2003). Mitochondrial biogenesis in mammals: The role of endogenous nitric oxide. *Science* **299,** 896–899.

Ohno, H., Zhu, G., Mohan, V. P., Chu, D., Kohno, S., Jacobs, W. R., Jr., and Chan, J. (2003). The effects of reactive nitrogen intermediates on gene expression in *Mycobacterium tuberculosis*. *Cell Microbiol.* **5,** 637–648.

Okamoto, T., Gohil, K., Finkelstein, E. I., Bove, P., Akaike, T., and van der Vliet, A. (2004). Multiple contributing roles for $NO^{\bullet}S2$ in LPS-induced acute airway inflammation in mice. *Am. J. Physiol. Lung Cell Mol. Physiol.* **286,** 198–209.

Paunesku, T., Mittal, S., Protic, M., Oryhon, J., Korolev, S. V., Joachimiak, A., and Woloschak, G. E. (2001). Proliferating cell nuclear antigen (PCNA): Ringmaster of the genome. *Int. J. Radiat. Biol.* **77,** 1007–1021.

Pilz, R. B., and Casteel, D. E. (2003). Regulation of gene expression by cyclic GMP. *Circ. Res.* **93,** 1034–1046.

Polverari, A., Molesini, B., Pezzotti, M., Buonaurio, R., Marte, M., and Delledonne, M. (2003). Nitric oxide–mediated transcriptional changes in *Arabidopsis thaliana*. *Mol. Plant Microbe. Interact.* **16,** 1094–1105.

Radi, R., Beckman, J. S., Bush, K. M., and Freeman, B. A. (1991). Peroxynitrite oxidation of sulfhydryls. The cytotoxic potential of superoxide and nitric oxide. *J. Biol. Chem.* **266,** 4244–4450.

Radisavljevic, Z. (2004). Inactivated tumor suppressor Rb by nitric oxide promotes mitosis in human breast cancer cells. *J. Cell Biochem.* **92,** 1–5.

Reynaert, N. L., Ckless, K., Korn, S. H., Vos, N., Guala, A. S., Wouters, E. F., Van Der Vliet, A., and Janssen-Heininger, Y. M. (2004). From the cover: Nitric oxide represses inhibitory κB kinase through S-nitrosylation. *Proc. Natl. Acad. Sci. USA* **101,** 8945–8950.

Sandau, K. B., Zhou, J., Kietzmann, T., and Brune, B. (2001). Regulation of the hypoxia-inducible factor 1alpha by the inflammatory mediators nitric oxide and tumor necrosis factor-alpha in contrast to desferroxamine and phenylarsine oxide. *J. Biol. Chem.* **276,** 39805–39811.

Schneiderhan, N., Budde, A., Zhang, Y., and Brune, B. (2003). Nitric oxide induces phosphorylation of p53 and impairs nuclear export. *Oncogene* **22,** 2857–2868.

Schopfer, F. J., Baker, P. R., and Freeman, B. A. (2003). $NO^{\bullet}$-dependent protein nitration: A cell signaling event or an oxidative inflammatory response? *Trends Biochem. Sci.* **281,** 646–654.

So, H. S., Park, R. K., Kim, M. S., Lee, S. R., Jung, B. H., Chung, S. Y., Jun, C. D., and Chung, H. T. (1998). Nitric oxide inhibits c-Jun N-terminal kinase 2 (JNK2) via S-nitrosylation. *Biochem. Biophys. Res. Commun.* **247,** 809–813.

Sommer, D., Coleman, S., Swanson, S. A., and Stemmer, P. M. (2002). Differential susceptibilities of serine/threonine phosphatases to oxidative and nitrosative stress. *Arch. Biochem. Biophys.* **404,** 271–278.

Stamler, J. S. (1994). Redox signaling: Nitrosylation and related target interactions of nitric oxide. *Cell* **78,** 931–936.

Takasaki, Y., Deng, J. S., and Tan, E. M. (1981). A nuclear antigen associated with cell proliferation and blast transformation. *J. Exp. Med.* **154,** 1899–1909.

Terry, C. M., Clikeman, J. A., Hoidal, J. R., and Callahan, K. S. (1999). TNF-alpha and IL-1alpha induce heme oxygenase-1 via protein kinase C, $Ca^{2+}$, and phospholipase A2 in endothelial cells. *Am. J. Physiol.* **276,** H1493–H1501.

Thomas, D. D., Miranda, K. M., Espey, M. G., Citrin, D., Jourd'heuil, D., Paolocci, N., Hewett, S. J., Colton, C. A., Grisham, M. B., Feelisch, M., and Wink, D. A. (2002). Guide for the use of nitric oxide (NO) donors as probes of the chemistry of NO and related redox species in biological systems. *Methods Enzymol.* **359,** 84–105.

Thomas, D. D., Espey, M. G., Ridnour, L. A., Hofseth, L. J., Mancardi, D., Harris, C. C., and Wink, D. A. (2004). Hypoxic inducible factor 1alpha, extracellular signal-regulated kinase, and p53 are regulated by distinct threshold concentrations of nitric oxide. *Proc. Natl. Acad. Sci. USA* **101,** 8894–8899.

Turko, I. V., and Murad, F. (2002). Protein nitration in cardiovascular diseases. *Pharmacol. Rev.* **54,** 619–634.

Turpaev, K., Bouton, C., Diet, A., Glatigny, A., and Drapier, J. C. (2005). Analysis of differentially expressed genes in nitric oxide-expressed human monocytic cells. *Free Radic. Biol. Med.* **38,** 1392–1400.

Vile, G. F., Basu-Modak, S., Waltner, C., and Tyrrell, R. M. (1994). Heme oxygenase 1 mediates an adaptive response to oxidative stress in human skin fibroblasts. *Proc. Natl. Acad. Sci. USA* **91,** 2607–2610.

Wang, E. Q., Lee, W. I., Brazeau, D., and Fung, H. L. (2002). cDNA microarray analysis of vascular gene expression after nitric oxide donor infusions in rats: Implications for nitrate tolerance mechanisms. *AAPS Pharm. Sci.* **4,** E10.

Wink, D. A., and Laval, J. (1994). The Fpg protein, a DNA repair enzyme, is inhibited by the biomediator nitric oxide *in vitro* and *in vivo*. *Carcinogenesis* **15,** 2125–2129.

Wink, D. A., and Mitchell, J. B. (1998). Chemical biology of nitric oxide: Insights into regulatory, cytotoxic, and cytoprotective mechanisms of nitric oxide. *Free Radic. Biol. Med.* **25,** 434–456.

Zamora, R., Vodovotz, Y., Aulak, K. S., Kim, P. K., Kane, J. M. III, Alarcon, L., Stuehr, D. J., and Billiar, T. R. (2002). A DNA microarray study of nitric oxide–induced genes in mouse hepatocytes: Implications for hepatic heme oxygenase-1 expression in ischemia/reperfusion. *Nitric Oxide* **7,** 165–186.

Zhou, X., Espey, M. G., Chen, J. X., Hofseth, L. J., Miranda, K. M., Hussain, S. P., Wink, D. A., and Harris, C. C. (2000). Inhibitory effects of nitric oxide and nitrosative stress on dopamine-beta-hydroxylase. *J. Biol. Chem.* **275,** 21241–21246.

# [28] NO Signaling in ARE-Mediated Gene Expression

*By* Eun Young Park and Sang Geon Kim

## Abstract

Nitric oxide (NO) and peroxynitrite, which serve as cell signal molecules, activate the antioxidant response element (ARE) for the induction of phase II antioxidant enzymes as an adaptive response. The reactive nitrogen species plays an essential role in Nrf2 activation and Nrf2 binding to the ARE present in the target genes. In this chapter, we describe the system by which the NO signaling pathway regulates ARE-mediated gene expression, which includes immunochemical assessment and gel shift analysis of Nrf2 activation.

## Introduction

Endogenous nitric oxide (NO) regulates a number of physiological responses including vascular tone, inflammatory responses, and gene transcription (Bogdan, 2001; Goodwin *et al.*, 1999; Knowles and Moncada, 1994). NO is derived from L-arginine via NO synthase (NOS) isoenzymes, and it is becoming evident that NOS activity is associated with human diseases and disorders (Domenico, 2004). Actions of NO involve stimulation of signal transduction pathways (Bogdan, 2001; Park *et al.*, 2000). Peroxynitrite, formed by the reaction between NO and superoxide, also serves as a cell signal molecule (Ortega and De Artinano, 2000). However, excess peroxynitrite produced from activated macrophages causes toxic and detrimental effects, such as DNA damage and protein oxidation. Therefore, reactive oxygen species, including NO and peroxynitrite, activate the antioxidant response element (ARE) for the induction of phase II antioxidant enzymes as part of adaptive responses (Kang *et al.*, 2002a). In this chapter, we describe the system by which the NO signaling pathway regulates ARE-mediated gene expression.

## Assessment of NO-Mediated Cell Signaling

### Chemicals

Chemical activators for the study of NO signaling include *S*-nitroso-*N*-acetyl-penicillamine (SNAP) and 3-morpholinosydnonimine (SIN-1). Similarly, chemical inhibitors may be useful to identify some of the cell

METHODS IN ENZYMOLOGY, VOL. 396
0076-6879/05 $35.00
DOI: 10.1016/S0076-6879(05)96028-X

signaling pathways (Lee *et al.*, 2001a; Park *et al.*, 2002). Wortmannin or LY294002 inhibits the activity of phosphoinositide 3-kinase (PI3-kinase). PD98059 serves as an inhibitory agent for the pathway of MKK1/ERK. SB203580 is a specific inhibitor of p38 mitogen-activated protein kinase (MAPK). However, it should be pointed out that PD98059, a chemical inhibitor of MKK1, promotes nuclear translocation of C/EBP$\beta$ and increases C/EBP$\beta$ binding to the C/EBP-binding consensus DNA sequence (Kang *et al.*, 2003). Activation of C/EBP$\beta$ by PD98059 leads to the induction of the C/EBP$\beta$ target genes, irrespective of its inhibition of MKK1/ERK1/2 activity. Thus, it is recommended that the cells transfected (preferably stable transfection) with the vector of MKK1 dominant-negative mutant [MKK1(−)] be used for exploration of the MKK1-ERK1/2 signaling pathway.

## Stable Transfection with Plasmid

In general, cells are transfected using Transfectam (Promega). Cells (e.g., H4IIE, Raw264.7, HeLa, and HEK293) are re-plated 24 h before transfection, usually at a density of $2 \times 10^6$ cells in a 10-cm diameter plastic dish. For use in a MKK1(−) vector, 20 $\mu$l of Transfectam is mixed with 10 $\mu$g of plasmid in 2.5 ml of minimal essential medium (MEM). Cells are transfected by addition of MEM containing each plasmid and Transfectam and are incubated at 37° in a humidified atmosphere of 5% $CO_2$ for 6 h. After addition of 6.25 ml of MEM containing 10% fetal calf serum, cells are incubated for an additional 48 h at 37°. To establish a stably transfected cell line, viable cells are subcultured at least five successive times in the medium containing 50 $\mu$g/ml of geneticin. For example, the colonies of the cells stably transfected with the plasmid of PI3-kinase p110 overexpression [p110(+)] or p85 overexpression [p85(+)] can be used to specify the role of PI3-kinase signaling from other overlapping cell signaling pathways such as PI4-kinase.

## Luciferase Reporter Gene Assay

To perform the dual-luciferase reporter assay system, cells ($7 \times 10^5$ cells/well) are re-plated in six-well plates overnight, serum starved for 12 h, and transiently transfected with an ARE-containing promoter-luciferase construct and pRL-SV plasmid (a plasmid that encodes for *Renilla* luciferase and is used to normalize transfection efficacy) in the presence of Lipofectamine Plus Reagent (Life Technologies) for 3 h. Transfected cells are incubated in Dulbecco's Modified Eagle Medium (DMEM) containing 1% fetal calf serum for 3 h and exposed to chemical activators in culture medium containing fetal calf serum for 18 h at 37°. Firefly and *Renilla* luciferase activities in cell lysates are measured using a luminometer. The activity of firefly luciferase is measured by adding Luciferase Assay Reagent II according to the manufacturer's instruction, and after quenching the reaction, the *Renilla* luciferase reaction is initiated by adding Stop & Glo reagent. The relative luciferase activity is calculated by normalizing firefly luciferase activity to that of *Renilla* luciferase.

## NO-Mediated Oxidative Stress

### Sulfur Amino Acid–Deficient Culture Medium

A previous study showed that sulfur amino acid deprivation (SAAD) from the incubation medium increases peroxide formation in the cells (Kang *et al.*, 2002a). NO or peroxynitrite, an oxygen species derived from NO, contributes to SAAD-induced oxidative stress. When H4IIE cells are pretreated with $N^G$-nitro-L-arginine methyl ester (L-NAME, 300 $\mu M$), an NOS inhibitor for 12 h, the rate of increase in DCF fluorescence by SAAD is about 25% decreased. Deficiency of L-arginine, an amino acid substrate required for NO production, also reduces SAAD-induced peroxide production. We found that the inhibition of prooxidant production by L-arginine deficiency was greater than that by L-NAME. In the cells loaded with dichlorofluorescein (DCFH), oxidation of the dye represents generalized oxidative stress, including the oxygen species $RO_2^\bullet$, $RO^\bullet$, $OH^\bullet$, HOCl, and $ONOO^\bullet$ (Halliwell and Gutteridge, 1998). DCFH is oxidized by peroxynitrite. Constitutive NOS (cNOS) is expressed in hepatocyte-derived cells lines. We showed that cNOS is involved in the prooxidant production in the cells under the condition of SAAD (Kang *et al.*, 2002a). The intensity of DCF fluorescence is 1.2- and 4-fold increased 10–30 min after exposure of the dye-loaded cells to SAAD medium.

SAAD disrupts a dynamic equilibrium of the GSH pool and limits the compensatory increase in GSH synthesis. SAAD that persistently increases peroxide formation causes formation of peroxynitrite from NO and superoxide. Production of intracellular peroxides can be monitored spectrofluorometrically using 2′,7′-dichlorofluorescein diacetate (DCFH-DA) as a fluorescent dye.

1. Cells are maintained in DMEM containing 10% fetal calf serum, 50 units/ml penicillin, and 50 $\mu$g/ml streptomycin at 37° in humidified atmosphere with 5% $CO_2$.
2. Sulfur amino acid–deprived MEM is reconstituted with Earle's balanced salt solution, vitamin mixture, and the amino acids except cystine and methionine. The monolaying cells are then cultured for

the desired time period in MEM with or without cystine and methionine.

3. The cells are suspended 12 h after serum deprivation and then DCFH-DA dissolved in ethanol is added at the final concentration of 10 $\mu M$ in the medium.

4. Oxidation of DCFH yields dichlorofluorescein. Fluorescence is monitored at the excitation and emission wavelengths of 485 and 530 nm, respectively, using a fluorescence plate reader.

## NO-Producing Chemicals

NO donors mimic the proper *in vivo* redox state and produce similar stoichiometric amounts of NO as that occurring in cells. SNAP has a short half-life, produces low concentrations of NO, and may yield reactive nitrogen species that mimic NO (Gbadegesin *et al.*, 1999). When cells are incubated in SAAD medium in the presence of SNAP, ARE-mediated gene induction is activated. Although SNAP (10–100 $\mu M$) alone does not increase ARE-mediated gene induction, SNAP in combination with SAAD increases the level of target gene transcript to a greater extent than that by SAAD alone. Therefore, NO is required for the ARE-mediated gene induction, but NO alone is not sufficient to stimulate induction of the gene (Kang *et al.*, 2002a). Peroxynitrite converted from NO in conjunction with other reactive oxygen species (e.g., superoxide) contributes to the ARE-mediated phase II enzyme induction.

SIN-1 is a chemical that produces equimolar NO and superoxide and thus induces peroxynitrite formation. SIN-1 under aerobic conditions decomposes NO˙ and superoxide and thus yields mostly (~76%) peroxynitrite (Castro *et al.*, 1996). We found that peroxynitrite increases the activities of PI3-kinase and Akt and activates Nrf2 (Kang *et al.*, 2002a). SIN-1 (300 $\mu M$) is active in translocating cytoplasmic Nrf2 to the nucleus from 3 to 12 h (Kang *et al.*, 2002a) and increases the band intensity of Nrf2-ARE binding complex 1–24 h after treatment. ARE activation is notably increased 3–6 h after SIN-1 treatment.

## Assessment of Nrf2 Activation

### Preparation of Nuclear and Cytoplasmic Fractions

Nuclear translocation of Nrf2 precedes Nrf2-mediated gene transcription (Kang *et al.*, 2002b; Zipper and Mulcahy, 2003). When cells are incubated in SAAD medium in the absence of L-arginine or in the presence of L-NAME (i.e., the conditions that inhibit NO production), Nrf2

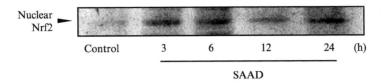

Control     3     6     12     24     (h)

SAAD

FIG. 1. Nuclear translocation of Nrf2 by sulfur amino acid deprivation (SAAD). Levels of nuclear Nrf2. The levels of Nrf2 were immunochemically assessed in the nuclear fractions of cells incubated in SAAD medium for 3–24h. All lanes contained 20μg of nuclear proteins.

activation by SAAD is inhibited. Whereas SAAD activates Nrf2 at 3–24 h (Fig. 1), L-arginine deficiency or L-NAME (300 $\mu M$, 2 h pretreatment) prevents an increase in the level of nuclear Nrf2 by SAAD. The level of cytosolic Nrf2 is reciprocally changed to that of nuclear Nrf2.

1. Nuclear extracts are prepared by differential centrifugations. The cells in dishes are washed with ice-cold phosphate-buffered saline (PBS), scraped, transferred to microtubes, and allowed to swell after the addition of 100 $\mu l$ hypotonic buffer containing 10 m$M$ HEPES (pH 7.9), 10 m$M$ KCl, 0.1 m$M$ EDTA, 2 m$M$ dithiothreitol (DTT), and 0.5 m$M$ phenylmethylsulfonylfluoride.

2. The lysates are incubated for 10–30 min in ice and centrifuged at 7200$g$ for 5 min at 4°. Supernatants are used as the cytoplasmic fractions for the assay of Nrf2. Preferably use fresh samples. The remaining samples may be stored at −70° until use.

3. The pellets containing crude nuclei are resuspended in 50 $\mu l$ of extraction buffer containing 20 m$M$ HEPES (pH 7.9), 400 m$M$ NaCl, 1 m$M$ EDTA, 10 m$M$ DTT, and 1 m$M$ phenylmethylsulfonylfluoride, and then are incubated for 30 min in ice. The samples are centrifuged at 15,800$g$ for 10 min to obtain supernatants containing nuclear fractions.

*Immunoblot Analysis of Nrf2*

The apparent molecular weight of cytoplasmic and nuclear proteins immunoreactive with anti-Nrf2 antibody varies in the literature (57–68 kDa) (Lee *et al.*, 2001b; Moi *et al.*, 1994; Zhang and Gordon, 2004). The apparent molecular weight of Nrf2, detected in the cytoplasmic proteins fractionated from cell lysates prepared with the buffer solution containing DTT (2 m$M$), is 57 kDa (Kang *et al.*, 2002b). In the H4IIE cells treated with *tert*-butylhydroquinone (*t*-BHQ, 6 h), 57 kDa Nrf2 is detected in the nuclear, but not in cytoplasmic, fraction. When the buffer solution lacking DTT is used for the preparation of cytoplasmic and nuclear fractions, anti-Nrf2 antibody (sc-100, Santa Cruz) detects two bands with the apparent molecular weight of 100 and 57 kDa in the cytoplasmic fraction of untreated cells. Anti-actin antibody recognizes both 100-kDa Nrf2 and 43-kDa actin in the subcellular fractions prepared under the non-reducing condition.

1. Cytosolic and nuclear Nrf2 is analyzed by sodium dodecylsulfate (SDS)–polyacrylamide gel electrophoresis (PAGE) and immunoblot analyses. The samples are fractionated by 7.5% gel electrophoresis and electrophoretically transferred to nitrocellulose paper.
2. The nitrocellulose paper is incubated with polyclonal rabbit anti-Nrf2 antibody (1:1000 dilution), followed by incubation with horseradish peroxidase–conjugated secondary antibody.
3. Immunoreactive protein is visualized through incubation with an ECL chemiluminescence detection kit. Equal loading of proteins is usually verified by actin immunoblotting with anti-actin antibody.
4. Change in the protein levels is determined via scanning densitometry of the immunoblot.

*Immunoprecipitation of Nrf2*

Interaction of Nrf2 with other protein(s) in the nucleus may be assessed with Nrf2 immunoprecipitates from nuclear extracts. For example, prooxidant increases the level of nuclear actin co-immunoprecipitated with anti-Nrf2 antibody in *t*-BHQ–treated cells.

1. Either nuclear fraction or total cell lysates (50 $\mu$g in 300 $\mu$l each) are incubated with polyclonal anti-Nrf2 antibody for 2 h at 4°.
2. The antigen–antibody complex is immunoprecipitated after incubation of the samples for 2 h at 4° with protein G–agarose. Immune complexes are solubilized in 2× Lammeli buffer and boiled for 5 min.
3. The proteins in the samples are resolved using 7.5% SDS-PAGE, transferred to nitrocellulose membranes, and immunoblotted with anti-Nrf2 antibody or the antibody of interest.
4. The blots are developed using an ECL chemiluminescence detection kit. BCIP and NBT can also be used for color development.

*Immunocytochemistry of Nrf2*

Cytoplasmic, perinuclear, and nuclear localization of Nrf2 is monitored by immunocytochemistry. Superposition of Nrf2 images with that of the target protein(s) of interest allows us to assess the possible interaction of Nrf2 with other intracellular protein(s). For example, Nrf2 binds with Keap1 (Itoh *et al.*, 1999).

1. Hepatocyte-derived cells (e.g., HepG2, HuH-7, H4IIE, and Hepa1c1c) are grown on Lab-TEK chamber slides and incubated in serum-deprived medium for 6 h.

2. For immunostaining, the cells are fixed in 100% methanol for 30 min and washed three times with PBS. After blocking in 5% bovine serum albumin (BSA) in PBS for 1 h at room temperature or overnight at 4°, the cells are incubated for 1 h with anti-Nrf2 antibody (1:100 dilution) in PBS containing 0.5% BSA.

3. The cells are incubated with fluorescein-isothiocyanate (FITC)–conjugated anti–immunoglobulin G (IgG) antibody (1:100) after serial washings with PBS. Counterstaining with propidium iodide (2 $\mu$g/ml) verifies the location and integrity of nuclei. The stained cells in samples are washed and microscopically examined.

### Gel Shift Assay

Gel shift assay is used to determine Nrf2 activation and to identify the components in protein–ARE complex. In general, the antibody alone specific for Nrf1 or Nrf2 reduces formation of the retarded band. Also, the presence of Nrf1, Nrf2, and small Maf antibodies induces immunodepletion or supershift of the retarded band (Kang *et al.*, 2002a).

1. A double-stranded DNA probe containing ARE consensus sequence (e.g., GSTA2 ARE-containing oligonucleotide, 5'-GATCATGGCATTGC ACTAGGTGACAAAGCA-3 '; NQO1 ARE-containing oligonucleotide, 5'-GATCCAGTCACAGT GACTCAGCAGAATCTG-3') is end-labeled with [$\gamma$-32p]ATP and T$_4$ polynucleotide kinase.

2. The gel shift reaction mixture contains 4 $\mu$l of 5× binding buffer containing 20% glycerol, 5 m$M$ MgCl$_2$, 250 m$M$ NaCl, 2.5 m$M$ EDTA, 2.5 m$M$ DTT, 0.25 mg/ml poly-dI-dC, 50 m$M$ Tris–Cl (pH 7.5), 5 $\mu$g of nuclear extracts, and sterile water in a total volume of 20 $\mu$l.

3. The reaction mixture is preincubated for 10 min. DNA-binding reactions are performed at room temperature for 30 min after the addition of 1-$\mu$l probe (10$^6$ cpm).

4. Competition experiment is performed by adding a 20-fold excess of unlabeled ARE oligonucleotide to the reaction mixture before the DNA-binding reaction. For supershift assay, the antibodies (2–6 $\mu$g each) are added to the reaction mixture after initial 20 min incubation and are additionally incubated for 2 h at 37°.

5. Samples are loaded onto 4% polyacrylamide gels at 100 V. The gels are removed, fixed, dried, and autoradiographed.

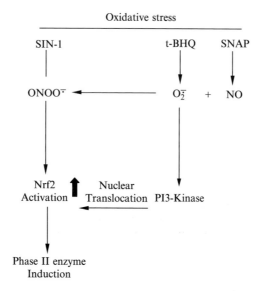

FIG. 2. A schematic diagram illustrating the pathway by which peroxynitrite induces phase II enzymes via Nrf2/antioxidant response element (ARE) activation. PI3-kinase, which regulates nuclear translocation of Nrf2, is stimulated by oxidative stress. Peroxynitrite is produced in cells by reaction of oxygen species with nitric oxide that is generated by constitutive nitric oxide synthase in hepatocytes.

## Conclusion

The cNOS, eNOS, and iNOS serve as the enzymes active in stimulating the cell signal pathway leading to phase II enzyme induction. We have described here the methods to study the cell signaling pathways of NO- or peroxynitrite-mediated activation of Nrf2/ARE. Peroxynitrite is produced as a consequence of interaction between NO and reactive oxygen species. The reactive nitrogen species plays an essential role in Nrf2 activation and Nrf2 binding to the ARE present in the target genes (Fig. 2).

## Acknowledgments

This work was supported by the Basic Sciences Research Program from Korea Research Foundation (FS0409-20040072), Ministry of Education, Republic of Korea.

## References

Bogdan, C. (2001). Nitric oxide and the regulation of gene expression. *Trends Cell Biol.* **11**, 66–75.

Castro, L., Alvarez, M. N., and Radi, R. (1996). Modulatory role of nitric oxide on superoxide-dependent luminol chemiluminescence. *Arch. Biochem. Biophys.* **333**, 179–188.

Domenico, R. (2004). Pharmacology of nitric oxide: Molecular mechanisms and therapeutic strategies. *Curr. Pharm. Res.* **10**, 1667–1676.

Gbadegesin, M., Vicini, S., Hewett, S. J., Wink, D. A., Espey, M., Pluta, R. M., and Colton, C. A. (1999). Hypoxia modulates nitric oxide–induced regulation of NMDA receptor currents and neuronal cell death. *Am. J. Physiol.* **277**, 673–683.

Goodwin, D. C., Landino, L. M., and Marnett, L. J. (1999). Effects of nitric oxide and nitric oxide–derived species on prostaglandin endoperoxide synthase and prostaglandin biosynthesis. *FASEB J.* **13**, 1121–1136.

Halliwell, B., and Gutteridge, J. M. C. (1998). "Free Radicals in Biology and Medicine," 2nd Ed. Oxford University Press, NY.

Itoh, K., Wakabayashi, N., Katoh, Y., Ishii, T., Igarashi, K., Engel, J. D., and Yamamoto, M. (1999). Keap1 represses nuclear activation of antioxidant responsive elements by Nrf2 through binding to the amino-terminal Neh2 domain. *Genes Dev.* **13**, 76–86.

Kang, K. W., Choi, S. H., and Kim, S. G. (2002a). Peroxynitrite activates NF-E2–related factor 2/antioxidant response element through the pathway of phosphatidylinositol 3-kinase: The role of nitric oxide synthase in rat glutathione S-transferase A2 induction. *Nitric Oxide* **7**, 244–253.

Kang, K. W., Lee, S. J., Park, J. W., and Kim, S. G. (2002b). Phosphatidylinositol 3-kinase regulates nuclear translocation of NF-E2–related factor 2 through actin rearrangement in response to oxidative stress. *Mol. Pharmacol.* **62**, 1001–1010.

Kang, K. W., Park, E. Y., and Kim, S. G. (2003). Activation of CCAAT/enhancer-binding protein beta by 2'-amino-3'-methoxyflavone (PD98059) leads to the induction of glutathione S-transferase A2. *Carcinogenesis* **24**, 475–482.

Knowles, R. G., and Moncada, S. (1994). Nitric oxide synthases in mammals. *Biochem. J.* **298**, 249–258.

Lee, S. E., Chung, W. J., Kwak, H. B., Chung, C. H., Kwack, K. B., Lee, Z. H., and Kim, H. H. (2001a). Tumor necrosis factor-alpha supports the survival of osteoclasts through the activation of Akt and ERK. *J. Biol. Chem.* **276**, 49343–49349.

Lee, J. M., Moehlenkamp, J. D., Hanson, J. M., and Johnson, J. A. (2001b). Nrf2-dependent activation of the antioxidant responsive element by tert-butylhydroquinone is independent of oxidative stress in IMR-32 human neuroblastoma cells. *Biochem. Biophys. Res. Commun.* **280**, 286–292.

Moi, P., Chan, K., Asunis, I., Cao, A., and Kan, Y. W. (1994). Isolation of NF-E2–related factor 2 (Nrf2), a NF-E2–like basic leucine zipper transcriptional activator that binds to the tandem NF-E2/AP1 repeat of the beta-globin locus control region. *Proc. Natl. Acad. Sci. USA* **91**, 9926–9930.

Ortega Mateo, A., and De Artinano, A. A. (2000). Nitric oxide reactivity and mechanisms involved in its biological effects. *Pharmacol. Res.* **42**, 421–427.

Park, H. S., Huh, S. H., Kim, M. S., Lee, S. H., and Choi, E. J. (2000). Nitric oxide negatively regulates c-Jun N-terminal kinase/stress–activated protein kinase by means of S-nitrosylation. *Proc. Natl. Acad. Sci. USA* **97**, 14382–14387.

Park, H. J., Kim, B. C., Kim, S. J., and Choi, K. S. (2002). Role of MAP kinases and their cross-talk in TGF-beta1–induced apoptosis in FaO rat hepatoma cell line. *Hepatology* **35**, 1360–1371.

Zhang, Y., and Gordon, G. B. (2004). A strategy for cancer prevention: Stimulation of the Nrf2-ARE signaling pathway. *Mol. Cancer Ther.* **3**, 885–893.

Zipper, L. M., and Mulcahy, R. T. (2003). Erk activation is required for Nrf2 nuclear localization during pyrrolidine dithiocarbamate induction of glutamate cysteine ligase modulatory gene expression in HepG2 cells. *Toxicol. Sci.* **73**, 124–134.

## [29]  Tyrosine Phosphorylation in Nitric Oxide–Mediated Signaling Events

*By* Hugo P. Monteiro, Carlos J. Rocha Oliveira, Marli F. Curcio, Miriam S. Moraes, and Roberto J. Arai

### Abstract

In this chapter, we provide an overview of nitric oxide (NO)–tyrosine phosphorylation signal transduction pathways, integrating them with the cyclic guanosine monophosphate (cGMP) and *S*-nitrosylation–mediated pathways that are triggered by NO. The second half of this chapter includes a description of the methods that our laboratory has used extensively to characterize the mechanisms involved in signaling events mediated by this pathway. These include assays for detecting protein tyrosine phosphorylation, tyrosine phosphorylation of the epidermal growth factor (EGF) receptor, phosphorylation of the ERK1/2 mitogen-activated protein (MAP) kinases, transfection of cells with modified forms of p21Ras, and an assay of p21Ras.

### Introduction

*Overview on the Participation of Protein Tyrosine Phosphorylation in Nitric Oxide–Mediated Signaling Events*

Nitric oxide (NO), a gaseous free radical, is generated through a five-electron oxidation of a guanidinium nitrogen on the amino acid L-arginine. This reaction is catalyzed by a family of enzymes, NO synthases (NOSs), which were originally subgrouped in three isoforms. Neuronal NOS (nNOS), endothelial NOS (eNOS), and inducible NOS (iNOS) display specific and independent physiological functions (Nathan, 1992). To carry out the tasks of each isoform, NO diffuses freely in biological systems and by doing this acts as a signaling radical.

NO was recognized as a unique signaling molecule shortly after three seminal observations: (1) the characterization of NO as one of the substances derived from the endothelium that controls the vascular tone (Furchgott and Zawadzki, 1980); (2) nitro vasodilators such as nitroglycerin and sodium nitroprusside (SNP) release NO to promote vascular relaxation (Gruetter *et al.*, 1980); (3) NO reacts with the iron-heme moiety of the soluble form of guanylyl cyclase in smooth muscle cells and other cell

METHODS IN ENZYMOLOGY, VOL. 396
0076-6879/05 $35.00
DOI: 10.1016/S0076-6879(05)96029-1

types, increasing the production of cyclic guanosine monophosphate (cGMP). Intracellular elevated levels of cGMP may eventually lead to relaxation of smooth muscle (Katsuki *et al.*, 1977). These findings provided an initial perspective from which NO could be considered as a highly diffusible second messenger that essentially signals through soluble guanylyl cyclase. However, since then this view has been radically modified with increasing experimental evidence indicating that NO also reacts with protein and nonprotein thiols (Stamler, 1994). The reaction between NO and thiols, named "*S*-nitrosylation," is remarkably specific, given the ubiquity of NO and the abundance of potential substrate thiols. Further, the number of cysteine residues in a protein will not determine its susceptibility to *S*-nitrosylation. A single cysteine residue in the presence of specific allosteric effectors such as $pO_2$, pH, and metal ions, located in an acid–base motif and in hydrophobic compartments, is a good candidate for nitrosylation (Hess *et al.*, 2001). Indeed, that is the case for all the proteins whose activities are regulated by NO, including hemoglobin, *N*-methyl-D-aspartate (NMDA) receptor–coupled channel, ryanodine receptor, Jun kinase, caspase-3, and the small guanosine triphosphatase (GTPase) p21Ras (Hess *et al.*, 2001). In the particular case of p21Ras, the original work by Lander *et al.* (1996) characterized cysteine 118 as a critical site for regulation of the GTPase by *S*-nitrosylation. *S*-nitrosylation of this residue triggers guanine nucleotide exchange and downstream signaling events that include recruitment of phosphatidylinositol 3-kinase and Raf-1 kinase (Deora *et al.*, 1998, 2000).

Protein phosphorylation was initially suggested as one possible mechanism through which NO stimulation of guanylyl cyclase could mediate signaling events. Elevation of the intracellular levels of cGMP leads to the activation of a group of serine/threonine kinases known as *protein kinases G* (PKG). PKG-mediated signaling events include inhibition of phospholipase C and stimulation of calcium removal from the cytoplasm (Lincoln *et al.*, 1996). In addition to PKG, other protein kinases have their activities modulated by NO. Early studies in our laboratory (Monteiro *et al.*, 1994; Peranovich *et al.*, 1995) indicate that NO released from two compounds, SNP and *S*-nitroso-*N*-acetyl-penicillamine (SNAP), stimulated tyrosine phosphorylation levels on a group of endogenous proteins in murine fibroblasts. The stable cGMP analogue, 8-Br-cGMP, mimicked the NO-mediated pattern of protein tyrosine phosphorylation, suggesting the participation of PKGs in the event. Four proteins, 126, 56, 43, and 40 kDa, had their phosphotyrosine content increased by the action of NO/cGMP. Later, these proteins were characterized as distal components of the epidermal growth factor (EGF)/integrins-mediated signal transduction pathway. The band corresponding to the 126-kDa protein was focal adhesion kinase (FAK). The 56-kDa protein was Src kinase, and the doublet 43- and 40-kDa protein corresponded to the extracellular-regulated ERK1/2 MAP kinases. Our findings led us to conclude that NO can signal through a cascade of reactions normally used by integrins and growth factors (Monteiro *et al.*, 2000).

Independent observations made by other groups corroborated our hypothesis and provided additional evidence on the role of NO/cGMP in tyrosine phosphorylation–mediated signaling events (Akhand *et al.*, 1999; Garcia-Benito *et al.*, 2000; Goligorsky *et al.*, 1999; Lander *et al.*, 1996; Rhoads *et al.*, 2004). In rat pancreatic acini, FAK and its immediate downstream substrate paxillin were phosphorylated on tyrosine upon stimulation with SNP and 8-Br-cGMP (Garcia-Benito *et al.*, 2000). In human umbilical vein endothelial cells, SNP and endogenously generated NO stimulated a transient increase on tyrosine phosphorylation of FAK and paxillin. NO-stimulated tyrosine phosphorylation involved the participation of guanylyl cyclase/cGMP and was directly associated with endothelial cell migration. Inhibition of eNOS activity and tyrosine phosphorylation significantly reduced the rate of endothelial cell migration. At low endogenous concentrations of NO, cell migration was favored, whereas high-output NO release obtained from NO donors eventually led to cell detachment. These observations emphasized the role of NO in the facilitation of the assembly and disassembly of focal adhesions complexes and its consequences on cell locomotion (Goligorsky *et al.*, 1999). These findings received further support from a study by Rhoads *et al.* (2004). The authors demonstrated an enhancement in NO synthesis coupled with stimulation of FAK phosphorylation in small intestinal cells supplemented with L-arginine. NO-stimulated tyrosine phosphorylation of FAK was directly related to small intestinal cell migration. Our findings on NO-stimulated tyrosine phosphorylation of Src kinase were confirmed and extended by a study performed by Akhand *et al.* (1999). Incubation of NIH3T3 murine fibroblasts with SNAP and immunoprecipitation of Src kinase revealed that SNAP promoted both *S*-nitrosylation and Src kinase autophosphorylation at tyrosine 416. In studies of signaling pathways used by integrins, Giancotti and Ruoslahti (1999) showed a connection between FAK/Src kinase to either p21Ras-ERK1/2 or p130Cas-c-Jun $NH_2$-terminal kinase (JNK), depending on the nature of the stimulus. Analogously, studies with NO, which stimulated signaling pathways used by integrins and growth factors, revealed a connection between FAK/Src kinase and p21Ras-ERK1/2 (Monteiro *et al.*, 2000; Rocha Oliveira *et al.*, 2003). Accordingly, early observations by Lander *et al.* (1996) described the NO-mediated activation of JNK, ERK, and p38 MAP kinase in human Jurkat T cells. NO-induced MAP kinase activation was effectively blocked by $\alpha$-hydroxyfarnesylphosphonic acid, an inhibitor of p21Ras. The participation of p21Ras in NO signal transduction was also suggested.

A point of convergence between different signaling cascades lies at p21Ras (Shields *et al.*, 2000). Indeed, our findings point to an integration of the different NO-mediated signaling cascades, based on p21Ras. Integration included the cGMP-guanylyl cyclase, *S*-nitrosylation, and the protein tyrosine phosphorylation-mediated signaling pathways. We demonstrated that NO/cGMP stimulation of p21Ras activity in RAECs is an essential requirement for stimulation of protein tyrosine phosphorylation and NO/cGMP–activated p21Ras mediates its effects through the Raf-1/MEK/ERK1/2 signaling pathway. The activation of ERK1/2 underlies the activation of the EGF receptor tyrosine kinase, which is essential for NO-stimulated protein tyrosine phosphorylation (Rocha Oliveira *et al.*, 2003).

## Methods

A great deal of information regarding the NO-mediated tyrosine phosphorylation signaling events has been obtained using Western blot analysis. Specific antibodies that recognize activated and nonactivated forms of different proteins that are part of the signaling pathways became invaluable tools in this endeavor. We have used various cell lines that include murine fibroblasts (3T3), RAECs, and human cervical carcinoma cell line (HeLa). Another experimental tool routinely used in our laboratory is the cellular transfection of negative dominant forms of p21Ras, a key element in the NO/tyrosine phosphorylation-signaling pathway. In this section, we describe some of the methods that our laboratory has used to elucidate the mechanisms involved in the signaling events mediated by this pathway.

### *Immunoblotting for Detection of Protein Tyrosine Phosphorylation*

The pattern of NO-stimulated tyrosine phosphorylation was originally obtained in murine fibroblasts, which were grown in Dulbecco's Modified Eagle Medium (DMEM) supplemented with 10% fetal bovine serum (FBS). Cultures were maintained in 100 mm plastic Petri culture dishes in a humidified atmosphere of air/$CO_2$ (95/5%) at 37°. Unless otherwise indicated, cells were starved in serum-free medium and incubated with the NO donor SNAP (100–500 $\mu M$) in the same conditions. After the incubation period (15–60 min at 37°), cells were solubilized in lysis buffer A (20 m$M$ HEPES, pH 7.5; 150 m$M$ NaCl; 10% glycerol; 1% Triton X-100; 1.5 m$M$ MgCl$_2$; 1 m$M$ EGTA) supplemented with protease inhibitors (1 $\mu$g/ml aprotinin; 1 $\mu$g/ml leupeptin; 1 m$M$ PMSF) and phosphatase

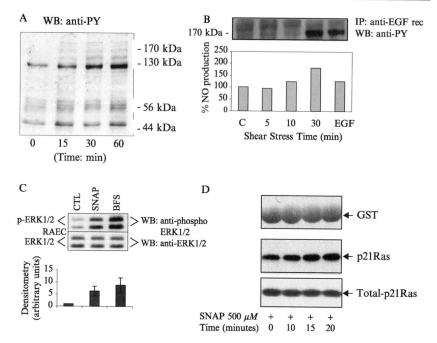

FIG. 1. (A) Murine fibroblasts were starved for 24 h and after that were incubated in the presence of 500 $\mu M$ S-nitroso-N-acetyl-penicillamine (SNAP) for increasing periods. Cells were lysed and the same amount of protein (60 $\mu g$/ml) was resolved by sodium dodecylsulfate (SDS) and blotted onto nitrocellulose sheets. Western blot analysis for tyrosine-phosphorylated proteins was performed as described in the section "Methods." (B) Rabbit aortic endothelial cells (RAECs) were exposed to shear stress (12 dynes/cm$^2$) for 5–30 min. Static controls were incubated with or without 100 ng/ml epidermal growth factor (EGF) (5 min at 37°). Production of nitric oxide (NO) was estimated using a chemiluminescence-based assay (bottom). Immunoprecipitation of the EGF receptor from cell lysates and Western blot analysis with the anti-PY antibody were performed as described in the section "Methods." (C) Serum-starved RAECs were treated with 100 $\mu M$ SNAP for 30 min at 37°. Western blot analysis for p-ERK1/2 (top) and total ERK1/2 (middle) are described in the "Methods" section. Normalized ERK1/2 (bottom) indicates the densitometric ratio of p-ERK1/2 to ERK1/2 and estimates the extent of activation of ERK1/2. (D) Activated p21Ras in HeLa cells treated with the NO donor SNAP was detected by precipitation with glutathione-S-transferase fusion protein Ras-binding domain (GST-RBD) followed by Western blotting with an anti-Ras antibody as described in the section "Methods" (top). Total p21Ras was identified by Western blotting of total cell lysate (middle). Also shown are (bottom) averaged data quantified by laser densitometry analysis of Western blots, expressed as relative intensity of the bands.

inhibitors (2 m$M$ Na$_3$VO$_4$; 50 m$M$ NaF; 10 m$M$ Na pyrophosphate). Total cell lysates (60 $\mu$g/lane) were resolved in 10% sodium dodecylsulfate–polyacrylamide gel electrophoresis (SDS-PAGE) and blotted onto nitrocellulose sheets. The nitrocellulose blots were saturated with TBS-T (10 m$M$ Tris pH 7.5; NaCl 150 m$M$; Tween 20 0.1%) and 0.1% bovine serum albumin (BSA) for 2 h at room temperature and were incubated overnight at 4° with anti-phosphotyrosine antibody (anti-PY) clone 4G10 (Upstate Biotechnology, Lake Placid, NY) diluted in TBS-T/0.1% BSA (1:2000). Proteins were visualized using a goat anti-mouse secondary antibody conjugated to HRP diluted in TBS-T (1:2000 and 1 h incubation at room temperature) and a chemiluminescence detection system (Fig. 1A).

NO-stimulated tyrosine phosphorylation of the EGF receptor was characterized by using immunoprecipitation of the receptor followed by Western blot with anti-PY. Upon treatment and cell lysis in supplemented lysis buffer A, lysates (2.0 mg protein) were incubated for 1 h with 4 $\mu$g of a mouse monoclonal anti-EGF receptor antibody (Sigma-Aldrich Co., St. Louis, MO) clone 29.1.1 preconjugated with protein A-Sepharose CL4B (15 mg/50% slurry in 20 m$M$ HEPES, pH 7.5). Immunoprecipitated receptors were blotted onto nitrocellulose sheets and probed with anti-PY, as described earlier (Fig. 1B).

## Immunoblotting for Detection of ERK1/2 and Phospho-ERK1/2 MAP Kinases

After cell treatments, lysis, and determination of protein concentrations, the nitrocellulose blots were saturated with TBS-T/5% skim milk for 2 h at room temperature. After three washes (15 min at room temperature) in TBS-T, membranes were incubated overnight at 4° with anti-phospho-p44/p42 MAP kinase antibody (Thr202/Tyr204) (Cell Signaling Technology, Beverly, MA) and diluted in TBS-T/5% BSA (1:2000). Proteins were visualized using a goat anti-rabbit secondary antibody conjugated to HRP diluted in TBS-T (1:3000 and 1 h incubation at room temperature) and a chemiluminescence detection system. Nitrocellulose membranes were stripped from the first antigen–antibody complex in stripping buffer (Tris–HCl, pH 6.7; 0.1% $\beta$-mercaptoethanol; 2% SDS) for 30 min, with agitation every 10 min at 60°. Stripped membranes were submitted to the same treatment and probed with anti-p44/p42 MAP kinase antibody (Cell Signaling Technology, Beverly, MA) diluted in TBS-T/5% BSA (1:2000). Because both antibodies are generated in rabbits, we adopt the same procedure to visualize the proteins. The ratio between the densitometric analysis of blots for both phospho-ERK1/2 and ERK1/2 MAP kinases revealed the extent of ERK1/2 MAP kinases activation (Fig. 1C).

## Transfection of Cell Cultures with Modified Forms of p21Ras

RAECs expressing the N17Ras protein (negative dominant mutant of p21Ras) under the control of the mouse mammary tumor virus (MMTV) promoter were obtained by transfection of 4 $\mu$g of pcDNAIII/pMMRasDN (Feig and Cooper, 1988) or empty pcDNAIII vector (control of the transfection) by the Lipofectin method (Invitrogen, Carlsbad, CA) according to the manufacturer's instructions. Cells stably transfected were selected in 400 $\mu$g/ml Geneticin (G418) containing F12 media supplemented with 10% FBS. After isolating the colonies, the transfected cell lines were kept in G418 (100 $\mu$g/ml). Clones were screened for the presence of N17Ras by Southern blot analysis, and the selected clones were tested for N17Ras transcription by Northern blot analysis.

HeLa cells expressing the p21Ras C118S, where cysteine 118 was replaced by a serine residue, were obtained by transfection of 4 $\mu$g of pcDNAIII/H-RasC118S (Lander et al., 1997) or pcDNAIII/H-Ras (control of the transfection) by the Lipofectamine method (Invitrogen, Carlsbad, CA) according to the manufacturer's instructions. Cells stably transfected were selected in 600 $\mu$g/ml Geneticin (G418) containing Minimum Essential Medium supplemented with 1 m$M$ sodium pyruvate and 10% FBS. After isolating the colonies, the transfected cell lines were kept in G418 (600 $\mu$g/ml). Transfected cells were characterized by expression levels and activity of p21Ras as described in the next section.

## Assay for p21Ras Activation

P21Ras activation was determined using the minimal Ras-binding domain (RBD) of Raf-1 (a51-131) glutathione-$S$-transferase (GST) fusion protein (GST-RBD), which binds very tightly to the GTP-associated form of p21Ras (Rooij and Bos, 1997). Bacteria expressing the GST-RBD fusion protein were induced with 1 m$M$ IPTG. Bacterial suspensions (OD ranging from 0.3 to 0.4 arbitrary units of absorbance) were sonicated on ice six times for 1 min in phosphate-buffered saline (PBS) containing 0.5 m$M$ dithiothreitol (DTT), 0.1 $\mu M$ aprotinin, 1 $\mu M$ leupeptin, and 1 m$M$ phenyl-methylsulfonylfluoride (PMSF). Bacterial lysates, stirred in 1% Triton X-100 and 10% glycerol for 30 min at 4°, were aliquoted and stored at −80°. The selected amount of crude GST-RBD was conjugated to glutathione-Sepharose beads at 4° overnight. Beads were centrifuged and washed three times with RIPA buffer (50 m$M$ Tris, pH 8.0; 150 m$M$ NaCl; 0.5% sodium deoxycholate; 1% NP-40; 0.1% SDS; 1 $\mu$g/ml aprotinin; 1 $\mu$g/ml leupeptin, and 1 m$M$ DTT). Beads were incubated with cell lysates, and bound material was assayed by Western blots. Blots were saturated in TBS-T/5% skim milk (2-h room temperature) and probed overnight (4°)

with a mouse monoclonal anti-pan Ras antibody (Calbiochem, San Diego, CA) diluted in TBS-T/5% BSA (1:2000). Proteins were visualized using a goat anti-mouse secondary antibody conjugated to HRP diluted in TBS-T (1:2000 and 1-h incubation at room temperature) and a chemiluminescence detection system. The remaining lysates (total protein 50 $\mu$g/ml) were probed with the same antibody to determine the levels of total endogenous p21Ras. The ratio between the intensity of the Ras signal in the GST-RBD pull down and that obtained from total p21Ras determined by densitometry is proportional to the activity of p21Ras (Fig. 1D).

## Acknowledgments

The work performed in our laboratory received the financial support from the following Brazilian Institutions: Fundação de Amparo a Pesquisa do Estado de S.Paulo (FAPESP; 00/12154-2), Conselho Nacional de Desenvolvimento Cientifico e Tecnológico (CNPq; 306814/2003-0), and Coordenadoria de Aperfeiçoamento de Pessoal de Nível Superior (CAPES).

## References

Akhand, A. A., Pu, M., Senga, T., Kato, M., Suzuki, H., Miyata, T., Hamaguchi, M., and Nakashima, I. (1999). Nitric oxide controls Src kinase activity through a sulphydryl group modification-mediated Tyr-527–independent and Tyr-416–linked mechanism. *J. Biol. Chem.* **274**, 25821–25826.

Deora, A. A., Win, T., Vanhaesebroeck, B., and Lander, H. M. (1998). A redox-triggered ras-effector interaction. Recruitment of phosphatidylinositol 3 kinase to Ras by redox stress. *J. Biol. Chem.* **273**, 29923–29928.

Deora, A. A., Hajjar, D. P., and Lander, H. M. (2000). Recruitment and activation of Raf-1 by nitric oxide–activated Ras. *Biochemistry* **39**, 9901–9908.

Feig, L. A., and Cooper, G. M. (1988). Inhibition of NIH3T3 cell proliferation by a mutant ras protein with preferential affinity for GDP. *Mol. Cell Biol.* **8**, 3235–3243.

Furchgott, R. F., and Zawadzki, J. Z. (1980). The obligatory role of endothelial cells in the relaxation of arterial smooth muscle by acetylcholine. *Nature* **288**, 373–376.

Garcia-Benito, M., San Roman, J. I., Lopez, M. A., Garcia-Marin, L. J., and Calvo, J. J. (2000). Nitric oxide stimulates tyrosine phosphorylation of p125 (FAK) and paxillin in rat pancreatic acini. *Biochem. Biophys. Res. Commun.* **274**, 635–640.

Giancotti, F. G., and Ruoslahti, E. (1999). Integrin signaling. *Science* **285**, 1028–1032.

Goligorsky, M. S., Abedi, H., Noiri, E., Takhtajan, A., Lense, S., Romanov, V., and Zachary, I. (1999). Nitric oxide modulation of focal adhesions in endothelial cells. *Am. J. Physiol.* **276**, C1271–C1281.

Gruetter, D. Y., Gruetter, C. A., Barry, B. K., Baricos, W. H., Hyman, A. L., Kadowitz, P. J., and Ignarro, L. J. (1980). Activation of coronary arterial guanylate cyclase by nitric oxide, nitroprusside, and nitrosoguanidine—Inhibition by calcium, lanthanum, and other cations, enhancement by thiols. *Biochem. Pharmacol.* **29**, 2943–2950.

Hess, D. T., Matsumoto, A., Nudelman, R., and Stamler, J. S. (2001). S-nitrosylation: Spectrum and specificity. *Nat. Cell. Biol.* **3**, E1–E3.

Katsuki, S., Arnold, W., Mittal, C. K., and Murad, F. (1977). Stimulation of guanylate cyclase by sodium nitroprusside, nitroglycerin and nitric oxide in various tissue preparations and comparison to the effects of sodium azide and hydroxylamine. *J. Cyclic Nucleotide Res.* **3,** 23–35.

Lander, H. M., Milbank, A. J., Tauras, J. M., Hajjar, D. P., Hempstead, B. L., Schwartz, G. D., and Kraemer, R. T. (1996). Redox regulation of cell signaling. *Nature* **381,** 380–381.

Lander, H. M., Hajjar, D. P., Hempstead, B. L., Mirza, U. A., Chait, B. T., Campbell, S., and Quilliam, L. A. (1997). A molecular redox switch on p21Ras. *J. Biol. Chem.* **272,** 4323–4326.

Lincoln, T. M., Cornwell, T. L., Komalavilas, P., and Boerth, N. (1996). Cyclic GMP–dependent protein kinase in nitric oxide signaling. *Methods Enzymol.* **269,** 149–166.

Monteiro, H. P., Peranovich, T. M. S., Fries, D. M., Stern, A., and Silva, A. M. (1994). Nitric oxide potentiates EGF-stimulated tyrosine kinase activity in 3T3 cells expressing human EGF receptors. *In* "Frontiers of Reactive Oxygen Species in Biology and Medicine" (K. Asada and T. Yoshikawa, eds.), pp. 215–218. Elsevier Science, Amsterdam.

Monteiro, H. P., Gruia-Gray, J., Peranovich, T. M. S., Barbosa de Oliveira, L. C., and Stern, A. (2000). Nitric oxide stimulates tyrosine phosphorylation of focal adhesion kinase, Src kinase, and mitogen-activated protein kinases in murine fibroblasts. *Free Radic. Biol. Med.* **28,** 174–182.

Nathan, C. (1992). Nitric oxide as a secretory product of mammalian cells. *FASEB J.* **6,** 3051–3064.

Peranovich, T. M. S., da Silva, A. M., Fries, D. M., Stern, A., and Monteiro, H. P. (1995). Nitric oxide stimulates tyrosine phosphorylation in murine fibroblasts in the absence and presence of epidermal growth factor. *Biochem. J.* **305,** 613–619.

Rocha Oliveira, C. J., Schindler, F., Ventura, A. M., Moraes, M. S., Arai, R. J., Debbas, V., Stern, A., and Monteiro, H. P. (2003). Nitric oxide and cGMP activate the Ras-MAP kinase pathway-stimulating protein tyrosine phosphorylation in rabbit aortic endothelial cells. *Free Radic. Biol. Med.* **35,** 381–396.

Rhoads, J. M., Chen, W., Gookin, J., Wu, G. Y., Fu, Q., Blikslager, A. T., Rippe, R. A., Argenzio, R. A., Cance, W. G., Weaver, E. M., and Romer, L. H. (2004). Arginine stimulates intestinal cell migration through a focal adhesion kinase dependent mechanism. *Gut* **53,** 514–522.

Rooij, J., and Bos, J. L. (1997). Minimal Ras-binding domain of Raf-1 can be used as an activation probe for Ras. *Oncogene* **14,** 623–625.

Shields, J. M., Pruitt, K., McFall, A., Shaub, A., and Der, C. J. (2000). Understanding Ras: "It ain't over till it's over." *Trends Cell. Biol.* **10,** 147–154.

Stamler, J. S. (1994). Redox signaling: Nitrosylation and related target interactions of nitric oxide. *Cell* **78,** 931–936.

# [30]   Identification and Evaluation of NO-Regulated Genes by Differential Analysis of Primary cDNA Library Expression (DAzLE)

*By* SUK J. HONG, VALINA L. DAWSON, and TED M. DAWSON

## Abstract

Nitric oxide (NO) has numerous physiological roles in the cell. One of the actions of NO is gene regulation through protein modification and signal transduction. In neurons, NO can be produced from neuronal NO synthase, which is activated by calcium following $N$-methyl-D-aspartate (NMDA) receptor activation. Differential analysis of cDNA library expression (DAzLE) was used to identify differentially expressed genes by NO. Fundamentally, this technique combines differential hybridization to isolate genes whose expression is differentially regulated with microarray to analyze the expression of the isolated genes. The expression of genes identified by the DAzLE method is verified further by quantitative real-time polymerase chain reaction (RT-PCR) and/or Northern blot analysis. The high selectivity and sensitivity of this technique for detecting differentially expressed gene transcripts enable the investigation and identification of a panel of genes that are regulated by NO.

## Introduction

Nitric oxide (NO) is a reactive free radical species that is a neuronal messenger. It can also mediate nitrosative and oxidative stress through nitrosation, nitration, or nitrosylation of various molecules, including lipid, DNA, and proteins. It is evident that NO plays many roles in the physiology and pathophysiology of the central nervous system (CNS), such as CNS development, long-term potentiation, trauma, ischemia, and neurodegenerative diseases (Contestabile *et al.*, 2003). Although physiological levels of NO subserve signaling and neuromodulatory roles, excessive creation of NO leads to neuronal death.

In neurons, NO produced from neuronal NO synthase (nNOS) can diffuse into glial cells or other neurons and act as a signaling molecule. As a downstream event of NO-mediated signaling, the modulation of gene expression by NO may play a key role in long-term changes of neurons, such as synaptic plasticity and neuronal development (Gibbs, 2003; Pilz and Casteel, 2003).

METHODS IN ENZYMOLOGY, VOL. 396                                   0076-6879/05 $35.00
        DOI: 10.1016/S0076-6879(05)96030-8

NO is known to affect gene regulation in cells, not only by activating NO-dependent signaling pathways (Schafer *et al.*, 2000), but also by direct posttranslational modification of transcription factors (Reynaert *et al.*, 2004). In this regard, it is not surprising that many transcription factors show redox sensitivity through cysteine residues in DNA binding domains under both nitrosative and oxidative stress (Marshall *et al.*, 2000).

The DAzLE method described in this chapter is designed to identify differentially regulated genes (Fig. 1). This method has several advantages: (1) Because DAzLE is based on the screening of a primary nonamplified cDNA library with the probes containing poly(dA/dT) tailless cDNAs that limit cross-hybridization between the 3′ end of the sequences, it can detect low-abundant transcripts and medium- to high-abundant transcripts with minimal redundancy in differentially expressed genes. (2) There is no limit in the number of genes for cDNA library differential screening, which

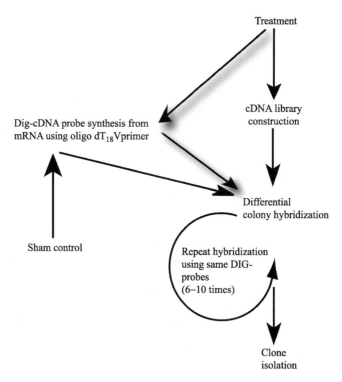

Fig. 1. Illustration of primary screening in differential analysis of cDNA library expression [differential analysis of cDNA library expression (DAzLE)].

allows one to investigate as many genes as possible. (3) A large number of genes from the primary screening can be analyzed using microarray simply and efficiently.

This technique can be applied not only to NO-mediated gene regulation (Li *et al.*, 2004), but also to other paradigms of gene modulation (Hong *et al.*, 2004).

Procedures

*Experimental Scheme (See Figure 2)*

Cortical Neuron Culture (14 DIV culture from E16 embryo)
→NADPH diaphorase staining (a marker of nNOS expression) is used to confirm the expression of nNOS and the maturation of the cortical neurons

⇓

50 $\mu M$ NMDA + 10 $\mu M$ glycine for 5 min to activate NMDA-induced signal transduction
→Six-h incubation → Collect total RNA from cells

⇓

mRNA purification using oligo(dT) column → cDNA library construction

⇓

cDNA library differential screening using cDNA probes derived from NMDA-treated cells and CSS-treated cells

⇓

Confirmation of the clones that are regulated by NMDA by Northern, reverse Northern blotting

⇓

Identification of the genes by DNA sequencing and genomics

⇓

Functional studies of regulated genes

*Cell Culture*

Primary cortical neurons are prepared from gestational day 16 fetal C57Bl mice and from nNOS-null C57Bl mice (Dawson *et al.*, 1993). The dissected cortical brain regions are incubated for 15 min in 0.027% trypsin/saline solution [5% phosphate-buffered saline (PBS), 40 m$M$ sucrose, 30 m$M$ glucose, 10 m$M$ HEPES, pH 7.4] and transferred to modified Eagle's medium (MEM) containing 10% horse serum, 10% fetal bovine serum, and 2 m$M$ glutamine. Cells are dissociated by trituration, counted, and

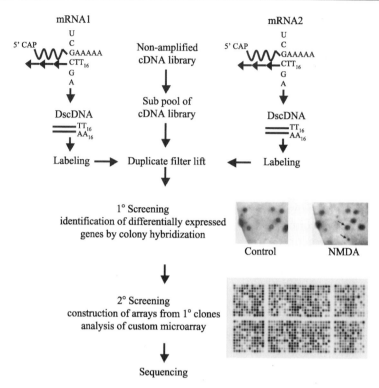

FIG. 2. Scheme displaying the use of differential analysis of cDNA library expression (DAzLE) in identification of $N$-methyl-D-aspartate receptor–mediated genes. $1.2 \times 10^6$ bacterial cDNA library clones were used for primary screening. There were 1152 clones that were identified from primary screening and arrayed in nitrocellulose membranes. The microarray membranes were used for secondary screening and used for isolation of genes regulated by nitric oxide. Reproduced with permission (Hong *et al.*, 2004).

plated in 24-well plates (Nunc) coated with poly-ornithine at a density of $4 \times 10^5$ cells/well. Four days after plating, the cells are treated with 6.7 $\mu$g/ml of 5-fluoro-2'-deoxyuridine for 3 days to inhibit proliferation of nonneuronal cells. Cells are maintained in MEM, 10% horse serum, 2 m$M$ glutamine and 25 m$M$ glucose in 7% $CO_2$ humidified, 37° atmosphere. The medium is changed twice a week. Mature neurons (>14 days in culture) are used. In the mature cultures, the percentage of neurons is approximately 70–90% of the total number of cells, as assessed by neuron-specific enolase (NSE) and glial fibrillary acidic protein (GFAP) immunocytochemical staining of neurons and astrocytes, respectively.

*Diaphorase Staining*

NADPH diaphorase staining is used to detect NOS-containing neurons in cultures. Cells are washed three times with PBS and fixed for 30 min at 4° in a 4% paraformaldehyde (PF), 0.1 $M$ phosphate buffer. The PF solution is washed away with Tris-buffered saline (TBS) containing 50 m$M$ Tris–HCI, 1.5% NaCl, pH 7.4. The reaction solution containing 1 m$M$ NADPH, 0.2 m$M$ nitroblue tetrazolium, 0.2% Triton X-100, 1.2 m$M$ sodium azide, 0.1 $M$ Tris–HCI, pH 7.2, is applied to the fixed cell cultures for 1 h at 37°. The reaction is terminated by washing away the reaction solution with TBS. All diaphorase-positive cells in each well are counted using an inverted microscope. Generally, mature cortical cultures contain more than 200 diaphorase-positive cells per well.

*Cell Treatment*

Mature neurons are washed with Tris-buffered control salt solution (CSS) containing 120 m$M$ NaCl, 5.4 m$M$ KCl, 1.8 m$M$ CaCl$_2$, 25 m$M$ Tris–HCl pH 7.4, and 20 m$M$ D-glucose. Next, 50 $\mu M$ NMDA and 10 $\mu M$ glycine in CSS is applied to the cells for 5 min to induce NMDA-dependent genes, and then the cells are washed and replaced with minimum essential medium containing 10% horse serum and incubated for 6 h in the incubator. Sham (control) treatment is performed as described earlier, except for 5-min treatment with CSS only. $N$-nitro-L-arginine (0.5 m$M$) is dissolved in CSS to make a stock solution. To inhibit NOS activity, $N$-nitro-L-Arg (100 $\mu M$) is added to neuronal cultures 15 min before NMDA treatment. Then, NMDA (50 $\mu M$), glycine (10 $\mu M$), and $N$-nitro-L-Arg (100 $\mu M$) are applied to the cultures for 5 min. Neuronal cultures are washed with culture medium three times. Using another set of neuronal cultures, NO donors (NOR3, 10 $\mu M$, Sigma; NOR3-NO$^-$, 10 $\mu M$) are added to culture medium to deliver NO directly to neurons.

*cDNA Library Construction*

The cDNA library should be large enough to contain representatives of all sequences of interest, some of which may be derived from low-abundance messenger RNAs (mRNAs). It includes a minimal number of clones that contain small (<500 bp) cDNA inserts, and it is composed of cDNA inserts that are near full-length copies of the mRNAs. Total RNA is extracted from neurons with TRIzol reagent (Invitrogen), and poly(A)$^+$ RNA is purified with oligo(dT) cellulose chromatography. A bacterial cDNA library from the mRNA of NMDA-treated neurons or NO-donors–treated neurons was constructed with CloneMiner cDNA library construction kit (Invitrogen).

## Colony Hybridization

To identify genes regulated by NO signaling pathway, bacteria containing the plasmid cDNA library are used for differential hybridization screening. Nylon membrane filters (137 mm) are laid on an agar plate (150 mm) taking care to avoid air bubbles between the membrane and agar surface. Up to 2000 bacterial colonies are applied onto the filters and incubated until they grow 0.1–0.2 mm in diameter. The template filter is peeled off and laid, colonies up, on a bed of sterile Whatman paper. A wetted sterile filter is held between two flat-bladed forceps and laid on the template filter. The sandwich is pressed firmly together with a velvet-covered replica-plating tool (or thick glass plate). The replica is peeled off the template and placed on a fresh agar plate (up to five filters). The replicas are incubated at 30–37° until the colonies develop to 0.5–1.0 mm. Filters are removed carefully from the plates and placed (colonies side up) on the prepared filter paper (on two layers) soaked with denaturation solution (0.5 N NaOH, 1.5 $M$ NaCl) for 15 min. Filters are then placed for 15 min onto the prepared filter paper soaked with neutralization solution (1.0 $M$ Tris–HCl, pH 7.5, 1.5 $M$ NaCl). Filters are transferred onto the prepared Whatman paper soaked with 2× SSC for 10 min and air-dried for 30 min. The transferred DNA is crosslinked with UV-crosslinker (1200 unit, two times). Then, each filter is placed on a clean piece of aluminum foil and treated with 2 ml of 2 mg/ml proteinase K. The solution is distributed evenly and incubated for 1–2 h at 37°. Using Whatman paper fully wetted with dH$_2$O, the filters are blotted between the Whatman paper and pressure is applied by passing over the area with a bottle. Cellular and agar debris are removed by gently pulling off the upper filter paper (the debris will stick to this filter paper). If necessary, the blotting step is repeated with a fresh piece of paper.

### Hybridization

1. Place up to three membrane discs in hybridization bag and add 100 ml prehybridization solution [DIG Easy Hyb Granules (Roche) can be used].
2. Prehybridize for 2 h in a hybridization oven at 42°.
3. Radioactive first-strand cDNA probe is prepared as described in Reverse Northern Blotting procedure. Denature the labeled probe by incubating at 68° for 10 min (reuse only). If the probe is synthesized freshly, boil it for 5 min and cool it rapidly on ice.
4. Fresh probe only: Mix the denatured probe with 20 ml hybridization solution, prewarmed to 42°.

5. Remove the prehybridization solution and add the hybridization solution.
6. Incubate for 2 h to overnight at 42°.
7. At the end of the hybridization, pour the hybridization solution into a 50-ml tube.
8. Wash the membranes twice for 5 min in ample 2× SSC, 0.1% SDS at room temperature with gentle agitation.
9. Transfer the membranes to 0.5× SSC, 0.1% SDS, and wash twice for 15 min at 68° with gentle agitation.

*Immunological Detection*

- Washing buffer: 0.1 *M* maleic acid, 0.15 *M* NaCl, pH 7.5, 0.3% Tween 20
- Maleic acid buffer: 0.1 *M* maleic acid, 0.15 *M* NaCl, pH 7.5
- Detection buffer: 0.1 *M* Tris–HCl, 0.1 *M* NaCl, pH 9.5
- Blocking solution: Dilute 10× blocking solution (Roche) 1:10 with maleic acid buffer (prepare fresh)
- Antibody solution: Centrifuge anti-digoxigenin-AP (Roche) for 5 min at 10,000 rpm in the original vial before each use, and pipette the proper volume carefully from the surface. Dilute anti-digoxigenin-AP 1:10,000 in blocking solution (3 $\mu$l in 30 ml).

1. After hybridization and stringency washes, rinse membranes briefly 1–5 min in washing buffer.
2. Incubate for 30 min in 100 ml blocking solution.
3. Incubate for 30 min in 30 ml antibody solution.
4. Wash 2 × 15 min in 100 ml washing buffer.
5. Equilibrate 2–5 min in 40 ml detection buffer.
6. Place membranes with DNA side up on a development folder (big X-ray film can be used).
7. Apply 1 ml CSPD ready-to-use (Roche) spread the substrate evenly and without air bubbles over the membrane (by covering Saran Wrap).
8. Incubate for 10 min at 15–25°.
9. Squeeze out excess liquid (do not dry the membrane).
10. Incubate the damp membrane for 10 min at 37° to enhance the luminescent reaction.
11. Expose to X-ray film for 25 min to 1 h at 15–25° (up to 48 h).
12. The bacterial colonies that show higher intensity on X-ray film with NMDA-treated neuronal probe are picked up, cultured in LB broth containing ampicillin, and preserved at −80° in 50% glycerol.

*Reverse Northern Blotting Procedure*

Plasmid DNAs from the positive bacterial clones are isolated, denatured, and spotted on a positively charged Nylon membrane (Amersham) with a 96-well vacuum manifold. The spotted DNA is crosslinked to the membrane with a UV crosslinker (Amersham Pharmacia). $^{32}$P-labeled first-strand cDNA is prepared by reverse transcription of total RNA. Next, 30 $\mu$g of total RNA is mixed with 4 $\mu$g of dT$_{18}$V, incubated 10 min at 70°, and cooled on ice for 5 min. The mixture is added with 50 m$M$ Tris–HCl (pH 8.3), 75 m$M$ KCl, 3 m$M$ MgCl$_2$, 10 m$M$ dithiothreitol (DTT), 0.5 m$M$ dATP, 0.5 m$M$ dGTP, 0.5 m$M$ dTTP, 0.02 8m$M$ dCTP, 100 $\mu$Ci $\alpha$-$^{32}$P-dCTP, and 200 units of Superscript RT II (Invitrogen) in a final 25-$\mu$l reaction solution. The reaction mixture is incubated at 42° for 1 h and at 42° for 30 min after addition of 200 units of Superscript RT II. The membrane is hybridized and washed as described earlier. The signal intensity of each spot is measured by ImageQuant software (Amersham Pharmacia), normalized, and compared between control and NMDA-treated neurons. Each image is overlaid with grids to compare signal intensities of individual spots.

*Microarray Construction and Analysis*

Because the gene expression of a large number can be easily monitored by microarray analysis, it is beneficial to make an array of differentially expressed genes. The NMDA-induced gene-enriched microarray is constructed by arraying polymerase chain reaction (PCR)–amplified cDNA clones at high density on a nylon membrane. 1152 bacterial clones are selected from differential screening. The plasmids are purified from 96-well bacterial cultures (Edge Biosystems), and the cDNA inserts are amplified by PCR. Each PCR product is verified by agarose gel electrophoresis, and each product is printed onto nylon membrane by an array robot. Thirty micrograms of total RNA is used to label cDNA probes by reverse transcription for hybridizing to the microarrays. Total RNA (30 $\mu$g) and 3 $\mu$g of oligo(dT)$_{18}$V are mixed (adjust final volume with RNase-free water to 13 $\mu$l), incubated at 70° for 10 min, cooled down to 42°, and then placed on ice. Reverse transcription reaction buffer [5× first-strand synthesis buffer (6 $\mu$l), 100 m$M$ DTT (2.5 $\mu$l), 33 m$M$ d(AGT)TP mix (0.6 $\mu$l), 100 $\mu M$ dCTP (1.5 $\mu$l), $\alpha$-$^{33}$P-dCTP (5 $\mu$l), and SuperscriptII (1.5 $\mu$l)] is added to RNA-oligo(dT) solution and incubated 42° for an hour. RNA is hydrolyzed at 65° for 30 min after adding 1 $\mu$l of 1% SDS, 1 $\mu$l of 0.5 $M$ EDTA, and 3 $\mu$l of 2 N NaOH. Sample is neutralized with 10 $\mu$l of 1 $M$ Tris–HCl (pH 7.5) and 3 $\mu$l of 2 N HCl. $^{33}$P-labeled cDNA is purified with ProbeQuant G-50 microspin column (Amersham). And then cDNA sample is boiled for 5 min and either directly added to the hybridization buffer or placed on ice before hybridization. [33]P-labeled cDNAs from sham-treated and NMDA-treated cortical neurons (or NO-donor–treated neuron) are used as the reference probe and the sample probe, respectively, in all hybridizations. Ten micrograms of polydeoxyadenylic acid and 20 $\mu$g of human CoT1 DNA (Invitrogen) are added to a DIG easy hybridization solution (Roche), and the microarray membrane is prehybridized at 42° for 1 h before the probe is added directly to the prehybridization solution. Denatured probe (95°, 5 min) is added to the solution and hybridized for 16–24 h. Unbound probe is removed by washing the membranes three times for 5 min each at 55° with 0.5× SSC solution containing 0.01% SDS and once for 5 min at room temperature with 0.06× SSC solution. The microarray membrane is exposed to a phosphoimage screen for 24 h. The screen is scanned in a phosphoimager at 50-micron resolution. Genes are selected as differentially expressed clones if their expression level deviated from that of sham-treated neurons by a factor of 2.5 in at least five of the samples from NMDA- (or NO-donor–) treated neurons or the standard deviation for the set of five values of z-ratios determined in the analysis of the time course of gene expression exceeded 0.8. The cDNAs displaying differential expression are selected, confirmed by Northern analysis or by RT-PCR. The sequences of differentially expressed genes are analyzed.

### Quantitative RT-PCR Analysis

*Preparation of Total RNA.* Total RNA (5 $\mu$g) is mixed with random hexamers (200 ng), incubated at 70° for 10 min and chilled on ice for 1 min. Reverse transcription buffer (20 m$M$ Tris–HCl, pH 8.4, 50 m$M$ KCl, 2.5 m$M$ MgCl$_2$, 10 m$M$ DTT, 500 $\mu M$ each dNTP) is added, and reaction mixture is incubated at 25° for 5 min. Superscript RT II (200 units) is added to the mixture and incubated for 10 min at 25°, for 50 min at 42°, and for 15 min at 70°. After the mixture is chilled on ice, 2 units of RNase H is added and further incubated for 20 min at 37°. The reaction is stopped by incubation for 15 min at 70°.

*QPCR (SYBR Green Method).* Primers (22mers, 50% GC content, PCR product size between 100 and 150 bp) are designed and synthesized. cDNA (1 $\mu$l) is mixed with 24 $\mu$l of reaction mixture (1× SYBR PCR buffer, 3 m$M$ MgCl$_2$, 200 $\mu M$ dNTPs, 0.5 units AMP Erase UNG, 100 n$M$ forward primer, 100 n$M$ reverse primer, and 1.25 U AmpliTaq Gold polymerase). PCR conditions are initial incubation at 50° for 2 min and 95° for 10 min, then 40 cycles of 95° for 15 s and 60° for 1 min. The change in fluorescence during PCR is measured in ABI Prism 7700 system.

References

Contestabile, A., Monti, B., and Ciani, E. (2003). Brain nitric oxide and its dual role in neurodegeneration/neuroprotection: Understanding molecular mechanisms to devise drug approaches. *Curr. Med. Chem.* **10,** 2147–2174.

Dawson, V. L., Dawson, T. M., Bartley, D. A., Uhl, G. R., and Snyder, S. H. (1993). Mechanisms of nitric oxide–mediated neurotoxicity in primary brain cultures. *J. Neurosci.* **13,** 2651–2661.

Gibbs, S. M. (2003). Regulation of neuronal proliferation and differentiation by nitric oxide. *Mol. Neurobiol.* **27,** 107–120.

Hong, S. J., Li, H., Becker, K. G., Dawson, V. L., and Dawson, T. M. (2004). Identification and analysis of plasticity-induced late-response genes. *Proc. Natl. Acad. Sci. USA* **101,** 2145–2150.

Li, H., Gu, X., Dawson, V. L., and Dawson, T. M. (2004). Identification of calcium- and nitric oxide–regulated genes by differential analysis of library expression (DAzLE). *Proc. Natl. Acad. Sci. USA* **101,** 647–652.

Marshall, H. E., Merchant, K., and Stamler, J. S. (2000). Nitrosation and oxidation in the regulation of gene expression. *FASEB J.* **14,** 1889–1900.

Pilz, R. B., and Casteel, D. E. (2003). Regulation of gene expression by cyclic GMP. *Circ. Res.* **93,** 1034–1046.

Reynaert, N. L., Ckless, K., Korn, S. H., Vos, N., Guala, A. S., Wouters, E. F., van der Vliet, A., and Janssen-Heininger, Y. M. (2004). Nitric oxide represses inhibitory kappaB kinase through *S*-nitrosylation. *Proc. Natl. Acad. Sci. USA* **101,** 8945–8950.

Schafer, U., Schneider, A., and Neugebauer, E. (2000). Identification of a nitric oxide-regulated zinc finger containing transcription factor using motif-directed differential display. *Biochim. Biophys. Acta* **1494,** 269–276.

# [31] Role of NO in Enhancing the Expression of HO-1 in LPS-Stimulated Macrophages

*By* KLAOKWAN SRISOOK, CHAEKYUN KIM, and YOUNG-NAM CHA

## Abstract

Macrophages serve as the first-line defense against invading pathogens by (a) overproducing $O_2^-$ via activation of NADPH-oxidase localized in its plasma membrane, (b) inducing the expression of inducible nitric oxide synthase (iNOS) and overproducing NO, and (c) generating highly toxic peroxynitrite (ONOO$^-$) to kill the invading pathogens without killing the macrophages themselves. Results show that this was due at least in part to the NO-derived induction of heme oxygenase-1 (HO-1) expression. The NO-derived induction of HO-1 caused (a) rapid elimination of toxic heme to inhibit lipid peroxidation and to prevent further induction of iNOS, (b) rapid production of bile pigment antioxidants to scavenge reactive

METHODS IN ENZYMOLOGY, VOL. 396
0076-6879/05 $35.00
DOI: 10.1016/S0076-6879(05)96031-X

oxygen ($O_2^-$) and nitrogen (NO) metabolites, and (c) rapid production of carbon monoxide (CO) to inhibit further production of $O_2^-$ and NO by blocking the activities of NADPH-oxidase and iNOS, respectively. Thus, the NO overproduced by the $O_2^-$-dependent induction of iNOS expression can scavenge $O_2^-$ to produce $ONOO^-$, first to kill the invading pathogens and second to enhance the HO-1 expression in macrophages. This allows the survival of host tissues from the injuries caused by inflammatory oxidative stress.

## Introduction

Macrophages serve as the first-line defense against invading pathogens by overproducing peroxynitrite ($ONOO^-$), the combination product of superoxide anion ($O_2^-$) and nitric oxide (NO) radicals (Huie and Padmaja, 1993). NADPH-oxidase (PHOX) and inducible NO synthase (iNOS) catalyze the overproduction of $O_2^-$ and NO radicals, respectively, in the stimulated macrophages (Forman and Torres, 2002; MacMicking et al., 1997). When macrophages are stimulated by invading pathogens or by inflammatory cytokines, activity of these heme-containing enzymes increases. Most typically, production of $O_2^-$ and NO in macrophages increases upon stimulation with bacterial lipopolysaccharide (LPS) (Srisook and Cha, 2004; Victor and De La Fuente, 2003). Although these $O_2^-$ and NO radicals are not sufficiently reactive to oxidize cellular GSH directly, they combine rapidly to generate the powerfully GSH-oxidizing $ONOO^-$ and cause damage, not only to invading pathogens but ironically also to macrophages themselves and surrounding host cells (Ferret et al., 2002; Wink et al., 1996). However, macrophages and most of the aerobic host cells can protect themselves by enhancing the expression of heme oxygenase-1 (HO-1; EC 1.14.99.3). HOs are involved in catalyzing the oxidative degradation of free heme, either newly synthesized or released from heme enzymes by the overproduced $O_2^-$ and NO, to liberate $Fe^{2+}$, biliverdin, and carbon monoxide (CO) (Maines, 1997). Both the liberated $Fe^{2+}$ and the released free heme can catalyze the production of highly toxic hydroxyl radical (HO*) from $H_2O_2$ via Fenton chemistry (Halliwell and Gutteridge, 1999). To prevent this, $Fe^{2+}$ is incorporated into ferritin and transported out (Balla et al., 1992; Cairo et al., 1995), and the released free heme is eliminated by HO activity. In most mammalian cells, biliverdin is converted to bilirubin by biliverdin reductase (Elbirt and Bonkovsky, 1999; Otterbein and Choi, 2000). Both biliverdin and bilirubin, the bile pigments, function as antioxidants scavenging and detoxifying reactive oxygen species (ROS; $H_2O_2$, HO*) and reactive nitrogen species (RNS; $ONOO^-$, $N_2O^-$) (Kaur et al., 2003; Mancuso et al., 2003; Stocker et al., 1987). Also, the CO produced by HO activity binds avidly to heme iron and can inhibit PHOX and iNOS activity, suppressing further production of $O_2^-$ and NO, respectively. Thus, the NO overproduced by iNOS in LPS-stimulated macrophages can serve not only to scavenge the $O_2^-$ but also to generate $ONOO^-$, the rapidly GSH-oxidizing product. By scavenging $O_2^-$ and by generating $ONOO^-$, NO may serve to protect cells from the $O_2^-$-derived toxicity as long as a high level of intracellular GSH can be maintained. As a consequence of this $ONOO^-$-dependent lowering of GSH level, expression of HO-1 is enhanced to eliminate the potentially toxic free-heme while rapidly producing the bile pigments and the CO, serving as antioxidants and as inhibitor of additional $O_2^-$ and NO production, respectively. Thus, the overproduction of NO caused initially by LPS-driven oxidative burst may then serve eventually to preserve cellular homeostasis by elevating HO-1 expression.

In this study, RAW 264.7 peritoneal macrophages were exposed to an NO donor, spermine NONOate (SPNO), to determine specifically the role of NO in enhancing HO-1 expression. SPNO was added to culture medium in increasing doses, and the increases of HO-1 messenger RNA (mRNA) expression and HO-1 protein accumulation were determined. These were compared with those enhanced by LPS, used as the positive control. In response to SPNO, HO-1 protein synthesis was increased in a dose-dependent manner without enhancing the iNOS expression (Fig. 1). In the LPS-treated cells, iNOS expression was induced markedly with overproduction of NO (Fig. 2A). This was accompanied by marked increase in HO-1 protein accumulation. This indicated that the overproduced NO resulting from LPS-derived induction of iNOS was causing HO-1 induction. In an effort to determine the effect of inhibiting this endogenous overproduction of NO on HO-1 induction, LPS-stimulated cells

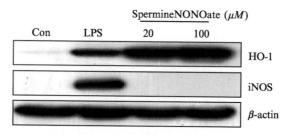

FIG. 1. Induction of HO-1 expression by nitric oxide (NO)–donor spermine-NONOate. Cells were treated with spermine-NONOate (20 $\mu M$, 100 $\mu M$) or 1 $\mu$g/ml lipopolysaccharide (LPS) and harvested at 12 h. Harvested cells were homogenized, and homogenates were analyzed to determine the contents of HO-1 and inducible NO synthase (iNOS) proteins employing immunoblot analysis as described in the section "Methods."

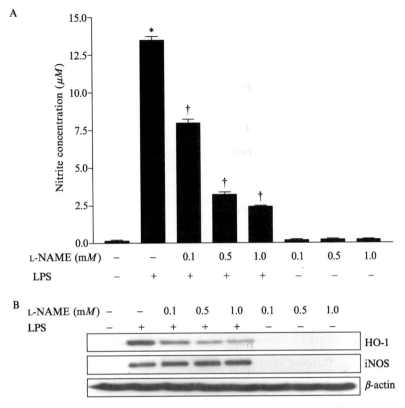

Fig. 2. Induction of HO-1 expression by nitric oxide (NO) produced endogenously. Cells were pretreated with varying doses of L-$N^G$-nitroarginine-methylester (L-NAME) (0.1, 0.5, and 1.0 mM) 30 min before addition of lipopolysaccharide (LPS) (1 µg/ml). Cells were harvested 12 h after the addition of LPS. (A) Accumulated nitrite concentrations present in the conditioned media were determined. Bar graph indicates mean ±SEM (N = 6). Using unpaired Student's $t$ test, asterisk (*) indicates $p < .001$ versus control media. Dagger ($^†$) indicates $p < .001$ versus the LPS-conditioned medium. (B) The immunoblot picture shows a representative result obtained from cells treated as described above.

were treated with a well-known competitive inhibitor of NOS activity [L-$N^G$-nitroarginine-methylester (L-NAME)]. L-NAME inhibited NO production in a dose-dependent manner (Fig. 2A) and inhibited the LPS-dependent induction of HO-1 expression, again, in a dose-dependent manner (Fig. 2B). Although not shown, HO activity determined by production of bilirubin correlated well with the SPNO-dependent increase and the L-NAME–dependent inhibition on the extent of HO-1 protein accumulation.

Combined, these results indicate that HO-1 expression and HO activity are increased by the NO overproduced in LPS-stimulated macrophages. Thus, macrophages can serve as the first-line defense against invading pathogens by (1) activating PHOX to overproduce $O_2^-$, (2) inducing iNOS to overproduce NO, and (3) generating highly toxic $ONOO^-$. This occurs without destroying the $ONOO^-$-generating macrophages. This may be due, at least in part, to the NO-derived induction of HO-1 expression for (1) rapid elimination of the toxic free-heme catalyzing the production of HO* from $H_2O_2$ via the Fenton reaction, (2) rapid production of bile pigment antioxidants, and (3) rapid inhibition on further production of $O_2^-$ and NO by using CO, the product of HO activity. In summary, stimulated macrophages overproducing $O_2^-$ can enhance NO production by inducing iNOS expression, perhaps to increase the $ONOO^-$ generation. This $ONOO^-$ is being used not only to kill the invading pathogens but also to deplete cellular GSH and to initiate enhanced HO-1 expression in the stimulated macrophages, all to defend the host from invading pathogens while aiding its own survival from oxidative injuries caused by pathogen-derived inflammatory responses.

## Materials and Methods

### Cell Culture

Murine peritoneal macrophage cell line RAW 264.7 was obtained from American Type Culture Collection (ATCC). Cells were cultured in Dulbecco's Modified Eagle Medium (DMEM) (Gibco/Invitrogen, Grand Island, NY) containing 100 U/ml of penicillin, 100 $\mu$g/ml of streptomycin, 4 m$M$ L-glutamine, 25 m$M$ D-glucose, 1 m$M$ sodium pyruvate, and 10% heat-inactivated characterized fetal bovine serum (FBS) (Hyclone Laboratories, Logan, UT) at 37° in humidified air containing 5% $CO_2$. The cells used in all experiments were from the same passage prepared from multiple vials of cells stored frozen in liquid $N_2$. Confluent cells in 100-mm plate were incubated for 10 min at 37° in cold Hanks' balanced salt solution (HBSS) without $Ca^{2+}$ and $Mg^{2+}$ and then scraped gently. The cells were pelleted, resuspended in a fresh medium, and then seeded into culture dish. After overnight growth, cells were treated with various experimental chemicals.

### Reverse Transcription Polymerase Chain Reaction

*Complementary DNA Synthesis.* The cell number used for analyzing mRNA expression was $1.5 \times 10^6$ and they were suspended in 3 ml medium and plated in 60-mm plates. Isolation of total RNA was carried out using the TRI reagent (Molecular Research Center, Cincinnati, OH) according to the manufacturer's instructions. Total RNA (500 ng) was reverse transcribed in a 30-$\mu$l reaction volume containing 3 units of avian myeloblastosis virus (AMV) reverse-transcriptase XL (TaKaRa, Shiga, Japan), 0.08 $\mu M$ of oligo(dT) primer, 0.7 m$M$ of each deoxyribonucleic triphosphate (dNTP), 5 m$M$ MgCl$_2$ and 3 $\mu$l of 10× concentrated reaction buffer. The reaction mixture was incubated at 42° for 30 min and at 99° for 5 min.

*Reverse Transcription Polymerase Chain Reaction.* The obtained complementary DNA (cDNA) was used in polymerase chain reaction (PCR). The sequences of primers used were 5′-AGA CTG GAT TTG GCT GGT CCC TCC-3′ (sense) and 5′-AGA ACT GAG GGT ACA TGC TGG AGC C-3′ (antisense) for iNOS; 5′-TGA AGG AGG CCA CCA AGG AGG-3′ (sense) and 5′-AGA GGT CAC CCA GGT AGC GGG-3′ (antisense) for HO-1; 5′-GTC GGT GTG AAC GGA TTT G-3′ (sense) and 5′-ACA AAC ATG GGG GCA TCA G-3′ (antisense) for glyceraldehyde phosphate dehydrogenase (GAPDH). The PCR mixture contained 200 $\mu M$ each of the dNTPs, 0.16 $\mu M$ each of the sense and antisense primers, 1 unit Taq polymerase (TaKaRa, Shiga, Japan), reaction buffer (10 m$M$ Tris–HCl, pH 8.3, 50 m$M$ KCl), 1.5 m$M$ MgCl$_2$, 3 $\mu$l of cDNA, and sterile distilled water to make up 30 $\mu$l. PCR was carried out as follows: one cycle of 1 min at 94°, 25 cycles each for 30 s at 94°, 30 s at 55°, and 1 min at 72°, and one final cycle of 10 min at 72°. The amplified products were mixed with 0.2 volume of 6× concentrated loading buffer (0.25% bromophenol blue, 0.25% xylene cyanol FF, 30% glycerol) and separated for 30 min at 100 V in a 1.5 % agarose gel made up in TBE buffer (89 m$M$ Tris–borate, 2 m$M$ EDTA, pH 8.0). The gels were stained with 1 $\mu$g/ml ethidium bromide solution for 10 min, destained in distilled water for 10 min with shaking, and then the DNA bands were visualized under ultraviolet (UV) light. Quantification of RT-PCR products was performed by densitometry for each band using Bio-Profil software version 99.04 (Vilber Lourmat Biotechnology, France). The image densities of PCR products for HO-1 and iNOS mRNAs were normalized with the density of GAPDH mRNA. Wearing gloves is recommended when working with RNA and ethidium bromide.

## Western Blot Analysis

*Preparation of Whole Cell Extracts.* For the preparation of proteins to be analyzed by Western blot assays, $2 \times 10^6$ cells plated in 3 ml of medium in 60-mm plates were used. Cells were washed with ice-cold phosphate-buffered saline (PBS) and then scraped in the presence of ice-cold lysis buffer containing 50 m$M$ HEPES (pH 7.5), 2 m$M$ EDTA, 50 m$M$ NaCl,

1 m$M$ MgCl$_2$, 1 m$M$ dithiothreitol (DTT), 0.3% Triton X-100, and a protease inhibitor mixture [10 $\mu$g/ml leupeptin, 10 $\mu$g/ml trypsin inhibitor, 2 $\mu$g/ml aprotinin, and 1 m$M$ phenylmethylsulfonylfluoride (PMSF)]. The cell lysate was sonicated twice on ice for 1 min with 1-s interval and at the output of 2 W employing the Vibracell ultrasonic processor. Protein concentrations were quantified using the BCA protein assay kit (Pierce, Rockford, IL).

*Protein Electrophoresis and Western Blotting.* Equal amounts of proteins obtained from cells were mixed with loading buffer [125 m$M$ Tris–HCl, pH 6.8, 5% 2-mercaptoethanol, 10% glycerol, 4% sodium dodecylsulfate (SDS), 0.01% bromophenol blue], heated for 10 min at 100°, and subjected to electrophoresis using 10% SDS polyacrylamide gels. Precision plus protein standards (Bio-Rad, Hercules, CA) were loaded for indication of molecular weights. Electrophoresis was performed for 2 h at 80 V in Tris–glycine electrophoresis buffer (192 m$M$ glycine, 25 m$M$ Tris, and 0.1% SDS) (Sambrook and Russell, 2001).

*Detection of iNOS and HO-1 Proteins.* Separated proteins were transferred onto polyvinylidene fluoride (PVDF) membrane (Millipore, Bedford, MA) by overnight 30-V electroblotting in transfer buffer (192 m$M$ glycine, 25 m$M$ Tris, and 10% methanol) kept at 4°, and the nonspecific bindings were blocked with TBS-T buffer [10 m$M$ Tris–HCl, pH 7.4, 100 m$M$ NaCl, and 0.1% (v/v) Tween 20] containing 5% nonfat dried milk at room temperature for 1 h or overnight at 4° with shaking. The membranes were then incubated at room temperature with anti-mouse HO-1 (Stressgen Biotechnologies, Victoria, Canada), anti-mouse iNOS (BD Transduction Laboratory, Lexington, KY), and anti-mouse $\beta$-actin antibodies (Sigma, St. Louis, MO) diluted at 1:500, 1:1000, and 1:5000, respectively. The duration of incubation employed to determine the contents of HO-1 and iNOS was 2 h, and 1 h for $\beta$-actin. The PVDF membranes were then washed five times with excess TBS-T buffer for 5 min with shaking. Then, the membranes were incubated for 1 h at room temperature with goat anti-mouse immunoglobulin G (IgG):horseradish peroxidase (HRP) antibody diluted at 1:1000 for detection of HO-1 and iNOS, and at 1:5000 for $\beta$-actin (BD Transduction Laboratory). After washing the membranes with excess TBS-T buffer five more times, the specific protein bands on PVDF membrane were visualized on X-ray film activated by chemiluminescence using Western blotting luminol reagent (Santa Cruz Biotechnology, Santa Cruz, CA). The intensities of each band signal were determined by densitometry using Bio-Profil software version 99.04 (Vilber Lourmat, France). The image densities of iNOS- and HO-1–specific bands were normalized, with the density of $\beta$-actin band used as the internal control to compare the amounts of iNOS and HO-1 protein accumulated in each sample.

*Nitrite Determination Using Griess Reaction as an Assay of NO
Production (iNOS Activity)*

Nitrite, a stable oxidation product of NO, was used as a measure of NOS activity. Nitrite present in the conditioned culture media was determined by a spectrophotometric assay based on the Griess reaction (Green *et al.*, 1982). For determination of nitrite concentration, 24-well plates were used, each well containing $1.5 \times 10^5$ cells cultured in 0.5 ml DMEM free of phenol red. One hundred micoliters of the conditioned medium was incubated with the same volume of Griess reagent [0.1% *N*-(1-naphtyl)-ethylenediamine and 1% sulfanilamide in 5% orthophosphoric acid] at room temperature for 10 min. The absorbance at 546 nm was measured using a microplate reader (Model PowerWaveX, Bio-Tek Instruments, Winooski, VT), and the nitrite concentration was determined from a standard curve generated with dilutions of sodium nitrite in DMEM free of phenol red.

*Heme Oxygenase Activity Assay*

*Preparation of Microsomal Extracts from Macrophages (Yet* et al., *2002).* RAW 264.7 cells ($\sim 5 \times 10^6$ cells/100-mm plate) were washed twice with cold pH 7.4 PBS and scraped in the presence of 1 ml of cold homogenization buffer (30 m$M$ Tris–HCl, pH 7.5, 0.25 $M$ sucrose, 0.15 $M$ NaCl containing 10 $\mu$g/ml leupeptin, 2 $\mu$g/ml aprotinin, 10 $\mu$g/ml trypsin inhibitor and 1 m$M$ phenylmethylsulfonyl fluoride (PMSF). Cells were sonicated on ice for 2 min with 5-s interval at 1 min and at the output of 2 W employing the Vibracell ultrasonic processor (Sonics & Materials, Newtown, CT). Cell lysates were centrifuged at 10,000$g$ for 15 min at 4°, and supernatants were collected and centrifuged again at 100,000$g$ for 1 h at 4° by employing an ultracentrifuge (model Optima LE-80K, Beckman Coulter). Finally, the microsomal pellet was resuspended in 100 m$M$ potassium phosphate buffer, pH 7.4, containing 2 m$M$ MgCl$_2$, 10 $\mu$g/ml leupeptin, 2 $\mu$g/ml aprotinin, 10 $\mu$g/ml trypsin inhibitor, and 1 m$M$ PMSF. Protein concentration of the collected microsomal fraction was measured using the BCA method according to manufacturer's instructions (Pierce, Rockford, IL).

*Preparation of Rat Liver Cytosol to be Used as the Source of Biliverdin Reductase (Yet* et al., *2002).* Rats were anesthetized by an i.p. injection of secobarbital (40 mg/kg body weight), and liver was perfused *in situ* through portal vein with cold PBS to flush out blood. The liver was cut free and washed twice with ice-cold PBS. Then, the liver was homogenized on ice in 3 volumes of homogenization buffer (30 m$M$ Tris–HCl, pH 7.5, 0.25 $M$ sucrose, and 0.15 $M$ NaCl containing the mixture of protease inhibitors) using a Polytron homogenizer (PT-3100, Kinematica, Switzerland). Homogenates were centrifuged at 10,000g for 15 min at 4°. The supernatant thus obtained was centrifuged again at 100,000g for 1 h at 4° using an ultracentrifuge. The supernatant was divided into many aliquots and stored at −80° until used. The frozen aliquots were thawed and used as the source of biliverdin reductase in the HO activity assay determining the amount of bilirubin produced.

*Measurement of Heme Oxygenase Activity.* The microsomal fraction (200 $\mu$l) obtained from variously treated macrophages was added to the reaction mixture (200 $\mu$l) containing 0.8 m$M$ NADPH, 2 m$M$ glucose-6-phosphate, 0.2 units glucose-6-phosphate dehydrogenase, 20 $\mu M$ hemin, 100 m$M$ potassium phosphate buffer, pH 7.4, and 2 mg of rat liver cytosol protein used as the source of biliverdin reductase. The samples were incubated for 1 h at 37° in a waterbath kept in dark, and the reaction was terminated by placing the samples on ice for at least 2 min. The amount of bilirubin formed was determined by calculation from the difference in absorbances between 464 and 530 nm (extinction coefficient, 40 m$M^{-1}$ cm$^{-1}$ for bilirubin). Heme oxygenase activity was expressed as nanomoles of bilirubin formed per milligram of protein per hour (Foresti *et al.*, 1997). NADPH solution should be prepared fresh on the day of use and protected from direct light.

## References

Balla, G., Jacob, H. S., Balla, J., Rosenberg, M., Nath, K., Apple, F., Eaton, J. W., and Vercellotti, G. M. (1992). Ferritin: A cytoprotective antioxidant strategem of endothelium. *J. Biol. Chem.* **267,** 18148.

Cairo, G., Tacchini, L., Pogliaghi, G., Anzon, E., Tomasi, A., and Bernelli-Zazzera, A. (1995). Induction of ferritin synthesis by oxidative stress. Transcriptional and post-transcriptional regulation by expansion of the free iron pool. *J. Biol. Chem.* **270,** 700.

Elbirt, K. H., and Bonkovsky, H. I. (1999). Heme oxygenase: Recent advances in understanding its regulation and role. *Proc. Assoc. Am. Phys.* **111,** 438.

Ferret, P. J., Soum, E., Negre, O., and Fradelizi, D. (2002). Auto-protective redox buffering systems in stimulated macrophages. *BMC Immunol.* **3,** 3.

Foresti, R., Clark, J. E., Green, C. J., and Motterlini, R. (1997). Thiol compounds interact with nitric oxide in regulating heme oxygenase-1 induction in endothelial cells. Involvement of superoxide and peroxynitrite anions. *J. Biol. Chem.* **272,** 18411.

Forman, H. J., and Torres, M. (2002). Reactive oxygen species and cell signaling: Respiratory burst in macrophage signaling. *Am. J Respir. Crit. Care Med.* **166,** S4.

Green, L. C., Wagner, D. A., Glogowski, J., Skipper, P. L., Wishnok, J. S., and Tannenbaum, S. R. (1982). Analysis of nitrate, nitrite, and [$^{15}$N]nitrate in biological fluids. *Anal. Biochem.* **126,** 131.

Halliwell, B., and Gutteridge, J. M. C. (1999). "Free Radicals in Biology and Medicine," 3rd Ed. Oxford Science Publications, New York.

Huie, R. E., and Padmaja, S. (1993). The reaction of NO with superoxide. *Free Radic. Res. Commun.* **18,** 195.

Kaur, H., Hughes, M. N., Green, C. J., Naughton, P., Foresti, R., and Motterlini, R. (2003). Interaction of bilirubin and biliverdin with reactive nitrogen species. *FEBS Lett.* **543,** 113.

MacMicking, J., Xie, Q. W., and Nathan, C. (1997). Nitric oxide and macrophage function. *Annu. Rev. Immunol.* **15,** 323.

Maines, M. D. (1997). The heme oxygenase system: A regulator of second messenger gases. *Annu. Rev. Pharmacol. Toxicol.* **37,** 517.

Mancuso, C., Bonsignore, A., Di Stasio, E., Mordente, A., and Motterlini, R. (2003). Bilirubin and S-nitrosothiols interaction: Evidence for a possible role of bilirubin as a scavenger of nitric oxide. *Biochem. Pharmacol.* **66,** 2355.

Otterbein, L. O., and Choi, A. M. K. (2000). Heme oxygenase: Colors of defense against cellular stress. *Am. J. Physiol. Lung Cell Mol. Physiol.* **279,** L1029.

Sambrook, J., and Russell, D. W. (2001). "Molecular Cloning: A Laboratory Manual," 3rd Ed. Cold Spring Harbor Laboratory Press, Cold Spring Harbor, New York.

Srisook, K., and Cha, Y. N. (2004). Biphasic induction of heme oxygenase-1 expression in macrophages stimulated with lipopolysaccharide. *Biochem. Pharmacol.* **68,** 1709.

Stocker, R., Yamamoto, Y., McDonagh, A. F., Glazer, A. N., and Ames, B. N. (1987). Bilirubin is an antioxidant of possible physiological importance. *Science* **235,** 1043.

Victor, V. M., and De La Fuente, M. (2003). Changes in the superoxide production and other macrophage functions could be related to the mortality of mice with endotoxin-induced oxidative stress. *Physiol. Res.* **52,** 101.

Wink, D. A., Hanbauer, I., Grisham, M. B., Laval, F., Nims, R. W., Laval, J., Cook, J., Pacelli, R., Liebmann, J., Krishna, M., Ford, P. C., and Mitchell, J. B. (1996). Chemical biology of nitric oxide: Regulation and protective and toxic mechanisms. *Curr. Topics Cell Regul.* **34,** 159.

Yet, S. F., Melo, L. G., Layne, M. D., and Perrella, M. A. (2002). Heme oxygenase 1 in regulation of inflammation and oxidative damage. *Methods Enzymol.* **353,** 163.

# [32]   G-Protein Signaling in iNOS Gene Expression

By SANG GEON KIM and CHANG HO LEE

## Abstract

Heterotrimeric G proteins are the molecular switches in the receptor-mediated transmembrane signaling system. Inducible nitric oxide synthase (iNOS) is inducible by a variety of inflammatory stimuli, which leads to vascular hyporeactivity. In this chapter, the system to study the cell signaling pathways downstream of the GPCR coupling to G proteins is described for the study of iNOS gene expression. The cellular signaling pathways by which ligand induces iNOS may serve as the pharmacological targets for preventing or treating vascular hyporeactivity.

METHODS IN ENZYMOLOGY, VOL. 396                                         0076-6879/05 $35.00
Copyright 2005, Elsevier Inc. All rights reserved.                        DOI: 10.1016/S0076-6879(05)96032-1

## Introduction

The G-protein–coupled receptors (GPCRs) consist of the largest family of transmembrane proteins. GPCRs as seven-transmembrane receptor proteins transduce signals from cell surface to intracellular effectors through various G proteins. Heterotrimeric G proteins, which are composed of $\alpha$, $\beta$, and $\gamma$ subunits, are considered the molecular switches in the receptor-mediated transmembrane signaling system. The agonist-occupied GPCRs interact with heterotrimeric G proteins and promote activation of $G\alpha$ proteins by catalyzing guanosine diphosphate (GDP)–guanosine triphosphate (GTP) exchange reaction, thereby causing dissociation of the heterotrimer into $\alpha$ and $\beta\gamma$ subunits (Clapham and Neer, 1997; Morris and Malbon, 1999). Both the $\alpha$ and the $\beta\gamma$ subunits can independently modulate the physiological activities of various cell signaling effectors. Many of the G-protein $\alpha$ subunits have been identified in mammalian cells. The $G\alpha$ proteins are classified into four subfamilies, namely $G\alpha_s$, $G\alpha_i$, $G\alpha_q$, and $G\alpha_{12}$, depending on their amino acid sequence homology (Hepler and Gilman, 1992; Wilkie *et al.*, 1992). Among the four subfamilies, the members of $G\alpha_{12}$ subunits are composed of $G\alpha_{12}$ and $G\alpha_{13}$ proteins that share more than 65% sequence homology. The $G\alpha_{12}$ subunits are mainly involved in the regulation of cell proliferation and differentiation processes via various intracellular signaling modules (Dhanasekaran *et al.*, 1998; Voyno-Yasenetskaya *et al.*, 1994; Xu *et al.*, 1993).

Vascular hyporeactivity is attributable to excess nitric oxide (NO) production, a key gaseous molecule inducing the collapse of the cardiovascular system (Szabo *et al.*, 1994). NO also affects the inflammatory processes (Szabo and Thiemermann, 1994; Wong and Billiar, 1995). Inducible NO synthase (iNOS) is expressed in endothelial cells, macrophages, or fibroblasts by various inflammatory stimuli, and it causes a large amount of NO to be produced, thereby affecting regulation of vascular hyporeactivity (Chesrown *et al.*, 1994; Salvucci *et al.*, 1998). In septic shock, an exhaustion of antithrombin and excess amount of free thrombin induce iNOS in macrophages, culminating in inflammation response that leads to vascular hyporeactivity via a large amount of NO production (Fig. 1).

Among the four $G\alpha$ subfamilies, $G\alpha_{12}$ members are activated by stimulation of thrombin, $TXA_2$, lysophosphatidic acid, and thyroid-stimulating hormone receptors (Fig. 1) (Chesrown *et al.*, 1994; Kang *et al.*, 2003; Salvucci *et al.*, 1998; Yamaguchi *et al.*, 2003). $G\alpha_{12}$ and $G\alpha_{13}$ can regulate various intracellular effectors or cellular responses such as platelet aggregation (Nieswandt *et al.*, 2002), actin-stress fiber formation, and apoptosis. GPCRs transmit signals through conformational changes upon ligand activation and interaction with G proteins. In this chapter, the system to study

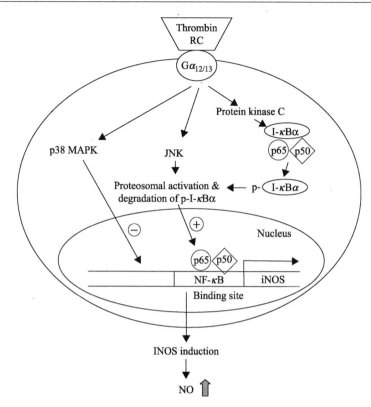

FIG. 1. The proposed cell signaling pathways for the induction of inducible nitric oxide synthase by thrombin.

the cell signaling pathways downstream the GPCR coupling to G proteins is described for the study of iNOS gene expression.

## Experimental Protocols Related to G-Protein Research

To investigate the role of the $G\alpha$ subunit in cell signaling pathway(s), $G\alpha$ protein may be overexpressed transiently or permanently in diverse types of cells by transfecting the complementary DNA (cDNA) encoding the $G\alpha$ protein of interest ($G\alpha$ wild type) or the guanosine triphosphatase (GTPase)–deficient $\alpha$ subunit (constitutively active $G\alpha$ subunit mutant) such as $G\alpha_q$(Q209L), $G\alpha_{11}$QL, $G\alpha_{14}$QL, $G\alpha_{15}$(Q212L), $G\alpha_{16}$(Q212L), $G\alpha_{12}$(Q229L), or $G\alpha_{13}$(Q226L), depending on the experimental purposes. Particularly when assessing the interaction between the GPCR and the G-protein $\alpha$ subunit, the cDNAs of wild-type $G\alpha$ subunit and the receptor of interest may also be co-expressed in the same target cells. Then, we assess ligand-stimulated biochemical changes, such as GDP–GTP exchange, accumulation of second messengers, protein phosphorylation, and gene expression, thereby identifying the cell signaling pathway regulated by the G protein of interest (Fig. 2).

## Transient Transfection

Although various transfection procedures are available for many types of cells, commercially available cationic lipid reagent is used for the expression of G-protein subunit. The optimized protocol described here

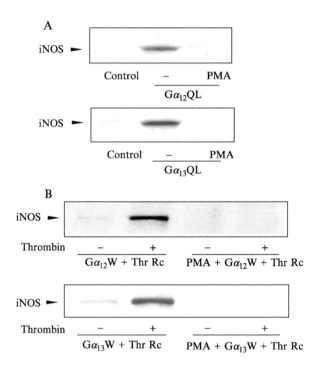

Fig. 2. Nuclear factor-$\kappa$B (NF-$\kappa$B)–mediated induction of inducible nitric oxide synthase (iNOS) by activated G$\alpha_{12/13}$. (A) Induction of iNOS by thrombin in the cells transfected with the plasmid encoding an activated mutant of G$\alpha_{12}$ (G$\alpha_{12}$QL) or G$\alpha_{13}$ (G$\alpha_{13}$QL). The control cells were transfected with the pCDNA plasmid. The cells that had been pretreated with phorbol 12-myristate 13-acetate (PMA) for 18 h were also transfected with each plasmid and further incubated for 18 h. (B) Induction of iNOS by thrombin in the cells transfected with the plasmid encoding the wild-type G$\alpha_{12}$ (G$\alpha_{12}$W) or G$\alpha_{13}$ (G$\alpha_{13}$W) with thrombin (Thr) receptor (Rc) co-transfection. Control cells were transfected with the pCDNA plasmid. The transfected cells were exposed to thrombin (10 units/ml) for 12 h.

yields relatively high transfection efficiency, which allows us to investigate the role of G-protein signaling in the iNOS expression.

1. Cells (e.g., COS-7, NIH3T3 fibroblast, HEK293, and Raw264.7) are maintained in Dulbecco's Modified Eagle Medium (DMEM) containing 10% fetal bovine serum (FBS), 50 units of penicillin/ml, and 50 $\mu$g of streptomycin/ml at 37° in a humidified atmosphere with 5% $CO_2$. The cells are plated at a density of $5 \times 10^6$/10 cm diameter dish and maintained in 80–90% confluency. Cells should be subjected to no more than 20 passages.

2. Cells are transfected by the method using LipofectAMINE reagent (Invitrogen). One microgram of the plasmid DNA and 3 $\mu$l of LipofectA-MINE reagent are mixed in 1 ml of antibiotic-free minimal essential medium (MEM) and then are added to the cells. Three hours later, culture medium is changed with serum-free MEM with antibiotics, and the cells are further incubated for 12–24 h to assess iNOS expression or for 1–3 h to monitor the activity of nuclear factor-$\kappa$B (NF-$\kappa$B) (a major transcription factor responsible for iNOS induction). The transfection efficiency is about 50% in the case of Raw264.7 cells if determined by the method of lacZ reporter transfection. Culture medium is changed with serum-free MEM with antibiotics and the cells are further incubated to assess iNOS expression or NF-$\kappa$B activation.

*Stable Transfection*

Cells may be stably transfected with the plasmid of interest to eliminate the effect of possible endotoxin contamination in the plasmid preparation. Cells are re-plated 24 h before transfection at a density of $2 \times 10^6$ cells in a 10 cm diameter dish. LipofectAMINE is mixed with 10 $\mu$g of a plasmid [e.g., JNK1 dominant negative mutant (JNK1[−])] in 2.5 ml of MEM. The cells are transfected by addition of MEM containing each plasmid and LipofectAMINE, and then incubated at 37° in a humidified atmosphere of 5% $CO_2$ for 6 h. After addition of 6.25 ml MEM with 10% FBS, the cells are incubated for an additional 48 h. Geneticin is added to select resistant colonies. A mixture of stably transfected clones or the cells derived from a single clone are used. Stable cell transfection should be verified by assaying the level of marker protein or its corresponding activity. For example, JNK1(−) stable cells should have no inducible JNK activity, which can be easily monitored by GST-c-Jun phosphorylation.

G-Protein Immunoblotting

1. The cells expressing recombinant G$\alpha_{12}$ or G$\alpha_{13}$ subunits are collected and washed with phosphate-buffered saline (PBS). The cells are lysed in 45 $\mu$l of RIPA buffer (1% nonidet P-40, 0.5% sodium deoxycholate, 0.1% sodium

dodecylsulfate (SDS), 2 m$M$ EDTA, 50 m$M$ NaF, 1 m$M$ phenylmethylsul-fonylfluoride, 4 $\mu$g/ml aprotinin, 2 $\mu$g/ml leupeptin, 200 $\mu M$ vanadate, and 0.1% $\beta$-mercaptoethanol).

2. To the lysates, 15 $\mu$l of 4× Lammeli buffer is added. The samples are boiled and subjected to SDS gel electrophoresis and Western blot analysis. G$\alpha_{12}$ protein is immunoblotted by reacting with an antibody (1:10,000) directed against the specific sequence (QENLKDIMLQ) of G$\alpha_{12}$. The expression of G$\alpha_{13}$ protein is similarly assessed by the antibody (1:10,000) against the specific sequence (HDNLKQLMLE) of G$\alpha_{13}$. The proteins are visualized with ECL chemiluminescence or West-Zol Western blot detection kit.

## [$^{35}$S]GTP$\gamma$S Binding Assay

Although immunoblotting with G-protein antibody provides a simple and specific method for identification of G$\alpha$ protein in subcellular fractions, immunoblot analysis alone may not be sufficient to reflect the activity of G$\alpha$ subunit. Agonist occupancy of GPCR promotes interaction of the receptor with heterotrimeric G proteins, which, in turn, facilitates the exchange of GDP to GTP. Thus, to provide evidence as to the functional coupling of GPCR to heterotrimeric G proteins, it is necessary to measure the extents of signal transduction downstream after receptor occupancy. Among the several ways of measuring the downstream events from proximal to distal to the agonist-induced receptor activation, [$^{35}$S]GTP$\gamma$S binding is considered an assay representing one of the earliest receptor-mediated events. This assay determines the extent of G-protein activation after agonist-induced stimulation of a GPCR by measuring the binding of nonhydrolyzable analogue of [$^{35}$S]GTP$\gamma$S to G$\alpha$ subunit (Harrison and Traynor, 2003).

The assay is relatively simple and is less subjected to amplification or regulation by other cellular processes that may occur when measuring the activities of downstream effector molecules. This assay has been used successfully in certain permeabilized cell types to evaluate GPCR-mediated G protein activity, and it is now most commonly performed by using the cell membrane preparation enriched with the endogenous or recombinant receptor of interest (Waelbroeck, 2001). The higher background signal of [$^{35}$S] GTP$\gamma$S binding that may be observed in G$\alpha$s- and Gq-coupled receptor activation is reduced by assaying the immunoprecipitate of [$^{35}$S]GTP$\gamma$S bound to the G$\alpha$ subunit of interest. In addition, G$\alpha$ subunit–specific antibodies are used to investigate the coupling specificity of certain G$\alpha$ subunits to the GPCR of interest by measuring the radioactivity of the [$^{35}$S]GTP$\gamma$S-G$\alpha$ complex. The scintillation proximity and newly developed fluorescence-based assays for the estimation of G-protein activation are now commercially available, which makes the assays more useful in monitoring the selectivity of various GPCRs to G-protein coupling.

The following experimental protocol provided in this context has been a widely used simple and easy-to-follow [$^{35}$S]GTP$\gamma$S binding assay method using the cell membrane preparation enriched with the endogenous or recombinant receptor of interest.

1. Add 10 $\mu$l of the 5× assay buffer containing 250 m$M$ HEPES buffer (pH 8.0), 5 m$M$ EDTA, 0.5% Lubrol PX, and 50 m$M$ MgSO$_4$ to a 5-ml polypropylene tube.
2. Add 1 $\mu$l of 10 m$M$ dithiothreitol (DTT).
3. Add the samples prepared from the membranes obtained from the cells expressing the receptor of interest to the tube, and add water to make the volume of 45 $\mu$l.
4. Add 5 $\mu$l of [$^{35}$S]GTP$\gamma$S (50 p$M$: $\sim$2500 cpm/pmol).
5. Incubate the samples for 30–45 min at 30°.
6. At the end of incubation, dilute each of the samples with 2 ml of the filtration buffer containing 20 m$M$ Tris–HCl (pH 8.0), 100 m$M$ NaCl, and 25 m$M$ MgCl$_2$.
7. Filter the reaction mixtures through BA 85 nitrocellulose filters in the Millipore manifold.
8. Wash the filters with a total volume of 12 ml of the filtration buffer.
9. Dry the filter, and count radioactivity in a liquid scintillation counter.
10. Convert the data to either the concentration of [$^{35}$S]GTP$\gamma$S bound/ mg membrane proteins or fold increases over the basal binding.

## NF-$\kappa$B–Mediated iNOS Induction

The expression of iNOS is controlled by the transcription factors, including NF-$\kappa$B. NF-$\kappa$B exists in the cytoplasm of unstimulated cells in a quiescent form bound to its inhibitor (Li and Verma, 2002). Binding of ligand to GPCR induces a series of cellular responses of NF-$\kappa$B activation, which include I-$\kappa$B$\alpha$ phosphorylation and degradation, nuclear translocation of the p65/p50 NF-$\kappa$B complex, and p65/p50 NF-$\kappa$B binding to DNA.

### Immunoblot Analysis

The levels of iNOS and subsequent intracellular events such as I-$\kappa$B$\alpha$ phosphorylation are assessed by immunoblot analysis.

1. To perform SDS–polyacrylamide gel electrophoresis (PAGE) and immunoblot analysis of iNOS protein, cells are lysed in the buffer

containing 20 mM Tris–Cl (pH 7.5), 1% Triton X-100, 137 mM sodium chloride, 10% glycerol, 2 mM EDTA, 1 mM sodium orthovanadate, 25 mM $\beta$-glycerophosphate, 2 mM sodium pyrophosphate, 1 mM PMSF, and 1 $\mu$g/ml leupeptin.

2. Cell lysates are centrifuged at 10,000g for 10 min to remove debris. The proteins are fractionated using a 7.5% separating gel to assess the level of iNOS. The fractionated proteins are electrophoretically transferred to nitrocellulose paper. iNOS is immunoblotted with anti-iNOS antibody.

3. I-$\kappa$B$\alpha$ phosphorylation precedes I-$\kappa$B$\alpha$ degradation. I-$\kappa$B$\alpha$ and its phosphorylated form are determined in cytosolic samples by SDS-PAGE and immunoblotting with anti-I-$\kappa$B$\alpha$ and anti-phosphorylated I-$\kappa$B$\alpha$ antibodies.

4. The secondary antibodies are horseradish peroxidase– or alkaline phosphatase–conjugated anti–immunoglobulin G (IgG) antibody. Nitrocellulose paper is developed using 5-bromo-4-chloro-3-indolylphosphate/4-nitroblue tetrazolium chloride or developed using ECL chemiluminescence system.

*Gel Retardation Assay*

Activation of NF-$\kappa$B is partly assessed by NF-$\kappa$B binding to the NF-$\kappa$B binding consensus oligonucleotide by gel shift analysis.

1. To prepare nuclear extracts, culture dishes are washed with ice-cold PBS. The dishes are then scraped and transferred to microtubes. Cells are allowed to swell by adding 100 $\mu$l of lysis buffer [10 mM HEPES (pH 7.9), 10 mM KCl, 0.1 mM EDTA, 0.5% Nonidet-P40, 1 mM DTT, and 0.5 mM PMSF]. Tubes are vortexed to disrupt cell membranes. The samples are incubated for 10 min on ice and centrifuged for 5 min at 4°.

2. The pellets containing crude nuclei are resuspended in 50 $\mu$l of the extraction buffer containing 20 mM HEPES (pH 7.9), 400 mM NaCl, 1 mM EDTA, 1 mM DTT, and 1 mM PMSF, and then are incubated for 30 min on ice. The samples are centrifuged at 15,800g for 10 min to obtain the supernatant containing nuclear extracts.

3. For determination of NF-$\kappa$B activation, the nuclear extracts prepared from the cells treated with a ligand are probed with a radiolabeled NF-$\kappa$B consensus oligonucleotide. The double-stranded DNA of NF-$\kappa$B (5′-AGTTGAGGGGACTTTCCCAGGC-3′) is end-labeled with [$\gamma$-$^{32}$P]ATP and T$_4$ polynucleotide kinase. The DNA probe is used for gel shift analysis.

4. The reaction mixture contains 2 $\mu$l of 5× binding buffer containing 20% glycerol, 5 m$M$ MgCl$_2$, 250 m$M$ NaCl, 2.5 m$M$ EDTA, 2.5 m$M$ DTT, 0.25 mg/ml (poly)dI-dC and 50 m$M$ Tris–Cl (pH 7.5), nuclear extracts (2–10 $\mu$g), and sterile water in a total volume of 10 $\mu$l. Incubations are initiated by addition of 1 $\mu$l probe (e.g., $10^6$ cpm) and continued for 20 min at room temperature.

5. The p50/p50 homodimer complex migrates slightly faster. In general, NF-$\kappa$B–DNA binding is visualized by increases in the band intensity of a slow migrating p65/p50 complex (e.g., 30 min to 3 h). A 20-fold excess of the NF-$\kappa$B probe abolishes the band retardation. In some experiments, an aliquot of nuclear extracts is incubated with 2 $\mu$g of highly specific anti-p65 and/or p50 antibody at room temperature for 1 h. The addition of both anti-p65 and anti-p50 antibodies causes a supershift with reduction in the band intensity of the p65/p50 complex.

6. Samples are loaded onto 4% polyacrylamide gels at 100 V. The gels are removed, fixed, dried, and autoradiographed.

*Immunocytochemistry of p65*

Because p65 is the major component of activated NF-$\kappa$B, translocation of p65 into the nucleus may be monitored in combination with gel shift analysis. Cells are treated with a ligand between 30 min and 3 h, fixed, and permeabilized. The p65 protein is located in the cytoplasm of untreated cells and moves into the nucleus after activation.

1. Cells are grown on Lab-TEK chamber slides and incubated in serum-deprived medium for 24 h.
2. For immunostaining, the cells are fixed in 100% methanol for 30 min and washed three times with PBS. After blocking in 5% bovine serum albumin (BSA) in PBS for 1 h at room temperature or overnight at 4°, cells are incubated for 1 h with anti-p65 antibody (1:100) in PBS containing 0.5% BSA.
3. The cells are incubated with fluorescein isothiocyanate–conjugated goat anti-rabbit IgG (1:100-1:1000) after serial washings with PBS.
4. Counterstaining with propidium iodide verifies the location and integrity of nuclei. Stained cells are washed and examined using a microscope.

*Cell Signaling for iNOS Induction by Thrombin*

We found that G$\alpha_{12}$QL or G$\alpha_{13}$QL (G$\alpha$12 or G$\alpha$13 activated mutant) expression notably increased NO production. Cells were transfected with the G$\alpha_{12}$W, G$\alpha_{13}$W, G$\alpha_{12}$QL, or G$\alpha_{13}$QL plasmid and then treated with an activator. The plasmid of $G\alpha_{15}QL$ ($G\alpha_{15}$ activated mutant) may be used as a negative control. We reported that JNK inhibition by the stable transfection with JNK1($-$) completely suppresses iNOS induction by thrombin, whereas p38 kinase inhibition by SB203580 enhances iNOS expression (Fig. 1) (Kang et al., 2003). The induction of iNOS by thrombin is regulated by the distinct and opposed functions of JNK and p38 kinase. In the case of the pathways of the $G\alpha_{12}$ protein–coupled receptors, protein kinase C–dependent phosphorylation is linked to $G\alpha_{12/13}$ activation (Kang et al., 2003).

## Concluding Remarks

Growth factors, cytokines, and proteases play an important role in the vascular responsibility by inducing iNOS and NO production via the pathway coupled with $G\alpha_{12/13}$. The cellular signaling pathways by which ligand induces iNOS may serve as the pharmacological targets for preventing or treating vascular hyporeactivity.

## Acknowledgments

This work was supported by the National Research Laboratory Program (2001), KISTEP, the Ministry of Science and Technology, Republic of Korea.

## References

Chesrown, S. E., Monnier, J., Visner, G., and Nick, H. S. (1994). Regulation of inducible nitric oxide synthase mRNA levels by LPS, INF-gamma, TGF-beta, and IL-10 in murine macrophage cell lines and rat peritoneal macrophages. Biochem. Biophys. Res. Commun. **200**, 126–134.

Clapham, D. E., and Neer, E. J. (1997). G protein beta gamma subunits. Annu. Rev. Pharmacol. Toxicol. **37**, 167–203.

Dhanasekaran, N., Tsim, S. T., Dermott, J. M., and Onesime, D. (1998). Regulation of cell proliferation by G proteins. Oncogene **17**, 1383–1394.

Harrison, C., and Traynor, J. R. (2003). The [$^{35}$S]GTPgammaS binding assay: Approaches and applications in pharmacology. Life Sciences **74**, 489–508.

Hepler, J. R., and Gilman, A. G. (1992). G proteins. Trends Biochem. Sci. **17**, 383–387.

Kang, K. W., Choi, S. Y., Cho, M. K., Lee, C. H., and Kim, S. G. (2003). Thrombin induces nitric-oxide synthase via Galpha$_{12/13}$-coupled protein kinase C–dependent I-kappaBalpha phosphorylation and JNK-mediated I-kappaBalpha degradation. J. Biol. Chem. **278**, 17368–17378.

Li, Q., and Verma, I. M. (2002). NF-$\kappa$B regulation in the immune system. Nat. Rev. Immunol. **2**, 25–34.

Morris, A. J., and Malbon, C. C. (1999). Physiological regulation of G protein–linked signaling. Physiol. Rev. **79**, 1373–1430.

Nieswandt, B., Schulte, V., Zywietz, A., Gratacap, M. P., and Offermanns, S. (2002). Costimulation of Gi- and G$_{12/13}$-mediated signaling pathways induces integrin alpha IIbbeta 3 activation in platelets. J. Biol. Chem. **277**, 39493–39498.

Salvucci, O., Kolb, J. P., Dugas, B., Dugas, N., and Chouaib, S. (1998). The induction of nitric oxide by interleukin-12 and tumor necrosis factor-alpha in human natural killer cells: Relationship with the regulation of lytic activity. *Blood* **92**, 2093–2102.

Szabo, C., and Thiemermann, C. (1994). Invited opinion: Role of nitric oxide in hemorrhagic, traumatic, and anaphylactic shock and thermal injury. *Shock* **2**, 145–155.

Szabo, C., Thiemermann, C., Wu, C. C., Perretti, M., and Vane, J. R. (1994). Attenuation of the induction of nitric oxide synthase by endogenous glucocorticoids accounts for endotoxin tolerance *in vivo*. *Proc. Natl. Acad. Sci. USA* **91**, 271–275.

Voyno-Yasenetskaya, T. A., Pace, A. M., and Bourne, H. R. (1994). Mutant alpha subunits of G12 and G13 proteins induce neoplastic transformation of Rat-1 fibroblasts. *Oncogene* **9**, 2559–2565.

Waelbroeck, M. (2001). Activation of guanosine $5'$-[$\gamma$-$^{35}$S] thiotriphosphate binding through $M_1$ muscarinic receptors in transfected Chinese hamster ovary cell membranes: 1. Mathematical analysis of catalytic G protein activation. *Mol. Pharmacol.* **59**, 875–885.

Wilkie, T. M., Gilbert, D. J., Olsen, A. S., Chen, X. N., Amatruda, T. T., Korenberg, J. R., Trask, B. J., De Jong, P., Reed, R. R., Simon, M. I., Jenkins, N. A., and Copeland, N. G. (1992). Evolution of the mammalian G protein alpha subunit multigene family. *Nat. Genet.* **1**, 85–91.

Wong, J. M., and Billiar, T. R. (1995). Regulation and function of inducible nitric oxide synthase during sepsis and acute inflammation. *Adv. Pharmacol.* **34**, 155–170.

Xu, N., Bradley, L., Ambdukar, I., and Gutkind, J. S. (1993). A mutant alpha subunit of G12 potentiates the eicosanoid pathway and is highly oncogenic in NIH 3T3 cells. *Proc. Natl. Acad. Sci. USA* **90**, 6741–6745.

Yamaguchi, Y., Katoh, H., and Negishi, M. (2003). N-terminal short sequences of alpha subunits of the $G_{12}$ family determine selective coupling to receptors. *J. Biol. Chem.* **278**, 14936–14939.

# [33]   Determination of Nitric Oxide–Donor Effects on Tissue Gene Expression *In Vivo* Using Low-Density Gene Arrays

*By* DOANH C. TRAN, DANIEL A. BRAZEAU, and HO-LEUNG FUNG

## Abstract

Gene array technology has been used to examine gene expression changes following drug treatments, including administration of nitric oxide (NO) donors. High-density arrays represent a powerful and popular method to analyze a large number of genes simultaneously. On the other hand, low-density arrays, available commercially at a lower cost, allow for the use of gene-specific primers, which reduces the risk of cross-hybridization among genes with similar sequence. For certain experiments in which the hypothesis is focused on a selected set of genes, use of low-density

METHODS IN ENZYMOLOGY, VOL. 396
0076-6879/05 $35.00
DOI: 10.1016/S0076-6879(05)96033-3

arrays might be more productive and cost-effective. Here, we describe our experience using low-density arrays to examine the effect of exposure to the NO-donor isobutyl nitrite on the expression of 23 cancer- and angiogenesis-related genes in mouse tissues. Detailed descriptions of data capture procedures, statistical tests, and confirmation studies using real-time quantitative (RTQ) reverse transcription polymerase chain reaction (RT-PCR) are presented. Three simple statistical methods, namely Student's $t$ test, significant analysis of microarrays (SAM), and permutation adjusted $t$ statistics (PATS), were applied on our gene array data, and their utilities were compared. All three methods yielded concordant results for the most significant genes, namely vascular endothelial growth factor (VEGF), VEGF receptor 3, Smad5, and Smad7. RT-PCR confirmed VEGF upregulation as observed via gene arrays. PATS appeared to be more robust than SAM in handling our small gene array data set. This statistical method, therefore, appears more suited for analyzing low-density gene array data. We conclude that low-density gene array is a useful screening method that can be performed with lower cost and less cumbersome data treatment.

## Introduction

Induction of endogenous nitric oxide (NO) production through stimulating the enzyme inducible NO synthase (iNOS, NOS II) is known, in several studies, to cause profound changes in cytokine regulation and gene expression (Jozkowicz *et al.*, 2001; Okamoto *et al.*, 2004; Speyer *et al.*, 2003; Zeidler *et al.*, 2004). However, few studies address these effects *in vivo* when exogenous NO donors are used either in therapy (e.g., nitroglycerin in the treatment of ischemic diseases) or as drugs of abuse (e.g., inhalant nitrites for their purported sexual effects). We have used gene arrays to explore whether *in vivo* nitroglycerin vascular tolerance is associated with significant changes in gene expression in the rat aorta (Wang *et al.*, 2002). We found that nitroglycerin treatment (to produce vascular tolerance) led to changes in expression of 290 vascular genes, whereas treatment with S-nitroso-N-acetylpenicillamine (SNAP), which did not cause tolerance, produced changes in the expression of only 41 genes. These results suggest the usefulness of the gene array technique for exploring genetic changes *in vivo* in experiments of pharmacological relevance.

In these earlier experiments, we used commercially available high-density arrays that contained several thousand genes. Many investigators currently adopt this approach, sometimes employing arrays with the entire genome of the species of interest. These high-density arrays represent a powerful tool in the initial search for gene expression changes or patterns of change throughout the genome. However, the substantial cost of high-density arrays often prohibits sufficient experimental replication, and statistical testing of the resultant data is consequently hindered or simply ignored. Experimental replication is particularly crucial because these arrays generate such an extensive body of data that, without suitable rigorous statistical examination, it is difficult to identify significantly altered genes from a substantial number of false positives and negatives.

In many biochemical and pharmacological experiments, the hypothesis is centered on examining the changes in the expression of a selected subset of genes. For example, the possible association of inhalant nitrite abuse with its tumorigenic activity (Soderberg, 1999) and Kaposi's sarcoma (Haverkos, 1990; Marmor *et al.*, 1982) would suggest a targeted testing of possible changes in cancer- and angiogenesis-related genes. In these instances, the alternative use of low-density complementary DNA (cDNA) arrays containing these genes is indicated. The technology for these lower density arrays is well established; the experiments are simple to perform, and data acquisition and analysis can be carried out without specialized equipment. These custom arrays are also commercially available at a lower cost, which permits the performance of sufficient experimental replicates to enable statistical testing. The inherent problem with cross-hybridizations (Evertsz *et al.*, 2001) in gene array experiments can be minimized with careful design of gene-specific primers used to generate labeled cDNAs, a task that is difficult and not practical with large arrays. For these reasons, we applied a commercially available array containing 23 genes focused on the cancer and angiogenesis pathways to assess the *in vivo* effects of the NO-donor isobutyl nitrite (ISBN) on their expression in liver and lung tissues (Tran *et al.*, 2003).

## Preparation of Tissue Samples for Gene Array Analysis (Tran *et al.*, 2003)

Male C57BL/6 mice were exposed to approximately 1400 parts per million (ppm) ISBN or regular breathing air (controls) for 4 h. Immediately at the end of exposure, mice were sacrificed, and liver and lung tissues were removed. To preserve RNA integrity, the tissues were removed quickly, immediately frozen in liquid nitrogen, and then transferred to storage at $-80°$. If flash freezing in liquid nitrogen is not feasible, an RNA preserving solution such as RNAlater (Ambion, Austin, TX) could be used.

Total RNA was isolated using a glass binding matrix (spin column)–based kit (Eppendorf, Westbury, NY). To maintain conditions that inhibit ribonuclease (RNase) activity, tissue samples were cut to size while frozen and ground up in the presence of cell lysis solution in a mortar and pestle cooled by liquid nitrogen. The frozen powder mixture of ground tissue and lysis buffer was then transferred to a centrifuge tube and allowed to thaw. It is essential that sufficient lysis solution be used to ensure complete RNase inhibition. The thawed mixture was then passed through a 22G needle several times to ensure complete cell disruption. Complete cell lysis reduces clogging of the spin column in subsequent steps and increases RNA yields. This lysis procedure generated high-quality total RNA preparations, as assessed by the presence of distinct 18S and 28S bands on denaturing formaldehyde-gel electrophoresis. Following spectrophotometric measurement at 260 nm to determine concentration and formaldehyde gel electrophoresis to determine integrity, the RNA was ready for gene array studies.

### Capture of Low-Density Gene Array Data

The manufacturer's protocol (SuperArray Bioscience Corporation, Frederick, MD) was followed for cDNA labeling, hybridization, and washing, except for the use of a more extended washing time (40 min/wash cycle) to ensure low background signals. Briefly, total RNA was reverse transcribed into cDNA using gene array–specific primers and [33]P-labeled dCTP, and hybridized to a gene array membrane overnight at 68°. Each membrane was washed and exposed to a phosphor screen for quantitation. Exposures were done in pairs (of treatment and control). An image of the array was obtained by scanning the phosphor screen in a Packard Cyclone system.

Each of the 23 genes is represented by two spots on the array membrane. The membrane also contains spots for two housekeeping genes, $\beta$-actin and glyceraldehyde-3-phosphate dehydrogenase (GAPDH), as well as two negative control spots. Using OptiQuant software (version 3.00), a grid of small circles (SC) was placed on top of the array. This procedure quantified the intensity of each spot, including any signal arising from background noise. To account for the background signal, we placed a grid of larger circles (LC) to quantify the total signal of each spot and the area surrounding it. The net signal of each spot was then calculated as follows:

$$Mean\ background\ signal\ (MBS) = \frac{Total\ signal\ of\ LC - Total\ signal\ of\ SC}{Area\ of\ LC - Area\ of\ SC}$$

$$(1)$$

$$Net\ spot\ signal\ per\ unit\ area\ (NS) = \frac{Total\ signal\ of\ SC}{Area\ of\ SC} - MBS \qquad (2)$$

The normalized gene signal was then calculated as follows, with negative results being set to zero:

$$Gene\ signal = \frac{NS_{geneX} - NS_{Negative\ control}}{NS_{\beta-actin} - NS_{Negative\ control}} \qquad (3)$$

This approach of background subtraction (i.e., using local surroundings in addition to the negative control spots) was chosen because of uneven background noise in some membranes, and it produced more consistent intensity values among data collected from different membranes.

To normalize for possible differences in the amounts of RNA used, RT efficiency, hybridization conditions, and so on, the signal for $\beta$-actin was used to normalize the gene signals on each corresponding membrane. The signal for this housekeeping gene was assumed to be constant. However, this assumption is not always valid (Schmittgen and Zakrajsek, 2000; Selvey *et al.*, 2001). We verified that $\beta$-actin was not changed after inhalant nitrite exposure under our exposure conditions via real-time quantitative (RTQ) reverse transcription polymerase chain reaction (RT-PCR).

### Statistical Treatment of Gene Array Data

The simplest approach to analyzing gene array data is to compare the ratio of gene expression with and without treatment using the fold-change method, with an arbitrary cutoff (e.g., twofold or threefold). This method was used often when lack of sufficient replication precluded any meaningful statistical analysis. In an attempt to improve its usefulness, a variation of the fold-change method, the pairwise fold-change method, implemented a requirement that such change be consistently observed between paired samples (Ly *et al.*, 2000). Examination of a data set revealed that these methods generated a 60–84% false-discovery rate (FDR) (Tusher *et al.*, 2001). Thus, the fold-change method, though simple to use, produces little meaningful information.

Although low-density focused arrays generate a smaller volume of data than their high-density counterparts, statistical analysis is still needed to confirm suspected changes and to minimize the presence of false positives and negatives. We have, therefore, compared the applicability of three relatively simple methods of gene array data analysis, viz., unpaired

Student's $t$ test, significant analysis of microarrays (SAM) (Tusher et al., 2001), and permutation adjusted $t$ statistic (PATS) (Wang et al., 2002).

The Student's $t$ test of two independent groups was used first to compare our study of ISBN exposure versus air control in mice. However, because there were 23 genes on the membrane and, therefore, requiring 23 separate $t$ tests, the probability of obtaining a false positive is very high. At an alpha level of 0.05, the probability of at least one gene being called significant by chance alone is 0.7 [$p = 1 - (0.95)^{23}$] (Dowdy et al., 2004). The normal correction for multiple tests such as the Bonferroni method is, however, too strict when applied to gene array data where a large number of comparisons are performed using limited sample size, often resulting in no significant genes (Tusher et al., 2001), as was found in the present case. In addition, the Bonferroni correction assumes that multiple tests are independent of one another. This assumption is questionable, given the known interdependence of many gene pathways. Other authors have devised alternative methods such as SAM and PATS to address this problem.

SAM is a method by Tusher et al. (2001), which first calculates a number called *relative difference* that is based on the mean variability of the repeated measurement of each gene, as well as a constant calculated based on the expression of all genes tested. Genes are called *significant* if the observed relative difference is different from the expected relative difference (a measure of random variability in the data calculated by averaging the relative difference of repeated permutations of the data set) by a user-adjustable value called *delta*. An FDR is then calculated by comparing the average number of genes falsely called significant as calculated from the list of all possible permutations to the total number called significant from the original set. The authors have distributed a computer program that automates this process (at http://www-stat-class.stanford.edu/SAM/SAMServlet). This method was shown to yield a lower FDR as compared to the fold-change or pairwise fold-change method.

The non-parametric permutation adjusted $t$ statistic approach (Wang et al., 2002) is also based on the resampling method of constructing all possible permutations of a data set. It compares the $t$ statistic of the observed grouping versus the distribution of $t$ statistics for all possible permutations of the data to assess the likelihood of the observed set occurring by chance alone. PATS considers each gene independently and, therefore, can be used for both large and small arrays. To apply this method with sufficient confidence ($p < .05$), at least four sets of arrays are required to produce greater than 20 unique permutations.

Both SAM and PATS were applied to our data set and their results compared.

## Results and Discussion

The lower cost of low-density arrays allowed us to generate eight sets of data for comparison. Using the standard unpaired $t$ test, four genes, namely VEGF, VEGF receptor 3, Smad5, and Smad7, were called significant from the liver samples at the $p < .05$ level. No gene was shown to be significantly changed in the lung. However, as mentioned earlier, the probability that a false positive might occur in the $t$-test analysis is 0.7, at alpha $= .05$. It should be noted that because of the relatively high level of variability in gene expression data, there is a risk of false-negative results as well. However, the larger number of replicates helps in reducing this risk.

Applying SAM to the identical data set yielded nine genes called significant at 8% FDR. The top four most significant genes were identical to those obtained from the $t$ test. Attempts to lower the FDR resulted in no significant genes being identified. We also tested various delta values yielding higher FDRs. The results were variable beyond the top four genes, as we noted that a lower FDR sometimes actually was associated with an increased number of significant genes reported. SAM was designed for and is expectedly more suitable for a larger data set than our 23-gene set. With our small set of genes, the FDR function was not smooth, and a small change in *delta* could lead to erratic results. For our small data set, therefore, SAM was not robust enough, so it was not used as a primary method to determine statistical significance.

The third method, PATS, worked equally well with larger data sets (Wang *et al.*, 2002) as well as ours because it constructs permutations for each individual gene and not the entire gene set. Because our data set consisted of eight treatments versus eight controls, there were 6435 unique permutations ($16!/(8!*8!)/2$). The probability that the observed set occurred by chance alone was calculated by dividing its $t$ statistic rank by 6435. We considered all genes with $p < .05$ as being significantly different. This method yielded the same four significant genes as reported by the Student $t$ test.

Once the number of genes is narrowed down to a few with high degree of certainty via statistical testing, the results could then be followed up using more robust experimental methods such as RTQ RT-PCR to confirm gene expression level. The upregulation of VEGF following nitrite exposure was confirmed using the RTQ RT-PCR method. RTQ RT-PCR showed that VEGF expression was increased by $2.43 \pm 1.27$ fold following ISBN exposure, which agreed well with the $2.33 \pm 1.20$ fold increase observed using gene array. Using the RTQ RT-PCR technique, we also did not find significant changes in $\beta$-actin and GAPDH expression levels. Our analysis, therefore, indicates that all three statistical methods are in agreement regarding the top four significant genes. SAM yielded a higher number of significant genes at an FDR of 8%, the lowest FDR we could achieve before no gene was deemed significant. The permutation-adjusted $t$-statistic approach appeared better suited for low-density array data sets, provided that there are at least four replicate arrays in each group. The computer codes for PATS written for Mathematica are available from D.A. Brazeau at (dbrazeau@buffalo.edu).

We believe that the use of low-density arrays may be considered in some experiments. The relatively simple and straightforward analysis of data allows scientists who are less familiar with gene array technology to perform experiments with lower costs and resources.

## References

Dowdy, S. M., Wearden, S., and Chilko, D. (2004). "Statistics for Research." Wiley, New York.

Evertsz, E. M., Au-Young, J., Ruvolo, M. V., Lim, A. C., and Reynolds, M. A. (2001). Hybridization cross-reactivity within homologous gene families on glass cDNA micro-arrays. *Biotechniques* **31,** 1182–1192.

Haverkos, H. W. (1990). Nitrite inhalant abuse and AIDS-related Kaposi's sarcoma. *J. Acquir. Immune Defic. Syndr.* **3,** S47–S50.

Jozkowicz, A., Cooke, J. P., Guevara, I., Huk, I., Funovics, P., Pachinger, O., Weidinger, F., and Dulak, J. (2001). Genetic augmentation of nitric oxide synthase increases the vascular generation of VEGF. *Cardiovasc. Res.* **51,** 773–783.

Ly, D. H., Lockhart, D. J., Lerner, R. A., and Schultz, P. G. (2000). Mitotic misregulation and human aging. *Science* **287,** 2486–2492.

Marmor, M., Friedman-Kien, A. E., Laubenstein, L., Byrum, R. D., William, D. C., D'Onofrio, S., and Dubin, N. (1982). Risk factors for Kaposi's sarcoma in homosexual men. *Lancet* **1,** 1083–1087.

Okamoto, T., Gohil, K., Finkelstein, E. I., Bove, P., Akaike, T., and van der Vliet, A. (2004). Multiple contributing roles for NOS2 in LPS-induced acute airway inflammation in mice. *Am. J. Physiol. Lung Cell. Mol. Physiol.* **286,** L198–L209.

Schmittgen, T. D., and Zakrajsek, B. A. (2000). Effect of experimental treatment on housekeeping gene expression: Validation by real-time, quantitative RT-PCR. *J. Biochem. Biophys. Methods* **46,** 69–81.

Selvey, S., Thompson, E. W., Matthaei, K., Lea, R. A., Irving, M. G., and Griffiths, L. R. (2001). Beta-actin—An unsuitable internal control for RT-PCR. *Mol. Cell Probes.* **15,** 307–311.

Soderberg, L. S. (1999). Increased tumor growth in mice exposed to inhaled isobutyl nitrite. *Toxicol. Lett.* **104,** 35–41.

Speyer, C. L., Neff, T. A., Warner, R. L., Guo, R. F., Sarma, J. V., Riedemann, N. C., Murphy, M. E., Murphy, H. S., and Ward, P. A. (2003). Regulatory effects of iNOS on acute lung inflammatory responses in mice. *Am. J. Pathol.* **163,** 2319–2328.

Tran, D. C., Yeh, K.-C., Brazeau, D. A., and Fung, H.-L. (2003). Inhalant nitrite exposure alters mouse hepatic angiogenic gene expression. *Biochem. Biophys. Res. Commun.* **310,** 439–445.

Tusher, V. G., Tibshirani, R., and Chu, G. (2001). Significance analysis of microarrays applied to the ionizing radiation response. *Proc. Natl. Acad. Sci. USA* **98,** 5116–5121.

Wang, E. Q., Lee, W. I., Brazeau, D., and Fung, H. L. (2002). cDNA microarray analysis of vascular gene expression after nitric oxide donor infusions in rats: Implications for nitrate tolerance mechanisms. *AAPS PharmSci.* **4,** E10.

Zeidler, P. C., Millecchia, L. M., and Castranova, V. (2004). Role of inducible nitric oxide synthase–derived nitric oxide in lipopolysaccharide plus interferon-gamma–induced pulmonary inflammation. *Toxicol. Applied Pharmacol.* **195,** 45–54.

# Section V

# Cell Biology and Physiology

# [34]   Cell H$_2$O$_2$ Steady-State Concentration and Mitochondrial Nitric Oxide

By MARIA CECILIA CARRERAS, MARIA CLARA FRANCO, DANIELA P. CONVERSO, PAOLA FINOCCHIETO, SOLEDAD GALLI, and JUAN JOSÉ PODEROSO

## Abstract

For many years, mitochondrial respiration was thought to follow an "all or nothing" paradigm supporting the notion that in the normal O$_2$ concentration range, respiration is mainly controlled by tissue demands. However, nitric oxide produced by cytosol or mitochondrial nitric oxide synthases adapts respiration to different physiologic conditions and increases the mitochondrial production of O$_2$ active species that contributes to NO clearance. Because mitochondrial NO utilization is sensitive to environmental or hormonal modulation, and because diffusible active species, like H$_2$O$_2$, are able to regulate genes related to proliferation, quiescence, and death, we surmised that the two mechanisms converge to elicit the different responses in cell physiology.

## O$_2^-$ and H$_2$O$_2$ as a Function of Oxygen Concentration

Many years ago, it was established that mitochondria produce oxygen-active species; in the physiological condition, 3–5% of oxygen is univalently reduced in normal respiration (Boveris and Cadenas, 1997). Formerly considered toxic bystanders of electron transfer reactions in mitochondria, superoxide anion (O$_2^-$) and its dismutation product H$_2$O$_2$ are now recognized as modulators of genes throughout oxidation of reactive groups in transcription factors, kinases, and phosphatases. Isolation of mitochondrial components led to the identification of five multimeric complexes (complexes I–V) embedded in the inner membrane. Nonenzymatic reactions of O$_2^-$ formation occur at complexes I and III, both depending on autoxidation of increased intermediary ubisemiquinone (UQH$^-$), at specific ubiquinone pools, as emerged from the use of rotenone (complex I inhibitor at PSSP protein) and antimycin (complex III inhibitor at cytochrome $b_H$) (Carreras *et al.*, 2004a).

METHODS IN ENZYMOLOGY, VOL. 396
0076-6879/05 $35.00
DOI: 10.1016/S0076-6879(05)96034-5

*Reactions 1, 2, and 3, $O_2^-$ dismutation to $H_2O_2$*

$$UQH^- + e^- \rightarrow UQ^{-\cdot} \qquad (1)$$

$$UQ^{-\cdot} + O_2 \rightarrow O_2^- + UQ \qquad (2)$$

$$O_2^- + O_2^- + 2H^+ \rightarrow H_2O_2 + O_2 \qquad (3)$$

Mitochondrial NO Utilization Conducts to $O_2^-$ Formation

Mitochondria continuously move in the transition between a quiet resting respiratory condition (state 4 $O_2$ uptake, 10–40 ng at O/min mg prot) and an active respiratory state in the presence of adenosine diphosphate (ADP) (state 3 $O_2$ uptake, 80–250 ng at O/min mg prot). In accordance with the very low cytochrome $c$ oxidase $K_M$ for oxygen (0.5–2.0 $\mu M$), for many years, mitochondrial respiration was thought to follow an "all or nothing" paradigm with the implicit notion that up to very low $O_2$ levels, respiration does not depend on $O_2$ availability and that electron transfer was rather modulated by redox pairs content. In the last years, however, it was extensively reported that nitric oxide (NO) reversibly binds to $Cu^{2+}$–B center of cytochrome oxidase and modulates electron transfer to $O_2$ and respiration in rat skeletal muscle, liver, and heart mitochondria (Brown, 1995; Poderoso *et al.*, 1996, 1999a,b). Similar effects were described in perfused rat heart (Poderoso *et al.*, 1998). NO-dependent inhibition of $O_2$ uptake is achieved at low and physiological NO concentrations; 50–100 n$M$ NO inhibits by a half cytochrome oxidase activity.

NO has become a significant contributor to cell signaling and behavior. From our perspective, the relationship between NO and oxygen-active species in mitochondria supports many of the programs involved in development, life, and death. In 1996, our group described that exposure of heart isolated mitochondria to NO markedly increases $O_2^-$ and $H_2O_2$ production rates (Poderoso *et al.*, 1996); subsequent studies proved this response to be a universal effect of NO on mitochondria.

Oxidant yields particularly significant results when it is reminded that NO inhibits cytochrome oxidase and reduces oxygen uptake. Thus, the matrix NO level determines two modes of mitochondrial $O_2$ reactive species production: at low NO and high $O_2$ uptake, and at high NO and low $O_2$ uptake. It is clear that in the first case, $d[O_2^-]/dt$ and derived $d[H_2O_2]/dt$ will mainly depend on $-d[O_2]/dt$, $[O_2]$ and [MnSOD]. However, in the second case, NO will increase $O_2^-$ by increasing the reduction level on the side of substrate and by other reactions, and it will react with $O_2^-$ to produce $ONOO^-$ in accordance with reactions 4 and 5; also $ONOO^-$ reacts with ubiquinol and increases further $O_2^-$ formation (Schöpfer *et al.*, 2000).

$$NO + UQH \rightarrow NO^- + UQ^{-\cdot} \tag{4}$$

$$NO + O_2^- \rightarrow ONOO^- \tag{5}$$

Therefore, in this context, $d[O_2^-]/dt$ and derived $d[H_2O_2]/dt$ will depend on matrix [NO] and [MnSOD]. At mitochondrial physiological [O$_2$] ($\sim$10–15 $\mu M$) and [NO] ($\sim$20–30 n$M$) levels, both responses are additive up to an NO concentration of about 0.1 $\mu M$, but afterwards (NO > 0.2–0.3 $\mu M$) further reduction of O$_2$ uptake increases NO-dependent species production but increases the ONOO$^-$/H$_2$O$_2$ ratio (Fig. 1A). From this perspective, it emerges that in the absence of NO with fully active complex activity, mitochondrial H$_2$O$_2$ production could hardly be upregulated, whereas it will be downregulated by decreasing transcriptional activity related to electron transfer chain components. Instead, in the presence of NO, O$_2^-$ and H$_2$O$_2$ release is modulated by matrix concentration, either increasing or decreasing at higher toxic NO levels. In the absence of a selective inhibition of electron transfer between complexes I and III, like that produced by genetic disorders, and in a range from physiological values to 10-fold higher concentration, NO is the unique metabolite that modulates mitochondrial oxygen species yield and function. In addition, NO decreases superoxide anion production, depending on oxidative

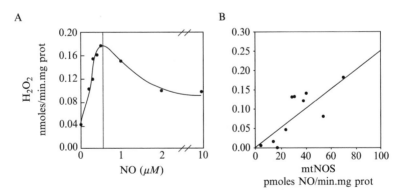

FIG. 1. (A) H$_2$O$_2$ production rate of isolated liver mitochondria as a function of nitric oxide (NO) concentration; mitochondria were exposed to NO solutions obtained by bubbling authentic NO gas; dashed line indicates critical NO concentration that favors further ONOO$^-$ formation. (B) In accordance with matrix NO level, activity of liver mitochondrial NO synthase (mtNOS) correlates with mitochondrial H$_2$O$_2$ production rate; the variations of mtNOS were obtained at different proliferation rates during liver development or in T$_4$-stimulated rats.

metabolism, and it binds $O_2^-$, which may be protective for mitochondria in terms of nonregulated protein or lipid oxidations.

## mtNOS, $H_2O_2$, and Peroxynitrite

The discovery of a differentiated NO synthase (NOS) in mitochondria (mtNOS) by independent groups (Ghafourifar and Richter, 1997; Giulivi et al., 1998) "affords unique organelle-based regulatory mechanisms for NO synthesis and has considerable implications for mitochondrial function" (Brookes, 2004).

Apparently, mtNOS content depends on the rate of nNOSα translocation to the organelles. Because mtNOS possesses characteristic posttranslational changes, like phosphorylation in Akt-sensitive Ser 1412 and N-terminal acylation, its content should depend on activation of Akt, which in turn increases the enzyme-specific activity (C. Giulivi, personal communication), and on the acylation rate that contributes to its insertion in the inner membrane. Additionally, mtNOS is subjected to selective modulation by thyroid status (Carreras et al., 2001), cold acclimation (Peralta et al., 2003), and brain plasticity (Riobó et al., 2002).

It was reported that the decrease of cytosolic proteins that anchor nNOS through PDZ domains, like dystrophin in heart muscle, increases enzyme translocation to mitochondria (Kanai et al., 2001). Thus, mtNOS content depends on nNOS expression, on the rate of posttranslational changes, and on the expression of proteins that anchor nNOS and modulate the intracellular traffic. It remains still undefined a putative role of chaperones heat shock protein-70 (Hsp70) and Hsp90 that contribute to nNOS folding and activity and are implicated in the mechanisms of translocation to mitochondria through interactions with outer membrane transporters (TOMs).

In our experience, there exist different situations that clearly promote transcriptional activity and nNOS translocation to mitochondria. In rat liver development (Carreras et al., 2004b), nNOS was present in cytosol but almost absent in fetal mitochondria and translocated to mitochondria during the first week after delivery. The increase of NO and $H_2O_2$ steady-state concentrations leads to the activation of kinases linked to cell cycles like p38 mitogen-activated protein kinase (MAPK) and to the inhibition of pro-proliferative kinases like ERK1/2, which contributes to a diminished cell proliferation rate and forces cells to enter in the quiescent adult condition. Similarly, in rat brain development, a postnatal increase of mtNOS and $H_2O_2$ is sustained by 1–2 wk in the transition between neuroblastic proliferation and the phase of cell arrest and apoptosis that characterizes structural synaptic plasticity (Riobó et al., 2002). In adult early stage (~90 days), mtNOS reaches a stable value, lower than that of the postnatal period. Preliminary experiments in our laboratory show that expression and activity of mtNOS and H$_2$O$_2$ of rat brain and liver increase during adult life and are maximal at 24–28 mo (maximal lifespan of the species). However, in other experiments, mtNOS activity was found to be decreased during aging (Boveris and Navarro, 2004).

On the other hand, we observed that hypothyroidism promotes a significant increase of nNOS messenger RNA (mRNA) and the translocation of nNOS to mitochondria associated with the formation of O$_2^-$ and ONOO$^-$ (unpublished data).

## The Calculation of Cell H$_2$O$_2$ Steady-State Concentration

As shown by Boveris and Cadenas (1997), cell H$_2$O$_2$ steady-state concentration ([H$_2$O$_2$]ss) depends on $d[H_2O_2]/dt$, on catabolizing enzymes, and on free diffusion of the species outside cells. The $d[H_2O_2]/dt$ is contributed by mitochondria and peroxisomes and, to a considerable lesser extent, by tissue-specific reactions. Because peroxisomes have the highest concentration of catabolizing catalase, mitochondria with very low levels of degrading enzymes are accepted as the main source of oxygen species. Considering the aforementioned effects, we can discriminate total mitochondrial yield as the sum of oxygen-dependent and NO-dependent $d[H_2O_2]/dt$:

$$\text{Total mt } d[H_2O_2]/dt = [d[H_2O_2]/dt]_{O2} + [d[H_2O_2]/dt]_{NO} \qquad (1)$$

Experimentally, $[d[H_2O_2]/dt]_{NO}$ has been found to be 30–40% of total mt $d[H_2O_2]/dt$ but remarkably is the modulable term of the equation. Depending on different tissues and species, H$_2$O$_2$ degrading enzymes include catalase, glutathione peroxidase (GPX), and thioredoxin peroxidases (TPXs). Thus, cell or tissue [H$_2$O$_2$]ss could be calculated as the ratio between H$_2$O$_2$ mitochondrial yield and enzyme activities, as estimated from known data or directly measured in the corresponding samples [Eq. (2)]:

$$[H_2O_2]ss = \frac{[[d[H_2O_2]/dt]_{O2} + [d[H_2O_2]/dt]_{NO}}{[cat]k_1 + [GPX]k_2 + [TPX]k_3} \qquad (2)$$

Measurement of $d[H_2O_2]/dt$

Detection of $H_2O_2$ is estimated by several assays linked to oxidation of a detector compound horseradish peroxidase (HRP) (Boveris et al., 1972). In the presence of $H_2O_2$, hydrogen donors are oxidized by HRP.

    a. $HRP + H_2O_2 \rightarrow HRP - H_2O_2$ [compound I]
    b. $HRP - H_2O_2 + AH_2 \rightarrow HRP + 2H_2O + A$

The amount of $H_2O_2$ produced is estimated by following the increase in fluorescent products from previously nonfluorescent hydrogen donors, such as 4-p-hydroxyphenyl acetic acid or by monitoring the decrease in fluorescence of initially fluorescent probes such as scopoletin. The same principle could be used for spectrophotometric assays, monitoring the oxidation of phenol red, but the method is less sensitive.

$H_2O_2$ production is continuously monitored with excitation and emission wavelengths at 315 and 425 nm, respectively, when the p-hydroxyphenyl acetic acid–HRP assay is used. The reaction medium consists of 50 m$M$ potassium phosphate buffer with 50 m$M$ L-valine, supplemented with mitochondrial substrates, 12.5 units/ml HRP, 250 $\mu M$ p-hydroxyphenyl acetic acid, 1 $\mu M$ Mn(III)TBAP (SOD mimetic), and 0.15 mg of mitochondrial protein per milliliter. L-Valine is added to inhibit arginase and prolong NO production, and SOD mimetic is used to make the dismutation rate of $O_2^-$ to $H_2O_2$ independent of variations of endogenous Mn-SOD. However, Mn(III)TBAP may have other effects on different experimental systems, because it reacts with $ONOO^-$ (Ferrer-Sueta et al., 2003). Also, $ONOO^-$ may oxidize pHPA (Ischiropoulos et al., 1996). In the presence of high NO concentrations, the assay could overestimate $H_2O_2$ yield by the inhibition of oxidation by NO, in part due to the reaction of NO with compound I and compound II, and in part to the reaction with the phenoxyl radical. These data suggest that the simultaneous generation of NO and peroxynitrite can interfere with the detection of $H_2O_2$.

As shown in Fig. 1B, mitochondrial production of $H_2O_2$ correlates with the modulation of mtNOS activity.

To discriminate $d[H_2O_2]/dt]_{O2}$ and $[d[H_2O_2]/dt]_{NO}$, purified mitochondria are incubated in the absence or the presence of NOS substrate L-arginine or NOS inhibitor L-N$^G$-monomethyl L-arginine, L-NMMA. Maximal $d[H_2O_2]/dt$ is obtained in state 4 oxygen uptake (in the absence of ADP acceptor), in the presence of the adequate substrates like 6 m$M$ malate plus 6 m$M$ glutamate (complex I substrate), or 10 m$M$ succinate (complex II substrate). NO-dependent $H_2O_2$ production was determined as the difference of $H_2O_2$ production rate in the presence of 100 $\mu M$ L-Arg alone or plus L-NMMA. The difference between the two determinations indicates the contribution of NO to total $d[H_2O_2]/dt$ as represented by the NOS-inhibited condition.

## Catalase Activity

Catalase activity in 7000$g$ supernatant is determined by measuring the decrease in the absorption at 240 nm. Reaction medium consists of 50 m$M$ potassium phosphate buffer, pH 6.8, and 10 m$M$ hydrogen peroxide, thereby determining the pseudo–first-order reaction constant ($k'$) of H$_2$O$_2$ decrease ($\epsilon_{240} = 41\ \mu M^{-1}\ cm^{-1}$) (Chance, 1954). Results are expressed as catalase content in pmol/mg protein ($k = 4.6 \times 10^7\ M^{-1}\ cm^{-1}$). Intracellular production of H$_2$O$_2$ can be estimated by using the catalase inhibitor, aminotriazole, which reacts with the compound I intermediate; thus, catalase is inhibited by aminotriazole only in the presence of H$_2$O$_2$. By determining the half-time of inactivation of catalase, steady-state H$_2$O$_2$ concentration can be approximated according to Eq. (3) (Royall et al., 1992):

$$[H_2O_2] = k_{cat}/k_1, \qquad (3)$$

where $0.5/t_{1/2} = -k_{cat}$ and $k_1 = 1.7 \times 10^7\ [mol/L]^{-1}\ s^{-1}$.

Under conditions in which GPX reductase or thioredoxins significantly contribute to H$_2$O$_2$ metabolism, calculations of H$_2$O$_2$ concentrations by this method will be underestimated. In the presence of NO, the measurement of catalase activity can be lower than expected by NO interaction with the heme group of the enzyme; catalase could be also nitrosylated at a high NO concentration (Foster and Stamler, 2004).

## Glutathione Peroxidase Activity

GPX activity is determined spectrophotometrically by the oxidation of NADPH at 340 nm ($\epsilon_{240} = 6.22\ mM^{-1}\ cm^{-1}$). Reaction medium consists of 100 m$M$ potassium phosphate buffer/EDTA 1 m$M$, pH 7.7, supplemented with 10 m$M$ ter-butyl hydroperoxide, 10 U/ml, glutathione reductase, 100 m$M$ GSH, 40 m$M$ sodium azide, and 10 m$M$ NADPH.

## Total Oxidants

Total oxidants may be estimated by the oxidation of 2′,7′-dichlorofluorescin (DCFH) to the fluorescent compound 2,7′-dichlorofluorescein (DCF). The diacetate form of DCFH (DCFH-DA) is taken up by cells, where intracellular esterases cleave the molecule to DCFH. In the presence of oxidants like H$_2$O$_2$, peroxynitrite (the product of NO and O$_2^-$

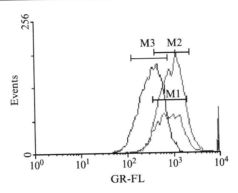

FIG. 2. Differential oxidation of 2′,7′-dichlorofluorescin (DCFH) of isolated hepatocytes in controls (M1), in the presence of 1 m$M$ L-Arg (M2) or plus 20/5 m$M$ L-NAME (M3), as assessed by flow cytometry.

reaction), hypochlorous acid, or lipid peroxides, DCFH is oxidized to DCF (Tarpey and Fridovich, 2001).

NO-dependent oxidant production can be estimated by preincubating $10^6$ cells in 1 ml PBS with 1 m$M$ L-Arg during 20 min or with 5 m$M$ NOS inhibitor L-NAME for 5 min, and then 20 min with L-Arg (Fig. 2).

The cellular fluorescence intensity is measured after 30 min of incubation with 5 $\mu M$ DCFH-DA using a flow cytometer (Herrera *et al.*, 2001). Propidium iodide (0.005%) is used to detect dead cells. For each analysis, 10,000 events are recorded.

### Intracellular and Intramitochondrial Superoxide Anion

The intracellular and intramitochondrial production of $O_2^-$ can be estimated by hydroethidine (HEt) oxidation to ethidium by flow cytometry (Fig. 3). HEt is cell permeant and can undergo two-electron oxidation to form the DNA-binding fluorophore ethidium bromide; in this reaction, however, fluorescence seems to depend on the formation of a fluorescent compound different to ethidium (Zhao *et al.*, 2003). The reaction is relatively specific for $O_2^-$, with minimal oxidation by $H_2O_2$, peroxynitrite, or hypochlorous acid.

NO-dependent $O_2^-$ is estimated by preincubating $10^6$ cells at similar conditions as described for DCFH assay and using 5 $\mu M$ HEt (Fig. 3). In a semiquantitative fashion, the kinetics of mitochondrial NO-dependent $O_2^-$ can be followed by co-localization of orange HEt and green 4-amino-5-methylamino-2′-7′-dichlorofluorescein diacetate (DAF-FM) by confocal microscopy. To detect submicromolar intracellular and intramitochondrial

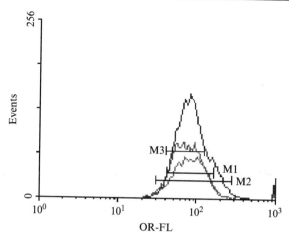

FIG. 3. Nitric oxide (NO)–dependent production of O$_2^-$ of isolated purified mitochondria as followed by hydroethidine (HEt) oxidation with flow cytometry in controls (M1) and in the presence of 1 m$M$ L-Arg alone (M2) or plus L-NAME 10/5 m$M$ (M3).

NO concentrations, 10$^6$ cells or 1 mg isolated and purified mitochondria are incubated in 1 ml PBS in the presence of 10 $\mu M$ DAF-FM for 30 min, and the fluorescence is measured by flow cytometry. DAF-FM diacetate is cell permeant and passively diffuses across cellular membranes, where it is deacetylated by intracellular esterases.

## Cell Cycle Regulation by H$_2$O$_2$

H$_2$O$_2$ regulates gene activity. Cyclins and cyclin-dependent kinases are subjected to regulation by H$_2$O$_2$ (Fig. 4A). Also, activation of MAPKs regulating cyclin expression is modified by H$_2$O$_2$ (Fig. 4B). We reported the correlation between selective MAPK activation, mitochondrial NO/[H$_2$O$_2$], and cyclin D1–3 expression that determine the cell entering from quiescent G$_0$ to G$_1$ (Carreras *et al.*, 2004b). Maximal cyclin expression is achieved at low or very low NO/H$_2$O$_2$ concentrations and correlates well with redox state (Fig. 4C). Accordingly, cyclin expression may be regulated in isolated cells by operating on H$_2$O$_2$ levels.

## Cell Proliferation and H$_2$O$_2$

At very low oxidative stress, cell proliferation is stimulated in normal or tumoral cells (Fig. 5A and B). However, normal proliferation must be carefully dissected from the tumorigenic activities. The redox-dependent tumorigenic mechanism requires persistent inflammation that is a current

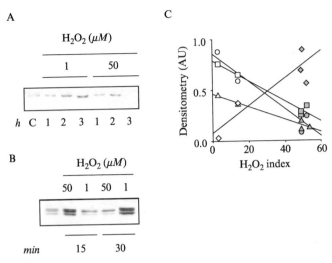

Fig. 4. (A) Kinetics of cyclin D1 expression in mammary gland LMM3 tumoral cell line at different $H_2O_2$ concentrations. (B) Differential effects of $H_2O_2$ on the kinetics of ERK1/2 activation in P2 hepatocytes. (C) Differential activation of mitogen-activated protein kinases (MAPKs) and cyclin D1 expression in isolated hepatocytes as a function of spontaneous $H_2O_2$ index ($[d[H_2O_2]/dt]_{NO}/[d[H_2O_2]/dt]_{O2/antimycin}) \times 100$. Symbols, □: P-ERK1; △: P-ERK2; ○: cyclin D1; ◇: P-p38MAPK (proliferating cells in white, quiescent cells in gray).

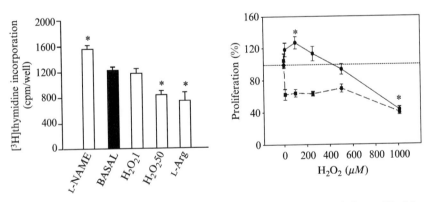

Fig. 5. (A) Proliferation rate of neonatal P2 hepatocytes is oppositely modified by subtracting $NO/H_2O_2$ with L-NAME or by adding L-Arg or $H_2O_2$. (B) Differential effects of $H_2O_2$ alone (circles) or plus catalase inhibitor ATZ (squares) on proliferation of mammary gland NMuMG cells; asterisk (*) denotes $p < .05$ respect to controls by analysis of variance. Catalase inhibitor determinations resulted different from control (no treatment), at any $H_2O_2$ concentration.

source of NO and H$_2$O$_2$ through the activity of macrophagic cells. The active species (including peroxynitrite) induce prolonged activation of proliferating pathways, like nuclear factor-$\kappa$B (NF-$\kappa$B), whereas immortalization and acquisition of tumoral features in growth and invasiveness depend on DNA mutations, mostly in connection with gene oxidation. In addition, both normal and tumoral cells are arrested or undergo apoptosis at high H$_2$O$_2$ levels (Fig. 5A and B). Previously, we reported that very low mtNOS activity and NO levels and H$_2$O$_2$ yield are a common finding to normal or tumoral cells (Galli $et$ $al.$, 2003); in different situations, mtNO/H$_2$O$_2$ should depend on the physiological modulation of mtNOS or on persistently decreased NOS/mtNOS activities, likely due to the existence of defective enzymes.

Proliferation Assays

Different methods have been developed to quantify cell proliferation, such as use of radioactive thymidine or BrdU (as a substitute for radioactive thymidine) to label DNA in live cells or the use of tetrazolium salts such as MTT that are reduced into colored formazan compounds. Absorbance can be read with an enzyme-linked immunosorbent assay (ELISA) plate reader at 492 nm. The biochemical procedure is based on the activity of mitochondria enzymes, which are inactivated shortly after cell death. This method was found to be efficient in assessing the viability of cells.

Cells are plated in 96-multiwell plastic dishes at the appropriate densities in culture media supplemented with 10% FCS. Effects of H$_2$O$_2$ can be monitored by supplementing cells with 0.1 $\mu M$ to 1 m$M$ H$_2$O$_2$ in culture media with 10% FCS. At 48 h of incubation, proliferation can be assessed with a non-radioactive cell proliferation assay (Galli $et$ $al.$, 2003) or by [$^3$H]-thymidine incorporation (Carreras $et$ $al.$, 2004b).

Redox modulation of cell proliferation can be assessed by controlling intracellular H$_2$O$_2$ steady-state by supplementation of the media with L-Arg (5 m$M$), L-NAME (5 m$M$), or H$_2$O$_2$, or by preincubating with catalase inhibitor 3-amino-1, 2, 4-triazole (ATZ) (5 m$M$) (Fig. 5A and B).

It is noteworthy that preincubation with L-NAME reflects pure O$_2$-dependent production of active species. In addition, intracellular H$_2$O$_2$ steady-state concentration may be adjusted, supplementing the media with glucose/glucose oxidase if antioxidant enzyme activities (catalase and GPX) are previously determined, as shown in Eq. (2).

## Effects of NO/H$_2$O$_2$ on Kinase Activation and Distribution

Cell responses are based on activation of different signaling pathways. As mentioned earlier, MAPK and Akt pathways are very sensitive to NO and redox conditions. In our experience, differential activation of MAPKs plays a significant role in normal development (Carreras *et al.*, 2004b). At low NO/H$_2$O$_2$ level, pro-proliferative ERK1/2 is phosphorylated, whereas at high levels, ERK1/2 activity is reduced, but pro-apoptotic p38MAPK and JNK activation augment. In addition, we observed that not only activation but subcellular distribution of ERK is affected by H$_2$O$_2$; exposing cells to 50 $\mu M$ associates with prolonged mitochondrial retention with a lower activity than that obtained at 1 $\mu M$ (Alonso *et al.*, 2004).

### Redox Activation of MAPKs and Phosphatases in Mitochondria

Measuring NO/H$_2$O$_2$ effects on kinase activities in cell fractions requires subcellular fractions, to immunoprecipitate the selected kinase and to measure upstream and downstream components of the pathways in the presence of NO or H$_2$O$_2$. For instance, modulatory effects of H$_2$O$_2$ on mitochondrial ERK1/2 and upstream MEK1/2 (Fig. 6A) can be analyzed in freshly isolated rat brain mitochondria (4 mg protein/ml) resuspended in buffer with a mix of antiproteases and antiphosphatases (ammonium molybdate, sodium orthovanadate, and sodium fluoride), and incubated by 5, 15, 30, and 60 min at 37° with 1–50 $\mu M$ H$_2$O$_2$ (Fig. 6) or stimulated to produce endogenous H$_2$O$_2$ with antimycin, and then compared with controls without

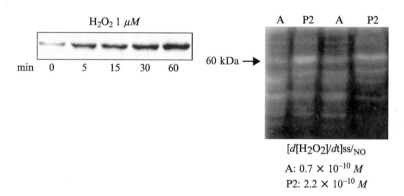

Fig. 6. (A) Time course of activation of constitutive mitochondrial MEK1/2 by H$_2$O$_2$. (B) Modulation of phosphatase activity in isolated brain mitochondria at two $[d[\text{H}_2\text{O}_2]/dt]\text{ss}/_{\text{NO}}$. (A) Adult; P2: neonatal rats. 60-kDa band corresponds to threonine–phosphatase PP1B.

H$_2$O$_2$. At the end of incubations, aliquots (25 $\mu$g proteins) are taken for ERK1/2 and MEK immune blotting, as described elsewhere. To test specific redox effects on mitochondrial ERK pathways, mitochondrial samples can be incubated with 1 and 10 $\mu M$ of U0126 (MEK1/2 inhibitor) in the presence or absence of H$_2$O$_2$.

### Mitochondrial Phosphatases (PTPs)

Phosphatases possess thiol groups and are very sensitive to redox changes. To test phosphatase activities (Fig. 6), mitochondria are washed two times with ice-cold PBS and are then lysed in buffer containing 50 m$M$ HEPES, pH 7.4, 150 m$M$ NaCl, 0.5% Nonidet P40, 1 m$M$ EDTA, 2 m$M$ EGTA, 20 g/ml leupeptin, 1 g/ml pepstatin, 200 KIE/ml aprotinin, and 1 m$M$ PMSF (protease inhibitors are added freshly). Sodium orthovanadate can also be included because it is a reversible PTP inhibitor and is effectively removed by sodium dodecylsulfate (SDS)–polyacrylamide gel electrophoreses (PAGE). Irreversible PTP inhibitors (such as Zn$^{2+}$, pervanadate, SH-alkylants) must, however, be omitted. The detection method is very sensitive, in that as little as 1–10 pg of some PTPs can be detected.

Measurement of phosphatases requires the utilization of a labeled substrate. To that purpose, 1 mg of poly(Glu4Tyr) is incubated at 25° with shaking overnight with 20 $\mu$g (~60 U) recombinant human pp60c-src (expressed as a GST-fusion protein in *Escherichia coli*), in 0.5 ml kinase buffer containing 50 m$M$ HEPES, pH 7.4, 2 m$M$ DTT, 0.1 m$M$ sodium orthovanadate, 10 m$M$ MgCl$_2$, 1 m$M$ MnCl$_2$, 0.2 m$M$ ATP, and 200 Ci of [$^{32}$P]ATP (3000 Ci/mmol) (Burridge and Nelson, 1995; Markova *et al.*, 2005). Precipitated $^{32}$P-poly(Glu4Tyr) is collected by centrifugation at 12,000$g$ in a Microfuge at 4° for 10 min, and the pellet is readily dissolved in 100 ml of 2M Tris–base; nonincorporated ATP is removed by passing the sample over a column of Sephadex G50, at pH 7.2. Aliquots of labeled poly (Glu4Tyr) are stored frozen at −80° and can be used for up to 8 weeks, except when silver staining of the gels is required.

### Gel Electrophoresis and Detection of PTPs

Detection of phosphatases is performed in 7.5, 10, or 12.5% SDS polyacrylamide gels casted by standard procedures, using a commercially available acrylamide–bisacrylamide stock (30:0.8), except that $^{32}$P-labeled poly(Glu4Tyr) is added to the polymerization mix at approximately 2 × 10$^5$ cpm/ml gel (background). Mitochondrial samples, samples of cell lysates, or immunoprecipitates are run under standard conditions. After electrophoresis, proteins in the gel must be renatured, and gels are incubated under shaking at room temperature for 1.5 h or overnight in 50–100 ml 50 m$M$ Tris–HCl, pH 8, and 20% isopropanol to remove SDS. On the next day, gels are stained with Coomassie by a standard procedure and dried and exposed to high-sensitivity films, using an intensifying screen at $-80°$, usually for several hours. Depending on the problem addressed, exposure times should be varied to achieve optimal detection. Because stringent quantification of the images is not possible anyway, we prefer to digitize the images by scanning the films at 300–400 dpi resolution.

## Concluding Remarks

Production and utilization of NO and $O_2$ active species are mutually dependent, particularly in the frame of *in vivo* signaling and metabolism in living cells. To us, the most clear example is given by the fact that mitochondria use NO to produce superoxide anion and $H_2O_2$. A controlled direct reaction between $O_2^-$ and NO leading to $H_2O_2$ and $ONOO^-$ formation has several physiological applications. First, it controls the NO steady-state concentration and consequently the activity of cytochrome oxidase and $O_2$ uptake. Second, it may modulate anaplerotic pathways by regulating the electron transfer chain and the Krebs cycle. Third, it may protect mitochondria from uncontrolled oxidation resulted form high electron transfer rates. Finally, $H_2O_2$ and $ONOO^-$ act as signal molecules that influence the balance among the different cascades and their distribution and activity to modify cell cycle and behavior. In part, the final effects could depend on the subcellular traffic of kinases to mitochondria and nucleus, as resulted from activities of upstream kinases and phosphatases. Considering the interdependence of nitrogen and oxygen active species, it is hard to experimentally dissect their respective effects in living cells; moreover, simultaneous detection of $N_2$ and $O_2$ species requires careful monitoring of methodological interferences as well as intracellular L-Arg and L-Arg transporters in cell membrane and mitochondria from the different studied tissues and species.

## References

Alonso, M., Melani, M., Jaitovich, A., Converso, D. P., Carreras, M. C., Medina, J. H., and Poderoso, J. J. (2004). Mitochondrial extracellular signal-regulated kinases (ERK1/2) are modulated during brain development. *J. Neurochem.* **89,** 248–256.

Boveris, A., and Cadenas, E. (1997). Cellular sources and steady-state levels of reactive oxygen species. *In* "Oxygen, Gene Expression, and Cellular Function" (L. Biadasz Clerch and D. J. Massaro, eds.), pp. 1–25. Marcel Dekker, New York.

Boveris, A., and Navarro, A. (2004). Rat brain and liver mitochondria develop oxidative stress and lose enzymatic activities on aging. *Am. J. Physiol. Regul. Integr. Comp. Physiol.* **287,** R1244–1249.

Boveris, A., Oshino, N., and Chance, B. (1972). The cellular production of hydrogen peroxide. *Biochem. J.* **128**, 617–630.

Brookes, P. S. (2004). Mitochondrial nitric oxide synthase. *Mitochondrion* **3**, 187–204.

Brown, G. C. (1995). Nitric oxide regulates mitochondrial respiration and cell functions by inhibiting cytochrome oxidase. *FEBS Lett.* **369**, 136–139.

Burridge, K., and Nelson, A. (1995). An in-gel assay for protein tyrosine phosphatase activity: Detection of widespread distribution in cells and tissues. *Anal. Biochem.* **232**, 56–64.

Carreras, M. C., Franco, M. C., Peralta, J. G., and Poderoso, J. J. (2004a). Nitric oxide, complex I, and the modulation of mitochondrial reactive species in biology and disease. *Mol. Aspects Med.* **25**, 125–139.

Carreras, M. C., Converso, D. P., Lorente, A. S., Barbich, M. R., Levisman, D. M., Jaitovich, A., Antico Arciuch, V. G., Galli, S., and Poderoso, J. J. (2004b). Mitochondrial nitric oxide synthase drives redox signals from proliferation and quiescence in rat liver development. *Hepatology* **40**, 157–166.

Carreras, M. C., Peralta, J. G., Converso, D. P., Finocchietto, P. V., Rebagliati, I., Zaninovich, A. A., and Poderoso, J. J. (2001). Modulation of liver mitochondrial NOS is implicated in thyroid-dependent regulation of O2 uptake. *Am. J. Physiol. Heart Circ. Physiol.* **281**, H2282–2288.

Chance, B. (1954). Special Methods, Catalase. *In* "Methods of Biochemical analysis" (D. Glick, ed.), pp. 408–424. Interscience, New York.

Ferrer-Sueta, G., Vitturi, D., Batinic-Haberle, I., Fridovich, I., and Radi, R. (2003). Reactions of manganese porphyrins with peroxynitrite and carbonate radical anion. *Biol. Chem.* **278**, 27432–27438.

Foster, M. W., and Stamler, J. S. (2004). New insights into protein S-nitrosylation. Mitochondria as a model system. *J. Biol. Chem.* **279**, 25891–25897.

Galli, S., Labato, M., Bal de Kier Joffè, E., Carreras, M. C., and Poderoso, J. J. (2003). Decreased mitochondrial nitric oxide synthase activity and hydrogen peroxide relate persistent tumoral proliferation to embryonic behavior. *Cancer Res.* **60**, 6370–6377.

Ghafourifar, P., and Richter, C. (1997). Nitric oxide synthase activity in mitochondria. *FEBS Lett.* **418**, 291–296.

Giulivi, C., Poderoso, J. J., and Boveris, A. (1998). Production of nitric oxide by mitochondria. *J. Biol. Chem.* **273**, 11038–11043.

Herrera, B., Alvarez, A. M., Sánchez, A., Fernández, M., Roncero, C., Benito, M., and Fabregat, I. (2001). Reactive oxygen species (ROS) mediates the mitochondrial-dependent apoptosis induced by transforming growth factor a in fetal hepatocytes. *FASEB J.* **15**, 741–751.

Ischiropoulos, H., Nelson, J., Duran, D., and Al-Mehdi, A. (1996). Reactions of nitric oxide and peroxynitrite with organic molecules and ferrihorseradish peroxidase: Interference with the determination of hydrogen peroxide. *Free Radic. Biol. Med.* **20**, 373–381.

Kanai, A. J., Pearce, L. L., Clemens, P. R., Birder, L. A., VanBibber, M. M., Choi, S. Y., de Groat, W. C., and Peterson, J. (2001). Identification of a neuronal nitric oxide synthase in isolated cardiac mitochondria using electrochemical detection. *Proc. Natl. Acad. Sci.* **98**, 14126–14131.

Markova, B., Gulati, P., Herrlich, P. A., and Bohmer, F. D. (2005). Investigation of protein-tyrosine phosphatases by in-gel assays. *Methods* **35**, 22–27.

Peralta, J. G., Finocchietto, P. V., Converso, D. P., Schöpfer, F., Carreras, M. C., and Poderoso, J. J. (2003). The modulation of mitochondrial nitric oxide synthase and energy expenditure in rat cold acclimation. *Am. J. Physiol. Heart Circ. Physiol.* **284**, H2375–2383.

Poderoso, J. J., Carreras, M. C., Lisdero, C. L., Riobó, N. A., Schöpfer, F., and Boveris, A. (1996). Nitric oxide inhibits electron transfer and increases superoxide radical production in rat heart mitochondria and submitochondrial particles. *Arch. Biochem. Biophys.* **328**, 85–92.

Poderoso, J. J., Carreras, M. C., Lisdero, C., Schöpfer, F., Riobó, N., Giulivi, C., Boveris, A., Boveris, A. A., and Cadenas, E. (1999a). The reaction of nitric oxide with ubiquinol: Kinetic properties and biological significance. *Free Radic. Biol. Med.* **26,** 925–935.

Poderoso, J. J., Lisdero, C., Schöpfer, F., Riobó, N., Carreras, M. C., Cadenas, E., and Boveris, A. (1999b). The regulation of mitochondrial oxygen uptake by redox reactions involving nitric oxide and ubiquinol. *J. Biol. Chem.* **274,** 37709–37716.

Poderoso, J. J., Peralta, J. G., Lisdero, C. L., Carreras, M. C., Radisic, M., Schöpfer, F., Cadenas, E., and Boveris, A. (1998). Nitric oxide regulates oxygen uptake and promotes hydrogen peroxide release by the isolated beating rat heart. *Am. J. Physiol. Cell. Physiol.* **274,** 112–119.

Riobó, N., Melani, M., Sanjuán, N., Carreras, M. C., Cadenas, E., and Poderoso, J. J. (2002). The modulation of mitochondrial nitric oxide synthase activity in rat brain development. *J. Biol. Chem.* **277,** 42447–42455.

Royall, J. A., Gwin, P. D., Parks, D. A., and Freeman, B. A. (1992). Responses of vascular endothelial oxidant metabolism to lipopolysaccharide and tumor necrosis factor-alpha. *Arch. Biochem. Biophys.* **294,** 686–694.

Schöpfer, F. J., Riobó, N. A., Carreras, M. C., Alvarez, B., Radi, R., Boveris, A., Cadenas, E., and Poderoso, J. J. (2000). Oxidation of ubiquinol by peroxynitrite: Implications for protection of mitochondria against nitrosative damage. *Biochem. J.* **349,** 35–42.

Tarpey, M. M., and Fridovich, I. (2001). Methods of detection of vascular reactive species: Nitric oxide, superoxide, hydrogen peroxide, and peroxynitrite. *Circ. Res.* **89,** 224–236.

Zhao, H., Kalivendi, S., Zhang, H., Joseph, J., Nithipatikom, K., Vasquez-Vivar, J., and Kalyanaraman, B. (2003). Superoxide reacts with hydroethidine but forms a fluorescent product that is distinctly different from ethidium: Potential implications in intracellular fluorescence detection of superoxide. *Free Radic. Biol. Med.* **34,** 1359–1368.

[35]  Cytotoxic and Cytoprotective Actions of $O_2^-$ and NO (ONOO$^-$) are Determined Both by Cellular GSH Level and HO Activity in Macrophages

*By* Klaokwan Srisook, Chaekyun Kim, and Young-Nam Cha

## Abstract

Survival of macrophages, which serve as the first-line defense against invading pathogens by invoking the overproduction of highly toxic peroxynitrite (ONOO$^-$), depends on their ability to maintain the intracellular GSH level and to induce the expression of heme oxygenase-1 (HO-1). The ONOO$^-$ is produced by macrophages stimulated by pathogens and is a powerful oxidant reacting directly with cellular GSH and proteins, killing both invading pathogens and macrophages themselves. However, macrophages can survive the toxicity of ONOO$^-$ by replenishing the depleted GSH level and by inducing HO-1 expression. In macrophages exposed to conditions overproducing $O_2^-$, NO, or ONOO$^-$, the cellular level of GSH

METHODS IN ENZYMOLOGY, VOL. 396
0076-6879/05 $35.00
DOI: 10.1016/S0076-6879(05)96035-7

decreased rapidly, and when excessive, cells died. However, in cells surviving the toxicity caused by lower doses of $O_2^-$, NO, or ONOO$^-$, the depleted intracellular GSH level was replenished, and HO-1 expression was increased, but not when they were coexposed to an inhibitor of HO-1 activity. Cells exposed to an inhibitor of GSH synthesis had greater induction of HO-1 expression and survived. However, cells exposed to an inhibitor of HO-1 activity died extensively and could not be revived by addition of N-acetylcysteine (NAC), a precursor of GSH synthesis. Thus, the dichotomous cytotoxic or cytoprotective effects of $O_2^-$, NO, or ONOO$^-$ in macrophages are determined both by cellular GSH level and by HO-1 activity.

## Introduction

Macrophages serve as the first-line defense against invading pathogens by overproducing peroxynitrite (ONOO$^-$), the combined product of superoxide anion ($O_2^-$) and nitric oxide (NO) radicals (Huie and Padmaja, 1993). NADPH-oxidase (PHOX) and inducible NO synthase (iNOS) catalyze the overproduction of $O_2^-$ and NO radicals, respectively, in the stimulated macrophages (Forman and Torres, 2002; MacMicking et al., 1997). When stimulated by invading pathogens or inflammatory cytokines, activity of these heme-containing enzymes producing the radicals is increased markedly, either by rapid activation of PHOX or by induction of iNOS. Simultaneous overproduction of $O_2^-$ and NO results in a rapid formation of ONOO$^-$ (6.7 $\times$ 10$^9$ M/s) (Huie and Padmaja, 1993), a severely GSH-oxidizing and protein-nitrating product, causing destruction of invading pathogens, but ironically, also the macrophages themselves and surrounding host tissues (Wink et al., 1996). However, the ONOO$^-$-stressed or GSH-depleted macrophages and host cells can protect themselves by enhancing the expression of heme oxygenase-1 (HO-1; EC 1.14.99.3) (Ishii et al., 2000; Li et al., 2002). At least three isoforms of HOs are present in most mammalian cells: the oxidative stress– or heme-inducible HO-1 and the constitutively expressed noninducible HO-2 and HO-3 (Maines et al., 1986; McCoubrey et al., 1997; Shibahara et al., 1993). Together, these HOs are involved in degrading the excess free heme while generating biliverdin (bilirubin) and carbon monoxide (CO) (Maines, 1997). Bile pigments (biliverdin and bilirubin) are known to scavenge and detoxify both reactive oxygen species (ROS) and reactive nitrogen species (RNS), thus serving as cytoprotective antioxidants (Kaur et al., 2003; Mancuso et al., 2003; Stocker et al., 1987). Furthermore, CO can bind avidly to the heme iron contained in heme enzymes like PHOX and iNOS, and inhibit additional production of $O_2^-$ and NO (Dulak and Jozkowicz, 2003; Slebos et al., 2003), respectively.

Thus, HO can provide cytoprotective effects against the injuries that can be caused by ROS and RNS.

Small amounts of $O_2^-$ and NO are generated even during normal cellular metabolism catalyzed by several heme-containing enzymes (i.e., PHOX, mitochondrial cytochromes, P-450, cNOS) (Connelly et al., 2003; Forman and Torres, 2001). Productions of these small amounts of $O_2^-$ and NO, and consequently the ONOO$^-$, are needed to support the constitutive levels of GSH and antioxidant enzymes like superoxide dismutases (SODs), glutathione peroxidase (GPX), and catalase, among others. Also, the free heme being synthesized constantly is needed for incorporation into heme enzymes like PHOX, cytochromes, and cNOS to support the normal production of physiological amounts of $O_2^-$ and NO (ONOO$^-$). Excess free heme being synthesized constantly is removed by the constitutively expressed HO-2 (Maines, 1997). Under normal conditions, although the production of these small amounts of $O_2^-$, NO, and free heme is, thus, cytoprotective, they are detoxified sufficiently by the existing cellular antioxidants, antioxidant enzymes, and HO-2, as mentioned earlier.

However, upon excessive production of $O_2^-$ [i.e., lipopolysaccharide (LPS)-stimulated oxidative burst and exposure to redox-cycling chemicals) and/or with inadequate detoxification, cells undergo oxidative stress, most notably by severe oxidation and depletion of intracellular GSH. Additionally, the overproduced ROS (and RNS) can bind the heme-containing enzymes directly and degrade the protein to release free heme (Droge, 2002). The released free heme can catalyze the Fenton reaction producing highly reactive HO$^•$ from $H_2O_2$ (Halliwell and Gutteridge, 1999; Jeney et al., 2002), and it must be eliminated by enhancing the HO activity. Thus, macrophages surviving from such oxidative stress respond by activating several redox-sensitive transcription factors like nuclear factor-$\kappa$B (NF-$\kappa$ B), AP-1, and Nrf2, which are involved in the transcriptional activation of the NO-producing iNOS (Kim et al., 1997), the GSH-synthesizing enzymes (Bea et al., 2003; Li et al., 2002), and the CO- and bile pigment–producing HO-1 (Alam et al., 2003; Camhi et al., 1998), respectively.

The overproduced $O_2^-$ and NO are not sufficiently reactive or toxic by themselves and do not oxidize or deplete GSH readily. However, when both are produced simultaneously and in excess, NO can rapidly scavenge $O_2^-$ to generate ONOO$^-$ at a diffusion-limited rate. In this connection, nearly 70% of $O_2^-$ produced in the LPS-stimulated macrophages with elevated iNOS is converted to ONOO$^-$. The overproduced ONOO$^-$ oxidizes GSH rapidly and depletes cellular GSH (Wink and Mitchell, 1998). In response to this ONOO$^-$-derived GSH oxidation and depletion, both the de novo GSH synthesis and HO-1 expression are increased in macrophages. GSH synthesis requires the supply of cysteine (Sato et al., 2001), and elevated HO-1 eliminates the released free heme rapidly while respectively producing antioxidants (bile pigments) and CO, the inhibitor of additional $O_2^-$ and NO production (see Chapter 31). Thus, in the macrophages surviving from $ONOO^-$ stress, upregulation of cystine transporter Xc (supplying cysteine needed for GSH synthesis) and HO-1 is readily observed (Srisook and Cha, 2004). Therefore, generation of $ONOO^-$ can be cytoprotective or cytotoxic, depending on the rate and amount of its production, and these are modulated by both the GSH level and HO activity.

In our study, RAW 264.7 peritoneal macrophages were exposed to menadione bisulfite (MBS), a redox-cycling chemical generating much endogenous $O_2^-$, and spermine NONOate (SPNO), an exogenous NO donor, each alone and in combination to generate $ONOO^-$, and we determined their ability to deplete GSH level and to induce HO-1 expression. In macrophages stimulated with increasing doses of MBS, a dose-dependent rapid depletion of cellular GSH level was observed (data not shown) with delayed upregulation of both HO-1 messenger RNA (mRNA) (data not shown) and protein expression (Fig. 1). This occurred without any increases in iNOS expression (Fig. 1) or NO production (data not shown). These results indicate that the $O_2^-$ overproduced by redox-cycling metabolism of MBS causes GSH depletion and HO-1 induction without endogenous production of NO (or $ONOO^-$). Alternatively, when the cells were treated with increasing doses of SPNO, again, a dose-dependent rapid depletion of cellular GSH level (2 h) was observed (data not shown) with

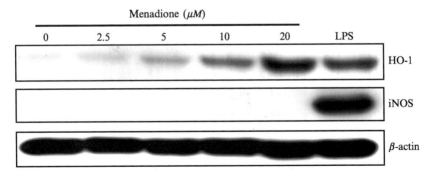

FIG. 1. Menadione induces HO-1 expression in a dose-dependent manner. RAW 264.7 macrophages were incubated with menadione bisulfite (MBS; 0–20 $\mu M$) or with 1 $\mu$g/ml lipopolysaccharide (LPS) for 12 h. Cells were homogenized and subjected to 10% sodium dodecylsulfate–polyacrylamide gel electrophoresis. The contents of HO-1 and inducible nitric oxide synthase (iNOS) proteins were determined by immunoblot analysis. The $\beta$-actin protein content was used to normalize the HO-1 and iNOS protein contents.

(A)

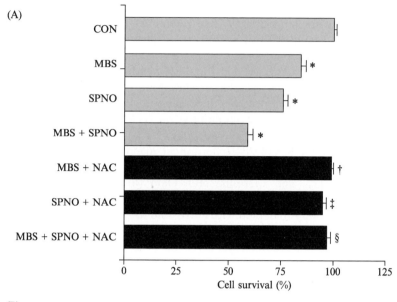

(B)

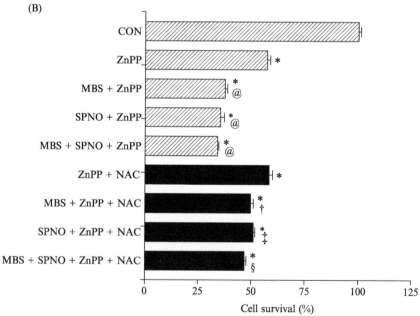

delayed upregulation of both HO-1 mRNA and protein expression (see Chapter 31). This suggests that NO alone enhances the HO-1 expression. In the cells treated with LPS, there was a biphasic induction of HO-1, occurring first with oxidative burst ($O_2^-$ production) and second with iNOS induction (NO production) or with $ONOO^-$ generation (Srisook and Cha, 2004). In the macrophages treated with MBS and SPNO to mimic the effect of generating $ONOO^-$, there was an immediate depletion of GSH level, accompanied by a large induction of HO-1 expression (data not shown).

Next, viability of macrophages treated in various combinations of MBS, SPNO, and $N$-acetylcysteine (NAC), a membrane-diffusible form of cysteine needed for GSH synthesis, was determined by employing the MTT test. Addition of NAC would prevent GSH depletion and, thus, block the HO-1 induction by inhibiting the activation of redox-sensitive transcription factors (i.e., NF-$\kappa$B, AP-1, and Nrf2) needed for inductions of iNOS and HO-1 expression. Whereas the cells treated with MBS ($O_2^-$), SPNO (NO), or their combination ($ONOO^-$) had markedly decreased survival, the cells treated additionally with NAC had full survival (Fig. 2A). These results indicated that GSH is needed for cytoprotection against the toxicity of $O_2^-$, NO and $ONOO^-$.

Next, to determine the necessity of HO activity for cell survival against the toxicity of $O_2^-$, NO, and $ONOO^-$, cells were treated with various combinations of MBS, SPNO, and zinc-protoporphyrin IX (ZnPP), an inhibitor of HO activity (Drummond, 1987). In the control macrophages treated with ZnPP alone, inhibiting the HO activity derived from the constitutively expressed HO-2, cell survival was decreased markedly. When the macrophages with elevated HO-1 expression (i.e., cells treated with MBS, SPNO, or their combination) were treated with ZnPP, greater

---

FIG. 2. Cell survival against the toxicity of $O_2^-$, nitric oxide (NO), and $ONOO^-$. Spermine NONOate (SPNO, 20 $\mu M$) or menadione bisulfite (MBS, 20 $\mu M$) was added to cells that have been preincubated with 2.5 m$M$ $N$-acetylcysteine (NAC) for 2 h (A) and with 10 $\mu M$ ZnPP for 30 min (B). Viability of cells harvested at 24 h after the addition of SPNO or MBS was determined using the MTT test. (A) Bar graph shows the mean ± SEM (N = 3). Using the unpaired Student's $t$ test, asterisk (*) indicates $p < .01$ compared to the viability of control cells. Dagger (†) indicates $p < .01$ compared with cells treated with MBS alone. Double dagger (‡) indicates $p < .01$ compared with cells treated with SPNO alone. Section (§) indicates $p < .001$ compared with cells treated with MBS plus SPNO. (B) Bar graph shows the mean ± SEM (N = 3). Asterisk (*) indicates $p < .001$ compared to the viability of control cells. The "at" sign (@) indicates $p < .001$ compared with cells treated with ZnPP. Dagger (†) indicates $p < .01$ compared with cells treated with MBS + ZnPP. Double dagger (‡) indicates $p < .01$ compared with cells treated with SPNO + ZnPP. Section (§) indicates $p < .001$ compared with cells treated with MBS + SPNO.

decreases in cell survival were observed (Fig. 2B). These decreased cell survivals caused by inhibition of either the constitutive or the inducible HO activity (ZnPP-treated cells) were not restored by addition of NAC. These results indicate that the HO activity, whether derived from constitutively expressed HO-2 or from stress-induced HO-1, is essential for cell survival from the toxicity of ROS and RNS. Combined, the obtained data suggest that either the cytoprotective or the cytotoxic effects (so-called the "double-edged sword" effects) of $O_2^-$, NO, and $ONOO^-$ are determined both by cellular GSH level and by HO activity in macrophages.

## Materials and Methods

### Cell Culture

Murine peritoneal macrophage cell line RAW 264.7 was obtained from American Type Culture Collection (ATCC). Cells were cultured in Dulbecco's Modified Eagle Medium (DMEM) (Gibco/Invitrogen, Grand Island, NY) containing 100 units/ml of penicillin, 100 $\mu$g/ml of streptomycin, 4 m$M$ L-glutamine, 25 m$M$ D-glucose, 1 m$M$ sodium pyruvate, and 10% heat-inactivated characterized fetal bovine serum (FBS) (Hyclone Laboratories, Logan, UT) at 37° in humidified air containing 5% $CO_2$. The cells used in all experiments were from the same passage obtained from multiple vials of cells frozen in liquid $N_2$. Confluent cells in a 100-mm plate were harvested by 10-min incubation at 37° in $Ca^{2+}$- and $Mg^{2+}$-free Hanks' balanced salt solution (HBSS) and then scraped gently using a cell scraper. The harvested cells were pelleted, resuspended in a fresh medium, and then seeded into culture dish. After an overnight growth, cells were treated with various experimental chemicals.

### Cell Survival by MTT Assay

Cell viability was evaluated by determining the mitochondrial function of living cells on the basis of their ability to reduce the yellow dye, tetrazolium salt 3-(4,5-dimethylthiazol-2-yl)-2,5-diphenyltetrazolium bromide (MTT) into blue formazan crystal, mainly by the functional mitochondrial dehydrogenases (Berridge and Tan, 1993). The formation of formazan is proportional to the number of functional mitochondria in living cells. Cells ($1.5 \times 10^5$ in 0.5 ml of 10% FBS-DMEM) were plated into 24-well plates and allowed to attach overnight; subsequently, cells were washed with warm HBSS. Ten microliters of 5 mg/ml MTT solution [dissolved in phosphate-buffered saline (PBS)] was added to each well for 2 h, and afterwards, the solution was aspirated and 0.5 ml of dimethylsulfoxide (DMSO) was put into each well to solubilize the blue formazan crystal product. The plate was returned to the 37° incubator for 5 min, and the solubilized formazan solution was transferred to another 96-well plate. The plates were then read using a microplate reader (PowerWaveX, Bio-Tek Instrument, Inc.), which measured the absorbance of solubilized formazan present in each well at 550 nm. This MTT cell survival assay is based on increased absorbance of the product formed by living cells and allows a relative determination of survival rates. Percentage of cell survival is expressed as (absorbance of treated well/absorbance of control well) × 100.

## Glutathione Assay

*Sample preparation*: RAW 264.7 macrophage cells ($\sim$1 × $10^6$ cells/ 60-mm plate) were washed twice with cold PBS (pH 7.4) and scraped after addition of 300 $\mu$l of 5% (w/v) 5-sulfosalicylic acid (5-SSA) and 0.2% (v/v) Triton X-100. Suspended cells were then freeze-thawed three times and finally sonicated on ice for 5 s at the output of 2 W by employing the Vibracell ultrasonic processor (Srisook and Cha, 2004) (Sonics & Materials, Newtown, CT). The amount of protein present in cell lysate was assayed using the BCA protein assay reagent kit (Pierce, Rockford, IL). After removing the protein residue by centrifugation at 20,000g for 15 min at 4°, the acidic supernatant fraction containing both the reduced (GSH) and oxidized glutathione (GSSG) was stored at −70° until assays for glutathione determinations could be performed.

*Total glutathione assay*: The total glutathione levels (reduced and oxidized glutathione) were determined using a method based on the enzymatic recycling assay employing glutathione reductase (GR) (Tietze, 1969). In the presence of NADPH, GR reduces the GSSG contained in the acidic supernatant to GSH, and the total reduced glutathione (GSH) is oxidized by 5,5′- dithio-bis-(2-nitrobenzoic acid) (DTNB) to yield GSSG and 5-thio-2-nitrobenzoic acid (TNB), a chromophore absorbing at 405 nm. Thus, the assay mixture (200 $\mu$l) contained 0.6 m*M* DTNB, 0.2 m*M* NADPH, 0.5 unit of GR, and 10 $\mu$l of the acidic supernatant in 0.1 M sodium phosphate buffer (pH 7.4) with 5 m*M* EDTA. The rate of TNB formation from DTNB is proportional to the total amount of glutathione, and this is followed spectrophotometrically for 6 min at 405 nm at 20-s intervals using a microplate reader (Model PowerWaveX, Bio-Tek Instruments, Winooski, VT) kept at 30°. A standard curve containing various known amounts of GSH (0.2–1.6 $\mu$*M*) dissolved in the same Triton X-100 and 5-SSA solution was generated to quantify the

GSH concentration present in various samples, and the results obtained with a reaction mixture containing GR, 5-SSA, and Triton X-100 without GSH was used as the blank. The results are expressed as nanomoles of total glutathione per milligram of cellular protein.

*Oxidized glutathione assay*: For the specific determination of oxidized glutathione (GSSG) levels, the thawed supernatant fraction (100 $\mu$l) was incubated first with 2 $\mu$l of undiluted 97% 2-vinylpyridine (2-VP, Aldrich, Milwaukee, WI) and undiluted 6 $\mu$l triethanolamine (TEA, Sigma-Aldrich, St. Louis, MO] for 1 h at room temperature. This was done to prevent the GSH present in the samples from reacting with DTNB (Griffith, 1980). Subsequently, the level of GSSG present in this derivatized sample was determined by employing the DTNB-GR recycling assays as described for determination of the total glutathione content. GSSG standard curve (0.02–0.16 $\mu M$) was prepared in the 5-SSA and Triton X-100 solution containing the same amount of 2-VP and TEA as the samples. The 2-VP should be stored at $-20°$. When the 2-VP becomes brown or viscous, it should be replaced with a new bottle or redistilled. To prevent autoxidation of GSH, TEA should be added to the side of reaction vial above the level of solution.

# References

Alam, J., Killeen, E., Gong, P., Naquin, R., Hu, B., Stewart, D., Ingelfinger, J. R., and Nath, K. A. (2003). Heme activates the heme oxygenase-1 gene in renal epithelial cells by stabilizing Nrf2. *Am. J. Physiol. Renal Physiol.* **284,** F743–752.

Bea, F., Hudson, F. N., Chait, A., Kavanagh, T. J., and Rosenfeld, M. E. (2003). Induction of glutathione synthesis in macrophages by oxidized low-density lipoproteins is mediated by consensus antioxidant response elements. *Circ. Res.* **92,** 386–393.

Berridge, M. V., and Tan, A. S. (1993). Characterization of the cellular reduction of 3-(4,5-dimethylthiazol-2-yl)-2,5-diphenyltetrazolium bromide (MTT): Subcellular localization, substrate dependence, and involvement of mitochondrial electron transport in MTT reduction. *Arch. Biochem. Biophys.* **303,** 474–482.

Camhi, S. L., Alam, J., Wiegand, G. W., Chin, B. Y., and Choi, A. M. K. (1998). Transcriptional activation of the HO-1 gene by lipopolysaccharide is mediated by 5' distal enhancers: Role of reactive oxygen intermediates and AP-1. *Am. J Respir. Cell Mol. Biol.* **18,** 226–234.

Connelly, L., Jacobs, A. T., Palacios-Callender, M., Moncada, S., and Hobbs, A. J. (2003). Macrophage endothelial nitric-oxide synthase autoregulates cellular activation and proinflammatory protein expression. *J. Biol. Chem.* **278,** 26480–26487.

Droge, W. (2002). Free radicals in the physiological control cell function. *Physiol. Rev.* **82,** 47–95.

Drummond, G. S. (1987). Control of heme metabolism by synthetic metalloporphyrins. *Ann. NY Acad. Sci.* **514,** 87–95.

Dulak, J., and Jozkowicz, A. (2003). Carbon monoxide – a "new" gaseous modulator of gene expression. *Acta Biochim. Pol.* **50,** 31–47.

Forman, H. J., and Torres, M. (2001). Redox signaling in macrophages. *Mol. Asp. Med.* **22,** 189–216.

Forman, H. J., and Torres, M. (2002). Reactive oxygen species and cell signaling: Respiratory burst in macrophage signaling. *Am. J. Respir. Crit. Care Med.* **166,** S4–8.

Griffith, O. W. (1980). Determination of glutathione and glutathione disulfide using glutathione reductase and 2-vinylpyridine. *Anal. Biochem.* **106,** 207–212.

Halliwell, B., and Gutteridge, J. M. C. (1999). "Free Radicals in Biology and Medicine," 3rd Ed. Oxford Science Publications, New York.

Huie, R. E., and Padmaja, S. (1993). The reaction of no with superoxide. *Free Radic. Res. Commun.* **18,** 195–199.

Ishii, T., Itoh, K., Takahashi, S., Sato, H., Yanagawa, T., Katoh, Y., Bannai, S., and Yamamoto, M. (2000). Transcription factor Nrf2 coordinately regulates a group of oxidative stress-inducible genes in macrophages. *J. Biol. Chem.* **275,** 16023–16029.

Jeney, V., Balla, J., Yachie, A., Varga, Z., Vercellotti, G. M., Eaton, J. W., and Balla, G. (2002). Pro-oxidant and cytotoxic effects of circulating heme. *Blood* **100,** 879–887.

Kaur, H., Hughes, M. N., Green, C. J., Naughton, P., Foresti, R., and Motterlini, R. (2003). Interaction of bilirubin and biliverdin with reactive nitrogen species. *FEBS Lett.* **543,** 113–119.

Kim, Y. M., Lee, B. S., Yi, K. Y., and Paik, S. G. (1997). Upstream NF-kappaB site is required for the maximal expression of mouse inducible nitric oxide synthase gene in interferon-gamma plus lipopolysaccharide-induced RAW 264.7 macrophages. *Biochem. Biophys. Res. Commun.* **236,** 655–660.

Li, N., Wang, M., Oberley, T. D., Sempf, J. M., and Nel, A. E. (2002). Comparison of the pro-oxidative and proinflammatory effects of organic diesel exhaust particle chemicals in bronchial epithelial cells and macrophages. *J. Immunol.* **169,** 4531–4541.

MacMicking, J., Xie, Q. W., and Nathan, C. (1997). Nitric oxide and macrophage function. *Annu. Rev. Immunol.* **15,** 323–350.

McCoubrey, W. K., Huang, T. J., and Maines, M. D. (1997). Isolation and characterization of a cDNA from the rat brain that encodes hemoprotein heme oxygenase-3. *Eur. J. Biochem.* **247,** 725–732.

Maines, M. D. (1997). The heme oxygenase system: A regulator of second messenger gases. *Annu. Rev. Pharmacol. Toxicol.* **37,** 517–554.

Maines, M. D., Trakshel, G. M., and Kutty, R. K. (1986). Characterization of two constitutive forms of rat liver microsomal heme oxygenase. Only one molecular species of the enzyme is inducible. *J. Biol. Chem.* **261,** 411–419.

Mancuso, C., Bonsignore, A., Di Stasio, E., Mordente, A., and Motterlini, R. (2003). Bilirubin and S-nitrosothiols interaction: Evidence for a possible role of bilirubin as a scavenger of nitric oxide. *Biochem. Pharmacol.* **66,** 2355–2363.

Sato, H., Kuriyama-Matsumura, K., Hashimoto, T., Sasaki, H., Wang, H., Ishii, T., Mann, G. E., and Bannai, S. (2001). Effect of oxygen on induction of the cystine transporter by bacterial lipopolysaccharide in mouse peritoneal macrophages. *J. Biol. Chem.* **276,** 10407–10412.

Shibahara, S., Yoshizawa, M., Suzuki, H., Takeda, K. K., and Meguro, K. (1993). Functional analysis of cDNAs for two types of human heme oxygenase and evidence for their separate regulation. *ENDO J. Biochem.* **113,** 214–218.

Slebos, D. J., Ryter, S. W., and Choi, A. M. (2003). Heme oxygenase-1 and carbon monoxide in pulmonary medicine. *Respir. Res.* **4,** 7.

Srisook, K., and Cha, Y. N. (2004). Biphasic induction of heme oxygenase-1 expression in macrophages stimulated with lipopolysaccharide. *Biochem. Pharmacol.* **68,** 1709–1720.

Stocker, R., Yamamoto, Y., McDonagh, A. F., Glazer, A. N., and Ames, B. N. (1987). Bilirubin is an antioxidant of possible physiological importance. *Science.* **235,** 1043–1046.

Tietze, F. (1969). Enzymic method for quantitative determination of nanogram amounts of total and oxidized glutathione: Applications to mammalian blood and other tissues. *Anal. Biochem.* **27,** 502–522.

Wink, D. A., Hanbauer, I., Grisham, M. B., Laval, F., Nims, R. W., Laval, J., Cook, J., Pacelli, R., Liebmann, J., Krishna, M., Ford, P. C., and Mitchell, J. B. (1996). Chemical biology of nitric oxide: Regulation and protective and toxic mechanisms. *Curr. Topics Cell Regul.* **34,** 159–187.

Wink, D. A., and Mitchell, J. B. (1998). Chemical biology of nitric oxide: Insights into regulatory, cytotoxic, and cytoprotective mechanisms of nitric oxide. *Free Radic. Biol. Med.* **25,** 434–456.

# [36] Determination of Mitochondrial Nitric Oxide Synthase Activity

*By* Pedram Ghafourifar, Melinda L. Asbury,
Sandeep S. Joshi, and Eric D. Kincaid

## Abstract

The main biological targets of nitric oxide (NO) are hemoproteins, thiols, and superoxide anion ($O_2^-$). Mitochondria possess several hemoproteins, thiol-containing molecules, and they are one of the prime cellular producers of $O_2^-$. Thus, these organelles remain one of the main biological targets for NO. Reports on the existence of a $Ca^{2+}$-sensitive mitochondrial NO synthase (mtNOS) have opened a new window in the field of NO and mitochondria research (Ghafourifar and Richter, 1997). mtNOS-derived NO reversibly decreases the activity of the mitochondrial hemoprotein, cytochrome *c* oxidase. This function of mtNOS regulates mitochondrial respiration and transmembrane potential ($\Delta\psi$). The NO generated by mtNOS reacts with mitochondrial thiol-containing proteins including caspase-3. Because the *S*-nitrosated caspase-3 remains apoptotically silent as long as it is located within the mitochondria, this function of mtNOS portrays an anti-apoptotic property for mtNOS. mtNOS-derived NO also reacts with $O_2^-$ to generate peroxynitrite. mtNOS-derived peroxynitrite induces oxidative stress and releases cytochrome *c* from the mitochondria, which represents a pro-apoptotic role for mtNOS. How mitochondria harmonize the reversible functions of mtNOS for mitochondrial respiration, its anti-apoptotic actions via *S*-nitrosation of caspase-3,

METHODS IN ENZYMOLOGY, VOL. 396
0076-6879/05 $35.00
DOI: 10.1016/S0076-6879(05)96036-9

versus the pro-apoptotic properties of peroxynitrite remains to be fully understood. However, intramitochondrial ionized $Ca^{2+}$ concentration ($[Ca^{2+}]_m$) and the status of mitochondrial reducing defense barriers seem to play crucial roles in orchestrating the functions of mtNOS for mitochondria and cells (Ghafourifar and Cadenas, 2005).

## Nitric Oxide and Nitric Oxide Synthases

The pioneering works of Moncada, which led to the discovery that the endothelium-derived relaxing factor is NO (Palmer $et\ al.$, 1987) changed our view of NO from being a toxic gas to that of extreme importance in biology (Koshland, 1992). NO is synthesized from L-arginine by NO synthase (NOS; EC 1.14.13.39) isozymes. This reaction is a two-step, five-electron oxidation of the terminal guanidino nitrogen of L-arginine with $N$-hydroxy-L-arginine as the intermediate. The reaction stoichiometrically consumes $O_2$ and produces L-citrulline as the final co-product. Three distinct isoforms of NOS have been well characterized in mammalian tissues, referred to as endothelial NOS (eNOS), neuronal NOS (nNOS), and inducible NOS (iNOS). All characterized NOS isozymes are heme-containing proteins that are dimeric under native conditions with a monomer molecular mass of about 126–160 kDa. nNOS and eNOS are typical $Ca^{2+}$/calmodulin-sensitive enzymes (e.g., eNOS and nNOS activities are regulated by cellular $Ca^{2+}$ oscillation). However, iNOS forms a tight complex with calmodulin at very low $Ca^{2+}$ concentrations, and alterations in cellular $Ca^{2+}$ status do not alter iNOS activity (Ghafourifar and Richter, 1999a).

NO exerts remarkable functions in various biological systems (Moncada $et\ al.$, 1991). For example, NO is one of the most important biological factors that maintain the blood pressure lowered and prevent platelet aggregation and adhesion. In the central nervous system, NO is the neurotransmitter for functions including memory formation. In the peripheral nervous system, NO is the modulator of nociception at the spinal cord and mediates the nonadrenergic and noncholinergic neurotransmission.

## Targets for NO

Biological functions of NO are mediated through its distinct reactions with hemoproteins, thiols, and $O_2^-$. Mitochondria contain several hemoproteins and thiols, and they remain one of the prime cellular producers of $O_2^-$. Thus, mitochondria are one of the prime biological targets for NO (Ghafourifar and Colton, 2003). The distinct reactions of NO with mitochondrial targets are briefly discussed in this chapter (Ghafourifar and Cadenas, 2005).

## Mitochondrial Targets of NO

The highly compartmentalized mitochondria provide distinct targets for reactions with NO (Scheme 1). Cytochrome $c$ oxidase, the terminal enzyme of the respiratory chain, is an abundant mitochondrial hemoprotein embedded in mitochondrial inner membrane. In rat liver mitochondria, cytochrome $c$ oxidase is a 204-kDa complex protein that constitutes 15–20% of the inner membrane proteins and consumes more than 90% of the oxygen used in the cells. The $O_2$ binding site of cytochrome $c$ oxidase is very specialized; however, NO exerts physicochemical properties very similar to $O_2$ that allow NO to compete with $O_2$ for the $O_2$ binding site. Thus, physiological concentrations of NO reversibly inhibit cytochrome $c$ oxidase and decrease $O_2$ consumption in a manner that resembles a pharmacological competitive antagonism between NO and $O_2$ (Ghafourifar *et al.*, 2000).

NO can react with a reduced thiol to produce a nitrosothiol. This reversible reaction is redox sensitive, and within mitochondria, it is preferred in the intermembrane space (Ghafourifar and Colton, 2003). Several mitochondrial apoptogenic proteins, including caspase-3, are located within this compartment. NO reacts with caspase-3 to produce $S$-nitrosated caspase-3 that is apoptotically silent. This anti-apoptotic property of NO protects the organelles against the proteolytic activity of the caspase and cells against apoptosis. Upon release from the mitochondria, the reduced environment of cytoplasm denitrosates and, thus, activates the caspase (Ghafourifar and Colton, 2003).

The reaction of NO with $O_2^-$ occurs with the nearly diffusion-controlled rate constant of $1.9 \times 10^{10} \ M \ s^{-1}$ (Kissner, 1997). This reaction results in the formation of peroxynitrite, a highly reactive NO-derived species. Within mitochondria, this reaction is preferred in the matrix space (Ghafourifar and Colton, 2003). Peroxynitrite causes oxidative damage to various mitochondrial respiratory chain components (Ghafourifar *et al.*, 1999b; Riobo *et al.*, 2001; Sharpe and Cooper, 1998; Szibor *et al.*, 2001), inactivates mitochondrial matrix enzymes such as manganese superoxide dismutase (MnSOD) (MacMillan-Crow *et al.*, 1998), and succinyl-coenzyme A (CoA):3-oxoacid CoA-transferase (SCOT) (Turko *et al.*, 2001), and it releases cytochrome $c$ from the mitochondria (Ghafourifar *et al.*, 1999e). Inactivation of MnSOD and SCOT plays a critical role in oxidative stress, and release of cytochrome $c$ from the mitochondria is one of the key events during apoptosis in many cells.

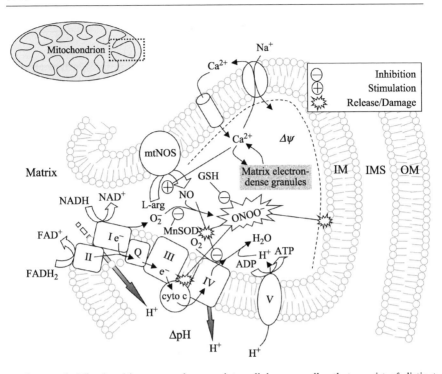

SCHEME 1. Mitochondria are membranous intracellular organelles that consist of distinct suborganelle compartments: The inner (IM) and the outer membrane (OM), the matrix, and the intermembrane space (IMS) (Scheme 1). The IM carries the oxidative phosphorylation system, which consists of four respiratory complexes (I–IV) that are embedded in the IM, coenzyme Q (ubiquinone; Q), and adenosine triphosphate (ATP) synthase that is often referred to as *complex V*. These complexes are functionally arranged in an electrochemical hierarchy based on their redox potentials. The respiratory chain provides a unique broad spectrum of redox potentials varying from $-280$ mV for complex I to 250 mV for complex IV. Electrons enter the chain through oxidation of NADH or $FADH_2$ and flow down the chain to complex IV to reduce $O_2$ to $H_2O$. Coupled to this electron flow, protons are extruded from the mitochondrial matrix into the IMS. Since the IM is impermeable to protons and protons can reenter the matrix only through the ATP synthase machinery, the proton extrusion establishes a transmembrane potential ($\Delta\psi$, negative inside) and an electrochemical gradient ($\Delta pH$, alkaline inside) across the coupling membrane. Mitochondrial transmembrane potential is much more negative than the cell membrane potential (e.g., fully energized isolated rat liver mitochondria built a $\Delta\psi$ up to $-200$ mV. The $\Delta\psi$ is the driving force for mitochondria to take up and retain $Ca^{2+}$. Mitochondria are one of the main cellular calcium-buffering organelles and can store relatively large quantities of calcium. However, intramitochondrial ionized calcium ($[Ca^{2+}]_m$) is maintained low by several mechanisms. Most importantly, mitochondria very rapidly precipitate the $[Ca^{2+}]_m$ to form nonionized calcium pools, the matrix electron-dense granules. The amount and content of these granules vary under different conditions; however, they mainly consist of calcium phosphate and hydroxy apatite $[Ca_3(PO_4)_2$ and $(Ca_3[PO_4]_2)_3 \cdot Ca(OH)_2]$. $Ca^{2+}$ can leave

Mitochondrial NOS

Between 1995 and 1996, several immunohistochemical studies suggested the presence of an NOS-like protein within mitochondria (Bates *et al.*, 1995, 1996; Frandsen *et al.*, 1996; Kobzik *et al.*, 1995). In 1997, the first report on the existence of a constitutively expressed and continuously active NOS in mitochondria (mtNOS), the association of mtNOS with the mitochondrial inner membrane, and the determination of mtNOS activity was published (Ghafourifar and Richter, 1997). It was shown that mtNOS is $Ca^{2+}$-sensitive [i.e., an increase in the intramitochondrial $Ca^{2+}$ concentration ($[Ca^{2+}]_m$) increases mtNOS activity], and that mtNOS exerts substantial control over mitochondrial respiration and transmembrane potential ($\Delta\psi$). These findings were confirmed by several other groups (Ghafourifar and Cadenas, 2005).

mtNOS and Mitochondrial Bioenergetics

By inhibiting cytochrome *c* oxidase activity, mtNOS-derived NO decreases $O_2$ consumption, $\Delta\psi$, and mitochondrial matrix pH (Ghafourifar and Richter, 1997, 1999c; Ghafourifar *et al.*, 1999e). Inhibition of the basal endogenous mtNOS activity increases basal mitochondrial $O_2$ consumption, and $\Delta\psi$ (Ghafourifar and Richter, 1997) causes mitochondrial matrix alkalinization and provides a resistance to the sudden drop of $\Delta\psi$ induced by elevation of $[Ca^{2+}]_m$ (Ghafourifar and Richter, 1999c; Ghafourifar *et al.*, 1999c). These findings suggest that mtNOS is continuously active and that it reversibly regulates mitochondrial respiration and respiration-dependent functions. mtNOS also exerts a feedback-regulatory function

---

mitochondria as a consequence of decreased $\Delta\psi$ (not shown) or in exchange with another cation such as $Na^+$. The latter mechanism is nonelectrogenic and occurs with preserved $\Delta\psi$.

Mitochondria possess a nitric oxide synthase (NOS), the mitochondrial NOS (mtNOS), which is associated with the IM and generates NO in a $Ca^{2+}$-sensitive fashion. mtNOS-derived NO competes with $O_2$ for its binding site at complex IV, cytochrome *c* oxidase. Thus, mtNOS reversibly regulates mitochondrial respiration and $\Delta\psi$.

Most of the $O_2$ consumed by mitochondria is reduced to $H_2O$ at complex IV; however, 2–5% of the $O_2$ is incompletely reduced by other respiratory chain complexes to generate superoxide anion ($O_2^-$). The mitochondrial respiratory chain is one of the prime cellular producers of $O_2^-$. Mitochondrial $O_2^-$ reacts with NO and produces the powerful oxidative species peroxynitrite ($ONOO^-$). mtNOS-derived $ONOO^-$ releases cytochrome *c*, increases peroxidation of mitochondrial membrane lipids, and oxidatively damages mitochondrial susceptible targets such as succinyl-C. A: 3-oxoacid C. A-transferase (SCOT) and manganese superoxide dismutose (MnSOD).

Mitochondrial reducing barriers such as MnSOD and reduced gluthathione (GSH) are located within the mitochondria matrix. MnSOD decreases the formation of $ONOO^-$ by using $O_2^-$, and GSH neutralizes the oxidative effects of $ONOO^-$ by converting the $ONOO^-$ to *S*-nitrosated species.

that protects mitochondria against $Ca^{2+}$ overload. Increased $[Ca^{2+}]_m$ stimulates mtNOS-derived NO formation (Arnaiz et al., 1999; Dedkova et al., 2003; Ghafourifar and Richter, 1997, Ghafourifar et al., 1999e, Kanai et al., 2001, 2004; Lores-Arnaiz et al., 2004) that causes $Ca^{2+}$ efflux from the mitochondria via at least two pathways: (1) passive $Ca^{2+}$ efflux as a result of decreased $\Delta\psi$ or (2) active $Ca^{2+}$ release through formation of peroxynitrite that stimulates the specific mitochondrial $Ca^{2+}$ release pathway with preserved $\Delta\psi$. The latter mechanism involves oxidation of mitochondrial pyridine nucleotides (Bringold et al., 2000).

## mtNOS, Apoptosis, and Oxidative Stress

Apoptosis is an evolutionarily conserved mechanism required for normal cell and tissue homeostasis (Kerr et al., 1972). Irregular apoptosis is one of the prime pathological mechanisms underlying numerous diseases, including cancer, ischemia/reperfusion-induced cardiac injury, and neurodegenerative diseases. Mitochondria play a key role in apoptosis (Ghafourifar et al., 2000; Green and Reed, 1998). Interaction with mitochondria is one of the early events in apoptosis induced by many apoptogenic factors, including NO and peroxynitrite. Mitochondria possess key pro- and anti-apoptotic proteins, such as Bax, caspases, and Bcl-2. Moreover, released cytochrome c is one of the most crucial apoptosis triggers in many cells. Accordingly, mitochondria are called the *switchboard of apoptosis* (Ghafourifar and Richter, 2001; Szibor et al., 2001).

A substantial number of reports indicate that NO induces apoptosis through mechanisms that involve formation of peroxynitrite (Ferrante et al., 1999; Keller et al., 1998; Leist et al., 1997). This form of apoptosis occurs with mitochondrial dysfunction and perturbed mitochondrial redox balance (Almeida et al., 1998; Keller et al., 1998). Between 2% and 5% of the electrons flowing through the respiratory chain *leak* out (Boveris and Cadenas, 2000; Chance et al., 1979). These electrons account for the fraction of the oxygen consumed by mitochondria to generate $O_2^-$ (Cadenas et al., 2001). NO reacts with $O_2^-$ to produce peroxynitrite with the nearly diffusion-controlled rate of $1.9 \times 10^{10}\ M\ s^{-1}$ (Kissner et al., 1997). NO and $O_2^-$ are produced in a close vicinity within mitochondria: $O_2^-$ is produced at the inner membrane during electron transfer, and NO is produced by the inner membrane associated mtNOS. Thus, it is very conceivable that mtNOS-derived NO forms peroxynitrite. Several groups have reported that mtNOS, indeed, generates peroxynitrite (Alvarez et al., 2003; Boveris et al., 2002; Bringold et al., 2000; Cadenas et al., 2001; Dedkova et al., 2003; Ghafourifar et al., 1999e; Kanai et al., 2004; Navarro, 2004; Poderoso et al.,

1999) and that a substantial amount of NO produced within mitochondria converts to peroxynitrite (Dedkova *et al.*, 2003; Kanai *et al.*, 2004; Lacza *et al.*, 2001). Accordingly, mtNOS has been called *peroxynitrite synthase* (Groves, 1999). Poderoso's group has shown that mitochondrial inner membrane produces peroxynitrite with the rate of $9.5 \times 10^{-8}$ $M$ $s^{-1}$ that accounts for 15% of the entire $O_2^-$ generated by the mitochondrial inner membrane (Cadenas *et al.*, 2001; Poderoso *et al.*, 1999). The first report on the generation of peroxynitrite by mtNOS demonstrated that mtNOS-derived peroxynitrite induced oxidative stress and released cytochrome *c* from the mitochondria in a manner prevented by the anti-apoptotic protein Bcl-2 (Ghafourifar *et al.*, 1999e). Boveris *et al.* (2002) have shown that mtNOS-derived peroxynitrite induces mitochondrial dysfunction and contractile failure in rat and human skeletal muscle. Heart mtNOS generates peroxynitrite that diminishes the oxidative phosphorylation capacity in mouse cardiomyocytes (Kanai *et al.*, 2004). Brain mtNOS also generates peroxynitrite that is involved in oxidative stress-related conditions (Alvarez *et al.*, 2003) including aging (Navarro, 2004). A negative correlation has been shown to exist between free radicals produced by rat brain mtNOS and brain development (Riobo *et al.*, 2001). A role for mtNOS-derived peroxynitrite in apoptotic cell death of SH-SY5Y neurons has also been reported (Dennis and Bennett, 2003).

## $Ca^{2+}$ Dependence of mtNOS

Some studies did not report mtNOS to be $Ca^{2+}$-sensitive (French *et al.*, 2001; Giulivi, 1998). Many buffers traditionally used to investigate mitochondria, including the buffers used in those studies, contain high concentrations of $Mg^{2+}$ ($\geq 1$ m$M$), a well-known mitochondrial $Ca^{2+}$ uptake blocker (McKean, 1991; Tsuda *et al.*, 1991; Votyakova *et al.*, 1993). We and others (Dedkova *et al.*, 2003; Ghafourifar *et al.*, 1999e; Kanai *et al.*, 2001, 2004) have shown that blockade of mitochondrial $Ca^{2+}$ uptake (e.g., by ruthenium red or by collapsing $\Delta\psi$) drastically decreases mtNOS activity. Moreover, Manzo-Avalos *et al.* (2002) have demonstrated that $Mg^{2+}$ also inhibits the activity of mtNOS in a dose-dependent manner. Therefore, using high concentrations of $Mg^{2+}$ decreases the mtNOS activity, causes mtNOS to appear $Ca^{2+}$-insensitive, and is undesirable in mtNOS research (Ghafourifar and Cadenas, 2005).

## What Isozyme Is mtNOS?

Two laboratories almost simultaneously reported that mitochondria from rat diaphragm skeletal muscle fibers (Kobzik *et al.*, 1995); rat non-synaptosomal brain (Bates *et al.*, 1995), and rat heart, skeletal muscle, and kidney (Bates *et al.*, 1996) cross-react with eNOS antibodies. Other reports also suggested association of eNOS with the mitochondria from rat heart (Reiner, 2001), brain, and liver (Lacza *et al.*, 2001). A systematic study by Gao *et al.* (2004) has addressed the association of eNOS with mitochondria and demonstrated that eNOS was only associated with the cytoplasmic face of the outer mitochondrial membrane. This study showed that even urea-denatured eNOS, which lacks association with the cytoplasmic membrane, associates with the mitochondrial outer membrane. The eNOS association with the mitochondrial outer membrane was fully abolished when mitochondria were treated with proteinase K, which removes proteins nonspecifically bound to the mitochondrial outer membrane. Moreover, deletion of a pentabasic peptide (residues 628–632) that did not affect the association of the eNOS with the cytoplasmic membrane did abolish the eNOS association to the mitochondrial outer membrane. These findings strongly argue that eNOS is not the mitochondrial *bona fide* NOS.

Frandsen *et al.* (1996) were the first to demonstrate that immunolabeling of cytochrome *c* oxidase, a mitochondrial inner membrane protein, solidly concurs with that of the nNOS. Several other laboratories have also observed the similarity of mtNOS with nNOS (Kanai *et al.*, 2001; Riobo *et al.*, 2001).

One study attempted to purify and characterize the amino acid sequence of rat liver mtNOS (Tatoyan and Giulivi, 1998). This study used one-step adenosine diphosphate (ADP)–affinity chromatography and obtained a protein from rat liver mitochondrial matrix that generated L-citrulline from L-arginine in a $Ca^{2+}$-independent manner (Tatoyan and Giulivi, 1998). Because the purified protein cross-reacted with an antibody against iNOS, which also generates L-citrulline from L-arginine in a $Ca^{2+}$-insensitive manner, the study concluded that mtNOS was iNOS. It needs to be mentioned that liver mitochondria are vastly engaged in citrulline formation via routes other than an NOS. Terrestrial ureotelic organisms including mammals deposit ammonia in liver mitochondria and convert the ammonia to urea in the urea cycle. The cycle, discovered by the pioneering work of Sir Hans Krebs in 1932 (Krebs, 1970), contains enzymes located in the cytoplasm and mitochondrial matrix. The mitochondrial component is the matrix enzyme carbamoyl-phosphate synthetase 1 (CPS-1; EC 2.7.2.5) that catalyzes the conversion of ammonia to carbamoyl phosphate, which further condenses with L-ornithine to produce L-citrulline. CPS-1 is an abundant matrix protein that constitutes about 20% of the total protein in rat liver mitochondrial matrix, and its activity is regulated by intramitochondrial $Mg^{2+}$ and adenosine triphosphate (ATP) (Powers, 1981; Rodríguez-Zavala *et al.*, 1997). Thus, it is likely that the mitochondrial matrix protein reported in that study (Tatoyan and Giulivi,

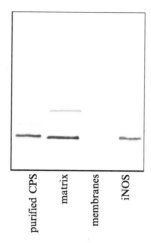

Fig. 1. Carbamoylphosphate synthetase-1 (CPS-1) and inducible nitric oxide synthase (iNOS) cross-reactivity. Cross-reactivity of mitochondrial abundant matrix protein CPS-1 and iNOS was tested. Purified CPS-1 (CPS-1; 40 ng), rat liver mitochondrial matrix (matrix; 250 ng), and membrane fraction (membranes; 250 ng) and recombinant murine iNOS (iNOS; 4.5 ng; Alexis Biochemicals, San Jose, CA) were run on 10% sodium dodecylsulfate polyacrylamide gel, blotted onto nitrocellulose membrane, and probed with an iNOS primary antibody (BD Transduction Biosciences, San Jose, CA; 1/2500 dilution) and anti-mouse secondary antibody (R&D Systems, Minneapolis, MN; 1/4000 dilution containing 0.1% nonfat milk) and visualized by Enhanced Chemiluminescence Plus (BD Transduction Biosciences, San Diego, CA).

1998) was CPS-1. We tested whether CPS-1 cross-reacts with iNOS antibodies by using purified CPS-1 (generous gifts from Dr. Carol Lusty, Columbia University, NY) and found that CPS-1 does cross-react with an iNOS antibody (Fig. 1).

Subsequent studies have addressed these discrepancies. Kanai *et al.* (2001) studied mtNOS in isolated intact mouse heart mitochondria and reported that the enzyme generates NO in response to elevation of $[Ca^{2+}]_m$. This study showed that inhibition of mitochondrial $Ca^{2+}$ uptake by ruthenium red or by collapsing the $\Delta\psi$ prevents heart mtNOS activity. This study also used knockout mice and demonstrated the absence of mtNOS activity in nNOS$^{-/-}$, but not in eNOS$^{-/-}$ or iNOS$^{-/-}$ mice, suggesting that heart mtNOS is related to nNOS. Riobo *et al.* (2001) demonstrated that rat brain mtNOS is an nNOS distinct from the 157-kDa cytoplasmic nNOS. This study showed that mtNOS is associated with the mitochondrial inner membrane and produces NO in a $Ca^{2+}$-sensitive manner (Riobo *et al.*, 2001). Other authors have also demonstrated that brain mtNOS is an nNOS antibody cross-reacting protein associated with the mitochondrial inner membrane and generates NO in a typical $Ca^{2+}$-sensitive manner (i.e., a calmodulin antagonist such as chlorpromazine inhibits the mtNOS activity) (Boveris *et al.*, 2002; Lores-Arnaiz *et al.*, 2004).

Taken together, most reports concur that mtNOS is associated with the mitochondrial inner membrane and generates NO in a $Ca^{2+}$-sensitive manner. However, the exact amino acid characterization of mtNOS remains to be further investigated.

### Determination of mtNOS Activity

Several limiting and confounding factors restrict *in vivo* studies. Less sophisticated models such as organelles, suborganelles, proteins, peptides, or simpler models are substantially used to uncover molecular mechanisms of physiological or pathological events. Isolated mitochondria maintain most of their features as within cells, yet numerous confounding factors that limit studying these organelles in cells or tissues are avoided. For example, isolated mitochondria contain almost all components of the oxidative phosphorylation, oxidative stress, and apoptosis machinery. Most of the mtNOS studies have been conducted using isolated mitochondria. Isolation of mitochondria from organs such as liver, heart, or brain has been discussed elsewhere. The following sections outline practical notes for studying mtNOS in isolated mitochondria.

### Isolation of Mitochondria

Perform the euthanasia by decapitation, and deplete the body of blood to limit exposure of mitochondria to NO reacting molecules including hemoglobin. Remove and place the organ (e.g., liver) in a dish on ice, and remove fat, ducts, vessels, connective tissues, and blood clots. Mince the tissue, and wash several times in ice-cold buffer to remove the remaining blood. It is important to keep the temperature low during homogenization steps. Our laboratory places the homogenizer in a *cold jacket* (a tube containing ice water that insulates the homogenizer). Over-homogenizing must be avoided because heat and the mechanical force produced during the homogenization strongly damage mitochondria and mtNOS. It is highly recommended that the homogenizer, pestle, centrifuge tubes, and all other containers used to isolate or handle the mitochondria not be washed with detergents. Detergents can gradually release from the glass, Teflon, or centrifuge tubes and dissociate membrane-associated proteins including mtNOS. During centrifugation steps, a red spot in low spin pellets may indicate contamination with blood. Gently remove the contaminating blood. If there is a fat layer floating above the supernatant of the high spin steps, remove it with a soft lint-free tissue. Discard the light brown fluffy layer lining the high spin pellets. This layer contains only minimal mitochondria and is highly contaminated with organelles other than mitochondria.

The purity of the mitochondrial preparation can be assessed by several methods. Our laboratory routinely measures the cytochrome $a$ content, which is done by a quick spectrophotometric assay. Measure the OD of a mitochondrial suspension (e.g., 1 mg/ml, at 605–630 nm before and after reduction with dithionite). Calculate the cytochrome $a$ concentration using extinction coefficient of $12 \text{ m}M^{-1} \text{ cm}^{-1}$.

Several assays are available to measure NOS activity. The rather unique structure and compartmentalization of mitochondria limit the use of some of these assays in mtNOS activity determination. Additionally, the activity of mtNOS, like many other mitochondrial enzymes, declines rapidly in isolated mitochondria. It is highly recommended to perform mtNOS-related experiments within 4–6 h of isolation of the mitochondria. In the following sections, we discuss three widely used mtNOS assays, namely photometry, amperometry, and radioassay.

*Photometric Determination of mtNOS Activity*

Spectrophotometry has been widely used to measure mtNOS activity using oxyhemoglobin (oxyHb) assay. Some laboratories have also used fluorophotometric NO probes to measure mtNOS activity.

*Oxyhemoglobin Assay.* NO stoichiometrically reacts with oxyHb to produce methemoglobin (metHb). The distinct optical behaviors of these species (Fig. 2) make the conversion of oxyHb to metHb a simple, practical, and rapid spectrophotometric NOS determination assay. This assay can be used to measure mtNOS activity; however, modifications need to be made and precautions need to be taken.

If isolated mitochondria are intact and not contaminated with the cytoplasmic components (e.g., cytosolic forms of NOS), normally a minimal metHb formation is observed. NO produced within mitochondria reacts with several agents, particularly $O_2^-$; therefore, amounts of NO that reach the extra mitochondria (detected by oxyHb) are smaller than the actual NO produced within mitochondria. Additionally, oxyHb cannot enter intact mitochondria because the mitochondrial inner membrane is impermeable to molecules larger than 100–150 daltons (oxyHb is 65 kDa). metHb formed by intact mitochondria may not be used as a means to quantify mtNOS activity and may pose artifacts including contamination with cytoplasmic forms of NOS. Some studies have suggested that oxymyoglobin (18 kDa) might be used to measure mtNOS activity in intact

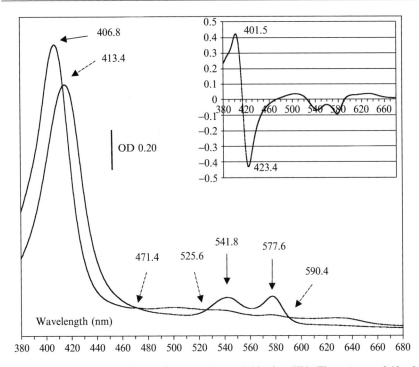

FIG. 2. Oxyhemoglobin (oxyHb) and methemoglobin (metHb). The spectra of 10 $\mu M$ bovine erythrocyte oxyHb and metHb dissolved in 100 $\mu M$ HEPES, pH 7.10, at 23° were recorded from 380 to 680 nm using an Aminco DW-2000 spectrophotometer. Peaks (solid lines) and isosbestic points (dashed arrows) are shown. The inset shows the difference spectrum.

mitochondria (Sarkela *et al.*, 2001). To use the oxyHb assay to measure mtNOS activity, we have used broken mitochondria (BM) (described later in this chapter). This preparation contains all the components of intact mitochondria, except membranes are ruptured to allow oxyHb to reach mtNOS that is associated with mitochondrial inner membrane. However, a reaction with $O_2^-$ generated by the mitochondrial inner membrane may supersede the reaction of mtNOS-derived NO with oxyHb. Thus, relatively high amounts of superoxide dismutase (SOD) (e.g., $\geq 1$ KU/ml) are highly recommended while using oxyHb assay to determine mtNOS activity.

PREPARATION OF BROKEN MITOCHONDRIA. Suspend the intact mitochondria in a hypoosmotic solution by adding 2–4 volumes of ice-chilled $H_2O$ containing a protease inhibitor cocktail: leupeptin, phenylmethylsulfonyl-fluoride (PMSF), pepstatin A, and aprotinin (10 $\mu M$ each). Mix the suspension, and apply a mild sonication (e.g., 100–150 W, 50% duty cycle,

75 s) to break the remaining intact mitochondria. To avoid oxidation of proteins and lipids during the sonification, water used in hypoosmotic shock  may be purged with $N_2$ for at least 15 min before addition to the mitochondrial suspension, and $N_2$ can be applied over the suspension during the sonification. Readjust the osmolality by adding pH-adjusted concentrated buffer (e.g., a 10× buffer) to the BM suspension. Centrifuge the suspension at 10,000g for 10 min at 4°. The supernatant is BM (Ghafourifar, 2002; Ghafourifar and Richter, 1997).

SPECTROPHOTOMETRIC DETERMINATION OF MTNOS ACTIVITY. Mitochondria contain all the substrates and cofactors that mtNOS requires, and elevation of $[Ca^{2+}]_m$ per se is sufficient to stimulate mtNOS activity in intact mitochondria (Ghafourifar and Richter, 1997; Ghafourifar et al., 1999e). However, dilution or oxidation of some of the mtNOS substrates or cofactors during the preparation of BM or submitochondrial particles (SMPs) may require the presence of the following substrates or cofactors in mtNOS assay medium:

1–10 $\mu$g/ml calmodulin: Prepare 1 mg/ml stock solution, aliquote, and store at −80°. Avoid freeze-thawing.

10–100 $\mu M$ L-arginine: Prepare a 1–100 m$M$ stock solution and store at −20°.

10–15 $\mu M$ tetrahydrobiopterin (BH$_4$): Prepare a 2 m$M$ stock solution in 10 m$M$ HCl immediately before the experiment. BH$_4$ undergoes rapid autoxidation upon dilution.

≥1 KU/ml SOD: Prepare 100–500 KU/ml stock solution. Aliquot and store at −20°. Avoid freeze-thawing.

PROCEDURE. The commercially available hemoglobin (horse heart is commonly used) is a mixture of oxyHb and metHb and cannot be used "as is" in oxyHb assay. Dissolve the hemoglobin in a buffer, fully reduce to oxyHb with dithionite, and remove the excess dithionite by purifying the oxyHb using column purification (our laboratory uses prepacked Sephadex columns). Measure the concentration of the purified oxyHb using $\varepsilon_{415\ nm}$ 131.0 m$M^{-1}$ cm$^{-1}$. Aliquot and store at −80°.

Add the mtNOS substrates and 4 $\mu M$ oxyHb to the cuvette, and record the optical density. Autoxidation of oxyHb in oxygenated aqueous solutions provides a spontaneous metHb formation that needs to be subtracted from the metHb produced by mtNOS-derived NO (Fig. 3). Add BM or SMP and continue recording the OD. Our laboratory generally uses up to 0.1 mg mitochondrial protein/ml. Several wavelengths can be used. For example, at 401 nm, the NO formation can be quantified using an $\varepsilon_{401(metHb-oxyHb)}$ of 49 m$M^{-1}$ cm$^{-1}$. If dual-wavelength spectroscopy is

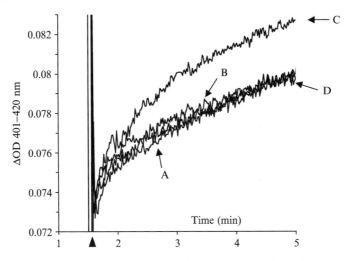

FIG. 3. Mitochondrial nitric oxide synthase (mtNOS) activity determination using oxy-hemoglobin (oxyHb) assay. mtNOS activity of rat liver broken mitochondria (0.03 mg/ml) was determined at 401–420 nm using a dual-beam dual-wavelengths Aminco DW-2000. Buffer (100 mM HEPES, pH 7.10) was added to the cuvette, and after reaching a steady reading, oxyHb was added (4 $\mu M$) (arrowhead). (A) Autoxidation of oxyHb in the absence of broken mitochondria. (B) Rat liver broken mitochondria (0.03 mg/ml) were present in the buffer before the addition of oxyHb. (C) Same as traces (B) while mtNOS activity was stimulated by $Ca^{2+}$ (100 $\mu M$), L-Arg (100 $\mu M$), and 1 KU Cu/Zn-SOD was present in the buffer before the addition of broken mitochondria. (D) Same as (C) with the addition of L-NMMA (100 $\mu M$).

possible, measurements at 401–420 nm with $\varepsilon_{401-420}$ (metHb-oxyHb) 100 mM$^{-1}$ cm$^{-1}$ provide a very sensitive measurement.

## Fluorophotometric Determination of mtNOS Activity

Some studies have used NO fluorescent probes such as 4,5-diamino-fluorescein diacetate (DAF-2-DA) and detected mtNOS activity in various cells (Dedkova *et al.*, 2003; Dennis and Bennett, 2003; Lacza *et al.*, 2003; Lopez-Figueroa *et al.*, 2000). However, it seems that experimental conditions may highly affect the fluorescent mtNOS assays. For example, one study used DAF fluorescent assay and failed to detect NO formation by isolated intact mouse brain mitochondria (Lacza *et al.*, 2003). The presence of 5 mM $Mg^{2+}$ in the NOS assay medium (as discussed earlier), along with experimental conditions involving depletion of the medium of $O_2$ and the low temperature used in this study, may account for the lack of NO formation by those mitochondria.

## Amperometric mtNOS Assay

Clark-type electrodes have been widely used to determine NOS activity. These electrodes are relatively easy to use, do not require handling radioactive material or rather sophisticated dual-beam spectroscopy, and are sensitive enough for most practical purposes. Several NO-sensitive electrodes that are commercially available perform under the same principle: NO diffuses through a semipermeable membrane and reaches the tip of the working electrode, where it becomes oxidized by the electrode to $NO_3$. This reaction generates an electric current proportional to the concentration of NO in the solution. The electrode is equipped with a sensitive amperometer that detects the current generated at the tip of the electrode. The commercially available NO sensors can be used to measure mtNOS activity; however, when BM or SMP is used, addition of NOS substrates and cofactors (discussed earlier) may be required. Figure 4 demonstrates a typical mtNOS activity assay using the NO electrode.

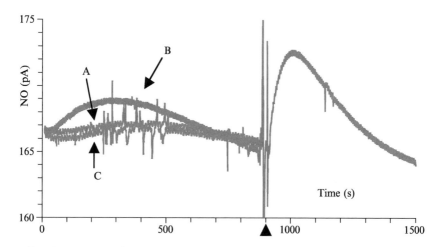

FIG. 4. Amperometric mitochondrial nitric oxide synthase (mtNOS) determination. Rat liver broken mitochondria (1 mg/ml) were added at time 0 to the NO chamber containing HEPES buffer (100 m$M$, pH 7.10) thermostated at 37°. The mtNOS activity was measured using an NO-sensitive electrode operated by an Apollo 4000 Free Radical Analyzer (World Precision Instruments, FL). (A) L-Arg (100 $\mu M$), tetrahydrobiopterin (10 $\mu M$), and Cu/Zn-SOD (2.7 KU/ml) were present in the buffer before the addition of broken mitochondria. (B) Same as (A) with addition of $Ca^{2+}$ (100 $\mu M$). (C) Same as (B) with addition of L-NMMA (100 $\mu M$). At the arrowhead, sodium nitrite (20 $\mu M$) was added to validate the test and provide a scale.

*mtNOS Activity Radioassay*

Determination of radiolabeled L-citrulline produced from radiolabeled L-arginine is a widely used NOS assay also used for NOS activity determination. To measure mtNOS activity in intact mitochondria, the citrulline assay is preferred over oxyHb assay. However, some points need careful attention.

As discussed, liver mitochondria are highly engaged in citrulline formation through the urea cycle. Although $K_m$ of L-arginine for NOS is 2–10 $\mu M$, whereas that of arginase is 8–10 m$M$, it is highly recommended that low amounts of radiolabeled L-arginine be used in mtNOS assay to avoid contribution of the urea cycle in the citrulline formed. Moreover, many buffers traditionally used to investigate mitochondrial functions contain high concentrations of $Mg^{2+}$ (discussed earlier). This cation is a well-known mitochondrial $Ca^{2+}$ uptake blocker, and blockade of mitochondrial $Ca^{2+}$ uptake decreases mtNOS activity. Additionally, $Mg^{2+}$ per se inhibits mtNOS activity. For example, in one study in which 5 m$M$ $Mg^{2+}$ was present in the citrulline assay medium, mtNOS exerted a very low $Ca^{2+}$-insensitive activity (French *et al.*, 2001). Therefore, presence of $Mg^{2+}$ in mtNOS assays including the citrulline assay is unfavorable.

*Procedure.* Supplement mitochondria or mitochondrial subfractions with L-[$^3$H]arginine (30,000–50,000 cpm) in the presence of NOS substrates (as discussed earlier) and incubate at 37°. Basal mtNOS activity can be determined by supplementing mitochondria with the radiolabeled L-[$^3$H] arginine without NOS substrates (Ghafourifar and Richter, 1997; Ghafourifar *et al.*, 1999e). At the end of the incubation period (generally ~10–30 min), terminate the mtNOS activity by adding ice-chilled stop solution containing 2 m$M$ EDTA, 1 m$M$ unlabeled L-citrulline, and 20 m$M$ sodium acetate buffer, pH 5.00.

Prepare the exchange resin (Dowex 50W × 8, mesh size 200–400, $H^+$ form) as described (Mayer *et al.*, 1994). Load the spin filters with 600–1000 $\mu$l of the resin, spin at 3000–5000$g$ for 2 min, and wash with 500 $\mu$l $H_2O$. Load the packed columns with the radiolabeled mitochondrial samples, and spin at 5000$g$ for 2–5 min. Rinse the resins two times with 200 $\mu$l $H_2O$, collect all the effluent in scintillation vials, add scintillation fluid, and determine the radioactivity. The radioactive L-arginine passing through the column after mock incubation (no mitochondrial material added) should not exceed 1% of the total radioactivity loaded on the column.

mtNOS and Mitochondrial Respiration

mtNOS-derived NO competes with $O_2$ and reversibly regulates mitochondrial respiration. In intact mitochondria, elevation of $[Ca^{2+}]_m$ is sufficient to stimulate mtNOS activity and decrease the respiration; however, in

BM or SMP, addition of other NOS substrates and cofactors (as discussed earlier) may be required.

*Procedure*

Add mitochondria or mitochondrial subfractions into the thermostated oxygen chamber containing the buffer, for example, 100 m$M$ HEPES. To stimulate mtNOS, incubate mitochondria with $Ca^{2+}$ 3–5 min before energizing them with respiratory substrates. To inhibit mtNOS activity in intact mitochondria, incubate the mitochondria with an NOS inhibitor 20–60 min on ice before the assay. Relatively high concentrations of NOS inhibitors may be required to inhibit mtNOS activity of intact mitochondria. Most commonly used NOS inhibitors are competitive L-arginine analogues and the intramitochondrial L-arginine concentration is in the millimolar range (Dolinska and Albrecht, 1998). Additionally, mitochondrial membranes are not permeable to most NOS inhibitors. Our laboratory generally

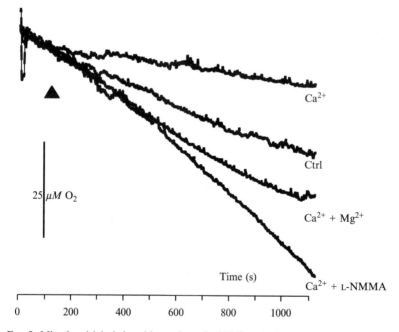

FIG. 5. Mitochondrial nitric oxide synthase (mtNOS) and mitochondrial $O_2$ consumption. Isolated intact beef heart mitochondria (1 mg/ml) (arrowhead) were added to the oxygen electrode chamber in the absence (Ctrl) or presence of 160 $\mu M$ $Ca^{2+}$ ($Ca^{2+}$) or 1 m$M$ $Mg^{2+}$ ($Mg^{2+}$) and respiration was measured using a Clark-type electrode (YSI, Yellow Springs, OH). To inhibit mtNOS activity, mitochondria were incubated with 1 m$M$ L-NMMA for 20 min on ice before the test (L-NMMA + $Ca^{2+}$).

incubates mitochondria with 100–300 $\mu M$ L-NMMA for 20–60 min on ice before the test. As discussed, presence of $Mg^{2+}$ in the buffer prevents mtNOS activation, so the mtNOS-derived NO-induced decrease in the respiration may not be observed (Fig. 5).

## Acknowledgment

This work was supported in part by the National Institute on Aging award AG023264-02, and in part by the Marshall University Center of Biomedical Research Excellence award 1 P20 RR020180 NCRR, NIH.

## References

Almeida, A., Heales, S. J. R., Bolanos, J. P., and Medina, J. M. (1998). Glutamate neurotoxicity is associated with nitric oxide–mediated mitochondrial dysfunction and glutathione depletion. *Brain. Res.* **790,** 209–216.

Alvarez, S., Valdez, L. B., Zaobornyj, T., and Boveris, A. (2003). Oxygen dependence of mitochondrial nitric oxide synthase activity. *Biochem. Biophys. Res. Commun.* **305,** 771–775.

Arnaiz, S. L., Coronel, M. F., and Boveris, A. (1999). Nitric oxide, superoxide, and hydrogen peroxide production in brain mitochondria after haloperidol treatment. *Nitric Oxide* **3,** 235–243.

Bates, T. E., Loesch, A., Burnstock, G., and Clark, J. B. (1995). Immunocytochemical evidence for a mitochondrially located nitric oxide synthase in brain and liver. *Biochem. Biophys. Res. Commun.* **213,** 896–900.

Bates, T. E., Loesch, A., Burnstock, G., and Clark, J. B. (1996). Mitochondrial nitric oxide synthase: A ubiquitous regulator of oxidative phosphorylation? *Biochem. Biophys. Res. Commun.* **218,** 40–44.

Boveris, A., Alvarez, S., and Navarro, A. (2002). The role of mitochondrial nitric oxide synthase in inflammation and septic shock. *Free Radic. Biol. Med.* **33,** 1186–1193.

Boveris, A., and Cadenas, E. (2000). Mitochondrial production of hydrogen peroxide regulation by nitric oxide and the role of ubisemiquinone. *IUBMB. Life.* **50,** 245–250.

Bringold, U., Ghafourifar, P., and Richter, C. (2000). Peroxynitrite formed by mitochondrial NO synthase promotes mitochondrial $Ca^{2+}$ release. *Free Radic. Biol. Med.* **29,** 343–348.

Cadenas, E., Poderoso, J. J., Antunes, F., and Boveris, A. (2001). Analysis of the pathways of nitric oxide utilization in mitochondria. *Free Radic. Res.* **33,** 747–756.

Chance, B., Sies, H., and Boveris, A. (1979). Hydroperoxide metabolism in mammalian organs. *Physiol. Rev.* **59,** 527–605.

Dedkova, E. N., Ji, X., Lipsius, S. L., and Blatter, L. A. (2003). Mitochondrial calcium uptake stimulates nitric oxide production in mitochondria of bovine vascular endothelial cells. *Am. J. Physiol. Cell. Physiol.* **286,** C406–C415.

Dennis, J., and Bennett, J. P., Jr. (2003). Interactions among nitric oxide and Bcl-family proteins after MPP+ exposure of SH-SY5Y neural cells I: MPP+ increases mitochondrial NO and Bax protein. *J. Neurosci. Res.* **72,** 76–88.

Dolinska, M., and Albrecht, J. (1998). L-Arginine uptake in rat cerebral mitochondria. *Neurochem. Int.* **33,** 233.

Ferrante, R. J., Hantraye, P., Brouillet, E., and Beal, M. F. (1999). Increased nitrotyrosine immunoreactivity in substantia nigra neurons in MPTP treated baboons is blocked by inhibition of neuronal nitric oxide synthase. *Brain. Res.* **823,** 177–182.

Frandsen, U., Lopez-Figueroa, M., and Hellsten, Y. (1996). Localization of nitric oxide synthase in human skeletal muscle. *Biochem. Biophys. Res. Commun.* **227,** 88–93.

French, S., Giulivi, C., and Balaban, R. S. (2001). Nitric oxide synthase in porcine heart mitochondria: evidence for low physiological activity. *Am. J. Physiol. Heart. Circ. Physiol.* **280,** H2863–H2867.

Gao, S., Chen, J., Brodsky, S. B., Huang, H., Adler, S., Lee, J. H., Dhadwal, N., Cohen-Gould, L., Gross, S. S., and Goligorsky, M. S. (2004). Docking of endothelial nitric oxide synthase (eNOS) to the mitochondrial outer membrane: A pentabasic amino acid sequence in the autoinhibitory domain of eNOS targets a proteinase K-cleavable peptide on the cytoplasmic face of mitochondria. *J. Biol. Chem.* **279,** 15968–15974.

Ghafourifar, P. (2002). *In* "Characterization of mitochondrial nitric oxide synthase: *Methods Enzymol.*" (E. Cadenas and L. Packer, eds.), Vol. 359, pp. 339–350. Academic Press, San Diego.

Ghafourifar, P., Bringold, U., Klein, S. D., and Richter, C. (2000). Mitochondrial nitric oxide synthase, oxidative stress and apoptosis. *Biol. Sign. Recep.* **10,** 57–65.

Ghafourifar, P., and Cadenas, E. (2005). Mitochondrial nitric oxide synthase. *Trends Pharmacol. Sci.* **3** (in press).

Ghafourifar, P., and Colton, C. A. (2003). Compartmentalized nitrosation and nitration in mitochondria. *Antioxid. Redox. Signal.* **5,** 349–354.

Ghafourifar, P., and Richter, C. (1997). Nitric oxide synthase activity in mitochondria. *FEBS. Lett.* **418,** 291–296.

Ghafourifar, P., and Richter, C. (1999a). Nitric oxide in mitochondria: Formation and consequences. *In* "From Symbiosis to Eukaryotism—ENDOCYTOBIOLOGY VII" (E. Wagner *et al.,* eds.), pp. 503–516. University of Geneva.

Ghafourifar, P., and Richter, C. (1999c). Mitochondrial nitric oxide synthase regulates mitochondrial matrix pH. *Biol. Chem.* **380,** 1025–1028.

Ghafourifar, P., and Richter, C. (2001). *In* "Mitochondrial Ubiquinone (Coenzyme Q10): Biochemical, Functional, Medical and Therapeutical Aspects in Human Health and Disease" (M. Ebadi, J. Marwah, and R. Chopra, eds.), Vol. 1, pp. 437–445. Prominent Press, AZ.

Ghafourifar, P., Richter, C., and Moncada, S. (1999b). Nitric oxide produces peroxynitrite at the level of cytochrome oxidase. *Acta. Physiol. Scand.* **167,** 63.

Ghafourifar, P., Schenk, U., Klein, S. D., and Richter, C. (1999c). Mitochondrial nitric-oxide synthase stimulation causes cytochrome c release from isolated mitochondria. Evidence for intramitochondrial peroxynitrite formation. *J. Biol. Chem.* **274,** 31185–31188.

Giulivi, C. (1998). Functional implications of nitric oxide produced by mitochondria in mitochondrial metabolism. *Biochem. J.* **332,** 673–679.

Green, D. R., and Reed, J. C. (1998). Mitochondria and apoptosis. *Science* **281,** 1309–1312.

Groves, J. T. (1999). Peroxynitrite: Reactive, invasive and enigmatic. *Curr. Opin. Chem. Biol.* **3,** 226–235.

Kanai, A. J., Pearce, L. L., Clemens, P. R., Birder, L. A., VanBibber, M. M., Choi, S. Y., de Groat, W. C., and Peterson, J. (2001). Identification of a neuronal nitric oxide synthase in isolated cardiac mitochondria using electrochemical detection. *Proc. Natl. Acad. Sci. USA* **98,** 14126–14131.

Kanai, A., Epperly, M., Pearce, L., Birder, L., Zeidel, M., Meyers, S., Greenberger, J., De Groat, W., Apodaca, G., and Peterson, J. (2004). Differing roles of mitochondrial nitric oxide synthase in cardiomyocytes and urothelial cells. *Am. J. Physiol. Heart. Circ. Physiol.* **286,** H13–H21.

Keller, J. N., Kindy, M. S., Holtsberg, F. W., St. Clair, D. K., Yen, H. C., Germeyer, A., Steiner, S. M., Bruce-Keller, A. J., Hutchins, J. B., and Mattson, M. P. (1998). Mitochondrial manganese superoxide dismutase prevents neural apoptosis and reduces ischemic brain injury: Suppression of peroxynitrite production, lipid peroxidation, and mitochondrial dysfunction. *J. Neurosci.* **18,** 687–697.

Kerr, J. F., Wyllie, A. H., and Currie, A. R. (1972). Apoptosis: A basic biological phenomenon with wide-ranging implications in tissue kinetics. *Br. J. Cancer.* **26,** 239–257.

Kissner, R., Nauser, T., Bugnon, P., Lye, P. G., and Koppenol, W. H. (1997). Formation and properties of peroxynitrite as studied by laser flash photolysis, high-pressure stopped-flow technique, and pulse radiolysis. *Chem. Res. Toxicol.* **10,** 1285–1292.

Kobzik, L., Stringer, B., Balligand, J. L., Reid, M, B., and Stamler, J. S. (1995). Endothelial type nitric oxide synthase in skeletal muscle fibers: Mitochondrial relationships. *Biochem. Biophys. Res. Commun.* **211,** 375–381.

Koshland, D. E., Jr. (1992). The molecule of the year. *Science* **258,** 1861.

Krebs, H. A. (1970). The history of the tricarboxylic acid cycle. *Perspect. Biol. Med.* **14,** 154–170.

Lacza, Z., Puskar, M., Figueroa, J. P., Zhang, J., Rajapakse, N., and Busija, D. W. (2001). Mitochondrial nitric oxide synthase is constitutively active and is functionally upregulated in hypoxia. *Free Radic. Biol. Med.* **31,** 1609–1615.

Lacza, Z., Snipes, J. A., Zhang, J., Horvath, E. M., Figueroa, J. P., Szabo, C., and Busija, D. W. (2003). Mitochondrial nitric oxide synthase is not eNOS, nNOS or iNOS. *Free Radic. Biol. Med.* **35,** 1217–1228.

Leist, M., Fava, E., Montecucco, C., and Nicotera, P. (1997). Peroxynitrite and nitric oxide donors induce neuronal apoptosis by eliciting autocrine excitotoxicity. *Eur. J. Neurosci.* **9,** 1488–1498.

Lopez-Figueroa, M. O., Coamano, C., Morano, M. I., Ronn, L. C., Akil, H., and Watson, S. J. (2000). Direct Evidence of nitric oxide presence within mitochondria. *Biochem. Biophys. Res. Commun.* **272**(1), 129–133.

Lores-Arnaiz, S., D'Amico, G., Czerniczyniec, A., Bustamante, J., and Boveris, A. (2004). Brain mitochondrial nitric oxide synthase: *in vitro* and *in vivo* inhibition by chlorpromazine. *Arch. Biochem. Biophys.* **430,** 170–177.

MacMillan-Crow, L. A., Crow, J. P., and Thompson, J. A. (1998). Peroxynitrite-mediated inactivation of manganese superoxide dismutase involves nitration and oxidation of critical tyrosine residues. *Biochemistry* **37,** 1613–1622.

Manzo-Avalos, S., Perez-Vazquez, V., Ramirez, J., Aguilera-Aguirre, L., Gonzalez-Hernandez, J. C., Clemente-Guerrero, M., Villalobos-Molina, R., and Saavedra-Molina, A. (2002). Regulation of the rate of synthesis of nitric oxide by Mg (2+) and hypoxia. Studies in rat heart mitochondria. *Amino Acids* **22,** 381–389.

Mayer, B., Klatt, P., Werner, E. R., and Schmidt, K. (1994). Molecular mechanisms of inhibition of porcine brain nitric oxide synthase by the antinociceptive drug 7-nitroindazole. *Neuropharmacology* **33,** 1253.

McKean, T. A. (1991). Calcium uptake by mitochondria isolated from muskrat and guinea pig hearts. *J. Exp. Biol.* **157,** 133–142.

Moncada, S., Palmer, R. M., and Higgs, E. A. (1991). Nitric oxide: physiology, pathophysiology, and pharmacology. *Pharmacol Rev.* **43,** 109–142.

Navarro, A. (2004). Mitochondrial enzyme activities as biochemical markers of aging. *Mol. Aspects. Med.* **25,** 37–48.

Palmer, R. M., Ferrige, A. G., and Moncada, S. (1987). Nitric oxide release accounts for the biological activity of endothelium-derived relaxing factor. *Nature* **327,** 524.

Poderoso, J. J., Lisdero, C., Schopfer, F., Riobo, N., Carreras, M. C., Cadenas, E., and Boveris, A. (1999). The regulation of mitochondrial oxygen uptake by redox reactions involving nitric oxide and ubiquinol. *J. Biol. Chem.* **274,** 37709–37716.

Powers, S. G. (1981). Regulation of rat liver carbamyl phosphate synthetase I. Inhibition by metal ions and activation by amino acids and other chelating agents. *J. Biol. Chem.* **256,** 11160–11165.

Reiner, M., Bloch, W., and Addicks, K. (2001). Functional interaction of caveolin-1 and eNOS in myocardial capillary endothelium revealed by immunoelectron microsopy. *J. Histochem. Cytochem.* **49**(12), 1605–1610.

Riobo, N. A., Clementi, E., Melani, M., Boveris, A., Cadenas, E., Moncada, S., and Poderoso, J. J. (2001). Nitric oxide inhibits mitochondrial NADH:ubiquinone reductase activity through peroxynitrite formation. *Biochem. J.* **359,** 139–145.

Rodríguez-Zavala, J. S., Saavedra-Molina, A., and Moreno-Sanchez, R. (1997). Effect of intramitochondrial $Mg^{2+}$ on citrulline synthesis in rat liver mitochondria. *Biochem. Mol. Biol. Int.* **41,** 179–187.

Sarkela, T. M., Berthiaume, J., Elfering, S., Gybina, A. A., and Giulivi, C. (2001). The modulation of oxygen radical production by nitric oxide in mitochondria. *J. Biol. Chem.* **276,** 6945–6949.

Sharpe, M. A., and Cooper, C. E. (1998). Interaction of peroxynitrite with mitochondrial cytochrome oxidase. Catalytic production of nitric oxide and irreversible inhibition of enzyme activity. *J. Biol. Chem.* **273,** 30961–30972.

Szibor, M., Richter, C., and Ghafourifar, P. (2001). Redox control of mitochondrial functions. *Antiox. Red. Sign.* **3,** 515–524.

Tatoyan, A., and Giulivi, C. (1998). Purification and characterization of a nitric-oxide synthase from rat liver mitochondria. *J. Biol. Chem.* **273**(18), 11044–11048.

Tsuda, T., Kogure, K., Nishioka, K., and Watanabe, T. (1991). Synergistic deleterious effect of micromolar Ca ions and free radicals on respiratory function of heart mitochondria at cytochrome C and its salvage trial. *Neuroscience* **44,** 335–341.

Turko, IV., Marcondes, S., and Murad, F. (2001). Diabetes-associated nitration of tyrosine and inactivation of succinyl-CoA:3-oxoacid CoA-transferase. *Am. J. Physiol. Heart. Circ. Physiol.* **281,** H2289–H2294.

Votyakova, T. V., Bazhenova, E. N., and Zvjagilskaya, R. A. (1993). Yeast mitochondrial calcium uptake: Regulation by polyamines and magnesium ions. *J. Bioenerg. Biomembr.* **25,** 569–574.

# [37] Functional Activity of Mitochondrial Nitric Oxide Synthase

*By* Laura B. Valdez, Tamara Zaobornyj, and Alberto Boveris

## Abstract

The functional activity of mitochondrial nitric oxide synthase (mtNOS) is determined by inhibiting $O_2$ uptake and by enhancing $H_2O_2$ production. The effect of mtNOS activity on mitochondrial $O_2$ uptake is assayed in state 3 respiration in two limit conditions of intramitochondrial NO: at its maximal and minimal levels. The first condition is achieved by supplementation with L-arginine and superoxide dismutase (SOD), and the second by

0076-6879/05 $35.00
DOI: 10.1016/S0076-6879(05)96037-0

addition of an NOS inhibitor and oxyhemoglobin. The difference between state 3 $O_2$ uptake in both conditions constitutes the mtNOS functional activity in the inhibition of cytochrome oxidase activity. The functional activity of mtNOS in enhancing mitochondrial $H_2O_2$ generation in state 4 is given by the NO inhibition of ubiquinol–cytochrome $c$ reductase activity. Simple determinations with the oxygen electrode or the measurement of mitochondrial $H_2O_2$ production can be used to assay the effects of physiological and pharmacological treatments on mtNOS activity.

Mitochondrial Nitric Oxide Production

Mitochondrial nitric oxide synthase (mtNOS) catalyzes the oxidation of L-arginine and NADPH by $O_2$ to yield L-citrulline and nitric oxide (NO), using $Ca^{2+}$ and calmodulin as cofactors. Considering the $Ca^{2+}$ requirement for enzymatic activity, mtNOS is a constitutive NOS (Giulivi, 1998; Tatoyan and Giulivi, 1998) that has been identified as the $\alpha$-isoform of neuronal NOS (nNOS-$\alpha$), with a myristoyl acylation different from endothelial NOS (eNOS) and phosphorylated in the C-terminal region (Elfering et al., 2002). NO production has been observed in mitochondria isolated from a series of mammalian organs: liver (Ghafourifar and Richter, 1997; Giulivi, 1998; Schild et al., 2003), brain (Carreras et al., 2002; Lores Arnaiz et al., 1999), thymus (Bustamante et al., 2000), kidney (Boveris et al., 2003a), diaphragm (Boveris et al., 2002a), and heart (Boveris et al., 2003b; Costa et al., 2002; French et al., 2001; Hotta et al., 1999; Kanai et al., 2001; Valdez et al., 2004).

The recognition of the mitochondrial NO production catalyzed by mtNOS and the effects of NO on mitochondrial electron transfer [i.e., the oxygen competitive inhibition of cytochrome oxidase (Brown and Cooper, 1994; Cleeter et al., 1994) and the inhibition of electron transfer between cytochromes $b$ and $c$ (Poderoso et al., 1996)] are consistent with a role of NO in the physiological regulation of mitochondrial energy production and signaling. The biological situations in which the regulation by NO of mitochondrial respiration are considered relevant include hypoxia, ischemia-reperfusion, inflammation, apoptosis, and aging. In the consideration of the effects of NO on mitochondrial functions, the production of NO by mtNOS is important by the diffusional vicinity of the NO source (mtNOS) and of NO targets: cytochrome oxidase and ubiquinol–cytochrome $c$ reductase. The ability of mtNOS activity to modulate by its product NO, mitochondrial $O_2$ uptake, and hydrogen peroxide ($H_2O_2$) production is known as *mtNOS functional activity* (Boveris et al., 2002b, 2003a,b) and is considered one of the major pathways by which NO exerts its role as an intracellular regulator in physiological, pathological, and pharmacological conditions.

Determination of mtNOS Functional Activity

## Mitochondria Isolation and Mitochondrial Membranes Preparation

Rat and mouse liver, heart, kidney, and brain mitochondria used to determine NO production, $H_2O_2$ production, and $O_2$ uptake are prepared by a classical procedure using 0.23 $M$ mannitol, 0.07 $M$ sucrose, 1 m$M$ EDTA, 10 m$M$ Tris–HCl, pH 7.4, as homogenization buffer. Renal cortex is separated from medulla and papilla and homogenized. The homogenate (1 g organ/9 ml homogenization medium) is centrifuged at 750$g$ for 10 min to discard nuclei and cell debris, and the supernatant is centrifuged at 7000$g$ for 10 min. The obtained pellet is washed and suspended in the same buffer and consists of intact coupled mitochondria able to carry out oxidative phosphorylation. Mitochondrial membranes are obtained by twice freezing and thawing and homogenization through a 15/10 hypodermic needle (Boveris *et al.*, 1972, 2002b; Lores Arnaiz *et al.*, 2004). Protein concentration is assayed by the Folin reagent using bovine serum albumin (BSA) as standard (Lowry *et al.*, 1951).

## Mitochondrial NOS Functional Activity: Inhibition of $O_2$ Consumption

Oxygen uptake is determined polarographically with a Clark-type electrode in a 1.5-ml chamber at 30°, in an air-saturated reaction medium consisting of 0.23 $M$ mannitol, 0.07 $M$ sucrose, 20 m$M$ Tris–HCl, pH 7.4, 1 m$M$ EDTA, 5 m$M$ phosphate, 4 m$M$ $MgCl_2$, and 1 mg of mitochondrial protein (Boveris *et al.*, 1999). With brain mitochondria, 0.2% BSA is added to the medium (Lores Arnaiz *et al.*, 2004), and with heart mitochondria $MgCl_2$ is omitted. Respiratory rates are determined with either 6 m$M$ malate and 6 m$M$ glutamate or 7 m$M$ succinate as substrates for complex I or complex II, respectively. State 3 active respiration is established by the addition of 0.5 m$M$ adenosine diphosphate (ADP). Oxygen uptake is expressed in nanograms-at O/min mg protein.

The effect of *mtNOS activity on mitochondrial $O_2$ uptake* is better assayed in state 3 respiration [the maximal physiological rate of $O_2$ uptake and adenosine triphosphate (ATP) synthesis or active respiration], achieved in the presence of excess respiratory substrate and ADP. Respiratory rates are determined in two limit conditions of intramitochondrial NO: at its maximal and minimal levels. The first condition is achieved by supplementation with 100–300 $\mu M$ L-arginine and 1 $\mu M$ SOD, and the second by addition of 1–2 m$M$ NOS inhibitor (L-NAME or L-NNA) and 20–25 $\mu M$ oxyhemoglobin ($HbO_2$). Figure 1 illustrates the alternative effects of L-arginine and L-NNA on the state 3 $O_2$ uptake of mouse brain mitochondria. Whereas L-arginine decreases respiration by 16%, L-NNA

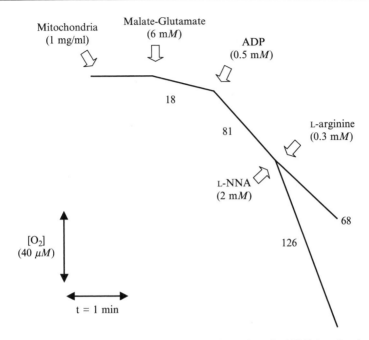

FIG. 1. Determination of mitochondrial nitric oxide synthase (mtNOS) functional activity in mouse brain mitochondria by measuring $O_2$ uptake. Mitochondria (1 mg protein/ml) were resuspended in 0.23 M mannitol, 0.07 M sucrose, 20 m$M$ Tris–HCl, pH 7.4, 1 m$M$ EDTA, 5 m$M$ phosphate, 0.2% bovine serum albumin (BSA), and 4 m$M$ MgCl$_2$. The effect of mtNOS activity on mitochondrial $O_2$ uptake is assayed by the determination of state 3 respiration. Respiratory rates are determined in two limit conditions of intramitochondrial NO: (1) in the presence of L-arginine and (b) in the presence of L-NNA. Numbers near traces indicate $O_2$ uptake in nanograms-at O/min mg protein. From Lores Arnaiz *et al.* (2004).

increases $O_2$ consumption by 55%. The summed effects, 58 ng-at O/min mg protein or 71%, determine the mtNOS functional activity in the inhibition of $O_2$ uptake. Supplementation of state 3 mitochondria with the mtNOS substrate L-arginine and with SOD decreases the respiration rate by 2–16%, depending on the organ (Fig. 1; Table I). It is understood that L-arginine and SOD have synergistic effects in maximizing the intramito-chondrial steady-state level of NO by providing the mtNOS substrate and by removing extramitochondrial $O_2^-$. Supplementation of the mitochondri-al preparation with an NOS inhibitor and HbO$_2$, a condition that minimizes intramitochondrial NO levels, increases the $O_2$ consumption by 8–55% in heart, kidney, liver, and brain mitochondria. These effects are explained by the continuous production of NO by mtNOS and the inhibition of

TABLE I
MITOCHONDRIAL NOS FUNCTIONAL ACTIVITY IN THE INHIBITION OF $O_2$ UPTAKE

| Conditions | Respiratory rate (ng-at O/min mg protein) | | | |
|---|---|---|---|---|
| | Heart | Liver | Kidney[a] | Brain[b] |
| Control | $234 \pm 11$ | $162 \pm 10$ | $201 \pm 7$ | $81 \pm 3$ |
| + L-arginine + SOD (a) | $199 \pm 9$ | $160 \pm 8$ | $193 \pm 5$ | $68 \pm 2$ |
| + NOS Inhibitor + HbO$_2$ (b) | $257 \pm 12$ | $195 \pm 12$ | $219 \pm 6$ | $126 \pm 6$ |
| mtNOS functional activity [(b) − (a)] | $58 \pm 15$ | $35 \pm 14$ | $26 \pm 8$ | $58 \pm 6$ |
| mtNOS functional activity [(b − a)/Control] × 100 | 25% | 22% | 13% | 71% |

[a] From Boveris et al. (2003a).
[b] From Lores Arnaiz et al. (2004).
Note: Malate (6 m$M$) and glutamate (6 m$M$) were used as substrates for heart, liver, and brain mitochondria, and succinate (7 m$M$) was used for kidney mitochondria. Experiments were carried out with mitochondria in metabolic state 3 [0.5 m$M$ adenosine diphosphate (ADP)]. Mitochondria were supplemented with 0.1–0.3 m$M$ L-arginine and 1 $\mu M$ superoxide dismutase (SOD) (a) or with a nitric oxide synthase (NOS) inhibitor and 20–25 $\mu M$ oxyhemoglobin (b). The NOS inhibitor was $N^{\omega}$- nitro-L-arginine (L-NNA, 2 m$M$) for heart, liver, and brain mitochondria, and $N^{\omega}$-nitro- L-arginine methyl ester (L-NAME, 1 m$M$) for kidney mitochondria.

cytochrome oxidase by NO. In this process, NO binds to the enzyme in its reduced form in a process that is competitive with oxygen (Brown and Cooper, 1994; Poderoso et al., 1996). Binding and inhibition are removed by washing or addition of HbO$_2$. The difference between the state 3 O$_2$ uptake with L-arginine and SOD (a) and with NOS inhibitor and HbO$_2$ (b) indicates the *mtNOS functional activity in the inhibition of cytochrome oxidase activity* that is conveniently expressed as a percentage fraction of state 3 respiration (Table I).

*Mitochondrial NOS Functional Activity: Enhancement of H$_2$O$_2$ Production*

Hydrogen peroxide production is determined in intact mitochondria by the scopoletin–horseradish peroxidase (HRP) method, following the decrease in fluorescence intensity at 365–450 nm ($\lambda$exc – $\lambda$em) at 30° (Boveris, 1984). The reaction medium consists of 0.23 $M$ mannitol, 0.07 m sucrose, 20 m$M$ Tris–HCl, pH 7.4, 0.8 $\mu M$ HRP, 1 $\mu M$ scopoletin, 0.3 $\mu M$ SOD, and 7 m$M$ succinate or 6 m$M$ malate and 6 m$M$ glutamate. A calibration curve is made using H$_2$O$_2$ (0.05–0.35 $\mu M$) as standard to express the fluorescence changes as nmol H$_2$O$_2$/min mg protein. To determine the functional activity of mtNOS in up regulating H$_2$O$_2$ production, the mitochondrial suspensions are supplemented with 100–300 $\mu M$ L-arginine or with an NOS inhibitor: L-NAME (1 m$M$) or L-NNA (2 m$M$).

Hydrogen peroxide production is physiologically maximal in mitochondrial state 4 (Boveris and Chance, 1973). The *regulatory activity of mtNOS on mitochondrial $H_2O_2$ generation* is assayed in mitochondrial state 4, the resting and nonphosphorylating mitochondrial state (controlled respiration), which is achieved in the presence of excess respiratory substrate and in the absence of ADP. Addition of L-arginine increases $H_2O_2$ production by 9–28% in kidney, brain, liver, and heart mitochondria, whereas the supplementation of the same preparation with NOS inhibitor decreases $H_2O_2$ generation by 3–55% (Table II). The difference in $H_2O_2$ production rate between the conditions of maximal and minimal NO levels is known as *the functional activity of mtNOS on the regulation $H_2O_2$ production.* The effects on $H_2O_2$ production are again explained by the intramitochondrial NO steady-state concentrations and, in this case, by the NO inhibition of ubiquinol–cytochrome $c$ reductase complex that enhances $H_2O_2$ production (Poderoso *et al.*, 1999). The functional activity of mtNOS in the enhancement of $H_2O_2$ production is conveniently expressed as a percentage or fraction of state 4 $H_2O_2$ production (Table II).

TABLE II

MITOCHONDRIAL NOS FUNCTIONAL ACTIVITY IN THE ENHANCEMENT OF $H_2O_2$ PRODUCTION

| Conditions | $H_2O_2$ production (nmol $H_2O_2$/min mg protein) | | | |
| --- | --- | --- | --- | --- |
| | Heart[a] | Liver | Kidney[b] | Brain[c] |
| Control | $0.70 \pm 0.04$ | $0.50 \pm 0.04$ | $0.44 \pm 0.03$ | $0.40 \pm 0.03$ |
| + L-arginine (a) | $0.90 \pm 0.03$ | $0.61 \pm 0.04$ | $0.48 \pm 0.03$ | $0.44 \pm 0.02$ |
| + NOS inhibitor (b) | $0.68 \pm 0.03$ | $0.47 \pm 0.03$ | $0.40 \pm 0.03$ | $0.18 \pm 0.02$ |
| + antimycin (c) | $1.05 \pm 0.05$ | $0.72 \pm 0.05$ | $0.70 \pm 0.05$ | – |
| mtNOS functional activity [(a) − (b)] | $0.22 \pm 0.05$ | $0.14 \pm 0.05$ | $0.08 \pm 0.04$ | $0.26 \pm 0.03$ |
| mtNOS functional activity [(a − b)/control] × 100 | 31% | 28% | 18% | 65% |

[a] From Boveris *et al.* (2003b).

[b] From Boveris *et al.* (2003a).

[c] From Lores Arnaiz *et al.* (2004).

*Note:* Experiments were carried out with mitochondria in metabolic state 4. Malate (6 m$M$) and glutamate (6 m$M$) were used as substrates for heart and liver mitochondria, and succinate (7 m$M$) was used for brain and kidney mitochondria. Mitochondria were supplemented with 0.1–0.3 m$M$ L-arginine (a) or with a nitric oxide (NOS) inhibitor (b). The NOS inhibitor was $N^\omega$-nitro-L-arginine (L-NNA, 2 m$M$) for heart, liver, and brain mitochondria, and $N^\omega$-nitro-L-arginine methyl ester (L-NAME, 1 m$M$) for kidney mitochondria.

Nitric Oxide Production

NO production is measured in intact mitochondria and mitochondrial membranes by following spectrophotometrically at 577–591 nm ($\varepsilon = 11.2$ m$M^{-1}$ cm$^{-1}$) in a Beckman DU 7400 diode array spectrophotometer the oxidation of oxyhemoglobin to methemoglobin, at 37° (Boveris *et al.*, 2002b; Carreras *et al.*, 1996; Murphy and Noack, 1994). The assay is performed with the samples supplemented with either L-arginine or a NOS inhibitor, and the difference in rates of methemoglobin formation gives the rates of NO production.

## Mitochondrial NOS Biochemical and Functional Activities

The reaction medium used to determine NO production by heart, liver, and kidney mitochondrial membranes (0.5–0.8 mg protein/ml) contains 50 m$M$ phosphate buffer, pH 7.4, 1 m$M$ L-arginine, 1 m$M$ CaCl$_2$, 100 $\mu M$ NADPH, 10 $\mu M$ dithiothreitol, 4 $\mu M$ CuZn-SOD (to avoid interference by O$_2^-$), 0.1 $\mu M$ catalase (to avoid oxyhemoglobin oxidation by H$_2$O$_2$), and 20 $\mu M$ oxyhemoglobin (in heme group). NO production by brain mitochondrial membranes is measured in the same reaction medium but at pH 5.8 and containing 50 $\mu M$ L-arginine. Controls are performed by the addition of NOS inhibitors (1 m$M$ or 2 m$M$ L-NNA) to give specificity to the oxyhemoglobin assay. The absorbance changes that are inhibitable by NOS inhibitors are due to NO formation.

Figure 2 shows the relationships between the biochemical and functional activities of mtNOS, measured in submitochondrial membranes and in intact mitochondria, respectively. NO production by heart, liver, kidney, and brain submitochondrial membranes is 0.69 $\pm$ 0.05, 1.37 $\pm$ 0.16, 0.86 $\pm$ 0.05, and 0.45 $\pm$ 0.05 nmol NO/min mg protein, respectively. The functional activity of mtNOS is determined both in inhibiting O$_2$ uptake and in enhancing H$_2$O$_2$ production. The marked difference in functional activity between brain mitochondria and liver, heart, and kidney mitochondria is understood as being due to the mitochondrial content of L-arginine, as it is shown for brain and liver mitochondria in Table III.

## Mitochondrial NO Release

Table IV shows the rates of NO release from liver, kidney, and heart intact mitochondria. The medium used to measure mitochondrial NO release consists of 0.23 $M$ mannitol, 0.07 $M$ sucrose, 20 m$M$ Tris–HCl, pH 7.4, 1 m$M$ EDTA, 5 m$M$ phosphate, 4 m$M$ MgCl$_2$, and 20 $\mu M$ oxyhemoglobin heme. Mitochondria (0.5 mg protein/ml) are energized by the addition of 6 m$M$ malate and 6 m$M$ glutamate or 7 m$M$ succinate, in the

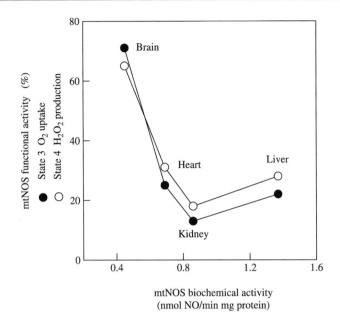

FIG. 2. Relationship of mitochondrial nitric oxide synthase (mtNOS) biochemical activity (in nmol NO/min mg protein) and mtNOS functional activity (in %) in mitochondria from different organs.

TABLE III
AMINO ACID CONTENT IN MOUSE BRAIN AND RAT LIVER MITOCHONDRIA

| | Content (nmol/mg protein) | |
| --- | --- | --- |
| Amino acid | Liver | Brain[a] |
| Arg | 0.62 ± 0.03 | 19.0 ± 0.5 |
| Asp | 0.38 ± 0.03 | 12.8 ± 0.9 |
| Glu | 1.68 ± 0.18 | 20.0 ± 2.0 |

[a] From Lores Arnaiz et al. (2004).

absence or in the presence of 0.5 m$M$ ADP. NO release from intact mitochondria is 40–50% lower in state 3 than in state 4, indicating that the state 4–state 3 transition is able to regulate mtNOS activity. Taking into account that the respiratory control is the ratio of $O_2$ uptake in state 3/state 4, the enzymatic control (EC) is defined as the ratio of NO production as state 4/state 3. Table IV shows that EC values are about 1.7–1.8 for mitochondria isolated from different organs.

TABLE IV
NO RELEASE AND $O_2$ UPTAKE BY LIVER, KIDNEY, AND HEART COUPLED MITOCHONDRIA

| | Oxygen uptake (ng-at O/min mg protein) | | | NO release (nmol/min mg protein) | | |
| --- | --- | --- | --- | --- | --- | --- |
| | State 4 | State 3 | $RC^a$ | State 4 | State 3 | $EC^b$ |
| Liver | | | | | | |
| Malate-glutamate | 24 ± 2 | 162 ± 10 | 6.8 | 1.32 ± 0.02 | 0.72 ± 0.1 | 1.8 |
| Succinate | 58 ± 5 | 274 ± 30 | 4.7 | 1.35 ± 0.05 | 0.73 ± 0.1 | 1.8 |
| Kidney | | | | | | |
| Malate-glutamate | 50 ± 4 | 180 ± 10 | 3.6 | 0.52 ± 0.05 | 0.31 ± 0.02 | 1.7 |
| Succinate | 79 ± 8 | 254 ± 27 | 3.2 | 0.73 ± 0.04 | 0.41 ± 0.05 | 1.8 |
| Heart | | | | | | |
| Malate-glutamate | 52 ± 6 | 234 ± 11 | 4.5 | – | – | – |
| Succinate | 106 ± 9 | 394 ± 35 | 3.7 | 0.62 ± 0.04 | 0.37 ± 0.06 | 1.7 |

[a] RC, respiratory control; RC = (state 3/state 4) $O_2$ consumption.
[b] EC, enzymatic control; EC = (state 4/state 3) NO production.

## Mitochondrial NOS Functional Activity in Pathological Situations and after Pharmacological Treatments

Table V lists a series of *in vivo* treatments that affect mtNOS activity. Both an increase and a decrease in mitochondrial NO production after pharmacological treatment or pathological situations were observed (Boveris *et al.*, 2003a,b; Lores Arnaiz *et al.*, 2004). The respiratory difference between the conditions with (a) L-arginine and with (b) the NOS inhibitor are taken as measurement of mtNOS activity as described earlier.

Treatment with the drug enalapril produced a marked increase (1.9–5.4 times) in NO production by heart, liver, and kidney mitochondria and an enhancement in mtNOS functional activity, measured through both the $O_2$ uptake (1.5–2.6 times) and the $H_2O_2$ production (1.9–3.5 times). Similarly, the endotoxic shock generated by lipopolysaccharide (LPS) administration produced a two-fold increase in the NO production of diaphragm mitochondria, which was reflected in a 1.5–2.6 times higher mtNOS functional activity. In contrast, *in vivo* treatment with chlorpromazine inhibited brain mtNOS activity and mtNOS functional activity, in both cases by 50%. A similar effect of decreased mtNOS activity has been observed in liver mitochondria after thyroxine administration.

To conclude, simple respiratory determinations with the oxygen electrode or through the measurement of mitochondrial $H_2O_2$ production can be used to assay effects of physiological or pharmacological treatments on mtNOS activity.

TABLE V

MITOCHONDRIAL NOS FUNCTIONAL ACTIVITY AND BIOCHEMICAL ACTIVITY (NO PRODUCTION) IN PATHOLOGICAL AND PHARMACOLOGICAL SITUATIONS

| Condition | Treatment and doses | Organ | NO production | mtNOS functional activity (Ratio referred to control activity) | |
| --- | --- | --- | --- | --- | --- |
| | | | | $O_2$ uptake | $H_2O_2$ production |
| Enalapril[a] | 30 mg/kg rat/day (i.p., 14 days) | Kidney | ↑ 5.4 | ↑ 2.5 | ↑ 3.5 |
| Enalapril[b] | 10 mg/kg mouse/day (oral, 14 days) | Heart | ↑ 1.9 | ↑ 1.5 | ↑ 1.9 |
| Enalapril[b] | 10 mg/kg mouse/day (oral, 14 days) | Liver | ↑ 1.9 | ↑ 2.6 | ↑ 2.6 |
| LPS | 10 mg/kg (i.p., single dose) | Liver | ↑ 0.65 | — | — |
| LPS | 10 mg/kg (i.p., single dose) | Diaphragm | ↑ 2.0 | ↑ 1.5 | ↑ 2.6 |
| Chlorpromazine[c] | 10 mg/kg mouse (i.p., single dose, 1 h) | Brain | ↓ 0.5 | ↓ 0.3 | ↓ 0.5 |
| Thyroxine ($T_4$)[d] | 0.2 mg/kg rat (subcutaneously, 14 days) | Liver | ↓ 0.5 | — | — |
| Hypothyroidism[d] | — | Liver | ↑ 3.6 | ↑ 5.0 | — |

[a] From Boveris et al. (2003a).
[b] From Boveris et al. (2003b).
[c] From Lores Arnaiz et al. (2004).
[d] From Carreras et al. (2001).

## Concluding Remarks

Mitochondrial NOS is a highly regulated mitochondrial enzyme, which in turn plays a regulatory role through mitochondrial NO steady-state levels that modulate $O_2$ uptake and $O_2^-$ and $H_2O_2$ production rates.

The mtNOS activity is biochemically modulated by mitochondrial metabolic states. The state 4–state 3 transition regulates NO release in coupled mitochondria, and the rates in state 3 are 40–50% lower than those in state 4. The continuous intramitochondrial generation of NO by mtNOS generates a NO steady-state concentration that exerts a regulatory role on mitochondrial $O_2$ uptake and $H_2O_2$ production. The functional activity of mtNOS in regulating mitochondrial respiration was observed in isolated mitochondria from heart, liver, kidney, and brain in state 4 and in state 3, and with malate-glutamate or succinate as substrates. This regulatory role extends from physiological levels of NO to pathologically increased NO levels.

## References

Boveris, A. (1984). Determination of the production of superoxide radicals and hydrogen peroxide in mitochondria. *Methods Enzymol.* **105,** 429–435.

Boveris, A., Alvarez, S., and Navarro, A. (2002a). The role of mitochondrial nitric oxide synthase in inflammation and septic shock. *Free Radic. Biol. Med.* **33,** 1186–1193.

Boveris, A., and Chance, B. (1973). The mitochondrial generation of hydrogen peroxide. *Biochem. J.* **134,** 707–716.

Boveris, A., Costa, L. E., Cadenas, E., and Poderoso, J. J. (1999). Regulation of mitochondrial respiration by adenosine diphosphate, oxygen and nitric oxide. *Methods Enzymol.* **301,** 188–198.

Boveris, A., D'Amico, G., Lores Arnaiz, S., and Costa, L. E. (2003b). Enalapril increases mitochondrial nitric oxide synthase activity in heart and liver. *Antioxid. Redox Signal.* **5,** 691–697.

Boveris, A., Lores Arnaiz, S., Bustamante, J., Alvarez, S., Valdez, L. B., Boveris, A. D., and Navarro, A. (2002b). Pharmacological regulation of mitochondrial nitric oxide synthase. *Methods Enzymol.* **359,** 328–339.

Boveris, A., Oshino, N., and Chance, B. (1972). The cellular production of hydrogen peroxide. *Biochem. J.* **128,** 617–630.

Boveris, A., Valdez, L. B., Alvarez, S., Zaobornyj, T., Boveris, A. D., and Navarro, A. (2003a). Kidney mitochondrial nitric oxide synthase. *Antioxid. Redox Signal.* **5,** 265–271.

Brown, G. C., and Cooper, C. E. (1994). Nanomolar concentrations of nitric oxide reversibly inhibit synaptosomal respiration by competing with oxygen at cytochrome oxidase. *FEBS Lett.* **356,** 295–298.

Bustamante, J., Bersier, G., Romero, M., Aron Badin, R., and Boveris, A. (2000). Nitric oxide production and mitochondrial dysfunction during rat thymocyte apoptosis. *Arch. Biochem. Biophys.* **376,** 239–247.

Carreras, M. C., Melani, M., Riobo, N., Converso, D. P., Gatto, E. M., and Poderoso, J. J. (2002). Neuronal nitric oxide synthases in brain and extraneural tissues. *Methods Enzymol.* **359,** 413–423.

Carreras, M. C., Peralta, J. G., Converso, D. P., Finocchietto, P. V., Rebagliati, I., Zaninovich, A. A., and Poderoso, J. J. (2001). Modulation of liver mitochondrial NOS is implicated in thyroid-dependent regulation of O(2) uptake. *Am. J. Physiol. Heart Circ. Physiol.* **281,** H2282–2288.

Carreras, M. C., Poderoso, J. J., Cadenas, E., and Boveris, A. (1996). Measurement of nitric oxide and hydrogen peroxide production from human neutrophils. *Methods Enzymol.* **269,** 65–75.

Cleeter, M. W., Cooper, J. M., Darley-Usmar, V. M., Moncada, S., and Schapira, A. H. (1994). Reversible inhibition of cytochrome c oxidase, the terminal enzyme of the mitochondrial respiratory chain, by nitric oxide. Implications for neurodegenerative diseases. *FEBS Lett.* **345,** 50–54.

Costa, L. E., La Padula, P., Lores Arnaiz, S., D'Amico, G., Boveris, A., Kurnjek, M. L., and Basso, N. (2002). Long-term angiotensin II inhibition increases mitochondrial nitric oxide synthase and not antioxidant enzyme activities in rat heart. *J. Hypertens.* **20,** 2487–2494.

Elfering, S. L., Sarkela, T. M., and Giulivi, C. (2002). Biochemistry of mitochondrial nitric-oxide synthase. *J. Biol. Chem.* **277,** 38079–38086.

French, S., Giulivi, C., and Balaban, R. S. (2001). Nitric oxide synthase in porcine heart mitochondria: Evidence for low physiological activity. *Am. J. Physiol. Heart Circ. Physiol.* **280,** H2863–2867.

Ghafourifar, P., and Richter, C. (1997). Nitric oxide synthase activity in mitochondria. *FEBS Lett.* **418,** 291–296.

Giulivi, C. (1998). Functional implications of nitric oxide produced by mitochondria in mitochondria metabolism. *Biochem. J.* **332,** 673–697.

Hotta, Y., Otsuka-Murakami, H., Fujita, M., Nakagawa, J., Yajima, M., Liu, W., Ishikawa, N., Kawai, N., Masumizu, T., and Kohno, M. (1999). Protective role of nitric oxide synthase against ischemia-reperfusion injury in guinea pig myocardial mitochondria. *Eur. J. Pharmacol.* **380,** 37–48.

Kanai, A. J., Pearce, L. L., Clemens, P. R., Birder, L. A., Van Bibber, M. M., Choi, S. Y., de Groat, W. C., and Peterson, J. (2001). Identification of a neuronal nitric oxide synthase in isolated cardiac mitochondria using electrochemical detection. *Proc. Natl. Acad. Sci. USA* **98,** 14126–14131.

Lores Arnaiz, S., Coronel, M. F., and Boveris, A. (1999). Nitric oxide, superoxide, and hydrogen peroxide production in brain mitochondria after haloperidol treatment. *Nitric Oxide Biol. Chem.* **3,** 235–243.

Lores Arnaiz, S., D'Amico, G., Czerniczyniec, A., Bustamante, J., and Boveris, A. (2004). Brain mitochondrial nitric oxide synthase: *In vitro* and *in vivo* inhibition by chlorpromazine. *Arch Biochem Biophys.* **430,** 170–177.

Lowry, O. H., Rosebrough, N. J., Farr, A. L., and Randall, R. J. (1951). Protein measurement with the Folin phenol reagent. *J. Biol. Chem.* **193,** 265–275.

Murphy, M. E., and Noack, E. (1994). Nitric oxide assay using hemoglobin method. *Methods Enzymol.* **233,** 240–250.

Poderoso, J. J., Carreras, M. C., Lisdero, C., Riobo, N., Schöpfer, F., and Boveris, A. (1996). Nitric oxide inhibits electron transfer and increases superoxide radical production in rat heart mitochondria and submitochondrial particles. *Arch. Biochem. Biophys.* **328,** 85–92.

Poderoso, J. J., Carreras, M. C., Schöpfer, F., Lisdero, C. L., Riobo, N. A., Giulivi, C., Boveris, A. D., Boveris, A., and Cadenas, E. (1999). The reaction of nitric oxide with ubiquinol: Kinetic properties and biological significance. *Free Radic. Biol. Med.* **26,** 925–935.

Schild, L., Reinheckel, T., Reiser, M., Horn, T. F., Wolf, G., and Augustin, W. (2003). Nitric oxide produced in rat liver mitochondria causes oxidative stress and impairment of respiration after transient hypoxia. *FASEB J.* **17,** 2194–2201.

Tatoyan, A., and Giulivi, C. (1998). Purification and characterization of a nitric-oxide synthase from rat liver mitochondria. *J. Biol. Chem.* **273,** 11044–11048.

Valdez, L. B., Zaobornyj, T., Alvarez, S., Bustamante, J., Costa, L. E., and Boveris, A. (2004). Heart mitochondrial nitric oxide synthase. Effects of hypoxia and aging. *Mol. Aspects Med.* **25,** 49–59.

[38] Tetrahydrobiopterin as Combined Electron/Proton Donor in Nitric Oxide Biosynthesis: Cryogenic UV–Vis and EPR Detection of Reaction Intermediates

*By* Antonius C. F. Gorren, Morten Sørlie, K. Kristoffer Andersson, Stéphane Marchal, Reinhard Lange, and Bernd Mayer

## Abstract

The role of tetrahydrobiopterin ($BH_4$) as a cofactor in nitric oxide synthase (NOS) has been the object of intense research in the last few years. It was found that in addition to its established effects on the NOS heme spin state, substrate affinity, and enzyme dimerization, $BH_4$ is required as a one-electron donor to oxyferrous $[Fe(II) \cdot O_2]$ heme that is formed as an intermediate in the catalytic cycle. Cryogenic spectroscopic techniques proved particularly useful in the identification of this role of $BH_4$ in NO synthesis. With these methods, the mechanism of fast reactions, such as the reaction of ferrous NOS with $O_2$, can be unraveled by lowering the reaction temperature to subzero values. This may not only reduce the rate to such an extent that the reaction can be followed on a time scale from seconds to minutes, but intermediates may be observed that do not accumulate at higher temperatures. Cryogenic ultraviolet–visible (UV–vis) and electron paramagnetic resonance spectroscopy have been applied to clarify why the $BH_4$ analogue 4-amino-tetrahydrobiopterin (4-amino-$BH_4$) is unable to support NO synthesis. In the course of these studies, evidence was gathered supporting a role for $BH_4$ as an obligate proton and electron donor. It is believed that the inhibitory action of 4-amino-$BH_4$ derives from an inability to serve as a proton donor, even though it is perfectly able to serve as an electron donor. In this chapter, the suitability, drawbacks, and advantages of cryogenic methods are discussed.

## Introduction

### Nitric Oxide Synthesis

The biosynthesis of nitric oxide (NO) is catalyzed by a small group of closely related enzymes, the NO synthases (NOSs) (EC 1.14.13.39) [see Pfeiffer *et al.* (1999), Stuehr (1999), and Alderton *et al.* (2001) for reviews]. There are three isoforms, called neuronal, endothelial, and inducible NOS (nNOS, eNOS, and iNOS), which are involved in neurotransmission,

METHODS IN ENZYMOLOGY, VOL. 396
0076-6879/05 $35.00
DOI: 10.1016/S0076-6879(05)96038-2

vasodilation, and the immune response, respectively. In accordance with their distinct functions, the isoforms differ in localization and regulation. Structurally and mechanistically, the isoforms are similar, albeit with subtle differences. All NOS isoforms catalyze the conversion of L-arginine to L-citrulline and NO in two consecutive reactions, with $N^G$-hydroxy-L-arginine (NHA) as an intermediate. Both reactions consume one molecule of $O_2$ and yield one molecule of $H_2O$, and both reactions require exogenous electrons (two and one equivalent for the first and second reaction, respectively) that are provided by NADPH. NOS is a homodimer, with each monomer consisting of an oxygenase and a reductase domain. Catalysis takes place in the oxygenase domain at a P450-type heme. The reductase domain shuttles electrons from NADPH to the heme via two flavin (one FAD and one FMN) moieties. Arg hydroxylation is thought to occur by the same mechanism as proposed for cytochrome P450-catalyzed mono-oxygenations. Accordingly, substrate binding and heme reduction are followed by $O_2$ binding. Subsequently, reduction of the oxyferrous complex by another electron generates the reduced oxyferrous complex (superoxy-ferrous/peroxy-ferric heme), which, after two protonation steps and O-O bond scission, is thought to yield a compound I type structure (oxyferryl heme plus a porphyrin centered radical cation). This compound is thought to rapidly hydroxylate the substrate by the so-called oxygen-rebound mechanism, resulting in ferric heme and (enzyme-bound) NHA. The second reaction proceeds similarly up to formation of the reduced oxyferrous species ($Fe(II) \cdot O_2^-$). Although no experimental evidence is yet available for the following steps in the cycle, the NHA reaction is thought to deviate from the mechanism proposed for the Arg reaction at that stage. Specifically, it is assumed that a compound I type species is not formed and that the preceding peroxy or hydroperoxy species reacts with NHA directly.

## The Role of Tetrahydrobiopterin

In addition to heme, FAD, FMN, and $Ca^{2+}$/calmodulin, NOS requires tetrahydrobiopterin ($BH_4$) as a cofactor [see Gorren and Mayer (2002), Wei *et al.* (2003a), and Werner *et al.* (2003) for reviews]. $BH_4$ binds to the oxygenase domain in close proximity to the heme, although not in the substrate-binding pocket. It is bound in the dimer interface, where it interacts with amino acid residues from both monomers. $BH_4$ has a range of structural and allosteric effects on NOS. As might be expected from the nature of the pterin-binding site, $BH_4$ stabilizes NOS dimerization. Likewise, because of the proximity of the pterin-binding site to the heme and the substrate-binding pocket, it is no surprise to find $BH_4$ affecting heme properties and substrate affinity. $BH_4$ causes a low-to-high spin shift of the heme; it increases the reduction potential of iNOS heme, and there is considerable cooperativity between binding of pterin and substrate. All of these effects, which vary in magnitude between isoforms, may modulate NOS activity. However, the role of $BH_4$ extends well beyond such effects, because without pterin, oxidation of NADPH becomes uncoupled from NO synthesis, and reduction of $O_2$ results in the formation of superoxide anion ($O_2^-$). Moreover, studies with pterin derivatives demonstrated that all pteridines that bound to NOS mimicked the structural and allosteric effects, whereas only the fully reduced (tetrahydro) species could sustain NO synthesis. These observations suggest that the key role of $BH_4$ involves direct participation in NOS catalysis as a redox-active agent. The elucidation of this aspect of the function of $BH_4$ in NO synthesis has largely relied on a combination of rapid kinetic techniques [rapid-scan/stopped-flow UV–vis spectroscopy, freeze-quench electron paramagnetic resonance (EPR) spectroscopy, rapid-quench-assisted reaction product analysis] and cryogenic spectroscopic methods.

## Experimental Procedures

### Cryoenzymology

In a previous contribution to this series, the methodology for the acquisition of cryogenic UV–vis optical spectra was already described (Gorren *et al.*, 2002). Reviews on and examples of this type of study can be found in Maurel *et al.* (1974), Douzou (1977), Andersson *et al.* (1979), Bonfils *et al.* (1980), Fink and Geeves (1979), Larroque *et al.* (1990), Travers and Barman (1995), and Bec *et al.* (1999). Essentially, the same method can be applied to obtain cryogenic EPR spectra, although such experiments are technically a bit more challenging because of the smaller sample volumes and higher enzyme concentrations and in our experience require some practice.

### Sample Preparation for Cryogenic EPR Experiments

Samples are prepared according to a similar procedure as applied for optical spectroscopy (Gorren *et al.*, 2002), except for the use of EPR tubes instead of optical cuvettes. Samples in 50% ethylene glycol, 50 m$M$ KP$_i$ (pH 7.5), containing 25–100 $\mu M$ NOS in a total volume of 200 $\mu$l, are made anaerobic by gassing with argon for 30–60 min at 0°. Subsequently, sodium dithionite is added at room temperature in 0.5 $\mu$l aliquots from an anaerobic stock solution (~30 m$M$) to reduce the enzyme. After each addition, the sample is stirred for 30–60 s with the tip of the needle, and the progress of reduction, which requires higher than stoichiometric amounts of dithionite (Du *et al.*, 2003; Hurshman and Marletta, 2002; Schmidt *et al.*, 2001), is checked by inspection of the shift of the optical Soret band from 396 to 412 nm. Samples are then cooled to $-30°$, and 250 $\mu$l $O_2$ is added from a syringe over approximately 5 s to initiate the reaction. By means of a thermocouple, it was checked that the temperature remained at $-30°$ during this procedure. After 20–30 s further incubation, the sample is flash-frozen in $-130°$ *n*-pentane and placed inside the EPR spectrophotometer for spectral acquisition.

We used a Bruker ESP300E EPR spectrometer equipped with an Oxford Instruments cryostat 900, operated at 9.65 GHz with 100-kHz modulation frequency, 1–10 $G$ modulation amplitude, 0.5–4.0 mW microwave power, and measuring temperatures between 4 and 100 K. Microwave power saturation data were collected between 0.3 and 200 mW.

## Results

### Identification of BH$_4$ as an Obligate Electron Donor to Fe(II)·O$_2$

Cryogenic methods have proven useful to study the effect of $BH_4$ on the oxidation of reduced NOS with $O_2$. Comparison of the intermediates observed in that reaction with or without Arg and in the presence or absence of $BH_4$, along with the observation of NHA formation under the same conditions only in the presence of Arg and $BH_4$, led to the proposal that $BH_4$ served as a one-electron donor to the $Fe(II) \cdot O_2$ complex in the reaction with Arg (Bec *et al.*, 1998). That hypothesis was confirmed by the observation of a trihydrobiopterin radical, both by rapid-freeze EPR and by cryogenic EPR (Du *et al.*, 2003; Hurshman *et al.*, 1999; Schmidt *et al.*, 2001; Wei *et al.*, 2001), and by demonstration that the kinetics of $Fe(II) \cdot O_2$ decay corresponded to those of pterin radical formation and NHA production (Wei *et al.*, 2001).

Cryogenic UV–vis spectroscopy also formed the basis for the proposal that $BH_4$ was an obligate electron donor in the second reaction cycle, with NHA as the substrate (Gorren *et al.*, 2000), even though the different electronic requirements of that reaction suggested that direct participation of the pterin might not be necessary, and despite the fact that no pterin radical was observed. This issue remained controversial until fairly recently, when it was demonstrated that the transient formation of a pterin radical does indeed accompany the disappearance of the oxyferrous complex (Wei *et al.*, 2003b). It is now widely accepted that $BH_4$ is the immediate electron donor to the oxyferrous complex with both substrates and that both cycles follow the same pathway at least until the formation of the $Fe(II) \cdot O_2^-$ complex.

### Putative Role of BH4 as an Obligate Proton Donor in NO Synthesis

We used cryogenic UV–vis and EPR spectroscopy to study the reactions of reduced eNOS oxygenase domain with $O_2$ in the presence of 4-amino-tetrahydrobiopterin (Sørlie et al., 2003). This compound differs from other inhibitory pteridines in having electrochemical properties similar to those of $BH_4$, suggesting that it should be able to substitute for $BH_4$ in NO synthesis (Gorren et al., 2001). Cryogenic EPR studies offered vital clues to the reaction mechanism in the presence of 4-amino-$BH_4$. In the reaction with Arg, single turnover at $-30°$ resulted in the formation of a trihydropteridine radical in the same high yield as observed in the presence of $BH_4$, which demonstrated that 4-amino-$BH_4$ is indeed able to reduce the oxyferrous complex (Sørlie et al., 2003). Contrary to the reaction with $BH_4$ as the cofactor, where ferric heme was formed alongside the pterin radical, the heme species accompanying the 4-amino-BH3˙ radical was EPR silent. Furthermore, microwave power saturation experiments indicated that the unusually fast spin relaxation of the BH3˙ radical, which has been ascribed to interaction with the nearby heme spin, is much slower for the 4-amino-BH3˙ radical, which suggests a diamagnetic heme. The corresponding optical spectrum, which exhibited a Soret maximum at 428 nm, was tentatively assigned to the oxyferrous complex (which is diamagnetic). Taken together, the intermediate formed in the reaction with Arg and 4-amino-$BH_4$ was identified as a $Fe(II)$-$O_2$...4-amino-BH3˙-$H^+$ species. Formation of this compound can be rationalized if the reaction is halted after electron transfer from the pterin to the heme. Dissociation of $O_2^-$ followed by binding of a second equivalent of $O_2$ to the ferrous heme would give rise to the observed intermediate. These results showed that electron transfer from 4-amino-$BH_4$ to the oxyferrous state is possible, but that this does not result in completion of the catalytic cycle. The proposed sequence of events is illustrated in Fig. 1, top.

In cryogenic EPR studies with NHA as the substrate, no 4-amino-BH3˙ radical was observed. Instead, a ferrous–NO complex was formed, with EPR parameters indicating the absence of a strong sixth ligand (Sørlie et al., 2003). The corresponding optical spectrum with a Soret maximum at 417–423 nm indicated that the heme was still 6-coordinate, but that the proximal ligand was no longer a thiolate anion. On the basis of the combined evidence, it was proposed that the intermediate in the reaction with NHA and 4-amino-$BH_4$ was probably a SH-$Fe(II)$-NO...4-amino-$BH_4$ species, in which the axial cysteine thiolate ligand is protonated. This

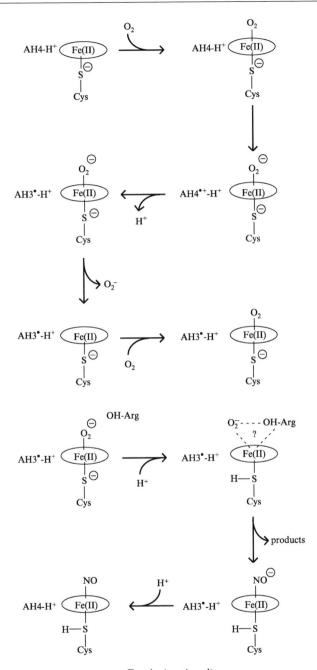

FIG. 1. *(continued)*

appears to imply that although 4-amino-$BH_4$ can reduce the oxyferrous complex, the catalytic cycle cannot proceed beyond that. In contrast with Arg, however, heme-bound $O_2^-$ is thought to react with bound substrate, resulting in the formation of citrulline and ferrous–NO heme, as well as in regeneration of reduced pteridine. Although the details of the mechanism, illustrated in Fig. 1, bottom, remain speculative, the observations show that as in the reaction with Arg, 4-amino-$BH_4$ reduces the oxyferrous complex but is unable to support completion of the catalytic cycle.

Inspection of the presumed mechanism of substrate oxygenation suggests that the sole reaction step subsequent to $Fe(II) \cdot O_2^-$ formation that is likely to be affected by the pteridine cofactor is the protonation of the complex that precedes O-O bond scission. Halting the reaction cycle at that specific reaction step would be consistent with the observed intermediates. It has been previously hypothesized that $BH_4$, but not 4-amino-$BH_4$, might be involved in protonation of the superoxy-ferrous complex (Crane *et al.*, 2000; Davydov *et al.*, 2002; Gorren and Mayer, 2002). The combined cryogenic EPR and optical studies discussed here provided strong, albeit largely circumstantial, evidence in favor of such a model, featuring $BH_4$ as an obligate combined electron/proton donor to the oxyferrous complex. According to this model, inhibition by 4-amino-$BH_4$ is due to its inability to serve as a proton donor, even though it is perfectly able to serve as an electron donor.

## Cryogenic $CO/O_2$ Exchange Studies

To evaluate the redox state of the intermediates in the reaction of reduced NOS with $O_2$ in cryogenic studies, one can make use of the fact that CO binds to the reduced heme only. The procedure is the same as

---

Fig. 1. Proposed reaction of reduced nitric oxide synthase (NOS) with $O_2$ in the presence of 4-amino-$BH_4$ at $-30°$. (A) Hypothetical reaction in the presence of Arg. After $O_2$ binding and electron transfer from the pteridine cofactor, the reaction cannot continue along the catalytic pathway, because proton transfer from 4-amino-$BH_4$ to heme-bound superoxide is blocked. Instead, a proton is lost to solvent. Dissociation of superoxide anion from the heme, perhaps facilitated by the negative charge on the proximal thiolate ligand, is followed by binding of a second equivalent of $O_2$, resulting in the intermediate observed spectroscopically. (B) Hypothetical reaction in the presence of $N^G$-hydroxy-L-arginine (NHA). The reaction is shown starting with the superoxy-ferrous complex that is formed after electron transfer from the pteridine. Instead of dissociating from the heme, bound superoxide may react with NHA in the heme pocket, resulting in a transient $Fe(II) \cdot NO^-$ complex. Electron transfer from that complex to the pteridine will yield the experimentally observed intermediate. Protonation of the proximal cysteinate ligand may be a consequence of the negatively charged distal ligands ($O_2^-$ or $NO^-$) in the intermediate states.

described for the reaction with $O_2$. Two 30-ml syringes filled with $O_2$ and CO, respectively, are precooled in a freezer. After preparation of the anaerobically reduced NOS sample in the spectrophotometer as described previously, precooled $O_2$ is added to the sample, immediately followed by addition of CO from the second syringe. Contrary to expectation, not all $Fe(II) \cdot O_2$ intermediates were able to form a $Fe(II) \cdot CO$ complex in the presence of CO at $-30°$. The ability of $O_2$/CO exchange correlated with the spectral properties of the oxy complexes. Partial CO complex formation occurred with intermediates exhibiting a Soret band at 420–421 nm, as indicated by the appearance of a Soret maximum at 444 nm under those conditions (Marchal $et$ $al.$, 2004a). In contrast, intermediates that displayed the remarkably red-shifted (428–432 nm) Soret maxima observed under some conditions showed no CO exchange at all but decayed to the ferric state instead. This was explained by a strong preference for the $Fe(III) \cdot O_2^-$ state over the $Fe(II) \cdot O_2$ resonance structure, resulting in $O_2^-$ dissociation from Fe(III) rather than $O_2$ dissociation from Fe(II).

The intermediates with 4-amino-$BH_4$ and its redox-inactive counterpart 4-amino-$BH_2$ are very similar and the red-shifted type. In agreement with this, no $Fe(II) \cdot CO$ complex was formed in the presence of 4-amino-$BH_2$ (Marchal $et$ $al.$, 2004b). In contrast, complete CO/$O_2$ exchange, with little or no Fe(III) formation, occurred in the presence of 4-amino-$BH_4$, which suggests the participation of 4-amino-$BH_4$ as an electron donor in the process, thus providing further support for the ability of 4-amino-$BH_4$ to reduce the oxyferrous complex (Marchal $et$ $al.$, 2004b).

## Comparison with Rapid Kinetic Methods

The most striking difference between rapid kinetic and cryogenic techniques is that standard rapid kinetic methods (rapid-scan/stopped-flow UV–vis spectroscopy; rapid-freeze EPR; freeze-quench product analysis) will usually yield results that are immediately relevant to the mechanism of catalysis. The same is not necessarily true for cryogenic methods because of the different experimental conditions (unphysiologically low temperature and the presence of a cryosolvent). Any kinetic information will also not be directly applicable to physiological conditions. It will, therefore, be necessary to confirm conclusions derived from cryogenic studies by additional methods. Indeed, the hypothesis that $BH_4$ is a one-electron donor in NO synthesis, proposed initially on the basis of cryogenic studies, was soon after confirmed by rapid kinetic techniques. However, the major disadvantage of cryogenic methods is also its greatest asset: Because of the different experimental conditions, intermediates may accumulate that are not detectable by standard rapid techniques. This may prove particularly useful in the case of NOS, where a whole series of intermediates presumed to be formed after the oxyferrous complex appear not to accumulate at ambient temperature.

The two faces of cryogenic spectroscopy are evident from results with NOS. Neither the 4-amino-$BH_3^\cdot$ radical observed in the reaction with Arg nor the ferrous NO complex obtained in the presence of NHA is observed in matching rapid kinetic studies (Hurshman *et al.*, 2003; Marchal *et al.*, 2004a). This apparent discrepancy calls for caution in the interpretation of the low-temperature data. One possible explanation involves the stability of the oxyferrous complex, which increases tremendously at lower temperatures; the apparent decay rate constants at 7 and $-30°$ are $5.4 \cdot 10^{-2}$ and $2.1 \cdot 10^{-4}$ $s^{-1}$, respectively (Gorren *et al.*, 2000; Marchal *et al.*, 2004a). Some observations suggest that electron transfer from $BH_4$ (or 4-amino-$BH_4$) to the heme is thermodynamically unfavorable and that the presence of substrate is required to pull the reaction forward (Gorren and Mayer, 2002). In that case, direct decay of the oxyferrous complex to Fe(III) and $O_2^-$ without oxidation of 4-amino-$BH_4$ is a viable option, and the chosen pathway will depend on the relative rate constants of $O_2^-$ formation from $Fe(II) \cdot O_2$ and $Fe(II) \cdot O_2^-$, as well as on the redox equilibrium between 4-amino-$BH_4$ and the oxyferrous complex. In view of the strong temperature dependence of the stability of the oxyferrous intermediate, it is conceivable that the pathway changes between cryogenic and ambient temperatures. This reconciles the cryogenic and rapid kinetic data and explains why at room temperature, no net 4-amino-$BH_4$ oxidation occurs. Irrespective of these considerations, the central observation that 4-amino-$BH_4$ can donate an electron to the heme and its implication for an additional role of $BH_4$ as a proton donor are probably not limited to cryogenic conditions, as they offer the only satisfactory explanation for the inhibitory properties of 4-amino-$BH_4$. However, final judgment on these issues must await confirmation or refutation by different methods.

## Acknowledgments

This work was supported by a grant from the Human Science Frontier Program (RGP0026/2001-M) and grant P15855 of the Fonds zur Förderung der Wissenschaftlichen Forschung in Österreich.

## References

Alderton, W. K., Cooper, C. E., and Knowles, R. G. (2001). Nitric oxide synthases: Structure, function and inhibition. *Biochem. J.* **357,** 593–615.
Andersson, K. K., Debey, P., and Balny, C. (1979). Subzero temperature studies of microsomal cytochrome P-450: O-dealkylation of 7-ethoxycoumarin coupled to single turnover. *FEBS Lett.* **102,** 117–120.

Bec, N., Gorren, A. C. F., Voelker, C., Mayer, B., and Lange, R. (1998). Reaction of neuronal nitric-oxide synthase with oxygen at low temperature. Evidence for reductive activation of the oxy-ferrous complex by tetrahydrobiopterin. *J. Biol. Chem.* **273,** 13502–13508.

Bec, N., Anzenbacher, P., Anzenbacherová, E., Gorren, A. C. F., Munro, A. W., and Lange, R. (1999). Spectral properties of the oxyferrous complex of the heme domain of cytochrome P450 BM-3 (CYP102). *Biochem. Biophys. Res. Commun.* **266,** 187–189.

Bonfils, C., Andersson, K. K., Maurel, P., and Debey, P. (1980). Cytochrome P-450 oxygen intermediates and reactivity at subzero temperatures. *J. Mol. Catal.* **7,** 299–308.

Crane, B. R., Arvai, A. S., Ghosh, S., Getzoff, E. D., Stuehr, D. J., and Tainer, J. A. (2000). Structures of the $N^\omega$-hydroxy-L-arginine complex of inducible nitric oxide synthase oxygenase dimer with active and inactive pterins. *Biochemistry* **39,** 4608–4621.

Davydov, R., Ledbetter-Rogers, A., Martásek, P., Larukhin, M., Sono, M., Dawson, J. H., Masters, B. S. S., and Hoffman, B. M. (2002). EPR and ENDOR characterization of intermediates in the cryoreduced oxy-nitric oxide synthase heme domain with bound L-arginine or $N^G$-hydroxyarginine. *Biochemistry* **41,** 10375–10381.

Douzou, P. (1977). "Cryobiochemistry—An Introduction." Academic Press, London.

Du, M., Yeh, H.-C., Berka, V., Wang, L.-H., and Tsai, A. (2003). Redox properties of human endothelial nitric-oxide synthase oxygenase and reductase domains purified from yeast expression system. *J. Biol. Chem.* **278,** 6002–6011.

Fink, A. L., and Geeves, M. A. (1979). Cryoenzymology: The study of enzyme catalysis at subzero temperatures. *Methods Enzymol.* **63,** 336–370.

Gorren, A. C. F., and Mayer, B. (2002). Tetrahydrobiopterin in nitric oxide synthesis: A novel biological role for pteridines. *Curr. Drug Metab.* **3,** 133–157.

Gorren, A. C. F., Bec, N., Schrammel, A., Werner, E. R., Lange, R., and Mayer, B. (2000). Low-temperature optical absorption spectra suggest a redox role for tetrahydrobiopterin in both steps of nitric oxide synthase catalysis. *Biochemistry* **39,** 11763–11770.

Gorren, A. C. F., Kungl, A. J., Schmidt, K., Werner, E. R., and Mayer, B. (2001). Electrochemistry of pterin cofactors and inhibitors of nitric oxide synthase. *Nitric Oxide* **5,** 176–186.

Gorren, A. C. F., Bec, N., Lange, R., and Mayer, B. (2002). Redox role for tetrahydrobiopterin in nitric oxide synthase catalysis: Low-temperature optical absorption spectral detection. *Methods Enzymol.* **353,** 114–121.

Hurshman, A. R., Krebs, C., Edmondson, D. E., Huynh, B. H., and Marletta, M. A. (1999). Formation of a pterin radical in the reaction of the heme domain of inducible nitric oxide synthase with oxygen. *Biochemistry* **38,** 15689–15696.

Hurshman, A. R., Krebs, C., Edmondson, D. E., and Marletta, M. A. (2003). Ability of tetrahydrobiopterin analogues to support catalysis by inducible nitric oxide synthase: Formation of a pterin radical is required for enzyme activity. *Biochemistry* **42,** 13287–13303.

Hurshman, A. R., and Marletta, M. A. (2002). Reactions catalyzed by the heme domain of inducible nitric oxide synthase: Evidence for the involvement of tetrahydrobiopterin in electron transfer. *Biochemistry* **41,** 3439–3456.

Larroque, C., Lange, R., Maurin, L., Bienvenue, A., and van Lier, J. E. (1990). On the nature of the cytochrome P450scc "ultimate oxidant": Characterization of a productive radical intermediate. *Arch. Biochem. Biophys.* **282,** 198–201.

Marchal, S., Gorren, A. C. F., Sørlie, M., Andersson, K. K., Mayer, B., and Lange, R. (2004a). Evidence of two distinct oxygen complexes of reduced endothelial nitric oxide synthase. *J. Biol. Chem.* **279,** 19824–19831.

Marchal, S., Lange, R., Sørlie, M., Andersson, K. K., Gorren, A. C. F., and Mayer, B. (2004b). CO exchange of the oxyferrous complexes of endothelial nitric-oxide synthase oxygenase domain in the presence of 4-amino-tetrahydrobiopterin. *J. Inorg. Biochem.* **98,** 1217–1222.

Maurel, P., Travers, F., and Douzou, P. (1974). Spectroscopic determinations of enzyme-catalyzed reactions at subzero temperatures. *Anal. Biochem.* **57,** 555–563.

Pfeiffer, S., Mayer, B., and Hemmens, B. (1999). Nitric oxide: Chemical puzzles posed by a biological messenger. *Angew. Chem. Int. Ed.* **38,** 1714–1731.

Schmidt, P. P., Lange, R., Gorren, A. C. F., Werner, E. R., Mayer, B., and Andersson, K. K. (2001). Formation of a protonated trihydrobiopterin radical cation in the first reaction cycle of neuronal and endothelial nitric oxide synthase detected by electron paramagnetic resonance spectroscopy. *J. Biol. Inorg. Chem.* **6,** 151–158.

Sørlie, M., Gorren, A. C. F., Marchal, S., Shimizu, T., Lange, R., Andersson, K. K., and Mayer, B. (2003). Single-turnover of nitric-oxide synthase in the presence of 4-amino-tetrahydrobiopterin. Proposed role for tetrahydrobiopterin as a proton donor. *J. Biol. Chem.* **278,** 48602–48610.

Stuehr, D. J. (1999). Mammalian nitric oxide synthases. *Biochim. Biophys. Acta* **1411,** 217–230.

Travers, F., and Barman, T. (1995). Cryoenzymology: How to practice kinetic and structural studies. *Biochimie* **77,** 937–948.

Wei, C.-C., Wang, Z.-Q., Wang, Q., Meade, A. L., Hemann, C., Hille, R., and Stuehr, D. J. (2001). Rapid kinetic studies link tetrahydrobiopterin radical formation to heme-dioxy reduction and arginine hydroxylation in inducible nitric-oxide synthase. *J. Biol. Chem.* **276,** 315–319.

Wei, C.-C., Crane, B. R., and Stuehr, D. J. (2003a). A tetrahydrobiopterin radical forms and then becomes reduced during $N^W$-hydroxy arginine oxidation by nitric-oxide synthase. *Chem. Rev.* **103,** 2365–2383.

Wei, C.-C., Wang, Z.-Q., Hemann, C., Hille, R., and Stuehr, D. J. (2003b). *J. Biol. Chem.* **278,** 46668–46673.

Werner, E. R., Gorren, A. C. F., Heller, R., Werner-Felmayer, G., and Mayer, B. (2003). Tetrahydrobiopterin and nitric oxide: Mechanistic and pharmacological aspects. *Exp. Biol. Med.* **228,** 1291–1302.

## [39]  Antisense-Mediated Knockdown of iNOS Expression in the Presence of Cytokines

*By* Karsten Hemmrich, Christoph V. Suschek, and Victoria Kolb-Bachofen

### Abstract

The impact of nitric oxide (NO) synthesized after activation by proinflammatory cytokines and/or bacterial products by an inducible NO synthase (iNOS) is still contradictory. Various methods to inhibit iNOS expression or activity have been established. A relatively new approach to inhibit iNOS-derived NO production is the antisense (AS) technique, which theoretically provides a specific and efficient method for inhibiting gene expression and function. This chapter focusses on the application of iNOS-specific AS-oligodeoxynucleotide (ODN) and highlights some of the pitfalls that must be considered to use this technique effectively.

### Introduction

Various methods to inhibit inducible nitric oxide synthase (iNOS) expression or activity are established that allow for elucidating a protective versus destructive role of high-output NOS during various stresses. A relatively new method to inhibit iNOS-derived NO production is the antisense (AS) technique, which theoretically provides a specific, rapid, and potentially high-throughput method for inhibiting gene expression (Stein, 2001). The concept of blocking the expression of a single gene by using AS-oligodeoxynucleotides (ODNs) is based on studies in the late 1960s proving that synthetic AS-ODNs indeed act in a sequence-specific manner. Today, the principal fields of AS-ODN application are the investigation of gene function by loss-of-function or decrease-of-function analyses and the development of AS drugs for therapeutic applications. In this chapter, we detail methods of how to apply iNOS-specific AS-ODNs and highlight some of the pitfalls of this technique.

### Design of Oligonucleotides

The AS technique offers an interesting approach, especially in the setting of closely related members of a gene family like iNOS and the two other NOSs. However, to ensure the specificity of inhibition, the design

METHODS IN ENZYMOLOGY, VOL. 396
0076-6879/05 $35.00
DOI: 10.1016/S0076-6879(05)96039-4

of the ODN is important. Table I gives several examples of AS-ODNs and control ODNs used for iNOS targeting. However, if you decide to create your own sequence, here are some points that should be considered:

- An ODN sequence resembling a sense sequence of the cellular genome may also bind transcription factors, thereby trapping nucleic acid–binding proteins, resulting in less specific inhibition of translation (Bielinska *et al.*, 1990).
- G-quartets have been reported to be antiproliferative, and they inhibit cell–cell and virus–cell interactions.
- CpG motifs, in which the CG residues are flanked by two purins on the 5'-end and two pyrimidines on the 3'-end, induce activation of Toll-like receptor 9–expressing cells (Bielinska *et al.*, 1990).
- Palindromic sequences of six or more bases induce $\alpha$- and $\gamma$-interferon production (Yamamoto *et al.*, 1994).

Concerning length of ODNs, it is important to know that the minimal length of a particular messenger RNA (mRNA) sequence specific for one species only out of the entire mRNA population is between 11 and 14 units (Ghosh and Cohen, 1992). Depending on temperature, ionic composition of the system, and base composition, an ODN of 14–20 bases offers optimal specific hybridization properties (Ghosh and Cohen, 1992).

Experimental Design

We describe the application of AS-ODN for rat aorta endothelial cells (ECs). For AS experiments, these cells are best cultured in six-well tissue culture plates in OPTIMEM I Reduced Serum Medium (Gibco, Catalogue No. 31985) [20% fetal calf serum (FCS)], seeded at a density of 200,000 cells/well and incubated overnight to allow attachment. The next day, culture supernatants are replaced by fresh serum-free OPTIMEM medium, and lipid-encapsulated AS-ODN and control-ODN are added. Cytokine activation is then performed, and inhibition of NO formation is confirmed on mRNA, protein, and nitrite level 24 h later.

*Procedure for ODN Incorporation*

ODNs are typically shipped as lyophilized DNA-Na salt (25 or 100 nmol). The pellet may stick to the cap of the tube, so start with a short spin in a microcentrifuge before opening. To produce a 100-$\mu M$ ODN solution, add 100 $\mu$l of $1 \times$ TE buffer (10 m$M$ Tris–HCl, 1 m$M$ EDTA, pH 8.0) per 10 nmol of ODN and vortex briefly. ODNs are then ready to use or should be stored at $-20°$.

TABLE I

ANTISENSE AND CONTROL SEQUENCES USED IN EXPERIMENTS AIMED AT iNOS KNOCKDOWN

| Cell type/animal | In vivo/ in vitro | Antisense sequences (ODN or PNA) | Controls: scrambled, nonsense, sense, missense sequences (ODN or PNA) | Reference |
| --- | --- | --- | --- | --- |
| Rat pulmonary artery smooth muscle cells | In vitro | 5-AAACTTCCA-GGGGCAAGC-3 | 5-GCTTGCCCCTGGAAGTTT-3 | Thomae et al., 1993 |
| Macrophages (bone marrow derived) | In vitro | 5-CTTCCAGGGGCAA-GCCATGTCTGAG-3 5-GGACTTGCAA-GTGAAATCC | 5-TCAGACATGGCTTGCCCCTGGAAG-3 5-CATCGGATTTCACTTGCAAGTCC-3 | Flesch et al., 1994 |
| Liver cells from male Wistar rats | In vitro | 5-GTGCTAATGCGG-AAGGTCATG-3 | 5-CATGACCTTCCGCATTAGCAC-3 sense | Kurose et al., 1996 |
| BSC-1 African green monkey kidney cells | In vitro | 5-ACAGGCCATCTCT-ATGGATTTACA-3 (bp 85-62 on human iNOS cDNA) | 5-TGTAAAGCCATACAGATGGCCTGT-3 (sense, bp 62-85) 5-TGTCCAATTAGCTCCGAGTCATAC-3 | Peresleni et al., 1996 |
| J774.1A mouse macrophage cell line | In vitro | | | Rothe et al., 1996 |
| Mouse mixed glial cell cultures from cerebral cortex of SJL/J-mice | In vitro | 5-CTAAGTTCAAAA-GCTGGGCAT-3 | 5-ATGCCCAGCTTTTGAACTTAG-3 5-AGCTAGTTACAGTGCAAGTCA-3 | Ding et al., 1996 |
| Mouse peritoneal macrophages C3H/HeN | In vitro | 5-TCCAGGGGCAA-GCCATGTCT-3 | 5-AGGTCCCCGTTCGGTACAGA-3 5-CTGCGAGTCGCACATTGAGC-3 5-TCTGTACCGAACGGGGACCT-3 5-TCTGGACCCAATGGGGACCT-3 5-TCCTGGGGCAAACCAGGTCT-3 | Arima et al., 1997 |

(continued)

TABLE I (continued)

| Cell type/animal | In vivo/ in vitro | Antisense sequences (ODN or PNA) | Controls: scrambled, nonsense, sense, missense sequences (ODN or PNA) | Reference |
|---|---|---|---|---|
| Macrophage- and T-cell depleted bone marrow cells | In vitro | 5-GGTGCTGCTTGTTAGG-AGGTCAAGTAAAGGGC-3 | 5-TGGCCCAGAAGGGGGTGCTGCATGCGGTGCAC-3 | Selleri et al., 1997 |
| RAW 264.7 murine macrophages | In vitro | 5-CCAGGGGCAAGCCATGTCT G-3 (bp 251-70); 5-CAAGCCATGTCTGAGACTT T-3 (bp 244-63); 5-GGGCAAGCCATG-3 (bp 254-70); 5-AAGCCATGTCTG-3 (bp 259-70); 5-AAGGGCA-3 (bp 253-259) | 5-GACGTGCGAGTCAGCACTGC-3 random; 5-CAGACATGGCTTGCCCCTGG-3 sense | Bilecki et al., 1997 |
| Mouse macrophage cell line J774.2 | In vitro | bp 2476-2969 | | Cartwright et al., 1997 |
| Female SJL mice | In vivo | 5-CAAGCCATGTC-TGAGACTTTG-3 | 5-CAAAGTCTCAGACATGGCTTG-3 | Ding et al., 1998 |
| Mouse macrophage cell line RAW 264.7 | In vitro | 5-AATTAAGCTTGCAG-CTAAGTATTAGAG-3 | 5-AATTAGATCTCACCTTGGTGAAGGGACTGAGC-3 | Giovine et al., 1998 |
| Rat peripheral blood natural killer (NK)-cells and spleen-NK cells | In vitro | 5-CTTCAGAGTCT-GCCCATTGCT-3 | 5-TCTCAGTGAGCCCTCATTCTG-3 | Cifone et al., 1999 |
| Human breast cancer cell line MCF-7 | In vitro | 5-AAATTTCCAA-GGACAGGC-3 | 5-GCCTGTCCTTGGAAATTT-3 | Binder et al., 1999 |
| Murine C3H 10T1/2 fibroblasts | In vitro | 5-GAACGGGGACCTTCA-3 (bp 260-74); 5-ACCGAGGGGCGTCGA-3 (bp 402-16); 5-GGTCGGCGGTGGTGGG-3 (bp 2480-94); 5-TTCTCCGACGGGGGGG-3 (bp 2686-700) | 5-CACTGTTGACTGGGG-3 (nonsense); 5-ATCGGACGCAGGCTA-3 (missense) | Lesoon-Wood et al., 1999 |

| Cell/organism | In vitro / In vivo | Sequence 1 | Sequence 2 | Reference |
|---|---|---|---|---|
| Murine endothelial cell line s-End-1 | In vitro | | | Cartwright et al., 2000 |
| A7r5 vascular smooth muscle cells | In vitro | 5-CAGGGGCAA-GCCATGTC-3 | 5-CACCGCCATGGCATCTG-3 | Ishigami et al., 2000 |
| Cultured human vascular endothelial cells | In vitro | | | Tanjoh et al., 2000 |
| C57BL/J6 mice | In vivo | 5-CACCTCCAACACAAGATC-3 | 5-CCTTCGTACCCTTTTTCC-3 | Dick et al., 2001 |
| C6 glioma cells | In vitro | | | Yin et al., 2001 |
| Male Sprague-Dawley rats | In vivo | 5-GGCAAGCCATGTCTG-3 | 5-ACCGACCGACGTGT-3 | Parmentier-Batteur et al., 2001 |
| Human colon carcinoma cell line HT-29 | In vitro | 5-CAGAAATTTCCAAGGACAGG CCAT-3 | | Chun et al., 2002 |
| Murine macrophages | In vitro | 5-Lys-CCTTTTCCTCTTTC-Gly-3 (PNA) | 5-Lys-CTTCTCCCTTTTTC-Gly-3 (PNA) | Chiarantini et al., 2002 |
| Lewis male rats | In vivo | 5-CTAAGCTCAAACGCTGGGCG T-3-NH$_2$ | 5-TGCGGGTCGCAAACTTGAATC-3-NH$_2$ | Voigt et al., 2002 |
| Rat aorta endothelial cells | In vitro | 5-TTTGCCTTATACTGTTCC-3 | 5-ACTACTACACTAGACTAC-3   5-ATATCCTTCCAGTACAG-3 | Hemmrich et al., 2003a,b |
| Murine osteoblastic MC3T3-E1 cells | In vitro | bp 52-264, GenBank M 84373 (Lyons et al., 1992) | | Abe et al., 2003 |
| Human fibroblasts | In vitro | 5-ACAGCTCAGTCCCTTCACCA A-3 | 5-TTGGTGAAGGGACTGAGAGCTGT-3 | Grasso et al., 2003; Renis et al., 2003 |
| Adult femal Sprague-Dawley rats | In vivo | 5-CTTCAGAGTCTGCCCATTGC T-3 | 5-TCTCAGTGAGCCCTCATTCTG-3 | Pearse et al., 2003 |
| Male Sprague-Dawley rats | In vivo | 5-GGCAAGCCATGTCTG-3 | 5-CGTCCCTATACGACC-3 | Steiner et al., 2004 |
| Mouse cholangiocytes 603 B cells (cell line) | In vitro | bp 207-641 | | Ishimura et al., 2004 |

To ensure ODN incorporation, lipid vehicles should be used. There are reports on ODN incorporation in the absence of transmembrane carriers (Arima *et al.*, 1997; Noiri *et al.*, 1996); however, most authors underline the necessity of or the improvement by uptake enhancers (Lappalainen *et al.*, 1997; Maus *et al.*, 1999). In terms of uptake and toxicity, best results were observed with Lipofectin Reagent (Gibco, Catalogue No. 18292-011). LipofectAMINE Reagent (Gibco, Catalogue No. 18324-012) is another frequently used lipid that also allows for good and long-lasting nuclear uptake of ODNs. However, LipofectAMINE reveals significant, concentration-dependent toxic effects and enhances iNOS expression in the presence of cytokines, with the net result of no inhibition of nitrite formation at the end of a respective experiment. We also examined FuGENE 6 Transfection Reagent (Roche, Catalogue No. 1 815 091) but found no ODN uptake under any condition tested. We, therefore, recommend the use of Lipofectin for iNOS-targeting AS experiments in primary cells. The following is a step-by-step procedure:

1. For preparation of lipid–ODN complexes, prepare the following solutions in sterile tubes:

Solution A1: 6 $\mu$l of the 100 $\mu M$ AS- and solution A2: control-ODN preparations are diluted separately into 100 $\mu$l OPTIMEM medium.
Solution B: For AS-ODN and control-ODN, two tubes are prepared and 6 $\mu$l of Lipofectin Reagent are diluted into 100 $\mu$l serum-free medium. If you decide to make lipid uptake enhancers yourself, it is of note that a Lipofectin solution of 1 $\mu$l ml$^{-1}$ has a final concentration of 0.75 $\mu M$ DOTMA and 0.68 $\mu M$ DOPE.

2. Solutions A and B are allowed to stand at room temperature for 40 min.

3. Solutions A and B are then combined, mixed by gentle pipetting, and incubated for another 15 min at room temperature to allow for micelle formation. *Note*: If ODNs are fluorescence labeled, for instance, with fluorescein-isothiocyanate (FITC) for tracking and confirmation of successful ODN uptake, light-protected tubes are recommended for preparing solution A and the mixture of solutions A and B.

4. In the meantime, supernatant of cells cultured overnight (8–12 h) should be replaced by 750 $\mu$l of fresh serum-free OPTIMEM medium for each well. *Note*: In contrast to the recommendations of the manufacturer, cells should not be washed before medium replacement because this may reduce the final extent of antisense inhibition.

We find that the omission of FCS during ODN uptake is crucial for effective ODN incorporation.

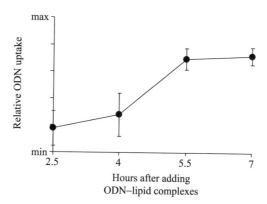

FIG. 1. Time-dependent uptake kinetics of oligodeoxynucleotides (ODNs) in endothelial cells. The kinetics of Lipofectin-mediated ODN uptake in endothelial cells were investigated during incubation in serum-reduced medium and addition of Lipofectin-encapsulated fluorescein-isothiocyanate (FITC)–labeled ODN. Data are from 18 individual experiments.

5. Next, the 200 $\mu$l mixture of solutions A and B is gently pipetted in small droplets onto cells, and cultures are further incubated at 37° in a $CO_2$ incubator.

6. Due to the uptake kinetics of Lipofectin-encapsulated ODN (Fig. 1), cell activation by cytokines [interleukin-1$\beta$ (IL-1$\beta$), tumor necrosis factor-$\alpha$ (TNF-$\alpha$), and interferon-$\gamma$ (IFN-$\gamma$), each at 1000 U ml$^{-1}$] should be performed 5.5 h after addition of lipid–ODN complexes. We analyzed mRNA, protein detection, and Griess assays 24 h after cytokine activation because rat cells show maximum activity at this point. Other species may have different kinetics. Cytokines should be premixed in a volume of 50 $\mu$l to give a final volume of 1 ml in every well after cytokine addition.

## Control for ODN Uptake

To microscopically ensure successful incorporation of ODN, we recommend use of AS-ODN with FITC labels (Biognostik, Göttingen, Germany). This allows for confirmation of ODN uptake and accumulation in the nucleus before adding cytokines (Fig. 2). It is reasonable to start monitoring ODN uptake 4–5 h after their addition because incorporation strongly increases at this time (Fig. 1). *Note:* In our experimental series, it was crucial to achieve an experimental condition that led to ODN accumulation in the nucleus. Conditions such as the presence of FCS during uptake, preventing nuclear transport, did not result in any inhibition.

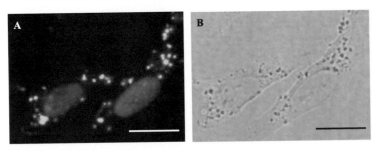

FIG. 2. Uptake of fluorescein-isothiocyanate (FITC)–labeled oligonucleotides (ODNs) into endothelial cells. (A) Culturing of endothelial cells in OPTIMEM serum-reduced medium in the presence of 6 $\mu$l Lipofectin in 1 ml OPTIMEM medium and ODN concentrations of 0.6 $\mu M$ leads to excellent uptake after 5.5 h. (B) Bright-field micrograph of the identical area. Bar in (A) and (B) = 50 $\mu$m.

## Analyses of mRNA Formation, Protein Expression, and Nitrite Production

Real-time polymerase chain reaction (PCR) has gained increasing importance. However, detailed experiments in our laboratory have shown that results are relatively identical compared to findings with conventional PCR. Because the newer technique is expensive and does not substantially increase information, we describe the conventional reverse transcription (RT) and PCR.

In detail, total cellular RNA (with 1 $\mu$g RNA/probe) can be prepared using the Omniscript RT Kit and RT carried out at 37° for 60 min with oligo(dT) (15mer) as primer. The cDNA (500 ng each) is used for PCR with primer ODN and amplification protocols as shown in Table II.

PCR products are subjected to electrophoresis on 1.8% agarose gels. Bands are visualized by ethidium bromide staining. Densitometric analysis of the visualized amplification products can be performed using the Kodak 1D software (Kodak, Stuttgart, Germany).

For Western blot analyses of the iNOS protein, cells are washed, scraped from the dishes, lysed, then transferred to a microcentrifuge tube, and boiled for 5 min in electrophoresis buffer. Proteins (30 $\mu$g/lane) are separated by electrophoresis in a 12% sodium dodecylsulfate (SDS)–polyacrylamide gel and transferred to nitrocellulose membranes. Further incubations are as follows: 2 h blocking buffer [2% bovine serum albumin (BSA), 5% nonfat milk powder, 0.1% Tween 20 in phosphate-buffered saline (PBS)], 1 h at 37° with a 1:2000 dilution of the monoclonal anti-iNOS antibody, and 1 h with a 1:2000 dilution of the secondary horseradish

TABLE II
LIST OF OLIGONUCLEOTIDES USED FOR iNOS OR GAPDH cDNA AMPLIFICATION

| Species/ product | Sequence[a] | Product size (bases) | Amplification conditions[b] | |
| --- | --- | --- | --- | --- |
| | | | Annealing | cycles |
| Rat iNOS | S ATGCCCGATGGCACCATCAGA | 394 | 60°, 30 s | 26 |
| | AS TCTCCAGGCCCATCCTCCTGC | | | |
| | AS AAGGCTTCCCCTGGAGAC | | | |
| Rat GAPDH | S CAACTACATGGTTTACATGTTCC | 416 | 60°, 30 s | 26 |
| | AS GGACTGTGGTCATGAGTCCT | | | |

[a] AS, antisense; S, sense.
[b] Polymerase chain reaction was started with 30 s at 94°, and amplification was always followed by a final incubation step at 72° for 10 min.

peroxidase–conjugated rabbit–anti-mouse–IgG antibody. Finally, blots are incubated for 5 min in ECL reagent (Pierce, Rockford, IL) and exposed to an autoradiographic film. To control for equal loading of total protein in all lanes, blots are also stained with a mouse anti–$\alpha$-tubulin antibody at a dilution of 1:2000.

For nitrite determination, cellular NO production is measured by quantifying the nitrite accumulation in culture supernatants of ECs using the diazotization reaction as modified by Wood et al. (1990) and $NaNO_2$ as standard.

*Antisense Inhibition on the mRNA, Protein, and Activity Level*

mRNA formation will only be partially inhibited ($\sim$20–30% inhibition). On the protein level, expression of iNOS will be strongly decreased by approximately 95% compared to controls. Nitrite levels, as determined 24 h after cytokine activation, usually showed a reduction of NO formation by around 60% compared to controls (*i.e.*, cytokines only).

It is of note that Lipofectin, if used as a control empty vehicle (*i.e.*, without the addition of ODNs), significantly increases the amounts of mRNA, protein, and nitrite in the presence of cytokines (Hemmrich et al., 2003a). This effect is not due to endotoxin contamination in the reagent and can be prevented by adding any kind of ODN. However, we recommend checking Lipofectin for endotoxin content by the limulus amebocyte lysate test before starting experiments.

# References

Abe, T., Hikiji, H., Shin, W. S., Koshikiya, N., Shima, S., Nakata, J., Susami, T., Takato, T., and Toyo-oka, T. (2003). Targeting of iNOS with antisense DNA plasmid reduces cytokine-induced inhibition of osteoblastic activity. *Am. J. Physiol. Endocrinol. Metab.* **285,** E614–E621.

Arima, H., Sakamoto, T., Aramaki, Y., Ishidate, K., and Tsuchiya, S. (1997). Specific inhibition of nitric oxide production in macrophages by phosphorothioate antisense oligonucleotides. *J. Pharm. Sci.* **86,** 1079–1084.

Bielinska, A., Shivdasani, R. A., Zhang, L. Q., and Nabel, G. J. (1990). Regulation of gene expression with double-stranded phosphorothioate oligonucleotides. *Science* **250,** 997–1000.

Bilecki, W., Okruszek, A., and Przewlocki, R. (1997). The effect of antisense oligodeoxynucleotides on nitric oxide secretion from macrophage-like cells. *Antisense Nucleic Acid Drug Dev.* **7,** 531–537.

Binder, C., Schulz, M., Hiddemann, W., and Oellerich, M. (1999). Induction of inducible nitric oxide synthase is an essential part of tumor necrosis factor-alpha-induced apoptosis in MCF-7 and other epithelial tumor cells. *Lab. Invest.* **79,** 1703–1712.

Cartwright, J. E., Johnstone, A. P., and Whitley, G. S. (1997). Inhibition of nitric oxide synthase by antisense techniques: Investigations of the roles of NO produced by murine macrophages. *Br. J. Pharmacol.* **120,** 146–152.

Cartwright, J. E., Johnstone, A. P., and Whitley, G. S. (2000). Endogenously produced nitric oxide inhibits endothelial cell growth as demonstrated using novel antisense cell lines. *Br. J. Pharmacol.* **131,** 131–137.

Chiarantini, L., Cerasi, A., Fraternale, A., Andreoni, F., Scari, S., Giovine, M., Clavarino, E., and Magnani, M. (2002). Inhibition of macrophage iNOS by selective targeting of antisense PNA. *Biochemistry* **41,** 8471–8477.

Chun, Y. J., Lee, S., Yang, S. A., Park, S., and Kim, M. Y. (2002). Modulation of CYP3A4 expression by ceramide in human colon carcinoma HT-29 cells. *Biochem. Biophys. Res. Commun.* **298,** 687–692.

Cifone, M. G., D'Alo, S., Parroni, R., Millimaggi, D., Biordi, L., Martinotti, S., and Santoni, A. (1999). Interleukin-2-activated rat natural killer cells express inducible nitric oxide synthase that contributes to cytotoxic function and interferon-gamma production. *Blood* **93,** 3876–3884.

Dick, J. M., Van Molle, W., Libert, C., and Lefebvre, R. A. (2001). Antisense knockdown of inducible nitric oxide synthase inhibits the relaxant effect of VIP in isolated smooth muscle cells of the mouse gastric fundus. *Br. J. Pharmacol.* **134,** 425–433.

Ding, M., Zhang, M., Wong, J. L., Voskuhl, R. R., and Ellison, G. W. (1996). Antisense blockade of inducible nitric oxide synthase in glial cells derived from adult SJL mice. *Neurosci. Lett.* **220,** 89–92.

Ding, M., Zhang, M., Wong, J. L., Rogers, N. E., Ignarro, L. J., and Voskuhl, R. R. (1998). Antisense knockdown of inducible nitric oxide synthase inhibits induction of experimental autoimmune encephalomyelitis in SJL/J mice. *J. Immunol.* **160,** 2560–2564.

Flesch, I. E., Hess, J. H., and Kaufmann, S. H. (1994). NADPH diaphorase staining suggests a transient and localized contribution of nitric oxide to host defence against an intracellular pathogen *in situ*. *Int. Immunol.* **6,** 1751–1757.

Ghosh, M. K., and Cohen, J. S. (1992). Oligodeoxynucleotides as antisense inhibitors of gene expression. *Prog. Nucleic Acid Res. Mol. Biol.* **42,** 79–126.

Giovine, M., Gasparini, A., Scarfi, S., Damonte, G., Sturla, L., Millo, E., Tonetti, M., and Benatti, U. (1998). Synthesis and characterization of a specific peptide nucleic acid that inhibits expression of inducible NO synthase. *FEBS Lett.* **426,** 33–36.

Grasso, S., Scifo, C., Cardile, V., Gulino, R., and Renis, M. (2003). Adaptive responses to the stress induced by hyperthermia or hydrogen peroxide in human fibroblasts. *Exp. Biol. Med. (Maywood)* **228**, 491–498.

Hemmrich, K., Suschek, C. V., Lerzynski, G., Schnorr, O., and Kolb-Bachofen, V. (2003a). Specific iNOS-targeted antisense knockdown in endothelial cells. *Am. J. Physiol. Cell Physiol.* **285**, C489–C498.

Hemmrich, K., Suschek, C. V., Lerzynski, G., and Kolb-Bachofen, V. (2003b). iNOS activity is essential for endothelial stress gene expression protecting against oxidative damage. *J. Applied Physiol.* **95**, 1937–1946.

Ishigami, M., Swertfeger, D. K., Hui, M. S., Granholm, N. A., and Hui, D. Y. (2000). Apolipoprotein E inhibition of vascular smooth muscle cell proliferation but not the inhibition of migration is mediated through activation of inducible nitric oxide synthase. *Arterioscler. Thromb. Vasc. Biol.* **20**, 1020–1026.

Ishimura, N., Bronk, S. F., and Gores, G. J. (2004). Inducible nitric oxide synthase upregulates cyclooxygenase-2 in mouse cholangiocytes promoting cell growth. *Am. J. Physiol. Gastrointest. Liver Physiol* **287**, G88–G95.

Kurose, I., Miura, S., Higuchi, H., Watanabe, N., Kamegaya, Y., Takaishi, M., Tomita, K., Fukumura, D., Kato, S., and Ishii, H. (1996). Increased nitric oxide synthase activity as a cause of mitochondrial dysfunction in rat hepatocytes: Roles for tumor necrosis factor alpha. *Hepatology* **24**, 1185–1192.

Lappalainen, K., Miettinen, R., Kellokoski, J., Jaaskelainen, I., and Syrjanen, S. (1997). Intracellular distribution of oligonucleotides delivered by cationic liposomes: Light and electron microscopic study. *J. Histochem. Cytochem.* **45**, 265–274.

Lesoon-Wood, L. A., Pierce, L. M., Lau, A. F., and Cooney, R. V. (1999). Enhancement of methylcholanthrene-induced neoplastic transformation in murine C3H 10T1/2 fibroblasts by antisense phosphorothioate oligodeoxynucleotide sequences. *Cancer Lett.* **147**, 163–173.

Lyons, C. R., Orloff, G. J., and Cunningham, J. M. (1992). Molecular cloning and functional expression of an inducible nitric oxide synthase from a murine macrophage cell line. *J. Biol. Chem.* **267**, 6370–6374.

Maus, U., Rosseau, S., Mandrakas, N., Schlingensiepen, R., Maus, R., Muth, H., Grimminger, F., Seeger, W., and Lohmeyer, J. (1999). Cationic lipids employed for antisense oligodeoxynucleotide transport may inhibit vascular cell adhesion molecule-1 expression in human endothelial cells: A word of caution. *Antisense Nucleic Acid Drug Dev.* **9**, 71–80.

Noiri, E., Peresleni, T., Miller, F., and Goligorsky, M. S. (1996). In vivo targeting of inducible NO synthase with oligodeoxynucleotides protects rat kidney against ischemia. *J. Clin. Invest.* **97**, 2377–2383.

Parmentier-Batteur, S., Bohme, G. A., Lerouet, D., Zhou-Ding, L., Beray, V., Margaill, I., and Plotkine, M. (2001). Antisense oligodeoxynucleotide to inducible nitric oxide synthase protects against transient focal cerebral ischemia-induced brain injury. *J. Cereb. Blood Flow Metab.* **21**, 15–21.

Pearse, D. D., Chatzipanteli, K., Marcillo, A. E., Bunge, M. B., and Dietrich, W. D. (2003). Comparison of iNOS inhibition by antisense and pharmacological inhibitors after spinal cord injury. *J. Neuropathol. Exp. Neurol.* **62**, 1096–1107.

Peresleni, T., Noiri, E., Bahou, W. F., and Goligorsky, M. S. (1996). Antisense oligodeoxynucleotides to inducible NO synthase rescue epithelial cells from oxidative stress injury. *Am. J. Physiol.* **270**, F971–F977.

Renis, M., Cardile, V., Grasso, S., Palumbo, M., and Scifo, C. (2003). Switching off HSP70 and i-NOS to study their role in normal and H2O2-stressed human fibroblasts. *Life Sci.* **74**, 757–769.

Rothe, H., Bosse, G., Fischer, H. G., and Kolb, H. (1996). Generation and characterization of inducible nitric oxide synthase deficient macrophage cell lines. *Biol. Chem. Hoppe Seyler* **377,** 227–231.

Selleri, C., Sato, T., Raiola, A. M., Rotoli, B., Young, N. S., and Maciejewski, J. P. (1997). Induction of nitric oxide synthase is involved in the mechanism of Fas-mediated apoptosis in haemopoietic cells. *Br. J. Haematol.* **99,** 481–489.

Stein, C. A. (2001). The experimental use of antisense oligonucleotides: a guide for the perplexed. *J. Clin. Invest.* **108,** 641–644.

Steiner, J., Rafols, D., Park, H. K., Katar, M. S., Rafols, J. A., and Petrov, T. (2004). Attenuation of iNOS mRNA exacerbates hypoperfusion and upregulates endothelin-1 expression in hippocampus and cortex after brain trauma. *Nitric Oxide* **10,** 162–169.

Tanjoh, K., Tomita, R., and Hayashi, N. (2000). Antisense oligodeoxynucleotides to human inducible nitric oxide synthase selectively inhibit induced nitric oxide production by human vascular endothelial cells: An experimental study. *Eur. J. Surg.* **166,** 882–887.

Thomae, K. R., Geller, D. A., Billiar, T. R., Davies, P., Pitt, B. R., Simmons, R. L., and Nakayama, D. K. (1993). Antisense oligodeoxynucleotide to inducible nitric oxide synthase inhibits nitric oxide synthesis in rat pulmonary artery smooth muscle cells in culture. *Surgery* **114,** 272–277.

Voigt, M., de Kozak, Y., Halhal, M., Courtois, Y., and Behar-Cohen, F. (2002). Downregulation of NOSII gene expression by iontophoresis of anti-sense oligonucleotide in endotoxin-induced uveitis. *Biochem. Biophys. Res. Commun.* **295,** 336–341.

Wood, K. S., Buga, G. M., Byrns, R. E., and Ignarro, L. J. (1990). *Biochem. Biophys. Res. Commun.* **170,** 80–88.

Yamamoto, T., Yamamoto, S., Kataoka, T., and Tokunaga, T. (1994). Ability of oligonucleotides with certain palindromes to induce interferon production and augment natural killer cell activity is associated with their base length. *Antisense Res. Dev.* **4,** 119–122.

Yin, J. H., Yang, D. I., Chou, H., Thompson, E. M., Xu, J., and Hsu, C. Y. (2001). Inducible nitric oxide synthase neutralizes carbamoylating potential of 1,3-bis(2-chloroethyl)-1-nitrosourea in c6 glioma cells. *J. Pharmacol. Exp. Ther.* **297,** 308–315.

# [40] Soluble Guanylyl Cyclase: The Nitric Oxide Receptor

*By* EMIL MARTIN, VLADIMIR BERKA,
AH-LIM TSAI, and FERID MURAD

## Abstract

Soluble guanylyl cyclase is recognized as the most sensitive physiologic receptor for nitric oxide. Binding of nitric oxide to the heme moiety of the cyclase induces its capacity to synthesize the second messenger cGMP. Although the changes in the state of the heme moiety upon exposure of enzyme to NO and its correlation to the stimulation of sGC catalytic activity are well documented, the exact mechanism of such coupling is not

METHODS IN ENZYMOLOGY, VOL. 396        0076-6879/05 $35.00

understood. Structure-functional studies are required to elucidate this process. In this chapter, we describe the method of expression and purification of recombinant human $\alpha_1/\beta_1$ isoform of sGC in insect cells, which can be a useful tool for such studies. Several approaches that enable characterization of the binding of NO to sGC heme moiety are also described.

## Introduction

Nitric oxide (NO) exercises multiple physiological functions, which include but are not limited to regulation of smooth muscle relaxation, platelet aggregation, and neurotransmission (Lucas et al., 2000). Soluble guanylyl cyclase (sGC) is one of the key proteins that play a crucial role in these processes. sGC is a heterodimeric protein composed of two subunits (Kamisaki et al., 1986). Two isoforms ($\alpha_1$ and $\alpha_2$) of the larger $\alpha$ subunit ($\sim$73–88 kDa depending on species) and two isoforms ($\beta_1$ and $\beta_2$) of the smaller $\beta$ subunit ($\sim$70 kDa) are found in vertebrates. So far only $\alpha_1/\beta_1$ and $\alpha_2/\beta_1$ heterodimers have been detected in vivo, with $\alpha_1/\beta_1$ being the most highly and ubiquitously expressed in vivo (Mergia et al., 2003; Zabel et al., 1998). C- terminal portions of $\alpha$ and $\beta$ subunits share extensive similarities among themselves and with catalytic domains of other guanylyl and adenylyl cyclases (Garbers, 2000). Under physiological conditions, only the heterodimer has catalytic activity.

In addition to the presence of both subunits, one protoporphyrin type IX ferrous heme moiety per heterodimer is required for NO activation. The heme moiety is coordinated by the histidine 105 residue of the $\beta$ subunit (Wedel et al., 1994; Zhao et al., 1998). Recent studies also suggested that Tyr135 and Arg139 residues of the $\beta$ subunit interact with the propionic groups of the heme moiety and are important for the stabilization of the protein heme interactions (Schmidt et al., 2004). Some structural studies indicated that the N-terminal domains of both subunits are necessary for successful and efficient incorporation of the heme moiety (Foerster et al., 1996) into a full-length heterodimer, whereas other studies suggest that N-terminal portion of $\alpha$ subunit is dispensable for this purpose (Koglin and Behrends, 2003). Because of the crucial role in binding heme and providing regulation by NO, the N-terminal regions of $\alpha$ and $\beta$ subunits are often referred to as the *regulatory domain of enzyme*. The catalytic domain of the enzyme is formed by the C-terminal portions of both subunits, and it converts guanosine triphosphate (GTP) into the second- messenger cyclic guanosine monophosphate (cGMP).

In the absence of any activators, the enzyme displays a relatively small activity with a turnover number of about 15–20 $min^{-1}$. Upon binding of the NO to the heme moiety, the catalytic activity of the enzyme is increased several hundred fold to a turnover rate of more than 1800 min$^{-1}$ due to a combined decrease in the $K_m$ for GTP and significant increase in the $V_{max}$. This highly activated state can be regarded as a high output state. The exact mechanism of this activation is not clearly understood. Various steady-state and kinetic spectroscopic studies (Makino *et al.*, 1999; Zhao *et al.*, 1999) demonstrated that the formation of nitrosyl heme results in the formation of a transient hexacoordinate complex, which is converted into a pentacoordinated nitrosyl heme as a result of the disruption of the coordinating bond with the histidine 105 of the $\beta$ subunit. Analysis of the kinetics of enzyme activation suggests that the enzyme is activated to its full potential only after the formation of pentacoordinated NO–heme complex (Zhao *et al.*, 1999). The sGC heme also readily binds carbon monoxide (CO) (Stone and Marletta, 1994). However, CO binding results in a hexacoordinate CO–heme complex, which correlates with only a slight (twofold to threefold) activation of the enzyme.

Although the correlation between the binding of NO to the heme moiety of sGC and the subsequent changes in the state of the heme and stimulation of sGC catalytic activity are well documented, there is little understanding of the exact nature of such coupling. Further structure-functional studies are required to elucidate the processes underlying this coupling. We describe the method of expression and purification of recombinant human $\alpha_1/\beta_1$ isoform of sGC in insect cells. Recombinant sGC, which preserved its functional properties, will be a useful tool for such studies. Several approaches that enable characterization of the binding of NO to sGC are also described.

## Expression of sGC in Sf9 Cells

### cDNA Isolation and Baculovirus Construction

The human sGC $\alpha_1$ and $\beta_1$ subunit's coding sequences were polymerase chain reaction (PCR)–amplified from the cDNA library from human cerebral cortex (Invitrogen) using the primers shown in Table I.

Primary PCR fragments are cloned into the pCR-Blunt plasmid according to the standard protocol suggested by the manufacturer to generate plasmids pCR-$\alpha$ and pCR-$\beta$. For purification purposes, several variant-containing hexahistidine tags were also generated. Using PCR, hexahistidine tags were introduced, either N-terminally directly following the methionine of the $\beta$ subunit or C-terminally in front of the stop codon of the $\alpha$ and $\beta$ subunit.

Secondary PCR fragments were inserted directionally into the EcoR I and Not I sites of pVL1392 plasmid to obtain pVL$\alpha$, pVL$\beta$, pVL$\beta^C$,

TABLE I
RT-PCR PRIMERS FOR HUMAN $\alpha_1$ AND $\beta_1$ SUBUNIT

| Subunit | 5' primer | 3' primer |
|---|---|---|
| $\alpha$ | CACCATGTTCTGCACGAGCTCAAGGATC | gcggccgcTCACTA↓GGGAAGTTTGGTGGAAGTC |
| $\beta$ | ACCATG↓TACGGATTTGTGAATCACGCC | gcggccgcAGATTCA↓GTACATCCTGCTTTGTTTCC |

Note: The first methionine triplet in 5' primer is indicated in bold fonts, while the stop codon is underlined. The arrow indicates the position of hexahistidine tag insertion (CATCACCATCACCATCAC for 5' and GTGATGGTGATGGTGATG for 3' primers). Sequences in italics indicate the Not I restriction site.

pVL$\beta^N$ plasmids (the superscripts indicate the position of the hexahistidine tag in the recombinant protein). Sf9 cells were transformed with these plasmids and BaculoGold DNA (Pharmingen) according to manufacturer's protocol to generate the recombinant baculoviruses. In another approach, the cDNA for the $\alpha$ subunit was cloned into the pAcHLT vector (Pharmingen), which fused with a 5-kDa N-terminal segment consisting of hexahistidine and a thrombin cleavage site. High-titer virus stocks were obtained from three amplification cycles using Sf9 cells cultured in monolayer according to a standard protocol. Viruses collected after the third amplification were used in further experiments. The second amplification viral stock was also stored at 4° in the dark and used to produce third-generation viruses, when necessary.

*Comments.* The efficiency of recombination (99.9%) permits the use of viral stocks directly without the plaque purification step. To ensure the quality of the working viral stocks, and to avoid the accumulation of mutation during multiple passages, recombinant viruses are regenerated from the original plasmid once in 12–18 mo.

### Expression of Recombinant sGC

Sf9 cells are cultured in Grace's insect medium supplemented with 10% (v/v) fetal calf serum (FCS), 1% (v/v) Pluronic 68, and 1% penicillin and streptomycin at 26.5–27.0°. Expression of sGC decreases significantly at temperatures higher than 27.5°. The cells are expanded in 500-ml cultures in 1-liter spinner flasks. The cells are either split or infected at a density of $2 \times 10^6$ cells/ml. The cells are not propagated for more than 3 wk. Cultures with excessive passages show great loss of cell viability and a significant decrease in protein yield. Typically, 2–3 liters of culture are co-transfected with both $\alpha$ and $\beta$ viruses at a multiplicity of infection of 5. The cells are cultured for 72 h after infection and collected by centrifugation at 800$g$. Higher centrifugation speeds should be avoided because it tends to damage the packed Sf9 cells. Typically, at this stage, the viability is about 90–95% with about $3 \times 10^6$ cells/ml.

*Comments.* If desired, the cells can be harvested 96 h after infection to increase the yield of protein. However, the viability of the cells should be strictly monitored because it could significantly decrease between 72 and 96 h, resulting in low yield of purified protein.

### Protein Purification

After the cells are harvested, the cell pellet is resuspended in 30–50 ml of buffer A [50 m$M$ TEA, pH 7.4, 10% glycerol, 4 m$M$ MgCl$_2$, and 2 m$M$ dithiothreitol (DTT) with 0.5 m$M$ EDTA, 0.5 m$M$ EGTA, 1 m$M$

phenylmethylsulfonylfluoride (PMSF), and 5 mg/ml each of pepstatin A, leupeptin, aprotinin, and chymostatin]. The cells are disrupted on ice by sonication on a W-225R Branson sonicator (Heat Systems-Ultrasonics) for three cycles, 10 pulses of 0.6 s each, at power setting of 4.5. The efficiency of sonication should be tested under the microscope to achieve a good purification yield. The lysate is then centrifuged at $10,000g$ for 10 min. The supernatant is then subjected to a $100,000g$ centrifugation for 90 min.

The $100,000g$ supernatant is applied at a rate of 1.5 ml/min onto a 30-ml DEAE-FF Superose column, which is equilibrated with buffer B (50 m$M$ TEA, pH 7.4, 10% glycerol, 4 m$M$ MgCl$_2$). The column is extensively washed after the sample loading to remove residual DTT and EDTA. The proteins are eluted with a 60-ml gradient of 0–500 m$M$ NaCl. Fractions containing NO-stimulated sGC activity, which are visibly yellow at this stage, are collected and used for the next purification step.

The crude sGC fraction collected from the DEAE- Sepharose is loaded onto a 20-ml Ni-agarose column with HisBind resin (Invitrogen) at a rate of 1 ml/min. The column is washed with 50 ml of buffer B containing 5 m$M$ imidazole, followed by an extensive wash with buffer B containing 50 m$M$ imidazole until a stable baseline is observed. The sGC enzyme is then eluted with buffer B containing 175 m$M$ imidazole. The yellow fractions are pooled together, added with 4 m$M$ DTT, and used for further purification. The enzyme is 90–95% pure at this stage but is not very stable because of the high concentration of imidazole. Figure 1A shows the Coomassie blue-stained sodium dodecylsulfate (SDS)–polyacrylamide gel of the Sf9 cell supernatant (Lane 1), sGC containing fractions of DEAE-FF Sepharose eluate (Lane 2), and pure sGC after Ni- chelating chromatography (Lane 3). This purification procedure works equally well for $\alpha\beta$ heterodimers containing one hexahistidine tag at the C-or N- terminus of the $\alpha$ subunit or the C-terminus of the $\beta$ subunit (Fig. 1B). Typically, 1.5–2.0 mg of sGC is obtained for each liter of Sf9 culture. Although the sample purity is similar, the yield for the heterodimer with the N-terminally tagged $\beta$ subunit is significantly lower (data not shown).

To remove the imidazole and further purify the enzyme, the sample is loaded onto a 10-ml HiTrap DEAE-FF Sepharose column (Amersham) equilibrated with buffer C (50 m$M$ TEA, pH 7.4, 10% glycerol, 4 m$M$ MgCl$_2$ and 5 $M$ DTT with 0.5 m$M$ of EDTA and EGTA), washed, and eluted with a gradient of 0–350 m$M$ NaCl. The yellow fractions are collected and, when necessary, concentrated on a YW-100 Centricon concentrator (Millipore).

As we reported earlier (Lee $et$ $al.$, 2000), heterodimer with $\beta$ subunit carrying the hexahistidine tag at the N-terminus is heme deficient and, thus, is insensitive to NO activation. All other variants carrying one hexahistidine tag display an activity of at least 14 $\mu$mol/min/mg protein

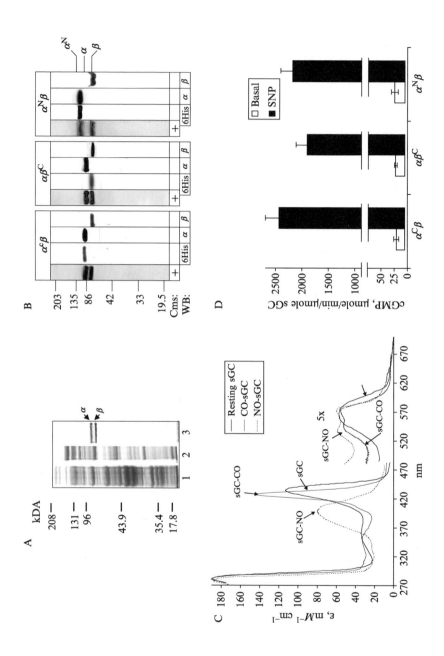

(a turnover rate of 1900/min) and more than 100-fold activation by NO-donor sodium nitroprusside (Fig. 1D). The heme content determined by pyridine hemochromogen using $\varepsilon_{557-540}$ value of 24.5 m$M^{-1}$ cm$^{-1}$ (de Duve, 1948) showed a stoichiometry of 0.6–0.9 heme/sGC heterodimer (data not shown). Data on enzyme activity and recovery in a representative purification from 6.3 × 10$^9$ Sf9 cells are summarized in Table II.

*Comments.* The purified enzyme is more stable at a concentration of more than 1 mg/ml. After addition of 5 m$M$ DTT, the sample can be stored without a significant ($\leq$30%) loss of original activity at $-80°$ for at least 6 mo. The preparation that is free of exogenously added thiols loses activity during storage within 2 wk.

### Binding of NO and CO to sGC

*Ultraviolet–visible (UV–vis) spectroscopic analysis* is the easiest approach to record the binding of gaseous ligand to sGC. As described previously (Gerzer *et al.*, 1981; Stone and Marletta, 1994) and represented in Fig. 1C, the resting enzyme shows a Soret band at 431 nm ($\varepsilon = 110$ m$M^{-1}$ cm$^{-1}$) and a broad $\alpha/\beta$ band around 550–590 nm ($\varepsilon = 11$ m$M^{-1}$ cm$^{-1}$). This spectrum is consistent with a high-spin 5- coordinated histidine-ligated heme (Antonini and Brunori, 1971). For most preparations, the ratio of 280/431 nm absorbance is 1.7, which is well in the range of reported ratio of 1.4 for the recombinant rat sGC (Brandish *et al.*, 1998), 1.8 for recombinant human sGC (Kosarikov *et al.*, 2001), or 2.4 for the bovine lung enzyme (Friebe *et al.*, 1997). To form the CO–sGC complex, 500 $\mu$l of sGC is placed in a 5-ml tightly sealed vial and depleted of oxygen by passing a stream of argon over the sample for 10 min. A stream of CO is then passed over the

---

FIG. 1. Purification of nitric oxide–sensitive sGC. (A) Coomassie staining of the lysate of Sf9 cells expressing sGC (Lane 1), sGC-containing DEAE-FF Sepharose eluate (Lane 2), and Ni-agarose eluate after 175 m$M$ imidazole wash (Lane 3). (B) DEAE-FF Sepharose/Ni-agarose purification generates high-purity preps for sGC with hexahistidine tag on the N- and C-terminus of $\alpha$ subunit or C-terminus of $\beta$ subunit. The lower mobility of the $\alpha^N$ subunit reflects the addition of 30 amino acid residues containing the thrombin cleavage site. Cms, Coomassie staining of 5 $\mu$g of purified enzyme; WB, Western blotting performed with Qiagen monoclonal antibodies raised against hexahistidine (6His), polyclonal antibodies raised against the residues 605–619 of the $\beta$ subunit or residues 634–647 of the $\alpha$ subunit. (C) Optical properties of the 4 $\mu$M sGC recorded on the HP8453 spectrophotometer. Solid line represents the ultraviolet–visible (UV–vis) spectra of untreated sGC, the dashed line is the spectra of sGC treated with CO, and the dotted line is the spectra of sGC treated with 50 $\mu$M of NO-donor DEA-NO. (D) Specific activity of various flagged sGC proteins represented as the turnover rate in $\mu$M cGMP/min/$\mu$M sGC. Activities of untreated (basal) or sodium nitroprusside-treated (SNP) enzyme are shown.

TABLE II
PURIFICATION OF RECOMBINANT HUMAN-SOLUBLE GUANYLYL CYCLASE

| Fraction | Volume (ml) | Protein (mg) | Total activity[a] (nmol/min) | Specific activity (nmol/min/mg) | Purification fold | Yeild (%) |
|---|---|---|---|---|---|---|
| 100,000g supernatant | 75 | 2625 | 126,000 | 48 | 1 | 100 |
| DEAE-FF Sepharose | 32 | 864 | 88,130 | 102 | 2.1 | 69.4 |
| Ni-agarose | 16 | 5 | 72,000 | 12,000[b] | 250 | 57.1 |
| DEAE-FF Sepharose | 7.2 | 4.8 | 86,400 | 18,000[b] | 375 | 68.5 |

[a] Only the activity in the presence of 100 $\mu M$ SNP is presented.
[b] The increase in specific activity in the last step reflects the inhibitory effect of 175 m$M$ imidazole, which is removed after DEAE-FF Sepharose.

sample for 5–10 s with constant shaking. The sample is transferred by a gas-tight syringe to a CO-filled sealed cuvette, and the UV–vis spectra are recorded in the 350–700 nm range. In a CO-bound form, sGC displays a Soret band at 423 nm ($\varepsilon = 143$ m$M^{-1}$ cm$^{-1}$) and a split of the broad $\alpha/\beta$ band into distinct 567 nm $\alpha$ ($\varepsilon = 11$ m$M^{-1}$ cm$^{-1}$) and 541 nm $\beta$ ($\varepsilon = 11$ m$M^{-1}$ cm$^{-1}$) bands. The measurements represented in Fig. 1C are in good agreement with previous results (Stone and Marletta, 1994) and are characteristic for a hexacoordinated CO–heme complex. To analyze the changes in spectroscopic properties of sGC upon NO binding, NO donors are used because under aerobic conditions, the affinity of sGC for NO is much higher than for oxygen. For this purpose, a 1-m$M$ stock solution of diethylamide adduct of NO (DEA/NO) donor is prepared in 10 m$M$ NaOH. As represented in Fig. 1C, addition of DEA/NO donor to sGC preparation results in the spectra typical for the formation of NO–heme complex. As described previously (Stone and Marletta, 1994) and shown in Fig. 1C, the NO–heme complex has a Soret peak at 399 nm ($\varepsilon = 80$ m$M^{-1}$ cm$^{-1}$) and 572 nm $\alpha$ ($\varepsilon = 12$ m$M^{-1}$ cm$^{-1}$) and 537 nm $\beta$ ($\varepsilon = 11.5$ m$M^{-1}$ cm$^{-1}$) bands, indicating the formation of a pentacoordinated NO–heme complex.

## EPR Analysis of sGC Heme

EPR spectra of resting ferrous sGC, oxidized sGC, and its ferrous–NO complex are shown in Fig. 2. To record the EPR spectra of the resting ferrous sGC heme, 300 $\mu$l of 6 $\mu M$ sGC is loaded in an EPR tube, frozen in dry ice/ethanol, and transferred into liquid nitrogen. The heme component, having fast relaxation, has to be measured at liquid helium temperature. The reduced sGC showed only small signals at $g = 4.3$, as nonspecific iron, and $g = 2$ region, as nonspecific organic radical, indicating a species with an even number of electrons (Fig. 2A, Spectrum a). To prepare a

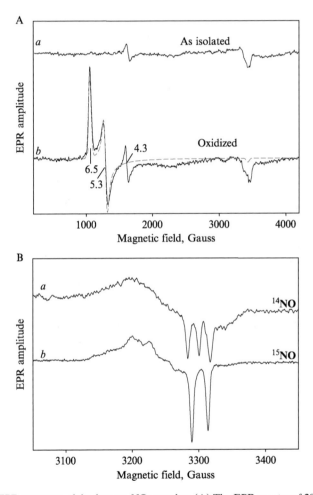

FIG. 2. EPR spectrum of the ferrous–NO complex. (A) The EPR spectra of 20 $\mu M$ resting ferrous sGC (a) and $K_3Fe(CN)_6$-oxidized ferric sGC (b) were recorded at liquid helium temperature (10 K) on a Brucker EMX EPR spectrometer under the following conditions: frequency, 9.61 gHz; power, 1 mW; modulation amplitude, 10 $G$; time constant, 0.33 s. (B) The EPR spectrum of the sGC NO–heme complex prepared anaerobically with [14]NO (a) and [15]NO (b) measured at liquid nitrogen temperature (120 K). EPR conditions are similar to those in (A) except frequency (9.28 gHz) and modulation amplitude (1 $G$).

sample of ferric-heme sGC, the sample is oxidized by titration with potassium ferricyanide $K_3Fe(CN)_6$. As the ferric-heme sGC has a Soret maximum at 393 nm, the oxidation of heme moiety is monitored by measuring the UV–vis spectra after each addition of potassium $K_3Fe(CN)_6$. $K_3Fe$ $(CN)_6$ is added until no increase in the 393-nm peak is observed. The $K_3Fe(CN)_6$-oxidized sGC showed a high-spin rhombic heme structure with anisotropic $g$ values at 6.5 and 5.3 and can be properly simulated by a rhombic heme center with 7.5% rhombicity (Fig. 2A, Spectrum b). These spectral characteristics are similar to earlier published data (Stone et al., 1996).

EPR spectrum of the fully converted ferrous–NO complex, in contrast, can be measured at liquid nitrogen temperature because of its much longer life time than the ferric sGC heme. To obtain this spectra, 20 $\mu M$ sGC heme was prepared anaerobically, as described earlier, and transferred into an anaerobic chamber (Coy Laboratory). Saturated solutions of purified $^{14}NO$ (Matherson) and $^{15}NO$ (Cambridge) were also prepared in a sealed vial, as described earlier, and transferred into an anaerobic chamber. Two hundred microliters of 20 $\mu M$ sGC were mixed in an EPR tube with 100 $\mu M$ of saturated (1 m$M$) NO solution and processed as described earlier. The ferrous–NO complex formed with either $^{14}NO$ or $^{15}NO$ exhibits EPR spectra predominantly for the 5-coordinate NO complex (Fig. 2B), highlighted by the nitrogen nuclear hyperfine features. The $^{14}NO$-ferrous sGC has three $g$ principal values at 2.08, 2.03, and 2.01 ($g_x/g_y/g_z$) with nuclear hyperfine splittings of 21, 18, and 17 $G$, respectively. On the other hand, EPR of $^{15}NO$-ferrous sGC has 2.08, 2.04, and 2.01 for the three $g$ principal values and 21, 21, and 22 $G$ for the corresponding nuclear hyperfine splittings. The conversion from triplet to doublet for the $g_z$ component is a manifestation for 5-coordinate NO–ferrous heme with NO as the distal ligand, and the proximal ligand is detached from heme iron upon NO binding. EPR analysis of our purified recombinant human sGC presented in Fig. 2 is in good agreement with previous EPR data of bovine and rat sGC (Makino et al., 1999; Stone et al., 1996).

*Monitoring the Kinetics of the Pentacoordinated Nitrosyl Heme Formation*

Previously reported time-resolved studies (Makino et al., 1999; Zhao et al., 1999) of the kinetics of NO binding to sGC heme and data presented here (Fig. 2) indicate that the formation of sGC pentacoordinated nitrosyl heme is a biphasic process. We describe here the method that permits monitoring of the dynamics of NO–heme complex formation.

*Preparation of NO Solutions.* For this analysis, pure NO gas (Matheson) is further purified by passing it through a *U*-shaped tube filled with dry KOH pellets. Ten milliliters of 50 m*M* TEA, pH 7.4, buffer is first made anaerobic by bubbling nitrogen gas for 10 min and then is saturated with purified NO (>99% purity) by bubbling for an additional 10 min to obtain a 1-m*M* solution of NO. The 60-$\mu M$ working solution is prepared by adding 600 $\mu$l of 1 m*M* NO stock solution in a Hamilton gas-tight syringe to 10 ml of $N_2$-saturated 50 m*M* TEA, pH 7.4, placed in a gas-tight syringe. During transfer, the gas phase in this syringe is kept at a minimum to avoid gas exchange and loss of NO. After injection of stock NO solution, the reaction syringe is sealed, and the working solution is used immediately for measurements.

*Preparation of Anaerobic sGC.* Protein samples in 50 m*M* TEA (pH 7.4) and 250 m*M* NaCl and 10% glycerol are made anaerobic in a tonometer by five cycles of vacuum (30 s)/argon (5 min) replacement using a glass anaerobic train.

*Time-Resolved Measurement of sGC NO–Heme Complex Formation.* These measurements are conducted using a Bio-SEQUENTIAL DX-18MV stopped- flow instrument (Applied Photophysics, Leatherhead, UK) equipped with a rapid scan photodiode array detector. Before measurements, the stopped-flow apparatus is made anaerobic by filling the instrument for 3–12 h with anaerobic 0.1 *M* pyrophosphate (pH 8.3) buffer containing saturated sodium dithionate. The 6-$\mu M$ anaerobic sGC solution is mixed with 60-$\mu M$ NO solutions at 24° in the stopped-flow instrument. The changes in the UV–vis spectra between 300 and 700 nm are recorded using the photodiode array detector (Applied Photophysics, Leatherhead, UK) at 400 scans/s. The time-dependent changes in the spectra are presented in Fig. 3A. As can be seen, a transient hexacoordinated NO-heme intermediate with absorbance maxima at 420 nm is formed within the dead time of the instrument. Previously reported estimations suggest a $k_{on} > 1.4 \times 10^8 \ M^{-1} \ s^{-1}$ for the initial binding step (Zhao *et al.*, 1999). The hexacoordinated NO complex is rapidly transformed into pentacoordinated NO complex with the maximum at 399 nm. Most of the reaction of 6-coordinate NO–heme complex formation occurred in the dead time of the stopped flow (~1.5 ms), and the observed changes in the 1-s reaction are essentially the conversion from 6-coordinate NO–heme complex to 5-coordinate NO–heme complex with isosbestic points at 406, 460, 530, and 560 nm. Global analysis using a single irreversible step reaction nicely

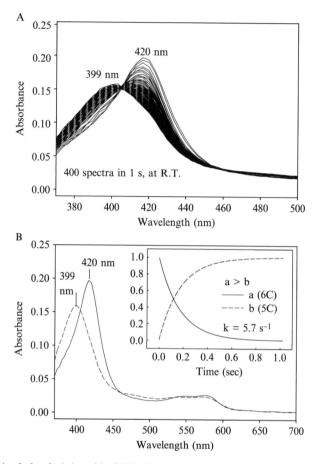

Fig. 3. Analysis of nitric oxide (NO) binding by a stopped-flow method. (Top) Time-resolved analysis of spectral changes of the ferrous sGC after mixing 6 $\mu M$ was with 60 $\mu M$ NO, and 400 spectra were recorded in 1 s at 24°. (Bottom) Global analysis of the data in the top panel to obtain resolved spectral species involved in the one-stage conversion from 6-c NO–heme intermediate (a) to 5-c NO–heme complex (b). (Inset) Resolved kinetics for $a$ and $b$ species. Rate is determined to be 5.7 s$^{-1}$.

fits the data and resolves the two spectroscopic components with a conversion rate of 5.7 s$^{-1}$ (Fig. 3B). It was also observed that this stage of reaction is linearly dependent on NO concentration, similar to the previous finding by Zhao *et al.* (1999). The meaning of this NO concentration dependence is not clear but gives mechanistic possibility for NO to displace the proximal histidine directly.

## Acknowledgments

This work was supported in part by HL64221 and GM61731 grants from the National Institutes of Health (NIH) (F. M.), the American Heart Association, Texas Affiliate grant 0465091Y (E. M.), and GM56818 from the NIH (A. -L. T.).

## References

Antonini, E., and Brunori, M. (1971). The derivatives of ferrous hemoglobin and myoglobin. *In* "Hemoglobin and Myoglobin in Their Reaction with Ligands" (Tatum and Neuberger, eds.). North-Holland Publishing Co., Amsterdam.

Brandish, P. E., Buechler, W., and Marletta, M. A. (1998). Regeneration of the ferrous heme of soluble guanylate cyclase from the nitric oxide complex: Acceleration by thiols and oxyhemoglobin. *Biochemistry* **37**, 16898–16907.

de Duve, C. (1948). A spectrophotomeric method for the simultaneous determination of myoglobin and hemoglobin in extracts of human muscle. *Acta Chem. Scand.* **2**, 264–289.

Foerster, J., Harteneck, C., Malkewitz, J., Schultz, G., and Koesling, D. (1996). A functional heme-binding site of soluble guanylyl cyclase requires intact N-termini of alpha 1 and beta 1 subunits. *Eur. J. Biochem.* **240**, 380–386.

Friebe, A., Wedel, B., Harteneck, C., Foerster, J., Schultz, G., and Koesling, D. (1997). Functions of conserved cysteines of soluble guanylyl cyclase. *Biochemistry* **36**, 1194–1198.

Garbers, D. L. (2000). The guanylyl cyclase receptors. *Zygote* **8**(Suppl. 1), S24–S25.

Gerzer, R., Bohme, E., Hofmann, F., and Schultz, G. (1981). Soluble guanylate cyclase purified from bovine lung contains heme and copper. *FEBS Lett.* **132**, 71–74.

Kamisaki, Y., Saheki, S., Nakane, M., Palmieri, J. A., Kuno, T., Chang, B. Y., Waldman, S. A., and Murad, F. (1986). Soluble guanylate cyclase from rat lung exists as a heterodimer. *J. Biol. Chem.* **261**, 7236–7241.

Koglin, M., and Behrends, S. (2003). A functional domain of the alpha1 subunit of soluble guanylyl cyclase is necessary for activation of the enzyme by nitric oxide and YC-1 but is not involved in heme binding. *J. Biol. Chem.* **278**, 12590–12597.

Kosarikov, D. N., Young, P., Uversky, V. N., and Gerber, N. C. (2001). Human soluble guanylate cyclase: Functional expression, purification and structural characterization. *Arch. Biochem. Biophys.* **388**, 185–197.

Lee, Y. C., Martin, E., and Murad, F. (2000). Human recombinant soluble guanylyl cyclase: Expression, purification, and regulation. *Proc. Natl. Acad. Sci. USA* **97**, 10763–10768.

Lucas, K. A., Pitari, G. M., Kazerounian, S., Ruiz-Stewart, I., Park, J., Schulz, S., Chepenik, K. P., and Waldman, S. A. (2000). Guanylyl cyclases and signaling by cyclic GMP. *Pharmacol. Rev.* **52**, 375–414.

Makino, R., Matsuda, H., Obayashi, E., Shiro, Y., Iizuka, T., and Hori, H. (1999). EPR characterization of axial bond in metal center of native and cobalt-substituted guanylate cyclase. *J. Biol. Chem.* **274**, 7714–7723.

Mergia, E., Russwurm, M., Zoidl, G., and Koesling, D. (2003). Major occurrence of the new alpha2beta1 isoform of NO-sensitive guanylyl cyclase in brain. *Cell Signal* **15**, 189–195.

Schmidt, P. M., Schramm, M., Schroder, H., Wunder, F., and Stasch, J. P. (2004). Identification of residues crucially involved in the binding of the heme moiety of soluble guanylate cyclase. *J. Biol. Chem.* **279**, 3025–3032.

Stone, J. R., and Marletta, M. A. (1994). Soluble guanylate cyclase from bovine lung: Activation with nitric oxide and carbon monoxide and spectral characterization of the ferrous and ferric states. *Biochemistry* **33,** 5636–5640.

Stone, J. R., Sands, R. H., Dunham, W. R., and Marletta, M. A. (1996). Spectral and ligand-binding properties of an unusual hemoprotein, the ferric form of soluble guanylate cyclase. *Biochemistry* **35,** 3258–3262.

Wedel, B., Humbert, P., Harteneck, C., Foerster, J., Malkewitz, J., Bohme, E., Schultz, G., and Koesling, D. (1994). Mutation of His-105 in the beta 1 subunit yields a nitric oxide-insensitive form of soluble guanylyl cyclase. *Proc. Natl. Acad. Sci. USA* **91,** 2592–2596.

Zabel, U., Weeger, M., La, M., and Schmidt, H. H. (1998). Human soluble guanylate cyclase: Functional expression and revised isoenzyme family. *Biochem. J.* **335**(Pt 1), 51–57.

Zhao, Y., Brandish, P. E., Ballou, D. P., and Marletta, M. A. (1999). A molecular basis for nitric oxide sensing by soluble guanylate cyclase. *Proc. Natl. Acad. Sci. USA* **96,** 14753–14758.

Zhao, Y., Schelvis, J. P., Babcock, G. T., and Marletta, M. A. (1998). Identification of histidine 105 in the beta1 subunit of soluble guanylate cyclase as the heme proximal ligand. *Biochemistry* **37,** 4502–4509.

# [41] Purification and Characterization of NO-Sensitive Guanylyl Cyclase

*By* MICHAEL RUSSWURM and DORIS KOESLING

## Abstract

Nitric oxide (NO)–sensitive guanylyl cyclase (GC) represents the receptor for the signaling molecule NO in mammals. The enzyme consists of two different subunits and contains a prosthetic heme group acting as an NO acceptor. Binding of NO to the heme accelerates the catalytic conversion from guanosine triphosphate (GTP) to cyclic guanosine monophosphate (cGMP) by 2.0–2.5 orders of magnitude. NO-sensitive GC represents a well-established drug target because the NO-releasing drugs used in the therapy of coronary heart disease act by stimulation of the enzyme. Furthermore, new NO-independent GC activators are under development. Characterization of the molecular mechanisms by which NO-releasing and NO-independent drugs control enzyme activity often requires very large amounts of enzyme, as do attempts to resolve the structure of GC. Because heterologous expression systems turned out to yield only small amounts of the enzyme, the purification from bovine lungs is still a convenient way to obtain the enzyme in large quantities. In this chapter, a fast method for purification of the enzyme from bovine lungs is described, and basic methods for characterization of the enzyme are summarized.

METHODS IN ENZYMOLOGY, VOL. 396                         0076-6879/05 $35.00
Copyright 2005, Elsevier Inc. All rights reserved.        DOI: 10.1016/S0076-6879(05)96041-2

Introduction

Nitric oxide (NO)–sensitive guanylyl cyclase (GC), one of the two cyclic guanosine monophosphate (cGMP)–forming cyclases, acts as the effector for the messenger molecule NO. NO-sensitive GC is an intracellular heterodimeric protein consisting of one $\alpha$ and one $\beta$ subunit and contains a prosthetic heme group (Denninger and Marletta, 1999; Friebe and Koesling, 2003; Russwurm and Koesling, 2004a; Wedel and Garbers, 2001). By binding to the prosthetic heme group, NO provokes a tremendous increase in catalytic activity [*i.e.*, formation of the second-messenger cGMP from guanosine triphosphate (GTP)]. Two $\alpha$ and two $\beta$ subunits ($\alpha_1$, $\alpha_2$, $\beta_1$, and $\beta_2$) have been cloned from mammals, but only $\alpha_1\beta_1$ and $\alpha_2\beta_1$ heterodimers have been demonstrated to form NO-sensitive cGMP-forming enzymes in expression systems. In contrast, the $\beta_2$ subunit has neither been expressed as a catalytically active enzyme, nor has it been detected on the protein level in tissues; the human $\beta_2$ subunit has been shown to contain a frameshift mutation (Behrends and Vehse, 2000). To date, only the $\alpha_1\beta_1$ isoform has been purified from native tissues; the $\alpha_2\beta_1$ enzyme can be expressed in and purified from heterologous systems. Characterization of the regulatory properties of GC is routinely performed with the $\alpha_1\beta_1$ enzyme, but the $\alpha_2\beta_1$ isoform has been demonstrated to show indistinguishable properties (Russwurm *et al.*, 1998). The two isoforms are characterized by a distinct subcellular and tissue-specific distribution; the $\alpha_1\beta_1$ is found in the cytosolic fractions of vascularized tissues, whereas the $\alpha_2\beta_1$ is preferentially expressed in the brain (Mergia *et al.*, 2003), where it is targeted to synaptic membranes by interaction with the synaptic protein PSD-95 (Russwurm *et al.*, 2001).

As early as 20 years ago, procedures for the purification of the NO-sensitive or soluble GC were described (Waldman and Murad, 1987). Shortly after the identification of NO-releasing substances as potent activators of the cytosolic cGMP-forming activity in the late 1970s (Arnold *et al.*, 1977; Kimura *et al.*, 1975), the NO-sensitive GC was purified as a heme-containing protein, and binding of NO to the prosthetic heme group was postulated as a mechanism of activation (Gerzer *et al.*, 1981). The physiological importance of the activation by NO became evident after identification of the endothelium-derived relaxing factor as NO (Furchgott and Zawadzki, 1980; Ignarro *et al.*, 1987; Palmer *et al.*, 1987). Starting with the enzyme purified from lung, the bovine and rat $\beta_1$ subunits were cloned and sequenced, which enabled the production of antibodies against synthetic peptides deduced from GC's sequence. Fortunately, one of those antibodies directed against the C-terminus of the $\beta_1$ subunit turned out to precipitate the native enzyme out of crude fractions. Using peptide antibodies for affinity columns during protein purification provides the possibility to elute the protein of interest with an excess of synthetic peptide, avoiding harsh conditions like changing to extreme pH values or adding urea. A purification scheme for GC based on this approach has been described in this series (Humbert *et al.*, 1991). However, during the last 15 years, the purification procedure has been significantly improved. Most importantly, the antibodies were coupled in an oriented manner to protein A Sepharose, instead of the unordered coupling to cyanogen bromide–activated Sepharose. All other steps of the purification have been newly developed, resulting in higher yields of purified protein with approximately 10-fold higher specific catalytic activities.

Assay for cGMP-Forming Activity

cGMP-forming activity of crude and purified fractions of GC is determined by conversion of $[\alpha\text{-}^{32}P]$GTP to $[^{32}P]$cGMP as described (Schultz and Böhme, 1984). In short, the enzyme is incubated with 500 $\mu M$ $[\alpha\text{-}^{32}P]$ GTP ($\sim$5 kBq), 3 m$M$ $MgCl_2$, 1 m$M$ cGMP, 3 m$M$ dithiothreitol (DTT), 0.5 mg/ml bovine serum albumin (BSA) in 50 m$M$ triethanolamine (TEA)/ HCl, pH 7.4, with or without addition of 100 $\mu M$ diethylamine NONOate (DEA/NO) in a total volume of 100 $\mu l$ for 10 min at 37°. Measurement of crude fractions may require addition of a GTP regenerating system (0.025 mg creatine phosphokinase/sample and 5 m$M$ creatine phosphate) and/or phosphodiesterase inhibitors. Reactions are stopped by addition of 450 $\mu l$ 120 m$M$ zinc acetate, and 5′ nucleotides are precipitated by addition of 450 $\mu l$ 120 m$M$ $Na_2CO_3$ and centrifugation (4 min, 14,000g, 21°). Subsequently, 800 $\mu l$ of the supernatant, together with 2 ml 0.1 $M$ perchloric acid, is applied to a column filled with 500 $\mu l$ neutral alumina equilibrated with 2 ml 0.1 $M$ perchloric acid. The columns are washed with 10 ml water and eluted with 4 ml 250 m$M$ sodium acetate directly into scintillation vials. Cerenkov radiation of the samples is measured in a $\beta$-counter, and the specific activity of the GC fractions is calculated using the following formula:

$$v(GC) = \frac{(C - C_0)}{C_t \times R} \times \frac{S}{P \times t},$$

where

v(GC)   specific activity of the enzyme (nmol cGMP $\times$ min$^{-1}$ $\times$ mg protein$^{-1}$)

C   counting rate (cpm) of the sample

$C_0$   counting rate (cpm) of a blank value (measured without GC in the incubation)

$C_t$ counting rate (cpm) of the total amount of GTP added to every sample

S  amount of substrate (GTP) used (nmol)

R  recovery rate of cGMP during precipitation and chromatography (usually 50%)

P  protein added per sample (mg)

t  incubation time (min)

The dynamic range of the assay is 0.015–0.5 nmol cGMP $\times$ min$^{-1}$ [*i.e.*, $\sim$0.025 $\mu$g purified enzyme (specific activity $\sim$20,000 nmol $\times$ min$^{-1}$ $\times$ mg) is used per reaction or 150 $\mu$g cytosolic protein (specific activity $\sim$3 nmol $\times$ min$^{-1}$ $\times$ mg)].

If fractions have to be measured very fast during the GC purification to decide which fractions have to be used in the next step, the chromatography step is omitted and the supernatant of the precipitation is measured directly. This will result in 10-fold higher blank counting rates ($\sim$1000 cpm compared to 100 cpm) and a recovery rate of 80%. During the purification, all fractions are measured in the presence of 100 $\mu M$ DEA/NO.

## Preparation of a Guanylyl Cyclase Antibody Affinity Column

Antibodies against the C-terminal sequence of the $\beta_1$ subunit (SRKNTGTEETEQDEN) are raised by standard methods. Before coupling, antisera are tested for their ability to precipitate the native enzyme by precipitation of GC activity with protein A Sepharose from the pooled fraction of the first chromatography step of the purification (see later discussion). Suitable antisera are selected for immobilization. To preserve the antigen-binding sites, the antibodies are bound to protein A Sepharose and subsequently crosslinked as follows.

One hundred milliliters of antiserum are adjusted to pH 8 by addition of 1/10 volume 1 $M$ Tris pH 8. Protein A Sepharose Fast Flow (100 ml) (Amersham Biosciences) is washed with 100 m$M$ Tris, pH 8 (1000 ml) and subsequently incubated with the antiserum in batch mode (1 h, room temperature, overhead rotator). The material is filled into a chromatography column and washed sequentially with 10 column volumes (CV) of (1) 100 m$M$ Tris, pH 8, (2) 10 m$M$ Tris pH 8, and (3) 200 m$M$ sodium borate, pH 9. The Sepharose is resuspended in 10 CV of 200 m$M$ sodium borate, pH 9, and 5 mg dimethylpimelimidate/ml buffer is added for crosslinking of immunoglobulin G (IgG) to protein A. The Sepharose is again incubated in batch mode (1 h, room temperature, overhead rotator) and subsequently washed with 3 CV 200 m$M$ ethanolamine, pH 8, to block remaining active sites. The Sepharose is then equilibrated with a sodium azide–containing storage buffer (see later discussion), packed into a suitable chromatography column, and used for the purification as described later in this chapter.

### Preparation of NO-Sensitive GC

A fast procedure for purification of NO-sensitive GC from bovine lung is described. After tissue homogenization and centrifugation, the supernatant is applied to an anion exchange chromatography using a step gradient, which removes 95% of contaminating proteins and reduces the volume by 90% (Table I). The eluate is bound to an immunoaffinity column and eluted by an excess of antigenic peptide, which removes nearly all contaminating proteins. The eluate is concentrated by anion exchange chromatography before a gel filtration chromatography. Thereafter, the pure GC is again concentrated on an anion exchange column. By using automated equipment, the whole procedure is usually completed within 30 h.

### Preparation of a Cleared Homogenate

All procedures are performed at 4°. Bovine lungs (5 kg) are minced and mixed with 25 liters of buffer A [75 m$M$ NaCl, 50 m$M$ TEA, pH 7.4 containing 1 m$M$ EDTA, 2 m$M$ DTT, 0.4 m$M$ phenylmethylsulfonylfluoride (PMSF), 0.2 m$M$ benzamidine] and homogenized using a Megatron MT 3000 equipped with an MTG 30/2 dispersing generator (Kinematica, Switzerland). The homogenate is centrifuged (14,000$g$, 15 min, 4°) in high-speed centrifuges with 6 × 500 ml rotors. As ultracentrifugation of 25 liters is hardly possible, the supernatants are roughly cleared by filtering through 604 filter paper (Ø 185 mm, Schleicher & Schuell) on a Büchner funnel.

TABLE I

SUMMARY OF THE PURIFICATION

| Fraction | Fraction vol. (ml) | Protein (mg/ml) | Protein (mg) | cGMP-forming activity (nmol/min) | Specific activity (nmol/min/mg) |
|---|---|---|---|---|---|
| Homogenate | 25,000 | 15 | 375,000 | 1,000,000 | 3 |
| Q Sepharose Big Beads | 2500 | 7 | 17,500 | 450,000 | 26 |
| Immunoaffinity | 500 | —[a] | —[a] | 87,000 | —[a] |
| Gel filtration | 11 | 0.4 | 5 | 84,000 | 16,970 |
| Source Q | 0.4 | 11 | 5 | 84,000 | 16,805 |

[a] Protein content not determined because of a large amount of antigenic peptide present.

The resulting filtrate (25 liters) contains approximately 400 g protein and 3 nmol $\times$ min$^{-1}$ $\times$ mg protein$^{-1}$ cGMP-forming activity.

### Preparation of an Enriched Fraction by Anion Exchange Chromatography

The filtrate is further processed in anion exchange chromatography. As column chromatography of the filtrate resulted in clogging of the column, the initial anion exchange step is performed in batch mode. Q Sepharose Big Beads (1.5 liters, Amersham Biosciences) equilibrated with buffer A is stirred for 1 h in the filtrate. Thereafter, the Sepharose is collected in a Büchner funnel ($\varnothing$ 240 mm) with a 30 $\mu m$ polyamide net. The Sepharose is repeatedly resuspended and washed with up to 15 liters buffer A until the color turns from red to yellow. The resuspended slurry is then poured into a Bioprocess Column ($\varnothing$ 100 mm, Amersham Biosciences). Subsequently, proteins are eluted in 250 ml fractions with 5 l buffer B (250 m$M$ NaCl, 50 m$M$ TEA, pH 7.4, containing 1 m$M$ EDTA, 2 m$M$ DTT, 0.4 m$M$ PMSF, 0.2 m$M$ benzamidine, 1 $\mu M$ pepstatin A). In the fractions, NO-stimulated cGMP-forming activity is determined as described earlier, and protein content is measured by the Warburg method [Protein concentration (mg/ml) $= 1.55 \times E_{280} - 0.76 \times E_{260}$]. Usually, cGMP-forming activity and total protein content correlate exactly; hence, if the chromatography is repeatedly performed under identical conditions, measurement of cGMP-forming activity can be omitted, and fractions are pooled according to protein content. The enriched fraction (2.5 liters) contains approximately 20 g of protein with a specific activity of 30 nmol cGMP $\times$ min$^{-1}$ $\times$ mg protein$^{-1}$. The Q Sepharose step, therefore, removes 95% of the total proteins, reduces the volume by 90%, and reaches a recovery of NO-stimulated cGMP-forming activity of 50%. The eluate of this chromatography is cloudy and thus must be sterile filtered using a Millipore All-Glass Filter Holder with an HVLP 0.45-$\mu m$ filter.

### Immunoaffinity Chromatography

The immunoaffinity column ($\varnothing$ 5 $\times$ 5 cm, 100 ml, flow rate 6.7 ml/min throughout the purification) is equilibrated with 2 CV of buffer B, and the Q Sepharose eluate is applied using a high-performance liquid chromatography (HPLC) pump. Subsequently, the column is washed with 8 CV of buffer C (150 m$M$ NaCl, 50 m$M$ TEA, pH 7.4, containing 1 m$M$ EDTA, 2 m$M$ DTT, 0.4 m$M$ PMSF, 0.2 m$M$ benzamidine, 1 $\mu M$ pepstatin A) followed by 2 CV of buffer D (50 m$M$ NaCl, 50 m$M$ TEA, pH 7.4, containing 1 m$M$ EDTA, 2 m$M$ DTT, 0.4 m$M$ PMSF, 0.2 m$M$ benzamidine, 1 $\mu M$

pepstatin A). Elution of GC is performed by circulating 100 ml of buffer D containing 0.2 mg/ml of the antigenic peptide for 2.5 h through the column, followed by rinsing with 150 ml of buffer D. The complete affinity chromatography step takes about 15 h and is performed overnight using a fully automated chromatography system to reduce the overall time required for the purification. The affinity column is regenerated with two CVs of each (1) 3 $M$ urea, (2) 0.5 $M$ NaCl, 0.1 $M$ sodium acetate, pH 4.5, and (3) 0.5 $M$ NaCl, 0.1 $M$ Tris, pH 8.5. Subsequently, the column is equilibrated with 10 CV of storage buffer (50 m$M$ NaCl, 50 m$M$ TEA, pH 7.4, containing 0.02% sodium azide) and stored at 4° until the next purification.

### Gel Filtration Chromatography

The eluate of the immunoaffinity chromatography (250 ml) contains a very dilute GC and therefore is subjected to an anion exchange step on a Source Q column (Amersham Biosciences, Ø 1 × 1.5 cm, flow rate 5 ml/min). The column is eluted with 4 ml buffer E (250 m$M$ NaCl, 50 m$M$ TEA, pH 7.4, containing 2 m$M$ DTT) directly onto a HiLoad Superdex 200 prep grade column (Amersham Biosciences, Ø 2.6 × 60 cm, flow rate 1.32 ml/min, running buffer F, 150 m$M$ NaCl, 50 m$M$ TEA, pH 7.4, containing 2 m$M$ DTT) connected in series. Elution is monitored at 280 nm (protein content) and 430 nm (Soret band). Fractions containing a Soret/protein coefficient of more than 90% are pooled, diluted to 1:4 with buffer D (without protease inhibitors and EDTA) to lower the salt concentration, and again concentrated on a Source Q column (Amersham Biosciences, Ø 0.5 × 5 cm, flow rate 1.5 ml/min; elution with 4 ml buffer E). Protein content is determined by measuring the absorbance at 280 nm (1 AU corresponds to 1 mg/ml GC). Fractions containing more than 1 mg/ml protein are snap frozen in liquid nitrogen and stored at −70° for several months without loss of activity; more dilute fractions are mixed with 1 volume glycerol before freezing at −70°. The purification usually yields 5 mg of GC at a concentration of approximately 10 mg/ml with specific NO-stimulated cGMP-forming activities of about 20,000 nmol × min$^{-1}$ × mg protein$^{-1}$.

### Characterization of NO-Sensitive GC

*Electrophoresis*

In sodium dodecylsulfate (SDS) gel electrophoresis under reducing conditions, the protein purified from bovine lung shows two bands of similar intensity at 70 and 74 kDa, corresponding to the $\beta_1$ and $\alpha_1$ subunits, respectively. Proteolytic cleavage, which sometimes occurs during the purification, leads to disappearance of the 74-kDa band and broadening of the 70-kDa band; obviously, a 4-kDa fragment is removed from the $\alpha_1$ subunit. However, this alters neither heme content nor catalytic activity of the enzyme. The enzyme purified from rat lung has been reported to show a 82-kDa $\alpha_1$ band and a 70-kDa $\beta_1$ band; the reason for the apparently larger $\alpha_1$ band remains unclear.

*UV–Vis Spectroscopy*

NO-sensitive GC as a hemoprotein shows a typical absorbance spectrum (Fig. 1) with a Soret peak at 431 nm (extinction coefficient 150 m$M^{-1}$ cm$^{-1}$), characteristic of a five coordinated ferrous heme iron. Binding of NO to the heme group can be followed spectrophotometrically. The absorbance maximum shifts to 399 nm, and the extinction coefficient decreases to 100 m$M^{-1}$ cm$^{-1}$, indicative of the formation of a five coordinated ferrous NO–heme complex (Fig. 1). It should be noted that binding of NO to the heme group and activation of the enzyme do not always coincide; that is, in the absence of the substrate $Mg^{2+}$ + GTP or the products $Mg^{2+}$ + cGMP + $PP_i$, binding of NO results in the formation of a nonactivated

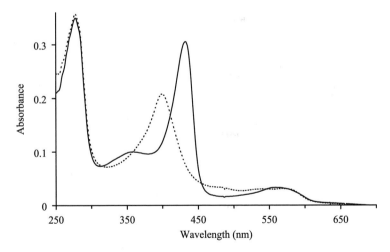

Fig. 1. Ultraviolet–visible (UV–vis) spectra of purified guanylyl cyclase (GC). UV–vis spectra of purified GC [0.3 mg/ml = 2 $\mu M$ in 3 m$M$ dithiothreitol (DTT), 50 m$M$ triethanolamine (TEA), pH 7.4) were recorded in a diode array detector (model 8453, Agilent Technologies) without (solid line) or with addition of 20 $\mu M$ Proli NONOate (Alexis) (dotted line).

state indistinguishable by spectroscopy from the active NO-bound state (Russwurm and Koesling, 2004b).

### Removal of the Prosthetic Heme Group

To distinguish between heme-dependent and heme-independent effects of activators or inhibitors of the enzyme, removal of the prosthetic heme group is required. Incubation of the enzyme with 0.5–2.0% Tween-20 for a few minutes at room temperature or 37° results in loss of the heme group, which should be judged by ultraviolet–visible (UV–vis) spectroscopy and measurement of NO stimulation (which should be lost). If removal of detergent is required (e.g., for spectroscopic analyses), anion exchange chromatography on DEAE Sepharose should be performed (binding of the enzyme in the presence of 50 mM TEA, pH 7.4, washing with 10 CV of the same buffer, and elution with 200 mM NaCl, 50 mM TEA, pH 7.4).

### Concluding Remarks

Purification of NO-sensitive GC has been significantly improved during recent years. The larger enzyme quantities obtained will allow a more thorough investigation of the molecular mechanisms of activation (i.e., how binding of NO to the prosthetic heme is translated into activation of catalysis). The extremely fast NO dissociation and the subsequent deactivation of the enzyme within seconds or possibly even faster represent the most intriguing feature of the enzyme, as all other heme-containing proteins exhibit far slower NO dissociation rates within minutes to hours (Bellamy et al., 2002; Kharitonov et al., 1997; Russwurm et al., 2002). The molecular determinants allowing the enzyme to repel NO faster than any other hemoprotein are still to be uncovered.

Furthermore, new drugs stimulating GC have been developed for the treatment of coronary heart disease and hypertension (Russwurm and Koesling, 2004a). The so-called NO sensitizers described to inhibit the enzyme's deactivation may work by stabilization of the active conformation or act by direct inhibition of NO dissociation. The heme-independent activators have been postulated to substitute for the heme moiety, although the underlying mechanism is unclear. Taken together, substantial progress has been made by the pharmaceutical industry to target the NO receptor GC for the treatment of vascular diseases. Using the purified enzyme, similar efforts should be made to unravel the molecular mechanisms by which those drugs exert their effects.

# References

Arnold, W. P., Mittal, C. K., Katsuki, S., and Murad, F. (1977). Nitric oxide activates guanylate cyclase and increases guanosine 3′:5′-cyclic monophosphate levels in various tissue preparations. *Proc. Natl. Acad. Sci. USA* **74,** 3203–3207.

Behrends, S., and Vehse, K. (2000). The $\beta_2$ subunit of soluble guanylyl cyclase contains a human-specific frameshift and is expressed in gastric carcinoma. *Biochem. Biophys. Res. Commun.* **271,** 64–69.

Bellamy, T. C., Griffiths, C., and Garthwaite, J. (2002). Differential sensitivity of guanylyl cyclase and mitochondrial respiration to nitric oxide measured using clamped concentrations. *J. Biol. Chem.* **277,** 31801–31807.

Denninger, J. W., and Marletta, M. A. (1999). Guanylate cyclase and the NO/cGMP signaling pathway. *Biochim. Biophys. Acta* **1411,** 334–350.

Friebe, A., and Koesling, D. (2003). Regulation of nitric oxide–sensitive guanylyl cyclase. *Circ. Res.* **93,** 96–105.

Furchgott, R. F., and Zawadzki, J. V. (1980). The obligatory role of endothelial cells in the relaxation of arterial smooth muscle by acetylcholine. *Nature* **288,** 373–376.

Gerzer, R., Hofmann, F., and Schultz, G. (1981). Purification of a soluble, sodium-nitroprusside–stimulated guanylate cyclase from bovine lung. *Eur. J. Biochem.* **116,** 479–486.

Humbert, P., Niroomand, F., Fischer, G., Mayer, B., Koesling, D., Hinsch, K. D., Schultz, G., and Böhme, E. (1991). Preparation of soluble guanylyl cyclase from bovine lung by immunoaffinity chromatography. *Methods Enzymol.* **195,** 384–391.

Ignarro, L. J., Buga, G. M., Wood, K. S., Byrns, R. E., and Chaudhuri, G. (1987). Endothelium-derived relaxing factor produced and released from artery and vein is nitric oxide. *Proc. Natl. Acad. Sci. USA* **84,** 9265–9269.

Kharitonov, V. G., Russwurm, M., Magde, D., Sharma, V. S., and Koesling, D. (1997). Dissociation of nitric oxide from soluble guanylate cyclase. *Biochem. Biophys. Res. Commun.* **239,** 284–286.

Kimura, H., Mittal, C. K., and Murad, F. (1975). Activation of guanylate cyclase from rat liver and other tissues by sodium azide. *J. Biol. Chem.* **250,** 8016–8022.

Mergia, E., Russwurm, M., Zoidl, G., and Koesling, D. (2003). Major occurrence of the new $\alpha_2\beta_1$ isoform of NO-sensitive guanylyl cyclase in brain. *Cell Signal* **15,** 189–195.

Palmer, R. M., Ferrige, A. G., and Moncada, S. (1987). Nitric oxide release accounts for the biological activity of endothelium-derived relaxing factor. *Nature* **327,** 524–526.

Russwurm, M., Behrends, S., Harteneck, C., and Koesling, D. (1998). Functional properties of a naturally occurring isoform of soluble guanylyl cyclase. *Biochem. J.* **335,** 125–130.

Russwurm, M., and Koesling, D. (2004a). Guanylyl cyclase: NO hits its target. *Biochem. Soc. Symp.* **71,** 51–63.

Russwurm, M., and Koesling, D. (2004b). NO activation of guanylyl cyclase. *EMBO J.* **23,** 4443–4450.

Schultz, G., and Böhme, E. (1984). Guanylate cyclase. *In* "Methods of Enzymatic Analysis" (J. Bergmeyer and M. Graßl, eds.), pp. 379–389. Verlag Chemie, Weinheim, Deerfield Beach, Basel.

Waldman, S. A., and Murad, F. (1987). Cyclic GMP synthesis and function. *Pharmacol. Rev.* **39,** 163–196.

Wedel, B., and Garbers, D. (2001). The guanylyl cyclase family at Y2K. *Annu. Rev. Physiol.* **63,** 215–233.

## [42]  The Measurement of Nitric Oxide Production by Cultured Endothelial Cells

*By* C. Michael Hart, Dean J. Kleinhenz, Sergey I. Dikalov, Beth M. Boulden, and Samuel C. Dudley, Jr.

Abstract

Nitric oxide (NO) produced by vascular endothelial cells (ECs) plays a critical role in normal vascular physiology. Important insights into mechanisms regulating the production of endothelial NO have been derived from *in vitro* studies employing cultured ECs. Although many techniques for the detection of NO have been described, many of these methodslack adequate sensitivity to detect the small amount of NO produced by cultured ECs. In this chapter, we describe three protocols that employ chemiluminescence, electron spin resonance, or electrochemical techniques to permit the reliable detection of EC NO production.

Introduction

The production of nitric oxide (NO) by vascular endothelial cells (ECs) plays a critical role in normal vascular physiology. Endothelial-derived NO reduces vascular tone, platelet activation, and aggregation, decreases stimulated vascular smooth muscle proliferation, and impairs leukocyte adherence. Endothelial dysfunction and impaired endothelial NO production constitute an early step in the pathogenesis of vascular disease. Within vascular endothelial cells, NO is produced constitutively from the amino acid, L-arginine, by the type III endothelial NO synthase (eNOS) isoform. Numerous investigators have substantially contributed to our understanding of the factors that regulate NO production by employing cultured EC models. However, reliably measuring the small amount of NO produced by eNOS within ECs can be challenging. Although a number of chemical- or fluorescence-based assays have been described that detect NO or its metabolites, limitations in the sensitivity and specificity of these assays frequently prevent their application to cultured EC models. Herein, we describe three techniques that we have successfully employed to reliably measure EC NO production.

Chemiluminescence Analysis of EC NO Production

The chemiluminescence method detects NO and its oxidation products, nitrite ($NO_2^-$) and nitrate ($NO_3^-$), in liquid culture media. The assay employs one of several commercially available chemiluminescence NO

METHODS IN ENZYMOLOGY, VOL. 396
0076-6879/05 $35.00
DOI: 10.1016/S0076-6879(05)96042-4

analyzers (such as Sievers NOA, Model 280, Boulder, CO). Liquid samples of culture media are injected into a reflux chamber that converts $NO_2^-$ and $NO_3^-$ into NO that is carried by inert gas into the analyzer. Before analysis, media samples are centrifuged at 2000$g$ for 1 min to remove any cellular debris. As illustrated in Fig. 1, 40-$\mu$l aliquots of the supernatant are aspirated into a Hamilton syringe and injected through an air-tight septum into a purge vessel containing 0.8% vanadium (III) chloride (VaCl$_3$) in 1 N HCl at 95° (A). These conditions convert $NO_2^-$ and $NO_3^-$ to NO in the purge vessel. The NO is then carried toward the NO analyzer (NOA) reaction cell in a steady stream of inert gas (argon or nitrogen). After leaving the purge vessel, the argon–NO mixture passes first through a cooled water-jacketed chamber that condenses evaporated VaCl$_3$ and returns it to the purge vessel (B). The sample/carrier gas mixture then bubbles through 1 N NaOH (C), followed by passage through an inline filter (D) to prevent acid vapors from reaching the NOA reaction cell. In the reaction cell (E), NO gas combines with ozone to produce NO in the excited state ($NO_2^*$), which upon decay emits light that passes through a red filter (F) and is detected at 600 nm by a photomultiplier tube (PMT) inside the NOA. The PMT generates a current proportional to the emitted light that is recorded in millivolts in the instrument software.

## Application to Cultured EC

EC monolayers are grown to confluence in standard serum-containing media. Before measurements of NO release, serum-containing culture media are decanted, and monolayers are washed with buffered saline at 37°. Serum-free culture media are preferred for measurements of NO release because higher protein concentrations in the injected culture media can cause foaming in the purge vessel. We generally use culture media that have lower background levels of $NO_2^-$ and $NO_3^-$, including Dulbecco's Modified Eagle Medium (DMEM) or Essential Basal Medium (EBM) (Fisher Scientific, Pittsburgh, PA). Serum-free medium at 37° is then added to each monolayer at a ratio of approximately 4 ml culture medium/3 −4 × $10^6$ cells. Experimental interventions are examined either by treating monolayers with reagents or an equivalent volume of vehicle before measurement of NO release or by including experimental reagents or vehicle in serum-free media that are collected and analyzed for NO release. Incubation in serum-free medium ± experimental interventions is then performed for designated intervals (typically 60 min) at 37° in a 5% $CO_2$/air incubator. The media (up to 1 ml from each monolayer) are then collected and frozen at −80° in plastic Eppendorf tubes. The cell monolayers are scraped into

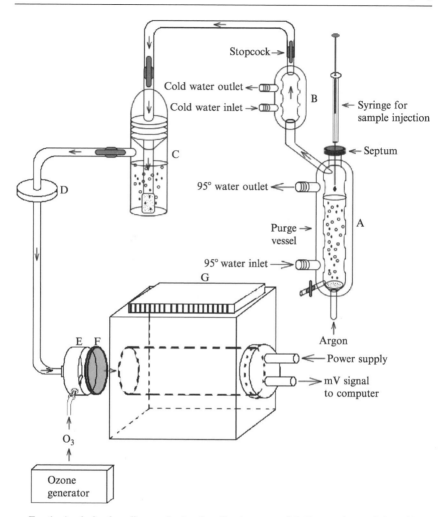

FIG. 1. Analysis of media samples by chemiluminescence. Media samples are injected into the purge vessel (A), which contains $VaCl_3$ at 95°. Sample nitric oxide (NO) is carried by argon out of the purge vessel and through the condenser (B), where any $VaCl_3$ is condensed by cold water and falls back down into the purge vessel. The NO-containing gas stream then passes through 1 N NaOH (C) to prevent HCl vapors from entering the reaction cell. After passing through an inline filter (D), NO enters the reaction cell (E), where it mixes with ozone to form $NO_2^*$. Upon decay, $NO_2^*$ emits light, which passes through a red filter (F) and is detected by the photomultiplier tube (PMT) (G), which is kept at $-15°$. This generates a current in millivolts proportional to the amount of NO in the sample, which is then sent to the software and quantified.

lysis buffer, and proteins are measured using the bicinchoninic acid protein assay (Pierce, Rockford, IL).

Figure 2 illustrates data generated with this technique in cultured porcine pulmonary artery ECs (PAECs) after treatment with the calcium ionophore, A23187, which increases intracellular calcium concentrations to activate eNOS and stimulates NO release. Confluent PAEC monolayers in 60-mm dishes were washed with warm Hanks' Balanced Salt Solution. Four milliliters of serum-free DMEM at 37° containing either 5 $\mu M$ A23187 or an equivalent volume of ethanol vehicle was then added to each dish and incubated for 1 h. The culture media and monolayers were then collected and processed as described earlier, and media samples were subjected to chemiluminescence analysis. Representative peaks generated by duplicate 40 $\mu l$ injections from vehicle- and A23187-treated PAECs are presented in Fig. 2A (inset). The area under the peak is proportional to NO in the sample. The software permits the user to define the baseline threshold for all samples, as well as the beginning and end of each peak, and it calculates the area under each sample curve (Fig. 2A). Sample NO concentrations are calculated by comparing the area of unknown sample peaks to the peak areas generated by the standard curve of known concentrations of $NaNO_3^-$ (Fig. 2B, inset). Preparation of the standards in the culture medium in which measurements are made permits accounting for $NO_2^-$ and $NO_3^-$ attributable to the culture medium itself. The measured concentration of NO is then expressed as picomole NO/min/mg protein. Basal EC NO production falls on the lower region of the standard curve (Fig. 2B, inset).

We have previously reported use of this chemiluminescence assay to detect basal and stimulated NO release in response to A23187 (Calnek *et al.*, 2003), insulin (Lynn *et al.*, 2004), acetylcholine, bradykinin, or histamine (Kleinhenz *et al.*, 2003) in various types of ECs including porcine pulmonary artery (Gupta *et al.*, 1998) and human aortic or human umbilical vein ECs (Calnek *et al.*, 2003). Advantages of this assay include its ability to detect small amounts of NO ($\sim$1 p$M$), the ability to measure NO from various types of media, the relatively small volume of media needed for analysis, and the low incidence of artifactual effects of treatment on the chemiluminescence signal. Disadvantages of this assay are that it requires specialized instrumentation, permits analysis of only a single sample at one time, may not permit reliable detection of EC NO production when NO release is decreased below basal rates of production, and is not facile at permitting analysis of rapid changes in NO release over brief intervals.

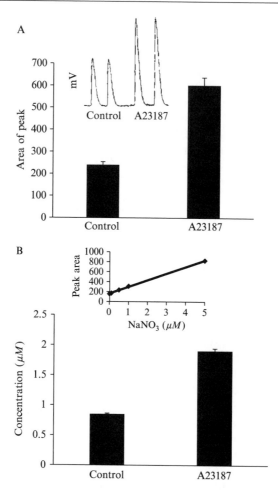

F_IG. 2. Endothelial cell (EC) nitric oxide (NO) release detected by the chemiluminescence technique. (A) ECs were treated with control media or media containing 5 $\mu M$ A23187 for 60 min. Duplicate samples from these ECs were injected into the NO analyzer (described in Fig. 1) generating peaks that are displayed in millivolts (inset) and averaged (±SD). (B) A standard curve is generated by injecting known amounts of $NaNO_3^-$ (inset), and the NO concentration of the samples is calculated by comparison to the standard curve.

## Electron Spin Resonance Techniques for Measuring EC NO Production

Electron spin resonance (ESR), one of the most direct and unambiguous methods for free radical detection, can be used for measurements of NO production by cultured EC. The short lifetime and quick relaxation of the NO molecule prevent the direct detection of NO with ESR. However, the use of spin traps that react with NO to produce stable paramagnetic molecules that have specific ESR spectra permits quantitative detection of NO. Colloid iron diethyldithiocarbamate, $Fe(DETC)_2$, is a reliable spin trap for NO detection in cultured ECs (Cai *et al.*, 2002; Dikalov *et al.*, 2002; Kleschyov *et al.*, 2002; Kuzkaya *et al.*, 2003).

*Preparation of Colloid Fe(DETC)$_2$ Stock Solution*

DETC (7.2 mg) and $FeSO_4$ $7H_2O$ (4.45 mg) (Sigma-Aldrich, St. Louis, MO) are each dissolved separately in 10 ml filtered, ice-cold, 0.9% NaCl that is first deoxygenated by bubbling with nitrogen. The cold DETC solution is then mixed 1:1 with $FeSO_4$ in a 10-ml plastic tube under constant nitrogen flow (avoid generating air bubbles). The resulting yellow-brownish $Fe(DETC)_2$ solution should be used immediately after preparation.

*Constructing a Calibration Curve*

The amount of NO detected in each sample can be quantified from the calibration curve for integral intensity of the ESR signal of $NO-Fe^{2+}(MGD)_2$ prepared with the NO donor, MAHMA-NONOate (1–20 $\mu M$) (Morley *et al.*, 1993). Briefly, $Fe(II)SO_4$ (1.4 mg) and sodium *N*-methyl-D-glucamine dithiocarbamate (MGD) (15 mg) are each dissolved in 1 ml deoxygenated 0.9% NaCl kept under nitrogen flow. The MGD and $FeSO_4$ are then mixed in a 1:1 ratio, and 0.2 ml of this mixture is added to 0.8 ml oxygen-free Krebs-HEPES buffer. Stock solutions of MAHMA-NONOate (2 m$M$) are dissolved in oxygen-free Krebs-HEPES buffer shortly before use. MAHMA-NONOate (final concentration 1–20 $\mu M$) is added to the Fe-MGD solution and incubated for 30 min at room temperature. The Fe-MGD solutions are then collected in 1-ml syringes, snap-frozen with liquid $N_2$, and subjected to ESR analysis.

*Application to Cultured ECs*

Post-confluent ECs in 100-mm plates are used for NO measurements with $Fe^{2+}(DETC)_2$. To test the role of superoxide or $H_2O_2$, selected ECs are incubated overnight with 100 U/ml PEG-SOD or PEG-catalase, respectively. ECs are then rinsed with ice-cold Krebs-HEPES buffer (99.01 m$M$ NaCl, 4.69 m$M$ KCl, 2.50 m$M$ $CaCl_2$, 1.20 m$M$ $MgSO_4$, 25 m$M$ $NaHCO_3$, 1.03 m$M$ $K_2HPO_4$, 20 m$M$ Na-HEPES, and 5.6 m$M$ D-glucose, pH 7.35) and treated for 5 min with reagents such as 100 $\mu M$ L-NAME, 10 $\mu M$ tetrahydrobiopterin, 5 $\mu M$ A23187, or 100 $\mu M$ $H_2O_2$ in 1.5ml Krebs-HEPES buffer (Cai *et al.*, 2002; Kuzkaya *et al.*, 2003).

Following these treatments, 0.5 ml $Fe^{2+}(DETC)_2$ colloid (prepared immediately before use) is spread throughout the plate by drop-wise addition to the EC monolayer surface without mixing. After incubation at 37° for 60 min, media are aspirated, and 0.6 ml Krebs-HEPES buffer is added to each plate. ECs are gently scraped with a rubber policeman and aspirated into 1-ml plastic syringes that are snap-frozen by immersion in liquid $N_2$. Because of its high lipophilicity, the $NO-Fe(DETC)_2$ complex is exclusively localized in the EC and not in the medium. The frozen samples can be stored at −80° and transported on dry ice. Before analysis, the frozen samples are removed by pushing the syringe plunger from the slightly warmed plastic syringe barrel. The samples are then chilled with liquid nitrogen and loaded into a finger dewar. The sample column is stabilized in the dewar with a cotton-tipped applicator, and the dewar is filled with liquid nitrogen and placed in a Bruker EMX ESR spectrometer for analysis. ESR spectrometer settings are as follows: microwave power, 10 mW; modulation frequency, 100 kHz; modulation amplitude, 5 G; field center, 3290 G; sweep width, 90 G; microwave frequency, 9.39 GHz; conversion time, 328 ms; time constant, 5.24 s; number of scans, 4; sweep time, 168 s. Data processing is performed by WIN-EPR software (Bruker BioSpin Corporation).

Nonstimulated bovine aortic ECs (BAECs) incubated with colloid Fe $(DETC)_2$ exhibited the triplet ESR signal characteristic of NO-Fe $(DETC)_2$ (g = 2.035; $A_N$ = 12.6 G), reflecting basal NO production (Fig. 3A). As expected, the calcium ionophore, A23187, sharply increased NO production, whereas the NOS inhibitor, L-NAME, completely blocked NO production. Scavenging of intracellular superoxide with PEG-SOD caused only a minor increase in the NO signal, whereas acute treatment with the redox cycling agent menadione (5 $\mu M$) completely blocked NO detection due to inactivation of NO by intracellular superoxide, an effect that was prevented by PEG-SOD (Fig. 3A). The amount of NO-Fe $(DETC)_2$ complex can be calculated from the calibration curve obtained by adding graded concentrations of MAHMA-NONOate to $Fe-MGD_2$ (Fig. 3B). Correlation of the integral intensity with the ESR amplitude of $NO-Fe(DETC)_2$ (Fig. 3C) provides the equation for calculating the amount of NO detected: [NO] = 0.3 [ESR amplitude] ($nM$). ESR data can be expressed as picomole per milligram of EC protein (Fig. 3A). Calculation of the amount of NO detected by analysis of the ESR amplitude is particularly useful in cases in which the ESR signal is weak or overlaps partially with other bioradicals (detection limit of $NO-Fe(DETC)_2$ ~10 n$M$). However, the line width of the $NO-Fe(DETC)_2$ ESR spectrum may vary considerably because of variations in the amount of $Fe(DETC)_2$ in membrane lipids and the amount of $Fe^{3+}$ in the $Fe(DETC)_2$ preparation. Thus, ESR amplitude can be used if the line widths of ESR spectra are the same.

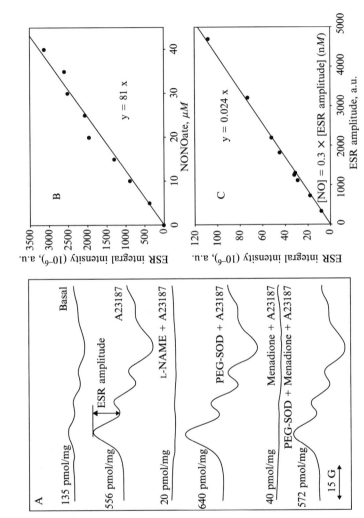

Fig. 3. Electron spin resonance (ESR) spectra of NO-Fe(DETC)$_2$ and calibration curves. (A) ESR spectra of NO-Fe(DETC)$_2$ in untreated bovine aortic ECs (BAECs) (basal) or treated with A23187, A23187 + L-NAME, PEG-SOD + A23187, Menadione + A23187, PEG-SOD + Menadione + A23187. (B) Calibration curve for NO-Fe(MGD)$_2$ prepared by MAHMA-NONOate. (C) Correlation of integral intensity with ESR amplitude of NO-Fe(DETC)$_2$.

Otherwise, integral intensities calculated by double integration of the ESR spectra provide more accurate calculation of NO concentration.

There are several advantages for EC NO detection with ESR and Fe $(DETC)_2$. $Fe(DETC)_2$ is specific for bioactive NO and does not detect nitrite or nitrate. In addition, the stability of the $NO\text{-}Fe(DETC)_2$ complex allows for measurement of the cumulative amount of bioactive NO produced over time. Limitations of this technique include the special handling required for $Fe(DETC)_2$ colloid to prevent oxidation. In addition, $NO\text{-}Fe^{2+}(DETC)_2$ can be oxidized by extracellular $H_2O_2$ or superoxide to form the ESR silent $NO\text{-}Fe^{3+}(DETC)_2$. However, our data show that bolus addition of 100 $\mu M$ $H_2O_2$ or the same amount of superoxide generated by xanthine oxidase decreased the $NO\text{-}Fe^{2+}(DETC)_2$ signal by 20%. Both $H_2O_2$ and superoxide caused line broadening of the ESR spectra because of accumulation $Fe^{3+}$ in the samples.

### Electrochemical Detection of EC NO Production

Electrochemical detection of NO has certain advantages, including temporal resolution, spatial resolution, and sensitivity with a detection limit sufficient for biological samples (Brovkovych et al., 1999; Malinski et al., 1996). In this technique, the voltage of an electrode is controlled to catalyze NO oxidation on the electrode surface. At a chosen sampling time, the current generated is a linear function of NO concentration at the electrode surface. Two common recording modes can be employed: cyclic voltametry and amperometry. In cyclic voltametry, voltage is varied, and current is measured. This mode is best suited for the identification of species, because the voltage corresponding to an increase in faradic current is determined by the half-cell potential (Fig. 4). In amperometry, voltage is held constant above the oxidation potential, and current is measured. This mode is used commonly to quantify species concentration. Electrodes are typically filaments made of carbon or platinum (Malinski et al., 1996; Xian et al., 2001). Electrodes are made more specific by coatings that either attract NO (Malinski and Taha, 2004) or exclude other oxidizable species (Friedemann et al., 1996). The following procedures describe techniques for measuring EC NO production using a carbon fiber electrode coated with Nafion and o-phenylenediamine (o-PD) exclusion coatings as previously reported (Friedemann et al., 1996). Although other electrodes are available, even commercially, we have chosen this electrode because of its small tip size, ease of fabrication, reliability, and low cost.

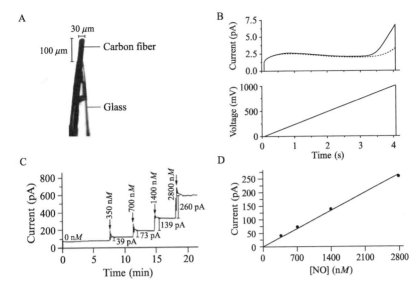

FIG. 4. Calibration of nitric oxide (NO) electrode. (A) Glass-encased, cylindrical carbon fiber electrode (30 $\mu$m × 100 $\mu$m exposed area). (B) Voltametry showing faradic current generated by NO. The dashed line shows current elicited by the voltage ramp in phosphate-buffered saline (PBS) only. With NO present in solution, an additional, faradic current representing the oxidation of NO can be seen at voltages above about +750 mV relative to an AgCl reference electrode. (C) Serial additions of a concentrated solution of NO are added and current recorded to create a calibration curve. (D) Using a cylindrical Nafion and o-PD coated carbon fiber electrode (30 $\mu$m × 100 $\mu$m exposed area), the detection limit is about 10 nM, and the response is linear up to 4 $\mu$M. The sensitivity from this calibration curve is 0.095 pA/nM, with $R^2 = 0.996$.

## Electrode Fabrication

Electrodes are fabricated as described previously (Friedemann et al., 1996) with a few simplifications. Suitable Nafion-coated carbon electrodes can be purchased commercially (World Precision Instruments, Inc., Sarasota, FL), but electrode integrity must be confirmed. An o-PD coating can be applied to increase NO specificity by polymerizing 5 mM o-PD on the electrode surface at +900 mV and 37° for 45 min. The o-PD solution is made fresh by dissolving one tablet o-PD (3 mg substrate/tablet, Sigma, St. Louis, MO) in 3.3 ml of 100 $\mu$M ascorbic acid in 0.1 M phosphate-buffered saline (PBS). After coating, the electrode is allowed to dry for 1 min, dipped in water, and then allowed to dry overnight. Electrodes are stored at 4° and can be used for several months after coating. Before each experiment, electrodes are hydrated by soaking in PBS for at least 1 h to increase baseline electrical stability. This form of exclusionary coating results in an electrode that is relatively selective for NO when compared with other biological species with oxidation potentials in this range, such as $H_2O_2$, superoxide, and ascorbic acid. To increase measurement accuracy in cells and tissue, inhibitors of NOS, such as $N_\omega$-nitro-L-arginine methyl ester (L-NAME) can be added during measurements, and the difference in current before and after exposure can be used to calculate NO concentration.

*Data Acquisition and Analysis*

Measurements are taken relative to a reference electrode (*e.g.*, AgCl), and noise reduction is accomplished by low-pass filtering. Electrodes should be calibrated before each experiment using dilutions of a saturated NO solution (Fig. 4). This solution is produced in a closed system by bubbling NO gas through deionized water that has been deoxygenated. Immediately after saturation, an NO solution at 20° has a concentration of 2 m*M* (Friedemann *et al.*, 1996; Mesáros, 1999). In a sealed system on ice, the solution NO concentration stabilizes at 1.4 m*M* over a 20-min period.

*Application to Cultured ECs*

Although electrodes are sufficiently sensitive to measure NO from single ECs (Malinski and Taha, 2004), we have generally measured EC NO production from cultured monolayers grown in 35-mm dishes at

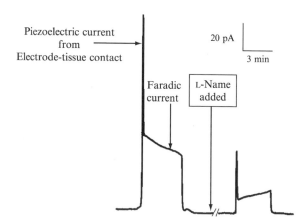

Fig. 5. Faradic current suppression by L-NAME. Charged at +900 mV, the electrode is advanced into contact with the tissue and withdrawn, generating a current spike and plateau. Repeating this procedure after incubation with 2 m*M* L-NAME shows suppression of most of the faradic current. The difference between the two plateau currents is assumed to represent nitric oxide (NO) concentration.

50–75% confluency (Cai *et al.*, 2003; Lynn *et al.*, 2004). To help minimize signal contamination, measurements are made with cells in 0.1 *M* PBS (pH 7.4) rather than in culture medium. With the dish of cells placed in a temperature controller (37°) on an inverted microscope stage, the electrode is positioned several micrometers away from the monolayer surface using a micromanipulator. The NO concentration is a function of distance from the cells, so the electrode must be placed at a consistent distance above the monolayer. Baseline current measurements are made with the electrode voltage held at +900 mV. The L-NAME suppressible signal is taken as indicative of the NO faradic current and can be converted to an NO concentration (Fig. 5).

In conclusion, electrochemistry can be used to reliably measure EC NO production. The method is desirable because of its spatial and temporal resolution. Although we have described methods for measuring EC NO production using Nafion and o-PD carbon fiber electrodes, these techniques can also be applied to other tissues. Alterations in the electrode design have allowed use of electrochemistry to detect other biologically relevant molecules, such as superoxide and $H_2O_2$ (Lacy *et al.*, 1998; McNeil *et al.*, 1995).

## References

Brovkovych, V., Stolarczyk, E., Oman, J., Tomboulian, P., and Malinski, T. (1999). Direct electrochemical measurement of nitric oxide in vascular endothelium. *J. Pharm. Biomed. Anal.* **19**, 135–143.

Cai, H., Li, Z., Dikalov, S., Holland, S. M., Hwang, J., Jo, H., Dudley, S. C., Jr., and Harrison, D. G. (2002). NAD(P)H oxidase–derived hydrogen peroxide mediates endothelial nitric oxide production in response to angiotensin II. *J. Biol. Chem.* **277**, 48311–48317.

Cai, H., Li, Z., Davis, M. E., Kanner, W., Harrison, D. G., and Dudley, S. C., Jr. (2003). Akt-dependent phosphorylation of serine 1179 and mitogen-activated protein kinase kinase/extracellular signal-regulated kinase 1/2 cooperatively mediate activation of the endothelial nitric-oxide synthase by hydrogen peroxide. *Mol. Pharmacol.* **63**, 325–331.

Calnek, D. S., Mazella, L., Roser, S., Roman, J., and Hart, C. M. (2003). Peroxisome proliferator-activated receptor gamma ligands increase release of nitric oxide from endothelial cells. *Arterioscler. Thromb. Vasc. Biol.* **23**, 52–57.

Dikalov, S., Landmesser, U., and Harrison, D. G. (2002). Geldanamycin leads to superoxide formation by enzymatic and non-enzymatic redox cycling: Implications for studies of Hsp90 and eNOS. *J. Biol. Chem.* **277**, 25480–25485.

Friedemann, M. N., Robinson, S. W., and Gerhardt, G. A. (1996). o-Phenylenediamine-modified carbon fiber electrodes for the detection of nitric oxide. *Anal. Chem.* **68**, 2621–2628.

Gupta, M. P., Steinberg, H. O., and Hart, C. M. (1998). $H_2O_2$ causes endothelial barrier dysfunction without disrupting the arginine-nitric oxide pathway. *Am. J. Physiol. Lung Cell. Mol. Physiol.* **18**, L508–L516.

Kleinhenz, D. J., Fan, X., Rubin, J., and Hart, C. M. (2003). Detection of endothelial nitric oxide release with the 2,3-diaminonapthalene assay. *Free Radic. Biol. Med.* **34**, 856–861.

Kleschyov, A. L., and Munzel, T. (2002). Advanced spin trapping of vascular nitric oxide using colloid iron diethyldithiocarbamate. *Methods Enzymol.* **359**, 42–51.

Kuzkaya, N., Weissmann, N., Harrison, D. G., and Dikalov, S. (2003). Interactions of peroxynitrite, tetrahydrobiopterin, ascorbic acid and thiols: Implications for uncoupling endothelial nitric oxide synthase. *J. Biol. Chem.* **278**, 22546–22554.

Lacy, F., O'Connor, D. T., and Schmid-Schonbein, G. W. (1998). Plasma hydrogen peroxide production in hypertensives and normotensive subjects at genetic risk of hypertension. *J. Hypertens.* **16**, 291–303.

Lynn, M., Rupnow, H., Kleinhenz, D., Kanner, W., Dudley, S., and Hart, C. M. (2004). Fatty acids differentially modulate insulin-stimulated endothelial nitric oxide production by an Akt-independent pathway. *J. Invest. Med.* **52**, 129–136.

Malinski, T., and Czuchajowski, L. (1996). Nitric oxide measurement by electrochemical methods. *In* "Methods in Nitric Oxide Research" (M. Feelisch and M. Stamler, eds.), pp. 319–339. John Wiley & Sons, West Sussex, England.

Malinski, T., and Taha, Z. (2004). Nitric oxide release from a single cell measured *in situ* by a porphyrinic-based microsensor. *Nature* **358**, 676–678.

McNeil, C. J., Athey, D., and Ho, W. O. (1995). Direct electron transfer bioelectronic interfaces: Application to clinical analysis. *Biosens. Bioelectron.* **10**, 75–83.

Mesáros, Š. (1999). Determination of nitric oxide saturated solution by amperometry on modified microelectrode. *Methods Enzymol.* **301**, 160–168.

Morley, D., and Keefer, L. (1993). Nitric oxide/nucleophile complexes: A unique class of nitric oxide–based vasodilators. *J. Cardiovasc. Pharmacol.* **22**, S3–S9.

Xian, Y., Liu, M., Cai, Q., Li, H., Lu, J., and Jin, L. (2001). Preparation of microporous aluminium anodic oxide film modified Pt nano array electrode and application in direct measurement of nitric oxide release from myocardial cells. *Analyst* **126**, 871–876.

# [43] Use of Microdialysis to Study Interstitial Nitric Oxide and Other Reactive Oxygen and Nitrogen Species in Skeletal Muscle

*By* MALCOLM J. JACKSON

## Abstract

Microdialysis techniques can be used to sample the interstitial space of tissues such as skeletal muscle. Analytical developments have allowed adaptations of these techniques to permit continuous monitoring of nitric oxide and a number of other reactive oxygen and nitrogen species in skeletal muscle extracellular space. Methods are described for assessment of interstitial nitrate and nitrite content, superoxide anion content, hydroxyl radical activity, and the content of relatively stable lipid radicals detectable using spin trapping and electron spin resonance techniques in skeletal muscle of rodents at rest and during contractile activity.

METHODS IN ENZYMOLOGY, VOL. 396
0076-6879/05 $35.00
DOI: 10.1016/S0076-6879(05)96043-6

Introduction

Reactive oxygen species (ROS) and reactive nitrogen species (RNS) including nitric oxide (NO), hydrogen peroxide, superoxide, and the hydroxyl radical have been implicated in the pathogenesis of many chronic disorders, but evaluation of their precise roles has been limited by the lack of appropriate techniques to directly monitor specific species in humans and animals *in vivo*. Current techniques rely almost entirely on indirect measurements of the end products of reaction of ROS with DNA, protein, or lipids (Halliwell and Gutteridge, 1989). ROSs play roles in normal physiological and pathogenetic processes and occur in all parts of the body. Relatively little is known about the roles of extracellular ROS, although neutrophils and other cells release substantial amounts of ROS into extracellular fluid (ECF) during normal activation. Nevertheless, it appears likely that extracellular ROSs may play a role in fundamental processes such as cell–cell signaling (Hancock, 1997; Jalkanen and Salmi, 2001) and aging (de Grey, 2000).

Analysis of the composition of the tissue extracellular (or interstitial) space is difficult because of lack of access to that body compartment. Microdialysis techniques provide a potential means of monitoring the composition of the tissue ECF that may not be available by more conventional means. The technique involves the positioning of a small probe containing a dialysis membrane into the tissue of interest. The membrane is perfused on the inside with a physiological medium at a very slow flow rate. The outside of the membrane is in contact with the tissue ECF. This allows water and low-molecular-weight compounds in the tissue ECF to perfuse across the dialysis membrane to be collected via the outlet port. Therefore, microdialysis allows continuous sampling of compounds in the ECF in close proximity to cells, providing accurate information on local chemical changes before such compounds are cleared and diluted in the circulatory system. The theory and application of the technique have been comprehensively reviewed by Benveniste and Huttemeier (1990).

Principles of Tissue Microdialysis and Potential for Measurements of
Extracellular ROS

A common comparison that is made in explaining the principles of microdialysis is that it mimics the passive functions of a small artificial blood vessel. The dialysis membrane is a hollow tubular membrane that is permeable to water and small solutes. The membrane is continuously perfused with a physiological solution that contains no substance of interest. This creates a concentration gradient along the membrane. The diffusion of substances will occur in the direction of the lowest concentration. There will potentially be bi-directional molecular and ionic traffic between the interior of the microdialysis probe and the surrounding tissue. Using this continuous flow through the membrane, the compound of interest diffuses into the perfusate and is collected for analysis. The constant flow results in incomplete equilibration of the dialysate with the ECF. This incomplete recovery is improved by the use of slow flow rates and relatively small dialysate volumes compared with area of the dialysis membrane (Ungerstedt, 1986). In practice, variations in recoveries are also minimized by the use of high-quality precision pumps to carefully deliver microliter volumes of perfusate over defined periods.

Attempts have been made to detect ROS in ECF by microdialysis since the beginning of the 1990s with early studies examining NO release (Balcioglu and Maher, 1993), spin trapping techniques to detect lipid radicals (Zini et al., 1992), and hydroxyl radical activity (Chiueh et al., 1992). Most of the initial studies reported data from rodent brain subjected to experimental models of pathologies, although subsequently the technique has been applied to the study of ECF in many other tissues including the spinal cord, cardiac tissue, skin, and skeletal muscle in several animal species and to a more limited extent in humans. In humans, studies of NO in the skin are relatively common (Clough et al., 1998; Katugampola et al., 2000), but ROS measurements by microdialysis techniques have also been undertaken in human skeletal muscle (Bangsbo, 1999) and in human brain during surgery (Marklund et al., 2000).

A number of approaches have been followed to examine ROS:

a. Direct detection of relatively stable ROS that cross the dialysis membrane and are collected via the outlet tube (e.g., $H_2O_2$) (Hyslop et al., 1995);

b. Analysis of stable metabolites of ROS that cross the dialysis membrane (e.g., nitrite and nitrate levels as a measure of NO generation) (Shinani et al., 1994);

c. Perfusion of microdialysis probes with "trapping" molecules that react with ROS to generate specific products that are collected and analyzed. Examples include the following:

   i. Detection of 2,3-dihydroxybenzoates formed from salicylate as an indicator of hydroxyl radical activity (Chiueh et al., 1992);
   ii. Perfusion with spin trapping agents followed by electron spin resonance analysis of the spin adducts (Zini et al., 1992);
   iii. Perfusion with hemoglobin to react with NO, followed by measurement of methemoglobin (Balcioglu and Maher, 1993);

iv. Perfusion with cytochrome *c* to examine its reduction by superoxide (McArdle *et al.*, 2001);

d. Measurement of low-molecular-weight pro-antioxidant and antioxidants that dialyze from the ECF and are collected via the outlet tube [*e.g.*, iron and ascorbate (Leveque *et al.*, 2003), uric acid (Marklund *et al.*, 2000), and glutathione (Sirsjo *et al.*, 1996)];

These techniques could also be used to obtain further information on ROS metabolism by examining the levels in the ECF of other low-molecular-weight substances that may be substrates for ROS-generating reactions [*e.g.*, hypoxanthine as a substrate for xanthine oxidase (Marklund *et al.*, 2000)] or contribute to the redox status of cells [*e.g.*, cysteine (Landolt *et al.*, 1991)].

## Use of Microdialysis to Measure Extracellular ROS in Skeletal Muscle

The increasing availability of microdialysis probes from commercial sources has greatly improved the range and reliability of probes for the study of skeletal muscle. Two main types of probe have been used: The "loop" dialysis membrane is effectively a short segment of a continuous tube, whereas "linear" probes have a concentric tube arrangement whereby the perfusion fluid enters through an inner tube, flows to its distal end, exits the tube, and enters the space between the inner tube and an outer dialysis membrane. The perfusion fluid then moves toward the proximal end of the probe, where the dialysis takes place. Various probes with different molecular weight cutoffs are available. The "linear" probes are widely used in comparison with other types. A schematic diagram of the placement of a probe in skeletal muscle is shown in Fig. 1. Examples of different types of probes can be found at http://www.bioanalytical.com/products/md/dlprob.html, http://www.microdialysis.se/probes.html, and http: //www.microbiotech.se.

We have experience only with the "linear" probes and have regularly used 10- or 4-mm membrane probes with 0.5-mm diameter and a molecular weight cutoff of 35,000 Da (e.g., from Metalant, Sweden). Similar probes are commercially available for use in a wide variety of animal species, and some have been specifically designed and approved for clinical use in humans.

## Experimental Procedures

### *Placement of the Probe*

Microdialysis probes are placed into suitable muscles of anesthetized rats or mice. In the rat, we have routinely examined the *tibialis anterior* muscle, but in the mouse the *gastrocnemius* muscle has been more

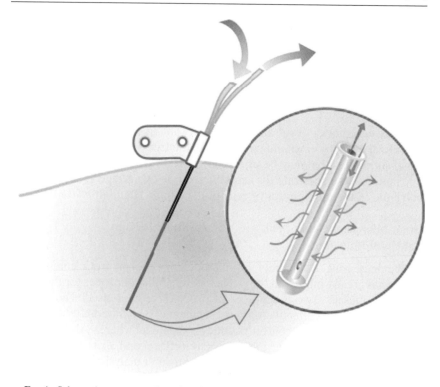

FIG. 1. Schematic representation of a microdialysis probe inserted into a limb muscle. The insert shows the potential for low-molecular-weight substances to diffuse in and out of the probe. Courtesy of CMA Microdialysis. (See color insert.)

frequently used to obtain the necessary muscle bulk. For placement of the probes (MAB 3.8.10, Metalant, Sweden), we use a splittable 22-gauge plastic introducer. It is technically feasible to place three or four probes in a single mouse *gastrocnemius,* facilitating multiple measurements from the same muscle with the probes remaining in the muscle of the anesthetized rodent for up to 4 h.

In humans, probes can be placed under local anesthetic. We have placed clinical probes in the *quadriceps* and *tibialis anterior* muscles, although other muscles could be examined.

### Detection of Hydroxyl Radical Activity in Muscle Extracellular Space in Mice

*Procedure.* Mice are anesthetized and probes (MAB 3.8.10, Metalant, Sweden) placed as described earlier. The microdialysis probe is perfused with 20 m$M$ salicylate in normal saline at a flow rate of 4 $\mu$l/min and allowed to stabilize for 30 min. Samples are then collected from the outlet tubes of the probe over sequential 15-min periods. The 2,3-dihydroxybenzoic acid (2,3-DHB) and 2,5-dihydroxybenzoic acid (2,5-DHB) generated from salicylate in the microdialysis fluids are measured as an index of reaction with hydroxyl radicals (Richmond *et al.*, 1981). 2,3-DHB and 2,5-DHB are measured by high-performance liquid chromatography (HPLC) with electrochemical detection, as described by Halliwell *et al.* (1988). Our HPLC system consists of a Rheodyne injector, HPLC pump (Gilson Model 303), Spherisorb 5 ODS column (HPLC technology): 25 × 4.6 mm with guard column and C-8 cartridge (BDH), and an electrochemical detector (Gilson Model 141). The HPLC eluant consists of 34 m$M$ sodium citrate, 27.7 m$M$ acetate buffer (pH 4.75) mixed with methanol 97.2:2.8 (v/v). Standard solutions of 2,3-DHB and 2,5-DHB are prepared in HPLC-grade water. Twenty microliters of samples or standards is eluted at a flow rate of 0.9 ml/min and monitored at +65 V with the electrochemical detector.

With the conditions and probes described, we have reported the formation of approximately 150 pmol 2,3-DHB/15 min from mouse skeletal muscle at rest (McArdle *et al.*, 2004). These values change if different length probes are used, if the molecular weight cutoff differs, or if the flow rate is varied. They are also likely to change with probes of different composition, although we have not examined this.

We have examined the effect of electrical stimulation of contraction on 2,3-DHB formation in the ECF of the mouse *gastrocnemius* muscle, and an increase of approximately 100% was observed (McArdle *et al.*, 2004) (Fig. 2).

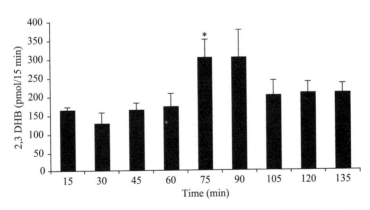

FIG. 2. Time course of production of 2,3-dihydroxybenzoate (DHB) from salicylate in microdialysates from the *gastrocnemius* muscle of anesthetized wild-type mice. The muscle was stimulated to undertake a protocol of demanding contractions via surface electrodes between 60 and 75 min. Redrawn, with permission, from McArdle *et al.* (2004).

*Comment and Limitations.* There are a number of general drawbacks with the microdialysis to assess ROS in skeletal muscle, and some specific points related to the technique used for assessment of extracellular hydroxyl radical activity should be considered:

• The cellular source of material detected in the microdialysates cannot be defined because the skeletal muscle ECF will be influenced by multiple cell types, such as endothelial cells and white cells, in addition to skeletal muscle cells.

• Insertion of the probe must cause some local trauma to the tissue, which might theoretically influence ROS measurements to a greater degree than other substances that are measured by microdialysis.

• Recoveries of ROS across the microdialysis membrane have not been defined and are much more difficult to quantify than for relatively stable molecules. Attempts to undertake such studies have been relatively unsuccessful (Pattwell *et al.*, 2001).

• Formation of 2,3-DHB from salicylate in biological systems has been claimed to occur specifically through hydroxyl radical–mediated hydroxylation, although *in vitro* studies have indicated that reaction with peroxynitrite also leads to hydroxylation of salicylate, and whether these reactions involve intermediary generation of hydroxyl radicals is unclear (see discussion in McArdle *et al.*, 2004). Inhibitor studies performed by our group are compatible with a major role for hydroxyl radical in forming 2,3-DHB from salicylate in muscle microdialysates (McArdle *et al.*, 2004).

### Detection of Superoxide Anion in Muscle Extracellular Space in Mice

*Procedure.* Mice are anesthetized and probes (MAB 3.8.10, Metalant, Sweden) placed as described earlier. The microdialysis probe is perfused with 50 $\mu M$ cytochrome $c$ in normal saline at a flow rate of 4 $\mu l/min$ and allowed to stabilize for 30 min. Samples are then collected from the outlet tubes of the probe over sequential 15-min periods. Reduction of cytochrome $c$ in the microdialysate is used as an index of superoxide radical in the microdialysate. Samples are analyzed using scanning visible spectrometry, and the superoxide content is calculated from the absorbance at 550 nm in comparison with the isobestic wavelengths at 542 and 560 nm. A molar extinction coefficient for reduced cytochrome $c$ of 21,000 is used for calculation of the superoxide anion concentration.

With the conditions and probes described, we have reported that levels of approximately 0.4 nmol superoxide/15 min can be detected in the ECF of mouse skeletal muscle at rest (McArdle *et al.*, 2004). Again, our experience demonstrates that these values change if different length probes are used, if the molecular weight cutoff differs, or if the flow rate is varied; and these values are also likely to change with probes of different composition, although we have not examined this. The values for superoxide tend to decrease over the initial sequential 15-min collection periods but eventually stabilize (McArdle *et al.*, 2004; Pattwell *et al.*, 2001). We have examined the effect of electrical stimulation of contraction on the reduction of cytochrome *c* in microdialysates from the ECF of the mouse *gastrocnemius* muscle, and an increase of approximately 60% was observed (McArdle *et al.*, 2004) (Fig. 3).

*Comment and Limitations.* The general comments concerning use of microdialysis to examine ROS in muscle ECF mentioned earlier also apply to superoxide measurements, but specifically the effect of trauma to tissues during probe insertion appears particularly important for superoxide detection. Levels that are detected fall during the first few sets of collections to eventually plateau (Fig. 3).

The lack of specificity of cytochrome *c* reduction as a measure of superoxide is a potentially important drawback with this approach. In our initial study, we observed that levels of cytochrome *c* in mouse microdialysates were reduced by about 50% on addition of purified superoxide dismutase (SOD) to the perfusate (McArdle *et al.*, 2001). It is likely that at least a proportion of the reduction of cytochrome *c* occurs outside the microdialysis probe because cytochrome *c* has a molecular weight of about

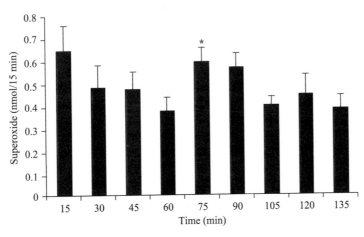

FIG. 3. Superoxide release monitored by reduction of cytochrome *c* in microdialysates from the *gastrocnemius* muscle of anesthetized wild-type mice. The muscle was stimulated to undertake a protocol of demanding contractions via surface electrodes between 60 and 75 min. Redrawn, with permission, from McArdle *et al.* (2004).

12 kDa, and the dialysis membrane cutoff is 35 kDa, so there will be substantial diffusion of the cytochrome $c$ out of the probe. In addition, because of the slow flow rate, there will be diffusion back into the probe. In that case, addition of high-molecular-weight purified SODs to the microdialysis fluid could not prevent the reduction of cytochrome $c$ by superoxide that occurs outside the probe. Our inhibitor data indicate that NO is unlikely to make a significant contribution to the reduction of cytochrome $c$ in this system. It has also been suggested that small molecules such as ascorbate or glutathione can reduce cytochrome $c$ *in vivo*, but in unpublished studies, we have observed that reduction of microdialysate cytochrome $c$ stopped immediately on the death of the mouse, suggesting a dependence of this reduction on metabolic activity that is not compatible with this hypothesis. Overall, therefore, the data are consistent with superoxide playing a substantial role in the reduction of cytochrome $c$ in microdialysates, but the lack of complete suppression of the reduction by exogenous SOD means that we cannot define precise levels from the current data (McArdle *et al.*, 2004).

### Detection of ROS in Muscle Extracellular Space Using Spin Traps with Electron Spin Resonance Detection

*Procedure.* Rodents are anesthetized and probes (MAB 3.8.10, Metalant, Sweden) placed as described earlier. The microdialysis probe is perfused with 140 mM $\alpha$-phenyl-*tert*-butylnitrone (PBN) in normal saline at a flow rate of 4 μl/min and allowed to stabilize for 30 min. Samples are then collected from the outlet tubes of the probe over sequential 15-min periods. Microdialysate samples are subsequently added to an equal volume of toluene, and the PBN adducts are extracted and stored at −80° until analyzed by electron spin resonance (ESR). Samples are degassed before analysis by X-band ESR.

Muscle microdialysis samples have been found to contain a PBN adduct that is similar or identical to that previously reported from spin trapping experiments with human and rat blood (Pattwell *et al.*, 2003). This ESR-detectable species has been described as a carbon or alkoxyl radical that may be a relatively stable intermediate, or end product, of lipid peroxidation. The level of this species has been shown to increase in rat muscle microdialysates after a period of ischemia and reperfusion (Pattwell *et al.*, 2003) (Fig. 4).

*Comment and Limitations.* PBN has been the spin trap of choice for microdialysis studies, but whether other spin trapping agents will produce greater information is unclear. Further studies in this area are required.

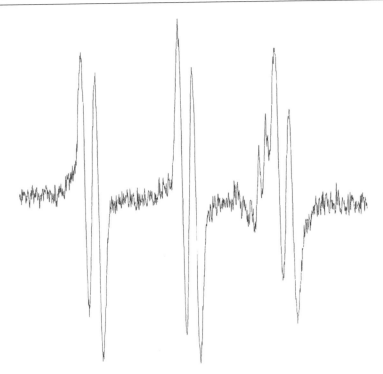

Fig. 4. Example of electron spin resonance spectra of α-phenyl-*tert*-butylnitrone (PBN)–trapped species in microdialysates from the *gastrocnemius* muscle of anesthetized wild-type rats. The muscle was subjected to a prolonged period of ischemia (4 h) followed by reperfusion for 1 h. Redrawn, with permission, from Pattwell *et al.* (2003).

## Detection of NO in Muscle Extracellular Space by Analysis of Nitrite and Nitrite Content

*Procedure.* Mice are anesthetized and probes (MAB 3.8.10, Metalant, Sweden) placed as described earlier. The microdialysis probe is perfused with normal saline at a flow rate of 4 μl/min and allowed to stabilize for 30 min. Samples are then collected from the outlet tubes of the probe over sequential 15-min periods. It has been reported that samples can be analyzed for nitrite and nitrate content by the Griess reaction (Shintani *et al.*, 1994), but to ensure sufficient sensitivity with the very small volumes of sample available (60 μl/15-min collection), we have used a commercial fluorometric assay (Cayman Chemical Co.) based on the method of Miles *et al.* (1995).

We have observed resting total nitrite and nitrate values of approximately 250 pmol/15 min in microdialysates from mouse skeletal muscle at rest with no major change after contractile activity (Vasilaki et al., in preparation).

*Comment and Limitations.* Recoveries of nitrate and nitrite across the microdialysis membrane can be readily evaluated because of the stable nature of these compounds, and hence, absolute values for NO generation can be calculated (Yamada and Nabeshima, 1997). However, there appear to be no comparative studies of the efficacy and utility of measurement of NO activity indirectly by examination of nitrite and nitrate content in microdialysates with direct detection of NO through, for instance, reaction with hemoglobin in the perfusate or amperometric approaches. Such studies are clearly required to validate the approach.

## Concluding Remarks

Microdialysis offers a potentially powerful approach to monitoring ROSs within tissues *in vivo*. Although there are still many reservations about the optimum techniques that should be used for specific ROSs, it seems that microdialysis can provide novel and reliable real-time evaluation of ROSs in tissue ECF. A wider utilization and thorough evaluation of the technique are likely to lead to a greater understanding of the roles of extracellular ROS in normal physiology and in disease states.

## References

Balcioglu, A., and Maher, T. J. (1993). Determination of kainic acid–induced release of nitric oxide using a novel hemoglobin trapping technique with microdialysis. *J. Neurochem.* **61,** 2311–2313.

Bangsbo, J. (1999). Vasoactive substances in the interstitium of contracting skeletal muscle examined by microdialysis. *Proc. Nutr. Soc.* **58,** 925–933.

Benveniste, H., and Huttemeier, P. C. (1990). Microdialysis—Theory and application. *Prog. Neurobiol.* **35,** 195–215.

Chiueh, C. C., Krishna, G., Tulsi, P., Obata, T., Lang, K., Huang, S. J., and Murphy, D. L. (1992). Intracranial microdialysis of salicylic acid to detect hydroxyl radical generation through dopamine autooxidation in the caudate nucleus: Effects of MPP+. *Free Radic. Biol. Med.* **13,** 581–583.

Clough, G. F., Bennett, A. R., and Church, M. K. (1998). Measurement of nitric oxide concentration in human skin *in vivo* using dermal microdialysis. *Exp. Physiol.* **83,** 431–434.

de Grey, A. D. (2000). The reductive hotspot hypothesis: An update. *Arch. Biochem. Biophys.* **373,** 295–301.

Halliwell, B., and Gutteridge, J. M. C. (1989). "Free Radical Biology and Medicine." Oxford University Press, Oxford.

Halliwell, B., Grootveld, M., and Gutteridge, J. M. (1988). Methods for the measurement of hydroxyl radicals in biomedical systems: Deoxyribose degradation and aromatic hydroxylation. *Methods Biochem. Anal.* **33,** 59–90.

Hancock, J. T. (1997). Superoxide, hydrogen peroxide and nitric oxide as signalling molecules: Their production and role in disease. *Br. J. Biomed. Sci.* **54,** 38–46.

Hyslop, P. A., Zhang, Z., Pearson, D. V., and Phebus, L. A. (1995). Measurement of striatal H2O2 by microdialysis following global forebrain ischemia and reperfusion in the rat: Correlation with the cytotoxic potential of H2O2 *in vitro. Brain Res.* **671,** 181–186.

Jalkanen, S., and Salmi, M. (2001). Cell surface monoamine oxidases: Enzymes in search of a function. *EMBO J.* **20,** 3893–3901.

Katugampola, R., Church, M. K., and Clough, G. F. (2000). The neurogenic vasodilator response to endothelin-1: A study in human skin *in vivo. Exp. Physiol.* **85,** 839–846.

Landolt, H., Langemann, H., Lutz, T., and Gratzl, O. (1991). Non-linear recovery of cysteine and glutathione in microdialysis. *In* "Monitoring Molecules in Neurosciences" (H. Rollema, B. H. C. Wersterink, and W.-J. Djirflout, eds.). Krips Repro, Holland.

Leveque, N., Robin, S., Makki, S., Muret, P., Rougier, A., and Humbert, P. (2003). Iron and ascorbic acid concentrations in human dermis with regard to age and body sites. *Gerontology* **49,** 117–122.

McArdle, A., Pattwell, D., Vasilaki, A., Griffiths, R. D., and Jackson, M. J. (2001). Contractile activity-induced oxidative stress: Cellular origin and adaptive responses. *Am. J. Physiol. Cell Physiol.* **280,** C621–C627.

McArdle, A., van der Meulen, J., Close, G. L., Pattwell, D., Van Remmen, H., Huang, T. T., Richardson, A. G., Epstein, C. J., Faulkner, J. A., and Jackson, M. J. (2004). Role of mitochondrial superoxide dismutase in contraction-induced generation of reactive oxygen species in skeletal muscle extracellular space. *Am. J. Physiol. Cell Physiol.* **286,** C1152–C1158.

Marklund, N., Ostman, B., Nalmo, L., Persson, L., and Hillered, L. (2000). Hypoxanthine, uric acid and allantoin as indicators of *in vivo* free radical reactions. Description of a HPLC method and human brain microdialysis data. *Acta Neurochir. (Wien)* **142,** 1135–1141.

Miles, A. M., Chen, Y., Owens, M. W., and Grisham, M. B. (1995). Fluorimetric determination of nitric oxide. *Methods* **7,** 40–47.

Pattwell, D., McArdle, A., Griffiths, R. D., and Jackson, M. J. (2001). Measurement of free radical production by *in vivo* microdialysis during ischemia/reperfusion injury to skeletal muscle. *Free Radic. Biol. Med.* **30,** 979–985.

Pattwell, D., Ashton, T., McArdle, A., Griffiths, R. D., and Jackson, M. J. (2003). Ischemia and reperfusion of skeletal muscle lead to the appearance of a stable lipid free radical in the circulation. *Am. J. Physiol. Heart Circ. Physiol.* **284,** H2400–H2404.

Richmond, R., Halliwell, B., Chauhan, J., and Darbre, A. (1981). Superoxide-dependent formation of hydroxyl radicals: Detection of hydroxyl radicals by the hydroxylation of aromatic compounds. *Anal. Biochem.* **118,** 328–335.

Shintani, F., Kanba, S., Nakaki, T., Sato, K., Yagi, G., Kato, R., and Asai, M. (1994). Measurement by *in vivo* brain microdialysis of nitric oxide release in the rat cerebellum. *J. Psychiatry Neurosci.* **19,** 217–221.

Sirsjo, A., Arstrand, K., Kagedal, B., Nylander, G., and Gidlof, A. (1996). *In situ* microdialysis for monitoring of extracellular glutathione levels in normal, ischemic and post-ischemic skeletal muscle. *Free Radic. Res.* **25,** 385–391.

Ungerstedt, U. (1986). Microdialysis—A new bioanalytical sampling technique. *Current Sep.* **7,** 43–46.

Yamada, K., and Nabeshima, T. (1997). Simultaneous measurement of nitrite and nitrate levels as indices of nitric oxide release in the cerebellum of conscious rats. *Neurochemistry* **68,** 1234–1243.

Zini, I., Tomasi, A., Grimaldi, R., Vannini, V., and Agnati, L. F. (1992). Detection of free radicals during brain ischemia and reperfusion by spin trapping and microdialysis. *Neurosci. Lett.* **138,** 279–282.

## [44]  Nitric Oxide, Proteasomal Function, and Iron Homeostasis—Implications in Aging and Neurodegenerative Diseases

By Srigiridhar Kotamraju, Shasi Kalivendi, Tiesong Shang, and
B. Kalyanaraman

Abstract

In this chapter, oxidant-induced transferrin receptor-mediated iron-signaling and apoptosis are described in endothelial and neuronal cells exposed to oxidants. The role of nitric oxide in the regulation of iron homeostasis and oxidant-induced apoptosis is described. The interrelationship between oxidative stress, iron-signaling, and nitric oxide-dependent proteasomal function provides a rational mechanism that connects both oxidative and nitrative modifications.

Nitric Oxide Stimulation of Proteasome Function:
A New Antioxidant Mechanism

Previous studies have attributed the antioxidant mechanism of •NO to its ability to react with peroxyl radicals (Hui and Padmaja, 1993). •NO is an effective chain-breaking antioxidant (Goss et al., 1997; Hogg et al., 1993). It has been reported to react with lipid peroxyl radicals at a nearly diffusion-controlled rate forming nitrated lipids (Goss et al., 1997; Hogg and Kalyanaraman, 1998; Hogg et al., 1993). •NO was reported to be 10,000-fold more potent than $\alpha$-tocopherol with respect to its ability to react with lipid peroxyl radicals (O'Donnell et al., 1997). •NO protected the endothelial cells against lipid hydroperoxide–induced cytotoxicity (Struck et al., 1995). It was proposed that •NO-mediated cytoprotection was due to scavenging of intracellular lipid peroxyl radicals (Struck et al., 1995). Thus, the antioxidant mechanism of •NO was generally attributed to its chemical effects (e.g., radical-scavenging mechanism).

Oxidized low-density lipoprotein (LDL)–mediated apoptosis in endothelial cells was shown to be abrogated by several •NO donors, including the NONOates (Kotamraju et al., 2001). It was previously reported that the antiapoptotic effect of •NO might be related to cellular regulation of iron signaling (Kotamraju et al., 2003). In that study, the cytoprotective effects of •NO were attributed to the upregulation of the proteasomal enzymes (Kotamraju et al., 2003). •NO-stimulated proteasomal activity was shown

METHODS IN ENZYMOLOGY, VOL. 396
0076-6879/05 $35.00
DOI: 10.1016/S0076-6879(05)96044-8

to be responsible for its cytoprotective and antioxidative effects in endothelial cells. These effects were attributed to the newly discovered antioxidant mechanism of •NO that is based on its biological signaling mechanism. The intriguing and complex relationship between the intracellular •NO, iron homeostasis, oxidative stress, and the cellular proteasomal function is shown in Fig. 1 (Kotamraju et al., 2004). Elucidating this connection may be important to the understanding of the proinflammatory and anti-inflammatory signaling mechanisms induced by oxidants and •NO. In this chapter, this unique aspect of •NO (i.e., ability to stimulate proteasomal function) is discussed in relation to its antioxidant and cytoprotective mechanisms.

Upregulation of Proteasomal Activity by •NO

Endogenous •NO was shown to be essential for maintaining the inherent proteasomal activity (Kotamraju et al., 2003). The proteolytic core of the 26S complex, the 20S proteasome, contains multiple peptidase activities, including the chymotrypsin-like activity (a serine endopeptidase that hydrolyzes peptide bonds present in the carboxyl group of the hydrophobic amino acids such as phenylalanine) and the trypsin-like activity (peptidase-mediated cleavage after basic side chains). Both chymotrypsin-like and trypsin-like activities were enhanced in the presence of •NO in endothelial cells.

Bovine aortic endothelial cells (BAECs) were obtained from Clonetics. Cells were usually obtained at the third passage, transferred to $75\text{-cm}^2$ filter

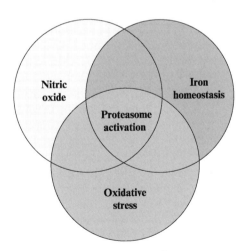

FIG. 1. Interrelationship between oxidative stress, •NO, intracellular iron, and proteasomal function. (See color insert.)

vent flasks, grown to confluence ($5.2 \times 10^6$ cells/75 cm$^2$) in Dulbecco's Modified Eagle Medium (DMEM) containing 10% fetal bovine serum (FBS), L-glutamine (4 mmol/L), penicillin (100 U/ml), and streptomycin (100 $\mu$g/ml), and then incubated at 37° in a humidified atmosphere of 5% $CO_2$ and 95% air. Cells were used between passages 4 and 14. On treatment day, the medium was replaced with DMEM containing 2% FBS, which had approximately 25–30 $\mu$g transferrin/ml.

Proteasome function assays consisted of measurements of 26S and 20S proteasome activities (Coux *et al.*, 1996; Pajonk *et al.*, 2002). The 26S proteasome function was measured as reported. Cells were washed with buffer I [50 mM Tris, pH 7.4/2 mM dithiothreitol (DTT)/5 mM MgCl$_2$/2 mM adenosine triphosphate (ATP)] and homogenized with buffer I containing 250 mM sucrose. Twenty micrograms of 10,000 $\times$ $g$ supernatant was diluted with buffer I to a final volume of 900 $\mu$l. The fluorogenic proteasome substrates SucLLVY-AMC (chymotrypsin-like) and Z-Leu-Leu-Lys-AMC (trypsin-like) were added in a final concentration of 80 $\mu$M. The proteolytic activity was measured by monitoring the release of the fluorescent group 7-amido-4-methylcoumarin (excitation 380 nm, emission 460 nm).

The 20S proteasome activity was determined according to the method developed by Grune *et al.* (1996). Cells were lysed in phosphate-buffered saline (PBS) containing 0.1% Triton X-100 and 0.5 mM DTT. The assay mixture contained 50 $\mu$l of buffer (50 mM Tris–HCl pH 7.8/20 mM KCl/5 mM MgCl$_2$/0.1 mM DTT, 250 $\mu$M SucLLVY-MCA) and 50 $\mu$l of cell lysate (15 $\mu$g of protein). After 30 min at 37°, the reaction was stopped by adding 1 ml of 0.2 M glycine buffer, pH 10, and the fluorescence of the liberated 7-amido-4-methylcoumarin was measured using the excitation and emission wavelengths at 365 and 460 nm, respectively.

BAECs were treated for 8 h with both L-NAME, a nonspecific inhibitor nitric oxide synthase (NOS) and D-NAME, an inactive structural analogue of L-NAME. As shown in Fig. 2, the trypsin-like activity of the 26S proteasome decreased in a dose-dependent manner in L-NAME–treated endothelial cells, but not in D-NAME–treated cells. Conversely, exogenously added •NO donors increased the proteasomal activity. In these studies, NONOates were used as •NO donors that released •NO slowly at a well-defined rate in the extracellular and intracellular milieu. NONOates release •NO as a result of a thermolytic decomposition in a 2:1 stoichiometry (i.e., two molecules of •NO are released from one molecule of the NONOate compound). DETA/NO, an extracellular •NO donor with a longer half-life (20 h), was used (Saavedra *et al.*, 2000). Because extracellular •NO is predominantly oxidized to $NO_2^-/NO_3^-$, relatively higher concentrations (50–100 $\mu$M) of DETA/NO were used. In contrast, the

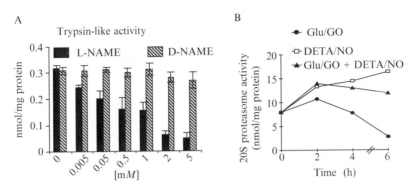

FIG. 2. (A) Effect of endogenous •NO on proteasomal activity in bovine aortic endothelial cells (BAECs). BAECs were pretreated with L-NAME, at the indicated concentrations for 8h, and the trypsin-like activity of the 26S proteasome was measured. (B) Effect of exogenous •NO donor on the proteasomal activity in BAECs. BAECs were pretreated with DETA/NO (100 $\mu M$) for 2h before the addition of Glu/GO (20 mu) for 6h. The 20S proteasome activity was measured as a function of time from 0 to 6h, using the fluorogenic peptide SucLLVY-MCA as a substrate.

cell-permeable esterase-specific pro-•NO donor that releases •NO inside the cell was used at a much lower concentration (1 $\mu M$) to elicit the same effects. These results show that •NO could stimulate and preserve the proteasomal activity in endothelial cells.

### Reversal of Antioxidant Effects of •NO by Proteasomal Inhibitors

To prove that the antioxidative effects of DETA/NO were due to stimulation of the cellular proteasomal activity, BAECs were pretreated with various inhibitors of proteasome [10 $\mu M$ lactacystin, 10 $\mu M$ MG-132, and 2 $\mu M$ epoxomycin (Sigma)]. These inhibitors by themselves did not induce significant oxidative stress or apoptosis in BAECs at the concentrations used.

Intracellular oxidative stress was measured by monitoring dichloro-fluorescein (DCF) formation (green fluorescence) (Tampo et al., 2003). BAECs were treated with $H_2O_2$ (1 $\mu M$/min) that was generated from glucose/glucose oxidase (Glu/GO). Before the addition of dichlorodihy-drofluorescein diacetate (DCFH/DA), cells were washed free of Glu/GO. DCFH/DA was added at a final concentration of 10 $\mu M$ and incubated for 20 min.

Cells were then washed twice with Dulbecco's PBS (DPBS). Fluorescence was monitored using a Nikon fluorescence microscope (excitation 488 nm, emission 610 nm) equipped with a fluorescein-isothiocyanate

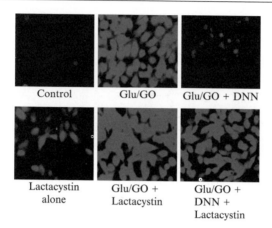

| Control | Glu/GO | Glu/GO + DNN |

| Lactacystin alone | Glu/GO + Lactacystin | Glu/GO + DNN + Lactacystin |

FIG. 3. Effect of proteasomal inhibitors on DCFH oxidation in BAECs treated with $H_2O_2$ and $^\bullet NO$. Cells were pretreated with the proteasome inhibitors and DETA/NO (100 $\mu M$) for 2h before treatment with Glu/GO (20 mu) (See color insert.)

(FITC) filter. Intracellular oxidant-induced activation of the active probe carboxy-DCFH to DCF was measured (Fig. 3). The DCF fluorescence intensity in oxidant-treated cells was nearly 1000-fold greater than in the control cells (Fig. 3). Glu/GO–mediated intracellular DCF fluorescence was abrogated in BAECs treated with DETA/NO that slowly released $^\bullet NO$ (7 n$M$/min). In the presence of lactacystin (Lac) or MG-132, DETA/ NO–mediated inhibition of DCF fluorescence was abolished (Fig. 3).

Another routinely used intracellular oxidative marker is the protein carbonyl (Levine *et al.*, 1994). Protein carbonyl levels were significantly elevated in Glu/GO–treated cells compared to the control cells. In the presence of DETA/NO, protein carbonyl levels were drastically decreased. Pretreatment with Lac significantly increased the protein carbonyl levels. These results are shown in Fig. 4. These findings point to a previously uncharacterized antioxidant role for $^\bullet NO$ (i.e., stimulator of proteasomal function). We then asked the question, how does enhancing the proteasomal activity by $^\bullet NO$ affect the antioxidant activity?

To address this question, the uptake of labeled iron ($^{55}Fe$) was measured as follows (Kotamraju *et al.*, 2002). BAECs were grown in DMEM containing 10% FBS until confluence. On the day of treatment, the medium was replaced with DMEM containing 2% FBS, and the cells were allowed to adjust to the medium conditions. Ferric chloride (0.2 $\mu$Ci of $^{55}Fe$) was added to the medium, and its levels were measured as a function of time. Cells were washed with DPBS, lysed with PBS containing 0.1% Triton X-100, and counted in a beta counter.

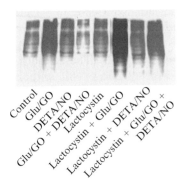

FIG. 4. Effect of proteasomal inhibitors on protein carbonyl formation in bovine aortic endothelial cells (BAECs) treated with $H_2O_2$ and $^{\bullet}NO$. Protein carbonyl levels were measured by Western blot analysis using a monoclonal rabbit nti-2,4-dinitrophenol antibody (Zymed).

BAECs were treated with Glu/GO with and without DETA/NO and proteasome inhibitors. $^{55}Fe$ uptake was nearly 2.3-fold higher in Glu/GO–treated cells than in control cells. In the presence of DETA/NO, this increase in $^{55}Fe$ uptake was inhibited. Lac or MG-132 treatment counteracted the effect of DETA/NO, as evidenced by the increase in $^{55}Fe$ uptake in Lac-treated cells. To probe the involvement of transferrin receptor (TfR) in oxidant-induced iron uptake, TfR levels were measured.

BAECs were washed with ice-cold PBS and resuspended in 100 $\mu$l of radio immunoprecipitation assay (RIPA) buffer [20 m$M$ Tris–HCl, pH 7.4, 2.5 m$M$ EDTA, 1% Triton X-100, 1% sodium deoxycholate, 1% sodium dodecylsulfate (SDS), 100 m$M$ NaCl, and 100 m$M$ NaF]. Following the addition of sodium vanadate, aprotinin, leupeptin, and pepstatin, as previously indicated (Matsunaga *et al.*, 2004), cells were homogenized, and the lysate was centrifuged for 15 min at 12,000$g$. Proteins were resolved on polyacrylamide gels and blotted onto nitrocellulose membranes. Membrane was washed twice with Tris-buffered saline (TBS) (140 m$M$ NaCl, 50 m$M$ Tris–HCl, pH 7.2) containing 1% Tween 20 before blocking the nonspecific binding with TBS containing 5% skim milk. Membrane was incubated with mouse anti-human transferrin receptor monoclonal antibody (1 $\mu$g/ml in TBS) (Zymed Laboratories) and 2% skim milk for 2 h at room temperature. Membrane was washed five times and detected with horseradish peroxidase–conjugated rabbit anti-mouse immunoglobulin G (IgG) (1:5,000) for 1.5 h at room temperature. The bands were detected by the enhanced chemiluminescence (ECL) method.

In the presence of Lac, TfR expression in cells treated with $H_2O_2$ and DETA/NO was enhanced. The increase in TfR expression by the proteasome inhibitor Lac suggests that TfR likely undergoes proteolytic degradation in the presence of $^\bullet$NO and $H_2O_2$. These results indicate a connection between $H_2O_2$-mediated iron signaling, $^\bullet$NO, and proteasomal activity.

## Reversal of Cytoprotective Effect of $^\bullet$NO by Proteasomal Inhibitors

The caspase-3 proteolytic activity was measured in BAEC treated with Glu/GO as a function of time. The caspase activity (caspase-3 assay kit, Clontech) in the $12,000g$ supernatant was measured in a spectrophotometer using DEVD-pNA (acetyl Asp-Glu-Val-Asp p-nitroanilide) as a substrate, according to the manufacturer's instructions provided with the assay kit. In Glu/GO-treated cells, the caspase-3 activity increased by nearly fivefold as compared to the control. In the presence of DETA/NO, which stimulated the proteosomal activity, the caspase-3 activation decreased in Glu/GO–treated cells. Proteasome inhibition with Lac caused an increase in caspase-3 activity in cells treated with Glu/GO and DETA/NO. These results suggest that the antiapoptotic effect of DETA/NO is due to stimulation of the proteasome.

## Implications in Age-Related Neurodegenerative Diseases

The role of oxidant-induced iron signaling has implications in aging and age-related diseases such as Parkinson's, Alzheimer's, and Friedreich's ataxia (Bishop et al., 2002; Killilea et al., 2003; Sipe et al., 2002). These diseases are characterized by defective respiratory chain components (e.g., complexes I and III) (Parker et al., 1989). As a result, there is increased formation of reactive oxygen species in mitochondria that leads to accumulation of iron levels and oxidative stress in mitochondria (Lan and Jiang, 1997). In Friedreich's ataxia, a neurodegenerative and cardiac disorder, there is deficiency in the mitochondrial chaperone protein frataxin (Bulteau et al., 2004). This disease is characterized by increased mitochondrial oxidative stress, iron accumulation, and diminished aconitase activity. NO production has been shown to be impaired in the substantia nigra of patients with Parkinson's disease (Kuiper et al., 1994). This was accompanied by a decrease in the activity of proteasomes. The antioxidative and cytoprotective effects of $^\bullet$NO were attributed to $^\bullet$NO-induced proteasomal activity (Kotamraju et al., 2003). Proteasomal inhibitors abrogated $^\bullet$NO-mediated cytoprotection and antioxidative effects. $^\bullet$NO as a stimulator of proteasomal function is still a nascent concept.

Many natural and synthetic drugs have been shown to stimulate •NO biosynthesis (Hattori *et al.*, 2002; Laufs *et al.*, 1998) and mitigate oxidative injury in endothelial cells. Clearly, the role of the •NO-stimulated proteolytic signaling mechanisms is vital to our understanding of the overall oxidative signaling mechanisms. Future investigations should undoubtedly focus on the following question: How does •NO stimulate the proteolytic signaling pathway? These efforts are currently underway in our laboratory.

## Acknowledgments

This work was supported by National Institutes of Health grants HL073056-01 and HL68769-01.

## References

Bishop, G. M., Robinson, S. R., Liu, Q., Perry, G., Atwood, C. S., and Smith, M. A. (2002). Iron: A pathological mediator of Alzheimer disease? *Dev. Neurosci.* **24**, 184–187.

Bulteau, A. L., O'Neill, H. A., Kennedy, M. C., Ikeda-Saito, M., Isaya, G., and Szweda, L. I. (2004). Frataxin acts as an iron chaperone protein to modulate mitochondrial aconitase activity. *Science* **305**, 242–245.

Coux, O., Tanaka, K., and Goldberg, A. L. (1996). Structure and functions of the 20S and 26S proteasomes. *Annu. Rev. Biochem.* **65**, 801–847.

Goss, S. P., Hogg, N., and Kalyanaraman, N. (1997). The effect of nitric oxide release rates on the oxidation of human low density lipoprotein. *J. Biol. Chem.* **272**, 21647–21653.

Grune, T., Reinheckel, T., and Davies, K. J. A. (1996). Degradation of oxidized proteins in K562 human hematopoietic cells by proteasome. *J. Biol. Chem.* **271**, 15504–15509.

Hattori, Y., Nakanishi, N., and Kasai, K. (2002). Statin enhances cytokine-mediated induction of nitric oxide synthesis in vascular smooth muscle cells. *Cardiovasc. Res.* **54**, 649–658.

Hogg, N., Kalyanaraman, B., Joseph, J., Struck, A., and Kalyanaraman, B. (1993). Inhibition of low-density lipoprotein oxidation by nitric oxide. Potential role in atherogenesis. *FEBS Lett.* **334**, 170–174.

Hogg, N., and Kalyanaraman, B. (1998). Nitric oxide and low-density lipoprotein oxidation. *Free Radic. Res.* **28**, 593–600.

Hui, R. E., and Padmaja, S. (1993). The reaction of no with superoxide. *Free Radic. Res. Commun.* **18**, 195–199.

Killilea, D. W., Atamna, H., Liao, C., and Ames, B. N. (2003). Iron accumulation during cellular senescence in human fibroblasts. *in vitro. Antioxid. Redox Signal* **5**, 507–516.

Kotamraju, S., Chitambar, C. R., Kalivendi, S. V., Joseph, J., and Kalyanaraman, B. (2002). Transferrin receptor-dependent iron uptake is responsible for doxorubicin-mediated apoptosis in endothelial cells: Role of oxidant-induced iron signaling in apoptosis. *J. Biol. Chem.* **277**, 17179–17187.

Kotamraju, S., Hogg, N., Joseph, J., Keefer, L. K., and Kalyanaraman, B. (2001). Inhibition of oxidized low-density lipoprotein-induced apoptosis in endothelial cells by nitric oxide. Peroxyl radical scavenging as an antiapoptotic mechanism. *J. Biol. Chem.* **276**, 17316–17323.

Kotamraju, S., Tampo, Y., Kalivendi, S. V., Joseph, J., Chitambar, C. R., and Kalyanaraman, B. (2004). Nitric oxide mitigates peroxide-induced iron-signaling, oxidative damage, and apoptosis in endothelial cells: Role of proteasomal function? *Arch. Biochem. Biophys.* **423**, 74–80.

Kotamraju, S., Tampo, Y., Keszler, A., Chitambar, C. R., Joseph, J., Haas, A. L., and Kalyanaraman, B. (2003). Nitric oxide inhibits H2O2-induced transferrin receptor-dependent apoptosis in endothelial cells: Role of ubiquitin-proteasome pathway. *Proc. Natl. Acad. Sci. USA* **100**, 10653–10658.

Kuiper, M. A., Visser, J. J., Bergmans, P. L., Scheltens, P., and Wolters, E. C. (1994). Decreased cerebrospinal fluid nitrate levels in Parkinson's disease, Alzheimer's disease and multiple system atrophy patients. *J. Neurol. Sci.* **121**, 46–49.

Lan, J., and Jiang, D. H. (1997). Excessive iron accumulation in the brain: A possible potential risk of neurodegeneration in Parkinson's disease. *J. Neural. Transm.* **104**, 649–660.

Laufs, U., La Fata, V., Plutzky, J., and Liao, J. K. (1998). Upregulation of endothelial nitric oxide synthase by HMG CoA reductase inhibitors. *Circulation* **97**, 1129–1135.

Levine, R. L., Williams, J. A., Stadtman, E. R., and Shacter, E. (1994). Carbonyl assays for determination of oxidatively modified proteins. *Methods Enzymol.* **233**, 346–357.

Matsunaga, T., Kotamraju, S., Kalivendi, S. V., Dhanasekaran, A., Joseph, J., and Kalyanaraman, B. (2004). Ceramide-induced intracellular oxidant formation, iron signaling, and apoptosis in endothelial cells: Protective role of endogenous nitric oxide. *J. Biol. Chem.* **279**, 28614–28624.

O'Donnell, V. B., Chumley, P. H., Hogg, N., Bloodsworth, A., Darley-Usmar, V., and Freeman, B. A. (1997). Nitric oxide inhibition of lipid peroxidation: Kinetics of reaction with lipid peroxyl radicals and comparison with alpha-tocopherol. *Biochemistry* **36**, 15216–15223.

Pajonk, F., Reiss, K., Sommer, A., and McBride, W. H. (2002). N-acetyl-L-cysteine inhibits 26S proteasome function: Implications for effects on NF-kappaB activation. *Free Radic. Biol. Med.* **32**, 536–543.

Parker, W. D., Jr., Boyson, S. J., and Parks, J. K. (1989). Abnormalities of the electron transport chain in idiopathic Parkinson's disease. *Ann. Neurol.* **26**, 719–723.

Saavedra, J. E., Shami, P. J., Wang, L. Y., Davies, K. M., Booth, M. N., Citro, M. L., and Keefer, L. K. (2000). Esterase-sensitive nitric oxide donors of the diazeniumdiolate family: *In vitro* antileukemic activity. *J. Med. Chem.* **43**, 261–269.

Sipe, J. C., Lee, P., and Beutler, E. (2002). Brain iron metabolism and neurodegenerative disorders. *Dev. Neurosci.* **24**, 188–196.

Struck, A. T., Hogg, N., Thomas, J. P., and Kalyanaraman, B. (1995). Nitric oxide donor compounds inhibit the toxicity of oxidized low-density lipoprotein to endothelial cells. *FEBS Lett.* **361**, 291–294.

Tampo, Y., Kotamraju, S., Chitambar, C. R., Kalivendi, S. V., Keszler, A., Joseph, J., and Kalyanaraman, B. (2003). Oxidative stress-induced iron signaling is responsible for peroxide-dependent oxidation of dichlorodihydrofluorescein in endothelial cells: Role of transferrin receptor-dependent iron uptake in apoptosis. *Circ. Res.* **92**, 56–63.

## [45]  Nitric Oxide Production by Primary Liver Cells Isolated from Amino Acid Diet–Fed Rats

By Yashige Kotake, Hideki Kishida, Dai Nakae, and Robert A. Floyd

### Abstract

Primary mixed liver cells were isolated from rats that had been fed an amino acid (AA) diet in which natural protein was replaced with a defined mixture of pure AAs. Nitric oxide (NO) production from these cells *in vitro* was monitored using a nitric oxide (NO)–selective fluorescent probe, diaminofluorescein, followed by flow cytometric analysis. High levels of NO fluorescence were seen in approximately half of liver cells isolated from rats fed an AA diet for 1–7 days, whereas there was baseline fluorescence in cells obtained from regular diet–fed rats. The apparent size of NO-producing cells was smaller than those not producing NO. The production of NO was inhibited when rats were treated with either inducible NO synthase (iNOS)– or endothelial NOS–specific inhibitor, and an inhibitor for iNOS induction during AA diet feeding. L-Arginine or L-glutamine (material for L-arginine biosynthesis) enriched diet showed the same NO augmentation as in AA diet. It is speculated that a high content of free L-arginine in AA diet may have caused enhanced NO production.

### Introduction

Amino acid (AA) diet is a diet that contains a mixture of L-amino acids in a defined ratio, as a substitute for natural protein (Rogers *et al.*, 1965). Although long-term consumption of AA diet in rats caused no major deleterious effects, when combined with a choline deficient diet, it significantly enhanced hepatocellular carcinoma formation (Nakae, 1999). Because AA diet is rich in free L-arginine, a substrate for nitric oxide (NO) synthase, NO production in the liver may be modulated by AA diet consumption. The objective of this study was to determine NO production in isolated primary liver cells in rats that had been fed AA diet for various durations. NO formation from isolated liver cells was determined by labeling NO with NO-specific fluorescence-labeling agent, followed by flow cytometric analysis.

METHODS IN ENZYMOLOGY, VOL. 396
0076-6879/05 $35.00
DOI: 10.1016/S0076-6879(05)96045-X

*Feeding Animals with Specific Diets*

Rats were treated strictly following the animal-use protocol approved by the institutional laboratory animal care and use committee in the Oklahoma Medical Research Foundation. Male Wistar rats (8–10 weeks old) were obtained from Charles River Laboratory (Indianapolis, IN). L-Amino acid diet, synthetic defined, was purchased from two vendors, ICN Biomedicals, Inc. (Irvine, CA), and Dyets, Inc. (Bethlehem, PA). The composition of L-amino acid in the two diets is similar; for example, each contains 1.3% L-arginine hydrochloride and 2.9% L-glutamic acid (starting material for L-arginine biosynthesis). The contents of L-arginine and L-glutamine in these AA diets are the same as choline-deficient L-amino acid–defined (CDAA) diet, which was used by Nakae (1999) to clarify issues regarding diet in a choline deficiency hepatocarcinogenesis model. Basal diet was a Purina 5005 rodent chow (Ralston Purina, St. Louis, MO). Small batches of L-arginine or L-glutamate–enriched diet were prepared in our laboratory by mixing the powdered basal diet with 10% by weight of the amino acid obtained from Sigma Chemical Co. (St. Louis, MO). Rats were fasted overnight (15 h), and then the specific diet was fed for 1 (24 h), 3, 7, 14, and 30 days. Basal diet control animals were also fasted overnight before continuing the feeding. Nitric oxide synthase (NOS) inhibitor, either $N$-nitro-L-arginine or aminoguanidine (Sigma, 150 mg/kg), was intraperitoneally administered twice, 12 h and 1 h before cell isolation. The inhibitor of inducible NOS (iNOS) induction $\alpha$-phenyl-*tert*-butylnitrone (PBN) (Sigma) was administered in the diet by adding at the rate of 0.3% w/w to the powdered AA diet.

*Primary Cell Isolation*

Primary liver cells were isolated following the method previously reported (Alpini *et al.*, 1994; Jeejeebhoy *et al.*, 1975). After rats were fed the diet for a specified period, they were anesthetized with isoflurane (Abbott Laboratories, Chicago, IL) using 95% oxygen–5% carbon dioxide as a carrier gas for a vaporizer (Surgevet/Anesco, Waukesha, WI). Under anesthesia, the abdomen was opened with a middle incision and the portal vein cannulated with polyethylene tubing (PE 205, Intramedic Becton-Dickenson). To remove blood from the liver, calcium-free Hanks' HEPES buffer was perfused into the cannulae with a peristaltic pump (Gilson Optima 2000, PerkinElmer, Boston, MA) at the flow rate of 30 ml/min for 10 min. Perfusate was allowed to drain from the opened inferior artery. The temperature of incoming perfusate was carefully maintained at $37 \pm 1°$ at the liver inlet by controlling the perfusate-reservoir temperature. The perfusate was switched to the same buffer containing 0.05% collagenase B

(Sigma) and allowed to perfuse with the same flow rate for 30 min. Well-digested liver became soft and flat and was readily dispersible in buffer with a plastic spatula. Cells were gently suspended in the buffer, and connecting tissues and cell debris were removed with a 220 nylon mesh filter. Cells were washed in buffer three times, and viability was determined using the Trypan blue exclusion method. A majority of liver cells isolated with the collagenase perfusion method is known to be hepatocytes (>90%) (Alpini *et al.*, 1994). The average size of hepatocyte (25 $\mu$m) is about 2.5 times larger than other liver cells, such as Kupffer cells and endothelial cells (Alpini *et al.*, 1994). Forward- and side-scattering dot-histograms obtained in flow cytometry indicated homogeneous size distribution (Fig. 1D), suggesting that the cell preparation was mainly hepatocytes.

*Flow Cytometry*

For flow cytometric analysis of NO production, isolated cells were suspended in fresh William's carbonate buffer (Sigma), and the NO-labeling agent diaminofluorescein acetate (DAFA, Sigma) (Itoh *et al.*, 2000; Nagano *et al.*, 2002) was added to 20 $\mu M$ and incubated for 20 min at 37°, and then kept on ice before subjecting them to analysis. Histograms were obtained using a FACScan flow cytometer (BD Biosciences, San Jose, CA), and fluorescence intensities were analyzed with BD CellQuest for greater than 10,000 gated cells.

Flow cytometry histograms indicated that liver cells that had been isolated from rats fed an AA diet for 1 day (or longer) increased NO production as compared to cells obtained from basal diet (regular rodent chow)–fed rats (Fig. 1). A summary of these experiments is as follows:

1. Approximately 50% of liver primary cells obtained from rats that had been fed AA diet 1 day or more produced DAFA-stainable NO (Fig. 1A). The average fluorescence intensity per cell in AA diet–fed rats was approximately four times higher than those on basal diet. NO production level has little dependence on the duration of AA diet feeding period up to 30 days (Fig. 3). The scattering histograms (Fig. 1B) indicate that the size of NO-producing cells was distinctively smaller than those not producing NO (Fig. 1B). Functional heterogeneity of hepatocytes from different lobular zone has been demonstrated (Alpini *et al.*, 1994). Periportal hepatocytes have smaller size, higher oxygen tension, and higher amino acid catabolism than pericentral hepatocytes. It is possible that NO-producing cells are distributed mainly in the periportal region. However, it has been demonstrated that the size of hepatocytes is readily influenced by administered drugs, such as pentobarbital (Willson *et al.*, 1984), so it is also possible that NO production could have caused the size change.

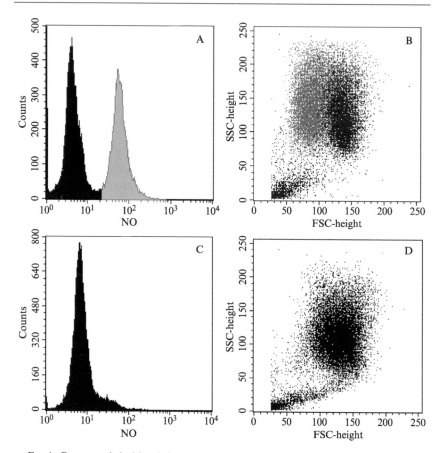

Fig. 1. Rats were fed with a defined L-amino acid (AA) diet or a basal diet for 3 days, and primary liver cells were isolated using *in situ* collagenase perfusion. Intact cells were then incubated with the nitric oxide (NO)–labeling compound diaminofluorescein acetate (DAFA) and subjected to flow cytometric analysis. (A) Flow cytometry histogram of liver cells isolated from a rat fed AA diet for 3 days, displaying NO fluorescence (log scale) versus counts (approximately equal to cell numbers). Lightly shaded peak is from cell population that produces more NO than that dark-shaded peak. (B) Dot plot displays forward scattering (FSC) height versus side scattering (SSC) height. Each dot corresponds to individual cells. Dots with light gray are from those NO-producing cells [i.e., lightly shaded peak in histogram (A)]. (C) Histogram obtained from liver cells isolated from basal diet–fed rats. (D) Dot plot for FSC versus SSC for liver cells isolated from basal diet–fed rats.

    2. L-Arginine– or L-glutamine–enriched diet (10 weight %) caused the appearance of NO-producing cell population (Fig. 2), but such NO-producing cells disappeared when the diet was switched from the enriched diet or an AA diet back to the basal diet for 1 day (Figs. 2 and 3).

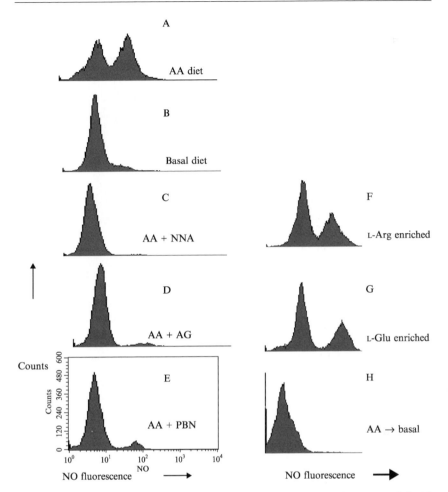

FIG. 2. Flow cytometry histograms obtained from liver cells isolated from rats fed with various diets and inhibitors. (A) Fed amino acid (AA) diet for 3 days; (B) fed basal diet; (C) fed AA diet for 3 days, and on the third day of feeding nitric oxide synthase (NOS) inhibitor $N$-nitroarginine (NNA) was administered i.p. 6 h and 30 min before cell isolation; (D) fed AA diet for 3 days, and on the third day of feeding NOS inhibitor, aminoguanidine (AG) was administered i.p. 6 h and 30 min before cell isolation; (E) fed AA diet supplemented with 0.3% α-phenyl-*tert*-butylnitrone (PBN) for 3 days; (F) fed basal diet supplemented with 10% L-arginine for 3 days; (G) fed basal diet supplemented with 10% Lglutamine for 3 days; (H) fed AA diet for 3 days then switched to basal diet feeding for 1 day.

3. The administration of NOS inhibitors, $N$-nitro-L-arginine (NNA) or aminoguanidine (AG) during AA diet feeding abolished NO production in these cells, indicating that NO was produced through L-arginine–dependent NOS pathways. NNA is considered an eNOS-specific inhibitor (specificity:

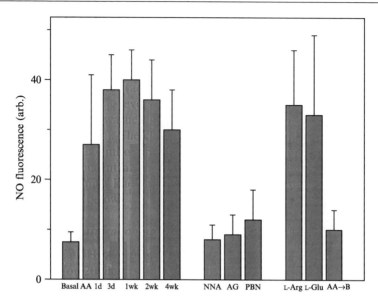

FIG. 3. Average fluorescence intensity per cell for liver cells isolated from rats on various diet regimens (three to five rats for each treatment). Rats were fed basal diet (Basal), or amino acid (AA) diet for 1 day (1d), 3 days (3d), 1 week (1wk), 2 weeks (2wk), and 4 weeks (4wk). Rats were fed AA diet for 3 days, and on the third day of feeding, nitric oxide synthase (NOS) inhibitor $N$-nitroarginine (NNA) or aminoguanidine (AG) was administered i.p. 6 h and 30 min before hepatocyte isolation, or fed AA diet supplemented with 0.3% α-phenyl-*tert*-butylnitrone (PBN) for 3 days. Rats were fed 10% L-arginine (L-Arg)– or L-glutamine (L-Glu)–supplemented basal diet for 3 days. Liver cells were isolated from rats fed AA diet for 3 days then switched to basal diet feeding for 1 day (AA→B).

enos/inos $= 30/1$) (Salerno *et al.*, 1997), and AG an iNOS-specific inhibitor (specificity: eNOS/iNOS $= 1/30$) (Southan *et al.*, 1996). In the present experiments, both NNA and AG inhibited NO formation to a similar extent (Figs. 2 and 3), suggesting that both eNOS and iNOS are involved in the NO production. Co-feeding of an inhibitor of iNOS induction PBN (0.3% w/w, 150 mg/kg/rat/day) (Kotake *et al.*, 1998) also suppressed NO formation (Figs. 2 and 3). We speculate that initial NO production through eNOS-mediated iNOS induction is the major source of NO in primary liver cells.

## Acknowledgments

We thank Dr. Lester A. Reinke and Mr. Danny Moore, University of Oklahoma Health Sciences Center, for the assistance in primary cell isolation. Support of this work was provided by National Institutes of Health grant CA82506.

References

Alpini, G., Phillips, J. O., Vroman, B., and LaRusso, N. F. (1994). Recent advances in the isolation of liver cells. *Hepatology* **20,** 494–514.

Itoh, Y., Ma, F. H., Hoshi, H., Oka, M., Noda, K., Ukai, Y., Kojima, H., Nagano, T., and Toda, N. (2000). Determination and bioimaging method for nitric oxide in biological specimens by diaminofluorescein fluorometry. *Anal. Biochem.* **287,** 203–209.

Jeejeebhoy, K. N., Ho, J., Greenberg, G. R., Philips, M. J., Bruce-Robertson, A., and Sodtke, U. (1975). Albumin, fibrinogen and transferrin synthesis in isolated rat hepatocyte suspensions. A model for the study of plasma protein synthesis. *Biochem. J.* **146,** 141–155.

Kotake, Y., Sang, H., Miyajima, T., and Wallis, G. L. (1998). Inhibition of NF-kappaB, iNOS mRNA, COX2 mRNA, and COX catalytic activity by phenyl-*N*-*tert*-butylnitrone (PBN). *Biochim. Biophys. Acta* **1448,** 77–84.

Nagano, T., and Yoshimura, T. (2002). Bioimaging of nitric oxide. *Chem. Rev.* **102,** 1235–1244.

Nakae, D. (1999). Endogenous liver carcinogenesis in the rat. *Pathol. Int.* **49,** 1028–1042.

Rogers, Q. R., and Harper, A. E. (1965). Amino acid diets and maximal growth in the rat. *J. Nutr.* **87,** 267–273.

Salerno, J. C., Martasek, P., Williams, R. F., and Masters, B. S. (1997). Substrate and substrate analog binding to endothelial nitric oxide synthase: Electron paramagnetic resonance as an isoform-specific probe of the binding mode of substrate analogs. *Biochemistry* **36,** 11821–11828.

Southan, G. J., and Szabo, C. (1996). Selective pharmacological inhibition of distinct nitric oxide synthase isoforms. *Biochem. Pharmacol.* **51,** 383–394.

Willson, R. A., Wormsley, S. B., and Muller-Eberhard, U. (1984). A comparison of hepatocyte size distribution in untreated and phenobarbital-treated rats as assessed by flow cytometry. *Dig. Dis. Sci.* **29,** 753–757.

# [46] Update on Nitric Oxide–Dependent Vasodilation in Human Subjects

*By* Craig J. McMackin and Joseph A. Vita

## Abstract

There currently is great interest in translating findings about the importance of nitric oxide (NO) in vascular biology to the clinical arena. The bioactivity of endothelium-derived NO can readily be assessed in human subjects as vasodilation of conduit arteries or increased flow, which reflects vasodilation of resistance vessels. This chapter provides an update on the available noninvasive methodology to assess endothelium-dependent vasodilation in human subjects.

METHODS IN ENZYMOLOGY, VOL. 396　　　　　　　　0076-6879/05 $35.00
DOI: 10.1016/S0076-6879(05)96046-1

Background

A vast body of work has emphasized the importance of endothelium-derived nitric oxide (NO) in vascular biology, and there currently is great interest in translating these findings to the clinical practice. Because it is a potent vasodilator, the bioavailability of endothelium-derived NO can readily be evaluated in human subjects by measuring changes in arterial diameter and blood flow. Human studies have shown that impaired endothelium-dependent vasodilation is associated with the presence of atherosclerosis and recognized cardiovascular disease risk factors. Many interventions that reduce cardiovascular risk also restore endothelium-dependent vasodilation toward normal. Importantly, prospective studies have shown that the presence of impaired endothelium-dependent vasodilation in the coronary or peripheral circulation identifies patients with increased risk for future cardiovascular disease events (Widlansky et al., 2003). In a review of methods for measurement of NO-dependent vasodilation in humans, Vita (2002) described invasive methods for the study of the coronary and peripheral circulations and use of two-dimensional ultrasound to study flow-mediated dilation of the brachial artery. Since that time, additional noninvasive approaches have emerged for study of NO-dependent control of vascular tone in the coronary circulation, central aorta, and peripheral microcirculation. In this chapter, we briefly mention advances in the previously described methods (Vita, 2002) and then describe these newer methods for assessment of NO-dependent vasodilation in humans (Table I).

Studies of the Coronary Circulation

*Invasive Studies*

A number of studies have examined NO-dependent vasodilation in the coronary circulation of patients undergoing cardiac catheterization. These studies assess changes in coronary artery diameter or coronary blood flow during intraarterial agonist infusion using quantitative coronary angiography and intracoronary Doppler, respectively (Vita, 2002). Though generally extremely safe, these studies have the potential to produce major complications such as coronary thrombosis or death. Because of their invasive nature, they are not well suited for repeated studies in the same individual or for the study of relatively low-risk populations. Despite these limitations, studies of the coronary circulation are the most clinically relevant for coronary artery disease. In particular, studies have shown that abnormalities of NO-dependent vasodilation in the coronary circulation are associated with increased risk for cardiovascular disease events (Halcox

TABLE I
NONINVASIVE METHODS FOR STUDY OF NO-DEPENDENT VASODILATION IN HUMAN SUBJECTS

| Method | Vascular bed | Stimulus for NO release | Advantages | Disadvantages |
|---|---|---|---|---|
| Vascular ultrasound Flow-mediated dilation (FMD) | Conduit brachial artery | Flow produced by reactive hyperemia | Well established<br>Demonstrated prognostic value<br>Correlates with coronary circulation | Operator dependent<br>Low signal/noise ratio<br>Lack of standardized methodology |
| Pulse-amplitude tonometry (PAT) | Small arteries in the fingertip | Flow produced by reactive hyperemia | Automatic<br>Rapid results<br>Correlates with coronary circulation | Clinical relevance not established<br>Minimal information about response to interventions |
| Venous occlusion Plethysmography | Forearm resistance vessels | Flow produced by reactive hyperemia | Stronger correlation with risk factors than FMD<br>Can be measured by several different techniques | Low reproducibility<br>Operator dependent |
| Arterial compliance Pulse wave analysis | Central aorta or brachial artery | None | More reproducible and stable over time<br>Low operator dependence | Only partially NO dependent<br>Relates to arterial structure and vascular tone |
| Transthoracic Doppler | Left anterior descending coronary artery | Adenosine or adenosine triphosphate | More clinically relevant circulation | Only partially NO dependent<br>Highly operator dependent<br>Lack of standardized methodology |

*et al.*, 2002; Schachinger *et al.*, 2000; Schindler *et al.*, 2003; Suwaidi *et al.*, 2000; Targonski *et al.*, 2003).

*Noninvasive Studies*

Given its clinical relevance, it would be desirable to obtain information about endothelium-dependent vasodilation of the coronary circulation in a noninvasive manner. Several studies have used transthoracic Doppler echocardiography to assess coronary blood flow reserve in the left anterior descending (LAD) circulation. This method involves obtaining Doppler flow signals of the distal LAD using an acoustic window near the midclavicular line in the fourth and fifth intercostals spaces. Signals are recorded at baseline and then after 2 min of intravenous adenosine triphosphate (ATP) infusion (140 $\mu$g/kg/min) to increase coronary blood flow. Changes in coronary blood flow are expressed as the ratio of ATP-induced to basal coronary flow velocity (Otsuka *et al.*, 2001). Coronary "flow reserve" measured in this manner is acutely impaired by passive cigarette smoking and improved by interventions known to improve endothelium-dependent vasodilation (Hirata *et al.*, 2004; Otsuka *et al.*, 2001).

The methodology is limited as a test of endothelial vasodilator function, however, because the response to ATP is only partially dependent on endothelium-derived NO. Furthermore, systemic ATP infusion lowers blood pressure and increases heart rate, which may alter coronary blood flow independently of endothelial function. The technique is highly operator dependent. No published studies have evaluated the effect of inhibitors of NO synthase (NOS) on the response. Thus, the technique has potential as a noninvasive method for assessing NO-dependent and NO-independent vasodilation in the human coronary circulation, but further studies are required before it can be generally accepted. A number of other noninvasive methodologies also show promise for examination of endothelium-dependent changes in coronary blood flow, including magnetic resonance imaging and positron emission tomography but currently are not in use for this purpose.

## Studies of the Arm and Hand

*Invasive Studies*

In light of the difficulty of studying the coronary circulation, many investigators have turned to the study of NO-dependent vasodilation in peripheral arteries. As reviewed by Vita (2002), studies of this type involve infusion of various vasoactive drugs into the brachial artery and measurement of vasodilation as changes in forearm blood flow using venous occlusion plethysmography or changes in radial artery diameter using high-resolution vascular ultrasound (Creager et al., 1990; Lieberman et al., 1996). The clinical relevance of these studies is predicated on the assumption that many cardiovascular disease risk factors are systemic in nature and have parallel effects in different vascular beds. This assumption is strongly supported by studies showing that an impaired blood flow response to acetylcholine and other endothelium-dependent vasodilators is associated with increased risk for cardiovascular disease events (Fichtlscherer et al., 2004; Heitzer et al., 2001; Perticone et al., 2001). Despite their clinical relevance, these studies require insertion of an arterial catheter, which reduces their applicability to the general population. Thus, the methodology remains extremely useful for studying selected populations and examining mechanisms of vascular dysfunction. However, there continues to be great interest in noninvasive methods to examine NO-dependent vasodilation in the periphery.

*Noninvasive Studies: Flow-Mediated Dilation*

A widely used noninvasive method to assess endothelial vasomotor function is brachial artery flow–mediated dilation as assessed by ultrasound (Corretti et al., 2002; Vita, 2002). In these studies, reactive hyperemia is induced by cuff occlusion of the arm, and changes in arterial diameter are measured using high-resolution ultrasound (Corretti et al., 2002). Flow-mediated dilation measured in this fashion depends on NO synthesis (Lieberman et al., 1996), correlates with endothelial vasomotor function in the coronary circulation (Anderson et al., 1995), and is reduced in the setting of traditional risk factors for coronary artery disease (Benjamin et al., 2004). In addition, impaired brachial artery flow–mediated dilation predicts short-term and long-term risk for cardiovascular disease events in patients with advanced atherosclerosis (Gokce et al., 2002) and in patients with hypertension (Modena et al., 2002).

Despite its clinical relevance, ultrasound-based studies have a number of limitations. The technique is technically demanding, and changes in brachial diameter produced by hyperemic flow (0.1–0.6 mm) are close to the limit of detection of ultrasound. Reproducibility depends greatly on image quality, and the technique requires time-consuming off-line image analysis. For these reasons, investigators have sought new techniques that are faster and simpler to perform.

*Noninvasive Studies: Pulse Amplitude Tonometry*

One emerging method is known as fingertip pulse amplitude tonometry (PAT). Studies are performed using a commercially available device (Endo-PAT 2000, Itamar Medical, Ltd.) that records the pulse amplitude in the fingertip at baseline and during reactive hyperemia. Hyperemia induces flow-mediated dilation within the fingertip and increases pulse amplitude. Simultaneous recordings are made from the contralateral finger and are used to adjust for changes in sympathetic tone and other systemic effects that might affect the signal during cuff occlusion and the hyperemic phase. Proprietary software provides further adjustment based on an empiric regression equation to account for baseline pulse amplitude, although the importance of making this adjustment remains unproven. The net response is expressed as the "reactive hyperemia PAT index." A preliminary study demonstrated that the increase in pulse amplitude is blocked, in part, by intraarterial infusion of monomethyl-L-arginine (L-NMMA), confirming that it depends in part on NO synthesis (Gerhard-Herman et al., 2002). Interestingly, the reactive hyperemia PAT index has been reported to correlate with brachial artery flow–mediated dilation in the arm and is inversely related to risk factors and the presence of coronary artery disease (Kuvin et al., 2003). The response also correlates with endothelial function in the coronary circulation (Bonetti et al., 2004). Finally, the response improves after enhanced external counter pulsation therapy, an intervention known to improve peripheral artery endothelial function (Bonetti et al., 2003).

   In our laboratory at Boston University School of Medicine, PAT and brachial ultrasound studies are done simultaneously using a single cuff occlusion to generate a period of reactive hyperemia, which stimulates flow-mediated dilation of both the conduit brachial artery and the small arteries in the finger. The PAT signals are recorded using thimble-shaped pneumatic probes that are placed on the index fingers of each hand. Patients lie supine with both wrists supported on foam blocks to allow the fingers to hang in an unsupported manner. The inflation pressure of the finger cuff is set to the diastolic pressure or 80 mm Hg (whichever is lower). Pulse recordings are made before cuff inflation and during the 1-min period beginning 1 min after 5-min cuff occlusion of the arm with the cuff placed on the upper arm. Figure 1 displays signals from a healthy subject and a subject with coronary artery disease. In a group of 252 unselected patients undergoing study of vascular function from our laboratory, the mean ($\pm$SD) reactive hyperemia PAT ratio was 2.2 $\pm$ 0.74 (range 1.23–5.69) with a highly skewed distribution. A prior study demonstrated that among patients referred for evaluation of chest pain, the reactive hyperemia PAT ratios were 1.31 $\pm$ 0.11 and 1.62 $\pm$ 0.47 for patients with and without exercise induced myocardial ischemia, respectively. We calculate that a sample size of 29 subjects per group would be required to detect a difference between groups of this magnitude with 80% power (alpha = 0.05) using log-transformed values for the reactive hyperemia/PAT ratio. These results suggest that clinically important differences between study groups

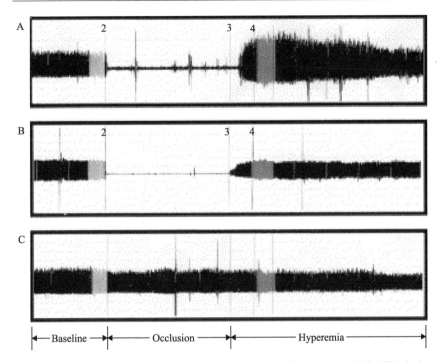

FIG. 1. Pulse amplitude recorded in the index finger with the Endo-PAT 2000 device (Itamar Medical, Ltd.) before, during, and after cuff occlusion of the arm, as described in the text. (A) The response from a healthy individual with no risk factors. (B) The very blunted response in an individual with coronary artery disease. (C) The response in the contralateral finger not subject to cuff occlusion, which demonstrates that the signal remains stable over time. The PAT ratio is calculated at baseline and between 1 and 2 min after cuff release. Reproduced, with permission, from Kuvin *et al.* (2003).

can be detected using this methodology in studies with samples sizes that are similar to those needed for study of brachial artery flow–mediated dilation (Vita, 2002). Overall, PAT appears to be a promising new methodology, but much work needs to be done to confirm its relation to other measures of NO-dependent vasodilation and to cardiovascular disease.

*Noninvasive Studies: Extent of Reactive Hyperemia*

Reactive hyperemia is the transient increase in limb blood flow that occurs after a period of limb occlusion and reflects ischemia-induced production of a variety of vasodilators, including adenosine and hydrogen ions that locally act on microvessels. A portion of the hyperemic response also depends on NO, possibly stimulated by local increases in shear stress during hyperemic flow. L-NNMA infusion blunts both the peak and the net hyperemic response in the forearm (Meredith *et al.*, 1996). Many investigators had suggested that reactive hyperemia is unaffected by cardiovascular disease. However, other studies have emphasized that reactive hyperemia is reduced in the setting of risk factors (Hayoz *et al.*, 1995; Higashi *et al.*, 2001; Mitchell *et al.*, 2004b) or coronary artery disease (Lieberman *et al.*, 1996), particularly the NO-dependent portion of the response (Higashi *et al.*, 2001). Reactive hyperemia also correlates inversely with systemic markers of inflammation, including C-reactive protein, interleukin-6, and the soluble form of intercellular adhesion molecule-1 (Vita *et al.*, 2004). Reactive hyperemia is the stimulus for brachial artery flow–mediated dilation, and we observed that a reduction in this stimulus accounts for much of the observed impairment in flow-mediated dilation observed in the setting of systemic risk factors (Mitchell *et al.*, 2004b). These findings suggest that noninvasive measures of flow can be used to assess reactive hyperemia as a clinically relevant correlate of endothelial vasomotor function.

We take two approaches to assessing reactive hyperemia. First, we use Doppler ultrasound to record flow signals from the brachial artery at

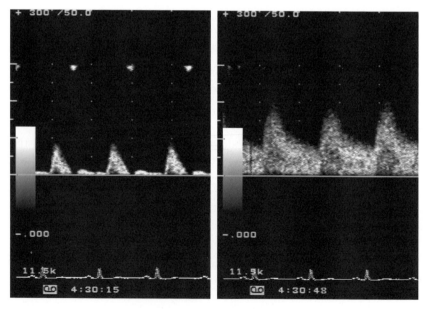

Fig. 2. Representative Doppler recordings from the brachial artery at baseline (left) and immediately after cuff release (right) reflecting hyperemic flow.

baseline and for 15 s after cuff release after 5-min occlusion of the upper arm (Vita, 2002). Typical flow signals are displayed in Fig. 2. The peak hyperemic response is typically observed within two or three beats after cuff release. Images are digitized on-line, and we measure the average flow velocity (area under the curve) for the peak cardiac cycle using one of several image analysis software packages (Brachial Analyzer, Medical Imaging Applications, Iowa City, IA). Table II presents reference values from a cohort of 503 healthy subjects studied in our laboratory. Many investigators express hyperemic flow as the ratio of peak to baseline flow, but a recent study suggests that the hyperemic flow velocity and hyperemic shear stress (calculated from the velocity, brachial artery diameter, and assumed values for blood viscosity) correlate most strongly with cardiovascular disease risk factors and prevalent cardiovascular disease (Mitchell *et al.*, 2004b).

A second method to assess reactive hyperemic uses venous occlusion plethysmography to measure forearm blood flow before and after cuff release (Higashi *et al.*, 2001). Blood flow measurements are made using a mercury-in-silastic strain gauge, upper arm and wrist cuffs, and a computerized plethysmograph (Hokanson, Inc.) (Vita, 2002). During these studies, the upper arm venous occlusion cuff is inflated to 40 mm Hg (or adjusted to optimize the tracing), and circulation to the hand is excluded by inflation of the wrist cuff to suprasystolic pressure before initiation of flow measurements. At least five measurements are made and averaged at baseline, and a recording is made every 20 s after cuff release for 2 min. Although this methodology has limited ability to "capture" the peak flow response, it provides a reproducible approach to examine the entire hyperemic response.

TABLE II

MEAN VALUES WITH 95% CONFIDENCE INTERVALS FOR BRACHIAL ARTERY FLOW[a]

| Age (y) | Sample size | Hyperemia volume flow ratio | Baseline flow velocity (cm/s) | Hyperemia flow velocity (cm/s) |
|---------|-------------|-----------------------------|-------------------------------|--------------------------------|
| <30 | n = 173 | 8.0 (6.0–10.1) | 12.0 (11.1–13.0) | 79.5 (75.2–83.7) |
| 30–39 | n = 125 | 7.3 (6.6–8.0) | 11.4 (10.3–12.5) | 78.0 (72.2–83.9) |
| 40–49 | n = 121 | 6.9 (6.1–7.6) | 11.9 (10.7–13.0) | 77.6 (72.0–82.3) |
| 50–59 | n = 56 | 6.7 (5.6–7.7) | 11.4 (9.9–13.0) | 74.1 (64.6–83.4) |
| ≥50 | n = 37 | 5.1 (3.3–6.9) | 12.2 (9.9–14.6) | 63.5 (50.1–76.2) |

[a] Displayed are mean values and 95% confidence intervals according to age. Results shown are for subjects without clinical history of coronary artery disease, peripheral vascular disease, diabetes mellitus, or hypertension.

*Noninvasive Studies: Pulse Wave Analysis of Arterial Stiffness*

There is great interest in examining arterial stiffness as a surrogate marker of atherosclerosis (Cohn *et al.*, 2004). A number of approaches can be used, including simple assessment of arterial pulse pressure measured by blood pressure cuff, pulse wave contour analysis assessed by tonometry, ultrasound visualization of arterial distensibility (calculated from the change in arterial diameter in relation to changes in blood pressure), and examination of pulse wave velocity. In regard to pulse wave velocity, a number of studies have shown that carotid-femoral pulse-wave velocity relates to cardiovascular disease risk factors and risk for future cardiovascular disease events (Cohn *et al.*, 2004). Whereas structural components of the arterial wall are major determinants of arterial stiffness, there is growing recognition that there also is a dynamic component of arterial stiffness that depends in part on arterial tone and endothelial release of NO. In support of the possibility, Wilkinson *et al.* (2002a,b) have observed that several measures of arterial stiffness are increased after systemic L-NMMA.

In our laboratory, we use applanation tonometry to assess vascular stiffness with a device developed at Cardiovascular Engineering, Inc. (Holliston, MA). Subjects lie quietly in a supine position, and pulse recordings are made from the carotid artery, brachial artery, radial artery, and femoral artery. Distances between recording sites are measured, and the pressures are calibrated using the brachial cuff pressure. Pulse recordings are gated using the electrocardiogram R-wave, and pulse wave velocity and the time of reflected waves are determined by blinded investigators. Reference values for a healthy, risk factor–free cohort were published this year (Mitchell *et al.*, 2004a). In some studies, we also made ultrasound recordings of flow and diameter of the left ventricular outflow tract, allowing us to calculate characteristic impendance, a variable that relates to stiffness of the proximal aorta (Mitchell *et al.*, 2002). One study demonstrated significant correlations between these measures of arterial stiffness and endothelial function (Nigam *et al.*, 2003), but further study will be required to define the precise contribute of endothelium-derived NO to arterial stiffness in different disease states.

Conclusions

The methodology for the study of NO-dependent vasodilation in intact humans continues to evolve. Many of the techniques are well established and have proven useful to study mechanisms of impaired NO bioavailability in atherosclerosis and related disease states and to evaluate potential therapies for these conditions. We suggest that some or all of these methods could be used clinically to assess cardiovascular risk or to guide risk-reduction therapy in individual patients. However, a great deal of work remains to be done to determine the clinical utility of these techniques.

## Acknowledgments

A Program Project Grant (HL60886), a Specialized Center of Research Grant (HL55993), and the Boston Medical Center General Clinical Research Center (M01RR00533) provided support for portions of this work.

## References

Anderson, T. J., Uehata, A., Gerhard, M. D., Meredith, I. T., Knab, S., Delagrange, D., Leiberman, E., Ganz, P., Creager, M. A., Yeung, A. C., and Selwyn, A. P. (1995). Close relation of endothelial function in the human coronary and peripheral circulations. *J. Am. Coll. Cardiol.* **26**, 1235–1241.

Benjamin, E. J., Larson, M. G., Keyes, M. J., Mitchell, G. F., Vasan, R. S., Keaney, J. F., Jr., Lehman, B., Fan, S., Osypiuk, E., and Vita, J. A. (2004). Clinical correlates and heritability of endothelial function in the community: The Framingham Heart Study. *Circulation* **109**, 613–619.

Bonetti, P. O., Barsness, G. W., Keelan, P. C., Schnell, T. I., Pumper, G. M., Holmes, D. R., Stuart, T. H., and Lerman, A. (2003). Enhanced external counterpulsation improves endothelial function in patients with coronary artery disease. *J. Am. Coll. Cardiol.* **41**, 370A (abstract).

Bonetti, P. O., Pumper, G. M., Higano, S. T., Holmes, D. R., Kuvin, J. T., and Lerman, A. (2004). Noninvasive identification of patients with early coronary atherosclerosis by assessment of digital reactive hyperemia. *J. Am. Coll. Cardiol.* **44**, 2137–2141.

Cohn, J. N., Quyyumi, A. A., Hollenberg, N. K., and Jamerson, K. A. (2004). Surrogate markers for cardiovascular disease: Functional markers. *Circulation* **109**, IV31–IV46.

Corretti, M. C., Anderson, T. J., Benjamin, E. J., Celermajer, D., Charbonneau, F., Creager, M. A., Deanfield, J., Drexler, H., Gerhard-Herman, M., Herrington, D., Vallance, P., Vita, J., and Vogel, R. (2002). Guidelines for the ultrasound assessment of endothelial-dependent flow-mediated vasodilation of the brachial artery. A report of the International Brachial Artery Reactivity Task Force. *J. Am. Coll. Cardiol.* **39**, 257–265.

Creager, M. A., Cooke, J. P., Mendelsohn, M. E., Gallagher, S. J., Coleman, S. M., Loscalzo, J., and Dzau, V. J. (1990). Impaired vasodilation of forearm resistance vessels in hypercholesterolemic humans. *J. Clin. Invest.* **86**, 228–234.

Fichtlscherer, S., Breuer, S., and Zeiher, A. M. (2004). Prognostic value of systemic endothelial dysfunction in patients with acute coronary syndromes: Further evidence for the existence of the "vulnerable" patient. *Circulation* **110**, 1926–1932.

Gerhard-Herman, M., Hurley, S., Mitra, D., Creager, M. A., and Ganz, P. (2002). Assessment of endothelial function (nitric oxide) at the tip of a finger. *Circulation* **106**, II–170 (abstract).

Gokce, N., Keaney, J. F., Jr., Menzoian, J. O., Watkins, M., Hunter, L., Duffy, S. J., and Vita, J. A. (2002). Risk stratification for postoperative cardiovascular events via noninvasive assessment of endothelial function. *Circulation* **105**, 1567–1572.

Halcox, J. P., Schenke, W. H., Zalos, G., Mincemoyer, R., Prasad, A., Waclawiw, M. A., Nour, K. R., and Quyyumi, A. A. (2002). Prognostic value of coronary vascular endothelial dysfunction. *Circulation* **106**, 653–658.

Hayoz, D., Weber, R., Rutschmann, B., Darioli, R., Burnier, M., Waeber, B., and Brunner, H. R. (1995). Postischemic blood flow response in hypercholesterolemic patients. *Hypertension* **26**, 497–502.

Heitzer, T., Schlinzig, T., Krohn, K., Meinertz, T., and Munzel, T. (2001). Endothelial dysfunction, oxidative stress, and risk of cardiovascular events in patients with coronary artery disease. *Circulation* **104**, 2673–2678.

Higashi, Y., Sasaki, S., Nakagawa, K., Matsuura, H., Kajiyama, G., and Oshima, T. (2001). A noninvasive measurement of reactive hyperemia that can be used to assess resistance artery endothelial function in humans. *Am. J. Cardiol.* **87**, 121–125.

Hirata, K., Shimada, K., Watanabe, H., Otsuka, R., Tokai, K., Yoshiyama, M., Homma, S., and Yoshikawa, J. (2004). Black tea increases coronary flow velocity reserve in healthy male subjects. *Am. J. Cardiol.* **93**, 1384–1388, A6.

Kuvin, J. T., Patel, A. R., Sliney, K. A., Pandian, G. P., Sheffy, J., Schnall, R. P., Karas, R. H., and Udelson, J. E. (2003). Assessment of peripheral vascular endothelial function with finger arterial pulse wave amplitude. *Am. Heart J.* **146**, 168–174.

Lieberman, E. H., Gerhard, M. D., Uehata, A., Selwyn, A. P., Ganz, P., Yeung, A. C., and Creager, M. A. (1996). Flow-induced vasodilation of the human brachial artery is impaired in patients <40 years of age with coronary artery disease. *Am. J. Cardiol.* **78**, 1210–1214.

Meredith, I. T., Currie, K. E., Anderson, T. J., Roddy, M. A., Ganz, P., and Creager, M. A. (1996). Postischemic vasodilation in human forearm is dependent on endothelium-derived nitric oxide. *AJP Heart Circ. Physiol.* **270**, H1435–H1440.

Mitchell, G. F., Izzo, J. L., Jr., Lacourciere, Y., Ouellet, J. P., Neutel, J., Qian, C., Kerwin, L. J., Block, A. J., and Pfeffer, M. A. (2002). Omapatrilat reduces pulse pressure and proximal aortic stiffness in patients with systolic hypertension: Results of the conduit hemodynamics of omapatrilat international research study. *Circulation* **105**, 2955–2961.

Mitchell, G. F., Parise, H., Benjamin, E. J., Larson, M. G., Keyes, M. J., Vita, J. A., Vasan, R. S., and Levy, D. (2004a). Changes in arterial stiffness and wave reflection with advancing age in healthy men and women: The Framingham Heart Study. *Hypertension* **43**, 1239–1245.

Mitchell, G. F., Parise, H., Vita, J. A., Larson, M. G., Warner, E., Keaney, J. F., Jr., Keyes, M. J., Levy, D., Vasan, R. S., and Benjamin, E. J. (2004b). Local shear stress and brachial artery flow-mediated dilation: The Framingham Heart Study. *Hypertension* **44**, 134–139.

Modena, M. G., Bonetti, L., Coppi, F., Bursi, F., and Rossi, R. (2002). Prognostic role of reversible endothelial dysfunction in hypertensive postmenopausal women. *J. Am. Coll. Cardiol.* **40**, 505–510.

Nigam, A., Mitchell, G. F., Lambert, J., and Tardif, J. C. (2003). Relation between conduit vessel stiffness (assessed by tonometry) and endothelial function (assessed by flow-mediated dilatation) in patients with and without coronary heart disease. *Am. J. Cardiol.* **92**, 395–399.

Otsuka, R., Watanabe, H., Hirata, K., Tokai, K., Muro, T., Yoshiyama, M., Takeuchi, K., and Yoshikawa, J. (2001). Acute effects of passive smoking on the coronary circulation in healthy young adults. *JAMA* **286**, 436–441.

Perticone, F., Ceravolo, R., Pujia, A., Ventura, G., Iacopino, S., Scozzafava, A., Ferraro, A., Chello, M., Mastroroberto, P., Verdecchia, P., and Schillaci, G. (2001). Prognostic significance of endothelial dysfunction in hypertensive patients. *Circulation* **104**, 191–196.

Schachinger, V., Britten, M. B., and Zeiher, A. M. (2000). Prognostic impact of coronary vasodilator dysfunction on adverse long-term outcome of coronary heart disease. *Circulation* **101,** 1899–1906.

Schindler, T. H., Hornig, B., Buser, P. T., Olschewski, M., Magosaki, N., Pfisterer, M., Nitzsche, E. U., Solzbach, U., and Just, H. (2003). Prognostic value of abnormal vasoreactivity of epicardial coronary arteries to sympathetic stimulation in patients with normal coronary angiograms. *Arterioscl. Thromb. Vasc. Biol.* **23,** 495–501.

Suwaidi, J. A., Hamasaki, S., Higano, S. T., Nishimura, R. A., Holmes, D. R., and Lerman, A. (2000). Long-term follow-up of patients with mild coronary artery disease and endothelial dysfunction. *Circulation* **101,** 948–954.

Targonski, P. V., Bonetti, P. O., Pumper, G. M., Higano, S. T., Holmes, D. R., Jr., and Lerman, A. (2003). Coronary endothelial dysfunction is associated with an increased risk of cerebrovascular events. *Circulation* **107,** 2805–2809.

Vita, J. A. (2002). Nitric oxide–dependent vasodilation in human subjects. *Methods Enzymol.* **359,** 186–200.

Vita, J. A., Keaney, J. F., Jr., Larson, M. G., Keyes, M. J., Massaro, J. M., Lipinska, I., Lehman, B., Fan, S., Osypiuk, E., Wilson, P. W. F., Vasan, R. S., Mitchell, G. F., and Benjamin, E. J. (2004). Brachial artery vasodilator function and systemic inflammation in the Framingham Offspring Study. *Circulation* **110,** 3604–3609.

Widlansky, M. E., Gokce, N., Keaney, J. F., Jr., and Vita, J. A. (2003). The clinical implications of endothelial dysfunction. *J. Am. Coll. Cardiol.* **42,** 1149–1160.

Wilkinson, I. B., MacCallum, H., Cockcroft, J. R., and Webb, D. J. (2002a). Inhibition of basal nitric oxide synthesis increases aortic augmentation index and pulse wave velocity *in vivo. Br. J. Clin. Pharmacol.* **53,** 189–192.

Wilkinson, I. B., Qasem, A., McEniery, C. M., Webb, D. J., Avolio, A. P., and Cockcroft, J. R. (2002b). Nitric oxide regulates local arterial distensibility *in vivo. Circulation* **105,** 213–217.

# [47] Assessing NO-Dependent Vasodilatation Using Vessel Bioassays at Defined Oxygen Tensions

*By* T. Scott Isbell, Jeffrey R. Koenitzer, Jack H. Crawford, C. R. White, David W. Kraus, and Rakesh P. Patel

## Abstract

Results from vessel bioassays have provided the foundation for much of our understanding of the mechanisms that control vascular homeostasis and blood flow. The seminal observations that led to the discovery that nitric oxide (NO) is a critical mediator of vascular relaxation were made with the use of such methodology, and many studies have used NO-dependent vessel relaxation as an experimental readout for understanding mechanisms that regulate vascular NO function. Studies have coupled controlling oxygen tensions within vessel bioassay chambers to begin to understand how oxygen—specifically hypoxia—regulate NO function, and

METHODS IN ENZYMOLOGY, VOL. 396
Copyright 2005, Elsevier Inc. All rights reserved.

0076-6879/05 $35.00
DOI: 10.1016/S0076-6879(05)96047-3

this context has identified red cells—specifically hemoglobin within—as critical modulators. Alone, vessel bioassays or measuring oxygen partial pressures ($pO_2$) is relatively straightforward, but the combination necessitates consideration of several factors. We use the example of deoxygenated red cells/hemoglobin-dependent potentiation of nitrite-dependent dilation to illustrate the salient factors that are critical to consider in designing and interpreting experiments aimed at understanding the interplay between oxygen and NO function in the vasculature.

## Introduction

### Vasoactivity of Nitric Oxide

The seminal studies that introduced nitric oxide (NO) to the biological arena and suggested that NO is a critical regulator of blood flow were related to its identification as an endothelium-derived relaxing factor (Ignarro et al., 1987; Palmer et al., 1987). Since then, and with the backing of almost 2 decades of research, it is now widely appreciated that NO regulates approximately 25% of basal blood flow in addition to a variety of processes important in vascular homeostasis (Quyyumi et al., 1995; Rees et al., 1989). The model that explains this function involves formation of NO from the endothelial isoform of NO synthase (eNOS), subsequent diffusion to the underlying smooth muscle, and activation of soluble guanylate cyclase (Ignarro, 2002). This initiates a signaling cascade that ultimately leads to vasodilation and increased blood flow. A central methodology used in developing this model is measuring tension in isolated vascular preparations treated with agonist and antagonists of NO-dependent signaling. Vessel bioassay studies have advanced our understanding of cardiovascular physiology and pharmacology in general, with recent studies extending the paradigm to include key regulatory functions for oxygen and red blood cells (RBCs) (Cosby et al., 2003; Crawford et al., 2003; James et al., 2004; Jia et al., 1996; McMahon et al., 2002; Pawloski et al., 2001; Wolzt et al., 1999).

### Hypoxia and Vascular Functions of NO

Oxygen and NO metabolism are intricately linked through multiple mechanisms ranging from oxygen being a substrate for NOS-dependent NO synthesis to NO regulating mitochondrial respiration and mediating tissue responses to hypoxia (Brookes et al., 2003; Edmunds et al., 2003; Gong et al., 2004; Hagen et al., 2003; Thomas et al., 2001). Physiological responses to tissue hypoxia are complex. For example, systemic hypoxia stimulates blood flow, a process vital in matching oxygen delivery to demand (Gonzalez-Alonso *et al.*, 2002). In contrast, hypoxia elicits vaso-constriction in the pulmonary circulation (Deem, 2004). Hypoxia is also a key factor in diseases associated with dysfunctional vascular NO metabolism (e.g., sepsis) (Crawford *et al.*, 2004). Vessel bioassay experiments have been instrumental in providing insights into the mechanisms through which oxygen modulates vascular hemodynamics, including activation of eNOS, modulation of tissue adenosine and potassium currents, and altered reactive oxygen species metabolism (Tune *et al.*, 2004).

More recent evidence suggests that hypoxia-induced vasodilatation occurs via circulating factors that respond to changing oxygen gradients by stimulating NO-dependent vasodilation. In this paradigm, either NO production within the hypoxic tissue is stimulated, or relatively stable precursor molecules are activated by hypoxia to release NO (Gladwin *et al.*, 2004). Within this framework, three mechanisms that incorporate the oxygen-sensing functions of the RBCs (specifically hemoglobin) have been proposed and include (1) adenosine triphosphate (ATP) release from the RBC and subsequent stimulation of purinergic signaling and eNOS activation (Ellsworth, 2004; Sprague *et al.*, 2003), and release of a vasodilatory stimulus from either (2) *S*-nitrosohemoglobin (Singel and Stamler, 2004) or (3) from nitrite reactions with deoxygenated RBCs (Cosby *et al.*, 2003; Nagababu *et al.*, 2003). This chapter is not intended to discuss these mechanisms (the reader is referred to review articles for a more detailed discussion) (Ellsworth, 2004; Gladwin *et al.*, 2004; Kim-Shapiro *et al.*, 2005; Singel and Stamler, 2004), but to outline general methods that allow coupling of techniques used in controlling oxygen concentrations to performing vessel bioassay studies from which the insights gained have been gleaned. Many texts have discussed the basic methodology associated with vessel bioassay studies, and these are, therefore, only briefly covered. This chapter focuses on the practical considerations associated with experimental design and interpretation when working with vessel bioassay systems that incorporate both oxygen control and the use of RBCs/hemoglobin.

## Vessel Bioassays at Defined Oxygen Partial Pressures

### Basic Principle of Vessel Bioassays

The classic approach for vessel bioassays is the measurement of changes in tension of a vessel isolated from a given vascular bed in response to the compound of interest. Typically, isolated vessels are cleaned of connective tissue and fat, segmented into individual ring segments (2–3 mm), and hung between two hooks connected to a force-displacement transducer in a tissue bath. Vessels are equilibrated at 37° in a Krebs-Henseleit (KH) buffer of the following composition (m$M$): NaCl 118; KCl 4.6; NaHCO$_3$ 27.2; KH$_2$PO$_4$ 1.2; MgSO$_4$ 1.2; CaCl$_2$ 1.75; Na$_2$ EDTA 0.03, and glucose 11.1 for 30 min. The KH is continuously perfused with a gas mixture, typically 21% or 95% O$_2$ and 5% CO$_2$, balanced with N$_2$. A passive tension is applied to the vessels (typically 2 g for rat thoracic aorta and 3 g for rabbit), and vessel viability and maximal contraction are then assessed by addition of a depolarizing dose of KCl (70 m$M$). Following a wash and re-equilibration period, vessels are pre-contracted to approximately 50% of maximal with a vasoconstrictive agent, typically 10$^{-7}$M phenylephrine (PE). Once a stable baseline is reached, the compound under investigation is added at increasing concentrations to assess its vasoactive effects. Percent relaxation and vasoconstriction are calculated by measuring the change in tension relative to submaximal contraction.

### Controlling Oxygen Partial Pressure

Oxygen partial pressure in vessel bioassay chambers can be varied by perfusing the buffer with specific gas mixtures containing 5% CO$_2$, the latter of which is essential for maintaining pH in bicarbonate-buffered systems. Exclusion of CO$_2$ can cause the calcium to precipitate out of solution and indicates a lack of control over pH. Most commercial vendors supply tanks with varying oxygen mixtures (0–95%), allowing for wide range for experimentation. To increase the variety of gas mixtures, the gas delivery system could include multiple mass flow controllers (several commercial suppliers are available), connecting a limited number of stock gas tanks to a common manifold. Such an arrangement will allow the user to select an infinite variety of gas mixtures to be delivered to the bioassay chambers.

In theory, at equilibrium with the perfused gas, the oxygen concentration in the buffered solution is the product of oxygen partial pressure (pO$_2$) and oxygen solubility. Solubility is a function of solvent composition (i.e., salinity) and the temperature. Temperature and salinity of the experimental media have an inverse effect on the solubility of oxygen; however, other constituents such as lipids can effectively increase solubility. Depending on the desired readout, oxygen can be expressed as partial pressure or converted to molarity using the correct solubility factor. For the purposes of experiments related to RBC and Hb-dependent effects, oxygen is expressed as pO$_2$ in millimeters of mercury (mm Hg) or torr. For a more thorough review of these principles, see Gnaiger and Forstner (1983).

In practice, however, the chamber bath pO$_2$ calculated from the gas mixture as described earlier provides only an approximation of the actual

$pO_2$, which can vary significantly because of several factors. First, all (to the best of our knowledge) vessel bioassay chambers are open systems that permit some gas exchange with the ambient atmosphere. This is a concern when performing experiments at a $pO_2$ that is different than the atmospheric oxygen pressure, especially at lower, physiologically relevant $pO_2$. This is best illustrated by the fact that perfusing with 95% $N_2$, 5% $CO_2$ gas, which is calculated to maintain the vessel bath at a $pO_2$ of 0 mm Hg, results in a measured $pO_2$ that varies between approximately 2 and 20 mm Hg, depending on gas flow rate. Other factors that affect the surface area of the solution–atmosphere interface (e.g., size and configuration of vessel bath) will also influence the bath $pO_2$. As shown in Fig. 1, gas is typically perfused through the solution from the bottom of the chamber at a rate that the experimenter sets to be sufficient to maintain mixing and solution homogeneity without damaging the vessel segment. In our hands, adjusting the gas perfusion rate only slightly can significantly affect solution $pO_2$ so that it can vary between 2 and 20 mm Hg when perfusing with 95% $N_2$, 5% $CO_2$. In addition to potential differences in gas flow between different experiments, perfusion into each individual vessel bath is usually regulated separately, giving rise to further variations in $pO_2$ in neighboring baths within a

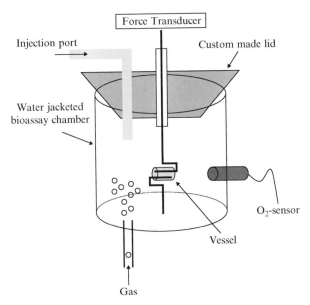

FIG. 1. Representative diagram of vessel bioassay chamber modified to accommodate an oxygen sensor and cap. For clarity buffer, wash, and water-jacket ports are not shown.

given experiment. In the context of working with hemoglobin, whose conformation state is regulated by the $pO_2$, which in turn may influence interactions with NO and its derivatives, this can result in different conformations and hence results. Other variables that can affect $pO_2$ include the length and composition of tubing through which solutions and gas pass. These concerns necessitate either installation of mass flow controllers or direct measurement of solution $pO_2$. The latter is discussed in more detail next.

## Measuring Oxygen Partial Pressure in Vessel Bath Solutions

### Polarographic Oxygen Sensors

Conducting experiments at defined $pO_2$ requires the ability to accurately and continuously monitor $pO_2$ in the bath solution. We use a Clark-style polarographic oxygen sensor (POS), which is inserted through the wall of the tissue bath (Fig. 1). It is assumed that the reader is familiar with the general principles of measuring $pO_2$ using an oxygen sensor, and thus it will not be discussed here. The placement of the sensor is important. As discussed later in this chapter, sealing the vessel bioassay chamber limits oxygen back-diffusion from the environment. To accommodate this, insertion of the sensor into the side wall is preferred, as opposed to insertion from the top. However, conventional bioassay chambers are not designed to accommodate this and therefore need to be adapted. We recommend that before modifying your chambers, the respective manufacturer be consulted to ensure compatibility of inserting an oxygen sensor into the chamber wall with the protocols for measuring vessel tension.

### Oxygen Sensor Calibration

Sensor calibrations should be performed under experimental conditions of media and temperature. An open bath system has the potential to gain or lose oxygen if perfused with a gas that has a $pO_2$ lower or higher than ambient, respectively. For this reason a two-point calibration with air, 21% $O_2$, and sodium dithionite, for 100% and 0% air saturation values, respectively, will limit calibration errors. Converting these percent saturations to partial pressures requires a measure of the barometric and water vapor pressure for the bath temperature. Because oxygen sensors respond linearly to $pO_2$, after such a calibration any perfusion gas above or below 21% $O_2$ will establish an equilibrium $pO_2$ in proportion to the air saturation $pO_2$.

## Experimental Setup

### *Open System*

Most applications of vessel tension measurements at specific $pO_2$ are performed in an open system. This offers the most ease in practical terms (for assembling vascular tissue in chambers, addition of reactants, etc.) but, as discussed earlier, it suffers from lack of fine control over $pO_2$, especially when working under hypoxic conditions. Typically, a range of oxygen levels can be achieved depending on perfused gas and bubbling rates, as discussed earlier. Table I serves as a guide for approximate solution $pO_2$ when perfusing KH buffer with defined gas mixtures at 37° using Radnoti 15-ml bioassay chambers.

### *"Closed" Systems*

To limit solution $pO_2$ variability associated with the open system, a closed system can be engineered by fitting chambers with a custom-made cap with ports to accommodate the tether from vessel to force transducer and the injection of reactants (see Fig. 1). Strictly speaking, this is not a closed system, because a gas phase still exists; however, sealing the

TABLE I

Predicted and Measured $pO_2$ Values in Krebs Henseleit Buffer, pH 7.4, 37° after Perfusion of Open Vessel Bioassay Chambers (15 ml Volume, Radnoti) with Gas Mixtures as Shown[a]

| Gas mixture | Predicted $pO_2$ (mm Hg) | Measured $pO_2$ (mm Hg) |
| --- | --- | --- |
| 95% $O_2$, 5% $CO_2$ | 660.25 | 609 |
| 21% $O_2$, 5% $CO_2$, 74% $N_2$ | 145.95 | 145.95 |
| 5% $O_2$, 5% $CO_2$, 90% $N_2$ | 34.75 | 62 |
| 1% $O_2$, 5% $CO_2$, 94% $N_2$ | 6.95 | 25 |
| 5% $CO_2$, 95% $N_2$ | 0 | 2–20 |

[a] The predicted values are calculated according to the equation $pO_2 = P_B - P_{H_2O} \times \%O_2$, where $P_B$ = barometric pressure, $P_{H_2O}$ = water vapor pressure (47 mm Hg), and $\%O_2$ = % oxygen in atmosphere (21%). Barometric pressure must be measured in the vicinity of the apparatus. The measured values represent an approximate as determined by oxygen sensors placed in chambers according to Fig. 1 and are included only to illustrate that differences between predicted and actual chamber $pO_2$ levels will exist. Furthermore, the actual $pO_2$ level will fluctuate depending on various factors as discussed in this chapter and thus should be measured directly for every experiment. The necessity for direct $pO_2$ measurement is highlighted by the range of $pO_2$ levels (2–20 mm Hg) that can be achieved by simply altering the rate of gas perfusion using 95% $N_2$, 5% $CO_2$.

chamber significantly decreases the surface area for back-diffusion, allowing finer control over lower solution $pO_2$. Again, direct measurement of $pO_2$ remains essential. The major practical limitation with this design is the care required in vessel preparation because the cap must allow an unimpeded connection between the vessel and force transducer.

## Simultaneous Measurement of Vessel Tension and Oxygen Partial Pressure

Thus far, the approaches discussed involve the standard protocol whereby the test compound is added at varying doses to the bioassay chamber at a fixed $pO_2$. Whereas this is useful for providing information on dose–response relationships, information on the oxygen dependence is limited to those $pO_2$ levels that can be attained by perfusion with gases of different oxygen compositions with detailed assessment of the oxygen dependence difficult to assess. This is illustrated by our interests in how RBCs stimulate nitrite-dependent vasodilation. Specifically, we and others have proposed that deoxygenated, but not oxygenated, RBCs convert nitrite into a vasodilator and that this process may contribute to hypoxic vasodilation (Cosby et al., 2003; Kim-Shapiro et al., 2005; Nagababu et al., 2003). Whether hemoglobin is oxygenated or deoxygenated is controlled by the hemoglobin $P_{50}$, which is the specific $pO_2$ at which hemoglobin is 50% saturated with oxygen and can vary between species, temperature, pH, and so on. Setting the $pO_2$ in bioassay chambers to a specific value is difficult (see earlier discussion). To directly address the question at what $pO_2$ do vessels start to dilate in response to different treatments, we adopted the following protocol in which vessel tension and $pO_2$ are monitored simultaneously while the $pO_2$ is changing continually. Specifically, vessels are precontracted, as in the conventional protocol discussed earlier, at a "high" oxygen tension (typically 95% or 21% $O_2$). Once stable, the reactants are added (e.g., RBCs and nitrite), and then the perfusion gas is switched to "low" oxygen gas mixture (95% $N_2$, 5% $CO_2$). Figure 2A shows representative data traces from such an experiment using rat thoracic aorta vessels under control (i.e., no treatment) conditions. At some $pO_2$ levels, vessels start to dilate spontaneously (discussed in more detail later), and this can be quantitated as shown in Fig. 2B. Figure 2C shows an alternative reaction profile that can be observed and differs slightly from that shown in Fig. 2A in that a relatively small ($\sim$1–3%) constriction is observed before dilation of vessel as the $pO_2$ decreases. The underlying nature of these differences is not clear but likely reflects inherent variances between vessels obtained from different animals. Figure 2D shows how the $pO_2$ at which dilation starts can be determined from traces shown in

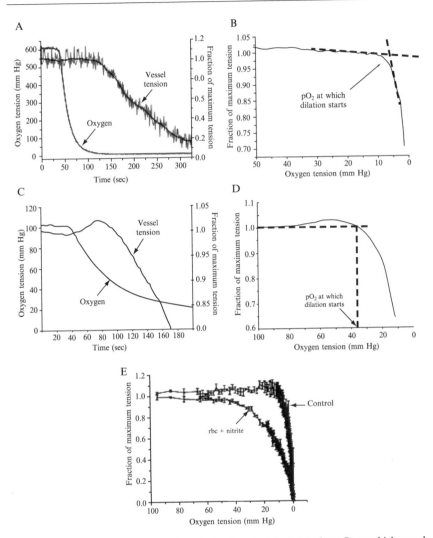

FIG. 2. Representative traces and methods of analysis to determine $pO_2$ at which vessels dilate when using protocol of simultaneous $pO_2$ and vessel tension measurement as vessel bath is deoxygenating. (A, C) Paired $pO_2$ and vessel tension traces from either control (i.e., no treatment) and treatment with rat red cells (0.3% hc) + nitrite (2 $\mu M$), respectively. Panels (A) and (C) also serve to illustrate the two types of responses observed. Specifically as the bath deoxygenates, at some $pO_2$, vessels start to relax (A) or relax after a small and transient (<3%) constriction (C). Note that in panel (A), vessel tension tracings are often accompanied by noise (gray line) when working with low oxygen tensions. In this case, a smoothing algorithm is applied and shown by a black line (Origin, Origin Lab Corporation). (B, D) Plots of $pO_2$ versus vessel tension [from data shown in panels (A) and (C), respectively]. The

Fig. 2C. Figure 2E shows cumulative data illustrating how the analysis (shown in Fig. 2B and D) was performed to yield insights into how RBCs potentiate nitrite-dependent vasodilation at $pO_2$ levels close to their $P_{50}$.

With control vessels, reoxygenation (after an initial deoxygenation) results in an increase in vessel tension due to the continued presence of contractile stimuli. Using rat thoracic aorta (note that different vascular preparations will have distinct oxygen sensitivities) and under control conditions, multiple cycles (five or more) of oxygenation and deoxygenation can be performed without compromising vessel dilation or contractile responses. If treatments are included, multiple cycles can be performed; however, reagent consumption and potential reactions with vascular tissue have to be considered. We recommend the first oxygenation–deoxygenation cycle be performed under control conditions to establish basal oxygen dependence and then reagents of interest added followed by a deoxygenation–reoxygenation cycle. Replicates can be obtained by using multiple vascular ring segments. Finally, these experiments can be performed in open or closed systems, but the latter will facilitate deoxygenation to lower $pO_2$. Note also that because a gas phase exists in both systems, the rate of deoxygenation is not linear. This is an important consideration for the conventional approach for performing vessel bioassays and necessitates that sufficient time be allowed for reaching the desired $pO_2$.

## Considerations and Limitations

### Critical $pO_2$

Because assessing changes in vessel tone is reliant on a stable baseline, it is critical to consider any effect of hypoxia on the basal tone. Moreover, at some concentrations, oxygen will become limiting for cytochrome $c$ oxidase/mitochondrial respiration and adenosine triphosphate (ATP) production. This value is referred to as the *critical $pO_2$*. Vascular contraction/dilation is ultimately dependent on ATP, so experiments performed at oxygen tensions below the critical $pO_2$ will be associated with changing baselines and will not accurately reflect any changes observed in response to test compounds. This underscores the importance of determining the critical $pO_2$ required to sustain viable vascular responses.

---

''threshold'' curves that are generated allow determination of the $pO_2$ at which vessels dilate (dashed lines). (E) Data obtained using this analysis to document that red cells + nitrite stimulate vasodilation at a higher $pO_2$ compared to control, and this value is similar to the hemoglobin $P_{50}$ (note that rat red cell $P_{50} = 35$–$40$ mm Hg). Reproduced, with permission, from Cosby *et al.* (2003).

The critical $pO_2$ is determined by monitoring the rates of oxygen consumption of the tissue, which are then plotted as a function of oxygen concentration and fitting to Michaelis-Menten equation. Whereas this is readily performed for cell suspensions, isolated organelles, and solubilized proteins in standard respirometer chambers, measuring rates of oxygen consumption in intact tissue is more difficult, largely because of the potential for tissue damage by the stirrer. To overcome this, we have used a stainless-steel screen stage (Fig. 3A) that is supported by an O-ring, which sits just above the stirrer in a respirometer designed for low oxygen experiments (Oxygraph, Oroboros, Innsbruck, Austria). Vessel segments can be mounted on the wire stage to ensure perfusion through the vessels while in the respirometer chamber. With the chamber sealed, the vessel oxygen consumption rate is determined as the slope of the declining chamber $pO_2$. Figure 3B shows an example of data obtained using this approach with rat thoracic aorta where critical $pO_2$ with these vessels is approximately 25 $\mu M$ under basal conditions, and it increases to approximately 35 $\mu M$ with PE precontracted vessels. These oxygen concentrations correspond to 18–30 mm Hg, which interestingly correlates with the $pO_2$ at which vessels start to spontaneously dilate (see Fig. 2B) and suggests that dilation in this system occurs as a result of loss of oxygen-dependent ATP production.

This raises an important cautionary note on performing experiments at low $pO_2$ levels, which may be close to or below the critical $pO_2$. As discussed, this situation can readily arise resulting in slowly changing baselines and introducing variability in experimental responses. We suggest that the critical $pO_2$ be determined for every vessel type used, and if possible, experiments should be designed so $pO_2$ levels above this are used. These effects of oxygen limitation again underscore the importance of direct measurement of $pO_2$ in each chamber under a given set of experimental conditions.

*Hemolysis*

Specifically in the context of RBCs, hemolysis is an important to measure for each experiment, and it can be assessed by centrifugation of experimental solution to pellet RBCs and determine hemoglobin in supernatant by standard spectroscopic techniques. Even a small degree of hemolysis can result in relatively high concentrations of cell-free hemoglobin that can efficiently scavenge NO and impact vessel tone. This can be prevented by inhibited NO production by the vessel using NOS inhibitors (e.g., L-NNA or L-NAME). Note that addition of these inhibitors will result in contraction alone and therefore should be added before test compounds

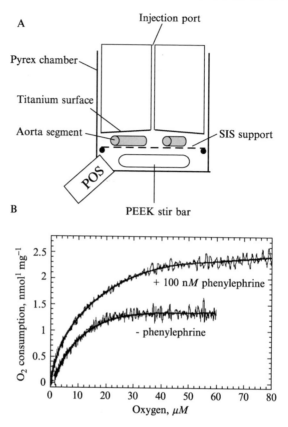

FIG. 3. Representative diagram of respirometer chamber used to determine the critical $pO_2$ of the vessel oxygen consumption rate. Geometry and materials are designed to ensure that the oxygen sensor, POS, responds rapidly to declining $pO_2$ and that oxygen egress from the materials is minimized (S/S, stainless steel; PEEK, polyether ether ketone). Vessels adhere to stainless-steel surface, presumably due to inherent surface tension, and using stirring speeds of 500 rpm in a 3-ml chamber does not move significantly. (B) Representative traces of oxygen consumption rate of rat aorta segments as a function of $pO_2$. Critical $pO_2$ is determined in vessels at rest and in those preconstricted with 100 n$M$ phenylephrine.

to obtain stable baselines. The extent of hemolysis is typically less than 0.5% but will increase with more vigorous bubbling.

### Oxygen Offloading from Red Cells

Another specific consideration with RBCs and hemoglobin is that they will bind or offload oxygen depending on the $pO_2$ and their $P_{50}$. Because the oxygen-binding capacity of hemoglobin is relatively high, deoxygenation beyond and below the $P_{50}$ can result in the release of significant oxygen into solution that would be measured by a sensor as an increase in solution concentration. Experimental media must, therefore, be allowed to reattain equilibrium after addition of RBCs or hemoglobin. For protocols in which deoxygenation is occurring as a function of time, the measured $pO_2$ by the sensor will reflect a complex balance of rates of oxygen offloading, rates of anoxic gas perfusion, and sensor response times. We find that with RBCs alone, the faster the rate of deoxygenation, the more pronounced the oxygen offloading phenomenon is, as determined by solution oxygen measurements.

*Foaming*

A limitation with perfusion of gas through the solution is that foaming results readily when proteins at high concentrations or solutions of relatively high hydrophobicity (e.g., RBCs and other membrane containing species) are required. Generally, the greater the degree of foaming, the greater the noise on tension measurements. This limits the maximum concentrations that can be used and, with RBCs, is typically 1% or less of hematocrit. Moreover, in open systems, foaming can create a new boundary at the solution–atmosphere interface that affects gas exchange, thus impacting oxygen in solution. The degree of foaming varies from chamber to chamber and adds to a list of factors that can lead to a lack of accurate control over $pO_2$. Commercially available "anti-foam" reagents can be used but must be controlled for any non-specific effects.

*Order of Preconstriction and Hypoxia*

For hypoxia experiments, two protocols can be employed. Either vessels can be precontracted first, followed by equilibration at desired $pO_2$, or vice versa. In each case, and depending on the degree, hypoxia will induce relaxation and development of a new baseline. In some laboratories, the basal tone is increased again by addition of more contractile stimulus. Alternatively, if a sufficient experimental window still exists (i.e., contractile tone) to observe dilation responses, test reagents can be added. In this case, comparison of dilatory stimuli at normoxia versus hypoxia must be normalized for the degree of change in baseline in response to hypoxia. This is important because hypoxia will potentiate nitrosovasodilator-dependent vasodilation and is illustrated by the example of *S*-nitrosohemoglobin, whose dilation, when normalized for hypoxia effects on vessel tone, does not show any oxygen dependence (Crawford *et al.*, 2003; Wolzt *et al.*, 1999).

*Position of Oxygen Sensor*

Placement of the sensor close to the vessel is critical, given that oxygen in solution may be heterogeneous, especially at the surface, where gas exchange is occurring and the bottom of the tissue bath, which may contain unstirred layers. If possible, we recommend placing the sensor at a height similar to where aortic rings are suspended.

## Summary

Vessel bioassays coupled with the ability to simultaneously measure $pO_2$ have proven to be critical in our understanding of mechanisms through which oxygen controls NO-dependent vasodilation. This is highlighted by concepts that propose a coupling of the oxygen-sensing function of hemoglobin to stimulating NO-dependent vasodilation. In this chapter, we described this approach and focused specifically on the factors that need to be considered when performing such experiments, with the most important, in our opinion, being direct measurement of $pO_2$ in the vessel bath of interest. Only when all factors are considered can accurate interpretation of the role of oxygen be made and implications for NO-dependent regulation of blood flow derived.

## Acknowledgments

This work was supported by National Institutes of Health (NIH) grants HL70146 (RPP) and AHA Grant in aid 0455296B (DWK), as well as an NIH cardiovascular training fellowship (TSI).

## References

Brookes, P. S., Kraus, D. W., Shiva, S., Doeller, J. E., Barone, M. C., Patel, R. P., Lancaster, J. R., Jr., and Darley-Usmar, V. (2003). Control of mitochondrial respiration by NO*, effects of low oxygen and respiratory state. *J. Biol. Chem.* **278**, 31603–31609.
Cosby, K., Partovi, K. S., Crawford, J. H., Patel, R. P., Reiter, C. D., Martyr, S., Yang, B. K., Waclawiw, M. A., Zalos, G., Xu, X., Huang, K. T., Shields, H., Kim-Shapiro, D. B., Schechter, A. N., Cannon, R. O., 3rd, and Gladwin, M. T. (2003). Nitrite reduction to nitric oxide by deoxyhemoglobin vasodilates the human circulation. *Nat. Med.* **9**, 1498–1505.
Crawford, J. H., Chacko, B. K., Pruitt, H. M., Piknova, B., Hogg, N., and Patel, R. P. (2004). Transduction of NO-bioactivity by the red blood cell in sepsis: Novel mechanisms of vasodilation during acute inflammatory disease. *Blood* **104**, 1375–1382.
Crawford, J. H., White, C. R., and Patel, R. P. (2003). Vasoactivity of S-nitrosohemoglobin: Role of oxygen, heme, and NO oxidation states. *Blood* **101**, 4408–4415.
Deem, S. (2004). Nitric oxide scavenging by hemoglobin regulates hypoxic pulmonary vasoconstriction. *Free Radic. Biol. Med.* **36**, 698–706.

Edmunds, N. J., Moncada, S., and Marshall, J. M. (2003). Does nitric oxide allow endothelial cells to sense hypoxia and mediate hypoxic vasodilatation? *In vivo* and *in vitro* studies. *J. Physiol.* **546**, 521–527.

Ellsworth, M. L. (2004). Red blood cell–derived ATP as a regulator of skeletal muscle perfusion. *Med. Sci. Sports Exerc.* **36**, 35–41.

Gladwin, M. T., Crawford, J. H., and Patel, R. P. (2004). The biochemistry of nitric oxide, nitrite, and hemoglobin: Role in blood flow regulation. *Free Radic. Biol. Med.* **36**, 707–717.

Gnaiger, E., and Forstner, H. (eds.) (1983). "Polarographic Oxygen Sensors, Aquatic and Physiological Applications." Springer-Verlag, New York.

Gong, L., Pitari, G. M., Schulz, S., and Waldman, S. A. (2004). Nitric oxide signaling: Systems integration of oxygen balance in defense of cell integrity. *Curr. Opin. Hematol.* **11**, 7–14.

Gonzalez-Alonso, J., Olsen, D. B., and Saltin, B. (2002). Erythrocyte and the regulation of human skeletal muscle blood flow and oxygen delivery: Role of circulating ATP. *Circ. Res.* **91**, 1046–1055.

Hagen, T., Taylor, C. T., Lam, F., and Moncada, S. (2003). Redistribution of intracellular oxygen in hypoxia by nitric oxide: Effect on HIF1alpha. *Science* **302**, 1975–1978.

Ignarro, L. J. (2002). Nitric oxide as a unique signaling molecule in the vascular system: A historical overview. *J. Physiol. Pharmacol.* **53**, 503–514.

Ignarro, L. J., Buga, G. M., Wood, K. S., Byrns, R. E., and Chaudhuri, G. (1987). Endothelium-derived relaxing factor produced and released from artery and vein is nitric oxide. *Proc. Natl. Acad. Sci. USA* **84**, 9265–9269.

James, P. E., Lang, D., Tufnell-Barret, T., Milsom, A. B., and Frenneaux, M. P. (2004). Vasorelaxation by red blood cells and impairment in diabetes: Reduced nitric oxide and oxygen delivery by glycated hemoglobin. *Circ. Res.* **94**, 976–983.

Jia, L., Bonaventura, C., Bonaventura, J., and Stamler, J. S. (1996). S-nitrosohaemoglobin: A dynamic activity of blood involved in vascular control. *Nature* **380**, 221–226.

Kim-Shapiro, D. B., Gladwin, M. T., Patel, R. P., and Hogg, N. (2005). The reaction between nitrite and hemoglobin: The role of nitrite in hemoglobin-mediated hypoxic vasodilation. *J. Inorg. Biochem.* **99**, 237–246.

McMahon, T. J., Moon, R. E., Luschinger, B. P., Carraway, M. S., Stone, A. E., Stolp, B. W., Gow, A. J., Pawloski, J. R., Watke, P., Singel, D. J., Piantadosi, C. A., and Stamler, J. S. (2002). Nitric oxide in the human respiratory cycle. *Nat. Med.* **8**, 711–717.

Nagababu, E., Ramasamy, S., Abernethy, D. R., and Rifkind, J. M. (2003). Active nitric oxide produced in the red cell under hypoxic conditions by deoxyhemoglobin-mediated nitrite reduction. *J. Biol. Chem.* **278**, 46349–46356.

Palmer, R. M., Ferrige, A. G., and Moncada, S. (1987). Nitric oxide release accounts for the biological activity of endothelium-derived relaxing factor. *Nature* **327**, 524–526.

Pawloski, J. R., Hess, D. T., and Stamler, J. S. (2001). Export by red blood cells of nitric oxide bioactivity. *Nature* **409**, 622–626.

Quyyumi, A. A., Dakak, N., Andrews, N. P., Husain, S., Arora, S., Gilligan, D. M., Panza, J. A., and Cannon, R. O., 3rd (1995). Nitric oxide activity in the human coronary circulation. Impact of risk factors for coronary atherosclerosis. *J. Clin. Invest.* **95**, 1747–1755.

Rees, D. D., Palmer, R. M., Hodson, H. F., and Moncada, S. (1989). A specific inhibitor of nitric oxide formation from L-arginine attenuates endothelium-dependent relaxation. *Br. J. Pharmacol.* **96**, 418–424.

Singel, D. J., and Stamler, J. S. (2004). Chemical physiology of blood flow regulation by red blood cells: Role of nitric oxide and S-nitrosohemoglobin. *Annu. Rev. Physiol.*

Sprague, R. S., Olearczyk, J. J., Spence, D. M., Stephenson, A. H., Sprung, R. W., and Lonigro, A. J. (2003). Extracellular ATP signaling in the rabbit lung: Erythrocytes as determinants of vascular resistance. *Am. J. Physiol. Heart Circ. Physiol.* **285,** H693–H700.

Thomas, D. D., Liu, X., Kantrow, S. P., and Lancaster, J. R., Jr. (2001). The biological lifetime of nitric oxide: Implications for the perivascular dynamics of NO and O2. *Proc. Natl. Acad. Sci. USA* **98,** 355–360.

Tune, J. D., Gorman, M. W., and Feigl, E. O. (2004). Matching coronary blood flow to myocardial oxygen consumption. *J. Appl. Physiol.* **97,** 404–415.

Wolzt, M., MacAllister, R. J., Davis, D., Feelisch, M., Moncada, S., Vallance, P., and Hobbs, A. J. (1999). Biochemical characterization of S-nitrosohemoglobin. Mechanisms underlying synthesis, no release, and biological activity. *J. Biol. Chem.* **274,** 28983–28990.

# [48] Nonenzymatic Nitric Oxide Formation during UVA Irradiation of Human Skin: Experimental Setups and Ways to Measure

*By* Christoph V. Suschek, Adnana Paunel, and
Victoria Kolb-Bachofen

## Abstract

Many of the local ultraviolet (UV)-induced responses, including erythema and edema formation, inflammation, premature aging, and immune suppression, can be influenced by nitric oxide synthase (NOS)–produced NO, which plays a pivotal role in cutaneous physiology. Besides enzyme-mediated NO production, UV radiation triggers an enzyme-independent NO formation in human skin. This occurs due to decomposition of photoreactive nitrogen oxides like nitrite and S-nitrosothiols, which are present in human skin at relatively high concentrations and lead to high-output formation of bioactive NO. This enzyme-independent NO formation opens a new field in cutaneous physiology and will extend our understanding of mechanisms contributing to skin aging, inflammation, and cancerogenesis but also functional protection. Therefore, it is of high interest to examine the chemical storage forms of these potential NO-generating agents in skin, the mechanisms and kinetics of their decomposition, and their biological relevance.

## Introduction

In human skin, several lines of evidence indicate that nitric oxide (NO) is involved in the control of wound-healing processes, in allergic skin manifestations, in microbicidal activity, antigen presentation, hair growth,

METHODS IN ENZYMOLOGY, VOL. 396
0076-6879/05 $35.00
DOI: 10.1016/S0076-6879(05)96048-5

in proliferation and differentiation of epidermal cells, and in the regulation of innate immune reactions and inflammatory responses. In addition, ultraviolet (UV)-induced processes such as erythema and edema formation, as well as melanogenesis, are also under NO-mediated control (Bruch-Gerharz *et al.*, 1998a,b). Moreover, it has been repeatedly shown that NO regulates the expression of a large number of genes, including protective stress response genes such as vascular endothelial growth factor (VEGF), heme oxygenase (HO)-1, and Bcl-2 (Ehrt *et al.*, 2001; Hemish *et al.*, 2003; Suschek *et al.*, 2003a). In addition, NO is an effective inhibitor of lipid peroxidation (Hogg and Kalyanaraman, 1999), and the coordinated action of NO on gene expression and preservation of membrane function effectively protects against either UVA- or reactive oxygen species (ROS)–induced apoptotic and necrotic cell death (Suschek *et al.*, 1999, 2001a).

It has been repeatedly shown that NO synthase (NOS)–dependent production of NO potentially occurs in all dermal cell types. Some of the NO molecules formed remain at or close to the point of their origin as nitroso compounds (RSNO or RNNO) and their oxidation products, nitrite and nitrate.

Both nitrite and RSNO compounds potentially represent an enzyme-independent source for NO either in an acidic milieu or during UVA exposure via decomposition or photolysis, respectively (Fischer and Warneck, 1996; Singh *et al.*, 1996). The nonenzymatic NO formation during UVA exposure potentially bridges the time gap between UV challenge and the enzymatic activity of inducible NOS (iNOS), which starts 8–10 h after UV irradiation (Kuhn *et al.*, 1998; Suschek *et al.*, 2001b), and represents a physiological response of human skin. Indeed, we show that during UVA challenge, nitrite represents a protective principle against ROS-induced cell death (Suschek *et al.*, 2003b).

With respect to the central biological relevance of NO in human cutaneous physiology and its protective role during exposure to UVA irradiation, it is of interest to determine the local concentrations of NO-derived products in human skin and to measure their decomposition and their ability to form NO upon UVA irradiation.

## Decision for an Applicative UVA Source

As mentioned, nitrogen oxide derivatives can decompose by the impact of UV light. Therefore, the choice of the proper light unit is decisive. Nitrite ions undergo photolysis, showing a maximum around 354 nm (Jankowski *et al.*, 2000). *S*-nitrosothiols (RSNO) have an absorption band peaking in the range of 330–340 nm of wavelength (Zhang *et al.*, 1996). Other compounds can also be added to the list for their well-established ability to undergo photolytic cleavage to NO in the UV or near-UV range. These include $N$-nitroso compounds (RNNO), which exhibit absorption maxima at 340–390 nm (Rao and Bhaskar, 1981); nitrosyl myoglobin (MbNO) with absorption maximum between 420 and 422 nm (Kharitonov et al., 1996); iron–sulfur nitrosyl complexes [such as Roussin's red salt (RRS)] with absorption maxima at 360–370 nm; and dinitrosyl–iron complexes, which exhibit an absorption maximum at 310 nm (Lobysheva et al., 1999). Furthermore, the intensities of light-induced NO formation strongly depend on the concentrations of the NO-releasing agents and the radiation power of the respective UV unit, which directly affects the decomposition efficiency. Thus, low NO formation rates (1–50 ppb) will result from low concentrations of the respective educts or may be due to a UV lamp with low performance. In both cases, a highly sensitive NO-detection system like the NO electrode or chemiluminescence detection (CLD) technique is required.

For UVA-induced NO formation in skin specimens or from skin homogenates, we use a 400 W mercury arc lamp (HPA 400W Cleo lamp), purchased from Phillips (Hamburg, Germany), emitting a UVA spectrum (320–420 nm) with a maximum of intensity between 340 and 380 nm (the lamp is used in a distance of 25 cm from the sample, corresponding to a radiant flux intensity of 34 mW/cm$^2$). For UVA-induced NO production in vitro from solutions containing photo-labile nitrogen oxides, we use a 4000-W mercury arc lamp from Sellas Medizinische Geräte (Gevelsberg, Germany), emitting a UVA-1 spectrum (340–410 nm) with a maximum of intensity at 366 nm (the lamp is used in a distance of 25 cm from the sample corresponding to a radiant flux intensity of 84 mW/cm$^2$), or a 2000-W mercury arc lamp from Sellas Medizinische Geräte (Gevelsberg, Germany) emitting a UVA spectrum (320–400 nm) with a maximum of intensity between 350 and 370 nm (the lamp is used in a distance of 25 cm from the sample corresponding to a radiant flux intensity of 70 mW/cm$^2$).

Preparation of Human Skin Samples

For detection of nonenzymatic NO formation during UVA exposure of human skin or of skin homogenate, or for determination of cutaneous nitrite, nitrate, $S$-, and $N$-nitroso compound concentrations, relatively bigger pieces of human skin specimens can be obtained from plastic surgery. Then commence as follows:

1. Immediately after extraction of the skin specimen, carefully remove the subdermal fat layer, and store the skin sample for transport at 4° in phosphate-buffered saline (PBS).

2. Upon arrival in the laboratory, flatten the skin specimen on a cold plastic foil–covered aluminium plate, place a plastic foil and then a second aluminium plate on top, and freeze this "sandwich" for 1 h at $-30°$.

3. Cut the frozen specimens into 10-mm squares using a sharp pair of scissors.

4. For immunohistochemical analysis of $S$-nitroso proteins, use a part of the still-frozen samples; embed it in Tissue-Tek (Reichert-Jung, Vienna, Austria), and snap-freeze in liquid nitrogen.

These samples can be stored in a freezer at $-30°$ for up to 3 wk without changes in concentration of the respective NO-derived metabolites or their NO-forming capacity during UVA challenge.

## Preparation of Human Skin Homogenate Supernatants

The best approach for determining the concentrations of cutaneous NO products or metabolites in human skin comprises the analysis of skin homogenate supernatants. To block degradation of the relatively unstable nitrogen oxide species nitrite, $N$-nitroso, or $S$-nitroso compounds during homogenization, care should be taken to avoid excessive light exposure, generation of heat, acidic pH, the presence of bivalent metal anions, and an oxidizing milieu. Therefore, the entire procedure should be carried out at $4°$ and direct light exposure should be avoided. Thiol alkylation by $N$-ethylmaleimide (NEM) and addition of ion chelators like EDTA are necessary for avoiding thiol trans-nitrosation or reduction of $S$-nitroso compounds, which could falsify the levels measured in the samples.

1. If available, use a cryomicrotom; cut the embedded 10-mm squares of skin specimens (see previous section) at $-25°$ into 20 $\mu$m thin sections with the plane of section parallel to the epidermis. Cut up to 2-mm deep into the dermis, equaling approximately 100 sections.

2. Determine the weight of the skin sections, and put sections in NEM-buffer [PBS containing 5 m$M$ (NEM), 2.5 m$M$ EDTA, protease inhibitor] in a 1:4 (w/v) dilution.

3. Homogenize for 2 min at $4°$ by using a dispersing system such as Ultra Turrax TP18/10 (Janke & Kunkel, Staufen, Germany).

4. Centrifuge the samples for 3 min at 1000$g$, and collect the supernatants.

5. Dilute the samples to obtain the protein content desired (preferably 5–20 mg/ml).

These homogenate supernatants can be used immediately for analysis or can be stored at $-30°$ for up to 2 wk.

When testing the impact of reduced thiols on nonenzymatic NO formation, skin homogenate supernatants can be prepared in the absence of the alkylation agent NEM, but these samples must be immediately used.

## Quantification of Nitrite, Nitrate, S-Nitroso, and N-Nitroso Compounds

To differentiate between compound classes without changing reaction solutions or conditions, samples can be pretreated with group-specific reagents before CLD analysis. Biological samples are typically divided into three aliquots: one used for direct injection (nitrite + nitroso compounds), one for preincubation with sulfanilamide (total nitroso species), and the third for preincubation with $HgCl_2$/sulfanilamide (mercury-resistant nitroso compounds). Again, as a general measure, samples are kept at 4° in the dark to avoid photolytic and thermolytic decomposition.

### Nitrite

Sulfanilamide is known to efficiently remove nitrite from solutions treated with strong acids (Cox and Frank, 1982; Marley et al., 2000). The amount of nitrite in a given sample can be quantified by subtraction of the peak areas of sample aliquots pretreated with sulfanilamide from that of untreated aliquots. For routine use, add to 500 $\mu$l of skin homogenate 55 $\mu$l of a 5% solution of sulfanilamide in 1 N HCl (final concentration 29 mmol/l), and incubate for 15 min at room temperature. Under these conditions, nitrite reacts with sulfanilamide to form a stable diazonium ion that will not be converted to NO by UVA.

### S-Nitrosothiols

The concentration of RSNOs in a given sample (500 $\mu$l) can be conveniently quantified by the difference between the detector signal obtained in the presence of sulfanilamide/$H^+$ (corresponding to total nitroso content) and the signal after pretreatment of the sample with 0.2% $HgCl_2$ (final concentration 7.3 mmol/l) and sulfanilamide/$H^+$ (final concentration 29 mmol/l as described earlier). Incubation with $HgCl_2$ results in cleavage of the S-NO bond without affecting recovery of nitrite or NO and forms the basis for the widely used Saville assay (1958). Time course studies with S-nitroso-glutathione (GSNO) and SNOAlb reveal that at physiological pH, a 20-min incubation period is required for complete S-NO cleavage. No difference is seen between sequential pretreatment with $HgCl_2$, followed by sulfanilamide/$H^+$ and co-incubation with both agents. Hence, a 30-min incubation with the combined $HgCl_2$/sulfanilamide/HCl reagent should be used for routine RSNO analysis.

*Other Nitroso Species*

The mercury-resistant part of the detector signal (i.e., the peak remaining after preincubation with $HgCl_2$/sulfanilamide/$H^+$) indicates the presence of RNNOs or metal nitrosyls, such as NO-heme species in the sample. Extensive further validation experiments with aqueous standards confirm that none of the species tested displays sensitivity higher than 5% toward mercury-induced cleavage under these conditions (Feelisch *et al.*, 2002).

*Nitrate*

Nitrate concentrations of the skin homogenates can be obtained by the nitrite concentrations obtained with vanadium chloride ($VCl_3$)–mediated reduction, followed by the Griess reaction and subtraction of the values found without nitrite reduction by the CLD analysis (Miranda *et al.*, 2001). Before using skin homogenate supernatants in the assay described later, deproteinize samples by addition of 1/10th volume of trichloroacetate (TCA) (10%) and centrifugation for 10 min at 10,000$g$. Nitrate measurements are performed at 37°. Typically, a nitrate standard solution (100 $\mu$l) is serially diluted (generally from 200 to 1.6 $\mu M$) in duplicate in a 96-well, flat-bottomed, polystyrene microtiter plate. The dilution medium is used as blank. After loading the plate with samples (100 $\mu$l), add 100 $\mu$l of the $VCl_3$ solution (saturated solution of 400 mg $VCl_3$ in 50 ml of 1 $M$ HCl) to each well, and rapidly add 50 $\mu$l of Griess reagent 1 (5% sulfanilamide in 1 N HCL) and 50 $\mu$l of Griess reagent 2 [0.1% $N$-(1-naphtyl)ethyleneamine dihydrochloride in $H_2O$]. The Griess solutions may be mixed immediately before application. Because reduction of nitrate to nitrite occurs in acidic solutions, the detection solution must be present during reduction, or the signal will be lost due to diffusion of nitrogen oxides into the gaseous phase. After incubation for 60–90 min, the absorbance at 540 nm is measured using a microplate reader.

Chemiluminescence for Detection of Nitrite, RSNO, and RNNO

The concentration of nitrite and various nitroso species can be determined after reductive cleavage by an iodide/triiodide-containing reaction mixture and subsequent determination of the NO released into the gas phase by its chemiluminescent reaction with ozone (CLD). NO reacts with $O_3$ to form nitrogen dioxide ($NO_2$); a proportion of the latter arises in an electronically excited state (NO$_2^*$), which, upon decay, emits light in the near-infrared region used for quantification by a photomultiplier (Clough and Thrush, 1967). Provided O$_3$ is present in excess and reaction conditions are kept constant, the intensity of light emitted is directly proportional to the NO concentration (Hampl *et al.*, 1996). The reaction mixture, consisting of 45 mmol/l potassium iodide (KI) and 10 mmol/l iodine (I$_2$) in glacial acetic acid, has to be kept at a constant temperature in a septum-sealed, water-jacketed reaction vessel, continuously bubbled with nitrogen (Feelisch *et al.*, 2002).

## UVA-Induced NO Formation from Human Skin

Irradiation of normal human skin specimens with UVA leads to cutaneous NO formation via decomposition of photo-labile NO equivalents. The transition of NO from skin tissue into the gas phase represents an indirect parameter of intracutaneous NO formation and can be detected by the CLD. Therefore, freshly isolated skin specimens should be UVA irradiated in a UVA light–permeable quartz glass–covered and sealed chamber that allows for continuous flux of helium gas and head-space gasses to the detection unit (Fig. 1). We use an 85-cm$^3$ chamber with an irradiation surface area of 16 cm$^2$, and a gas flux of 100 ml/min. The NO amounts detected can be calculated as micromole NO/min $\times$ cm$^2$ of skin. For the examination of UV-independent effects on NO release, such as temperature, a UV filter (opening above 420 nm) can be used.

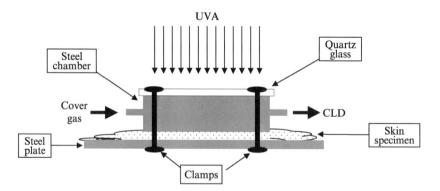

FIG. 1. For the detection of UVA-induced nitric oxide formation from UVA-irradiated skin specimens, a UVA-light-permeable quartz glass covered and sealed chamber can be used, allowing for continuous flux of head space gasses to the CLD detection unit.

Detection of UVA-Induced NO Release from Skin
Homogenate Supernatants

The following procedure is representative for measurements routinely performed in our laboratory:

1. Dilute freshly prepared skin homogenate supernatants in PBS buffer (containing 5 m$M$ N-ethyl-maleimide and 2.5 m$M$ EDTA) to a final concentration of 2 mg protein/ml.

2. In the presence of a silicon-based defoamer, irradiate this solution by UVA in a quartz glass cylinder, permanently exhausting of the gases for detection by the CLD method (Fig. 2). For example, we routinely measure the NO formation from 15-ml skin homogenate supernatant in a 120-cm$^3$ glass cylinder with 3.3-cm diameter and a gas flux of 100 ml/min. The NO amounts detected can be calculated as pM/min × mg protein.

3. Differential reduction or deprivation of nitrogen oxide species from samples allows for examination of the individual molecules contributing to UVA-induced NO formation via photodecomposition. For this, five aliquots are needed, and these have to be treated as follows:

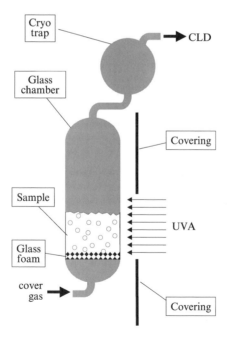

FIG. 2. Nitric oxide formation from UVA-irradiated skin homogenate supernatants can be quantified by using a UVA-light-permeable quartz glass cylinder permanently exhausting of the gases for detection by the CLD method.

a. NO formed during UVA irradiation of untreated skin homogenate supernatants represents decomposition of all photo-labile nitrogen oxide species.

b. Add 1/10th of volume of a 5% solution of sulfanilamide in 1 N HCl to the sample (final concentration 29 mmol/l), and incubate for 15 min at room temperature. The NO formed during subsequent UVA irradiation of those treated skin homogenate supernatants is due to decomposition of all photo-labile nitrogen oxide species except nitrite.

c. Incubate the sample with 0.2% $HgCl_2$ (final concentration 7.3 mmol/l) and sulfanilamide/$H^+$ (final concentration 29 mmol/l, as described earlier) for 20 min at room temperature. The NO formed during subsequent UVA irradiation is due to decomposition of all photo-labile nitrogen oxide species except nitrite and RSNO.

d. Incubate supernatants with 3 U nitrate reductase for 60 min at 37°. Then, add 0.2% $HgCl_2$ (final concentration 7.3 mmol/l) and sulfanilamide/$H^+$ (final concentration 29 mmol/l, as described earlier), and incubate for 20 min at room temperature. The NO formed during subsequent UVA irradiation is due to decomposition of all photo-labile nitrogen oxide species except nitrite, nitrate, and RSNO.

e. Incubate supernatants with 3 U nitrate reductase for 60 min at 37°. Then, add 1/10th volume of a 5% solution of sulfanilamide in 1 N HCl to the sample (final concentration 29 mmol/l), and incubate for 15 min at room temperature. The NO formed during subsequent UVA irradiation is due to all photo-labile nitrogen oxide species except nitrite and nitrate.

The impact of the individual nitrogen oxide species on UVA-induced NO formation from skin homogenate supernatants can be calculated by subtraction of the NO amounts detected (pM NO/min × mg protein).

### Detection of S-Nitroso Proteins by Immunohistochemistry

The concentration of RSNOs in skin tissue reflects the appearance and action of active NO. As shown earlier, a quantitative detection of photo-labile NO derivatives in the skin can be elegantly achieved by the CLD technique. Unfortunately, this method cannot provide any information on the localization or distribution of S-nitroso proteins in the skin. This can be easily achieved by immunochemical staining methods using specific antisera. Routinely, we examine S-nitroso protein formation on cryostat sections (7 μm of thickness), obtained from the skin specimens embedded and snap frozen as described earlier. For these purposes, we fix samples by glutaraldehyde (0.2% in TBS, pH 7.0) for 15 min at 4° in a moist chamber, and the S-nitroso proteins are detected by using the previously de-scribed rabbit anti–S-nitrosocysteine antiserum (Mnaimneh *et al.*, 1997, 1999) (Calbiochem, Luzern, Switzerland) used in a 1:100 dilution in TBS (supplemented with 3% low-fat milk powder and 0.5% Tween 20, pH 7.4). As a negative control, we use a nonrelevant rabbit antiserum (Calbiochem).

## References

Bruch-Gerharz, D., Ruzicka, T., and Kolb-Bachofen, V. (1998a). Nitric oxide in human skin: Current status and future prospects. *J. Invest. Dermatol.* **110**, 1.

Bruch-Gerharz, D., Ruzicka, T., and Kolb-Bachofen, V. (1998b). Nitric oxide and its implications in skin homeostasis and disease - a review. *Arch. Dermatol. Res.* **290**, 643.

Clough, P. N., and Thrush, B. A. (1967). Mechanism of chemiluminescent reaction between nitric oxide and ozone. *Trans. Faraday Soc.* **63**, 915.

Cox, R. D., and Frank, C. W. (1982). Determination of nitrate and nitrite in blood and urine by chemiluminescence. *J. Anal. Toxicol.* **6**, 148.

Ehrt, S., Schnappinger, D., Bekiranov, S., Drenkow, J., Shi, S., Gingeras, T. R., Gaasterland, T., Schoolnik, G., and Nathan, C. (2001). Reprogramming of the macrophage transcriptome in response to interferon-gamma and Mycobacterium tuberculosis: Signaling roles of nitric oxide synthase-2 and phagocyte oxidase. *J. Exp. Med.* **194**, 1123.

Feelisch, M., Rassaf, T., Mnaimneh, S., Singh, N., Bryan, N. S., Jourd'Heuil, D., and Kelm, M. (2002). Concomitant S-, N-, and heme-nitros(yl)ation in biological tissues and fluids: Implications for the fate of NO *in vivo*. *FASEB J.* **16**, 1775.

Fischer, M., and Warneck, P. (1996). Photodecomposition of nitrite and undissociated nitrous acid in aqueous solution. *J. Phys. Chem.* **100**, 18749.

Hampl, V., Walters, C. L., and Archer, S. L. (1996). Determination of nitric oxide by the chemiluminiscence reaction with ozone. *In* "Methods in Nitric Oxide Research" (M. Feelisch and J. S. Stamler, eds.), p. 309. John Wiley & Sons, Chichester, UK.

Hemish, J., Nakaya, N., Mittal, V., and Enikolopov, G. (2003). Nitric oxide activates diverse signaling pathways to regulate gene expression. *J. Biol. Chem.* **278**, 42321.

Hogg, N., and Kalyanaraman, B. (1999). Nitric oxide and lipid peroxidation. *Biochim. Biophys. Acta* **1411**, 378.

Jankowski, J. J., Kieber, D. J., Mopper, K., and Neale, P. J. (2000). Development and intercalibration of ultraviolet solar actinometers. *Photochem. Photobiol.* **71**, 431.

Kharitonov, V. G., Bonaventura, J., and Sharma, V. S. (1996). Interactions of nitric oxide with heme proteins using UV-VIS spectroscopy. *In* "Methods in Nitric Oxide Research" (M. Feelisch and J. S. Stamler, eds.), pp. 39–45. Wiley, Chichester, U.K.

Kuhn, A., Fehsel, K., Lehmann, P., Krutmann, J., Ruzicka, T., and Kolb-Bachofen, V. (1998). Aberrant timing in epidermal expression of inducible nitric oxide synthase after UV irradiation in cutaneous lupus erythematosus. *J. Invest. Dermatol.* **111**, 149.

Lobysheva, V. A., II, Serezhenkov, and Vanin, A. F. (1999). Interaction of peroxynitrite and hydrogen peroxide with dinitrosyl iron complexes containing thiol ligands *in vitro*. *Biochemistry (Mosc)* **64**, 153.

Marley, R., Feelisch, M., Holt, S., and Moore, K. (2000). A chemiluminescense-based assay for S-nitrosoalbumin and other plasma S-nitrosothiols. *Free Radic. Res.* **32**, 1.

Miranda, K. M., Espey, M. G., and Wink, D. A. (2001). A rapid, simple spectrophotometric method for simultaneous detection of nitrate and nitrite. *Nitric Oxide* **5**, 62.

Mnaimneh, S., Geffard, M., Veyret, B., and Vincendeau, P. (1997). Albumin nitrosylated by activated macrophages possesses antiparasitic effects neutralized by anti-NO-acetylated-cysteine antibodies. *J. Immunol.* **158,** 308.

Mnaimneh, S., Geffard, M., Veyret, B., and Vincendeau, P. (1999). Detection of nitrosylated epitopes in Trypanosoma brucei gambiense by polyclonal and monoclonal anti-conjugated-NO-cysteine antibodies. *C. R. Acad. Sci. III* **322,** 311.

Rao, C. N. R., and Bhaskar, K. R. (1981). Photolytic cleavage of *N*-nitroso compounds. *In* "The Chemistry of the Nitro and Nitroso Groups" (H. Feuer, ed.), pp. 153–154. Krieger, Huntington, New York.

Saville, B. (1958). A scheme for the colorimetric determination of microgram amounts of thiols. *Analyst* **83,** 670.

Singh, R. J., Hogg, N., Joseph, J., and Kalyanaraman, B. (1996). Mechanism of nitric oxide release from S-nitrosothiols. *J. Biol. Chem.* **271,** 18596.

Suschek, C. V., Briviba, K., Bruch-Gerharz, D., Sies, H., Kroncke, K. D., and Kolb-Bachofen, V. (2001a). Even after UVA-exposure will nitric oxide protect cells from reactive oxygen intermediate-mediated apoptosis and necrosis. *Cell Death Differ.* **8,** 515.

Suschek, C. V., Bruch-Gerharz, D., Kleinert, H., Forstermann, U., and Kolb-Bachofen, V. (2001b). Ultraviolet A1 radiation induces nitric oxide synthase-2 expression in human skin endothelial cells in the absence of proinflammatory cytokines. *J. Invest. Dermatol.* **117,** 1200.

Suschek, C., Krischel, V. V., Bruch-Gerharz, D., Berendji, D., Krutmann, J., Kroncke, K. D., and Kolb-Bachofen, V. (1999). Nitric oxide fully protects against UVA-induced apoptosis in tight correlation with Bcl-2 up-regulation. *J. Biol. Chem.* **274,** 6130.

Suschek, C. V., Schnorr, O., Hemmrich, K., Aust, O., Klotz, L. O., Sies, H., and Kolb-Bachofen, V. (2003a). Critical role of L-arginine in endothelial cell survival during oxidative stress. *Circulation* **107,** 2607.

Suschek, C. V., Schroeder, P., Aust, O., Sies, H., Mahotka, C., Horstjann, M., Ganser, H., Murtz, M., Hering, P., Schnorr, O., Kroncke, K. D., and Kolb-Bachofen, V. (2003b). The presence of nitrite during UVA irradiation protects from apoptosis. *FASEB J.* **17,** 2342.

Zhang, Y. Y., Xu, A. M., Nomen, M., Walsh, M., Keaney, J. F., Jr., and Loscalzo, J. (1996). Nitrosation of tryptophan residue(s) in serum albumin and model dipeptides. Biochemical characterization and bioactivity. *J. Biol. Chem.* **271,** 14271.

[49]  Nitric Oxide and Cell Signaling: *In Vivo* Evaluation
      of NO-Dependent Apoptosis by MRI and Not
      NMR Techniques

*By* Sonsoles Hortelano, Miriam Zeini,
Paqui G. Través, and Lisardo Boscá

Abstract

Apoptosis plays a key role in many pathological circumstances, such as neurodegenerative diseases. In these processes, the involvement of nitric oxide (NO) has been well established, and the ability of NO to exert cellular damage due to its reactive oxidative properties is perhaps the primary neurotoxic mechanism. The caspase 3 activation has recently been observed in stroke, spinal cord trauma, head injury, and Alzheimer's disease. Although numerous techniques have been described to evaluate apoptosis, these approaches involve invasive techniques and cannot provide detailed information about apoptosis *in vivo*. In this chapter, we describe the use of functional magnetic resonance imaging (fMRI) as a non-invasive technique to detect apoptosis *in vivo*. fMRI techniques can detect apoptosis at early stages in the process, allowing the onset in intact biological systems, providing a useful tool for monitoring apoptosis progression.

Apoptosis constitutes an essential event in several physiopathological processes such as development, selection in the immune system, neurodegenerative diseases, and host defense against pathogens and tumor cells (Green, 1998; Nagata, 1997; Thompson, 1995). Various stimuli can induce apoptosis, either specifically (*i.e.*, through cell type–specific receptors) or broadly by activating common mitochondrial pathways involved in apoptotic commitment; among them, the role of NO in apoptotic cell death has been well established. Interestingly, NO, produced from L-arginine and molecular oxygen in a reaction catalyzed by three different NO synthase (NOS) isoenzymes, can prevent or induce apoptosis, depending on its concentration and the cellular redox state. In macrophages, chondrocytes, $\beta$-pancreatic cells, fibroblasts, neurons, glial cells, or hepatocytes, high or moderate levels of NO induce apoptosis through the release of mitochondrial apoptogenic factors and the activation of caspases.

The neurotoxicity of NO in the central nervous system has been extensively studied, and the involvement of NO in a number of neurological disorders including neurodegenerative diseases (Dawson, 1994; Meldrum and Garthwaite, 1990; Zhang and Steiner, 1995) is well established. The

METHODS IN ENZYMOLOGY, VOL. 396                    0076-6879/05 $35.00
DOI: 10.1016/S0076-6879(05)96049-7

presence of stimuli that lead to low but sustained or overproduction of NO will likely cause neuronal damage through cell death of specific neurons. Indeed, the ability of NO to exert cellular damage due to its reactive oxidative properties is perhaps the primary neurotoxic mechanism. Cumulating evidence strongly suggests that apoptosis contributes to neuronal death in various neurodegenerative pathologies. Activation of the cysteine protease caspase-3 appears to be a key event in the execution of apoptosis in the central nervous system. In addition to NO, a high concentration of NO and peroxynitrite, a reaction product of NO with superoxide anion, can promote apoptotic death in neuronal cells through the indirect activation of caspases. Consistent with the proposal that apoptosis plays a central role in neurodegenerative diseases, caspase-3 activation has been observed in stroke, spinal cord trauma, head injury, and Alzheimer's disease (Mattson et al., 2000; Takuma et al., 2004). Indeed, peptide-based caspase inhibitors prevent neuronal loss in animal models of head injury and stroke, suggesting that these compounds may be the forerunners of nonpeptide small molecules that halt the apoptotic process implicated in these neurodegenerative disorders.

Caspases, synthesized as relatively inactive zymogens, must undergo a process of activation during apoptosis. Based on their order of activation, caspases are classified into two families: the initiator caspases, which include caspase-2, -8, -9, and -10; and the effector caspases, which include caspase-3, -6, and -7 (Shi, 2002). In response to upstream apoptotic stimuli, the initiator caspases undergo an autocatalytic processing and activation (Shi, 2002). Once activated, an initiator caspase specifically cleaves and hence activates an effector caspase zymogen. The effector caspases are responsible for the proteolytic cleavage of a broad spectrum of cellular targets, leading ultimately to cell death. The known cellular substrates include structural components (such as actin and nuclear lamin), regulatory proteins (such as DNA-dependent protein kinase), inhibitors of deoxyribonucleases (such as DFF45 or ICAD), and other proapoptotic proteins and caspases.

## Alternative Methods to Evaluate Apoptosis in Intact Animals

The underlying mechanisms for the selective death of neurons are the object of intensive study; however, most of the methods used to evaluate apoptosis involve invasive techniques. For this reason, an experimental approach to monitor the state of neurons under in vivo conditions could provide a follow-up of the factors that cause selective neuron degeneration, as well as the potential therapeutic intervention to evaluate the pharmacological actions of new drugs intended to delay the progression of the disease. In this context, the use of functional magnetic resonance imaging (fMRI) as a noninvasive technique is paramount in neuroscience for studying neural activities orchestrated during a particular behavior.

## Materials and Methods

### Animals

Adult male Sprague-Dawley rats, weighing 200–225 g (Charles River Laboratories, Inc., Wilmington, MA) were maintained under conditions of controlled temperature ($23 \pm 1°$) and lighting (lights on 8–20 h), with food and water provided *ad libitum*. All experiments included in the study were performed following the highest standards of animal care, monitoring health care, and minimizing pain and suffering, in accordance with National and International Laws for the Care and Use of Laboratory Animals. Groups of five animals were used in each condition.

### Surgical Preparation of the Animals and MRI Acquisition System

Animals were anesthetized with xylazine (20 mg/kg, AnaSed Ben Venue Laboratories, Bedford, OH) and ketamine (100 mg/kg, Phoenix Scientific, Inc., St. Joseph, MD), both injected intraperitoneally. The rat was placed supine in the stereotaxic unit, and 5 m$M$ GSNO in phosphate-buffered saline (PBS) $\pm 100$ $\mu M$ z-VAD (5 $\mu$l, final volume) was administered stereotaxically into the striatum of the left cerebral hemisphere. The coordinates used were anteroposterior (AP) of $+1.5$ mm, mediolateral (ML) of 2.5 mm, and dorsoventral (DV) of 4 mm. For controls, the same volume of PBS (5 $\mu$l) was injected into the striatum of the right cerebral hemisphere. After treatment, the animal was immediately removed from the stereotaxic instrument and placed in a prone position to restore more physiological conditions. The rat was then placed on a warm plaque to maintain its normal temperature after surgery. At 5 h after treatment, lesions produced by GSNO were monitored by MRI. The anesthetized animal was transferred to the MRI scanner room and fixed in place in a custom-made holder. MRI experiments were performed on a 4.7 T horizontal bore magnet (Biospec BMT 47/40, Bruker Instruments, Germany). T2-weighted measurements were performed before treatment and 5 h after GSNO injection.

### Tissue Preparation, Immunohistochemistry, and Determination of Apoptosis

Animals were sacrificed after MRI analysis with $CO_2$, and tissues were collected in a 30% sucrose solution in PBS (w/v), and left overnight at 4°. Samples were then frozen in 2-methylbutane at $-80°$. Serial 8-$\mu$m thick sections were cut with a microtome and placed on gelatinized glass. The tissue sections were fixed with 4% paraformaldehyde (pH 7.4 in PBS), for 20 min at room temperature. Then the preparations were washed with PBS for 30 min and permeabilized for 5 min with cold permeabilization solution (0.1% Triton X-100 in 0.1% sodium citrate).

Apoptotic cell death was then determined and quantified using TUNEL staining. After washing again with PBS, tissue sections were incubated with 50 μl of TUNEL reaction mixture for 1 h at 37°. This labeling solution included terminal deoxynucleotidyltransferase and nucleotides in reaction buffer, including fluorescein-isothiocyanate (FITC)–dUTP for labeling.

Multiple measurements were performed at the same time. Activated caspase-3 was also detected in these tissues. Sections were stained with a phycoerythrin (PE)-conjugated antibody against caspase-3 for 1 h at the same time as the TUNEL reaction mixture. After washing several times with PBS, nuclei were visualized by incubation with Hoescht.

*T2 Measurements*

Figure 1 shows T2 maps from representative control and GSNO ± z-VAD–treated animals. GSNO-treated animals exhibited a hyperintense area in the T2 maps obtained 5 h after treatment, whereas the animals treated with GSNO and z-VAD exhibited a small lesion area. The injection of PBS in the right hemisphere had no effect on the T2 maps, indicating that the rich-T2 area obtained in GSNO-treated animals was due to the

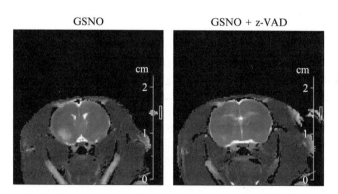

Fig. 1. T2 measurement. Animals were anesthetized with xylazine (20 mg/kg) and ketamine (100 mg/kg) and were placed supine in the stereotaxic unit. GSNO (5 mM) in phosphate-buffered saline (PBS) ± 100 μM z-VAD (5 μl) were injected stereotaxically into the striatum of the left cerebral hemisphere. For controls, the same volume of PBS (5 μl) was injected into the striatum of right cerebral hemisphere. T2 maps of coronal sections from the GSNO ± z-VAD treated animals showed the T2 hyperintensity in the damaged area obtained at 5 h after treatment.

toxic effect of NO, and not to the edema that usually appears immediately after injection. The observation that z-VAD inhibited NO toxicity suggests that caspases were involved in the NO-dependent apoptosis. The size of the areas with high T2 signals in the left hemisphere was measured (Table I), as well as the values of the T2 intensity in the left hemisphere with respect to the control area in the right hemisphere. The data obtained show that GSNO increased T2 intensity by 25%, whereas z-VAD prevented this effect.

### NO-Dependent Apoptosis and Involvement of Caspases

Apoptosis and caspase-3 activation were measured in tissue sections after TUNEL staining and incubation at the same time with specific antibodies against active caspase-3. Data for the tissue sections obtained after MRI experiments are shown in Fig. 2. The TUNEL and caspase-3 staining in tissue sections from GSNO-treated animals corresponded closely with the hyperintense area on the T2 maps.

Taken together, these data show a model of NO-dependent apoptosis in the brain and offer the possibility of an *in vivo* evaluation of cell death in

TABLE I
SIZE OF RELATIVE AREAS WITH HIGH T2 INTENSITY AND VALUES OF T2

|  | Control | GSNO | GSNO + z-VAD |
| --- | --- | --- | --- |
| Area | 0.001 cm$^2$ | 0.0758 cm$^2$ | 0.038 cm$^2$ |
| T2 intensity | 67.26 ± 5.03 ms | 89.28 ± 13.80 ms | 69.43 ± 3.94 ms |

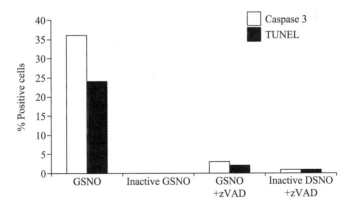

FIG. 2. Measurement of apoptosis and caspase-3 activation. Tissue sections were stained using TUNEL technique following the protocols provided by the supplier. Sections were stained with a phycoerythrin-conjugated antibody against caspase-3 at the same time as the TUNEL reaction mixture. Nuclei were stained with Hoescht 33258.

whole tissues, such as brain, using fMRI technology. The availability of more potent magnetic fields could benefit this type of study for other organs in which apoptosis is a major pathological issue, such as the heart under myocardial infarction.

### References

Dawson, D. A. (1994). Nitric oxide and focal cerebral ischemia: Multiplicity of actions and diverse outcome. *Cerebrovasc. Brain Metab. Rev.* **6**, 299–324.

Green, D. R. (1998). Apoptotic pathways: The roads to ruin. *Cell* **94**, 695–698.

Mattson, M. P., Culmsee, C., and Yu, Z. F. (2000). Apoptotic and antiapoptotic mechanisms in stroke. *Cell Tissue Res.* **301**, 173–187.

Meldrum, B., and Garthwaite, J. (1990). Excitatory amino acid neurotoxicity and neurodegenerative disease. *Trends Pharmacol. Sci.* **11**, 379–387.

Nagata, S. (1997). Apoptosis by death factor. *Cell* **88**, 355–365.

Shi, Y. (2002). Mechanisms of caspase activation and inhibition during apoptosis. *Mol. Cell* **9**, 459–470.

Takuma, H., Tomiyama, T., Kuida, K., and Mori, H. (2004). Amyloid beta peptide-induced cerebral neuronal loss is mediated by caspase-3 *in vivo. J. Neuropathol. Exp. Neurol.* **63**, 255–261.

Thompson, C. B. (1995). Apoptosis in the pathogenesis and treatment of disease. *Science* **267**, 1456–1462.

Zhang, J., and Steiner, J. P. (1995). Nitric oxide synthase, immunophilins and poly(ADP-ribose) synthetase: Novel targets for the development of neuroprotective drugs. *Neurol. Res.* **17**, 285–288.

# [50]    Microelectrode for *In Vivo* Real-Time Detection of NO

*By* Tayfun Dalbasti and Emrah Kilinc

### Abstract

Nitric oxide (NO) is gaining importance with its diverse spectrum of clinic effects. However, there is still a need for an ideal sensor to monitor its concentration in tissue. An ideal sensor should not interfere with the ongoing physiological process, while making fast, reliable, and repeatable measurements. We have designed a microelectrode for electrochemical NO measurement from tissue with relatively low interference and reliable results upon calibration. Details of electrode preparation and calibration procedure are explained along with an experiment to monitor effects of photodynamic therapy.

METHODS IN ENZYMOLOGY, VOL. 396              0076-6879/05 $35.00
        DOI: 10.1016/S0076-6879(05)96050-3

Nitric Oxide

Nitric oxide, an endogenously synthesized free radical of great physiological and pathophysiological importance, displays various biological functions after the induction of the corresponding NO synthase (NOS). Neuronal NOS (nNOS)–derived NO acts as a neuronal signal in the central nervous system; inducible NOS (iNOS)–derived NO contributes to the pathology of many clinical conditions, and endothelial-derived NOS (eNOS)–originated NO regulates vascular tone in the endothelial cell systems.

It was biologically studied in detail and first characterized *in vivo* as an endothelial-derived relaxing factor (EDRF) in 1983 by Furchgott (1983) and in 1987 by Palmer *et al.* The biosynthesis of NO has been reviewed by various authors (Butler and Williams, 1993; Moncada *et al.*, 1991; Stamler, 1994).

*Background of Detection*

Measuring NO in biological media is extremely hard because of its very low concentration values (nanomolar levels) and supershort lifetime (nanoseconds) (Dalbasti *et al.*, 2002).

Most of the techniques detecting NO release use indirect methods based on the determination of secondary species or adducts (e.g., nitrites and L-citrulline).

Current techniques for direct measurements of NO include some spectroscopic methods (such as chemiluminescence, UV–vis spectroscopy, etc.) and electrochemical methods. Several types of electrochemical electrodes for NO detection have been reported in the literature (Pallini *et al.*, 1998).

*Electrochemical Detection Principle*

NO is usually oxidized at the surface of the working electrode at a positive potential of approximately 800–900 mV (vs. Ag/AgCl reference), resulting in the generation of a small redox current. This oxidation redox potential may vary in accordance to the electrode material used or the catalytic compound used for surface modification. The overall electrode reaction for aqueous NO oxidation may be summarized as follows:

$$NO - e^- \xrightarrow{\sim +900 \ mV} NO^+$$

Because $NO^+$ is a relatively strong Lewis acid, in the presence of $OH^-$, it is converted into nitrite ($NO_2^-$):

$$NO^+ + OH^- \rightarrow HNO_2$$

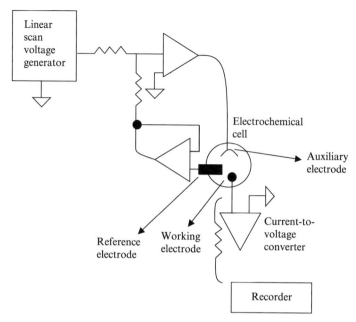

SCHEME 1. A typical three-electrode potentiostat.

The oxidation potential of $NO_2^-$ is very close to that of NO in aqueous solution; therefore it can be further oxidized into nitrate ($NO_3^-$). The oxidation current flowing between the working and reference electrodes is directly proportional to the concentration of NO oxidized as described earlier. This technique of applying such a constant redox potential and sampling the response current versus time is known as *amperometry* (chronoamperometry as well). Thus, the redox current proportional to NO concentration is measured amperometrically using a NO meter or a conventional potentiostat (Zhang and Broderick, 2000). NO meters and potentiostats use three electrode systems for sampling response current, which is illustrated and simplified in Scheme 1.

## A Sample Experiment: Electrode Construction and Application for *In Vivo* Real-Time Detection

There is a demand for an ideal NO detection method that is fast, durable, repeatable, specific, sensitive, and biocompatible, and that should not interfere with the tissue and physiological processes. So far, no electrode design fulfills these requirements. Electrochemical electrode designs are most close to these specifications. Following the instructions below, constructing a working NO measurement electrode is the main aim of this chapter. Please keep in mind that *this electrode design is only for approved experimental animal studies, and none of the procedures described hereafter are tested for human subjects.*

## Chemicals and Materials Required

Pt–Ir Alloy Wire
Teflon-insulated, multistranded Pt–Ir alloy wire (total diameter 130 $\mu$m) (10% Ir–90% Pt), Medwire (http://www.sigmundcohn.com/english/medwire.html)
Nafion
A synthetic perfluorinated polymer, containing sulfonic and carboxylic acid functions, solution containing 5% in a mixture of lower aliphatic alcohols and water; highly toxic and flammable; Fluka 70160
o-Phenylenediamine (OPD)
CAS [95-54-5], brownish-yellow crystals; slightly soluble in water, freely soluble in alcohol, chloroform, ether; may darken in storage; FW 108.14 g/mol, Sigma P 2903
$\alpha$-Aminolevulinic acid hydrochloride (ALA)
CAS [5451-09-2], freely soluble in water, FW 167.6 g/mol, Sigma A 3785
Potassium phosphate monobasic ($KH_2PO_4$) (potassium dihydrogen-phosphate)
CAS [7778-77-0], FW 136.1 g/mol, Sigma Sigma P 5379
Potassium phosphate dibasic ($K_2HPO_4$) (dipotassium hydrogen phosphate)
CAS [7758-11-4], FW 174.2 g/mol, Sigma P 8281
Sulfuric acid ($H_2SO_4$)
CAS [7664-93-9], 95–98%, FW 98.08 g/mol, Aldrich 32,050-1
Potassium nitrite ($KNO_2$)
CAS [7758-09-0], FW 85.10 g/mol, Sigma P 7391
Nitrogen gas ($N_2$) (Extra Pure)
Local source: Habas sinai ve Tibbi Gazlar Istihsal Endüstrisi A.S. Istanbul/Turkey [Phone: +90 (216) 452 56 00 Fax: +90 (216) 452 25 70, www.habas.com.tr]
Bunsen burner

## Preparation of Solutions

NO stock solution may easily be prepared by bubbling NO gas through deoxygenated 0.05 M phosphate buffer (1.36 g/L $KH_2PO_4$ and 6.96 g/L $K_2HPO_4$, pH = 7.4) for 30 min. Such an NO-bubbled buffer will have a concentration of about 2 m*M* (1.95 mmol/L) NO at saturation (Mesaros *et al.*, 1997; Young, 1981). The buffer to be bubbled with NO should be freshly prepared and heated to boiling just before cooling in an ice bath immediately. This boiling and cooling procedure is necessary to de-gas the solution (oxygen free). Boiling may preferably be performed in an ultrasonic cleaner, which will help increase the de-gas performance. The $O_2$-free buffer may then be bubbled with pure $N_2$ for 30 min to help it remain $O_2$ free. The NO gas used to saturate the buffer may either be supplied in commercial tanks (Matheson Tri-Gas, www.matheson-trigas.com/mathportal) or be prepared simultaneously in the lab. The production of NO gas in the lab is simpler and cheaper than the commercial choice, *but maximum attention should be paid during the process, as NO immediately on contact with air is converted to the extremely poisonous nitrogen dioxide ($NO_2$)*!! (Budavari, 2001).

Under adequate ventilation, NO gas may be produced in leak-free glassware or solution wash bottles. For this procedure in a leak-free glass chamber, 125 mol of 6 M $H_2SO_4$ is added in portions onto 50 g of $KNO_2$. In a highly acidic media, $KNO_2$ is decomposed, and equivocal NO is formed, which is simultaneously transferred through a leak-free connection into the buffer solution sitting in an ice bath, kept under $N_2$ atmosphere. When this bubbling procedure is performed for 30 min, the buffer is saturated with NO in a concentration of 1.95 m*M*, as indicated earlier, which is stable for 48 h at 4°.

### Electrode Fabrication

Working electrodes (microelectrodes) may be prepared from Teflon-insulated Pt–Ir alloy wire (Pt 90%–Ir 10%, 130 $\mu$m total diameter) by removing 3 mm of the Teflon coating (from the tip of the wire) using ordinary flame (Bunsen burner).

Pt–Ir alloy may be preferred because of the reported electrocatalytic effect of iridium in redox reactions (Wang *et al.*, 1996), and pure Pt is more fragile than the Pt–Ir alloy, which makes it harder to apply to real tissue samples.

Electrodes should be preconditioned in 0.5 *M* $H_2SO_4$ solution by applying a constant potential of 1.9 V for 30 s, followed by a cyclic voltametry between the ranges −0.25 and +1.1 V with a scan rate of 100 mV/s for 10 min. Preconditioning is necessary to remove any impurity left on the wire after the ordinary flame removal of the Teflon layer. Electrode fouling is another problem faced with non-pretreated electrodes, in which response current decays rapidly and is lost in seconds.

Prepared electrodes should be dip-coated four times (four layers) with 1/10th diluted Nafion solution, and each dried at 200° for 4 min as indicated (Friedemann *et al.*, 1996; Kilinc *et al.*, 2002). Nafion layers formed on the wire surface are responsible for repelling the negatively charged species and ions that might interfere during *in vivo* applications. Such interferences may include highly electroactive species, such as ascorbic acid, uric acid, and various drug metabolites.

Finally, electropolymerization of 50 m$M$ OPD (in buffer) should be performed on the Nafion-coated electrodes by applying 0.650 V (vs. Ag/AgCl reference) for 10 min. This additional layer of OPD on Nafion will improve the conductance of the electrode and help the Nafion layer in repelling the species of interference. Electrodes should be calibrated individually before use.

*Electrode Calibration*

A series of standard NO solutions may be prepared by diluting this NO-saturated stock for use in electrode calibration. Increasing the final concentration of NO with 10 nm increments in a 10 mL standard electrochemical cell may help in obtaining a calibration plot linear enough to work in a range of *in vivo* NO levels (Dalbasti *et al.*, 2002; Friedemann *et al.*, 1996; Kilinc *et al.*, 2002).

*NO Detection*

Albino rats weighing around 250 g may be used for *in vivo* recordings. A midline skin incision over the scalp may be performed to expose bregma and lambdoid sutures (under anesthesia). An approximately 4.0-mm diameter burr hole is necessary over the cerebellar hemisphere (taking great care to keep the dura intact). The dura may be opened under microscopic control.

The reference and auxiliary electrodes have to be placed under the skin close to the working electrode. In the control group of recordings, the Pt–Ir working electrode should carefully be placed in the cortex, and steady state recordings must be obtained.

In the study group, 50 $\mu$l, 5 mg/ml ALA solution in 0.05 $M$ phosphate buffer (pH 7.4) should be applied topically with a syringe (100 $\mu$l total volume) attached to a 26-gauge needle (Fig. 1).

After waiting 10 min for ALA diffusion in tissue, the intracortical electrode was placed, and simultaneous current-time recordings were obtained ($\sim$0.90 V vs. Ag/AgCl) until a steady state current recording was observed. This level must be recorded as a baseline, and then

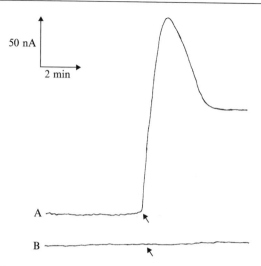

Fig. 1. Typical amperogram showing the *in vivo* nitric oxide (NO) release in rat cerebellum in the presence (5 mg/ml) (A) and absence (B) of ALA during ALA-mediated PDT. Arrows indicate the time of application of the 150-W excitation light.

broad-spectrum light (150-W halogen lamp) was applied from a 5-cm distance to the surface while recording was continued to show the generation of NO during ALA-mediated PDT.

In the study group, in contrast to the control group (where ALA is not administered), an amperogram similar to Fig. 1A should be obtained. In the control group, because NO is not released in the tissue, no increase in the response current should be observed (Fig. 1B).

## Commercial Electrochemical NO Measurement Systems

Although alternative direct and indirect measurement techniques are present in literature for NO detection, for commercially available systems, only a few electrochemical choices are present. However, with some basic electronics knowledge, one could design a simple measurement system to be used with NO electrodes. As mentioned earlier, the idea is to apply constant voltage to the electrode and measure the current flowing through. Because electrode impedance is high (several mega ohms) and measured current is very low (pico-nano amperes), special precautions should be taken in component selection and circuit design. The details of the companies providing commercial hardware and software for NO detection are summarized in Table I.

TABLE I
COMPANIES PROVIDING SYSTEMS FOR NO DETECTION

| Company | Address | Web | Products | Contact |
|---|---|---|---|---|
| World Precision Instruments, Inc. | 175 Sarasota Center Boulevard, Sarasota, FL 34240 | www.wpiinc.com | Free Radical Detection Nitric Oxide Detection Oxygen Detection Neurotransmitter Detection pH Meters and Electrodes | Phone: (941) 371-1003 Fax: (941) 377-5428 marketing@wpiinc.com |
| Innovative Instruments, Inc. | Innovative Instruments, Inc., 8533 Queen Brooks Ct., Tampa, FL 33637 | www.2in.com/ | Nitric Oxide Detection Oxygen Detection Reference Electrodes pH Sensors Nitrate Reductor | Phone: (813) 727-0676 Fax: (813) 914-8686 ztaha@worldnet.att.net |
| Diamond General Development Corporation | 333 Parkland Plaza, Ann Arbor, MI 48103-6202 | www.diamondgeneral.com | Nitric Oxide Detection Oxygen Detection Amplifier/Display Meters Calibration Cells | Phone: (734) 332-0200 Toll Free: (800) 678-9856 Fax: (734) 332-4775 mail@diamondgeneral.com |
| Inter Medical Co., Ltd. | 40-4, 3-chome, Imaike, Chikusa, NAGOYA, 464-0850, Japan | www1.sphere.ne.jp/intermed | Nitric Oxide Detection Oxygen Detection Electrode for Spine Functions | Phone: +81 52 731 8000 Fax: +81 52 731 5050 intermed-2@mbs. sphere.ne.jp |

## References

Budavari, S. (2001). "The Merck Index," 13th Ed. Merck Research Laboratories, Whitehouse Station, NJ.

Butler, A. R., and Williams, D. L. H. (1993). The physiological role of nitric oxide. *Chem. Soc. Rev.* **22**(4), 233–241.

Dalbasti, T., Cagli, S., Kilinc, E., Oktar, N., and Ozsoz, M. (2002). Online electrochemical monitoring of nitric oxide during photodynamic therapy. *Nitric Oxide* **7**, 301–305.

Friedemann, M. N., Robinson, S. W., and Gerhardt, G. A. (1996). o-Phenylenediamine-modified carbon fiber electrodes for the detection of nitric oxide. *Anal. Chem.* **68**, 2621–2628.

Furchgott, R. (1983). Role of endothelium in response of vascular smooth muscle. *Circulation Res.* **53**(5), 557–573.

Kilinc, E., Yetik, G., Dalbasti, T., and Ozsoz, M. (2002). Comparison of electrochemical detection of acetylcholine-induced nitric oxide release (NO) and contractile force measurement of rabbit isolated carotid artery endothelium. *J. Pharmaceut. Biomed. Analysis* **28**, 345–354.

Mesaros, S., Grunfeld, S., Mesarosova, A., Bustin, D., and Malinski, T. (1997). Determination of nitric oxide saturated (stock) solution by chronoamperometry on a porphyrin microelectrode. *Anal. Chim. Acta* **339**, 265–270.

Moncada, S., Palmer, R. M. J., and Higgs, E. A. (1991). Nitric oxide: Physiology, pathophysiology, and pharmacology. *Pharmacol. Rev.* **43**, 109–142.

Pallini, M., Curulli, A., Amine, A., and Palleschi, G. (1998). Amperometric nitric oxide sensors: A comparative study. *Electroanalysis* **10**(15), 1010–1016.

Palmer, R. M. J., Ferrige, A. G., and Moncada, S. (1987). Nitric oxide release accounts for the biological activity of endothelium-derived relaxing factor. *Nature* **327**, 524–526.

Stamler, J. S. (1994). Redox signaling: Nitrosylation and related target interactions of nitric oxide. *Cell* **78**, 927–936.

Wang, J., Rivas, G., and Chicharro, M. (1996). Iridium-dispersed carbon paste enzyme electrodes. *Electroanalysis* **8**(5), 434–437.

Young, C. L. (ed.) (1981). "Oxides of Nitrogen," Solubility Data Series, Vol. 8. IUPAC, Pergamon Press, Oxford.

Zhang, X., and Broderick, M. (2000). Amperometric detection of nitric oxide. *Mod. Asp. Immunobiol.* **1**(4), 160–165.

# [51] Real-Time Detection of Nitric Oxide Isotopes in Lung Function Tests

*By* H. Heller, R. Gäbler, and K.-D. Schuster

## Abstract

In lung function tests, the determination of the pulmonary diffusing capacity (D) using the single-breath method is a commonly applied technique. The calculation of D is performed on the basis of accurate measurements of indicator gas concentrations. In this chapter, we demonstrate the

METHODS IN ENZYMOLOGY, VOL. 396
Copyright 2005, Elsevier Inc. All rights reserved.
0076-6879/05 $35.00
DOI: 10.1016/S0076-6879(05)96051-5

appropriateness of the stable nitric oxide (NO) isotopes $^{14}$NO and $^{15}$NO in revealing reliable data of D. We performed studies on animals ($^{14}$NO) by using respiratory mass spectrometry (M3) and on humans ($^{15}$NO) by applying laser magnetic resonance spectroscopy (LMRS). The equipment was characterized by sufficient detection limits of 70 parts/billion at [$^{14}$NO] = 0.001% (M3) and 40 parts/billion at [$^{15}$NO] = 0.002 % (LMRS), respectively.

Lastly, we were able to show that D-values for $^{14}$NO indeed reveal the entire diffusive properties of the alveolar-capillary membrane and that $^{15}$NO is a useful indicator gas for reflecting disturbances of pulmonary gas exchange.

## Introduction

In lung function tests, the determination of the pulmonary diffusing capacity (D) is one of the commonly used techniques for evaluating the efficiency of alveolar-capillary gas uptake. This involves an inhalation of indicator gas, where each breath is held for a couple of seconds and an alveolar probe is then sampled from the end-expiratory portion of the exhaled gas mixture (single-breath method). The calculation of D is based on accurate measurements of inspiratory and end-expiratory indicator gas concentrations. The less indicator gas remains within alveolar space at the end of breath-holding, the more disappears into pulmonary capillary blood, thus pointing to sufficient alveolar-capillary gas uptake. Therefore, the validity of such an evaluation crucially depends on the accuracy in measuring gas concentrations.

The aim of our work was to assess the appropriateness of the stable nitric oxide (NO) isotopes $^{14}$NO and $^{15}$NO in revealing reliable data of D by performing studies on animals and humans. We analyzed $^{14}$NO by employing respiratory mass spectrometry and applied laser magnetic resonance spectroscopy (LMRS) to detect $^{15}$NO.

## Respiratory Mass Spectrometry: Animal Studies ($^{14}$NO)

A respiratory mass spectrometer is an apparatus that is used for the continuous separation and quantitative measurement of gas components in inspiratory and expiratory gas mixtures according to their different mass/charge ratios (m/z). We used a variable-collector magnetic sector respiratory mass spectrometer (M3, Varian MAT, Bremen, Germany), which was improved and modified to increase its sensitivity and long-time stability (Schuster et al., 1979).

In summary, the signal/noise ratio was enhanced by employing low-noise electrometer amplifiers and by reducing mechanical vibrations caused by the vacuum rotary pumps. M3 is designed with an electron-bombardment ion source. To further improve sensitivity, the ion current was increased by making the ion source tighter against the vacuum chamber and by inserting a dosage valve between the inlet system and the connected rotary pump. Resolving power was enhanced by narrowing the ion source slit. Long-time stability was extended by controlling the temperature of the vacuum chamber and the peak position. To put the latter modification into effect, the ion orbits were stabilized using a feedback circuitry operated between nitrogen ion current and accelerating voltage (response time: 30 ms). However, the most important adaptation was the application of a reference gas technique. Here, dried sample gas was repeatedly compared with a reference gas, which was mixed by a gas control unit to ensure the same main components: nitrogen, oxygen, carbon dioxide, and argon. Thus, the sample and the reference gas differed only in respect of their $^{14}NO$ content. This method allowed us to diminish drift errors and cross-talk effects. We set the $^{14}NO$ collector (m/z = 30) at a maximal distance from the ion source to gain an optimum resolving power for the $^{14}NO$ analyses. As we were conducting our tests on rabbits, approximately 50 ml of sample gas volume was available for analyses. We, therefore, had to minimize the gas consumption of the mass spectrometer. We reduced the sampling rate to 5 ml/min by installing a 2-m heated steel inlet capillary. All in all, the aforementioned measures revealed corresponding detection limits of between 70 parts/billion at $[^{14}NO] = 0.001\%$ and 500 parts/billion at $[^{14}NO] = 0.08\%$.

We performed a series of animal studies to check the suitability of $^{14}NO$ for producing reliable data on the diffusive properties of the alveolar-capillary membrane using the modified M3. For this purpose, we executed computerized single-breath experiments on anesthetized and artificially ventilated rabbits. After a maximum deflation, the animals were inflated with $^{14}NO$ containing test gas $[(^{14}NO) = 0.001–0.08\%]$: breath was held for 2–12 s, and the total expired gas was sampled by deflating the lungs through a spiral stainless-steel tube [3.5 mm inside diameter (ID), length 5 m]. The gas mixture stored within the tube was analyzed using mass spectrometry, beginning with the last sampled (end-expiratory) portion.

As the most important outcome of our *in vivo* studies, we found that measurements of pulmonary $^{14}NO$ uptake in rabbits indeed reveal the entire diffusive properties of the alveolar-capillary membrane (Heller and Schuster, 1998), and that the pulmonary $^{14}NO$ diffusing capacity is not influenced by potential biochemical reactions of $^{14}NO$ with pulmonary tissues (Heller and Schuster, 1997).

## Laser Magnetic Resonance Spectroscopy: Human Studies ($^{15}$NO)

The LMRS method is well established for the spectroscopy of paramagnetic molecules such as NO (Heller *et al.*, 2004; McKellar, 1981; Muertz *et al.*, 1999). The most common procedure used in experiments of this type is frequency modulation. To detect free radicals, the magnetic moment of these molecules may be exploited to provide frequency modulation of the absorption of the species through the Zeeman effect. This extreme sensitivity is achieved by polarization detection, which is one of the most sensitive methods in the mid-infrared wavelength region. The advantage of this technique is that NO is detected selectively and with extreme sensitivity. Therefore, the NO measurement is not interfered with by any other gas, allowing various NO isotopes (like $^{14}$NO and $^{15}$NO) to be distinguished. This advantage makes the technique unique for differentiating between endogenously produced $^{14}$NO and exogenously applied $^{15}$NO (Heller *et al.*, 2002).

To understand the signal generation of molecules with a magnetic moment, we may look at the simplest molecular transition, a Q (1/2)-transition instead of the Q (3/2)-transition of NO (Ganser *et al.*, 2003). The molecular transition occurs between the two states that are doubly degenerate and have equal magnetic moments. Applying an external magnetic field results in the so-called *Zeeman splitting of the molecular states*. A straightforward sinusoidal Zeeman modulation, in combination with a linearly polarized beam, does not produce a net signal at zero external field. The alternating modulation field defines the axis of quantization for the two circular components of the laser light. The magnetic substates coincide when the flux density $B_0$ is zero. The allowed transitions in the usual experimental setup are $\Delta M = +1$ or $\Delta M = -1$. When the field is scanned through zero, one type shows increasing and the other one decreasing frequency. For linearly polarized light, both components emit signals of opposite sign and therefore totally cancel each other out to zero. To obtain a non-zero signal, an asymmetry for the two coinciding transitions has to be introduced. An external magnetic field that is used for Zeeman tuning of the molecular transitions automatically increases the degeneracy. One magnetic substate (either $\Delta M = +1$ or $\Delta M = -1$) for the transition is selected, and a non-zero signal of the LMRS results.

The central part of the apparatus is a carbon monoxide laser with internal wavelength selection. The laser is stabilized on the gain maximum of a particular laser line by means of a standard frequency-modulation technique. Because the CO laser is only stepwise tunable, an external magnetic field has to be applied to tune the molecular NO transition (Zeeman tuning) into complete resonance with the laser frequency. The laser output is deflected into the bore of a magnet (solenoid, flux density: 0.1490 T). The total length of the detection zone is 200 mm. This is the homogenous part of the magnetic field; it is surrounded by a small coil for Zeeman modulation (at 8 kHz). The active volume of the inner glass cell is around 10 ml with a working pressure of about 25 mbar. The polarization of the emerging beam is analyzed by a Rochon prism that is almost crossed with respect to the polarization of the incoming laser beam. The infrared radiation is detected by liquid nitrogen cooled InSb detector and fed into a lock-in amplifier. The signal output is processed by a personal computer.

For lung function tests, the normal LMRS setup was modified to meet the requirements of clinical studies. The liquid nitrogen cooled carbon monoxide laser was replaced by an ethanol cooled sealed-off laser. Once the needed temperature in the laser and the cooling bath was reached, it took less than 20 min to start and calibrate the apparatus. The gas flow was adjusted at 40 ml/min to match the measurement time. Because the concentration is determined relatively by using the LMRS, calibration of the spectrometer is needed from time to time, with a defined gas mixture of $^{15}$NO calibration gas (20 parts/million $^{15}$NO in nitrogen) plus $^{15}$NO-free synthetic air at constant total gas pressure. The resulting detection limit of the LMRS was 40 parts/billion at $[^{15}NO] = 0.002\%$.

During the clinical studies and before every single-breath maneuver, the test gas had to be mixed (calibration gas, $^{15}$NO-free air). To minimize the reaction of $^{15}$NO with oxygen, $^{15}$NO was added last to the test gas. The remaining $^{15}$NO in the gas tubing (200 ml) from the mixing process was first measured and then served as a calibration of the concentration. Immediately after the single-breath experiment, the exhaled gas was analyzed using the LMRS.

In our human studies, we showed that pulmonary $^{15}$NO diffusing capacity significantly depended on mean alveolar volume and individual body height (Heller et al., 2003), and that in patients suffering from interstitial lung disease, values of D for $^{15}$NO were significantly reduced in five patients, reflecting impaired alveolar-capillary oxygen exchange in each case (Heller et al., 2004).

References

Ganser, H., Urban, W., and Brown, J. M. (2003). The sensitive detection of NO by Faraday modulation spectroscopy with a quantum cascade laser. Mol. Phys. **101**(4–5), 545–550.

Heller, H., and Schuster, K.-D. (1997). Single-breath diffusing capacity of NO independent of inspiratory NO concentration in rabbits. Am. J. Physiol. **273**, R2055–R2058.

Heller, H., and Schuster, K.-D. (1998). Nitric oxide used to test pulmonary gas exchange in rabbits. *Pflügers Arch.* **437**, 94–97.

Heller, H., Brandt, S., Gäbler, R., and Schuster, K.-D. (2002). The use of [15]N-labeled nitric oxide in lung function tests. *Nitric Oxide Biol. Ch.* **6**(1), 98–100.

Heller, H., Gäbler, R., Brandt, S., Jentsch, A., Granitza, K., Eixmann, B., Breitbach, T., Franz, C., Utkin, Y., Urban, W., and Schuster, K.-D. (2003). Pulmonary [15]NO uptake in man. *Pflügers Arch.* **446**, 256–260.

Heller, H., Korbmacher, N., Gaebler, R., Brandt, S., Breitbach, T., Juergens, U., Grohe, C., and Schuster, K.-D. (2004). Pulmonary [15]NO uptake in interstitial lung disease. *Nitric Oxide Biol. Ch.* **10**(4), 229–232.

McKellar, A. R. W. (1981). Mid-infrared laser magnetic resonance spectroscopy. *Faraday Discuss. Chem. Soc.* **71**, 63.

Muertz, P., Menzel, L., Bloch, W., Hess, A., Michel, O., and Urban, W. (1999). LMR spectroscopy: A new sensitive method for on-line recording of nitric oxide in breath. *J. Appl. Physiol.* **86**(3), 1075–1080.

Schuster, K.-D., Pflug, K.-P., Förstel, H., and Pichotka, J. P. (1979). Adaptation of respiratory mass spectrometry to continuous recording of abundance ratios of stable oxygen isotopes. *In* "Recent Developments in Mass Spectrometry in Biochemistry and Medicine" (A. Frigerio, ed.), Vol. 2, pp. 451–462. Plenum Publishing, New York.

# [52]   ESR Techniques for the Detection of Nitric Oxide *In Vivo* and in Tissues

*By* SERGEY DIKALOV and BRUNO FINK

## Abstract

Plasma levels of nitrite/nitrate may not accurately reflect endothelial nitric oxide synthase (eNOS) function because of interference by dietary nitrates. Nitrosyl hemoglobin (HbNO), a metabolic product of nitric oxide (NO•), may better correlate with bioavailable NO•, but it may depend on the activity of different NOS isoforms and may be affected by dietary nitrite/nitrate. This work examined the correlation between vascular endothelial NO• release and blood levels of HbNO. We measured HbNO in mouse blood using electron spin resonance (ESR) spectrometry, and we quantified vascular production of NO• using colloid Fe(DETC)$_2$ and ESR. C57Blk/6 mice who were fed a high-nitrate diet had levels of plasma HbNO increased 10-fold, whereas those fed a low-nitrate diet had decreased HbNO levels from $0.58 \pm 0.02$ to $0.48 \pm 0.01 \ \mu M$. Therefore, a low-nitrate diet is essential when using HbNO as a marker of eNOS activity. Treatment with L-NAME and the eNOS-specific inhibitor L-NIO halved HbNO formation, which reflects the complete inhibition of NO• release by aorta endothelium. Treatment of mice with the selective inducible NOS (iNOS)

METHODS IN ENZYMOLOGY, VOL. 396
0076-6879/05 $35.00
DOI: 10.1016/S0076-6879(05)96052-7

inhibitor, 1400W, or the selective neuronal NOS (nNOS) inhibitor $N$-AANG did not alter either blood HbNO levels or vascular NO$^\bullet$. The relationship between HbNO and NO$^\bullet$ production by the endothelium (0.23 $\mu M$ HbNO to 5.27 $\mu M$/h of NO$^\bullet$/mg of dry weight aorta) was found to be identical for both C57Blk/6 mice and mice with vascular smooth muscle–targeted expression of p22phox associated with strong increase in eNOS activity. These results support the important role of eNOS in the formation of circulating HbNO, whereas iNOS and nNOS do not contribute to HbNO formation under normal conditions. These data suggest that HbNO can be used as a noninvasive marker of endothelial NO$^\bullet$ production *in vivo*.

## Introduction

In the past decade, a growing interest in nitric oxide (NO$^\bullet$) called for reliable and sensitive techniques for its quantification both *in vitro* and *in vivo*. Developing the *in vivo* technique was the most challenging demand. Most *in vivo* studies investigating endothelium function relied on nitrite/nitrate measurements in blood plasma (Minamino *et al.*, 1997). Nitrite and nitrate are metabolic products of NO$^\bullet$ and are used to quantify NO$^\bullet$ by biotransformation of nitrates (Hasegawa *et al.*, 1999; Minamiyama *et al.*, 1999; Salvemini *et al.*, 1992). Unfortunately, nitrite/nitrate plasma levels are strongly affected by the dietary consumption of nitrite/nitrate, which is difficult to minimize, even in laboratory conditions. Moreover, the nitrite/nitrate level does not reflect bioactive amount of NO$^\bullet$ because the inactivation of NO$^\bullet$ by superoxide and other oxidants leads to the formation of nitrite/nitrate (Pfeiffer *et al.*, 1997). A method for *in vivo* NO$^\bullet$ detection based on the formation of nitrosyl hemoglobin (HbNO) in the reaction of deoxyhemoglobin (Hb) with NO$^\bullet$ has been described (Hall *et al.*, 1996). Therefore, to overcome the limitations of nitrite/nitrate assay, we used electron spin resonance (ESR) to analyze HbNO in frozen blood (Jaszewski *et al.*, 2003; Landmesser *et al.*, 2003).

The process of NO$^\bullet$ transfer into erythrocytes is of critical biological importance because it controls plasma NO$^\bullet$ bioavailability and diffusional distance of endothelial-derived NO$^\bullet$. It has been suggested that NO$^\bullet$ under physiological conditions is consumed rather than conserved by reaction with oxyhemoglobin (Joshi *et al.*, 2002). Indeed, rate constants imply that most of the NO$^\bullet$ will react with oxyhemoglobin-producing nitrate and methemoglobin (Scheme 1), whereas only a minor fraction will form HbNO detectable by ESR (Henry *et al.*, 1997). Our experiments with mouse blood show that only 10% of NO$^\bullet$ forms HbNO, whereas most of NO$^\bullet$ is oxidized to nitrite.

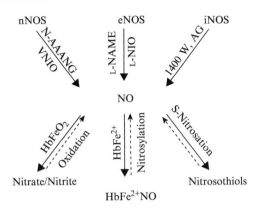

SCHEME. 1. Possible pathways for HbNO formation.

NO• is a well-known ligand of deoxygenated hemoglobin (Keilin *et al.*, 1937). Paramagnetic properties of HbNO have been studied with the use of ESR (Ingram *et al.*, 1955), and *in vitro* and *in vivo* ESR studies demonstrate the presence of HbNO *in vivo* during inflammation, drug metabolism, and treatment with statins (Glover *et al.*, 1999; Kosaka *et al.*, 1994; Ongini *et al.*, 2004; Takahashi *et al.*, 1998). The lifetime of HbNO ranges from 12 min to 20 h, with an average lifetime of 4 h, so it can be accumulated in the blood in substantial concentrations (Henry *et al.*, 1997). Although an electrochemical assay for measurement of HbNO was developed (Palmerini *et al.*, 2004), ESR remains the most direct and unambiguous method for HbNO measurement. Development of high-sensitivity ESR spectrometers has made possible the quantification of 5-coordinate HbNO in blood (Hall *et al.*, 1996; Jaszewski *et al.*, 2003). A growing body of evidence supports the idea that HbNO can be used as a marker of bioavailable NO• *in vivo*. ESR analysis of 5-coordinate HbNO permits the highly specific detection of low levels of HbNO during hypertension and other diseases (Datta *et al.*, 2004; Glover *et al.*, 1999).

NO• production *in vivo* can also be analyzed *ex vivo* in various tissues. Previously, NO• production in rabbit and mice aorta has been measured by colloid iron diethyldithiocarbamate, Fe(DETC)$_2$ and ESR spectroscopy (Khoo *et al.*, 2004; Kleschyov *et al.*, 2002). This chapter details the ESR techniques for measuring blood HbNO and NO• production in tissues with colloid Fe(DETC)$_2$. It shows the association of blood HbNO with NO• production in the tissue.

Materials and Procedures

$FeSO_4$, DETC, and A23187 were obtained from Sigma-Aldrich (St. Louis, MO). L-NAME, L-NIO, 1400W, and N-AANG were purchased from EMD Biosciences (San Diego, CA). The modified Krebs-HEPES buffer (KHB) for vessel studies was composed of 99.01 mmol/L NaCl, 4.69 mmol/L KCl, 2.50 mmol/L $CaCl_2$, 1.20 mmol/L $MgSO_4$, 25 mmol/L $NaHCO_3$, 1.03 mmol/L $K_2HPO_4$, 20 mmol/L Na-HEPES, and 5.6 mmol/L D-glucose, pH 7.35. All other chemicals were obtained from Sigma-Aldrich (St. Louis, MO) in the highest grade available.

*Animals*

C57Blk/6 (wild-type) and endothelial NOS (eNOS) knockout (KO) mice were obtained from Jackson Laboratories (Bar Harbor, ME). Studies were performed on 12- to 18-week-old male mice. Mice overexpressing the NADPH oxidase p22$^{phox}$ subunit in vascular smooth muscle cells (VSMC) (Tg$^{p22smc}$ mice) were created by cloning the p22$^{phox}$ complementary DNA (cDNA) downstream of these cells (Laude *et al.*, 2004). Investigation of *in vivo* effects of selective eNOS, iNOS, and nNOS inhibitors on HbNO formation and NO$^{\bullet}$ production in the aorta of C57Blk/6, p22$^{phox}$-overexpressed, and eNOS KO mice was performed using subcutaneous injection of L-NIO [L-$N^5$-(1-imunoethyl)ornitine], 1400 W [N-(3-amino-methyl)benzylacetamidine], or N-AANG [(4S)-N-(4-amino-5[aminoethyl] aminopentyl)-$N'$-nitroguanidine] (1 mg/kg in 0.9% NaCl) twice a day. On the day of study, mice were injected with 100 U heparin/25 g of body weight intraperitoneal 5 min before euthanization with $CO_2$. The renal and iliac arteries were cut, and the aorta was flushed gently two times with 1 ml of cold KHB. The aorta was rapidly removed and dissected free of adherent tissues. During preparation, the vessels were maintained in chilled KHB (6°) using a thermo-stabilized cold plate (Noxygen Science Transfer & Diagnostics, GmbH, Elzach, Germany).

*Preparation of Blood for Measurements of HbNO*

Heparinized blood (0.7 ml) was collected by a 26G needle from the right ventricle of the $CO_2$ euthanized mice. The blood was then frozen in liquid nitrogen in 1-ml syringes and kept at −80° before measurements. The frozen blood was transferred from syringes into ESR dewar flasks filled with liquid nitrogen. ESR measurements of HbNO were obtained at the temperature of liquid nitrogen (77°K). The amount of detected NO$^{\bullet}$ was determined from the calibration curve for intensity of the ESR signal of erythrocytes treated with known concentrations of nitrite (1–25 $\mu M$) and $Na_2S_2O_4$ (20 m$M$). Erythrocytes were prepared by immediate centrifugation at 3000 rpm for 5 min. Erythrocytes were separated from plasma and resuspended in deoxygenated phosphate-buffered saline (PBS) buffer.

The three-line hyperfine spectrum of the 5-coordinate complex of NO$^\bullet$ with hemoglobin (Hall *et al.*, 1996; Jaszewski *et al.*, 2003) was recorded with an X-band EMX ESR spectrometer (Bruker Instruments, Inc., Billerica, MA) using a high-sensitivity SHQ microwave cavity in a finger dewar filled with liquid nitrogen. ESR spectrometer settings were as follows: microwave power, 10 mW; modulation frequency, 100 kHz; modulation amplitude, 5 G; field center, 3320 G; sweep width, 320 G; microwave frequency, 9.39 GHz; conversion time, 655 ms; time constant, 5.24 s; number of scans, 2; sweep time, 336 s.

## Preparation of Colloid Fe(DETC)$_2$

To prepare 0.8 m$M$ Fe$^{2+}$(DETC)$_2$ stock solution, DETC (7.2 mg) and FeSO$_4$ 7H$_2$O (4.45 mg) were separately dissolved under nitrogen flow in two volumes (10 ml) of filtered and deoxygenated ice-cold 0.9% NaCl. A cold solution of DETC was mixed with FeSO$_4$ in the ratio of 1:1 in either a microcentrifuge tube (1.5 ml) without air bubbles or a 10-ml plastic tube in the nitrogen flow. The formed 0.8 m$M$ Fe(DETC)$_2$ colloid solution was yellow-brown and was used immediately after preparation.

## Incubation of Aorta with Colloid Fe(DETC)$_2$

Blood vessels were carefully and thoroughly cleaned of adhering fat and cut into 2-mm rings. An ice-cold KHB was used for cleaning and storing vessels. Four aortic rings were placed in 1.5 ml of KHB on a 12-well plate. Incubation of aortic rings was started by the addition of 0.5 ml colloid Fe(DETC)$_2$ to tissue samples with 1.5 ml KHB in each well. Samples can be treated with the various stimuli for NO$^\bullet$ production (H$_2$O$_2$, A23187) before addition of Fe(DETC)$_2$. In some studies, aortic segments were preincubated with PEG-SOD (100 U/ml), PEG-catalase (100 U/ml) for 4 h, or tetrahydrobiopterin (20 $\mu M$), apocynin (50 $\mu M$) for 15 min before treatment with Fe(DETC)$_2$. Aortic segments of vessels were incubated at 37° for 60 min in the presence of 200 $\mu M$ colloid Fe(DETC)$_2$. Because of its high lipophilicity, the formed NO-Fe(DETC)$_2$ complex was exclusively localized in the vascular tissue and not in the medium. The vessels were then placed in the center of a 1-ml syringe with 0.6 ml KHB and snap-frozen in the liquid nitrogen. Samples were stored for up to 6 mo at −80° before ESR analysis. The syringe was briefly rubbed between the palms, and a frozen sample column was removed from the syringe by gentle push from the slightly warmed plastic. Sample was chilled with liquid nitrogen and loaded into a finger dewar, fixed by cotton ball, filled with liquid nitrogen on the top, and then analyzed with a Bruker EMX ESR spectrometer. The ESR spectrometer settings were as follows: microwave power, 10 mW; modulation frequency, 100 kHz; modulation amplitude, 5 G; field center, 3290 G; sweep width, 90 G; microwave frequency, 9.39 GHz; conversion time, 328 ms; time constant, 5.24 s; number of scans, 4; sweep time, 168 s. The amount of detected $NO^{\bullet}$ was determined from the calibration curve for integral intensity of the ESR signal of NO-$Fe^{2+}(MGD)_2$ prepared at various concentrations (1–20 $\mu M$) of the NO-donor MAHMA-NONOate (Morley et al., 1993).

### Results and Discussion

*Electron Spin Resonance Analysis of HbNO.* Nitrosyl hemoglobin (HbNO) is stable and can be measured in the blood by ESR (Fig. 1). The ESR spectrum of mouse blood consists of three major components: ceruloplasmin, free radical component, and nitrosyl hemoglobin (Fig. 1). The amount of HbNO was quantified after subtracting ceruloplasmin and the free radical components from the ESR spectrum of the blood (Fig. 1D). HbNO can be analyzed either in the blood (Fig. 1A–D) or in the isolated erythrocytes (Fig. 1E). Unlike other $NO^{\bullet}$ assays, the formation of HbNO is proportional to the bioavailable amount of $NO^{\bullet}$ (Ongini et al., 2004). This allowed us to study the effect of hypertension on HbNO content in the red blood cells, showing a significant decrease in HbNO in DOCA-salt–treated mice compared with control mice (Landmesser et al., 2003).

ESR spectroscopy provides convenient quantification of HbNO by double integration of ESR spectrum or by analysis of ESR amplitude (Fig. 1E), which has the advantage of better resolution of HbNO, even in the ESR spectrum of the blood (Fig. 1A and D). Quantification of HbNO demonstrates that only 10% of $NO^{\bullet}$ produces HbNO in the aerobic samples (Fig. 1E). Therefore, HbNO calibration probes must be prepared in air-free blood treated with $Na_2S_2O_4$.

*Effect of Dietary Supplementation of Nitrates.* It has been reported that nitrite can be reduced to NO by deoxyhemoglobin and cause vasodilation in the human circulation under hypoxic condition (Cosby et al., 2003). Various intracellular compartments, such as endoplasmic reticulum, and mitochondria are also involved in bioconversion of nitrates (Kozlov et al., 2003a). Therefore, we tested the effect of dietary supplementation of nitrate with the drinking water (100 mg/L) of animals fed by conventional labor diet #5001 containing 33 mg/kg nitrate and 5 mg/kg nitrite, compared with basic low nitrate/nitrite diet containing 12 mg/kg nitrate and 3 mg/kg nitrite. Our results demonstrated a significant 20% decrease (from 0.58

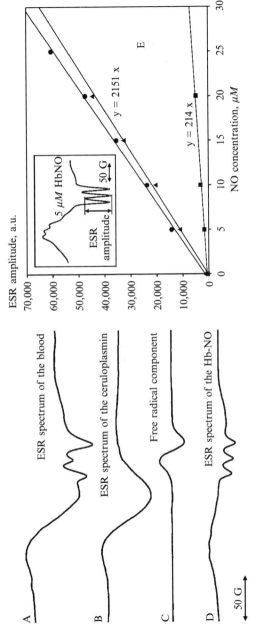

FIG. 1. Measurements of nitrosyl hemoglobin (HbNO) in mouse blood by subtraction of electron spin resonance (ESR) spectra of ceruloplasmin and the free radical component from the ESR spectrum of the blood (A–D). Calibration curve for HbNO (E). Insert shows ESR spectrum of HbNO in erythrocytes. ESR signal HbNO was determined after incubation of washed erythrocytes with nitrite (●), MAHMA NONOate (▲) in the presence of 20 mM of $Na_2S_2O_4$ or in the absence of $Na_2S_2O_4$ (■).

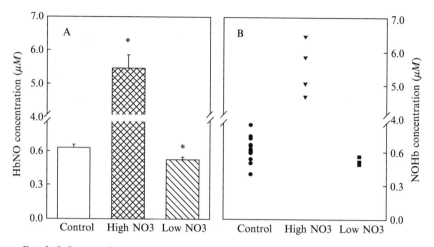

FIG. 2. Influence of dietary nitrite/nitrate on nitrosyl hemoglobin (HbNO) formation in mice. C57Blk/6 mice were investigated after feeding with a conventional diet containing 33 mg/kg nitrate and 5 mg/kg nitrite (Control), with 100 mg/L nitrate in the drinking water (high $NO_3$), or with a basic low nitrite/nitrate diet containing 12 mg/kg nitrate and 3 mg/kg nitrite (low $NO_3$). Data presented as mean $\pm$SEM (A) and as scatter plot (B).

$\pm0.02$ to $0.48$ $\pm0.01$ $\mu M$) in HbNO in animals fed the low-nitrite diet, whereas supplementation with the nitrate in the drinking water led to a 10-fold increase in HbNO content (Fig. 2A). These data suggest minimal interference of plasma nitrate with HbNO formation; however, a controlled low nitrate/nitrite diet was very useful for consistent HbNO data (Fig. 2B).

*Detection of NO in Aorta.* Colloid Fe(DETC)$_2$ has been used to study NO$^\bullet$ production in mouse aorta (Khoo *et al.*, 2004; Laude *et al.*, 2004). Similar to HbNO, Fe(DETC)$_2$ detects only bioactive NO$^\bullet$, does not interfere with nitrite/nitrate, and does not inhibit vascular Cu/Zn SOD (Kleschyov *et al.*, 2002). The NO-Fe(DETC)$_2$ was found to be exclusively associated with blood vessels, suggesting a "one-way delivery" of the trap into the cellular hydrophobic compartments. A time-dependent linear increase of the ESR signal was observed during 2-h incubation, which implies high availability of Fe(DETC)$_2$ and stability of NO-Fe(DETC)$_2$ complex.

Stimulation of the vascular tissue with Ca-ionophore (A-23187) sharply increased NO$^\bullet$ production in the aorta, showing characteristic triplet EPR signal (g = 2.035; $A_N$ = 12.6 G) of NO-Fe(DETC)$_2$ (Fig. 3A and B). Treatment of the mice with the irreversible NOS inhibitor, L-NAME, or the addition of L-NAME *in vitro* completely blocked NO$^\bullet$ production (Fig. 3C and D). The use of a highly selective reversible eNOS inhibitor

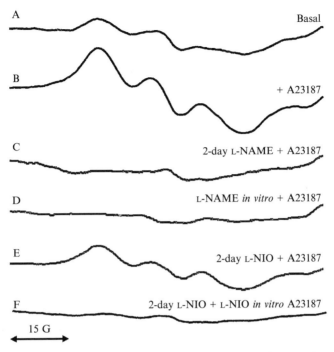

A                                                    Basal

B                                        + A23187

C                                   2-day L-NAME + A23187

D                                   L-NAME *in vitro* + A23187

E                                   2-day L-NIO + A23187

F                    2-day L-NIO + L-NIO *in vitro* A23187

15 G
←——→

FIG. 3. Electron spin resonance (ESR) spectra of NO-Fe(DETC)$_2$ in aortic segments of C57Blk/6 mice under nonstimulated (basal) conditions (A); after stimulation with 10 $\mu M$ A-23187 (B); *in vivo* treatment of C57Blk/6 mice with L-NAME (100 mg/L in drinking water) and additional stimulation with A-23187 (C); following *in vitro* treatment with 100 $\mu M$ L-NAME and stimulation with A-23187 (D); after a 2-day treatment with selective endothelial nitric oxide synthase (eNOS) inhibitor L-NIO and stimulated with A-23187 (E); after additional pretreatment with 10 $\mu M$ L-NIO (F).

L-NIO led to incomplete inhibition of NO$^\bullet$ production (Fig. 3E) in the aorta, due to the washing out of L-NIO from the tissue during aorta preparation. Indeed, *in vitro* addition of L-NIO completely blocked NO$^\bullet$ production (Fig. 3F). Therefore, we used *in vitro* supplementation of reversible NOS inhibitors (L-NIO, N-AANG, 1400W) in the concentration of 10 $\mu M$ to test their effects on aorta NO$^\bullet$ production.

*Association of Blood HbNO with Vascular NO$^\bullet$ Production.* To determine the association of circulating HbNO with the production of NO$^\bullet$ in endothelium, we treated the mice with either L-NAME or the eNOS-specific inhibitor L-NIO and measured HbNO concentration in the blood and NO$^\bullet$ production in the aorta endothelium. Of note is the fact that IC$_{50}$ for targeted enzyme can be a thousand-fold lower than IC$_{50}$ for nontargeted enzyme. For example, 1400 W inhibits iNOS with K$_d$ = 7 n$M$, which

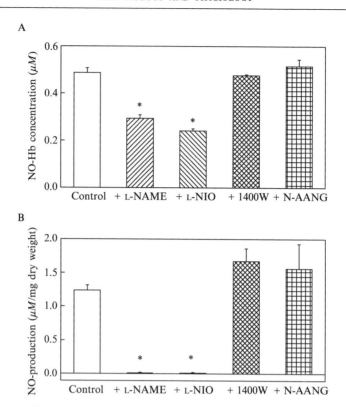

FIG. 4. (A) Nitrosyl hemoglobin (HbNO) concentration in the blood of C57Blk/6 mice after a 2-day treatment (subcutaneous 1 mg/kg, twice a day) with highly selective inhibitors of endothelial nitric oxide synthase (eNOS)–L-NIO, nNOS–N-AANG, iNOS–1400 W, or with a nonselective inhibitor of NOS isoforms–L-NAME (100 mg/L in drinking water); (B) NO• production in aortic sections of the same animals.

is 5000-fold lower than one of eNOS (Parmentier *et al.*, 1999). This treatment decreased HbNO formation by half (Fig. 4A). Furthermore, both L-NAME and L-NIO completely inhibited the release of NO• by aorta endothelium (Fig. 4B). The treatment with selective iNOS inhibitor 1400 W or selective nNOS inhibitor *N*-AANG did not alter either blood HbNO levels or vascular NO• production (Fig. 4A and B).

Finally, we determined the relationship between HbNO and endothelial NO• production as a ratio of L-NAME inhibitable HbNO (Fig. 5A) to NO• production in aorta (Fig. 5B). This ratio (0.23 $\mu M$ HbNO to 5.37 $\mu M$/h of NO•/mg dry weight aorta) was found to be identical for both C57Blk/6 mice and mice with a vascular smooth muscle–targeted expression of p22[phox] (Khatri *et al.*, 2004), in which vascular NO• production was

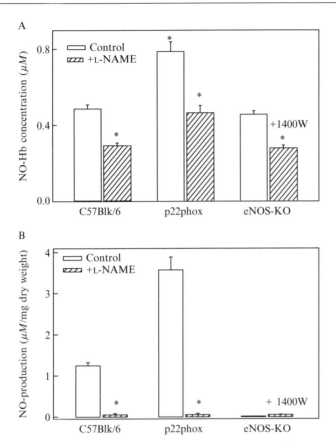

FIG. 5. Nitrosyl hemoglobin (HbNO) concentration in blood (A) and NO• production in aortic sections (B) of C57Blk/6, p22phox, or endothelial nitric oxide synthase (eNOS) knockout mice after a 2-day treatment with NOS inhibitor L-NAME or with inducible NOS inhibitor 1400 W.

markedly increased due to $H_2O_2$-mediated increase in eNOS expression (Laude *et al.*, 2004).

Of note is the fact that the HbNO level in eNOS KO mice was similar to that of the C57Blk/6 mice (Fig. 5A). However, the HbNO in eNOS KO mice was inhibited by the iNOS inhibitor 1400W, which did not affect HbNO in the C57Blk/6 mice (Figs. 4 and 5). These data imply that eNOS KO mice have increased activity of iNOS, making iNOS a source of HbNO, which is likely a compensatory effect of the loss of eNOS in these transgenic mice. Our results suggest that HbNO can be used not only as an index of *in vivo* NO• production by eNOS under normal conditions but also as an evidence of increased activity of iNOS under inflammation.

*Pathways for HbNO Formation.* Our results describe the role of eNOS and iNOS in the HbNO formation. Preliminary data suggest that expression of nNOS in the vasculature of transgenic mice may significantly contribute to HbNO production. nNOS and iNOS are not significant sources of NO• in the vasculature under normal physiological conditions. Inflammation, however, does greatly increases NO• production by iNOS, which may result in a dramatic increase in HbNO (Kozlov *et al.*, 2003b).

Our experiments show that NOS inhibitors do not completely block HbNO formation. It is possible that a part of the residual amount of HbNO can be derived from either endogenous or exogenous of nitrite/nitrate (low-nitrite diet had 12 mg/kg nitrate and 3 mg/kg nitrite). Alternatively, NOS inhibition may increase *in vivo* reduction of exogenous and endogenous nitrite to NO•, which was demonstrated in hypoxic tissues (Gladwin *et al.*, 2004). In our analysis, we considered the residual amount of HbNO as an NOS-independent background, which was the same in both the C57Blk/6 and the eNOS KO mice (Fig. 5A).

## Conclusion

It is clear that HbNO is strongly associated with vascular NO• production. Our results support an important role of eNOS in the formation of HbNO. Under normal physiological conditions, iNOS and nNOS do not contribute to HbNO formation. Although a low-nitrate diet is essential for analysis of HbNO, the fact that HbNO reflects bioavailable levels of vascular NO• makes HbNO useful as a noninvasive marker of endothelial function.

## Acknowledgments

This work was supported by NIH grants R0-1 HL39006, PO-1 HL58000, and AHA 6D6 0430201N.

## References

Cosby, K., Partovi, K. S., Crawford, J. H., Patel, R. P., Reiter, C. D., Martyr, S., Yang, B. K., Waclawiw, M. A., Zalos, G., Xu, X., Huang, K. T., Shields, H., Kim-Shapiro, D. B., Schechter, A. N., Cannon, R. O., 3rd, and Gladwin, M. T. (2003). Nitrite reduction to nitric oxide by deoxyhemoglobin vasodilates the human circulation. *Nat. Med.* **9,** 1498–1505.

Datta, B., Tufnell-Barrett, T., Bleasdale, R. A., Jones, C. J., Beeton, I., Paul, V., Frenneaux, M., and James, P. (2004). Red blood cell nitric oxide as an endocrine vasoregulator: A potential role in congestive heart failure. *Circulation* **109,** 1339–1342.

Gladwin, M. T. (2004). Haldane, hot dogs, halitosis, and hypoxic vasodilation: The emerging biology of the nitrite anion. *J. Clin. Invest.* **113,** 19–21.

Glover, R. E., Ivy, E. D., Orringer, E. P., Maeda, H., and Mason, R. P. (1999). Detection of nitrosyl hemoglobin in venous blood in the treatment of sickle cell anemia with hydroxyurea. *Mol. Pharmacol.* **55,** 1006–1010.

Hall, D. M., and Buettner, G. R. (1996). *In vivo* spin trapping of nitric oxide by heme: electron paramagnetic resonance detection ex vivo. *In* "Methods in Enzymology" (L. Packer, ed.), Vol. 268, pp. 188–192. Elsevier Inc., San Diego.

Hasegawa, K., Taniguchi, T., Takakura, K., Goto, Y., and Muramatsu, I. (1999). Possible involvement of nitroglycerin converting step in nitroglycerin tolerance. *Life Sci.* **64,** 2199–2206.

Henry, Y. A. (1997). EPR characterization of nitric oxide binding to hemoglobin. *In* "Nitric Oxide Research from Chemistry to Biology: EPR Spectroscopy of Nitrosylated Compounds." (Y. A. Henry and B. R. G. Ducastel, eds.), pp. 59–86. R. G. Landes Company, Austin, TX.

Ingram, D. J. E., and Bennett, J. E. (1955). Paramagnetic resonance in phtalocyane, haemoglogin, and other organic derivatives. *Discuss. Faraday Soc.* **19,** 140–146.

Keilin, D., and Hartree, E. F. (1937). Reaction of nitric oxide with haemoglobin and methaemoglobin. *Nature* **139,** 548.

Khatri, J. J., Johnson, C., Magid, R., Lessner, S. M., Laude, K. M., Dikalov, S. I., Harrison, D. G., Sung, H. J., Rong, Y., and Galis, Z. S. (2004). Vascular oxidant stress enhances progression and angiogenesis of experimental atheroma. *Circulation* **109,** 520–525.

Kleschyov, A. L., and Munzel, T. (2002). Advanced spin trapping of vascular nitric oxide using colloid iron diethyldithiocarbamate. *In* "Methods in Enzymology" (E. Cadenas and L. Packer, eds.), Vol. 359, pp. 42–51. Elsevier Inc., San Diego.

Kosaka, H., Sakaguchi, H., Sawai, Y., Kumura, E., Seiyama, A., Chen, S. S., and Shiga, T. (1994). Effect of interferon-gamma on nitric oxide hemoglobin production in endotoxin-treated rats and its synergism with interleukin 1 or tumor necrosis factor. *Life Sci.* **54,** 1523–1529.

Khoo, J. P., Alp, N. J., Bendall, J. K., Kawashima, S., Yokoyama, M., Zhang, Y. H., Casadei, B., and Channon, K. M. (2004). EPR quantification of vascular nitric oxide production in genetically modified mouse models. *Nitric Oxide* **10,** 156–161.

Kozlov, A. V., Dietrich, B., and Nohl, H. (2003a). Various intracellular compartments cooperate in the release of nitric oxide from glycerol trinitrate in liver. *Br. J. Pharmacol.* **139,** 989–997.

Kozlov, A. V., Szalay, L., Umar, F., Fink, B., Kropik, K., Nohl, H., Redl, H., and Bahrami, S. (2003b). EPR analysis reveals three tissues responding to endotoxin by increased formation of reactive oxygen and nitrogen species. *Free Radic. Biol. Med.* **34,** 1555–1562.

Landmesser, U., Dikalov, S., Price, S. R., McCann, L., Fukai, T., Holland, S. M., Mitch, W. E., and Harrison, D. G. (2003). Oxidation of tetrahydrobiopterin leads to uncoupling of endothelial cell nitric oxide synthase in hypertension. *J. Clin. Invest.* **111,** 1201–1209.

Laude, K., Cai, H., Fink, B., Hoch, N., Weber, D. S., McCann, L., Kojda, G., Fukai, T., Dikalov, S., Ramasamy, S., Gamez, G., Griendling, K. K., and Harrison, D. G. (2004). Hemodynamic and biochemical adaptations to vascular smooth muscle overexpression of p22$^{phox}$ in mice. *Hypertension* **44,** In Press.

Morley, D., and Keefer, L. (1993). Nitric oxide/nucleophile complexes: A unique class of nitric oxide–based vasodilators. *J. Cardiovasc. Pharmacol.* **22,** S3–S9.

Minamino, T., Kitakaze, M., Sato, H., Asanuma, H., Funaya, H., Koretsune, Y., and Hori, M. (1997). Plasma levels of nitrite/nitrate and platelet cGMP levels are decreased in patients with atrial fibrillation. *Arterioscler. Thromb. Vasc. Biol.* **17,** 3191–3195.

Minamiyama, Y., Takemura, S., Akiyama, T., Imaoka, S., Inoue, M., Funae, Y., and Okada, S. (1999). Isoforms of cytochrome P450 on organic nitrate-derived nitric oxide release in human heart vessels. *FEBS Lett.* **452,** 165–169.

Jaszewski, A. R., Fann, Y. C., Chen, Y. R., Sato, K., Corbett, J., and Mason, R. P. (2003). EPR spectroscopy studies on the structural transition of nitrosyl hemoglobin in the arterial-venous cycle of DEANO-treated rats as it relates to the proposed nitrosyl hemoglobin/nitrosothiol hemoglobin exchange. *Free Radic. Biol. Med.* **35,** 444–451.

Joshi, M. S., Ferguson, T. B., Jr, Han, T. H., Hyduke, D. R., Liao, J. C., Rassaf, T., Bryan, N., Feelisch, M., and Lancaster, J. R., Jr. (2002). Nitric oxide is consumed, rather than conserved, by reaction with oxyhemoglobin under physiological conditions. *Proc. Natl. Acad. Sci. USA* **99,** 10341–10346.

Ongini, E., Impagnatiello, F., Bonazzi, A., Guzzetta, M., Govoni, M., Monopoli, A., Del Soldato, P., and Ignarro, L. J. (2004). Nitric oxide (NO)–releasing statin derivatives, a class of drugs showing enhanced antiproliferative and antiinflammatory properties. *Proc. Natl. Acad. Sci. USA* **101,** 8497–8502.

Parmentier, S., Bohme, G. A., Lerouet, D., Damour, D., Stutzmann, J. M., Margaill, I., and Plotkine, M. (1999). Selective inhibition of inducible nitric oxide synthase prevents ischemic brain injury. *Br. J. Pharmacol.* **127,** 546–552.

Palmerini, C. A., Arienti, G., and Palombari, R. (2004). Electrochemical assay for determining nitrosyl derivatives of human hemoglobin: Nitrosylhemoglobin and S-nitrosylhemoglobin. *Anal. Biochem.* **330,** 306–310.

Pfeiffer, S., Gorren, A. C., Schmidt, K., Werner, E. R., Hansert, B., Bohle, D. S., and Mayer, B. (1997). Metabolic fate of peroxynitrite in aqueous solution. Reaction with nitric oxide and pH-dependent decomposition to nitrite and oxygen in a 2:1 stoichiometry. *J. Biol. Chem.* **272,** 3465–3470.

Salvemini, D., Pistelli, A., Mollace, V., Anggard, E., and Vane, J. (1992). The metabolism of glyceryl trinitrate to nitric oxide in the macrophage cell line J774 and its induction by *Escherichia coli* lipopolysaccharide. *Biochem. Pharmacol.* **44,** 17–24.

Takahashi, Y., Kobayashi, H., Tanaka, N., Sato, T., Takizawa, N., and Tomita, T. (1998). Nitrosyl hemoglobin in blood of normoxic and hypoxic sheep during nitric oxide inhalation. *Am. J. Physiol.* **274,** 349–357.

# Author Index

# Subject Index

## A

Ace1, inhibition by nitric oxide and nitroxyl, 307–309

Aging
neurodegenerative disease and oxidant-induced iron signaling, 532–533
protein nitrotyrosine identification
mass spectrometry
ionization techniques, 167–168
proteolytic fragment generation, 166–167
tandem mass spectrometry for peptide sequencing, 168–169
overview, 160, 162
proteomic studies in models, 160–161, 168–169
separation of proteins
affinity enrichment after biotin tagging, 163–164
immunoprecipitation, 163
isoelectric focusing, 165
multidimensional high-performance liquid chromatography, 164–165
two-dimensional gel electrophoresis, 162

Angeli's salt, quantum mechanical computations
decomposition mechanisms, 35, 37
reduction potential computation, 40–41

Apoptosis
neurons
metallothionein protection, 283–284, 286, 292–293
SIN-1 induction, 283
nitric oxide induction mechanisms, 429–430
nitric oxide inhibition in endothelial cells, 526
nitric oxide studies in rats
functional magnetic resonance imaging
animals, 581
instrumentation, 581
overview, 580–581
surgical preparation, 581
T2 measurements, 582–583
immunocytochemistry of caspases, 581–583
TUNEL assay, 582

## B

BH4, *see* Tetrahydrobiopterin
Biotin switch assay, *see* S-Nitrosylation

## C

Chemiluminescence, *see* Nitric oxide; S-Nitrosothiols; S-Nitrosylation
Confocal microscopy, *see* Fluorescence resonance energy transfer
Coronary circulation, *see* Vasodilation, nitric oxide

## D

DAzLE, *see* Differential analysis of cDNA library expression
Differential analysis of cDNA library expression
advantages, 360–361
nitric oxide-mediated genes
cDNA library construction, 363
cell culture, 361–362
cell treatment, 363
colony hybridization, 364–365
diapharose staining, 363
immunological detection, 365
microarray construction and analysis, 366–367
quantitative reverse transcriptase–polymerase chain reaction, 367
reverse northern blotting, 366
principles, 360–361

649

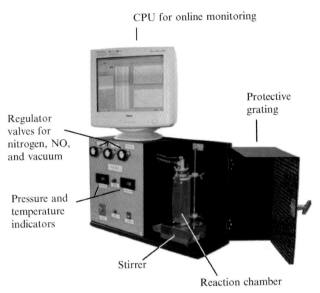

KONTER *ET AL.*, CHAPTER 2, FIG. 1. The NOtizer for the preparation of diazeniumdiolates.

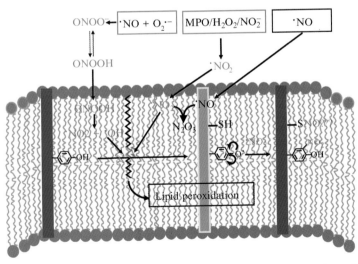

ZHANG *ET AL.*, CHAPTER 18, SCHEME 1. Proposed Mechanism for Transmembrane Tyrosine Nitration and Oxidation.

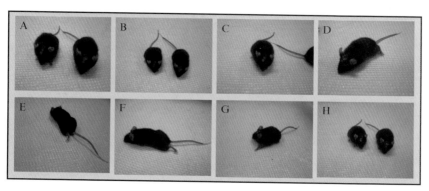

EBADI *ET AL.*, CHAPTER 24, FIG. 1. (A) Postural irregularity. (B) Drooping body posture. (C) Body tremors. (D) Neck muscle rigidity. (E) Unilateral hind-limb extension. (F) Walking difficulty. (G) Stiff tail. (H) Reduced mobility. SIN-1 (100 ng) accentuated 6-OHDA hemiparkinsonism as illustrated in (E–H). Note that pictures were taken 6 h after intranigral 6-OH-DA and/or SIN-1 microinjection. SN-pc 6-OH-DA: 500 ng + SIN-1 (100 ng)/500 nl in PBS pH 7.5 m 0.15 *M*, 0.01% ascorbic acid.

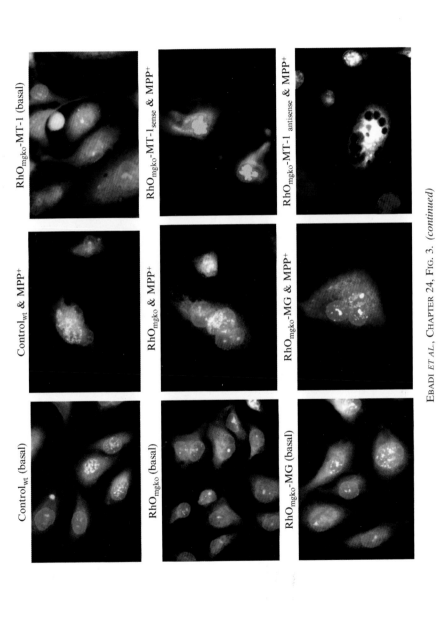

RhO$_{mgko}$-MT-1 (basal)

RhO$_{mgko}$-MT-1$_{sense}$ & MPP$^+$

RhO$_{mgko}$-MT-1$_{antisense}$ & MPP$^+$

Control$_{wt}$ & MPP$^+$

RhO$_{mgko}$ & MPP$^+$

RhO$_{mgko}$-MG & MPP$^+$

Control$_{wt}$ (basal)

RhO$_{mgko}$ (basal)

RhO$_{mgko}$-MG (basal)

EBADI *ET AL.*, CHAPTER 24, FIG. 3. *(continued)*

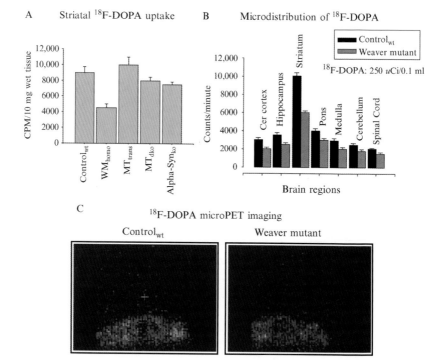

A Striatal $^{18}$F-DOPA uptake

B Microdistribution of $^{18}$F-DOPA

C $^{18}$F-DOPA microPET imaging

Control$_{wt}$          Weaver mutant

EBADI *ET AL.*, CHAPTER 24, FIG. 5. (A) Histogram demonstrating significantly ($p < .05$) reduced striatal $^{18}$F-DOPA uptake in WM$_{homo}$ mice as compared to control$_{wt}$, MT$_{trans}$, MT$_{dko}$, and $\alpha$-Syn$_{ko}$ mice. Data are mean $\pm$SD of five determinations in each group. (B) Histogram demonstrating $^{18}$F-DOPA microdistribution in the central nervous system (CNS) of control$_{wt}$ and weaver mutant mice. The data mean $\pm$ SD of eight determinations in each experimental group. (C) High-resolution microPET imaging illustrating significantly increased $^{18}$F-DOPA uptake in the CNS of control$_{wt}$ mouse (left panel) as compared to WM$_{homo}$ mice (right panel). The radioactivity is de-delocalized in the kidneys of WM$_{homo}$ mouse.

EBADI *ET AL.*, CHAPTER 24, FIG. 3. Multiple fluorochrome analysis of MPP$^+$ (100 $\mu M$) apoptosis in SK-N-SH neurons. Overnight treatment of MPP$^+$ induced DNA condensation in control$_{wt}$ neurons and nuclear DNA fragmentation in RhO$_{mgko}$ neurons. Transfection of RhO$_{mgko}$ neurons with either complex-1 gene or MT-1$_{sense}$ oligonucleotides attenuated MPP$^+$ apoptosis, whereas transfection with MT-1$_{antisense}$ oligonucleotides augmented blebbing, plasma membrane perforations, DNA fragmentation, and condensation. Digital fluorescence images were captured by SpotLite digital camera and analyzed by Image Pro software. (Fluorochromes: Blue: DAPI nuclear DNA stain; Red: JC-1 mitochondrial $\Delta\psi$ marker, Green: Fluorescein isothiocyanate) (RNA, and protein stain).

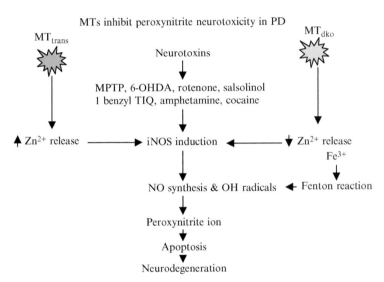

MTs inhibit peroxynitrite neurotoxicity in PD

$MT_{trans}$

$MT_{dko}$

Neurotoxins

MPTP, 6-OHDA, rotenone, salsolinol
1 benzyl TIQ, amphetamine, cocaine

$\uparrow$ $Zn^{2+}$ release $\longrightarrow$ iNOS induction $\longleftarrow$ $\downarrow$ $Zn^{2+}$ release
$Fe^{3+}$

NO synthesis & OH radicals $\longleftarrow$ Fenton reaction

Peroxynitrite ion

Apoptosis

Neurodegeneration

EBADI *ET AL.*, CHAPTER 24, FIG. 6. A proposed model of MT-mediated inhibition of peroxynitrite neurotoxicity. MTs release zinc ions during MPTP or other THIQ-induced oxidative and nitrative stress. Zinc inhibits inducible nitric oxide synthase (iNOS) and thus prevents the synthesis of peroxynitrite ions in $MT_{trans}$ mice. In $MT_{dko}$ mice, due to significantly reduced zinc in the cell, iNOS is induced and NO is synthesized which participates in peroxynitrite ion synthesis by reacting with OH radicals generated from $Fe^{3+}$-mediated Fenton reaction. Furthermore, zinc deficiency may enhance $Fe^{3+}$-mediated neurotoxicity in $MT_{dko}$ mice.

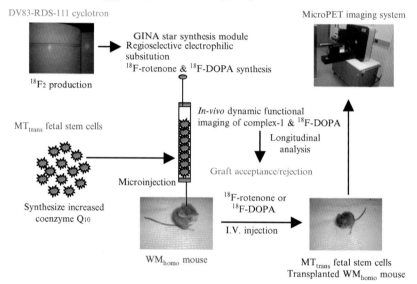

MT$_{trans}$ cell replacement therapy in genetic model of PD

DV83-RDS-111 cyclotron

MicroPET imaging system

GINA star synthesis module
Regioselective electrophilic
subsitution
$^{18}$F-rotenone & $^{18}$F-DOPA synthesis

$^{18}$F$_2$ production

MT$_{trans}$ fetal stem cells

*In-vivo* dynamic functional
imaging of complex-1 & $^{18}$F-DOPA

Longitudinal
analysis

Graft acceptance/rejection

Microinjection

Synthesize increased
coenzyme Q$_{10}$

$^{18}$F-rotenone or
$^{18}$F-DOPA

I.V. injection

WM$_{homo}$ mouse

MT$_{trans}$ fetal stem cells
Transplanted WM$_{homo}$ mouse

EBADI *ET AL.*, CHAPTER 24, FIG. 7. MT$_{trans}$ fetal stem cells are genetically resistant to dihydroxyphenylacetaldehyde (DOPAL) apoptosis, so they can be transplanted in the striatal region of WM$_{homo}$ mice exhibiting progressive dopaminergic degeneration and parkinsonism. The outcome of the grafts is evaluated by $^{18}$F-DOPA and $^{18}$F-rotenone imaging with high-resolution micro–positron emission tomography (PET) scanning. The procedure to prepare fetal stem cells is described in detail in Sharma and Ebadi (2003).

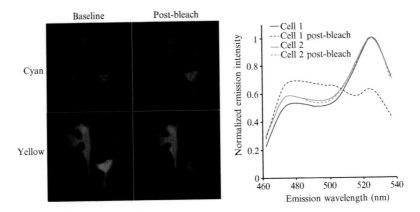

St. Croix *et al.*, Chapter 26, Fig. 2. Confirmation of energy transfer using acceptor photobleaching of the fluorescence resonance energy transfer–metallothionein (FRET-MT) construct. Lung endothelial cells were infected with an adenoviral vector encoding the fluorescent FRET-MT reporter molecule and were imaged 24 h later. FRET was detected in real time, using full spectral confocal imaging. The images show the separation of the two emitted signals (cyan and yellow) following spectral unmixing based on individual calibration spectra for each protein. The graph shows the spectral report provided by the Zeiss software. After selective photobleaching of cell 1, the donor (cyan) was unquenched, resulting in a pronounced increase in the peak emission intensity (~485 nm) indicative of positive FRET. In contrast, there were no changes in the emission intensity of the unbleached cell 2.

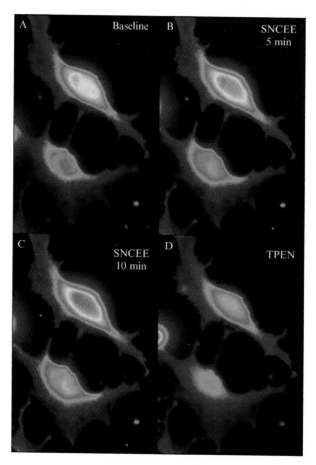

St. Croix *et al.*, Chapter 26, Fig. 4. FluoZin-3 fluorescence showed a TPEN-chelatable, time-dependent increase in response to application of L-SNCEE (50 $\mu M$), reaching a plateau at 10 min with a final mean increase of 88 ± 16% (n = 3).

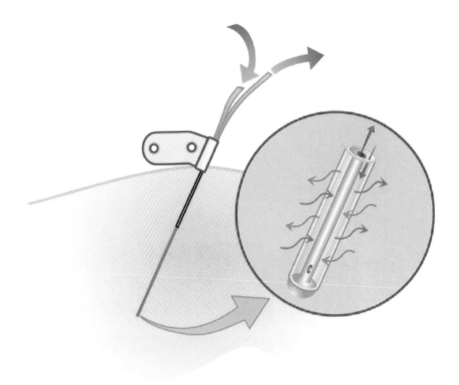

JACKSON, CHAPTER 43, FIG. 1. Schematic representation of a microdialysis probe inserted into a limb muscle. The insert shows the potential for low-molecular-weight substances to diffuse in and out of the probe. Courtesy of CMA Microdialysis.

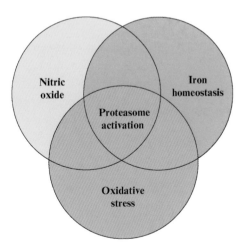

KOTAMRAJU *ET AL.*, CHAPTER 44, FIG. 1. Interrelationship between oxidative stress, •NO, intracellular iron, and proteasomal function.

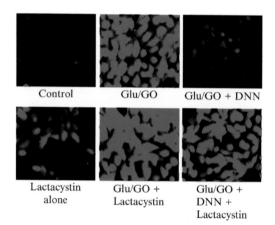

Control         Glu/GO        Glu/GO + DNN

Lactacystin     Glu/GO +       Glu/GO +
alone         Lactacystin     DNN +
                                   Lactacystin

KOTAMRAJU *ET AL.*, CHAPTER 44, FIG. 3. Effect of proteasomal inhibitors on DCFH oxidation in BAECs treated with $H_2O_2$ and $^{\bullet}NO$. Cells were pretreated with the proteasome inhibitors and DETA/NO for 2 h before treatment with Glu/GO.